Methods in Enzymology

Volume LVII

Bioluminescence and Chemiluminescence

EDITED BY

Marlene A. DeLuca
DEPARTMENT OF CHEMISTRY
UNIVERSITY OF CALIFORNIA AT SAN DIEGO
LA JOLLA, CALIFORNIA

ACADEMIC PRESS New York San Francisco London 1978
A Subsidiary of Harcourt Brace Jovanovich, Publishers

ACADEMIC PRESS, INC.
111 Fifth Avenue, New York, New York 10003

United Kingdom Edition published by
ACADEMIC PRESS, INC. (LONDON) LTD.
24/28 Oval Road, London NW1 7DX

Library of Congress Cataloging in Publication Data
Main entry under title:

Bioluminescence and chemiluminescence.

(Methods in enzymology ; v. 57)
Includes bibliographical references and index.
1. Bioluminescence. 2. Chemiluminescence.
I. DeLuca, Marlene. II. Title.
QP601.M49 vol. 57 [QH641] 574.1'925'08s
[574.1'9125]
ISBN 0-12-181957-4 78-13442

PRINTED IN THE UNITED STATES OF AMERICA

Methods in Enzymology

Volume LVII

BIOLUMINESCENCE AND CHEMILUMINESCENCE

METHODS IN ENZYMOLOGY

EDITORS-IN-CHIEF

Sidney P. Colowick   Nathan O. Kaplan

Section II. Bacterial Luciferase

Section III. *Renilla reniformis* Luciferase

Table of Contents

Section I. Firefly Luciferase

Section IV. Aequorin

Section V. *Cypridina*

Section VI. Earthworm Bioluminescence

Section VII. *Pholas dactylus*

Section VIII. Chemiluminescent Techniques

Section IX. Instrumentation and Methods

Contributors to Volume LVII

Article numbers are in parentheses following the names of contributors.
Affiliations listed are current.

DAVID G. ALLEN (31), *Department of Physiology, University College London, London, England*

JAMES MICHAEL ANDERSON (28, 41), *Department of Biochemistry, Boyd Graduate Studies Research Center, University of Georgia, Athens, Georgia 30602*

THOMAS O. BALDWIN (14, 22), *University of Illinois at Urbana-Champaign, Urbana, Illinois 61801*

MARGARETA BALTSCHEFFSKY (5), *Department of Biochemistry, Arrhenius Laboratory, University of Stockholm, S-104 05 Stockholm, Sweden*

JAMES E. BECVAR (21), *Department of Chemistry, University of Texas, El Paso, Texas 79968*

JOHN R. BLINKS (31), *Department of Pharmacology, Mayo Foundation, Rochester, Minnesota 55901*

ROBERT C. BOGUSLASKI (12, 37), *Ames Research and Development Laboratories, and Chemistry Department, Corporate Research, Miles Laboratories, Inc., Elkhart, Indiana 46514*

LEMUEL J. BOWIE (2), *Laboratory Service, Veterans Administration Hospital, San Diego, California 92161*

ROBERT T. BUCKLER (37), *Ames Research and Development Laboratories, and Chemistry Department, Corporate Research, Miles Laboratories, Inc., Elkhart, Indiana 46514*

ROBERT J. CARRICO (12, 37), *Ames Research and Development Laboratories, and Chemistry Department, Corporate Research, Miles Laboratories, Inc., Elkhart, Indiana 46514*

B. CHANCE (25), *Johnson Research Foundation, Department of Biochemistry and Biophysics, University of Pennsylvania School of Medicine, Philadelphia, Pennsylvania 19104*

EMMETT W. CHAPPELLE (7, 43), *Goddard Space Flight Center, NASA, Greenbelt, Maryland 20771*

THOMAS W. CLINE (16), *Department of Biology, Princeton University, Princeton, New Jersey 08540*

MILTON J. CORMIER (11, 27, 28, 29), *Department of Biochemistry, Boyd Graduate Studies Research Center, University of Georgia, Athens, Georgia 30602*

MARLENE DELUCA (1, 23), *Department of Chemistry, University of California at San Diego, La Jolla, California 92093*

JODY W. DEMING (7, 43), *Department of Microbiology, University of Maryland, College Park, Maryland 20742*

GEORGE J. FAINI (41), *Department of Biochemistry, Boyd Graduate Studies Research Center, University of Georgia, Athens, Georgia 30602*

LARRY R. FAULKNER (40), *Department of Chemistry, University of Illinois at Urbana-Champaign, Urbana, Illinois 61801*

RICHARD FERTEL (10), *Department of Pharmacology, The Ohio State University College of Medicine, Columbus, Ohio 43210*

CHARLES R. HACKENBROCK (4), *Laboratories for Cell Biology, Department of Anatomy, University of North Carolina at Chapel Hill, Chapel Hill, North Carolina 27514*

GARY C. HARRER (31), *Department of Pharmacology, Mayo Foundation, Rochester, Minnesota 55901*

J. W. HASTINGS (13, 14, 20, 21), *The Biological Laboratories, Harvard University, Cambridge, Massachusetts 02138*

OSMUND HOLM-HANSEN (8), *Institute of Marine Resources, Scripps Institution of Oceanography, University of California at San Diego, La Jolla, California 92093*

KAZUO HORI (28), *Department of Biochemistry, Boyd Graduate Studies Research Center, University of Georgia, Athens, Georgia 30602*

EDWARD JABLONSKI (23), *Department of Chemistry, University of California at San Diego, La Jolla, California 92093*

BRIAN R. JEWELL (31), *Department of Physiology, University College London, London, England*

FRANK H. JOHNSON (30, 32), *Department of Biology, Princeton University, Princeton, New Jersey 08540*

R. DON JOHNSON (12), *Ames Research and Development Laboratories, and Chemistry Department, Corporate Research, Miles Laboratories, Inc., Elkhart, Illinois 46514*

DAVID M. KARL (8, 9), *Department of Oceanography, University of Hawaii, Honolulu, Hawaii 96822*

PRASAD KOKA (26), *Bioluminescence Laboratory, Department of Biochemistry, University of Georgia, Athens, Georgia 30602*

NEAL LANGERMAN (42), *Department of Chemistry and Biochemistry, Utah State University, Logan, Utah 84322*

JOHN LEE (26), *Bioluminescence Laboratory, Department of Biochemistry, Boyd Graduate Studies Research Center, University of Georgia, Athens, Georgia 30602*

JOHN J. LEMASTERS (4), *Laboratories for Cell Biology, Department of Anatomy, University of North Carolina at Chapel Hill, Chapel Hill, North Carolina 27514*

ARNE LUNDIN (5, 6), *National Defence Research Institute, S-172 04 Sundbyberg, Sweden*

W. D. MCELROY (1), *Department of Biology, University of California at San Diego, La Jolla, California 92093*

JOHN C. MATTHEWS (11), *Department of Biochemistry, University of Georgia, Athens, Georgia 30602*

PATRICK H. MATTINGLY (31), *United States Public Health Service Hospital, Seattle, Washington 98114*

EDWARD A. MEIGHEN (18), *Department of Biochemistry, McGill University, Montreal, Quebec, Canada H3G 1Y6*

A. M. MICHELSON (35), *Institut de Biologie Physico-Chimique, 75005 Paris, France*

MICHAEL G. MULKERRIN (34), *Bioluminescence Laboratory, Department of Biochemistry, Boyd Graduate Studies Research Center, University of Georgia, Athens, Georgia 30602*

KENNETH H. NEALSON (15), *Marine Biology Research Division, Scripps Institution of Oceanography, University of California at San Diego, La Jolla, California 92093*

DAVID A. NIBLEY (43), *Boeing Company, Houston, Texas 77058*

MIRIAM Z. NICOLI (14), *University of Illinois at Urbana-Champaign, Urbana, Illinois 61801*

R. OSHINO (25), *Biochemistry Laboratory, Kobe Yamate Women's College Ikuta-ku, Kobe, 650 Japan*

GRACE L. PICCIOLO (7, 43), *Food and Drug Administration, Bureau of Medical Devices Laboratory, Silver Spring, Maryland 20910*

HANS N. RASMUSSEN (3), *Institute of Biological Chemistry, DK-2100 ϕ Copenhagen, Denmark*

DAVID F. ROSWELL (36), *Department of Chemistry, Loyola College, Baltimore, Maryland 21210*

HARTMUT R. SCHROEDER (37), *Ames Research and Development Laboratories, and Chemistry Department, Corporate Research, Miles Laboratories, Inc., Elkhart, Indiana 46514*

W. RUDOLF SEITZ (38), *Department of Chemistry, University of New Hampshire, Durham, New Hampshire 03824*

H. H. SELIGER (44), *McCollum-Pratt Institute and Department of Biology, The Johns Hopkins University, Baltimore, Maryland 21218*

OSAMU SHIMOMURA (30, 32), *Department of Biology, Princeton University, Princeton, New Jersey 08540*

PHILIP E. STANLEY (19, 24), *Department of Clinical Pharmacology, The Queen Elizabeth Hospital, Woodville, South Australia 5011*

MICHAEL A. TRUSH (39), *Departments of Pharmacology and Toxicology, West Virginia University Medical Center, Morgantown, West Virginia 26506*

FREDERICK I. TSUJI (33), *Molecular Biology Section, National Science Foundation, Washington, D.C. 20550*

SHIAO-CHUN TU (17), *Department of Biophysical Sciences, University of Houston, Houston, Texas 77004*

S. ULITZUR (20), *Department of Food Engineering and Biotechnology, Technion-Israel Institute for Technology, Haifa, Israel*

KNOX VAN DYKE (39), *Departments of Pharmacology and Toxicology, West Virginia University Medical Center, Morgantown, West Virginia 26506*

MENNO VAN LEEUWEN (31), *Department of Physiology, University of Amsterdam, Amsterdam, The Netherlands*

JOHN E. WAMPLER (34, 41), *Bioluminescence Laboratory, Department of Biochemistry, Boyd Graduate Studies Research Center, University of Georgia, Athens, Georgia 30602*

WILLIAM W. WARD (29), *Department of Biochemistry and Microbiology, Cook College, Rutgers University, New Brunswick, New Jersey 08903*

BENJAMIN WEISS (10), *Department of Pharmacology, Medical College of Pennsylvania, Philadelphia, Pennsylvania 19129*

EMIL H. WHITE (36), *Department of Chemistry, The Johns Hopkins University, Baltimore, Maryland 21218*

MARK E. WILSON (39), *Departments of Pharmacology and Toxicology, West Virginia University Medical Center, Morgantown, West Virginia 26506*

Preface

Historically, studies of bioluminescence and chemiluminescence were of a scholarly nature directed toward an understanding of mechanisms involved in the creation of excited states of molecules. During the past ten years, however, it has become apparent that bio- and chemiluminescence are powerful tools for assaying a variety of biologically important molecules. The applications have become so numerous and diverse that it seemed timely to devote a volume in the *Methods in Enzymology* series to these techniques.

The introduction of modern electronic instruments has made it possible to measure light emission with great precision. Thus ordinary chemical or enzyme-catalyzed reactions coupled to light-emitting systems can be used to determine a variety of biologically important molecules with a sensitivity that cannot be achieved by other methods.

Since this is the first time that a Methods volume has been devoted to bio- and chemiluminescence, I have deviated somewhat from the format used in other volumes. An introductory chapter for each different bioluminescent organism has been included for the benefit of those workers who are not familiar with the subject.

The first section is devoted to firefly luciferase. In the presence of excess luciferin and Mg^{2+} the light intensity is proportional to added ATP. The reaction has been used to detect ATP directly or in coupled reactions which either produce or use ATP.

Bacterial luciferase has been useful for the determination of NADH and NADPH. It is possible, in a coupled reaction, to determine the appearance or disappearance of these reduced nucleotides with great precision. The other luciferases discussed are unique in their requirements and are useful for various assays.

The discussion of chemiluminescence is far from complete, but an effort has been made to cover those areas that are related to biologically interesting molecules or processes.

The final section is devoted to instrumentation. Since the use of luminescent assays is critically dependent on the precision of light measurement, I feel this is a very important section. While most investigators will not build their own instrument, it should be useful to have an evaluation of the available commercial instruments. The section on calibration of phototubes is essential in order to have consistency of measurements among various laboratories.

The application of bio- and chemiluminescence techniques to the determination of compounds of biological interest is relatively recent, and one can expect many new and useful techniques to be introduced during the next few years. I hope that this volume of Methods will stimulate such developments.

I would like to thank all of the authors for their help in the initial planning of this volume and for their cooperation in supplying manuscripts. The people at Academic Press have been most helpful and I am grateful to them for having made my job much easier.

MARLENE DELUCA

METHODS IN ENZYMOLOGY

EDITED BY

Sidney P. Colowick and Nathan O. Kaplan

VANDERBILT UNIVERSITY
SCHOOL OF MEDICINE
NASHVILLE, TENNESSEE

DEPARTMENT OF CHEMISTRY
UNIVERSITY OF CALIFORNIA
AT SAN DIEGO
LA JOLLA, CALIFORNIA

METHODS IN ENZYMOLOGY

EDITORS-IN-CHIEF

Sidney P. Colowick   Nathan O. Kaplan

VOLUME XX. Nucleic Acids and Protein Synthesis (Part C)
Edited by KIVIE MOLDAVE AND LAWRENCE GROSSMAN

VOLUME XXI. Nucleic Acids (Part D)
Edited by LAWRENCE GROSSMAN AND KIVIE MOLDAVE

VOLUME XXII. Enzyme Purification and Related Techniques
Edited by WILLIAM B. JAKOBY

VOLUME XXIII. Photosynthesis (Part A)
Edited by ANTHONY SAN PIETRO

VOLUME XXIV. Photosynthesis and Nitrogen Fixation (Part B)
Edited by ANTHONY SAN PIETRO

VOLUME XXV. Enzyme Structure (Part B)
Edited by C. H. W. HIRS AND SERGE N. TIMASHEFF

VOLUME XXVI. Enzyme Structure (Part C)
Edited by C. H. W. HIRS AND SERGE N. TIMASHEFF

VOLUME XXVII. Enzyme Structure (Part D)
Edited by C. H. W. HIRS AND SERGE N. TIMASHEFF

VOLUME XXVIII. Complex Carbohydrates (Part B)
Edited by VICTOR GINSBURG

VOLUME XXIX. Nucleic Acids and Protein Synthesis (Part E)
Edited by LAWRENCE GROSSMAN AND KIVIE MOLDAVE

VOLUME XXX. Nucleic Acids and Protein Synthesis (Part F)
Edited by KIVIE MOLDAVE AND LAWRENCE GROSSMAN

VOLUME XXXI. Biomembranes (Part A)
Edited by SIDNEY FLEISCHER AND LESTER PACKER

VOLUME XXXII. Biomembranes (Part B)
Edited by SIDNEY FLEISCHER AND LESTER PACKER

VOLUME XXXIII. Cumulative Subject Index Volumes I–XXX
Edited by MARTHA G. DENNIS AND EDWARD A. DENNIS

VOLUME XXXIV. Affinity Techniques (Enzyme Purification: Part B)
Edited by WILLIAM B. JAKOBY AND MEIR WILCHEK

VOLUME XXXV. Lipids (Part B)
Edited by JOHN M. LOWENSTEIN

VOLUME XXXVI. Hormone Action (Part A: Steroid Hormones)
Edited by BERT W. O'MALLEY AND JOEL G. HARDMAN

VOLUME XXXVII. Hormone Action (Part B: Peptide Hormones)
Edited by BERT W. O'MALLEY AND JOEL G. HARDMAN

VOLUME XXXVIII. Hormone Action (Part C: Cyclic Nucleotides)
Edited by JOEL G. HARDMAN AND BERT W. O'MALLEY

VOLUME XXXIX. Hormone Action (Part D: Isolated Cells, Tissues, and Organ Systems)
Edited by JOEL G. HARDMAN AND BERT W. O'MALLEY

VOLUME XL. Hormone Action (Part E: Nuclear Structure and Function)
Edited by BERT W. O'MALLEY AND JOEL G. HARDMAN

VOLUME XLI. Carbohydrate Metabolism (Part B)
Edited by W. A. WOOD

VOLUME XLII. Carbohydrate Metabolism (Part C)
Edited by W. A. WOOD

VOLUME XLIII. Antibiotics
Edited by JOHN H. HASH

VOLUME XLIV. Immobilized Enzymes
Edited by KLAUS MOSBACH

VOLUME XLV. Proteolytic Enzymes (Part B)
Edited by LASZLO LORAND

VOLUME XLVI. Affinity Labeling
Edited by WILLIAM B. JAKOBY AND MEIR WILCHEK

VOLUME XLVII. Enzyme Structure (Part E)
Edited by C. H. W. HIRS AND SERGE N. TIMASHEFF

Volume XLVIII. Enzyme Structure (Part F)
Edited by C. H. W. Hirs and Serge N. Timasheff

Volume XLIX. Enzyme Structure (Part G)
Edited by C. H. W. Hirs and Serge N. Timasheff

Volume L. Complex Carbohydrates (Part C)
Edited by Victor Ginsburg

Volume LI. Purine and Pyrimidine Nucleotide Metabolism
Edited by Patricia A. Hoffee and Mary Ellen Jones

Volume LII. Biomembranes (Part C: Biological Oxidations)
Edited by Sidney Fleischer and Lester Packer

Volume LIII. Biomembranes (Part D: Biological Oxidations)
Edited by Sidney Fleischer and Lester Packer

Volume LIV. Biomembranes (Part E: Biological Oxidations)
Edited by Sidney Fleischer and Lester Packer

Volume LV. Biomembranes (Part F: Bioenergetics) (in preparation)
Edited by Sidney Fleischer and Lester Packer

Volume LVI. Biomembranes (Part G: Bioenergetics) (in preparation)
Edited by Sidney Fleischer and Lester Packer

Volume LVII. Bioluminescence and Chemiluminescence
Edited by Marlene DeLuca

Volume LVIII. Cell Culture (in preparation)
Edited by William B. Jakoby and Ira H. Pastan

Volume LIX. Nucleic Acids and Protein Synthesis (Part G, in preparation)
Edited by Kivie Moldave and Lawrence Grossman

Volume LX. Nucleic Acids and Protein Synthesis (Part H, in preparation)
Edited by Kivie Moldave and Lawrence Grossman

VOLUME 61. Enzyme Structure (Part H, in preparation)
Edited by C. H. W. HIRS AND SERGE TIMASHEFF

VOLUME 62. Vitamins and Coenzymes (Part D, in preparation)
Edited by DONALD B. MCCORMICK AND LEMUEL D. WRIGHT

Section I

Firefly Luciferase

[1] Purification and Properties of Firefly Luciferase

By MARLENE DELUCA and W. D. MCELROY

Firefly luciferase in the presence of luciferin (D-LH_2), ATP-Mg, and molecular oxygen catalyzes the production of light according to the following reactions:

$$\text{D-LH}_2 + \text{ATP} \overset{\text{Mg}^{2+}\text{E}}{\rightleftharpoons} \text{E}\cdot\text{LH}_2\text{AMP} + \text{PP}_i \quad (1)$$

$$\text{E}\cdot\text{LH}_2\text{AMP} + \text{O}_2 \rightarrow \text{E}\cdot\text{P} + \text{AMP} + \text{CO}_2 + \text{light} \quad (2)$$

Reaction (1) is the activation of D-LH_2 resulting in the formation of an enzyme-bound luciferyl adenylate (E·LH_2AMP). Reaction (2) requires stoichiometric amounts of molecular oxygen and leads to the formation of enzyme-bound excited product which subsequently decomposes to give off light and a tightly enzyme-bound product.[1,2] The structure of luciferin, the product oxyluciferin, and dehydroluciferin are shown in Fig. 1. A quantum yield has been demonstrated to be near unity.[3] Plant *et al.*[4] using ^{14}C-carboxyl-labeled luciferin in the presence of excess enzyme, ATP, and O_2 demonstrated the quantitative liberation of CO_2 from the luciferin during the reaction. The nature of the intermediates in the oxidative step is still not completely resolved. Alternative mechanisms have been discussed in two recent reviews.[1,2]

The color of the emitted light shows a peak at around 560 nm. This emission can be affected by temperature, pH, and metal ions.[5] At low pH or in the presence of lead, mercury, or other heavy metals, the emission peak is shifted to the red, showing an emission around 615 nm (see Fig. 2 for the effect of zinc). Thus, when one uses this system for the assay of ATP, it is critical to run internal controls to make sure that the light color has not been shifted. Because of the greater sensitivity of phototubes in the blue in contrast to the red, a shift to the red will give an apparent lower level of ATP. A dark-adapted eye can readily discern the difference in color between the control and one emitting at longer wavelengths. Under most conditions this phenomenon is rarely encountered.

[1] W. D. McElroy and M. DeLuca, *in* "Chemiluminescence and Bioluminescence" (M. J. Cormier, D. M. Hercules, and J. Lee, eds.), p. 285. Plenum, New York, 1973.
[2] M. DeLuca, *Adv. Enzymol.* **44,** 37 (1976).
[3] H. Seliger and W. D. McElroy, *Arch. Biochem. Biophys.* **88,** 136 (1960).
[4] P. Plant, E. H. White, and W. D. McElroy, *Biochem. Biophys. Res. Commun.* **31,** 98 (1968).
[5] H. Seliger and W. D. McElroy, *Proc. Natl. Acad. Sci. U.S.A.* **52,** 75 (1964).

ISBN 0-12-181957-4

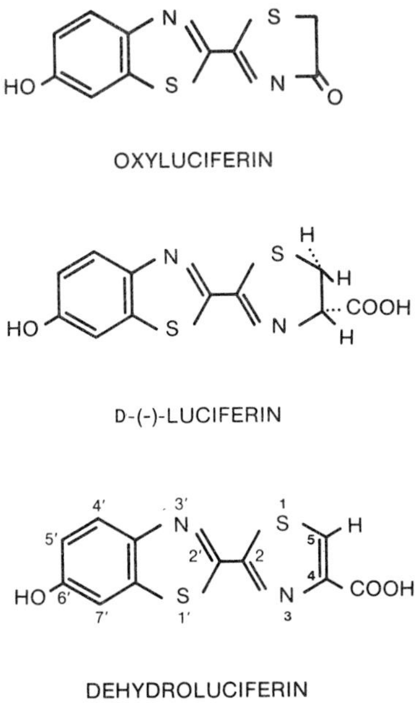

OXYLUCIFERIN

D-(-)-LUCIFERIN

DEHYDROLUCIFERIN

FIG. 1. Structures of oxyluciferin, D-(—)-luciferin, and dehydroluciferin.

One of the products of the first reaction, namely inorganic pyrophosphate, is a potent inhibitor of the light reaction, since it tends to prevent the formation of LH_2AMP. It is important to make sure that inorganic pyrophosphate is not present in significant amounts. The crude firefly extract contains a very powerful inorganic pyrophosphatase so the incubation of the crude extract at room temperature with added magnesium ion for a few minutes will reduce the concentration of inorganic pyrophosphate as well as residual ATP, which is often present in most preparations. Crude firefly preparations known to contain ATP should be allowed to stand at room temperature for an hour or two before partial purification is attempted. The pH should be maintained at approximately 7.5 and Mg^{2+} should be added to a final concentration of about 0.1 m*M*. Under these conditions residual nucleotide triphosphate will be rapidly broken down to inorganic phosphate and the corresponding nucleotide.

In crude firefly extracts and some luciferin preparations, dehydroluciferin is usually present. It is a potent inhibitor of the light reaction. It reacts with ATP and the enzyme according to the reaction

$$\text{L} + \text{ATP} \xrightleftharpoons{\text{Mg}^{2+}\text{E}} \text{L-AMP} + \text{PP}_i$$

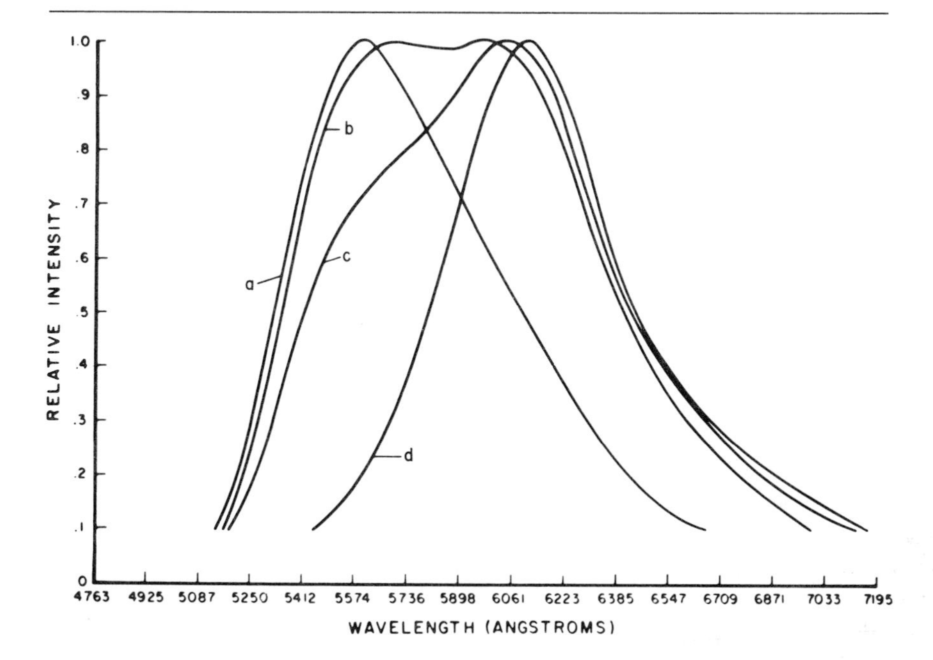

FIG. 2. Effect of zinc on the bioluminescent spectra of mixtures containing *Photinus pyralis* enzyme and synthetic luciferin. Curves: a, Control, $NaZn^{2+}$, pH 7.6; b, 0.13 m*M* $ZnCl_2$; c, 0.4 m*M* $ZnCl_2$; d, 2.3 m*M* $ZnCl_2$. From H. Seliger and W. D. McElroy, *Proc. Natl. Acad. Sci. U.S.A.* **52,** 75 (1964).

The dehydroluciferyl adenylate remains tightly bound to the enzyme. It is this inhibited enzyme that can be released by pyrophosphate through a reversal of the reaction and explains why the secondary addition of pyrophosphate in the presence of excess ATP and luciferin gives a stimulated light emission whose duration depends on the amount of pyrophosphate added. Gradual inhibition returns as the pyrophosphate is hydrolyzed by the pyrophosphatase.

The specificity of the enzyme for ATP has served as the basis for the development of many sensitive assays. In principle, the enzyme can be used to measure total ATP in a sample, or it can be coupled to ATP producing or consuming reactions. The sensitivity is such that as little as 10^{-14} mol of ATP can be accurately measured. As will be discussed in detail in a later section, the crude preparations contain a number of coupling enzymes so that if measurements are not made rapidly, other polyphosphates and other types of reactions can contribute to the rephosphorylation of adenylic acid or ADP to form ATP, and thus will produce light to varying degrees depending upon the concentration of ATP and other polyphosphates. It is for this reason that it is desirable wherever possible to use the flash-height technique, i.e., measuring the peak emission upon injecting ATP within a few seconds, thus eliminating interfering reactions in determination of ATP.

We will describe here the method of collecting and drying fireflies in order to prepare an appropriate acetone powder. We will then describe the preparations and properties of a crude and partially purified luciferase and finally the preparation of crystalline firefly luciferase. Since a large number of different procedures have been used by a number of workers, it is impossible to summarize all of this in detail. We will therefore describe in some detail the methods that we have developed in our laboratory for the preparation of these materials. They seem to be satisfactory for a number of different uses and have been tested by a number of investigators.

Preparation of the Crude Extract

Live fireflies can be dehydrated in a continuous vacuum with calcium chloride within 24 hr. Care should be taken not to overload the desiccator. Better material can be obtained if the fireflies are frozen with Dry Ice and dehydrated in the frozen state over a 24-hr period. It is absolutely critical that fireflies be well dried because the presence of water leads to a rapid inactivation of the luciferase. The one disadvantage of drying the fireflies alive in a vacuum over calcium chloride is that it does lead to the destruction of some of the luciferin and the formation of some dehydroluciferin. The dried lanterns should be removed from the whole body of the firefly by using tweezers to separate the lantern from the body. This can be done rapidly since the lanterns tend to separate readily on contact with the tweezers. Such lanterns can be stored in the dried state and kept in the deep freeze for years without loss of activity.

Because of the presence of a high lipid content, it is desirable to make an acetone powder that removes not only lipids but certain other inhibitors.

We will describe the preparation of the crude firefly luciferase using the same proportions as we normally use in the preparation of the crystalline enzyme. We usually start with 60 g of very dry acetone powder that has been very carefully ground to make sure that it is extremely fine and homogeneous. This represents approximately 12,000–13,000 fireflies if they are *Photinus pyralis*. Usually 75–100 g of dry firefly lanterns are ground in the cold to a fine powder with a mortar and pestle. This dried powder is then added to approximately 2 liters of cold acetone and gently swirled and allowed to stand in the cold for 30 min to 1 hr. The material is filtered on a Büchner funnel, washed once or twice with 200–300 ml of cold acetone, and then filtered until dry and powdery. Usually 15 min of aspiration is adequate in the cold. The filtered powder is then placed in a desiccator with appropriate drying agents, evacuated with a pump, and stored overnight. This dried powder will maintain its activity for several months in the dried state, but it has been our experience that, because of the tendency to take up water, it is best to prepare the acetone powder just prior to use.

For the crude enzyme preparation, 60 g of acetone powder are ground with 300 ml of 10% ammonium sulfate plus 1 m*M* EDTA at pH 8.0. It is critical in this grinding process that the pH be maintained in the alkaline state. On extraction the solution tends to become acid, and this leads to a rapid inactivation of the luciferase. By adding 1 *N* NaOH during the grinding process, the pH can be maintained at or about 7.5. After thoroughly grinding in the mortar and pestle, place the preparation in a blender and gently extract for 10–15 min. Frothing should be avoided because the enzyme can be readily denatured by surface inactivation. After this extraction, the suspension is spun in a refrigerated centrifuge for approximately 20–30 min at 7000 *g*. Although this extract contains well over 90% of the luciferase, the extraction should be repeated once more with 150 ml of 10% ammonium sulfate and 1 m*M* EDTA at pH 8.0. Combine the supernatants and centrifuge at 20,000 *g* in a refrigerated centrifuge for at least 30 min. The supernatant can now be removed and placed in plastic tubes and frozen. This crude extract will maintain good luciferase activity for months. Repeated thawing and freezing should be avoided. This same procedure can be used for any number of fireflies.

Hopkins[6] has studied a number of different ways of extracting luciferase from the acetone powder and has found varying results, although the initial specific activities in the crude extracts were the same when they were extracted at pH 8 in the presence of 1 m*M* EDTA whenever he used (a) water, (b) 0.75 *M* glycylglycine, (c) 0.5 *M* Trizma, (d) 0.4 *M* ammonium sulfate, or (e) 0.1 *M* potassium phosphate.

The stability of the enzyme in these crude solutions did vary considerably. At about 4° the enzyme was most stable in the phosphate and ammonium sulfate solutions. At room temperatures the differences were very pronounced. In ammonium sulfate solution, 50% of the enzyme was lost in 3–4 hr, but no activity was lost in phosphate in this same time period.

This crude preparation of firefly luciferase can be used for ATP assays, but it is essential in most cases to use the flash-height technique. This is particularly true for assaying ATP in crude extracts of material that contains polynucleotide triphosphate in great excess of ATP. A number of workers have used modifications for the preparation of a partially purified luciferase that removes much of the adenylate kinase, transphospho-

[6] T. Hopkins, Ph.D. thesis, Johns Hopkins University, Baltimore, Maryland, 1968.

rylases, and ADP. This is usually done by fractionation with enzyme-grade ammonium sulfate. A very active luciferase preparation can be made by taking the crude extract and adding ammonium sulfate to approximately 35% saturation, maintaining the pH at 7.5 with sodium hydroxide. After standing in the cold for 10–15 min, the preparation is centrifuged and the precipitate is discarded. The supernatant is now brought to approximately 70–75% ammonium sulfate saturation, again maintaining the pH at 7.5; after standing for 15 min in the cold, the solution is centrifuged and the precipitate is dissolved in 10% ammonium sulfate with 1 m*M* EDTA at pH 8.0. This is a very active preparation, but it requires the addition of luciferin in order to obtain maximum activity. Significant quantities of luciferin are present in all the supernatants mentioned above. It can be readily recovered and added back to the luciferase preparation. The recovery technique is discussed later.

Nielsen and Rasmussen,[7] Chappelle and Picciolo,[8] and others have demonstrated that one can obtain a fairly pure and highly specific luciferase by using Sephadex columns. This is extremely valuable because one can use small amounts of material and obtain highly purified luciferase where this is essential.

Preparation of Crystalline Luciferase

The following description of the preparation of the crystalline luciferase is taken from Green and McElroy.[9] The crude extract is made in a blender using 300 ml of 1 m*M* EDTA with 60 g of acetone powder. The pH of the solution should be monitored with a pH meter and concentrated NaOH added in order to maintain the pH around 7.5. Blend without foaming for about 3 min, centrifuge, and extract once more with the same volume of EDTA. Combine the two supernatants and centrifuge for 30 min in the refrigerated centrifuge at 7000 *g*. The crude extract is first treated with calcium phosphate gel. A first treatment is with a low concentration of phosphate gel, which removes a small amount of enzyme and a large amount of impurities, and a second higher concentration of phosphate gel, which absorbs all the enzyme from solution. Of the calcium phosphate gel,[10] 400 ml is centrifuged at 3500 *g* for 10 min and the supernatant is discarded.

[7] R. Nielsen and H. Rasmussen, *Acta Chem. Scand.* **22,** 1757 (1968).

[8] E. W. Chappelle and G. L. Picciolo, eds. "Analytical Applications of Bioluminescence and Chemiluminescence" (NASA Publ. No. SP-388). National Aeronautics and Space Administration, Washington, D.C., 1975.

[9] A. A. Green and W. D. McElroy, *Biochim. Biophys. Acta* **20,** 170 (1956).

[10] Preparation of gel: Dissolve 89 g of $CaCl_2 \cdot 2H_2O$ in 1 liter of H_2O and pour into an 18-liter carboy. Fill the carboy two-thirds full of tap water. Dissolve 152 g of $Na_3PO_4 \cdot 12H_2O$ in 1

Approximately 600 ml of the crude extract is added to the gel and mixed for approximately 3 min in a blender, taking care to prevent foaming. After thorough mixing, the gel is spun down at 3500 *g* in a refrigerated centrifuge, and the supernatant is stored in the cold. A liter of the calcium phosphate gel is spun at 3500 *g* for approximately 10 min, and the supernatant is discarded. The second gel is mixed in a blender for 3 min with the supernatant from the first gel treatment and allowed to stand for approximately 10 min in the cold, after which the preparation is centrifuged for approximately 10 min at 3500 *g*. The supernatant does contain a fair amount of luciferin and can be extracted with ethyl acetate.[11] The gel is washed with 300 ml of 1 m*M* EDTA, pH 8. After mixing for 3 min, spin at 3500 *g* and save the gel. This is an important step, since this alkaline wash removes a large amount of bound fluorescent material and essentially all the remaining luciferin.

The gel is now eluted with 300 ml of 20% saturated ammonium sulfate with 1 m*M* EDTA at pH 8. This is usually done in the blender for about 10 min, and then the mixture is spun for 10 min at 3500 *g*. This should be extracted twice with 300 ml and the supernatants combined. The combined eluates, which contain essentially all the luciferase, are fractionated with ammonium sulfate. Because of coprecipitation problems, the ammonium sulfate has to be added stepwise. This usually involves adding solid ammonium sulfate to raise the saturation from 20% to 40%, maintaining the pH at 7.5. After standing for 10–15 min in the cold, the precipitate is removed by centrifugation and discarded. The supernatant is now taken to 50% saturation, and after standing for 10 min the precipitate is removed by centrifugation and saved. The supernatant is taken to 60% ammonium sulfate saturation, and the precipitate is removed by centrifugation. The precipitates are dissolved in a minimum volume of 10% ammonium sulfate

liter of water and add slowly, with constant stirring, to the carboy. Adjust the pH to approximately 7.5 with 9 ml of concentrated HCl. Check the pH with appropriate paper. Allow to stand overnight. The gel will slowly settle. Remove as much water as possible from the top with a plastic or rubber tube, and refill with tap water. This should be repeated approximately seven times with tap water. As the washing continues, the gel tends to settle rapidly, so it is not necessary to leave overnight. Around the 5th, 6th and 7th wash the gel should settle in a few hours with a very sharp boundary. A final three washes should be done with distilled water. After the last wash, the gel is transferred to a 6-liter Erlenmeyer flask and allowed to settle overnight. After removal of excess water, the gel is ready for use. Concentration of gel should be about 16 mg/ml dry weight.

[11] The pH of the supernatant is adjusted to approximately 3.5 with HCl and is extracted gently three times with 100-ml portions of redistilled ethyl acetate. Then 50 ml of water are added to the combined ethyl acetate extracts, and the latter is removed under vacuum. The pH of the luciferin should be adjusted to about 6.5. Alkaline pH levels should be avoided, since luciferin will slowly oxidize to dehydroluciferin. Also the luciferin should be maintained in a red flask or a tube with foil around it, since light will also give rise to photooxidation.

PURIFICATION OF LUCIFERASE

Fraction	Total light units $\times 10^7$	Light units/mg protein
Crude extract	6.1	3,000
Supernatant of first gel	5.2	3,500
Elution of second gel	3.8	6,000
50–60% $(NH_4)_2SO_4$ fraction	2	11,000
First crystallization	1.8	22,000
Second crystallization	1.5	26,000
Third crystallization	1.3	27,500
Fourth crystallization	1.2	28,000
Fifth crystallization	1.2	28,000

containing 1 m*M* EDTA at pH 8. The precipitate will dissolve in 5–6 ml of the ammonium sulfate solution. Crystals are usually obtained from the 50–60% ammonium sulfate saturation fraction by dialysis and occasionally from the 40–50% fraction.

Crystallization of the enzyme depends upon the fact that it is a euglobulin, i.e., a water-insoluble protein. Crystallization takes place upon dialysis against a solution of low ionic strength. The dialyzing solution contains 1 m*M* EDTA, 10 m*M* NaCl, and 2 m*M* Na_2HPO_4, the final pH being adjusted to 7.3. The enzyme is placed in a dialyzing bag and tied, leaving some air space for expansion, and placed in 1 liter of the medium in the cold. After dialysis overnight, crystallization takes place. In some cases it is desirable to change the solution and dialyze for another 24 hr. After overnight dialysis, the crystals are centrifuged in the cold and dissolved in 0.4 *M* ammonium sulfate and 1 m*M* EDTA at pH 7.8. Recrystallization can be effected by redialyzing against a liter of fresh solution. Since the enzyme is extremely sensitive to oxidizing agents and heavy-metal ions, it is essential that the dialyzing membrane be carefully washed. The membrane is usually boiled in a solution of 10% sodium bicarbonate, 1 m*M* EDTA, and 1 m*M* βME, and further washed membranes can be stored in a solution of 1 m*M* βME.

A protocol for one series of crystallizations is given in the table.[12] The specific activity has increased about 10 times over the original crude extract. The crystals are homogeneous upon analytical ultracentrifugation and electrophoresis. Furthermore, the contaminating enzyme, pyrophosphatase, has been reduced from a very high level to essentially zero. The crystalline luciferase can be precipitated with ammonium sulfate in the presence of 1 m*M* EDTA at pH 7.5 and stored at 4°. For some reason the

[12] W. D. McElroy, this series, Vol. 6, p. 445.

crystalline enzyme is not too stable in the frozen state, and thus it is better to keep it as a slurry (10 mg of protein per milliliter) and dilute the enzyme appropriately when one wants to use it for enzymic assay. The slurry has been kept for several months at 4° without loss of activity. It is important, however, to keep in mind that, if the stored material is not carefully capped, ammonia will distill off and the pH will drop. It is absolutely essential that the pH be maintained above 7 by the addition of small quantities of dilute sodium hydroxide.

The recovery of crystalline luciferase is quite good with the above procedure. On the average we recover approximately 45% of the luciferase after 4–5 times crystallization. Often there is some loss in the first crystallization, but very little in subsequent steps. The quantity of luciferase that can be obtained from 60 g of acetone powder does vary, but on the average we recover approximately 70 mg of crystalline enzyme.

The extinction coefficient (OD_{280}, 1 cm) with 1 mg of protein per milliliter is 0.75. Although the molecular weight of the enzymically active unit is 50,000, only one MgATP is bound per MW 100,000. The evidence suggests the presence of two different subunits of MW 50,000 each; however, it has not been possible to separate the two, using a variety of techniques.[1,2]

Kinetics of the Light Reaction

The kinetics of the light reaction using crude enzyme are quite different from those observed with the crystalline luciferase. The difference is due primarily to a number of contaminating enzymes and the presence of inorganic pyrophosphatase and dehydroluciferin in the crude extracts. We will discuss in some detail the nature of these differences in the section that follows.

We have routinely used the following system for the assay of either the luciferase, ATP, or luciferin. The following proportions are used in a typical ATP assay: 2.1 ml of glycylglycine buffer, 25 m*M*, pH 7.5; 0.1 ml of 0.1 *M* magnesium sulfate; 0.01 ml of enzyme (0.10 mg of protein per milliliter); 0.2 ml of sample or standard ATP. Luciferase at this dilution should be used immediately, since it slowly loses activity even at 0°. It is convenient to inject the sample of ATP into the reaction mixture using a 0.25-ml syringe. The flash height that results is directly proportional to the ATP concentration at 10 μM final concentration or below. Delay in reaching the maximum is indicative of the presence of ADP and myokinase or other nucleotide triphosphates.

Using a crude enzyme, there is enough attached or bound luciferin to give adequate light for ATP determinations. However, these preparations can be greatly stimulated by the addition of luciferin to a final concentration of 50 μM. For example, in one crude commercial preparation where one obtained approximately 50 light units when used according to their instructions, it was possible to increase the light activity to over 2000 light units by the addition of luciferin and raising the magnesium to a concentration of 5 mM. In the assay for ATP, it is essential to have excess magnesium in order to convert all the ATP into magnesium ATP, which is the true substrate for the luciferase activating step. If possible, one should use the flash-height measurement as the technique for assaying for ATP, since it gives a higher degree of specificity and a greater sensitivity and accuracy than any other procedure for measuring ATP. Recently, DeLuca and McElroy[13] demonstrated that upon the injection of ATP in a reaction mixture, as described above, at 25° there is a 25-msec lag before light emission occurs and, furthermore, there is an additional 300 msec before the light reaches its peak output. Thus the actual injection of the ATP is not too critical in terms of observing the initial flash height, although the injection should be vigorous enough to give adequate mixing in front of the phototube.

The flash height observed on injecting ATP into a mixture containing crystalline luciferase with excess luciferin is shown in Fig. 3.[14] There is a rapid rise to a peak and then a relatively fast initial decay with a second slower decay over a long period of time. It is evident that there is some degree of product inhibition and that the enzyme can turn over slowly in the presence of ATP and excess luciferin. The nature of this product inhibition has been discussed in previous reviews in great detail.[1,2] In contrast to this type of flash, when one uses the crude extract and injects ATP there is a rapid flash and a very rapid decay to a very low level of luminescence. This rapid flash and decay to the low level is due to at least two major factors: (1) the activation of dehydroluciferin, which is present in the crude extract and acts as a very potent inhibitor of the luciferase; (2) the presence of inorganic pyrophosphatase, which destroys the pyrophosphate in the system and thus tends to allow the inhibiting reaction of dehydroluciferin to become much more favored. These results have been presented and discussed in some detail in previous publications.[14]

The rate of the light decrease in the crude extract is also proportional to the ATP concentration. The higher the ATP, the faster the decay and the greater the inhibition. Under appropriate conditions the light intensity after 3 min of mixing is proportional to ATP, but the low concentrations of ATP are brighter than the high concentrations. Thus if one is not using the flash-height technique, care must be taken when one makes readings. Some

[13] M. DeLuca and W. D. McElroy, *Biochemistry* **13,** 921 (1974).

[14] W. D. McElroy and H. Seliger, *in* "Light and Life" (W. D. McElroy and B. Glass, eds.), p. 219. Johns Hopkins Press, Baltimore, Maryland, 1961.

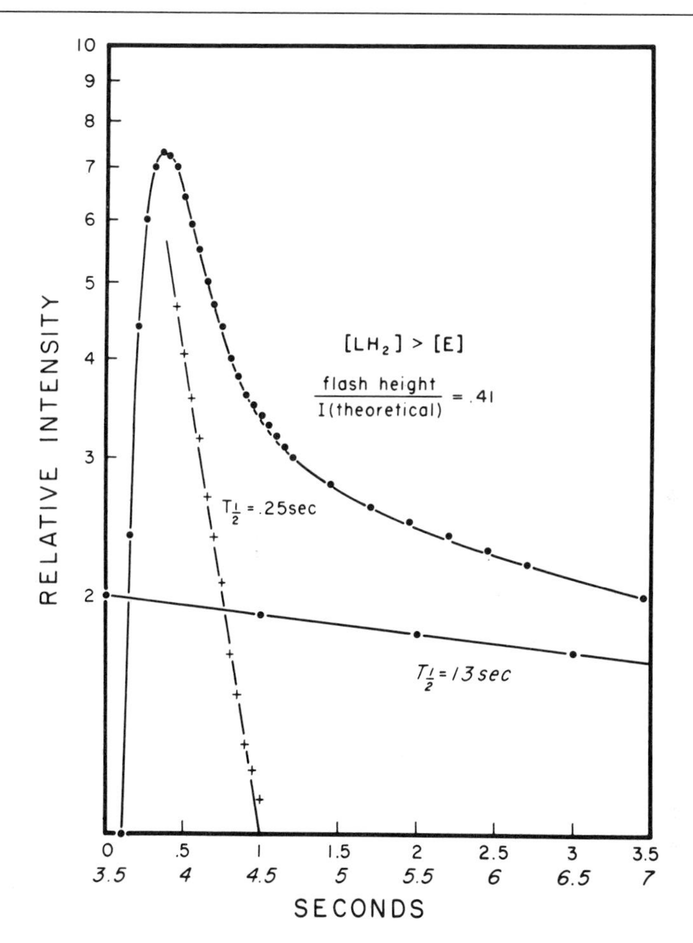

FIG. 3. Flash height and decay of *in vitro* firefly reaction. LH_2 and ATP are in great excess of enzyme. From W. D. McElroy and H. Seliger, *in* "Light and Life" (W. D. McElroy and B. Glass, eds.), p. 22. Johns Hopkins Press, Baltimore, Maryland, 1961.

of these effects can be greatly reduced by using highly diluted luciferase preparations, but care should be taken not to leave the diluted enzyme standing, for it is rapidly inactivated in highly dilute solution.

Effects of Arsenate

Because the rapid flash and decay using the crude extracts makes it difficult to determine ATP if one does not have a fast recorder and a good photometer detecting system, Strehler and Totter[15] were led to add various

[15] B. Strehler and J. Totter, *Arch. Biochem. Biophys.* **40,** 28 (1952).

inhibitors that would prevent the flash from occurring. They observed that arsenate was an effective inhibitor that did lower the flash height and tended to prolong the light emission for a given amount of ATP. However, addition of arsenate greatly lowers the sensitivity of the detecting system and should be avoided wherever possible. There are several commercial instruments now available that allow one to use the flash height, and under these circumstances, arsenate is not a recommended ingredient for the reaction mixture for the reasons given above. Using either crude or partially purified luciferase and under carefully designed conditions, it is possible to assay ATP down to at least 10^{-14} mol. By actually using quantum counting techniques, i.e., cooling the phototube, it is possible to get to even greater sensitivity if it is desired.

Specificity

As has been noted in many previous publications the assay is specific for ATP. However, precursors to ATP will also produce light provided the appropriate enzymes for ATP formation are present. The crystalline enzyme will respond only to ATP, although dATP will give about 5% of the activity of ATP. However, as indicated above, the crude enzyme does contain a number of compounds capable of phosphorylating ADP to form ATP; thus care should be taken in the analysis of the flash-height curve. However, in the presence of mixtures of ADP to ATP of 95 to 5, it is possible to determine ATP flash height with an error no greater than 5%. A number of articles in the present volume as well as many others that have been published and reviewed elsewhere have demonstrated that it is possible to couple a host of enzyme systems that either destroy or generate ATP to the firefly system. Consequently these will not be discussed in the present review.

Precautions

The light-emitting reaction is extremely sensitive to high concentrations of salt. Thus it is essential to dilute crude extracts appropriately. Usually, if the salt concentration is no greater than 0.1 m*M* there is no interference with the light reaction. As indicated earlier, heavy metals do change the color of the light from a yellow-green to a red. Thus if a phototube with low red sensitivity is used, an apparent inhibition can be observed that is actually due to a shift in color of the light.

Luciferase is a sulfhydryl-containing enzyme and requires two essential sulfhydryls for catalytic activity. Thus any reagents interacting with sulfhydryl groups tend to inhibit the luciferase reaction. However, most of these can be detected for appropriate corrections by using internal controls. The pH optimum is such that one should run the reaction around pH 7.5. However, even pH levels as high as 8 can be used. At pH levels below 8, a red light occurs instead of the usual yellow-green, thus giving rise to an apparent inhibition.

The activity of firefly luciferase is very sensitive to specific anion inhibition.[16] The order of effectiveness of inhibition by the anions, $SCN^- > I^- \sim NO_3^- > Br^- > Cl$, followed their position in the Hofmeister series. Extracts containing such anions must be diluted appropriately.

[16] J. L. Denburg and W. D. McElroy, *Arch. Biochem. Biophys.* **141,** 668 (1970).

[2] Synthesis of Firefly Luciferin and Structural Analogs

By LEMUEL J. BOWIE

Luciferin Synthesis

Firefly luciferin, D-(—)-2-(6′-hydroxy-2′-benzothiazolyl)-Δ^2-thiazoline-4-carboxylic acid, was first isolated in pure form from the American firefly *Photinus pyralis* in 1957 by Bitler and McElroy.[1] Approximately 9 mg of crystalline luciferin from approximately 15,000 fireflies was used to perform the initial characterization of this molecule. Luciferin was subsequently chemically synthesized by White *et al.*[2] and further characterized.[3] The structure of luciferin (LH_2) and two useful structural analogs, dehydroluciferin (L) and dehydroluciferol (LOH) are shown in Fig. 1.[4] Owing to the lability of the thiazoline portion of the luciferin molecule, synthetic approaches have generally involved the synthesis of a suitable benzothiazolyl derivative followed by condensation with cysteine to form the thiazoline ring of the luciferin molecule or derivative directly. In contrast, dehydroluciferin and dehydroluciferol have a very stable thiazolyl ring, which has made them very useful as close structural analogs of luciferin in mechanistic studies of firefly luciferase.[5,6]

[1] B. Bitler and W. D. McElroy, *Arch. Biochem. Biophys.* **72,** 358 (1957).
[2] E. H. White, F. McCapra, G. F. Field, and W. D. McElroy, *J. Am. Chem. Soc.* **83,** 2402 (1961).
[3] E. H. White, F. McCapra, and G. F. Field, *J. Am. Chem. Soc.* **85,** 337 (1963).
[4] G. E. Blank, J. Pletcher, and M. Sax, *Biochem. Biophys. Res. Commun.* **42,** 583 (1971).
[5] J. L. Denburg, R. T. Lee, and W. D. McElroy, *Arch. Biochem. Biophys.* **134,** 381 (1969).
[6] L. J. Bowie, V. Horak, and M. DeLuca, *Biochemistry* **12,** 1845 (1973).

ISBN 0-12-181957-4

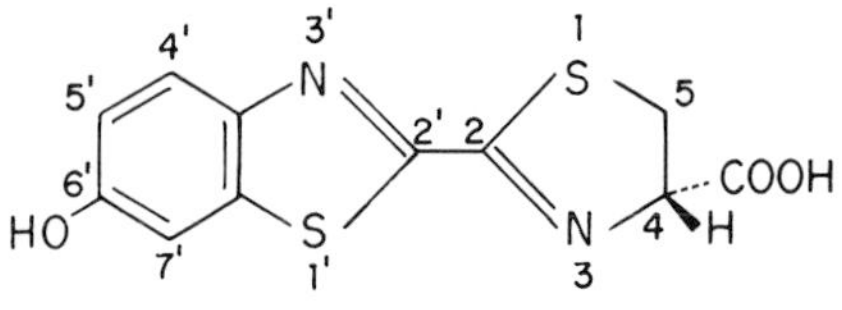

Fig. 1. Structures of luciferin (LH_2), dehydroluciferin (L), and dehydroluciferol (LOH). Rings are oriented with the N=C—C=N system transplanar as has been demonstrated for luciferin by X-ray analysis. From G. E. Blank, J. Pletcher, and M. Sax, *Biochem. Biophys. Res. Commun.* **42,** 583 (1971).

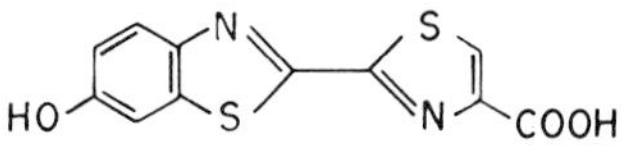

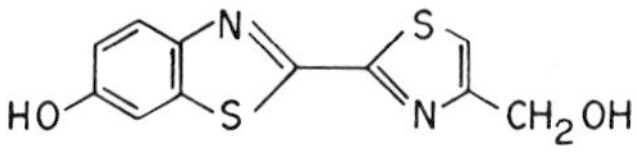

Figure 2 summarizes the three major synthetic paths that have been reported for luciferin. The key intermediate in all these schemes is 2-cyano-6-methoxybenzothiazole. Upon removal of the methyl group at position 6, the molecule is condensed with D-cysteine to give luciferin. Method 1 was used for the first chemical synthesis by White *et al.*[2] but was soon replaced by the much shorter approach (method 2)[7] utilizing a commercially available benzothiazole derivative. For large-scale preparations of luciferin, however, method 3 by Seto *et al.*[8] proved to be a very convenient and useful route. Therefore, in the following detailed discussion of the preparation of luciferin, this approach with minor modifications will be described.

Procedure

The method of Seto *et al.*[8] involves the reaction of carbamoylthiocarbonylthioacetic acid (I), prepared fresh and without isolation, with *p*-anisidine (II) to form 4-methoxythiooxanilamide (III). This product is then cyclized to form 2-carbamoyl-6-methoxybenzothiazole (IV), followed by conversion to the corresponding nitrile (V) as seen in Fig. 3. The rest of the synthesis proceeds as outlined in Fig. 2.

Preparation of 4-Methoxythiooxanilamide (III)

Since carbamoylthiocarbonylthioacetic acid is fairly unstable, it is prepared and used in alkaline solution without isolation by first dissolving

[7] E. H. White, H. Wörther, G. F. Field, and W. D. McElroy, *J. Org. Chem.* **30,** 2344 (1965).
[8] S. Seto, K. Ogura, and Y. Nishiyama, *Bull. Chem. Soc. Jpn.* **36,** 332 (1963).

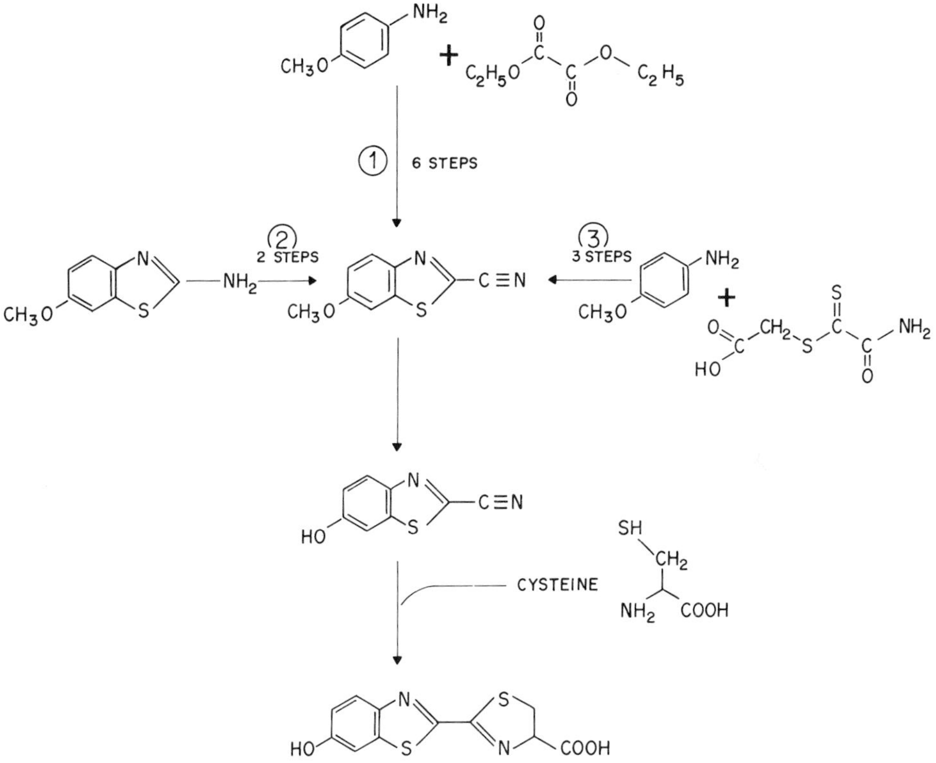

FIG. 2. Synthetic approaches to the synthesis of luciferin: (1) E. H. White, F. McCapra, G. F. Field, and W. D. McElroy, *J. Am. Chem. Soc.* **85,** 337 (1963); (2) E. H. White, H. Wörther, G. F. Field, and W. D. McElroy, *J. Org. Chem.* **30,** 2244 (1965); (3) S. Seto, K. Ogura, and Y. Nishiyama, *Bull. Chem. Soc. Jpn.* **36,** 332 (1963).

potassium hydroxide (300 g) in methanol (2000 ml). The solution is then divided into two halves: one half to be saturated with hydrogen sulfide and the other to be saturated with nitrogen. The half to be saturated with hydrogen sulfide is rapidly bubbled for approximately 3 hr with H_2S that has been dried by passage through calcium chloride. Owing to its odoriferous and toxic nature, this gas should be connected to a suitable trap. Two liters of 10% (w/v) lead acetate is very useful. As H_2S is initially bubbled through the alcoholic potassium hydroxide, the solution becomes light green and bubbles slowly escape from the trap without any decoloration of the lead acetate solution. Concomitantly, a slow stream of nitrogen is passed through the other flask. After approximately 1.5 hr, a black precipitate begins to form in the trap and the methanol–potassium hydroxide solution gradually turns yellow-green. A trichloroacetamide solution is prepared at this time by adding 185 g to 1000 ml of methanol and set aside.

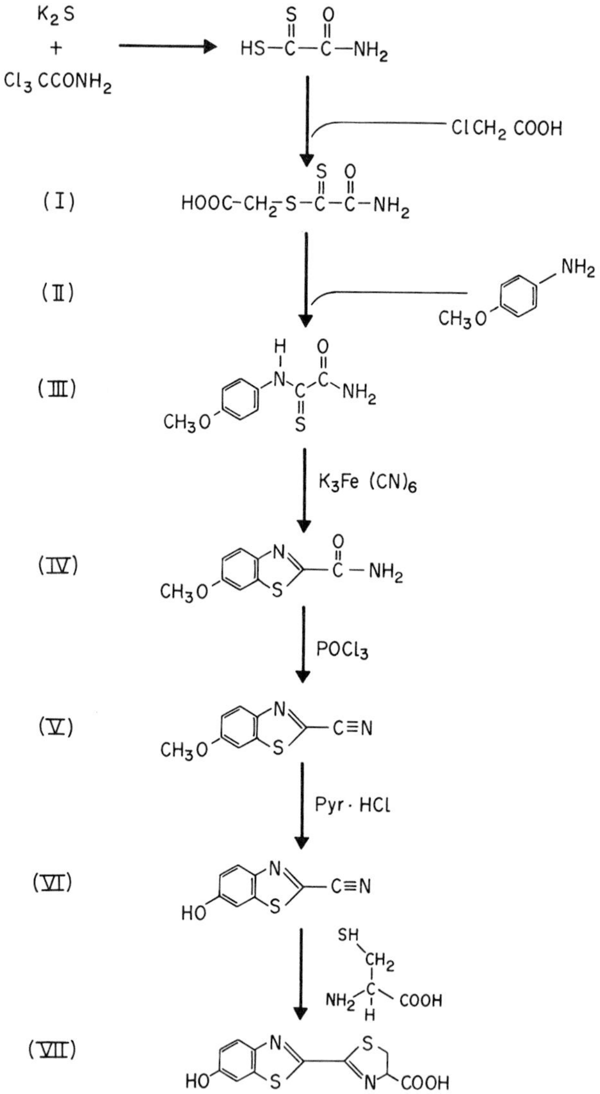

K_2S + Cl_3CCONH_2 → HS—C(=S)—C(=O)—NH_2

↓ $ClCH_2COOH$

(I) HOOC–CH_2–S–C(=S)–C(=O)–NH_2

(II) ↓ CH_3O–C_6H_4–NH_2

(III) CH_3O–C_6H_4–N(H)–C(=S)–C(=O)–NH_2

↓ $K_3Fe(CN)_6$

(IV) CH_3O–benzothiazole–C(=O)–NH_2

↓ $POCl_3$

(V) CH_3O–benzothiazole–C≡N

↓ Pyr · HCl

(VI) HO–benzothiazole–C≡N

↓ SH–CH_2–C(H)(NH_2)–COOH

(VII) HO–benzothiazole–thiazoline–COOH

FIG. 3. Reaction scheme for the synthesis of luciferin according to the method of S. Seto, K. Ogura, and Y. Nishiyama, *Bull. Chem. Soc. Jpn.* **36,** 332 (1963).

The first two flasks are mixed in a 6-liter Erlenmeyer flask and equilibrated with nitrogen for a short time (5 min). The trichloroacetamide solution is then added slowly with stirring while continuing to pass nitrogen through the mixture. As the solution gradually turns red, the temperature rises, and an ice bath should be used to maintain the temperature at ambient or below. Trichloroacetamide is again added slowly with cooling. This process is repeated until all the trichloroacetamide is added. The solution is then deep red but progressively turns reddish brown over a period of 30 min. A solution of monochloroacetic acid (160 g) in water (1000 ml) is neutralized with solid potassium carbonate and added to the above mixture. Upon vigorous shaking a deep red color forms. A layer of inorganic salt is allowed to form for approximately 5 min and removed by filtration. The filtrate is immediately added to *p*-anisidine (100 g) in 50% (v/v) aqueous methanol (1000 ml). (Note: Reagent grade *p*-anisidine must be used; it should be tan to near white, not brown.) As nitrogen bubbling is continued, yellow crystals begin to form. After approximately 30 min the flask is stoppered, and crystallization is allowed to continue overnight at 4°. The yellow crystals are removed by filtration. The yield of 4-methoxythiooxanilamide (III) is approximately (95–125 g) with a melting point of 183°–185°. Recrystallization of this product is not necessary for subsequent cyclization.

Preparation of 2-Carbamoyl-6-methoxybenzothiazole (IV)

4-Methoxythiooxanilamide (50 g) is dissolved in sodium hydroxide (100 g in 2000 ml) and added dropwise, using a separatory funnel, to a stirred solution of potassium ferricyanide (1 lb, or approximately 450 g) in 1 liter of water at room temperature. Fumes that are possibly toxic are generated during this reaction, requiring the use of a fume hood. A fine precipitate forms immediately, but the reaction is allowed to proceed with stirring for 1 hr. The precipitate (approximately 30 g) is then removed by filtration. Repeating the above procedure with an additional 40 g of product (III) yields an additional (20–25 g) of 2-carbamoyl-6-methoxybenzothiazole (IV). The precipitates are combined and recrystallized from methanol, yielding (50–55 g) of product (IV), with a melting point of 254°–256° (decomp.).

Preparation of 2-Cyano-6-methoxybenzothiazole (V)

The above product (50 g) is refluxed as a suspension in phosphorus oxychloride (500 ml) for 90 min. At the end of this time the excess phosphorus oxychloride is distilled off under reduced pressure. Care should be taken to remove all excess $POCl_3$ since it can interact with water with the evolution of heat. The product is cooled, and small aliquots are poured onto crushed ice (no excess liquid present) and mixed to ensure good yields. The resultant precipitate is filtered and extracted with benzene. This extract is filtered, and petroleum ether is added slowly to give a colorless precipitate. This yields approximately 30 g of 2-cyano-6-methoxybenzothiazole (V). Sublimation of the precipitate gives colorless needles (m.p. 128°–130°).

Preparation of 2-Cyano-6-hydroxybenzothiazole (VI)

Removal of the methyl group at position 6 is accomplished in a pyridine hydrochloride melt at 200°–210°. Fresh pyridine hydrochloride is most easily prepared by passing HCl gas over the surface of pyridine in a flask equipped with a reflux condenser. A controlled flow of hydrogen chloride is conveniently generated by allowing concentrated sulfuric acid to drip into a flask of ammonium chloride. The entire apparatus is shown in Fig. 4. The hydrogen chloride is first dried by passage through concentrated H_2SO_4. The safety bottle trap is included as a precaution against "sucking back" of contents of the reaction vessel (1-liter three-necked flask).

Hydrogen chloride generated in this manner is passed over the surface

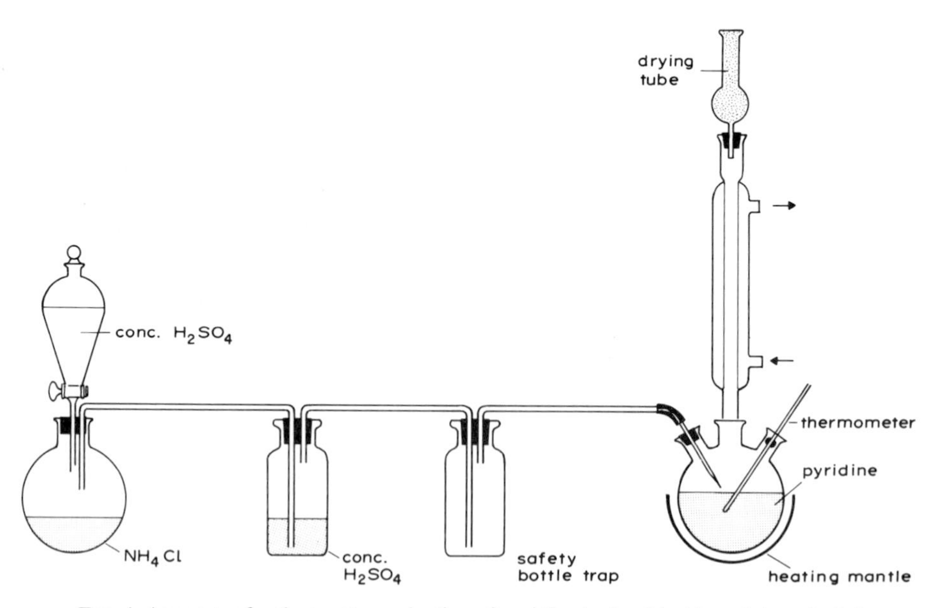

FIG. 4. Apparatus for the *in situ* production of pyridine hydrochloride and demethylation of 2-cyano-6-methoxybenzothiazole.

of pyridine (250 ml). The pyridine is heated under reflux until the reaction temperature reaches 180°. The solution at this time becomes somewhat yellow and pasty. Product (V) (20 g) is then added, and HCl is again passed over the surface. The temperature is increased to 200°–210° and held there for 3 hr. Hydrogen chloride flow is then stopped, the reaction mixture is allowed to cool, and finally the flask is placed on an ice bath. Small aliquots of the product, a red paste, are removed and mixed with crushed ice (2000 ml) and sodium carbonate (10 g). The residual material in the flask is washed out with small amounts of cold water into the sodium carbonate–ice mixture. The pH of the final mixture should be in the range of 6.5–7.0. The product is removed by filtration and dried in air. Traces of starting material (V) are extracted by boiling for 1 hr in benzene (1000 ml) and filtering while hot. It is important to remove all traces of 2-cyano-6-methoxybenzothiazole (V) at this stage since the methoxyluciferin formed by its reaction with cysteine is somewhat difficult to separate from luciferin. The residue after extraction is recrystallized from methanol to yield a fine, light brown solid with a melting point of 209°–211°. The yield of product (VI) may vary at this stage but should be in the range of 6–10 g. Owing to the importance of eliminating small traces of any starting material from this final product, it is necessary to confirm purity by ultraviolet absorption spectroscopy or thin-layer chromatography. The absorption spectrum of compound (V) in dilute alkali has a maximum in the region around 320 nm while compound (VI) has an absorption maximum at approximately 390 nm. Therefore, the absorption spectrum of the product (VI) in dilute base should show no shoulder in the region between 320 and 360 nm. Alternatively, the product can be chromatographed on silica gel plates (Baker Chemical Co.) using chloroform/ethyl acetate (5:1) as the solvent system. The relative migration can be monitored by visualizing with a long wavelength ultraviolet lamp. Under these conditions product (VI) has a relative mobility of 0.51. Any traces of material which chromatograph with an R_f of 0.75 should be removed by further extraction with hot benzene and recrystallization of the residue from methanol.

Preparation of D-*(−)-2-(6′-Hydroxy-2′-benzothiazolyl)-Δ^2-thiazolinecarboxylic Acid—"Firefly Luciferin"*

D-Cysteine hydrochloride (5 g) and potassium carbonate (4.0 g) are dissolved in 125 ml of water. The solution is adjusted to pH 7.5 if necessary and is thoroughly equilibrated with nitrogen (approximately 1 hr). Concomitantly, 2-cyano-6-hydroxybenzothiazole (5.0 g) is dissolved in methanol (150 ml) in a low-actinic glass flask (or a flask that has been masked to exclude light) and equilibrated with nitrogen for the same period of time. The solution of D-cysteine is then added to the flask containing the benzothiazole derivative and allowed to react at room temperature with stirring and under nitrogen stream for an additional 1.5 hr. The solution is adjusted to pH 6–7 with dilute hydrochloric acid, and the precipitate is collected by filtration. After redissolving in warm methanol, the solution is concentrated under vacuum until precipitation commences. The solution is then removed, reequilibrated with nitrogen for a short time, and allowed to precipitate overnight at 4° in a stoppered flask protected from light. The product is isolated as fine, pale yellow needles (6–7 g) with a melting point of 200°–204° (decomp.). Owing to the extreme lability of the product, it should be stored dry, under nitrogen atmosphere, in sealed or tightly stoppered tubes and protected from light by the use of aluminum foil, tape, etc.

Storage and Handling of Luciferin

Luciferin is a pale-yellow solid that recrystallizes with difficulty and sublimes with decarboxylation and decomposition. In aqueous solutions, it is sensitive to extremes in pH, especially in the presence of oxygen and light. Racemization occurs rapidly in certain solvents (e.g., 7% per hour at 4° in aqueous pyridine). Alkaline solutions, in the presence of oxygen, can give rise to dehydroluciferin. For these reasons, if crystalline luciferin is to be stored for extended periods of time, it should be well desiccated in light-tight containers and under nitrogen atmosphere. Under these conditions luciferin can be stored indefinitely without oxidation, racemization, or photodecomposition. Aqueous solutions at near neutral pH values can be stored at 4° safely for periods of 1–2 months if protected from light, and preferably under nitrogen atmosphere.

Spectral Properties of Luciferin

Luciferin is a highly fluorescent molecule, exhibiting a quantum yield of 0.62 in aqueous solutions at a pH 11.[9] Owing to the presence of the 6′-hydroxyl group, luciferin can undergo both ground-state and excited-state ionization to form the phenolate ion. The lowest energy absorption for the ground-state phenol form occurs at 327 nm, and the corresponding absorption for the ground-state phenolate ion occurs at 385 nm, as shown in Fig. 5A. Irrespective of the ground-state species that is excited (in aqueous solutions), fluorescence emission occurs primarily from a single species,

[9] R. A. Morton, T. A. Hopkins, and H. H. Seliger, *Biochemistry* **8**, 1598 (1969).

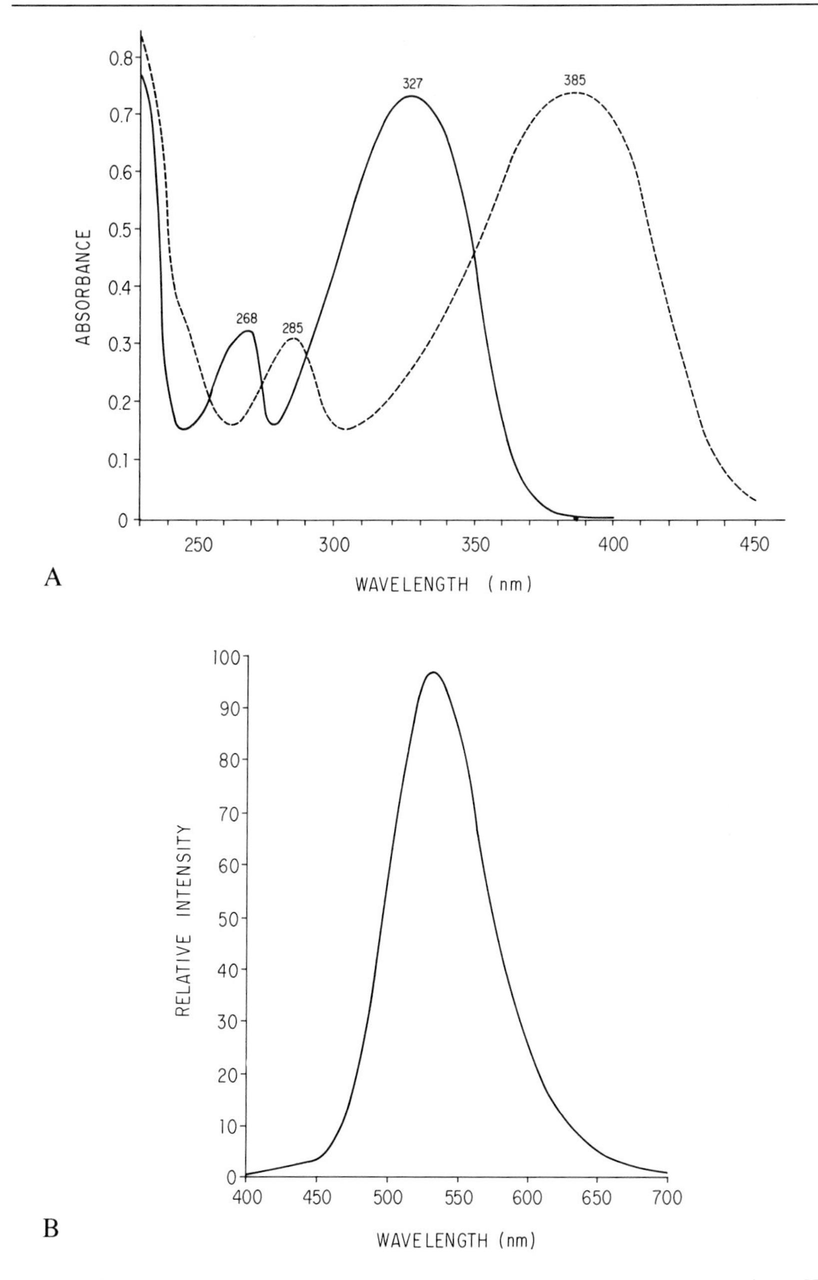

FIG. 5. (A) Normalized absorption spectra of luciferin at pH 4 (——) and at pH 11 (---). (B) Fluorescence emission spectrum (uncorrected) for luciferin at pH 4. The normalized emission spectrum for luciferin at pH 9 (not shown) is superimposable.

the excited-state phenolate (emission maximum approximately 537 nm), as seen in Fig. 5B. This results from ionization of the phenol in the excited state to form excited-state phenolate prior to fluorescence emission. In nonpolar solvents, where excited-state ionization is inhibited, emission occurs predominantly from the excited-state phenol (emission maximum at approximately 420 nm[9]). These properties of luciferin and some of its analogs have been very useful in studying the luciferase active site microenvironment.[6,9,10]

Criteria for Purity

The single most definitive test for structure and purity is examination of *in vitro* bioluminescence quantum yields and peak shapes for the yellow green (565 nm) emission. If the quantum yield is less than 0.7, the luciferin should be purified by recrystallization from methanol. This approach although described in some detail[11] is not simple to perform, however, and requires the use of purified luciferase. As an alternative, a combination of thin-layer chromatography and absorption and fluorescence spectroscopy can be used to assess the general purity of the product. Luciferin migrates as a single spot with an R_f of 0.31 on silica gel plates using a solvent system of ethyl acetate and methanol (10: 1). Paper chromatography using Whatman No. 3 and development with a solvent system of 95% ethanol and 1 *M* ammonium acetate, pH 7.5 (7:3) gives rise to a single spot at an R_f of 0.55. The relative migrations can easily be visualized by observation of the developed chromatograms with a long-wavelength (365 nm) ultraviolet lamp, since luciferin and many of its derivatives are highly fluorescent. Chromatography on silica gel plates which incorporate a fluorescent indicator is even more useful, since this allows visualization, by noting areas of fluorescent quenching, of contaminants that are not fluorescent but have ultraviolet absorption.

Another convenient technique for estimating purity is to examine absorption spectra of aqueous solutions of luciferin. At neutral pH, the ratio of the absorbance at 327 nm to that at 268 nm should be 2.3 or greater; and at alkaline pH, the ratio of absorbance at 385 nm and 285 nm should be 2.5. If a small amount of pure luciferin is available, fluorescence excitation and emission spectra of pure versus synthetic luciferin can be compared and provides another convenient means of estimating purity. If pure luciferin is not available, fluorescence excitation and emission spectra can be corrected for lamp output and phototube sensitivity and compared to pub-

[10] M. DeLuca, L. Brand, T. A. Cebula, H. H. Seliger, and A. F. Makula, *J. Biol. Chem.* **246,** 6702 (1971).

[11] H. H. Seliger and W. D. McElroy, *Arch. Biochem. Biophys.* **88,** 136 (1960).

lished spectra,[9] although this procedure is somewhat more difficult and error prone.

Table I summarizes both the spectral and chromatographic properties of luciferin, its analogs, and synthetic precursors. This information can be useful in the preliminary characterization of these products.

Synthesis of Related Analogs

Dehydroluciferin

Although dehydroluciferin can be easily prepared by the oxidation of alkaline aqueous solutions of luciferin by the use of potassium ferricyanide or molecular oxygen,[3] the subsequent isolation and purification of large quantities is somewhat difficult. Therefore, dehydroluciferin can be most readily prepared by direct thiazole synthesis.[3] 2-Cyano-6-methoxybenzothiazole (V) is converted to the corresponding thioamide, 2-thiocarboxamido-6-methoxybenzothiazole, by treatment with hydrogen sulfide, pyridine, and triethylamine. This thioamide readily condenses with methyl bromopyruvate to form the dimethyl derivative of dehydroluciferin. Treatment of this derivative with boiling, concentrated hydrobromic acid then yields dehydroluciferin.

Dehydroluciferol

Dehydroluciferol is prepared by reduction of the carboxylic acid group of dehydroluciferin. Owing to the limited solubility of dehydroluciferin in nonpolar solvents, reduction is conveniently accomplished after esterification of the carboxylic acid group and protection of the phenolic hydroxyl group.[6] Esterification also serves to facilitate the reduction and allows the use of less vigorous and more selective reducing agents. Dehydroluciferin is esterified by refluxing in acidified absolute ethanol. The dehydroluciferyl ethyl ester thus formed is allowed to react with dihydropyran (under nitrogen) to give the tetrahydropyranyl ether derivative, which is subsequently reduced with $LiAlH_4$ or $NaAlH_2(OCH_2CH_2OCH_3)_2$ in benzene. Hydrolysis is then achieved rapidly under mild conditions (e.g., 0.1 *N* HCl, 5 min) to give dehydroluciferol.

Luciferyl and Dehydroluciferyl Adenylate

The synthesis of the adenylates of luciferin and dehydroluciferin is accomplished by condensation in aqueous pyridine in the presence of dicyclohexyl carbodiimide according to the method of Morton *et al.*[9] with

TABLE I
SPECTRAL AND CHROMATOGRAPHIC PROPERTIES OF LUCIFERIN AND ANALOGS

Compound		Absorption			Relative migration[a] (R_f)		
		Lowest energy band (nm)	Molar absorptivity (liter/mol cm)	Fluorescence λ max (nm)	Silica gel, solvent A	Silica gel, solvent B	Paper, solvent C
Luciferin	acid	327	1.82×10^4	537	0.00	0.31	0.55
	base	385	1.82×10^4	537			
Dehydroluciferin	acid	348	2.06×10^4	555	0.00	0.34	0.33
	base	396	2.06×10^4	555			
Dehydroluciferol	acid	353	2.06×10^4	555	0.05	0.60	—
	base	398	2.06×10^4	555			
(III)		350	1.00×10^4	—	0.45	0.64	—
(IV)		315	1.58×10^4	—	0.33	0.60	—
(V)		320	1.74×10^4	—	0.72	0.68	—
(VI)		323	1.55×10^4	—	0.42	0.68	—
Luciferyl adenylate	acid	336	1.51×10^4	548	—	—	—
	base	(unstable)	—	—	—	—	—
Dehydroluciferyl adenylate	acid	354	—	—	—	—	—
	base	407	—	—	—	—	—

[a] Chromatographic solvent systems used: solvent A—5:1 chloroform/ethyl acetate; solvent B—10:1 ethyl acetate/methanol; Solvent C—7:3 95% ethanol/1 *M* ammonium acetate, pH 7.5.

minor modifications. Luciferin (10 mg) and adenylic acid (30 mg) are dissolved in 1 ml of dry pyridine containing 0.15 ml of 0.5 *N* HCl. Dicyclohexylcarbodiimide (400 mg) in 1 ml of dry pyridine is added, and the reaction is allowed to proceed for 90 min at 4°. The product is then precipitated by the addition of 6 volumes of cold acetone (−20°), filtered, and washed with cold acetone until all traces of pyridine are removed. The precipitate is then dissolved in 10 m*M* sodium acetate containing 40 m*M* sodium chloride, pH 4.5, and the solution is placed on a 1 cm × 30 cm Sephadex G-25 (fine) column after removal of the undissolved residue. Elution with the same buffer gives rise to adenylic acid, followed by

TABLE II
STRUCTURAL ANALOGS OF LUCIFERIN

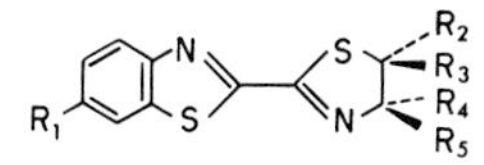

Compound	R1	R2	R3	R4	R5	References[b]
L-(+)-Luciferin	OH	H	H	H	COOH	3
Luciferyl adenylate	OH	H	H	CO-AMP[a]	H	12
6′-Aminoluciferin	NH_2	H	H	COOH	H	7
6′-Methoxyluciferin	OCH_3	H	H	COOH	H	7
Decarboxyluciferin	OH	H	H	H	H	7
Deshydroxyluciferin	H	H	H	COOH	H	7
5,5-Dimethylluciferin	H	CH_3	CH_3	COOH	H	7
cis-5-Methylluciferin	OH	CH_3	H	COOH	H	13
trans-5-Methylluciferin	OH	H	CH_3	COOH	H	13
5,5-Dimethylluciferyl adenylate	OH	CH_3	CH_3	CO-AMP[a]	H	14
5,5-Dimethyloxyluciferin	OH	CH_3	CH_3	O (as ketone)	—	15

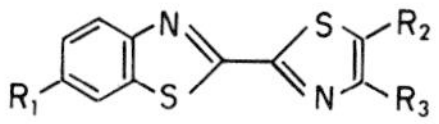

Compound	R1	R2	R3	References[b]
Oxyluciferin	OH	H	OH	15
Oxyluciferin diacetate	CH_3CO	H	CH_3CO	15
Dehydroluciferyl adenylate	OH	H	CO-AMP[a]	12

[a] CO-AMP, mixed anhydride formed from the carboxylic acid group of luciferin (or derivative) and the phosphoric acid group of adenosine 5′-phosphate.

[b] Numbers refer to text footnotes.

luciferyl adenylate, and finally free luciferin in the eluates. Owing to the small amounts of product in the eluates, luciferyl adenylate and luciferin can be most conveniently assayed utilizing the bioluminescent reaction with luciferase or luciferase plus ATP-Mg, respectively. Because of the instability of luciferyl adenylate (50% hydrolysis in 24 hr), it should be used as soon as possible. Dehydroluciferyl adenylate can be synthesized and purified in an analogous manner.

Other Analogs

Many additional analogs have been synthesized and have proved to be very useful in probing the molecular mechanism of firefly bioluminescence. A summary of the most useful of these is given in Table II[3,7,12–15] along with references that include descriptions of their synthesis. Of necessity, certain luciferyl analogs have been omitted along with a number of benzothiazole derivatives. However, descriptions of their preparation and properties can generally be found among the references listed, notably references cited in footnotes 5, 7, and 12.

[12] W. C. Rhodes and W. D. McElroy, *J. Biol. Chem.* **233,** 1528 (1958).
[13] E. H. White, E. Rapaport, T. A. Hopkins, and H. H. Seliger, *J. Am. Chem. Soc.* **91,** 2178 (1969).
[14] T. A. Hopkins, H. H. Seliger, E. H. White, and M. W. Cass, *J. Am. Chem. Soc.* **89,** 7148 (1967).
[15] N. Suzuki, M. Sato, K. Nishikawa, and T. Goto, *Tetrahedron Lett.* **53,** 4683 (1969).

[3] Preparation of Partially Purified Firefly Luciferase Suitable for Coupled Assays

By HANS N. RASMUSSEN

The purification of extracts of firefly tails for use in the analysis of ATP requires methods that can be applied to rather small amounts of material. The fractionation requirements are, however, well defined: (1) removal of ATP and other substances of low molecular weight to lower the blank values, and (2) removal of interfering enzymes to decrease the systematic errors of the analysis.

The substances of low molecular weight are easily removed by column gel filtration on low-porosity Sephadex gels.[1] Luciferin and dehydro-

[1] R. Nielsen and H. N. Rasmussen, *Acta Chem. Scand.* **22,** 1757 (1968).

ISBN 0-12-181957-4

luciferin are, furthermore, strongly adsorbed on this material from neutral buffers, and they may be separated from each other and from other extract components. The two compounds are completely desorbed in distilled water.

The ATP analysis has numerous systematic errors. A number of these are caused by the presence of some kinases, inorganic pyrophosphatase, and other enzymes in the firefly extract. The kinases, in particular nucleosidediphosphate kinase (EC 2.7.4.6) and adenylate kinase (EC 2.7.4.3), limit the specificity of the analysis by catalyzing ATP formation from other nucleotides. The inorganic pyrophosphatase (EC 3.6.1.1) activity of the firefly extract causes a rapid decline of the rate of the light reaction. The decline of the flash is furthermore influenced by enzymes that catalyze removal of ATP from the reaction mixture. The rate of decline is affected by several factors. Of most significant analytical influence is probably the inhibitory action of even low concentrations of ADP, which causes inhibition of the inorganic pyrophosphatase.[2] The enzyme-mediated interferences become serious analytical limitations if the ATP analysis is used in coupled assays where the ADP:ATP ratio may be very high, or if the light intensity is not measured immediately after the mixing of sample and firefly extract.

Luciferase may be purified from small amounts of firefly extract by column chromatography, either by gel filtration on high-porosity Sephadex gels[1] or by ion-exchange chromatography. Bény and Dolivo[3] introduced the use of DEAE-Sephadex for this chromatography. The method described below, single-solvent chromatography on DEAE-Sepharose, is a modification of this method.

The present procedure for purification of the light-emitting system consists of gel filtration for removal of compounds of low molecular weight, followed by purification of luciferase by ion-exchange chromatography. The procedure is preferable for routine preparations because a reasonably good purification is obtained in minimal time and under rather uncritical conditions.

Assay Methods

In the present purification method, the column fractionations are followed sufficiently well by visual observation of fluorescence and by assay of luciferase. All important fractions from the gel filtration exhibit fluorescence in ultraviolet light, and only luciferase need be localized in the

[2] H. N. Rasmussen and R. Nielsen, *Acta Chem. Scand.* **22,** 1745 (1968).

[3] M. Bény and M. Dolivo, *FEBS Lett.* **70,** 167 (1976).

fractions from the DEAE-Sepharose chromatography. The other enzymes of interest are well separated from the luciferase, and assays of these are not needed for routine purposes. They may be assayed by the methods described in this series, or, in the cases of nucleosidediphosphate kinase and adenylate kinase, by measuring, with the luciferase method, the rate of formation of ATP from the proper nucleotides.[1]

Luciferase is assayed according to the principle used by McElroy,[4] namely, by measuring the light emission of the enzyme sample mixed with buffer containing magnesium ions, luciferin, and ATP. The light emission decreases after the mixing, hence it is necessary to measure either at the moment of mixing or at some fixed time after the mixing. The assay conditions are not critical, but they depend on the equipment used for the light measurement. A suitable reaction mixture is composed of a 25-μl sample to which is added 850 μl of buffer containing luciferin from 0.5 mg of firefly tails and 1 nmol of ATP. The buffer consists of glycine (50 m*M*), Na_2HAsO_4 (10 m*M*), Na_2EDTA (1 m*M*), and $MgSO_4$ (5 m*M*), pH 7.7. The enzyme may be detected visually in the dark by adding, to 50 μl of sample, 100 μl of the same buffer containing luciferin from 0.5 mg of firefly tails and 100 nmol of ATP.

The recovery of light-emitting system in the purification procedure is measured on the basis of the initial light intensity (the flash height) obtained when enzyme preparation and luciferin are mixed with ATP. The activities of nucleosidediphosphate kinase and adenylate kinase are measured by the initial rates of increase of light emission due to GTP + ADP, and ADP, respectively. The enzyme preparations are diluted with the glycine–arsenate buffer described above to a concentration corresponding to 1 mg of firefly tails per milliliter. This rather arbitrary concentration has proved to be useful in the instrument used for the light measurement. With this instrument,[2] the enzyme–luciferin solution (850 μl) is injected pneumatically into the cuvette containing the test solution (max. 200 μl). The optical geometry is very favorable, and the photomultiplier is operated at a nominal sensitivity of 5 amp/lumen.

Purification Procedure

The following procedure is suited for purification of extract from 200 mg of firefly tails, sufficient for 200 analyses. It requires 1 day's work, followed by chromatography overnight. All operations are preferably carried out in a cold room.

Extraction. Desiccated firefly tails are ground dry in a cold porcelain

[4] M. DeLuca and W. D. McElroy, this volume [1].

mortar with a little acid-washed sand. Extraction buffer (20 ml per gram of tails) is added gradually with continued grinding. This buffer is 100 m*M* glycylglycine with 1 m*M* Na_2EDTA and 10 m*M* $MgSO_4$, pH 7.7. The activity of the firefly tails and the efficiency of the grinding may be judged from the light emitted by the suspension.

The suspension is centrifuged, e.g., 15,000 rpm for 30 min, and the clear supernatant is isolated with minimal contamination from the upper lipid layer. One gram of firefly tails is usually extracted in a single operation. The extract is stored frozen. The luciferase is very stable under these conditions, but refreezing of thawed extracts should be avoided.

Gel Filtration. Sephadex G-25, fine (Pharmacia, Uppsala, Sweden), is swollen for at least 3 hr at room temperature in the appropriate buffer. The gel is packed by sedimentation in a column 1.5 cm in diameter and ca. 30 cm long. A filter paper disk is placed on the gel surface. Any neutral buffer may be used for the fractionation; if the gel filtration is to be followed by ion-exchange chromatography, the Tris–EDTA buffer is used (see below); if the gel-filtered enzyme is to be used directly for analysis, a glycine-arsenate buffer is preferable. This buffer (50 m*M* glycine, 10 m*M* Na_2HAsO_4, 1 m*M* Na_2EDTA, pH 7.7) is advantageous because it is very resistant to microbial contamination, which otherwise may increase the blank value of the ATP analysis.

The firefly extract (max. 4 ml) is centrifuged, e.g., 8000 rpm in 15 min, and applied on the column, which then is eluted with buffer at a rate of about 15 ml/hr and with fraction volumes of 1–1.5 ml. The constancy of the flow rate is not critical. During the elution, four fluorescent bands are formed on the column (Fig. 1). These are, from the left: a faint, yellow band (luciferase), a blue band (unknown compound, eluted behind ATP), a green-yellow band (luciferin), and a yellow band (dehydroluciferin). The proteins are eluted from the column after about 2 hr in a volume of about 8 ml. The luciferase is easily detectable in the fractions by the yellow fluorescence. When the luciferase is eluted from the column, the flow rate is increased to about 50 ml/hr until the blue fluorescent compound has left the column. The elution is then continued with distilled water as eluent at a rate of about 25 ml/hr and with fraction volumes of about 3 ml. Luciferin and dehydroluciferin are eluted together in a total volume of about 12 ml.

The column is reequilibrated by elution with at least two bed volumes of buffer.

Ion-Exchange Chromatography. DEAE-Sepharose CL-6B (Pharmacia) is transferred to the sulfate form, packed, and equilibrated by a procedure based on the recommendations given by Himmelhoch.[5] About

[5] S. R. Himmelhoch, this series, Vol. 22 [26].

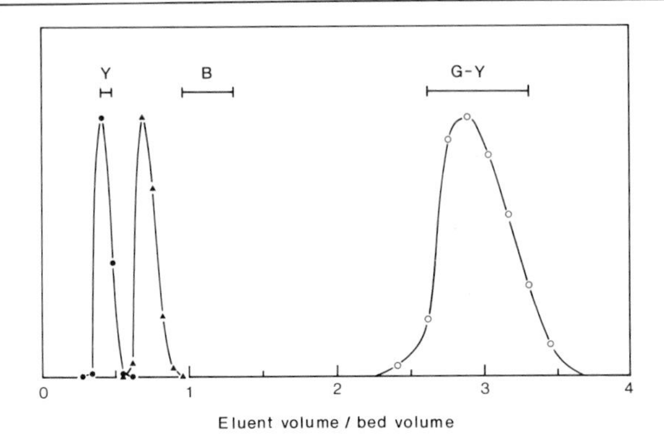

FIG. 1. Fractionation of crude enzyme extract on Sephadex G-25, fine. Column diameter, 1.6 cm; height, 36 cm. The column was eluted with glycine–arsenate buffer at a rate of 10 ml/hr. Fraction volume was 5 ml. The fluorescent fractions are indicated by the bars (Y, yellow; B, blue; G-Y, green-yellow). The peaks are normalized: ●, luciferase; ▲, ATP; ○, luciferin. Dehydroluciferin is not shown in the figure; it was eluted with maximum at 5.0 bed volumes.

18 ml of the DEAE-Sepharose suspension supplied are filtered on a glass filter funnel, and the ion-exchanger is washed with distilled water. It is suspended in a solution of 100 m*M* Tris, 1 m*M* Na_2EDTA, and 500 m*M* $MgSO_4$, the pH of which is adjusted to 7.7 with 2 *M* sulfuric acid. The pH of the suspension is controlled and, if necessary, readjusted with 2 *M* Tris. The suspension is filtered and the ion-exchanger is washed with distilled water before it is transferred to Tris-EDTA buffer (see below). After control and readjustment of pH, the material is packed by sedimentation in a column 0.9 cm in diameter and ca. 18 cm long. This column is eluted with Tris-EDTA buffer at a rate of about 50 ml/hr until the effluent has reached the pH value of the influent (3–5 bed volumes).

The Tris-EDTA buffer is prepared by adjusting the pH of 35 m*M* Tris and 1 m*M* Na_2EDTA to 7.70 with 2 *M* sulfuric acid. Prior to neutralization, the solution is equilibrated to the temperature of the cold room used for the fractionation. The pH meter is calibrated with standard buffer of the same temperature.

The gel-filtered luciferase fractions are applied to the column, which is eluted with Tris–EDTA buffer at a rate of about 10 ml/hr and with fraction volumes of about 2 ml. The fractions containing more than about 25% of the peak luciferase activity are pooled, supplemented with purified luciferin, and stored frozen. For analysis, this purified luciferase–luciferin preparation is diluted with glycine–arsenate buffer (50 m*M* glycine, 10 m*M* Na_2HAsO_4, 1 m*M* Na_2EDTA, pH 7.7) to a concentration corresponding to

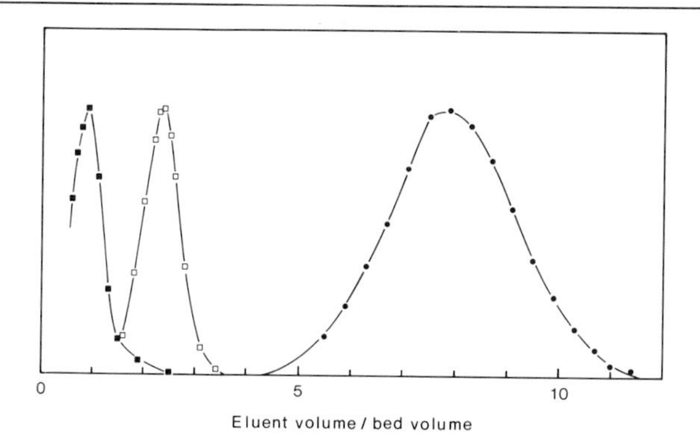

FIG. 2. Fractionation of firefly enzymes on DEAE-Sepharose CL-6B. Column diameter, 0.9 cm; height, 17 cm. The protein fractions (6.8 ml) from the Sephadex G-25 gel filtration were applied on the column, which was eluted with Tris–EDTA buffer, pH 7.70, at a rate of 8 ml/hr. Fraction volume was 1 ml. The peaks are normalized: ■, nucleoside diphosphate kinase; □, adenylate kinase; ●, luciferase.

1 mg of firefly tails per milliliter, and magnesium sulfate is added to a final concentration of 5 m*M*.

The ion-exchange material is most conveniently regenerated by the procedure described above.

Comments on the Purification Procedure

Table I shows data from a typical purification. Based on the initial light intensity per mol of ATP, the recovery of light-emitting system is about 50%, and the specific activity is increased 6–7 times. The blank values of the two purified preparations listed are equivalent to about 10^{-14} mol of ATP.

The luciferase is not saturated with luciferin under the conditions of analysis. Doubling of the luciferin concentration increases the initial light intensity ca. 30%. Dehydroluciferin present in the crude enzyme extract and in the purified luciferin inhibits the enzyme about 25%. Luciferin may be separated from dehydroluciferin, simply by continuing the gel filtration without water elution (cf. Fig. 1). The disadvantages of this fractionation, compared with that described, are that a longer fractionation time is required and less-concentrated fractions are obtained.

The flash curves are of different shapes for the preparations listed. The preparation purified by ion exchange showed very slow decline of the light signal. This is in accordance with the assumption that the major part of the decline is caused by inorganic pyrophosphatase, which is not detected in

TABLE I

PURIFICATION OF FIREFLY LIGHT-EMITTING SYSTEM

Preparation	Total volume (ml)	Total protein[a] (mg)	Signal[b] with 10^{-11} mol of ATP: Flash height: Amperes/mol	Flash height: Amperes/(mol × μg protein)	Signal in % of flash height after: 5 Sec	15 Sec	ATP formation[b] (mol/sec) from 3×10^{-9} mol: ADP	ADP + GTP
Crude extract	4.0	17.7	580	6.6	64	48	2.8×10^{-12}	2.4×10^{-10}
Sephadex G-25 fractions	7.5	13.7	467	6.8	67	58	1.7×10^{-12}	2.4×10^{-10}
DEAE fractions	47	1.3	284	44	98	95	1.1×10^{-14}	3.4×10^{-12}

[a] Determined after W. Schaffner and C. Weissmann, *Anal. Biochem.* **56,** 502 (1973).

[b] See the section Assay Methods.

the fractions from the ion-exchange chromatography. With crude enzyme extract and Sephadex G-25 purified enzyme, the shapes of the flash curves furthermore depend on the amount of ATP analyzed.[2] This is not the case with the preparation purified by ion exchange, which shows a different flash only at the highest amount of ATP analyzed (10^{-9} mol).

Significant amounts of nucleosidediphosphate kinase and adenylate kinase activity, 1.4% and 0.6%, respectively, appears in the luciferase fractions, although the enzymes are well separated on the DEAE-Sepharose column (Fig. 2). Variation of a number of conditions has not improved the purification. The phenomenon may be explained as a result of strong, atypical binding of a part of the two enzymes to the ion-exchange material. The chromatography has also been carried out by starting at lower ionic strength or higher pH; under these conditions the two kinases are eluted almost as shown in Fig. 2. The luciferase is then eluted either by a stepwise increase in ionic strength or by a linear gradient. In both instances, several percent of the kinase activities are eluted in a peak closely associated with the luciferase peak. The effect is also observed with DEAE-Sephadex A-50.

Alternative and Related Techniques

Luciferase may be purified in small amounts by ammonium sulfate fractionation at pH 8. The major part of the enzyme precipitates between 57% and 65% saturation.[6] The degree of precipitation is, however, somewhat influenced by the protein concentration, and the actual conditions should be found by increasing the ammonium sulfate concentration in small increments. The technique is favorably combined with the Sephadex G-25 fractionation described above. A 4-fold decrease in adenylate kinase activity with little loss of luciferase activity may be obtained. This gives an enzyme preparation sufficiently pure for some coupled assays.

The fractionation of the firefly light-emitting system on various Sephadex gels is described by Nielsen and Rasmussen.[1] A purification of luciferase, almost as good as that described above, was obtained after 2 days' chromatography. Momsen[7] crystallized the enzyme by dialyzing the fractions from this gel filtration against 5 m*M* Na_2HAsO_4 and 1 m*M* Na_2EDTA, pH 6.3, at 4° under nitrogen and in the dark. A preparation with very low adenylate kinase activity was obtained.

The adsorption of luciferin and dehydroluciferin to low-porosity Sephadex gels is probably unique for this material. The data of Schram *et*

[6] W. D. McElroy and J. Coulombre, *J. Cell. Comp. Physiol.* **39,** 475 (1952).
[7] G. Momsen, *Anal. Biochem.* **82,** 493 (1977).

al.[8] show that luciferin is not adsorbed to BioGel P-10. Hence it is not separated from ATP on this material. The adsorption to Sephadex G-10 is greatly increased at pH 3.[9] This has been used for concentrating and purifying luciferin from analytical reaction mixtures containing extract of 1 g of firefly tails per liter. Only little of the luciferin is used up in the analytical reaction, and the reaction mixtures may be stored frozen for years. The method involves adjustment to pH 3.0 with formic acid, filtration, application on a Sephadex G-10 column (2.5 × 20 cm for 1 liter), and elution of luciferin with neutral buffer. The eluate is applied on a smaller G-10 column after adjustment to pH 3.0, and the luciferin is eluted with distilled water, pH of which is adjusted to 9 with NaOH.

The ion-exchange chromatographic method of Bény and Dolivo[3] requires 3 days' chromatography. In the publication, the purification is evaluated in a manner different from the present one, but it is probably of the same order of magnitude.

The stepwise elution procedure mentioned in the section Comments on the Purification Procedure gives the luciferase in a very small volume. It may be advantageous if a less efficient purification, in particular with respect to nucleosidediphosphate kinase, can be tolerated. The conditions of chromatography are similar to those described, except that the Tris-EDTA buffer is adjusted to pH 8.00. After about 10 bed volumes, elution is continued with 100 m*M* Tris, 1 m*M* Na_2EDTA, and 20 m*M* $MgSO_4$, pH adjusted to 8.0. The luciferase is eluted in a volume of about 7 ml, approximately one bed volume after the shift of influent.

[8] E. Schram, R. Cortenbosch, E. Gerlo, and H. Roosens, *in* "Organic Scintillators and Liquid Scintillation Counting" (D. L. Horrocks and C.-T. Peng, eds.), p. 127. Academic Press, New York, 1971.

[9] G. Momsen, personal communication, 1976.

[4] Firefly Luciferase Assay for ATP Production by Mitochondria

By JOHN J. LEMASTERS and CHARLES R. HACKENBROCK

By virtue of the great sensitivity of the photomultiplier tube to light, the firefly luciferase assay has long been employed when extremely small quantities of adenosine triphosphate (ATP) need to be measured.[1] In this chapter we describe an extension of the luminescence assay by which ATP can be measured not only sensitively and quantitatively, but continuously

[1] B. L. Strehler, *Methods Biochem. Anal.* **16,** 99 (1968).

ISBN 0-12-181957-4

as well, in suspensions of metabolically active mitochondrial and submitochondrial membranes.[2,3]

Principle

The luminescence assay for ATP is based on the following luciferase-catalyzed reactions:[4]

$$E + LH_2 + ATP \overset{Mg^{2+}}{\longleftrightarrow} E \cdot LH_2AMP + PP_i \quad (1)$$

$$E \cdot LH_2AMP + O_2 \rightarrow E + CO_2 + AMP + \text{oxyluciferin} + \text{light} \quad (2)$$

The initial activation step, Reaction (1), is the formation of enzyme-bound luciferyl adenylate ($E \cdot LH_2AMP$) and pyrophosphate (PP_i) from luciferin (LH_2) and ATP. Reaction (1) is reversible and requires a divalent cation such as Mg^{2+}. The luciferyl adenylate–enzyme complex then reacts irreversibly with molecular oxygen [Reaction (2)] to produce a quantum of light, AMP, CO_2, and oxyluciferin, the decarboxyketo derivative of luciferin. Luminescence is inhibited noncompetitively by a product of Reaction (2), presumably by oxyluciferin.[5] This product inhibition is the major factor to be contended with in employing firefly luciferase luminescence as a continuous measure of ATP concentration.

Materials

Luciferase and Luciferin

Firefly luciferase and synthetic luciferin of high purity and specific activity are available commercially (DuPont Corp., Instrument Division, Wilmington, Delaware) in premixed form as vials of lyophilized powder. The contents of one vial are dissolved in 3 ml of 0.1 *M* sodium phosphate or potassium phosphate buffer, pH 7.4, to yield a solution containing approximately 710 μM luciferin and 1000 units of luciferase per milliliter. A more precise determination of luciferin concentration can be made spectrophotometrically at 327 nm using a molar extinction coefficient of 18,000.[6] One unit of enzyme as furnished by DuPont equals 1.6–1.9 μg of protein. The enzyme solution is stable for several days if stored at 0° to 4° and shielded from light. Crude aqueous extracts of luciferase and luciferin

[2] J. J. Lemasters and C. R. Hackenbrock, *Biochem. Biophys. Res. Commun.* **55,** 1262 (1973).
[3] J. J. Lemasters and C. R. Hackenbrock, *Eur. J. Biochem.* **67,** 1 (1976).
[4] M. DeLuca, *Adv. Enzymol.* **44,** 37 (1976).
[5] J. J. Lemasters and C. R. Hackenbrock, *Biochemistry* **16,** 445 (1977).
[6] R. A. Morton, T. A. Hopkins, and H. H. Seliger, *Biochemistry* **8,** 1598 (1969).

may be prepared by the homogenization of commercially available desiccated firefly tails. However, such extracts are generally not satisfactory for use with mitochondria, since they contain extraneous enzyme activities, such as adenylate kinase, pyrophosphatase, and nucleoside diphosphokinase.[1] Moreover, an unidentified factor in these extracts decreases the coupling of mitochondrial oxidative phosphorylation and reduces ATP:O ratios and respiratory control ratios by as much as 50%. Commercially available extracts of firefly tails in arsenate buffer are also to be avoided, since arsenate is an uncoupler of substrate level phosphorylation in mitochondria.

Adenine Nucleotides

Aqueous stock solutions of AMP, ADP, and ATP are made in 0.1 *M* and adjusted to pH 7.4 with 1 *N* NaOH or 1 *N* HCl. Their concentrations are quantitated spectrophotometrically at 259 nm using a molar extinction coefficient of 15,400. These stock solutions may be stored frozen for several months with little deterioration. On the day of the experiment, working solutions of 10 m*M* and 1 m*M* are prepared as needed.

Mitochondrial Incubation Medium

For routine work a simple incubation medium containing 150 m*M* sucrose, 5 m*M* $MgCl_2$, 10 m*M* KP_i buffer, pH 7.4, 5 m*M* sodium succinate, and 5 μM rotenone will support mitochondrial respiration and oxidative phosphorylation.

Mitochondrial Fractions

Rat liver mitochondria are prepared by homogenization and differential centrifugation in 0.25 *M* sucrose to a stock concentration of 50–100 mg of protein per milliliter.[7] Inner membrane vesicles are prepared by sonication of the mitoplast (inner membrane-matrix) fraction[8] to a stock concentration of 20–50 mg of protein per milliliter.[2] Techniques for isolation and fractionation of mitochondria are discussed elsewhere.[9]

Instrumentation

Photometers from a variety of spectrophotometers and fluorometers can be adapted to the measurement of luminescence. We employ the

[7] W. C. Schneider, *J. Biol. Chem.* **176,** 259 (1948).
[8] C. Schnaitman and J. W. Greenawalt, *J. Cell Biol.* **38,** 158 (1968).
[9] This series, Volumes 10 and 31.

photometer of a Brice-Phoenix light-scattering photometer (VirTis Co., Gardiner, New York). Sensitivity can often be greatly increased by removing any barriers that shield the phototube from scattered light and by moving the sample chamber as close as possible to the phototube. It is essential for accurate quantitation of the luminescence signal during a continuous recording that the reaction chamber be constantly and uniformly stirred by a magnetic bar or other device. Luminescence can be measured from an ordinary 1-cm light path cuvette or, as is our usual practice, from a glass, water-jacketed cell (Gilson Medical Electronics, Model OX-705, Middleton, Wisconsin) fitted with a Clark electrode (Yellow Springs Instrument Co., Yellow Springs, Ohio) for polarographic measurement of oxygen.[10] Light-tight additions to the reaction chamber are made with microliter syringes through a rubber stopper fitted into the lid of the photometer compartment. Rapid mixing is carried out in an Aminco-Morrow stopped-flow apparatus (American Instrument Co., Silver Springs, Maryland); in which case luminescence is detected by the photometer of an Aminco-Chance spectrophotometer.

Procedure

Theory

Quantitation of ATP concentration during a continuous luminescence recording requires an understanding of luciferase reaction kinetics, the essential features of which are illustrated in Fig. 1. Here, an aliquot of ATP, [S], is added to a solution of firefly luciferase and luciferin. The result is a rapid development of luminescence whose intensity maximum, or flash height, is proportional to ATP concentration in accordance with the Michaelis–Menten equation. The intensity maximum lasts only a few seconds and is followed by an exponential decay that may persist for several hours. This decay is due to product inhibition. It is not due to consumption of ATP or luciferin by the luminescence reaction, since the absolute rate of the reaction (which is proportional to the intensity of luminescence) is negligible compared to the concentrations of the reactants. Luminescence remains sensitive to ATP concentration. When a second aliquot of ATP, [X], is added, luminescence abruptly increases and then resumes a decay pattern. Since product inhibition of luminescence is noncompetitive with respect to ATP,[5] the following equation applies:

$$\frac{V_{\mathrm{s}}}{V_{\mathrm{s+x}}} = \frac{1}{[\mathrm{S}] + [\mathrm{X}]} \left(\frac{\mathrm{K_s}\,[\mathrm{S}]}{\mathrm{K_s} + [\mathrm{S}]} \right) + \frac{[\mathrm{S}]}{\mathrm{K_s} + [\mathrm{S}]} \qquad (3)$$

[10] R. W. Estabrook, this series, Vol. 10, p. 41.

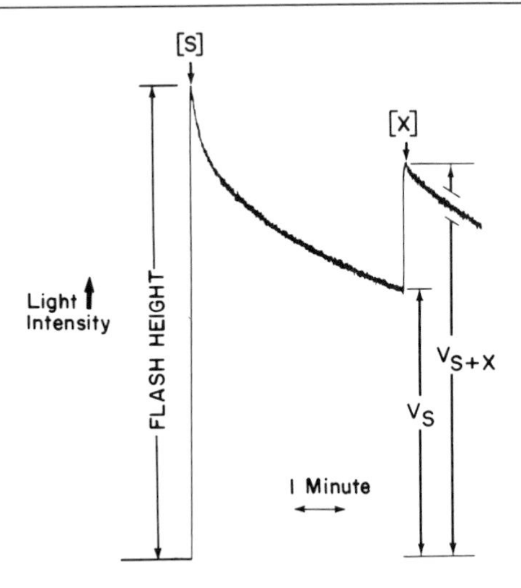

FIG. 1. Luminescence response of luciferase enzyme system to ATP. ATP, 239 μM, is added at [S] and again 4 min later at [X]. Reaction medium is 10 units of luciferase per milliliter, 7.1 μM luciferin, 5 mM $MgCl_2$, 70 mM sucrose, 220 mM mannitol, 2 mM HEPES, 7.5 mM K_2HPO_4, 5 mM sodium succinate, 0.5 mM EDTA, 6.25 μM rotenone, 1 mg of bovine serum albumin per milliliter, pH 7.4, 23°. From J. J. Lemasters and C. R. Hackenbrock, *Biochem. Biophys. Res. Commun.* **55,** 1262 (1973).

where V_s and V_{s+x} represent, respectively, the intensity of luminescence just before and just after the addition of [X]; [S] and [X] are the concentrations of the respective aliquots of ATP; and K_s is the Michaelis constant.

Estimation of K_s

Before Eq. (3) can be applied, K_s must first be determined. Although flash-height data can be employed to estimate K_s, it is much easier to make the determination following successive additions of ATP to the luciferase reaction medium, as shown in Fig. 1. Equation (3) is rearranged and solved for K_s:

$$K_s = \frac{[S]\,(1 - V_s/V_{s+x})}{V_s/V_{s+x} - [S]/([S] + [X])} \qquad (4)$$

With this expression it is then possible to calculate K_s in a single experiment from values of V_s and V_{s+x} and the concentrations of S and X ([S] and [X]). For greatest precision [S] and [X] should equal the expected value of K_s (200–300 μM). Since K_s is sensitive to a number of variables, its measurement should take place in a complete medium, including mitochondria. The omission of mitochondria alone may significantly alter the apparent value of K_s. When using loosely coupled mitochondrial membrane fractions, it is prudent to add oligomycin to the reaction medium in order to prevent ATP hydrolysis during the determination of K_s.

When K_s is determined, V_{max}, the maximum velocity of the reaction, may also be calculated.

$$V_{max} = V_o (K_s/[S] + 1) \quad (5)$$

where V_0 is the flash height following the first aliquot of ATP, [S].

Oxidative Phosphorylation

Figure 2 illustrates the response of luciferase to ATP synthesis during oxidative phosphorylation. When mitochondria are first added to the luciferase-containing medium, there is a slow rate of respiration and a small, but significant increase in luminescence. After 2 min, active oxidative phosphorylation is initiated by addition of a mixture of AMP and ADP.

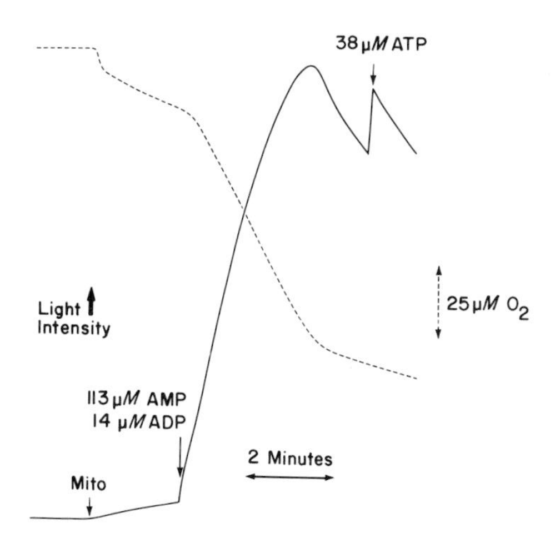

FIG. 2. Oxidative phosphorylation of AMP and ADP by intact mitochondria. The solid line represents luminescence, and the dotted line shows oxygen concentration. Reaction medium is 155 mM sucrose, 5 mM $MgCl_2$, 5 mM sodium succinate, 5 μM rotenone, 10 units of luciferase per milliliter, 7.1 μM luciferin, 1 mg of protein per milliliter of mitochondria (Mito), 11 mM KH_2PO_4–K_2HPO_4 buffer, pH 7.4, 23°. AMP, ADP, and ATP are added in the indicated amounts. From J. J. Lemasters and C. R. Hackenbrock, *Biochem. Biophys. Res. Commun.* **55,** 1262 (1973).

ATP synthesis is immediately demonstrated by a steadily rising luminescence signal and is accompanied by an accelerated rate of respiration. After another few minutes, luminescence begins to decline and respiration returns to its original slow rate. This indicates that all the added adenine nucleotide has been consumed by conversion to ATP.

In order to calculate ATP concentration at the completion of active oxidative phosphorylation, ATP standard is added. The calculation is made by rearranging Eq. (3) and solving for [S]:

$$[\mathrm{S}] = \frac{-(K_\mathrm{s} + [\mathrm{X}])}{2} + \sqrt{\frac{V_\mathrm{s}/V_{\mathrm{s+x}}\,(K_\mathrm{s} - [\mathrm{X}])^2 - (K_\mathrm{s} + [\mathrm{X}])^2}{4\,(V_\mathrm{s}/V_{\mathrm{s+x}} - 1)}} \tag{6}$$

V_s and V_{s+x} here represent the intensities of luminescence just before and just after the addition of [X], the ATP standard. This calculation applied to the example of Fig. 2 confirms our conclusion that all the added adenine nucleotide has been converted to ATP.

Isolated mitochondria are rich in adenylate kinase, an enzyme that reversibly catalyzes the conversion of two ADP molecules to one molecule each of ATP and AMP. Therefore, in Fig. 2 oxidative phosphorylation is initiated by both AMP and ADP in a ratio calculated to be in adenylate kinase equilibrium with ATP already present. Since the adenylate kinase equilibrium constant is approximately 1,[11] concentrations of AMP and ADP are chosen such that

$$[\mathrm{ADP}]^2/[\mathrm{AMP}] \simeq [\mathrm{ATP}]_{\mathrm{endogenous}} \tag{7}$$

where $[\mathrm{ATP}]_{\mathrm{endogenous}}$ is ATP present at the time the exogenous adenine nucleotide is added. Endogenous ATP concentration must be determined in a parallel experiment and is 1–3 μM under the conditions used in Fig. 2. The purpose of adding ADP and AMP in this fashion is to prevent any increase in ATP concentration from occurring that is the result of adenylate kinase activity alone. The importance of this maneuver is illustrated in Fig. 3, where ADP initiates oxidative phosphorylation. In this case the luminescence signal is markedly biphasic—the initial rapid phase being due to adenylate kinase activity, and the succeeding slower phase being representative of oxidative phosphorylation.

It is often desirable to know the initial rate of ATP generation, $d[\mathrm{S}]/dt$. This can be calculated from the initial rate of luminescence increase, dV/dt. The Michaelis–Menten equation is differentiated to obtain:

$$d[\mathrm{S}]/dt = (dV/dt) \cdot K_\mathrm{s} V_{\max}/(V_{\max} - V)^2 \tag{8}$$

[11] L. Noda, *in* "The Enzymes" (P. D. Boyer, ed.), Vol. VIII, p. 279. Academic Press, New York, 1973.

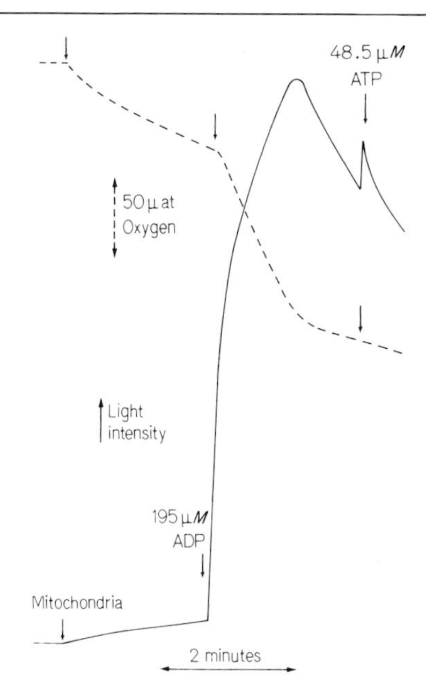

FIG. 3. Oxidative phosphorylation of ADP by intact mitochondria. Reaction medium is as described for Fig. 2. From J. J. Lemasters and C. R. Hackenbrock, *Eur. J. Biochem.* **67,** 1 (1976).

where V is the intensity of luminescence. When $V = 0$, the equation reduces to:

$$d[\mathrm{S}]/dt = (dV/dt) \cdot K_s/V_{\max} \quad (9)$$

This expression presupposes that the luciferase–luciferin medium is unreacted; otherwise product inhibition would reduce the effective value of $V_{\max}$.

Equation (9) may be used to estimate the initial rate of ATP synthesis catalyzed by sonicated inner membrane vesicles (Fig. 4). This membrane fraction lacks adenylate kinase, and ADP alone may be used to initiate oxidative phosphorylation. Also, there is no endogenous ATP formed by the inner membrane vesicles, and thus luciferase is essentially unreacted when ADP is added. From the slope of the luminescence recording, dV/dt, immediately following the ADP addition, the initial rate of ATP synthesis, $d[\mathrm{S}]/dt$, can be calculated with Eq. (9). In this instance d[S]/dt is 1.4 μmol of ATP per second per gram of protein.

If luciferase and luciferin have already commenced reacting, $d[\mathrm{S}]/dt$ may still be determined with Eq. (8) provided that ATP concentration, [S], is known and dV/dt is large in comparison to the product inhibition-

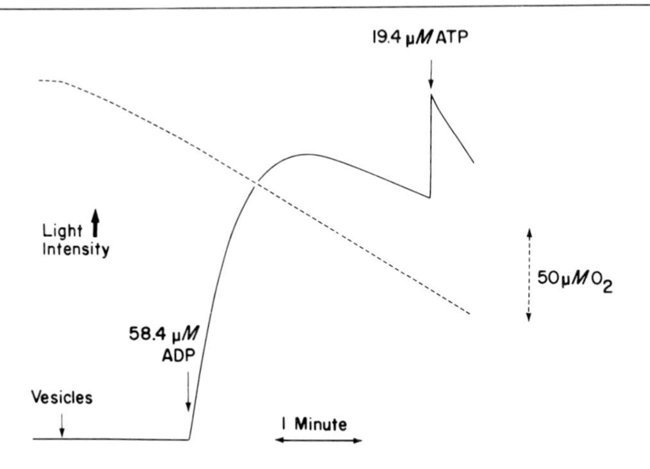

FIG. 4. Oxidative phosphorylation by sonicated inner membrane vesicles. Reaction medium is as described for Fig. 2 with the exception that mitochondria are replaced by vesicles, 0.5 mg of protein per milliliter. From J. J. Lemasters and C. R. Hackenbrock, *Biochem. Biophys. Res. Commun.* **55,** 1262 (1973).

mediated decline in luminescence. V_{max} must be redetermined to allow for product inhibition:

$$V_{max} = \frac{V(K_s + [S])}{[S]} \tag{10}$$

This value for V_{max} is then substituted into Eq. (8). Because of the greater number of variables involved, this calculation [Eqs. (8) and (10)] is subject to a greater degree of experimental error and should be used with appropriate caution.

Equation (3) may be rearranged and solved for one more variable, [X]:

$$[X] = [S]\left(\frac{K_s}{V_s/V_{s+x}\,(K_s + [S]) - [S]} - 1\right) \tag{11}$$

This expression may then be employed when the ATP equivalence of some abrupt change in luminescence is desired. Figure 5 illustrates its application. Mitochondria are added to the basic reaction medium as before. After 2 min of incubation, KCN, a respiratory inhibitor, is added. A few seconds later, after the KCN has taken effect, a mixture of AMP and ADP is added. A suitable control (not shown) demonstrates that the resulting abrupt increase in luminescence is not due to adenylate kinase. Rather it appears to be due to ATP synthesis associated with the discharge of the energized state of the mitochondria. Equation (11) permits quantitation of this abrupt increase in luminescence. V_s and V_{s+x} in Eq. (11) represent, respectively,

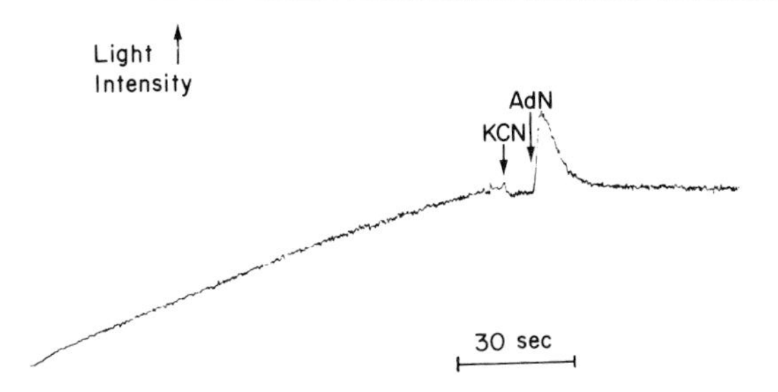

FIG. 5. Adenine nucleotide (AdN)-stimulated ATP production after KCN inhibition of respiration in mitochondria. Reaction medium is as described for Fig. 2. KCN (when added) is 2 m*M*. AdN (when added) is composed of 58 μM AMP and 15 μM ADP. The recording begins with the addition of mitochondria.

the intensities of luminescence just before and just after the abrupt change in question. [S] is ATP concentration before the abrupt change and must be determined in a parallel experiment. In the example shown in Fig. 5 the sharp rise in luminescence is equivalent to 1.4 μmol of ATP per gram of protein.

In most of our experiments we employ luciferase and luciferin at concentrations of 10 units/ml and 7.1 μM, respectively. These concentrations may be varied to suit differing needs. As shown in Fig. 6, large concentrations provide a stronger luminescence signal and greater sensitivity at low ATP concentrations but also cause a faster rate of product inhibition. At lower concentrations the signal loses strength and sensitivity, but product inhibition proceeds more slowly. In general, it is best to employ luciferase and luciferin concentrations that are as low as adequate sensitivity will permit.

Rapid Kinetics

The continuous nature of the luminescence signal lends itself to the examination of the rapid kinetics of ATP formation. However, the time resolution of this application is limited by the rapid kinetic characteristics of the luminescence reaction itself which, as shown in Fig. 7, has three components.[3,12] First is a lag time of 40 ± 5 msec after the rapid mixing of ATP with luciferase and luciferin, during which virtually no light is produced. Second is an increase in luminescence that rises to a maximum, the flash height, with a half-time of 190 ± 10 msec. Last, and not apparent on the time scale of Fig. 7, is a slow decay of luminescence because of product

[12] M. DeLuca and W. D. McElroy, *Biochemistry* **13**, 921 (1974).

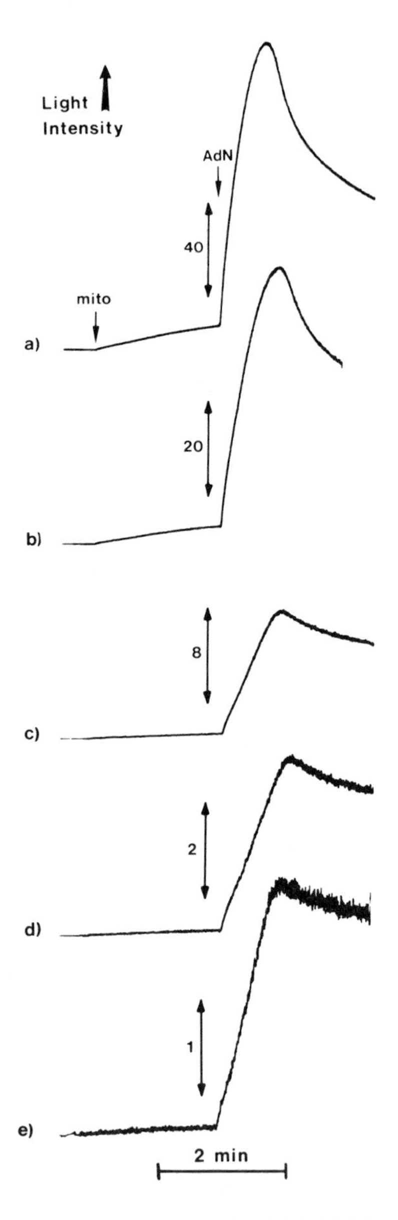

FIG. 6. The luminescence response to mitochondrial ATP formation measured over a 20-fold range of luciferase and luciferin concentration. The vertical scale, which is different for each recording, is indicated in equivalent arbitrary luminescence units. Luciferase and luciferin concentrations are, respectively: (a) 50 units/ml and 35.5 μM; (b) 25 units/ml and 17.8 μM; (c) 10 units/ml and 7.1 μM; (d) 5 units/ml and 3.6 μM; (e) 2.5 units/ml and 1.8 μM. Reaction medium is 100 mM sucrose, 5 mM $MgCl_2$, 5 mM sodium succinate, 1 mg of protein per milliliter of mitochondria, 10 mM KPO_4 buffer, 15 mM Tris buffer, pH 7.4, 23°. Adenine nucleotide (AdN), when added, is composed of 106 μM AMP and 28 μM ADP. From J. J. Lemasters and C. R. Hackenbrock, *Eur. J. Biochem.* **67**, 1 (1976).

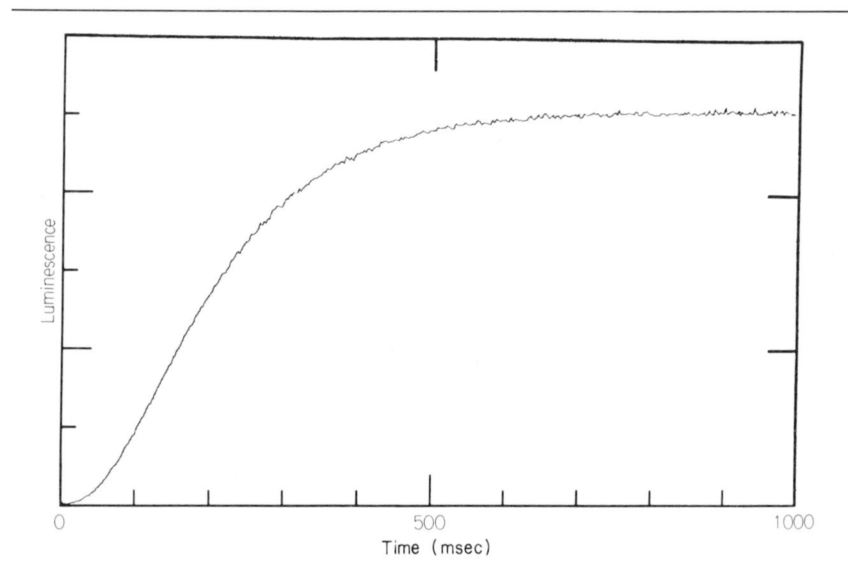

FIG. 7. Rapid-reaction kinetics of luciferase luminescence. ATP is rapidly mixed with luciferase and luciferin in a stopped-flow apparatus. Mixed reaction medium is 5 μM ATP, 125 units of luciferase per milliliter, 89 μM luciferin, 170 mM sucrose, 5 mM $MgSO_4$, 5 mM sodium succinate, 5 μM rotenone, 10 mM KH_2PO_4–K_2HPO_4 buffer, 25 mM Tris-HEPES buffer, pH 7.4, 23°. From J. J. Lemasters and C. R. Hackenbrock, *Eur. J. Biochem.* **67,** 1 (1976).

inhibition. The lag time and half-maximal rise time are essentially constant over a concentration range of 0.5 to 250 μM ATP.

The rapid kinetics of ATP synthesis initiated by a pulse of oxygen to reduced, anaerobic inner membrane vesicles is shown in Fig. 8A. One syringe of a stopped-flow apparatus contains anaerobic vesicles and ADP, and the other syringe contains an oxygenated solution of luciferin and luciferase. The contents of the two syringes are rapidly mixed, and the resulting luminescence signal increases in a linear fashion. A lag time of 40–60 msec precedes the development of any luminescence. Since this lag time approximates that of the luminescence reaction itself, it can be concluded that ATP synthesis must begin within 20 msec or less of mixing. The ensuing linearity of the luminescence signal shows that a maximal rate of ATP synthesis is rapidly achieved and then maintained. Since ATP concentration is low in these experiments (<5 μM), luminescence is linearly proportional to ATP concentration, and product inhibition is negligible over the first few seconds of the reaction.

In Fig. 8B conditions are altered so that an initial burst of ATP synthesis occurs immediately after mixing followed by a slower, steady-state rate.

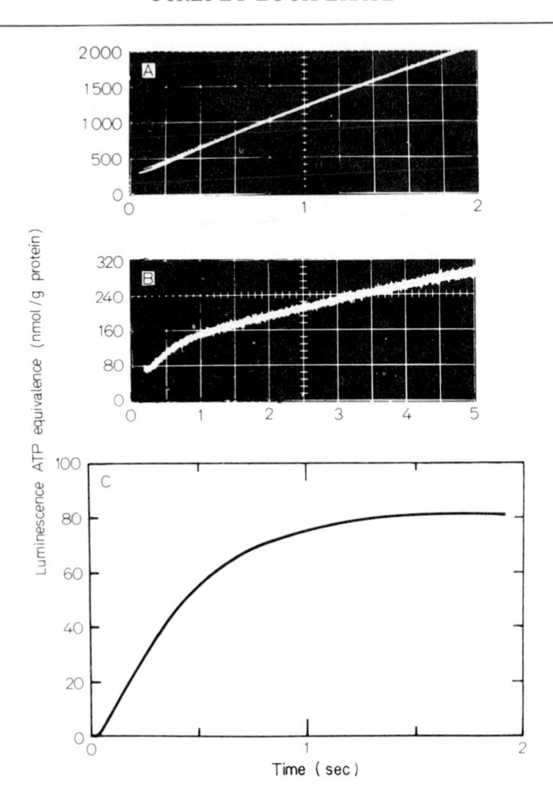

FIG. 8. Rapid kinetics of oxygen-pulsed ATP synthesis by sonicated inner membrane vesicles. Luminescence is measured vertically in units equivalent to 1 nmol of ATP per gram of protein. In (A), oxygen is pulsed to reduced vesicles with succinate as respiratory substrate. The recording is the result of mixing an anaerobic suspension containing 170 m*M* sucrose, 5 m*M* $MgSO_4$, 5 m*M* sodium succinate, 5 μM rotenone, 97.4 μM ADP, 4 mg. of protein per milliliter of vesicles, 10 m*M* KH_2PO_4–K_2HPO_4 buffer, 25 m*M* Tris–HEPES buffer, pH 7.4, with equal parts of an aerobic solution containing 170 m*M* sucrose, 5 m*M* $MgSO_4$, 5 m*M* sodium succinate, 5 μM rotenone, 250 units of luciferase per milliliter, 177.5 μM luciferin, 10 m*M* KH_2PO_4–K_2HPO_4 buffer, 25 m*M* Tris–HEPES buffer, pH 7.4, 23°. In (B), oxygen is pulsed to antimycin-inhibited and tetramethylphenylenediamine-bypassed sonicated inner membrane vesicles. Antimycin blocks site II of the respiratory chain. Tetramethylphenylenediamine is added in sufficient quantity to restore respiration to 20% of its original rate by bypassing the antimycin-inhibited site. The recording is the result of mixing an anaerobic suspension containing 162.5 m*M* sucrose, 5 m*M* $MgSO_4$, 5 m*M* sodium succinate, 5 μM tetramethylphenylenediamine, 0.25 μg of antimycin per milliliter, 97.4 μM ADP, 2.5 mg of protein per milliliter of vesicles, 10 m*M* KH_2PO_4–K_2HPO_4 buffer, 25 m*M* Tris–HEPES buffer, pH 7.4, with equal parts of an aerobic solution containing 162.5 m*M* sucrose, 5 m*M* sodium succinate, 5 m*M* $MgSO_4$, 0.25 μg of antimycin per milliliter, 5 μM tetramethylphenylenediamine, 250 units of luciferase per milliliter, 177.5 μM luciferin, 10 m*M* KH_2PO_4–K_2HPO_4 buffer, 25 m*M* Tris–HEPES buffer, pH 7.4, 23°. (C) A derived plot showing the luminescence produced by the initial burst of ATP synthesis in (B). From J. J. Lemasters and C. R. Hackenbrock, *Eur. J. Biochem.* **67,** 1 (1976).

This pattern of ATP synthesis is clearly evident in the luminescence recording. By subtracting the linear portion from the overall recording, we can generate a derived signal that represents the initial burst of ATP synthesis. The magnitude of the derived signal is equivalent to 0.08 μmol of ATP per gram of protein, and its half-maximal rise time is 300 msec. Since the half rise time of the luminescence reaction itself is 190 msec, the difference between these two values, approximately 100 msec, may be considered an order of magnitude value for the half-time of the initial burst of ATP synthesis.

General Considerations

If an enzyme obeys classical Michaelis–Menten kinetics, then at low substrate concentrations ($<< K_s$) reaction velocity and substrate concentration are linearly related. Luciferase is no exception to this rule even during a continuous reaction, since product inhibition does not alter the apparent K_s of luminescence. Therefore, at concentrations below about 25 μM, Eqs. (6), (8), and (11) need not be used, and ATP concentration may be calculated on the assumption that it is linearly proportional to luminescence. ATP standards must still be employed, however, since the proportionality constant between ATP concentration and luminescence is constantly changing as a consequence of product inhibition.

Several practical considerations deserve mention. Since oxygen is an absolute requirement for luminescence, care must be taken when working with mitochondria to assure that oxygen concentration is maintained at saturating levels with respect to the luciferase ($\geq$ 50 μM O_2). Attention must also be paid to the ionic composition of the reaction medium. Luminescence decreases with increasing ionic strength. Monovalent anions are especially deleterious in this regard.[13] If, for example, 75 mM NaCl were to replace sucrose in Fig. 2, luminescence would then be almost impossible to measure. Some salt should be included in a mitochondrial reaction medium, however, so that active transport by mitochrondria will not significantly change bulk ion concentrations.

Other substances are potent inhibitors of luminescence, including anesthetic agents[14,15] and the fluorescent probes anilinonaphthalene sulfonate (ANS) and toluidinonaphthalene sulfonate (TNS).[16] In addition, uncouplers of oxidative phosphorylation, such as dinitrophenol and carbonyl cyanide *m*-chlorophenyl hydrazone (CCCP), potently inhibit the luciferase reac-

[13] J. L. Denburg and W. D. McElroy, *Arch. Biochem. Biophys.* **141,** 668 (1970).
[14] I. Ueda and H. Kamaya, *Anesthesiology* **38,** 425 (1973).
[15] I. Ueda, H. Kamaya, and H. Eyring, *Proc. Natl. Acad. Sci. U.S.A.* **73,** 481 (1976).
[16] M. DeLuca, *Biochemistry* **8,** 160 (1969).

tion.[17] This type of specific inhibition is usually obvious, as these compounds when added to the reaction medium cause an abrupt decrease in luminescence that cannot be accounted for by enzymic hydrolysis of ATP.

Firefly luciferase is extremely specific for ATP as substrate in the luminescence reaction. No other naturally occurring nucleoside triphosphate has significant activity. AMP and ADP at high concentrations can act as competitive inhibitors of ATP,[18] but we have not encountered this inhibition as a significant problem at the concentrations of ADP and AMP used in our experiments. The rapidity and relatively simple instrumental requirements of the luciferase assay make it suitable for routine use and for use in other biological systems where ATP is of importance.

[17] J. J. Lemasters and C. R. Hackenbrock, unpublished observations.
[18] R. T. Lee, J. L. Denburg, and W. D. McElroy, *Arch. Biochem. Biophys.* **141,** 38 (1970).

[5] Measurement of Photophosphorylation and ATPase Using Purified Firefly Luciferase

By ARNE LUNDIN and MARGARETA BALTSCHEFFSKY

Firefly luciferase has been previously used for measurement of electron transport-linked ATP formation in chloroplasts and mitochondria.[1-6] The use of crude luciferase reagents and insufficient knowledge of the parameters determining the intensity of the bioluminescence have made quantitative evaluation of the data difficult. Based on knowledge of the kinetics of product inhibition during firefly luciferase luminescence,[7] a quantitative method by which the ATP concentration may be continuously monitored in mitochondria and submitochondrial vesicles has been described by Lemasters and Hackenbrock.[5,6]

A firefly luciferase method for continuous monitoring of low concentrations of ATP making corrections for product inhibition unnecessary has

[1] B. L. Strehler and J. R. Totter, *Arch. Biochem. Biophys.* **40,** 28 (1952).
[2] B. L. Strehler and D. D. Hendley, *in* "Light and Life" (W. D. McElroy and B. Glass, eds.), p. 601. Johns Hopkins Press, Baltimore, Maryland, 1961.
[3] B. L. Strehler, *Methods Biochem. Anal.* **16,** 99 (1968).
[4] G. P. Kukarskikh, T. Y. Krendeleva, and A. B. Rubin, *Biophysics* **17,** 88 (1972).
[5] J. J. Lemasters and C. R. Hackenbrock, *Biochem. Biophys. Res. Commun.* **55,** 1262 (1973).
[6] J. J. Lemasters and C. R. Hackenbrock, *Eur. J. Biochem.* **67,** 1 (1976).
[7] J. J. Lemasters and C. R. Hackenbrock, *Biochemistry* **16,** 445 (1977).

ISBN 0-12-181957-4

been developed by Lundin *et al.*[8-11] In this method reaction conditions are chosen to reduce product inhibition to a negligible level and to avoid luciferase inactivation or a measurable consumption of ATP in the luciferase reaction.[8] Under such conditions the intensity of the bioluminescence is proportional to the ATP concentration and changes only as a consequence of a changing ATP concentration.[8] In the linear range, i.e., at ATP concentrations $\leq 1 \mu M$,[8] the ATP concentration may thus be monitored simply by measuring the light intensity. At the end of the experiment the ratio between ATP concentration and light intensity is determined by the addition of internal ATP standard. Since this ratio is valid during the entire experiment, elaborate calibrations are eliminated.

A luciferase preparation suitable for continuous monitoring of ATP may be obtained by gel filtration essentially as described by Nielsen and Rasmussen.[8,12] A more purified enzyme may be needed in some applications and may be prepared by isoelectric focusing as described by Lundin *et al.*[10,11] With both methods it may be advisable to start the purification with ammonium sulfate precipitation.[11] The analytical usefulness of the continuous monitoring technique and the purified luciferase preparations described above has been proved in the assay of several enzymes and metabolites.[8-11]

The present paper describes the use of purified luciferase reagent for continuous monitoring of formation and breakdown of ATP in photophosphorylating systems. The measurement of ATP formation in *Rhodospirillum rubrum* chromatophores has been previously described by Lundin *et al.*[9] Photophosphorylation may be measured both under continuous illumination and after single short flashes, causing only one or a few turnovers of the cyclic electron-transport chain. This opens up new possibilities for obtaining a detailed picture of the events involved in photophosphorylation.

The ATPase activity may be measured at low ATP concentrations with the purified luciferase reagent.[13,14] In *Rhodospirillum rubrum* chromatophores a flash-induced stimulation of the ATPase activity was found with this method.[13,14] Since the stimulation could be attributed to a

[8] A. Lundin, A. Rickardsson, and A. Thore, *Anal. Biochem.* **75**, 611 (1976).

[9] A. Lundin, A. Thore, and M. Baltscheffsky, *FEBS Lett.* **79**, 73 (1977).

[10] A. Lundin, A. Rickardsson, and A. Thore, *in* "Proceedings of the Second Biannual ATP Methodology Symposium" (G. A. Borun, ed.), pp. 205–218. SAI Technology Company, San Diego, California, 1977.

[11] A. Lundin, this volume [6].

[12] R. Nielsen and H. Rasmussen, *Acta Chem. Scand.* **22**, 1757 (1968).

[13] M. Baltscheffsky and A. Lundin, Abstract presented at International Symposium on Membrane Bioenergetics, Spetsai, Greece, 1977.

[14] M. Baltscheffsky and A. Lundin, Abstract presented at 4th International Congress on Photosynthesis, Reading, England, 1977.

change in K_m rather than in V_{max}, it would have been difficult to detect at the high ATP concentrations used in the conventional assays of ATPase activity.

The specificity,[15] sensitivity, and convenience of the luciferase assay of ATP make the method described in the present communication a very attractive alternative to the "pH-method"[16] and the "^{32}P-method."[17] As described above, the continuous monitoring of ATP by purified luciferase has already proved to be a valuable analytical tool in the study of bacterial photophosphorylation, and this will undoubtedly also be the case in other types of phosphorylating systems.

Assay Method

Principle. The formation or breakdown of ATP in the photophosphorylating system is monitored by the light emitted in the luciferase reaction. There are certain restrictions on the reaction conditions put by the luciferase system.[8] The ATP concentration may not exceed 1 μM, pH should be in the interval 6–8, and the luciferase concentration should be chosen to result in an adequate sensitivity with a negligible analytical interference by product inhibition or ATP consumption in the luciferase reaction.

The measurement of photophosphorylation in *Rhodospirillum rubrum* chromatophores after single flashes is illustrated in Fig. 1. The reaction mixture contained chromatophores, ADP, inorganic phosphate, and luciferase reagent and was illuminated with a single 1-msec flash resulting in the first increase of the bioluminescence due to ATP formation by the chromatophores. The second increase of the bioluminescence was obtained after the addition of internal ATP standard. In both cases the half-time of the rise of the bioluminescence was 200 msec, which is similar to the value reported for the luciferase reaction by DeLuca and McElroy.[18] Thus it is not possible to follow the kinetics of the ATP formation unless the rate of the phosphorylating reaction is decreased as compared to the luciferase reaction. However, the yield of ATP after single flashes is conveniently determined and found to correspond to 1 ATP/189–330 bacteriochlorophylls.[9]

The monitoring of steady-state photophosphorylation in *R. rubrum* chromatophores is illustrated in Fig. 2. Results on the rate of ATP forma-

[15] W. D. McElroy and A. A. Green, *Arch. Biochem. Biophys.* **64,** 257 (1956).
[16] M. Nishimura, T. Ito, and B. Chance, *Biochim. Biophys. Acta* **59,** 177 (1962).
[17] S. Saphon, J. B. Jackson, and H. T. Witt, *Biochim. Biophys. Acta* **408,** 67 (1975).
[18] M. DeLuca and W. D. McElroy, *Biochemistry* **13,** 921 (1974).

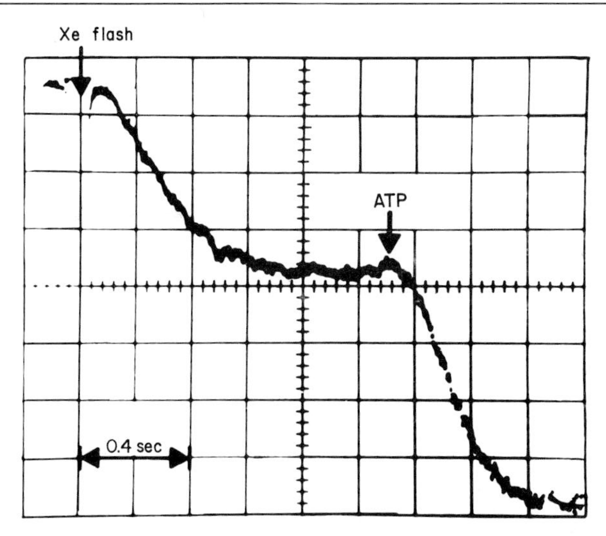

FIG. 1. Measurement of photophosphorylation after a single flash. The reaction mixture contained chromatophores from *Rhodospirillum rubrum* corresponding to 39 μM bacteriochlorophyll, 20 μM ADP, 2 mM inorganic phosphate, 0.1 M glycylglycine, pH 7.75, and luciferase reagent. The first increase (downward) of the bioluminescence represents the ATP formed after a 1-msec Xe flash filtered through triple layers of Wratten 88 A gelatin filter. The second increase is obtained at the addition of internal ATP standard in a final concentration of 0.24 μM. Reproduced from A. Lundin, A. Thore, and M. Baltscheffsky, *FEBS Lett.* **79,** 73 (1977).

tion obtained with the luciferase technique agree well with those obtained with the "pH-method."[16] The sensitivity of the luciferase technique allows the use of very dilute chromatophore suspensions with a negligible "self shadowing." In contrast to radiometric methods, the luciferase method is continuous and may be used to obtain true initial velocity kinetics.

Appropriate controls, including the use of uncouplers, ionophores, and electron-transport inhibitors, have shown that the method described actually measures photophosphorylation.[9] It may be of importance to realize that the method measures only ATP accessible to luciferase; i.e., ATP bound to or enclosed in the photophosphorylating particles is not measured.

The measurement of ATPase activity in *R. rubrum* chromatophores is illustrated in Fig. 3. In the experiment the initial ATP concentration was 0.1 μM, and after measurement in the dark the ATPase activity was stimulated by a single flash. The duration of the flash-induced stimulation of ATPase activity is approximately 40 sec and is shortened in the presence of FCCP.[13,14] The stimulated ATPase has a low K_m and is dependent on the presence of ADP and/or inorganic phosphate.

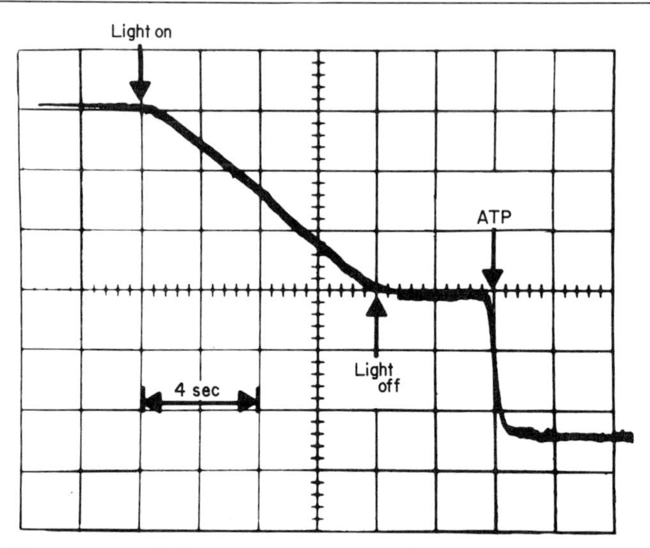

FIG. 2. Measurement of photophosphorylation during continuous illumination. Experimental conditions as in Fig. 1, but with a chromatophore concentration corresponding to 0.39 μM bacteriochlorophyll. Saturating illumination was obtained from a 20-W halogen lamp through a 875-nm interference filter. The linear increase (downward) of the bioluminescence is due to steady-state photophosphorylation. After illumination, internal ATP standard in a final concentration of 0.11 μM was added for calibration of the experiment. Reproduced from A. Lundin, A. Thore, and M. Baltscheffsky, *FEBS Lett.* **79,** 73 (1977).

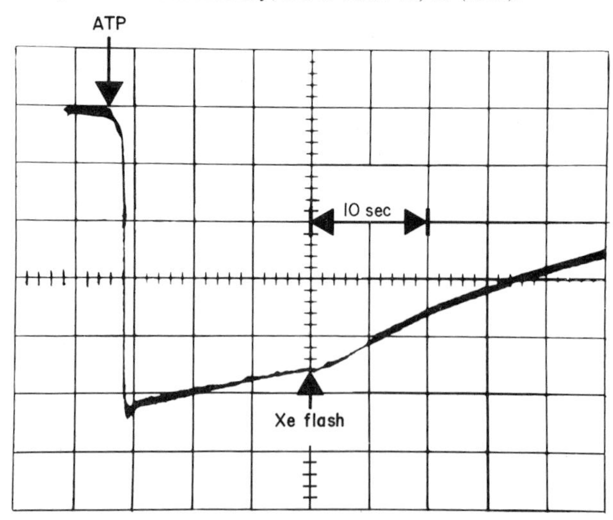

FIG. 3. Measurement of ATPase activity before and after activation with a single flash. Experimental conditions as in Fig. 1, but with 0.1 μM ATP and no ADP. The addition of ATP results in a sudden increase (downward) of the bioluminescence followed by a slow decrease due to ATPase activity. The 1-msec Xe flash resulted in an almost 3-fold activation of the ATPase.

Reagents. The pH value of all reagents is adjusted to pH 7.75 with sodium hydroxide or acetic acid.

Luciferase reagent: purified luciferase corresponding to a protein concentration of 3–15 μg/ml, 0.42 m*M* luciferin, 30 m*M* magnesium acetate, and 0.3% bovine serum albumin
Buffer: 0.15 *M* glycylglycine
Inorganic phosphate: 0.1 *M* sodium phosphate
ADP: 1 m*M* ADP purified by ion-exchange chromatography[11] and containing 1 m*M* EDTA
Internal ATP standard: 20 μ*M* ATP containing 20 m*M* Tris-acetate buffer and 1 m*M* EDTA

Assay Procedure. Luciferase reagent (0.5 ml) is added to a mixture of photophosphorylating system and buffer adjusted to result in measurable activities and a final volume of 1.5 ml. Photophosphorylation is measured after addition of ADP (30 μl) and inorganic phosphate (30 μl). ATPase activity is measured after addition of ATP in a final concentration of ≤ 1 μ*M*.[8] Experiments may be calibrated by the addition of internal ATP standard (10 μl).

Instrumentation. The measurements described in the present paper were done using a cuvette chamber from a Johnson Foundation dual-beam spectrophotometer with the measuring light turned off.[9] A generally applicable instrument should have the features listed below:

1. Bioluminescence should be measured with a photomultiplier with a high signal-to-noise ratio for firefly bioluminescence, e.g., EMI 9789, using a stabilized high-voltage supply. The signal from the photomultiplier may be read out on a storage oscilloscope or a rapid recorder.

2. The cross-illumination should be at 90° angle from the photomultiplier. Analytical interference from scattered cross-illumination may be reduced by filtering the light both at the inlet and in front of the photomultiplier provided cross-illumination and bioluminescence have different spectral distributions.

3. The equipment should be supplied with stirring facilities and a light-tight channel to allow rapid microliter additions to the cuvette in the measuring position.

4. The measuring position should be thermostatically controlled, since both the phosphorylating and the bioluminescent systems are temperature dependent.

Analytical Performance. The main advantage of the technique is the high sensitivity, allowing the measurement of very low amounts of photosynthetically produced ATP and the assay of ATPase activity at low ATP concentrations. With a modified spectrophotometer, a limit of 10 n*M* was obtained by turning off the measuring light,[9] and with simple equipment a detection limit of 10 p*M* ATP is easily obtained.[19] Increased sensitivity may also be obtained by increasing the luciferase concentration until the bioluminescence is affected by product inhibition and breakdown of ATP in the luciferase reaction.

Firefly luciferase is highly specific, and no other naturally occurring ribonucleoside triphosphate may substitute for ATP.[15] Nucleotide converting enzymes, e.g., adenylate kinase contaminating the photophosphorylating system, may, however, result in analytical interference. Owing to the sensitivity of the luciferase-coupled assay these problems may often be reduced by changing reactions conditions, e.g., using lower substrate concentrations.

The bioluminescent reaction has a full response time of a few hundred milliseconds, and the bioluminescent signal is stable for several minutes.[8] Thus ATP concentration may be continuously monitored in most short-time as well as long-time experiments. Since sampling is avoided and calibration is done with internal ATP standard, the accuracy and convenience of the technique is excellent, with a variation of duplicates always within a few percent.

The advantages of the luciferase-coupled assay makes it a valuable analytical tool that undoubtedly will contribute to the understanding of the basic events in electron transport coupled phosphorylation.

Acknowledgments

The authors thank Miss Gunilla Eriksson and Mrs. Anne Rickardsson for purifying luciferase and ADP. This work was supported by the Swedish Board for Technical Development and by a grant from the Swedish Natural Science Research Council (to M. B.).

[19] A. Lundin and A. Thore, *Anal. Biochem.* **66,** 47 (1975).

[6] Determination of Creatine Kinase Isoenzymes in Human Serum by an Immunological Method Using Purified Firefly Luciferase

By ARNE LUNDIN

Creatine kinase (CK) consists of two subunits, M and B. It appears in different tissues as the isoenzymes MM (muscle type), BB (brain type), and MB (hybrid type). In serum, increased total creatine kinase level is found after several types of damage to muscle tissue and is mainly due to increased levels of the MM isoenzyme. However, in serum from patients

ISBN 0-12-181957-4

suffering from acute myocardial infarction the MB fraction of the total creatine kinase activity is significant (6–25% at maximum total creatine kinase activity[1]). Increased serum level of the BB isoenzyme is found only in rare cases.[2-4] Determination of creatine kinase in serum, the MB isoenzyme in particular, is therefore of great value in the diagnosis of myocardial infarction.[2,3,5,6]

Several techniques for determination of the isoenzyme pattern of creatine kinase in serum have been developed. The isoenzymes may be electrophoretically separated on cellulose acetate,[7] polyacrylamide,[8] agar,[9] or agarose[10] and assayed either directly in the gel[8-10] or after elution of the creatine kinase bands.[7] Separation may also be achieved by ion-exchange chromatography on microcolumns of DEAE-Sephadex[11,12] or DEAE-cellulose.[13,14] The assay is usually performed spectrophotometrically according to Rosalki,[15] but fluorometry may also be used.[16]

Selective inactivation of isoenzymes of creatine kinase by precipitating antibodies has been described by Jockers-Wretou and Pfleiderer.[17] A more rapid method using an inhibiting antibody directed against the M subunit of human creatine kinase, leaving the B activity unaffected, has been described by Neumeier *et al.*[1] The MB activity may thus be calculated directly from remaining B activity, since BB activity can be found in serum only in rare cases.[2-4]

The assay of total creatine kinase activity using firefly luciferase was

[1] D. Neumeier, W. Prellwitz, U. Würzburg, M. Brundobler, M. Olbermann, H.-J. Just, M. Knedel, and H. Lang, *Clin. Chim. Acta* **73,** 445 (1976).

[2] L. Ljungdahl, S. Hofvendahl, W. Gerhardt, and J. Börjesson, *Clin. Chim. Acta* **78,** 43 (1977).

[3] M. Knedel, H. Lang, U. Würzburg, H. Schönborn, and H. P. Schuster, *Dtsch. Med. Wochenschr.* **101,** 983 (1976).

[4] E. K. Zsigmond, W. H. Starkweather, G. S. Duboff, and K. A. Flynn, *Anesth. Analg. (Paris)* **51,** 827 (1972).

[5] R. Roberts, K. S. Gowda, P. A. Ludbrood, and B. E. Sobel, *Am. J. Cardiol.* **36,** 433 (1975).

[6] A. Konttinen and H. Somer, *Am. J. Cardiol.* **29,** 817 (1972).

[7] R. Roberts, P. D. Henry, S. A. G. J. Witteveen, and B. E. Sobel, *Am. J. Cardiol.* **33,** 650 (1974).

[8] A. R. Smith, *Clin. Chim. Acta* **39,** 351 (1972).

[9] C. R. Roe, L. E. Limbird, G. S. Wagner, and S. T. Nerenberg, *J. Lab. Clin. Med.* **80,** 577 (1972).

[10] A. L. Sherwin, G. R. Siber, and M. M. Elhilali, *Clin. Chim. Acta* **17,** 245 (1967).

[11] D. W. Mercer, *Clin. Chem.* **20,** 36 (1974).

[12] D. W. Mercer and M. A. Varat, *Clin. Chem.* **21,** 1088 (1975).

[13] D. A. Nealon and A. R. Henderson, *Clin. Chem.* **21,** 392 (1975).

[14] D. A. Nealon and A. R. Henderson, *Clin. Chem.* **21,** 1663 (1975).

[15] S. B. Rosalki, *J. Lab. Clin. Med.* **69,** 696 (1967).

[16] J. A. S. Rokos, S. B. Rosalki, and D. Tarlow, *Clin. Chem.* **18,** 193 (1972).

[17] E. Jockers-Wretou and G. Pfleiderer, *Clin. Chim. Acta* **58,** 223 (1975).

first described by Strehler and Totter.[18] Witteveen *et al.*[19] developed a method by which the amount of MB could be determined in a mixture of MM and MB by assay of the enzyme activity at high and at low substrate concentrations using spectrophotometry and bioluminescence, respectively. This was possible owing to a slight difference in K_m for the two isoenzymes.

Injection of luciferase reagent into ATP solution generally results in a flash of light. The decay of the light is due to product inhibition, inactivation of luciferase, or consumption of ATP in the luciferase reaction or other enzymic reactions. Using a purified luciferase preparation, Lundin *et al.*[20] have defined conditions resulting in an almost constant light emission with a negligible decay rate at ATP concentrations less than 1 μM. Under these conditions continuous measurement of light emission may be used for nondestructive monitoring of ATP converting reactions, since the emission will be proportional to the ATP concentration in each instant. The application of this technique for the assay of several metabolites and enzymes including creatine kinase has been described.[20]

The immunological assay of creatine kinase isoenzymes by the bioluminescent method described in the present paper combines the sensitivity of the luciferase method[20] with the specificity of the immunological method of Neumeier *et al.*[1] In contrast to methods involving concentration of the serum or separation of the isoenzymes by electrophoresis or ion-exchange chromatography, the bioluminescent method is rapid and well suited for routine work.

Assay Method

Principle. The assay is based on continuous monitoring of the ATP formed in the creatine kinase reaction using the firefly system.[20]

$$\text{ADP} + \text{creatine phosphate} \xrightarrow{\text{CK}} \text{ATP} + \text{creatine}$$

$$\text{ATP} + \text{luciferin}_{\text{red}} + \text{O}_2 \xrightarrow{\text{luciferase}} \text{AMP} + \text{PP}_\text{i} + \text{luciferin}_{\text{ox}} + \text{CO}_2 + \text{light}$$

Specific determination of the MB isoenzyme in serum is achieved by immunoinhibition of M-subunit activity[1] or by separation of the isoenzymes by ion-exchange chromatography.[11,12]

Figure 1 shows a model experiment designed to illustrate the effect on

[18] B. L. Strehler and J. R. Totter, *Arch. Biochem. Biophys.* **40,** 28 (1952).

[19] S. A. G. J. Witteveen, B. E. Sobel, and M. DeLuca, *Proc. Natl. Acad. Sci. U.S.A.* **71,** 1384 (1974).

[20] A. Lundin, A. Rickardsson, and A. Thore, *Anal. Biochem.* **75,** 611 (1976).

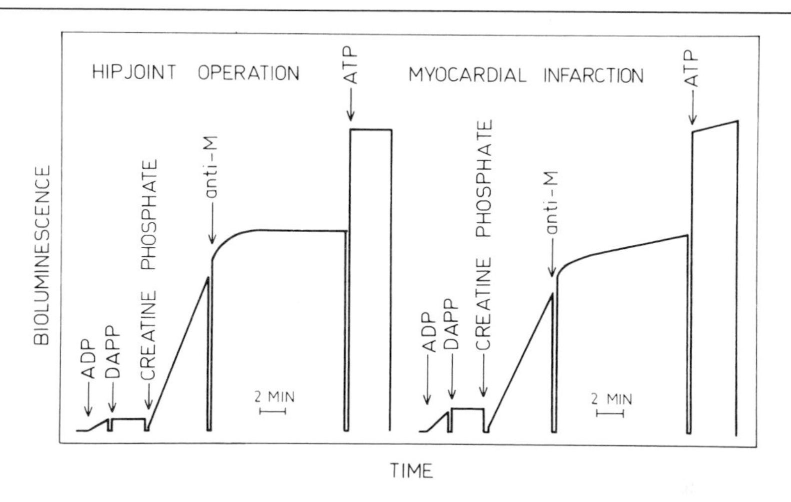

FIG. 1. The effect of substrates and inhibitors on ATP formation in the immunoinhibition assay of creatine kinase isoenzymes measured by the bioluminescent method. Serum samples (2.5 μl) were obtained from a patient operated in the hip joint [containing essentially only isoenzyme MM (250 U/liter)] and from a patient suffering from acute myocardial infarction containing MM (160 U/liter) and MB (40 U/liter). Further details are given in the text.

the ATP formation of substrates and inhibitors used in the assay of creatine kinase isoenzymes by the immunological method described in the present paper. The ATP formation after the addition of 10 μM ADP is due to adenylate kinase in the serum. The analytical interference by adenylate kinase is reduced by using a low ADP concentration and by the addition of the specific adenylate kinase inhibitor diadenosine pentaphosphate[21] (DAPP; 0.1 μM). Creatine phosphate (10 mM) inhibits luciferase to one-third of the original activity and initiates the creatine kinase activity. Addition of M-subunit inhibiting antibody inhibits the creatine kinase activity almost completely in the serum from the patient operated in the hip joint (only MM isoenzyme), but in the serum from the patient suffering from myocardial infarction 10% of the activity remains, indicating the presence of MB isoenzyme. The experiment is finished by the addition of internal ATP standard (0.2 μM) to make possible the calculation of ATP concentration from light-intensity data.

The low concentrations of ADP (10 μM) and creatine phosphate (10 mM) used in the present assay results in a lower rate of the creatine kinase reaction as compared to the spectrophotometric assay. If the bioluminescent assay is run at room temperature and the spectrophotometric assay

[21] P. Feldhaus, T. Fröhlich, R. S. Goody, M. Isakov, and R. H. Schirmer, *Eur. J. Biochem.* **57,** 197 (1975).

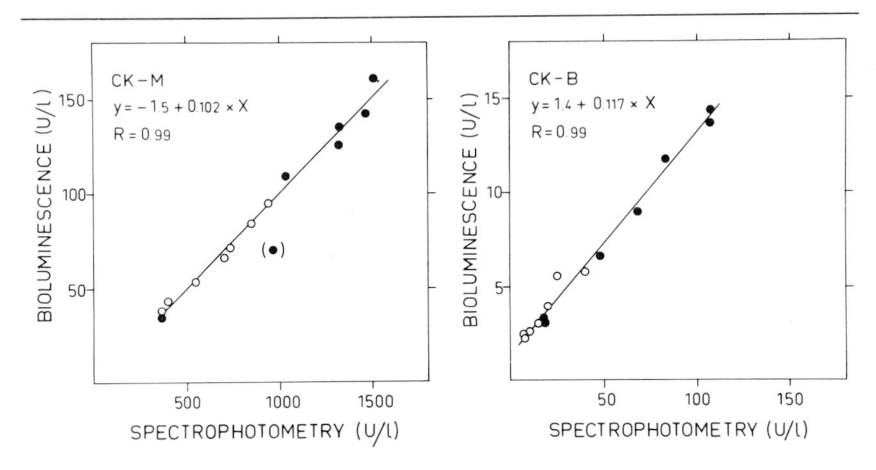

FIG. 2. Linear regression analysis of the correlation between the spectrophotometric and bioluminescent assays of creatine kinase (CK) isoenzymes. Serum samples were obtained with 6-hr intervals from two patients (○ and ●) suffering from acute myocardial infarction.

at 37°, the conversion factor between spectrophotometric units and bioluminescence units is approximately 0.11, as shown in Fig. 2.

The spectrophotometric assay of creatine kinase gives values on the activity below zero in some serum samples. This has been previously described[1,22] and may be due to analytical interference with the auxiliary reactions. Luciferase is very specific for ATP, and in the bioluminescent assay of creatine kinase luciferase is the only auxiliary enzyme. Interference resulting in negative creatine kinase values has not been observed with the bioluminescent assay. This is also indicated by the equation for the linear regression analysis of the MB determination as shown in Fig. 2.

Reagents. The pH value of all reagents is adjusted to 6.7 with acetic acid.

Reagent A: approximately 20 μg/ml luciferase (purified as described below), 0.47 m*M* D-luciferin, 33 m*M* magnesium acetate, and 0.33% bovine serum albumin

Reagent B: 0.1 *M* imidazole acetate, 1 m*M* EDTA, and 15 m*M* *N*-acetylcysteine

Reagent C: 0.5 m*M* ADP (purified as described below), 50 μ*M* diadenosine pentaphosphate, 1 m*M* EDTA, and 0.1 *M* imidazole acetate

Reagent D: creatine phosphate, 0.5 *M*

[22] R. G. Mueller, K. Neville, D. D. Emerson, and G. E. Lang, *Clin. Chem.* **21,** 268 (1975).

Reagent E: 20 μM ATP,[23] 1 mM EDTA, and 0.1 M imidazole acetate
Reagent F: Anti-human M subunit-inhibiting antibody produced in goats by multiple injections of purified, reactivated MM isoenzyme emulsified in complete Freund's adjuvant.[1] Purified and lyophilized antiserum[24] corresponding to a complete inhibition of 10 U of MM isoenzyme (37°) is dissolved in 2 ml of distilled water.

Reagents may be kept in the refrigerator for at least a week. Before the assay, reagents A and B are allowed to attain room temperature.

Purification and Assay of ADP. Commercial ADP generally contains approximately 1% ATP and is therefore purified by ion-exchange chromatography on Dowex (1-X8, 200–400 mesh)[25] before it is used in the assay. ADP (50 mg) is dissolved in distilled water (25 ml) and adjusted to pH 10. The solution is applied to a column (0.5 × 2.5 cm) and contaminating adenosine, adenine, and AMP are eluted (1 ml/min) with 3 mM HCl (100 ml). ADP is eluted (1 ml/min) with 10 mM HCl containing 20 mM NaCl. Peak fractions are pooled and adjusted to pH 6.7 with tris(hydroxymethyl)-aminomethane.

The ADP concentration of the pool is a few millimolar and may be determined after conversion of ADP to ATP as described by Lundin *et al.*[20] ADP is diluted 10,000-fold in 0.1 M Tris · acetate, pH 7.75, containing 2 mM EDTA. Diluted ADP (0.65 ml) is added to a mixture of reagent A (0.3 ml), 1 M potassium acetate (25 μl), and 20 mM phosphoenolpyruvate (5 μl). The increase of the bioluminescence obtained at the addition of pyruvate kinase (10 units in 10 μl) is proportional to the ADP concentration in the reaction mixture. The assay is calibrated by the addition of internal ATP standard, reagent E (10 μl). After correction for reagent blank (no added ADP) the ADP concentration may be calculated.

Purification of Luciferase. Several methods for purification of firefly luciferase have been published,[26-28] and some are described in the present volume. The method described here is based on ammonium sulfate precipitation followed by isoelectric focusing. The method has been worked out for crude firefly lantern extract obtained from Sigma, FLE-50, and the

[23] May be assayed spectrophotometrically in a coupled reaction using phosphoglyceric phosphokinase and glyceraldehyde phosphate dehydrogenase according to Sigma Technical Bulletin No. 366-UV.

[24] For the present work, M subunit-inhibiting antibody was generously supplied by Biochemical Research, E. Merck, Darmstadt, F.R.G.

[25] W. E. Cohn and C. E. Carter, *J. Am. Chem. Soc.* **72,** 4273 (1950).

[26] A. A. Green and W. D. McElroy, *Biochim. Biophys. Acta* **20,** 170 (1956).

[27] R. Nielsen and H. Rasmussen, *Acta Chem. Scand.* **22,** 1757 (1968).

[28] M. Bény and M. Dolivo, *FEBS Lett.* **70,** 167 (1976).

initial purification step may have to be somewhat modified with crude extracts obtained from other sources.

Crude lantern extract corresponding to 250 mg of dried lantern is dissolved in 22 ml of distilled water. Particulate matter is removed by centrifugation. 5.51 g of ammonium sulfate are added to 20 ml of supernatant; after 10 min at room temperature with slow stirring, the precipitate is removed by centrifugation. Ammonium sulfate, 1.45 g, is added to the supernatant; after 10 min with slow stirring, the precipitate is collected by centrifugation. The pellet is dissolved in 3 ml of distilled water, 44 ml of 40% sucrose, 2 ml of 40% ampholine, pH 5–7, 0.5 ml of 0.1 *M* EDTA, and 0.5 ml of 0.1 *M* dithiothreitol. This solution is used as dense solution (50 ml), and 1.6% ampholine in distilled water (50 ml) as light solution, in a density gradient made in a LKB 8100-10 (110 ml column) or similar electrofocusing equipment. A voltage of 600 V (max 2–3 W) is applied for approximately 60 hr. Luciferase is focused in one small and one large band with isoelectric points 5.8 ± 0.1 and 6.3 ± 0.1 at 5°, respectively. Fractions from the large band with activity higher than half the activity of the peak fraction are pooled. The neutralized luciferase preparation is stored at −20° in the presence of 0.1% bovine serum albumin.

The overall yield of the procedure is 65 ± 10%. With respect to specific activity and contamination by adenylate kinase, nucleoside diphosphokinase, and pyrophosphatase, the luciferase preparation obtained by isoelectric focusing is significantly better than that obtained by gel chromatography.[29]

Separation of Isoenzymes by Ion-Exchange Chromatography. Separation of creatine kinase isoenzymes by ion-exchange chromatography on microcolumns may be used as an alternative or control to the less laborious and more sensitive immunological method described in the present paper.

Serum (1 ml) is applied to a DEAE-Sephadex A-50 column (0.8 × 3 cm) and eluted with seven 1-ml portions of buffer I (50 m*M* Tris · HCl, 50 m*M* NaCl, pH 7.4) followed by five 1-ml portions of buffer II (50 m*M* Tris · HCl, 300 m*M* NaCl, pH 7.4). Fractions (1 ml) are collected in dithiothreitol (final concentration 10 m*M*) and assayed either separately or in pools of fractions 1–4 (MM) and 9–12 (MB).

The identity of the MM and MB fractions has been confirmed by the immunoinhibition method as shown in Fig. 3. The fractions were assayed spectrophotometrically[1] with and without M-subunit-inhibiting antibody. No significant B activity was found in the MM fractions or in the serum

[29] A. Lundin, A. Rickardsson, and A. Thore, *in* "Proceedings of the Second Biannual ATP Methodology Symposium" (G. A. Borun, ed.), pp. 205–218. SAI Technology Company, San Diego, California, 1977.

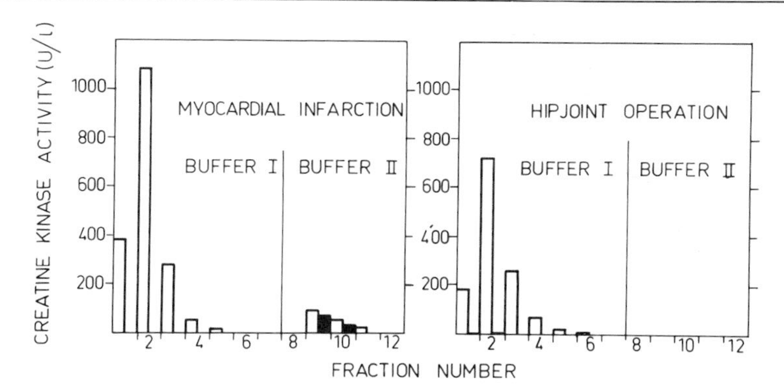

FIG. 3. Separation of serum creatine kinase isoenzymes by ion-exchange chromatography. Total creatine kinase activity (open bars) and MB activity (filled bars) were determined spectrophotometrically in the absence and in the presence of M-subunit inhibiting antibody, respectively. The serum obtained from a patient suffering from acute myocardial infarction contained (results after separation within brackets) 1796 (1822) U of MM isoenzyme per liter and 114 (120) U of MB isoenzyme per liter. The serum from the patient operated in the hip joint contained 1128 (1226) U of MM isoenzyme per liter, but no detectable MB isoenzyme.

from the patient operated in the hip joint. In the serum from the patient suffering from acute myocardial infarction, the MB fractions showed B activity as well as total creatine kinase activity. The separation of the isoenzymes by ion-exchange chromatography as described above results in a complete recovery of applied creatine kinase activity and is highly reproducible.

Assay of Creatine Kinase. The assay of creatine kinase in a series of serum samples is started by the determination of reagent blank values followed by the assay of a control serum with known activity. These assays are performed as shown in Fig. 4A and B with addition of internal ATP standard and separate determination of adenylate kinase background (Fig. 4A). Serum samples are then assayed as shown in Fig. 4C using the internal standard values (ΔI_{ADP} and ΔI_{CP}) obtained with the same sample volume of the control serum. The activity obtained with the control serum is used to calculate a conversion factor between results obtained by bioluminescence and spectrophotometry. The following steps are included in the assay (cf. Fig. 4):

Preincubation: Serum sample ($\leq$50 μl) is added to a mixture of reagent A (0.3 ml) and reagent B (adjusted to result in a final volume of 1 ml). For the immunological assay of MB isoenzyme, reagent F (20 μl) is also added. Incubate 15 min at room temperature.

Addition of ADP: Reagent C (20 μl) is added and the rate of the rise of the bioluminescence, $(dI/dt)_{ADP}$, is measured for 0.5–1 min.

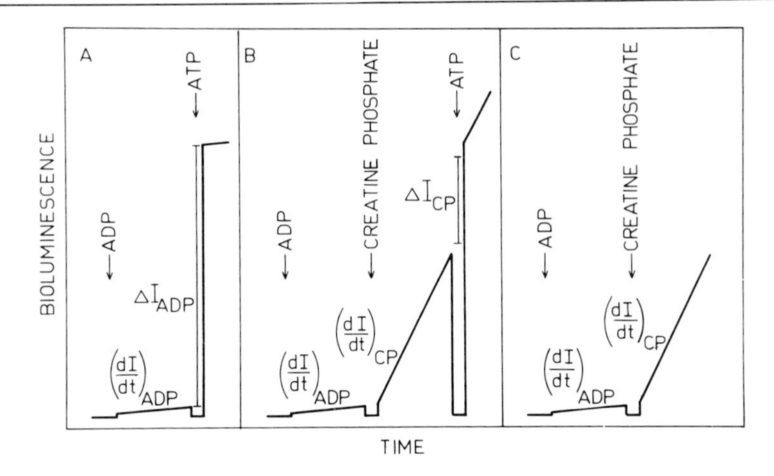

FIG. 4. Types of assays included in the determination of creatine kinase by the bioluminescent method. (A) Assay of background adenylate kinase activity with internal ATP standard. (B) Assay of adenylate kinase and creatine kinase with internal ATP standard for the creatine kinase assay. (C) Assay of adenylate kinase and creatine kinase without internal ATP standard.

Addition of creatine phosphate: Reagent D (20 μl) is added, and the rate of the rise of the bioluminescence, $(dI/dt)_{CP}$, is measured for 0.5–1 min.

Addition of internal ATP standard: Reagent E (10 μl) is added, resulting in an increase of the ATP concentration, C_{Stand}, and a simultaneous increase of the bioluminescence, ΔI_{ADP} or ΔI_{CP}. The total ATP concentration after addition of internal standard may not exceed 1 μM.[20]

The creatine kinase activity in the reaction mixture (CK) is calculated by Eq. (1). The correction for adenylate kinase activity is less than in the spectrophotometric assay[30] and is important only in the assay of isoenzyme MB in very low activities (cf. below).

$$\mathrm{CK} = \left[\frac{(dI/dt)_{CP}}{\Delta I_{CP}} - \frac{(dI/dt)_{ADP}}{\Delta I_{ADP}}\right] C_{Stand} \quad (1)$$

The creatine kinase activity in the serum is obtained after correction for (1) the CK value of the reagent blank, (2) the dilution of the serum in the assay, and (3) the lower reaction rate in the bioluminescent assay as compared to spectrophotometric methods.

Sensitivity and Reproducibility. The sensitivity of the bioluminescent assay is influenced by reagent blank and by serum adenylate kinase. With

[30] W. Gerhardt, L. Ljungdahl, J. Börjesson, S. Hofvendahl, and B. Hedenäs, *Clin. Chim. Acta* **78**, 29 (1977).

50-μl samples, the reagent blank corresponds to less than 1 U/liter.[31] Diadenosine pentaphosphate inhibits serum adenylate kinase to a level corresponding to less than 1 U/liter in 95% of the sera; in the remaining sera, it is less than 3% of the CK activity, with some very rare exceptions. This means that even without correction for adenylate kinase, the sensitivity of the bioluminescent assay is better than the spectrophotometric assay. With correction for reagent blank and remaining adenylate kinase activity, creatine kinase activities above 1 U/liter may be determined with an intraserial coefficient of variation of 3–14% depending on the creatine kinase activity. With thermostatically controlled equipment a considerable improvement of the reproducibility might be expected.

Acknowledgments

The author thanks Miss Inger Styrelius for performing the assays of creatine kinase; Miss Gunilla Eriksson, Miss Karin Mellström, and Mrs. Anne Rickardsson for performing the purification of luciferase and ADP; Drs. Willy Gerhardt (Helsingborg hospital), Rolf Nordlander and Olle Nyquist (Huddinge hospital), Erik Orinius (Karolinska hospital), and Anders Thore (National Defense Research Institute) for valuable discussions. This work was supported by the Swedish Board for Technical Development and by the Swedish National Association against Heart and Chest Diseases.

[31] Units are defined as micromoles of ATP formed per minute in the spectrophotometric assay at 37° and are approximately equal to 0.11 unit in the bioluminescent assay at room temperature.

[7] Determination of Bacterial Content in Fluids

By EMMETT W. CHAPPELLE, GRACE L. PICCIOLO, and JODY W. DEMING

Principles

An investigator using the firefly luciferase ATP assay to determine bacterial concentrations in fluid samples should be aware of those factors that will influence the accuracy of the measurement. These include the degree to which intracellular bacterial ATP varies in a given fluid environment, the possibility that "dead" cells may contribute to the final ATP measurement, the ability of the chosen extractant to provide maximum extraction of bacterial ATP and minimal inhibition of the luciferase enzyme reaction, the differentiation of bacterial ATP from other sources of ATP if present, and the sensitivity of the assay, which may not be adequate without sample concentration. These factors are addressed briefly before presenting assay methods as developed by this laboratory.

Variation in Intracellular ATP. The utilization of the firefly luciferase ATP assay to determine the number of bacteria in a fluid sample requires using a predetermined ATP per cell value. The value used in this laboratory for the ATP content of an average bacterium cultured in broth (trypticase soy) is 2.5×10^{-10} μg of ATP per organism (colony-forming unit). This is an average value ranging from 0.28 to 8.9×10^{-10} μg for 19 species tested,[1] and can be expected to vary over the growth cycle, decreasing as the population ages.[2-4]

A number of investigators have observed decreases in the ATP concentration in going from early log phase into the stationary phase.[5,6] In certain cases this decrease can approach an order of magnitude. Variances may also depend on the culturing environment; for example, *Escherichia coli* at the same growth phase when cultured in urine showed a 2-fold increase in ATP over *E. coli* cultured in nutrient broth.[6]

Thus, for each type of sample fluid, average bacterial ATP content and variation over the growth cycle should be determined by culturing in filter-sterilized sample fluid species of bacteria normally found in that environment. This information will help to establish the limits of accuracy of converting total ATP content to bacterial titer for a given application. Environmental conditions known to affect ATP content include dramatic changes in temperature, pH, nutrient supply, and oxygen tension.[7-11]

Interference by ATP in "Dead" Cells. In addition to viable cells, defined as colony-forming units, the ATP assay measures actively metabolizing cells present in the sample that simply fail to form colonies on a particular agar medium owing to growth requirements, inhibitory factors, etc. For example, when *E. coli* is treated with an antibiotic that inhibits DNA replication, i.e., nalidixic acid, plate counts indicate no viable cells

[1] E. W. Chappelle and G. V. Levin, *Biochem. Med.* **2,** 41 (1968).

[2] E. Freese, P. Fortnagel, and R. Schmitt, *in* "Spores IV" (L. L. Campbell, ed.), p. 82. American Society for Microbiology, Bethesda, Maryland, 1969.

[3] W. Klofat, G. L. Picciolo, E. W. Chappelle, and E. Freese, *J. Biol. Chem.* **224,** 3270 (1969).

[4] R. D. Hamilton and O. Holm-Hansen, *Limnol. Oceanogr.* **12**(2), 319 (1967).

[5] C. C. Lee, R. F. Harris, J. D. H. Williams, J. K. Syers, and D. E. Armstrong, *Soil Sci. Soc. Am., Proc.* **35,** 86 (1971).

[6] V. N. Bush, G. L. Picciolo, and E. W. Chappelle, *in* "Analytical Applications of Bioluminescence and Chemiluminescence" (E. W. Chappelle and G. L. Picciolo, eds.), p. 35 (NASA, Publ. No. SP-388). National Aeronautics and Space Administration, Washington, D.C., 1975.

[7] W. W. Forrest, *J. Bacteriol.* **90,** 1013 (1965).

[8] H. A. Cole, J. M. Wimpenny, and D. E. Hughes, *Biochim. Biophys. Acta* **143,** 445 (1967).

[9] R. E. Strange, H. E. Wade, and F. A. Dark, *Nature (London)* **199,** 55 (1963).

[10] R. B. Bailey and L. W. Parks, *J. Bacteriol.* **111,** 542 (1972).

[11] O. Holm-Hansen, *in* "Estuarine Microbial Ecology" (L. H. Stevenson and R. R. Colwell, eds.), p. 73. Univ. of South Carolina Press, Columbia, South Carolina, 1973.

after 24 hr. An ATP assay, however, indicates by high intracellular ATP levels that a full complement of intact, actively metabolizing bacteria remain in the sample.[12]

In order for an ATP assay to be a valid measurement of live bacteria (both viable and actively metabolizing cells), the contribution of bacterial ATP from "dead" cells should be negligible. Holm-Hansen and others have reported that ATP does not exist in measurable quantities in association with nonliving cells.[4,13-15] However, we believe that this may not always be the case.

In an actively metabolizing cell, the rate of ATP breakdown (decrease in steady-state concentration) is the net result of enzymic synthesis and hydrolysis. Enzymic hydrolysis of ATP is known to be stimulated if the cell is altered chemically by an uncoupler of oxidative phosphorylation, i.e., 2,4-dinitrophenol. Other modifications of the cell, including physical damage, can also affect ATPase activity. One of the authors has observed an enzymic involvement in ATP hydrolysis after a bacterial suspension was subjected to ultrasonic oscillation, the rate of ATP breakdown being markedly reduced when the ultrasonic treatment was performed at approximately 100° or at an acidic pH.[16] Where chemical hydrolysis is the only mechanism for ATP destruction, the half-life of ATP may vary from a few minutes to several hours, again depending upon such factors as pH, temperature, and the nature of associated cations. Whether or not "dead" cells contribute measurable quantities of ATP to a sample depends not only on the manner in which the cells are damaged or killed, but also on the nature of their fluid environment.

When a culture of *E. coli* is allowed to grow over an extended time period, i.e., 15 days, and the stationary phase extends into the death phase, viable cell counts decrease rapidly, while intracellular ATP levels diminish only gradually.[16] This would indicate that, in situations where cell death is the result of nutrient limitations or end-product inhibition, ATP concentrations do not undergo the drastic changes seen in other types of cell death.

At this time we can present no solution to the problem of differentiating live and "dead" bacteria in a given sample, a problem that requires consid-

[12] G. L. Picciolo, E. W. Chappelle, H. Vellend, S. Tuttle, C. G. Schrock, J. W. Deming, M. Barza, and L. Weinstein, "Application of Firefly Luciferase Assay for Adenosine Triphosphate (ATP) to Antimicrobial Drug Sensitivity Testing" (NASA, Publ. No. TM-D8439). National Aeronautics and Space Administration, Washington, D.C., 1977.

[13] O. Holm-Hansen and C. R. Booth, *Limnol. Oceanogr.* **11,** 510 (1966).

[14] O. Holm-Hansen, *Plant Cell Physiol.* **11,** 689 (1970).

[15] J. W. Patterson, P. L. Brezonik, and H. D. Putnam, *Environ. Sci. Technol.* **4,** 569 (1970).

[16] E. W. Chappelle, G. L. Picciolo, H. Okrend, R. R. Thomas, J. W. Deming, and D. A. Nibley, *in* "Proceedings of the Second Biannual ATP Methodology Symposium," (G. A. Borun, ed.), p. 611. SAI Technology Company, San Diego, California, 1977.

erable and rigorous investigation. However, we feel that interference in the ATP assay by "dead" bacteria in a sample whose nature and history is undefined should be considered as a possible source of error.

Extraction of ATP from Bacteria. The ideal extracting agent for bacterial ATP should provide complete release of all ATP with minimal ATP hydrolysis and minimal inhibition of the luciferase reaction. Numerous agents have been tested and compared for extraction efficiency, including ionizing and nonionizing organic solvents, inorganic acids, and boiling buffers.[17,18]

Although several methods have proved adequate in terms of efficiency, inorganic acids (particularly nitric acid, which forms no interfering precipitate upon neutralization) are recommended for bacteria because of superior extraction efficiency, ease of handling, immediate extraction, and pH inactivation of any naturally occurring (or added) ATPases. Extraction by boiling denatures hydrolyzing enzymes, but bacterial cells submerged in boiling tris(hydroxymethyl)aminomethane(Tris), for example, are not immediately exposed to 100°, and some ATP hydrolysis may occur.[19] Because an acidic pH will decrease luciferase activity, the extracted sample must be neutralized to pH 7.75 coincident with reacting with the luciferase preparation.

Interference by Nonbacterial ATP in the Sample. Nonbacterial ATP in the sample, including intracellular ATP from eukaryotic cells and free soluble ATP,[20,21] must be removed prior to bacterial ATP extraction and assay. Physical separation of bacterial cells from other cell types through differential filtration or centrifugation may result in an underestimate of bacterial titer, because bacteria are known to adhere to larger particles. A more quantitative approach is selective chemical lysis of nonbacterial cells followed by the addition of an ATPase to hydrolyze the freed ATP and any other soluble ATP present. In physiological fluids such as urine and blood, which contain both free soluble ATP and mammalian cells, these functions can be performed, respectively by Triton X-100, a nonionic detergent, and

[17] E. A. Knust, E. W. Chappelle, and G. L. Picciolo, *in* "Analytical Applications of Bioluminescence and Chemiluminescence" (E. W. Chappelle and G. L. Picciolo, eds.), p. 27 (NASA, Publ. No. SP-388). National Aeronautics and Space Administration, Washington, D.C., 1975.

[18] A. Lundin and A. Thore, *Appl. Microbiol.* **30**(5), 713 (1975).

[19] D. M. Karl and P. A. LaRock, *J. Fish. Res. Board Can.* **32,** 599 (1975).

[20] G. L. Picciolo, E. W. Chappelle, J. W. Deming, M. A. McGarry, D. A. Nibley, H. Okrend, and R. R. Thomas, *in* "Handbook on Infectious Disease Methodology" (D. Amsterdam, ed.), Vol. 1. Dekker, New York, in press.

[21] R. Hodson and F. Azam, *in* "Proceedings of the Second Biannual ATP Methodology Symposium" (G. A. Borun, ed.), p. 127. SAI Technology Company, San Diego, California, 1977.

apyrase, an ATP-hydrolyzing enzyme extracted from potatoes. [Triton X-100 (0.1%) is known to lyse *Streptococcus*, however, and to inhibit subsequent growth of some bacteria.] About 5% of the nonbacterial ATP may remain bound to macromolecules or cellular debris. By lowering the pH to 4.25 with an acidic solution such as malate buffer, this residual ATP is desorbed, making it susceptible to attack by the apyrase, which remains marginally active at that pH. The addition of nitric acid simultaneously inactivates the apyrase as it extracts bacterial ATP for assay.

Sensitivity and Calculations. With the use of a purified luciferase enzyme preparation, the lower limit of bacterial detection by the ATP assay is about 10^5 cells per milliliter (*E. coli* cells cultured in Trypticase soy broth and washed in sterile, distilled water or 0.9% NaCl). Sensitivity is limited by the light response of the "blank," produced by the reaction of filter-sterilized sample fluid, processed in the same manner as the bacteria-laden sample, with the luciferase enzyme preparation. Sample-minus-blank light response can be related directly to the response obtained from a known concentration of chemical ATP to establish the ATP content of the sample. Chemical ATP, added with the extractant to simulate bacterial ATP release, represents an extraction standard.

For many applications, adequate sensitivity can be achieved only by concentrating the bacteria into a reduced sample volume prior to ATP extraction. Both centrifugation and filtration allow bacterial concentration with the added advantages of washing the bacterial cells and removing soluble factors that otherwise might inhibit the luciferase reaction. Filtration excludes particulate matter from the final extract if bacterial ATP is extracted on the filter and collected as a filtrate. However, not all sample fluids, particularly physiological specimens, lend well to rapid filtration in the volumes required for adequate sensitivity. Filtration can be facilitated by slight shifts in pH or temperature to alter component solubilities, by the addition of 0.001% Triton X-100 to all reagents added prior to extraction, and by the use of rhozyme, a mixture of proteases extracted from *Aspergillis oryzae*, which degrades glycoproteins and mammalian cell membranes found in urine specimens.

Assay Methods

Instrumentation. Several photometric instruments specifically designed to measure low light levels as encountered in the firefly luciferase ATP assay are available commercially. These photometers, other suitable instrumentation, and an automated ATP assay procedure are discussed in detail in this volume [43].

Reagents. All reagents should be prepared in sterile, distilled, deionized water (SDDW) from chemicals of the highest purity available. Although disposable plasticware is preferable, glassware acid-cleaned and thoroughly rinsed in distilled water to remove residual ATP can be used. Potassium salts are particularly inhibitory to the luciferase reaction; therefore, pH adjustments of all reagents should be made with NaOH and HCl.

Adenosine triphosphate (ATP), 1.0 mg per milliliter of stock solution prepared in SDDW and stored at $-20°$ until needed; diluted and used as a standard

Apyrase, 10 mg of 30 m*M* $CaCl_2$ per milliliter (Ca^{2+} required as a cofactor at an acting concentration of 5 m*M*); a very stable and extremely active ATP-hydrolyzing enzyme with ATPase, ADPase, and AMPase activity (Sigma Chemical Company, St. Louis, Missouri)

Cleland's reagent (dithiothreitol), 1 m*M* solution in Tris buffer used for luciferase enzyme preparation; stabilizes purified luciferase by protecting sulfhydryl groups; maintains maximum luciferase activity for 8 hr with refrigeration

Luciferase enzyme preparation (extracted and purified from firefly lanterns with synthetic luciferin added[22] or available commercially from DuPont, Inc., Wilmington, Delaware), lyophilized powder rehydrated in 0.25 *M* Tris, pH 8.2, 10 m*M* $MgSO_4$, and 1 m*M* Cleland's reagent 30–45 min before use

Magnesium sulfate ($MgSO_4$), 10 m*M* solution in Tris buffer used for luciferase enzyme preparation; required cofactor in the luciferase reaction

Malate buffer, 50 m*M* malic acid, pH 4.25, prepared in SDDW; used to desorb membrane-bound ATP for subsequent hydrolysis

Nitric acid (HNO_3), acting concentration of 0.1 *N*; inactivates apyrase and extracts bacterial ATP

Rhozyme-41, mixture of proteases extracted from *Aspergillis oryzae*, 8% solution in SDDW prepared 24 hr before use; facilitates filtration of physiological fluids, degrades glycoproteins and mammalian cell membranes

Sodium sulfate ($NaSO_4$), 85 m*M* solution in SDDW; alternative diluent to SDDW for diluting nitric acid extract; adjusts ionic strength of the injected sample

Tris(hydroxymethyl)aminomethane (Tris), 0.25 *M* solution at pH 8.2 (with 10 m*M* $MgSO_4$ and 1 m*M* Cleland's reagent); rehydration

[22] D. A. Nibley, G. L. Picciolo, E. W. Chappelle, and M. A. McGarry, *in* "Proceedings of the Second Biannual ATP Methodology Symposium" (G. A. Borun, ed.), p. 441. SAI Technology Company, San Diego, California, 1977.

buffer for lyophilized luciferase–luciferin powder; maintains reaction pH optimum when sample contains nitric acid

Triton X-100 (TX), nonionic detergent, acting concentration of 0.12% selectively ruptures mammalian cells; 0.001% facilitates rapid sample and reagent filtration

Nonconcentrated Procedure. This procedure, which actually effects a 10-fold dilution of the sample, is recommended for analysis of pure cultures or samples containing high levels of bacterial ATP and little if any nonbacterial ATP. The method is simple, rapid, and very useful in laboratory studies monitoring the effects of chemical agents on bacteria; for example, the susceptibility of bacteria to antibiotics.[12]

One-tenth milliliter of 1.5 *N* HNO_3 is added to a 0.5-ml sample in a sterile, polypropylene tube for extraction of bacterial ATP. The extract is diluted with 4.4 ml of SDDW and mixed well. After injection of 0.1 ml of the diluted extract into 0.1 ml of luciferase enzyme preparation, the peak light response is recorded. (The merits of peak versus integrated light signal are addressed in this volume [43].)

To make sure that no soluble ATP contributes to the measurement of intracellular bacterial ATP, 0.1 ml of apyrase solution can be added to the sample prior to acid extraction. (If mammalian cells are present, the apyrase solution should contain 0.6% TX.) Fifteen minutes should be allowed for complete hydrolysis of soluble ATP before adding nitric acid.

Centrifugation Procedure. This procedure is specifically designed to detect significant bacterial infections in clinical urine samples that also may contain up to 10^7 leukocytes per milliliter. For other types of samples that require bacterial concentration (a 25-fold concentration is provided by this procedure), but do not challenge the system with such tremendous stores of nonbacterial ATP, some of the following steps should be considered optional.

Two-tenths milliliter of a 6% solution of TX is added to a 10-ml sample in a sterile polypropylene centrifuge tube. After centrifugation at 10,400 *g* (RCF) for 15 min at 20°, the supernatant is decanted. Any remaining soluble ATP is hydrolyzed by adding 1.0 ml of apyrase solution to the pellet and mixing well. After addition of 5.0 ml of 0.9% NaCl as a wash, the apyrase is allowed to hydrolyze for 15 min. One milliliter of 0.5 *M* malic acid, pH 4.25, is added, the sample is centrifuged again at 10,400 *g* (RCF) for 15 min at 20°, and the supernatant is decanted.

The only source of ATP now remaining in the pellet is intracellular bacterial ATP, which is extracted by resuspending the pellet in 0.2 ml of 0.1 *N* HNO_3 (mixed well). The extract is diluted with 0.2 ml of SDDW, and 0.1 ml of the final volume is assayed with 0.1 ml of the luciferase enzyme preparation.

Simple Filtration Procedure. If the sample fluid passes readily through a filter with minimal damage to the bacteria, and the sample contains fewer than 10^5 cells per milliliter, the following concentration procedure is appropriate.

A 10-ml sample (volume variable depending on desired concentration) is passed through a 0.22-μm membrane filter and washed with 10 ml of 0.9% NaCl. (A desirable membrane characteristic is minimal volume retention.) A 0.25-ml volume of 0.1 *N* HNO_3 is applied to the membrane surface (25 mm diameter) for bacterial ATP extraction. Before the extract is collected as filtrate, an equivalent volume of $NaSO_4$ is added to the filter as diluent. One-tenth milliliter of the filtrate is assayed by injection into 0.1 ml of luciferase enzyme preparation.

Soluble ATP and/or mammalian cell ATP can be removed in this case by pretreating the 10-ml sample with 0.2 ml of 6% TX and 1.0 ml of apyrase solution, allowing 15 min for ATP hydrolysis, and filtering and washing with NaCl as described above.

Complex Filtration Procedure. This procedure is modeled after the centrifugation procedure, and incorporates steps to facilitate filtration of physiological fluids, such as urine. The procedure is not recommended for routine use, however, because some clinical urine specimens resist effective filtration despite attempts to minimize clogging.

Sample pretreatment to minimize clogging includes the addition of 0.2 ml of 6% TX, 1.0 ml of 1.5 *M* Tris, pH 8.3, and 1.0 ml of 8% rhozyme to a 10-ml sample in a test tube. Standing time of 5 min is allowed for mammalian cell lysis and degradation of urine glycoproteins, etc. The sample is then passed through an 8-μm cellulose ester prefilter membrane overlaying a 0.4-μm bacteriological polycarbonate membrane.

After the filters are washed with 10 ml of malate buffer, 1.0 ml of apyrase solution is applied for 15 min. A second wash with 10 ml of Na_2SO_4 prevents buffering by malic acid and apyrase retained in the filter matrix.

For final extraction of bacterial ATP, 0.4 ml of 0.1 *N* HNO_3 is applied to the filters, followed by a 0.4-ml diluent and collection by filtration. One-tenth milliliter is assayed as described previously.

[8] Biomass and Adenylate Energy Charge Determination in Microbial Cell Extracts and Environmental Samples

By OSMUND HOLM-HANSEN and DAVID M. KARL

In both laboratory and field studies it is often desirable to determine the total amount of living cellular material (biomass) as well as the activity or metabolic state of the cells. Conventional methods (i.e., fresh or dry weight determinations, rate of increase of cell numbers, etc.) usually cannot be used owing to (1) lack of sensitivity in the analytical procedures, (2) the heterogeneous assemblage of cellular materials, (3) the presence of dead cells, and (4) the presence of detrital material not associated with the living cells. Estimation of cellular biomass by measurement of adenosine triphosphate (ATP) is generally not limited by any of these considerations.

The rationale for using ATP to estimate biomass is the ubiquitous distribution of ATP in all living cells, the rapid loss of ATP from dead cells, and the fairly uniform concentration of ATP in the protoplasm of all microbial cells. Data on ATP concentrations can thus be extrapolated to biomass parameters, such as cellular organic carbon or dry or fresh weight.[1-3]

In many studies it is desirable to determine the biochemical activities or metabolic states of the cells in addition to total biomass. One promising approach to the investigation of the energetic state of cells is to determine the concentrations and ratios of the adenine ribonucleotides (ATP, ADP, and AMP), which are involved in stoichiometric energy transduction between most metabolic reactions, and thus are of great significance with respect to regulation of cellular metabolism. Atkinson and his colleagues have defined adenylate energy charge (EC) as

$$\mathrm{EC} = \frac{[\mathrm{ATP}] + 1/2[\mathrm{ADP}]}{[\mathrm{ATP}] + [\mathrm{ADP}] + [\mathrm{AMP}]} \tag{1}$$

whereby the EC is a measure of the metabolically available energy stored in the adenine nucleotides. Since cells in normal growth conditions have control systems that maintain the concentrations of these nucleotides within narrow range limits, marked decreases in the EC may reflect en-

[1] O. Holm-Hansen and C. R. Booth, *Limnol. Oceanogr.* **11,** 510 (1966).

[2] O. Holm-Hansen, *Plant Cell Physiol.* **11,** 689 (1970).

[3] O. Holm-Hansen, *in* "Estuarine Microbial Ecology" (L. H. Stevenson and R. R. Colwell, eds.), p. 73. Univ. of South Carolina Press, Columbia, South Carolina, 1973.

ISBN 0-12-181957-4

vironmental stress or growth conditions that interfere with the normal functioning of the organism.

In this chapter we describe the methodology for quantitative determinations of ATP, ADP, and AMP in cell extracts and environmental samples, by use of the firefly luciferin–luciferase bioluminescent reaction. Difficulties and sources of error are discussed in regard to the analytical phase as well as the preliminary steps of sampling and extraction of the nucleotides from living cells.

Assay Principles

$$\mathrm{ATP} + \mathrm{LH_2} + \mathrm{O_2} \xrightarrow{\mathrm{FL}} \mathrm{AMP} + \mathrm{PP_i} + \mathrm{CO_2} + \mathrm{P} + h\nu \quad (2)$$

$$\mathrm{ADP} + \mathrm{PEP} \xrightarrow{\mathrm{PK}} \mathrm{ATP} + \text{pyruvate} \quad (3)$$

$$\mathrm{AMP} + \mathrm{ATP} \xrightleftharpoons{\mathrm{AK}} 2\ \mathrm{ADP} \quad (4)$$

where FL = firefly luciferase (EC 2.7. .), PK = pyruvate kinase (EC 2.7.1.40), AK = adenylate kinase (EC 2.7.4.3), LH_2 = reduced luciferin, P = product, PP_i = inorganic pyrophosphate, PEP = phosphoenolpyruvate.

Firefly luciferase is specific for ATP,[4,5] and hence it is not necessary to separate ATP from other nucleotides in the extracts. The light-emission reaction catalyzed by luciferase [Eq. (2)] is more complicated than shown above (see this volume [1]); as there is one quantum of visible light emitted per ATP molecule hydrolyzed, quantification of the light flux enables one to calculate the ATP concentration in the extract.

To determine ADP and AMP, these nucleotides must first be converted to ATP, which is then measured as in Eq. (2). To determine ADP, subsamples of the extract are incubated with excess PEP and pyruvate kinase [see Eq. (3)]. The ATP formed in this reaction is measured as in Eq. (2) and, after subtraction of the initial concentration of ATP, is equivalent to the original concentration of ADP in the sample extract. The concentration of AMP in another subsample is likewise determined by converting the AMP to ADP via the adenylate kinase reaction [Eq. (4)], and simultaneous conversion of the ADP formed to ATP via the pyruvate kinase reaction [Eq. (3)].

The most practical and convenient way to relate the measured light flux to ATP concentrations is to use a series of internal ATP standards. A variety of light-detecting instruments can thus be used, as it is necessary

[4] A. Green and W. D. McElroy, *Biochim. Biophys. Acta* **20,** 170 (1956).
[5] M. DeLuca, *Adv. Enzymol.* **44,** 37 (1976).

only to measure relative changes in light emission rather than absolute energy fluxes.

Reagents and Equipment

Reagents. Unless otherwise indicated, catalog numbers or information refer to items sold by Sigma Chemical Company.[6]

- Firefly lantern extract (FLE-50)
- Sodium arsenate buffer, 100 m*M*, pH 7.4
- ATP, ADP, and AMP, sodium salts, 2 μM solutions in Tris buffer, dispensed into small aliquots (3–5 ml), and stored frozen (−20°)
- Magnesium sulfate solution, 40 m*M*
- Tris(hydroxy)aminomethane−HCl buffer, 20 m*M*, pH 7.75
- Potassium phosphate buffer, 225 m*M*, pH 7.4
- Magnesium chloride solution, 45 m*M*
- D-Luciferin, optional (L 9504)
- Pyruvate kinase, 350–500 units per milligram of protein (P 1506)
- Adenylate kinase, 1000–1500 units per milligram of protein (M 3003)
- Phosphoenolpyruvate (P 7002)

Equipment

- Light-detection instrument: Any one of a number of commercially available general instruments, such as fluorometers, spectrophotometers or liquid scintillation counters, are suitable for measuring light emission. However, in order to obtain efficient and reliable data, with maximum sensitivity, a specially designed ATP photometer is often required. Instruments specifically designed for ATP analyses are marketed by the American Instrument Company (Chem-Glow photometer), E. I. DuPont, Inc. (Luminescence Biometer), and SAI Technology Company (ATP Photometer). A detailed description and comparison of these three instruments can be found in this volume [43].
- Strip-chart recorder with adjustable ranges from 10 mV to 10 V
- Water baths: one at 30°, one at 100°
- Glassware: All glassware used for nucleotide analyses should be thoroughly cleaned in order to eliminate the possibility of contamination; if possible, disposable glassware should be used: (a) Pyrex glass flasks, 50 and 250 ml, (b) vials, 18 × 40 mm (Kimble, No. 74400), and (c) glass culture tubes, 12 × 75 mm (Scientific Products, No. T-1285-3).

[6] References to Sigma Chemical Company or to other manufacturers do not represent any endorsement of these sources, but are cited only for the convenience of the reader.

Pipettes: (a) 5- and 10-ml glass or disposable plastic, (b) automatic pipette 0.1 to 1.0 ml, with disposable tips, (c) automatic pipettes of 50-μl and 200-μl volume, with disposable tips

Automatic injector: For use in conjunction with the light-detection instruments. Must be capable of rapid and reproducible injection of sample into the enzyme preparation; disposable tips are recommended.

Assay Methods

The procedures described below assume that the investigator has a nucleotide-containing extract that does not contain any inhibitors that would interfere with the reactions shown in Eqs. (2)–(4); later sections will discuss extract preparation and sources of analytical interference.

Luciferase Enzyme Preparation

Each vial of lyophilized firefly lantern extract (50 mg/vial) is reconstituted by adding 5 ml of distilled water, 10 ml of 100 m*M* sodium arsenate buffer (pH 7.4), and 10 ml of 40 m*M* $MgSO_4$. The enzyme mixture is allowed to stand for 3–4 hr at room temperature (23°–25°) in order to reduce the level of background light emission. After aging, the insoluble residue is removed by passage through a filter or by centrifugation. The clear filtrate or supernatant is transferred to a clean flask, and the enzyme preparation is now ready for use. Since considerable variation occurs between individual vials of firefly lantern extract, the contents of several vials should be pooled whenever large volumes of enzyme are required.

Standard Nucleotide Solutions

Internal standards are used to calculate absolute concentrations of ATP, ADP, and AMP in sample extracts; a fresh set of standards must be prepared for each day's work. A small tube of the frozen nucleotide is thawed and sequentially diluted with Tris buffer to give a series of solutions ranging in concentration from 200 n*M* to 1 n*M*. All thawed standards are discarded immediately after use. These precautions are necessary to minimize hydrolysis of the nucleotides during storage and also to minimize bacterial contamination of working solutions. There should be no significant loss of nucleotides in these standard solutions during an 8-hr period.

Light Measurement

There are several ways to quantitate the amount of ATP when using the luciferase bioluminescent reaction. The two methods most commonly used are (1) to record the peak height of the light emission, which should occur within 3 sec after injection of the sample, and (2) to integrate the light flux for a set time period. The major advantages of the integrated mode are (a) increased sensitivity, (b) ease and reliability of mixing, and (c) no need to depend upon a peak-height response, which may be difficult or impossible to measure with certain instruments. Reliability of the peak-height analysis depends upon a very rapid and complete mixing of all reactants. This is best accomplished by use of an automatic pipette, which ensures consistent mixing velocities for all samples. The peak-height mode of analysis offers the advantages of speed of assay and minimum interference from other enzymes or substrates that may affect the rate of the luciferase-catalyzed reaction.[7,8] This is particularly relevant when determining AMP concentrations, as adenylate kinase [Eq. (4)] and PEP are heat stable and cause a continual synthesis of ATP when injected into the luciferase preparation. For this reason, all work described in this chapter will utilize the peak-height mode of light measurement. As our work has been done with the ATP Photometer and Automatic Pipette produced by SAI Technology Co. (San Diego, California), volumes of enzyme and sample as well as size of the reaction vials to be used refer to that instrument.

ATP Determination

A glass vial (18 × 40 mm) containing 1.0 ml of the luciferase enzyme preparation is placed in the sample holder of the ATP photometer. The automatic pipette, containing 0.2 ml of the sample to be analyzed, is placed on top of the sample holder and the background light emission of the enzyme preparation is recorded by pushing the assay switch on the photometer. After clearing the digital display, the release mechanism on the automatic pipette is pushed, causing injection of the sample into the enzyme preparation and starting the timing circuits. The maximum peak height is obtained as a digital readout on the photometer. Although it is not necessary, it is informative to connect the ATP photometer to a strip-chart recorder in order to monitor the kinetics of the light-emission reaction. This is particularly useful in recognizing possible interferences with the luciferase reaction (effects of inhibitors, other nucleotides, etc.) as well as any instrument malfunction (e.g., leaky pipette). With any one batch of samples (up to about 100), a series of standards is generally run before and after the samples. The standard ATP solutions should have the same ionic composition as the sample solutions to ensure that the reactivity of ATP is similar in all determinations. Calculations of ATP concentrations are described in a later section.

[7] H. Rasmussen and R. Nielsen, *Acta Chem. Scand.* **22,** 1745 (1968).

[8] A. Lundin and A. Thore, *Anal. Biochem.* **66,** 47 (1975).

Enzymic Conversion of ADP and AMP to ATP

Although detailed procedures have already been described for adenine nucleotide determinations in plant tissues[9] and bacterial cell extracts,[10] several modifications were necessary in order to adapt these techniques for the relatively low levels of adenine nucleotides in most environmental samples.[11]

Each batch of samples to be analyzed for ATP, ADP, and AMP must be prepared and processed together with (a) a Tris-buffered reagent blank, (b) a set of 6–8 ATP standards ranging from 1 n*M* to 200 n*M* and (c) a series of solutions containing various ratios of ATP, ADP, and AMP. For each test sample, 200 μl of the Tris-buffered extract are pipetted into each of 4 disposable glass culture tubes (12 $\times$ 75 mm) labeled A (for ATP), B (for ATP + ADP), C and D (for ATP + ADP + AMP). The following additions are then made to these tubes:

Tubes A: Add 50 μl of a solution containing $MgCl_2$ (15 m*M*) and potassium phosphate buffer (75 m*M*, pH 7.4).

Tubes B: Add 50 μl of a solution containing $MgCl_2$ and potassium phosphate buffer (as above) in addition to phosphoenolpyruvate (0.5 m*M*) and pyruvate kinase (20 μg).

Tubes C: Add 50 μl of a solution containing $MgCl_2$, potassium phosphate buffer, PEP, and PK (as above), in addition to adenylate kinase (25 μg).

Tubes D: Add 50 μl of a solution similar to that added to C tubes, except that the solution also contains 10 ng of ATP.

Blanks, ATP standards, and the standard mixtures of ATP, ADP, and AMP are all treated as tubes A–D above. The reason for having the D tubes is that the efficiency of the adenylate kinase reaction is affected by the total adenine nucleotide concentration in the sample,[11] and inclusion of a series of D tubes (with increased concentrations of ATP) will indicate the efficiency of the adenylate kinase reaction in the sample extracts.

All the tubes are incubated for 30 min at 30°, placed into a boiling water bath (100°) for 2 min, and allowed to adjust to room temperature (~30 min). The samples are now ready for ATP analyses according to the methods described above.

Data Reduction and Calculation of the EC Ratio

Net light emission is determined by subtracting the appropriate blank value from each of the total light-emission determinations. Standard curves

[9] A. Pradet, *Physiol. Veg.* **5,** 209 (1967).

[10] A. G. Chapman, L. Fall, and D. E. Atkinson, *J. Bacteriol.* **108,** 1072 (1971).

[11] D. M. Karl and O. Holm-Hansen, *Mar. Biol.*, in press (1978).

are prepared by plotting net peak light emission on the ordinate versus ATP concentration on the abscissa, for each of the three sets of standard data. From these curves, the ATP concentration in each of the three reaction mixtures (ATP, ATP + ADP, ATP + ADP + AMP) is determined, and the amounts of ATP enzymically produced from ADP and AMP are calculated as differences between these measured values. The adenylate EC ratio is determined using the formulation of Ball and Atkinson[12] to reduce the propagation of errors.

If the integral mode of light detection is used, calculation of ATP concentrations becomes more difficult for the following reasons: (1) The addition of phosphoenolpyruvate to crude luciferase preparations results in a significant level of light emission, which increases with time after injection, owing to the presence of pyruvate kinase and ADP in the enzyme preparations. (2) Since adenylate kinase is a heat-stable protein, it remains active following the heat deactivation period and competes with luciferase for the ATP in solution. (3) The plot of ATP versus integrated light flux does not result in a linear plot as expected, especially at ATP concentrations greater than or equal to 20 n*M*. Although graphical analyses of integrated light values have been used to calculate ATP concentrations, greater precision is achieved through the use of a computer program, such as the one described by Booth.[13]

As discussed previously, peak height analyses may be conducted without any of the above considerations, and therefore are strongly recommended for adenine nucleotide determinations.

Discussion

Sampling of Cells and Extraction of Nucleotides

As ATP is rapidly turned over in growing cells,[14] and rapid and dramatic variations result from changes in environmental conditions,[15] it is essential that the cells be extracted as quickly as possible after sample collection. The most direct method of sampling bacterial or algal cultures is through direct injection of the sample into the extraction medium. If this procedure is utilized, it is imperative to measure the levels of soluble (cell-free) nucleotides in the culture medium as well, and to correct the original adenine nucleotide determinations for these dissolved concentrations. The most convenient method for determining cell-free adenine nucleotides is to

[12] W. J. Ball and D. E. Atkinson, *J. Bacteriol.* **121,** 975 (1975).

[13] C. R. Booth, *in* "Proceedings of ATP Methodology Seminar" (G. A. Borun, ed.), p. 104. SAI Technology Company, San Diego, California, 1975.

[14] D. E. F. Harrison and P. K. Maitra, *Biochem. J.* **112,** 647 (1969).

[15] J. Coombs, P. J. Halicki, O. Holm-Hansen, and B. E. Volcani, *Exp. Cell Res.* **47,** 315 (1967).

filter a portion of the culture medium through a membrane filter (pore dimensions dependent upon mean cell diameter) and to extract and assay an equivalent volume of the filtrate.

With most environmental samples, however, a preliminary concentration of the cellular material is required before the extraction procedure. The most practical and effective method for cell concentration is through vacuum filtration. The choice of filter is dictated by the size spectrum of cells. For marine bacterial cells, microfine glass fiber filters (Reeve Angel, 984H), or membrane filters such as Millipore (GS) or Nucleopore (0.2 μm), are most commonly used. For phytoplankton and larger unicells, Whatman (GF/C) glass fiber filters are effective. Immediately after the final amount of liquid has passed through the filter, the vacuum should be released and the filter should be quickly immersed into the extraction medium. It should be mentioned that vacuum filtration has been shown to have a detrimental effect on the concentration of ATP, and on the relative concentrations of ATP, ADP, and AMP recovered from naturally occurring microbial populations[11]; therefore, caution should be taken whenever preliminary concentration of the samples is required. The effect of vacuum filtration can be demonstrated by filtering increasing volumes of a particular water sample and then normalizing the results to a particular volume (i.e., normalize all samples to nanogram of ATP liter^{-1}). As the sample volume is increased, the concentration of ATP liter^{-1} and the value of the adenylate energy charge both decrease. As a compromise, minimal sample volumes should be extracted in order to reduce the effect of vacuum filtration.[11]

Azam and Hodson[16] have described an additional method for ATP determinations from dilute suspensions of microorganisms. This procedure involves the addition of H_2SO_4 directly to the water sample, followed by adsorption of the nucleotides onto activated charcoal and subsequent elution with ammoniacal ethanol. The samples are then evaporated to dryness and reconstituted in Tris buffer (pH 7.7, 20 m*M*). It is necessary to correct the measured ATP concentrations for the concentration of dissolved ATP in each sample. An equal volume of cell-free (0.2 μm filtrate) liquid must also be extracted as above. Since the efficiency of ATP binding to and elution from activated charcoal is highly variable, it is necessary to correct all the determinations using an ATP internal standard. Furthermore, since the binding and elution characteristics of the various adenine nucleotides are not identical,[11,17] and also owing to the possibility of charcoal-catalyzed ATP hydrolysis, this procedure is not recommended for energy charge determinations.

[16] F. Azam and R. E. Hodson, *in* "Proceedings of the Second Biannual ATP Methodology Symposium" (G. A. Borun, ed.), p. 109. SAI Technology Company, San Diego, California, 1977.

[17] D. M. Ireland and D. C. B. Mills, *Biochem. J.* **99,** 283 (1966).

A number of different extraction media and methodologies are currently in use for adenine nucleotide analyses. These include: boiling buffers,[1,18-20] sulfuric acid,[21] perchloric acid,[22] nitric acid,[23] trichloroacetic acid,[24] *n*-butanol,[25] chloroform,[26] and *n*-bromosuccinimide.[27] It has been our experience that excellent results are obtained with boiling Tris buffer (pH 7.7, 20 m*M*) for aqueous suspensions of microbial cells, and with cold sulfuric acid for multicellular organisms and soils or sediments. When using boiling Tris buffer, it is essential that the temperature be maintained at 100° to deactivate the ATPases. Moreover, when samples are directly injected into the boiling Tris, it is necessary to keep the injection volume low relative to the extractant volume (i.e., $\leqslant 2\%$) to minimize temperature drops. After continued heating for 4–5 min, the samples are removed from the water bath, restored to the original volume with distilled water, and stored frozen (−20°) until time of analysis.

When extracting multicellular organisms or soil or sediment samples, boiling buffers generally result in low ATP recoveries. This is due in part to the effects of thermal gradients associated with fluid–solid mixtures[28] and in part to loss of ATP through adsorption at neutral or alkaline pH. These difficulties can be averted through the use of cold sulfuric acid as the extraction medium.[21,28] After an extraction period of 15–20 min at 4°, the nucleotides are separated from the solid phase by centrifugation or vacuum filtration, and the pH of the acid extracts is adjusted to 7.7 with sodium hydroxide. The neutralized extracts are stored frozen (−20°) until time of analysis. Since both Na^+ and SO_4^{2-} ions inhibit the luciferase reaction, the ATP standards must be prepared in the same extraction and neutralization fluids.

Purity of Luciferase Preparation

Considerable variation exists among laboratories concerning the purity, preparation, handling, and use of the firefly luciferin–luciferase prepa-

[18] E. Beutler and M. C. Baluda, *Blood* **23,** 688 (1964).

[19] S. Lin and H. P. Cohen, *Anal. Biochem.* **24,** 531 (1968).

[20] K. Bancroft, E. A. Paul, and W. J. Wiebe, *Limnol. Oceanogr.* **21,** 473 (1976).

[21] C. C. Lee, R. F. Harris, J. D. Williams, D. E. Armstrong, and J. K. Syers, *Soil Sci. Soc. Am., Proc.* **35,** 82 (1971).

[22] R. E. Strange, H. E. Wade, and F. A. Dark, *Nature (London)* **199,** 55 (1963).

[23] J. W. Deming, M. W. McGarry, E. W. Chappelle, and G. L. Picciolo, *in* "Proceedings of the Second Biannual ATP Methodology Symposium" (G. A. Borun, ed.), p. 465. SAI Technology Company, San Diego, California, 1977.

[24] E. Beutler and K. Mathai, *Blood* **30,** 311 (1967).

[25] E. W. Chappelle and G. V. Levin, *Biochem. Med.* **2,** 41 (1968).

[26] A. M. Dhople and J. H. Hanks, *Appl. Microbiol.* **26,** 399 (1973).

[27] N. H. MacLeod, E. W. Chappelle, and A. M. Crawford, *Nature (London)* **223,** 267 (1969).

[28] D. M. Karl and P. A. LaRock, *J. Fish. Res. Board Can.* **32,** 599 (1975).

rations. In general, the specific application of the methods will dictate the choice of the reagents. Crude reagents are less sensitive and less specific than purified preparations, but they are also considerably less expensive. Both crude and partially purified enzyme preparations are available commercially; a number of laboratory techniques are also available for purification of the firefly reagents. These include: $(NH_4)_2SO_4$ precipitation followed by adsorption onto calcium phosphate gels,[29] Sephadex G-100 chromatography,[7] DEAE-cellulose chromatography,[30] and isoelectric focusing.[31] The techniques described in this chapter can be used with either crude or partially purified reagents.

Reaction Specificities

Firefly Luciferase. Although crystalline firefly luciferase is specific for ATP,[4,5] a number of ribose and deoxyribose nucleotides, especially guanosine triphosphate (GTP), will stimulate light emission in most commercial luciferase preparations (crude, as well as "purified"). The degree of analytical interference will depend upon the particular enzyme preparation used (see previous section), the method used to detect light emission (peak vs integral), and the [ATP]/[NTP] ratio of the cell extract. Since the time course of light emission resulting from non-adenine nucleotide triphosphates is slower than the ATP-dependent light-emission kinetics (see this volume [9]), peak height measurements will substantially reduce this source of interference.

Pyruvate Kinase. The pyruvate kinase reaction [Eq. (3)] is relatively nonspecific and will catalyze the transfer of ~P from PEP to GDP, IDP, UDP, and CDP, as well as ADP.[32] The corresponding nucleotide triphosphates that are formed can then serve as substrates for the nucleoside diphosphate kinase reaction, resulting in light emission in impure luciferase preparations.

Adenylate Kinase. The reaction given in Eq. (4) is very specific for AMP as the phosphoryl acceptor, but relatively nonspecific for ATP as the phosphoryl donor[33] (i.e., AMP + NTP $\rightleftharpoons$ NDP + ADP).

[29] W. D. McElroy, this series, Vol. 2, p. 854, 1955.

[30] I. Bekhor, J. K. Pollard, and W. Drell, *in* "Proceedings of the Second Biannual ATP Methodology Symposium" (G. A. Borun, ed.), p. 27. SAI Technology Company, San Diego, California, 1977.

[31] A. Lundin, A. Rickardsson, and A. Thore, *in* "Proceedings of the Second Biannual ATP Methodology Symposium" (G. A. Borun, ed.), p. 205. SAI Technology Company, San Diego, California, 1977.

[32] F. J. Kayne, *in* "The Enzymes" (P. D. Boyer, ed.), 3rd ed., Vol. 8, p. 353. Academic Press, New York, 1973.

[33] L. Noda, *in* "The Enzymes" (P. D. Boyer, ed.), 3rd ed., Vol. 8, p. 279. Academic Press, New York, 1973.

Analytical Interferences

In addition to the problems discussed above in the section entitled Sampling of Cells and Extraction of Nucleotides, other sources of analytical interference may be encountered during the enzymic determination of ATP. These include: (1) the presence of inorganic and organic ions in the sample extracts, resulting in loss of ATP in solution (i.e., through chelation) or in decreased luciferase activity, (2) the presence of humic acid-like substances in the sample extracts that impart a yellow color to the solution, thereby resulting in attenuation of the emitted light, (3) turbidity of the final extracts resulting in light scattering and absorption, (4) the presence of a high concentration of inorganic particulate material in the final extracts resulting in loss of ATP through adsorption, and (5) the presence of contaminating enzymes, in either the sample extracts or the luciferase preparation, that compete with luciferase for the ATP in solution (e.g., ATPase or adenylate kinase) or that result in the production of ATP through transphosphorylase reactions (e.g., nucleoside diphosphate kinase or pyruvate kinase).

Most of the above sources of error can be detected, and corrected for, through the use of an ATP internal standard, as discussed by Strehler.[34] The internal standard may be added in the form of an ATP salt solution, as live or lyophilized bacterial cells, or as radiolabeled ATP. To minimize the effects of ionic interference, it is imperative that the standard ATP solutions be prepared in an ionic medium identical to that of the samples. Peak-height measurements significantly decrease the analytical interference due to the presence of non-adenine nucleotide triphosphates and therefore, are strongly recommended. If available, a strip-chart recorder or oscilloscope should be interfaced with the photometer in order to detect deviations from the standard ATP-dependent light-emission reaction kinetics.

Sensitivity and Reproducibility

The sensitivity of this assay procedure is governed to a large extent by the instrumentation used to detect light emission, and also by the purity of the luciferase preparation. Using crude commercially available luciferase preparations available from Sigma Chemical Company (FLE-50) prepared as described in this chapter, and a commercial ATP photometer (SAI Technology Co.), the lower limit of ATP detection (i.e., >200 counts above background light emission) is about 1 n*M* ATP. If additional sensitivity is required, exogenous luciferin may be added to the enzyme prepa-

[34] B. L. Strehler, *Methods Biochem. Anal.* **16,** 99 (1968).

rations, thereby enabling the detection of 1 pM.[35] When extremely sensitive determinations of ATP, ADP, and AMP are to be conducted, it may be necessary to reduce the level of adenine nucleotides contained within the pyruvate kinase and adenylate kinase enzyme preparations. This is most easily achieved via equilibrium dialysis or gel filtration techniques.

The lower limit of ADP detection is governed by the rate of the pyruvate kinase reaction. Using the enzyme and substrate concentrations described in this report, the conversion of ADP to ATP is complete at least down to a concentration of 1 nM ADP. At ADP concentrations of < 1 nM, the amount of pyruvate kinase used per sample should be increased.

The rate-limiting step for the conversion of AMP to ATP via the coupled adenylate kinase/pyruvate kinase reaction is the initial conversion of AMP to ADP (i.e., AMP + ATP $\rightleftharpoons$ 2 ADP). The apparent K_m of adenylate kinase for AMP is a function of the concentration of ATP in solution; therefore, the overall reaction rate is dependent upon the concentration of ATP in the reaction mixture.[11] If a sufficient concentration of ATP is added to the extract, 1 nM AMP may easily be detected. At lower concentrations, the light emission resulting from the conversion of AMP to ATP becomes a small portion of the total light emission, owing to the requirement for exogenous ATP.

The reproducibility of the peak assay procedure as routinely performed in our laboratories is ± 1–2% of the mean ($n = 8$) throughout the entire range of ATP standards.

Relationship of ATP Concentrations to Biomass

In many ecological studies it is often convenient to express ATP levels in terms of cellular organic carbon, which in turn can be extrapolated to wet-weight or dry-weight determinations, approximate caloric content, etc. There is a considerable body of data in the literature indicating that the ATP concentration in unicellular algae and bacteria is a good measure of the total cellular organic carbon.[2,36] For most microbial cells, the ratio of organic carbon to ATP averages about 250 (i.e., ATP = 0.4% of the cellular organic carbon). In contrast, multicellular organisms, such as crustaceans, bivalves, and annelids, appear to have more ATP per unit carbon, thereby resulting in carbon to ATP ratios of less than 100 (i.e., ATP $\geq$1% of the cellular organic carbon).[37-41] Therefore, in order to obtain the most reliable estimates of biomass in environmental samples, it is essential to remove the

[35] D. M. Karl and O. Holm-Hansen, *Anal. Biochem.* **75,** 100 (1976).
[36] R. D. Hamilton and O. Holm-Hansen, *Limnol. Oceanogr.* **12,** 19 (1967).
[37] N. Balch, *Limnol. Oceanogr.* **17,** 906 (1972).
[38] H. Goerke and W. Ernst, *in* "Proceedings of 9th European Marine Biology Symposium" (H. Barnes, ed.), p. 683. Univ. Press, Aberdeen, Scotland, 1975.

larger organisms (i.e., prefiltration for water, sieving for sediments) prior to the extraction procedure, or to determine their biomass independently and correct the total biomass calculation for these measured values. For pure cultures of bacteria and algae, ATP determinations can be used as a rapid, sensitive, and accurate method of monitoring changes in growth and cell viability (see this volume [7]).

Acknowledgments

The authors thank Dr. A. F. Carlucci for helpful comments and criticism during the preparation of this chapter. The methods described in this report were developed under ERDA Contract EY-76-C-03-0010, PA 20.

[39] V. Bämstedt and H. R. Skjoldal, *Sarsia* **60,** 63 (1976).

[40] A. D. Ansell, *J. Exp. Mar. Biol. Ecol.* **28,** 269 (1977).

[41] D. M. Karl, J. A. Haugness, L. Campbell, and O. Holm-Hansen, *J. Exp. Mar. Biol. Ecol.*, in press.

[9] Determination of GTP, GDP, and GMP in Cell and Tissue Extracts

By DAVID M. KARL

Over the past two decades, an enormous amount of research effort has been directed toward determining the concentrations, and elucidating the physiological roles, of adenine nucleotides (ATP, ADP, and AMP) within cells and tissues; however, relatively little effort has been invested in obtaining comparable information on the additional intracellular purine and pyrimidine nucleotides. In recent years, guanosine triphosphate (GTP) has been shown to be an essential factor for the initiation, the aminoacyl tRNA binding, and the translocation processes of protein biosynthesis.[1] Routine quantitative analyses of guanine nucleotides (GTP, GDP, and GMP) in cell and tissue extracts have generally been neglected owing to the lack of a rapid, sensitive, and specific assay procedure comparable to the firefly luciferase bioluminescent assay for the determination of ATP. Several methods have been described for isolating GTP from cell and tissue nucleotide mixtures, including various separation procedures such as paper[2] thin-layer ion-exchange,[3,4] and high-pressure liquid[5] chromatography.

[1] F. Lucas-Lenard and F. Lipmann, *Annu. Rev. Biochem.* **40,** 409 (1971).

[2] G. Zweig and J. R. Whitaker, "Paper Chromatography and Electrophoresis," Vol. 1. Academic Press, New York, 1967.

[3] K. Randerath, "Thin-Layer Chromatography." Academic Press, New York, 1966.

[4] M. S. P. Manandhar and K. Van Dyke, *Anal. Biochem.* **60,** 122 (1974).

[5] P. R. Brown, *J. Chromatogr.* **52,** 257 (1970).

ISBN 0-12-181957-4

This chapter describes a simple, sensitive, and reproducible method for determining the concentrations of GTP, GDP, and GMP in extracts of biological materials, without prior chromatographic separations. This procedure is based upon the selective enzymic degradation of specific nucleotide triphosphates (NTP's) within cell extracts, and the fact that crude firefly luciferase preparations contain transphosphorylase enzymes and ADP, which together result in the quantitative production of ATP (and therefore light) in response to the addition of a wide variety of ribose and deoxyribose nucleotide triphosphates.[6,7] In addition to a complete description of the experimental procedures and equipment necessary for the determination of GTP, GDP, and GMP, a brief discussion of the pertinent analytical principles, assay limitations, and possible sources of interference is presented.

Assay Principles

$$\text{D-glucose} + \text{ATP} \underset{}{\overset{\text{HK}}{\rightleftharpoons}} \text{D-glucose-6-P} + \text{ADP} + \text{H}^+ \qquad (1)$$

$$\text{D-glucose-6-P} + \text{NADP}^+ \xrightarrow{\text{G6P-DH}} \text{gluconate-6-P} + \text{NADPH} + \text{H}^+$$

$$\alpha\text{-D-glucose-1-P} + \text{UTP} \xrightarrow{\text{UDPG-PP}} \text{UDP-glucose} + \text{PP}_i \qquad (2)$$

$$\text{GMP} + \text{ATP} \overset{\text{GK}}{\rightleftharpoons} + \text{GDP} + \text{ADP} \qquad (3)$$

$$\text{GDP} + \text{phosphoenolpyruvate} \xrightarrow{\text{PK}} \text{pyruvate} + \text{GTP} \qquad (4)$$

$$\text{ADP} + \text{GTP} \overset{\text{NDPK}}{\rightleftharpoons} \text{GDP} + \text{ATP} \qquad (5)$$

$$\text{ATP} + \text{luciferin} \xrightarrow{\text{L}} \text{product} + \text{PP}_i + \textit{light} \qquad (6)$$

where HK = hexokinase (EC 2.7.1.1), G6P-DH = glucose-6-phosphate dehydrogenase (EC 1.1.1.49), UDPG-PP = UDP-glucose pyrophosphorylase (EC 2.7.7.9), GK = guanosine-5′-monophosphate kinase (EC 2.7.4.8), PK = pyruvate kinase (EC 2.7.1.40), NDPK = nucleoside-5′-diphosphate kinase (EC 2.7.4.6), and L = firefly luciferase (EC 2.7. .)

Equipment and Reagents

Equipment

Glassware: All glassware used for nucleotide analyses should be thoroughly cleaned in order to eliminate the possibility of contamination; if possible, disposable glassware should be used. (a) Pyrex

[6] B. L. Strehler, *Methods Biochem. Anal.* **16,** 99 (1968).
[7] M. S. P. Manandhar and K. Van Dyke, *Microchem. J.* **19,** 42 (1974).

glass flasks, 50 and 250 ml, (b) glass culture tubes, 12 × 75 mm (Kimble), (c) glass vials, 18 × 45 mm (Scientific Products)

Pipettes: (a) 5- and 10-ml glass or disposable plastic; (b) automatic pipettes of 50-μl and 200-μl volume, with disposable tips; (c) automatic pipette 0.1–1.0 ml, with disposable tips

Incubation equipment: a water bath to maintain a uniform temperature (30°) during the period of the enzymic reactions

Boiling water bath

Instrument for quantitative light detection: Any one of a number of commercially available fluorometers, spectrophotometers, liquid scintillation counters, or ATP photometers is suitable for this purpose.[6,8]

Strip-chart recorder or oscilloscope

Automatic pipette: This is essential for peak-height measurements in order to ensure reproducible sample injection velocities.

Reagents

Potassium phosphate buffer, 225 m*M*, pH 7.4

$MgCl_2$ solution, 45 m*M*

D-Glucose/α-D-glucose-1-P solution, 1.5 m*M* each (Sigma Chemical Co.)

Phosphoenolpyruvate solution, 1.5 m*M* (Sigma Chemical Co.)

$NADP^+$ solution, 1.5 m*M* (Sigma Chemical Co.)

Tris(hydroxymethyl)aminomethane-HCl buffer (Tris buffer), 20 m*M*, pH 7.75 (Sigma Chemical Co.)

Sodium salts of GTP, GDP, GMP, ATP, ADP, and UTP (Sigma Chemical Co.)

$MgSO_4$ solution, 40 m*M*

Sodium arsenate buffer, 100 m*M*, pH 7.4

D-Luciferin, synthetic (Sigma Chemical Co.); optional

Firefly lantern extract (Sigma Chemical Co., catalog No. FLE-50)

HK/G6P-DH, crystalline, from yeast, 2:1 activity ratio (Sigma Chemical Co.)

PK, crystalline, from rabbit muscle, 350–500 units per milligram of protein (Sigma Chemical Co.)

GK, from hog brain, 10 units per milligram of protein (Boehringer Mannheim)

UDPG-PP, from beef liver, 100 units per milligram of protein (Boehringer Mannheim)

NDPK, from yeast (Sigma Chemical Co.); optional

[8] Picciolo *et al.*, this volume [43].

Methods

Sample Collection and Nucleotide Extraction

Extraction of guanine nucleotides from the appropriate sample material is the first step in the quantitative assay procedure. A number of different extraction media and methodologies are currently available for nucleotide determinations; these include boiling buffers,[9–12] sulfuric acid,[13] perchloric acid,[14] trichloroacetic acid,[15] *n*-butanol,[16] chloroform,[17] and *n*-bromosuccinimide.[18] The procedure finally selected for routine nucleotide extractions will depend upon the nature of the biological material to be analyzed (i.e., bacterial cells, mammalian tissue, etc.). Several different extraction techniques should be tested, using the biological material in question, to evaluate such analytical properties as the rate of nucleotide release, ionic interference of the extraction fluid(s), residual enzyme activity, chemical stability of the nucleotides, ease of operation, reproducibility, and the absolute levels of GTP resulting in the final cell extracts.

CAUTION: Since the rate of nucleotide triphosphate turnover in growing cells is extremely fast (i.e., for bacteria, 1–2 sec), it is essential that the biological material be extracted immediately after collection. Centrifugation[19] and vacuum filtration[20] may result in a decrease in the level of NTPs within the cell extracts.

Luciferase Enzyme Preparations

Each vial of lyophilized firefly lantern extract (50 mg/vial) is reconstituted by adding 5 ml of distilled water, 10 ml of 100 m*M* sodium arsenate buffer (pH 7.4), 10 ml of $MgSO_4$ (40 m*M*), and 2.5 μg of ADP. The enzyme mixture is allowed to stand for 3–4 hr at room temperature (23°–25°) in order to reduce the level of background light emission. After aging, the insoluble residue is removed by vacuum filtration or centrifugation. The

[9] E. Beutler and M. C. Baluda, *Blood* **23,** 688 (1964).
[10] O. Holm-Hansen and C. R. Booth, *Limnol. Oceanogr.* **11,** 510 (1966).
[11] S. Lin and H. P. Cohen, *Anal. Biochem.* **24,** 531 (1968).
[12] K. Bancroft, E. A. Paul, and W. J. Wiebe, *Limnol. Oceanogr.* **21,** 473 (1976).
[13] C. C. Lee, R. F. Harris, J. D. Williams, D. E. Armstrong, and J. K. Syers, *Soil Sci. Soc. Am., Proc.* **35,** 82 (1971).
[14] R. E. Strange, H. E. Wade, and F. A. Dark, *Nature (London)* **199,** 55 (1963).
[15] E. Beutler and K. Mathai, *Blood* **30,** 311 (1967).
[16] E. W. Chappelle and G. V. Levin, *Biochem. Med.* **2,** 41 (1968).
[17] A. M. Dhople and J. H. Hanks, *Appl. Microbiol.* **26,** 399 (1973).
[18] N. H. MacLoed, E. W. Chappelle, and A. M. Crawford, *Nature (London)* **223,** 267 (1969).
[19] H. A. Cole, J. W. T. Wimpenny, and D. E. Hughes, *Biochim. Biophys. Acta* **143,** 445 (1967).
[20] O. Holm-Hansen and D. M. Karl, this volume [8].

clear filtrate or supernatant is transferred to a clean flask, and the enzyme preparation is now ready for use. Since considerable variation occurs between individual vials of firefly lantern extract, the contents of several vials should be pooled whenever large volumes of enzyme are required.

Standard Nucleotide Solutions

For each firefly luciferase preparation (1 or more vials), a set of nucleotide standards must be analyzed. Stock GTP, GDP, and GMP solutions (2 μM) are prepared in Tris buffer and stored frozen ($-20°$) in 5-ml aliquots. When required, a single vial is thawed and working solutions (1 nM to 100 nM) are made up by dilution of the stock with Tris buffer. A separate set of standard solutions is prepared for each assay and discarded after a single day's use. This procedure will eliminate any problems due to GTP hydrolysis.

GTP Determinations

Two-hundred microliters are pipetted into a disposable glass culture tube (12 × 75 mm), labeled and placed into a test tube rack. A series of GTP standard solutions covering the full range of expected experimental values, as well as reagent blanks, are also processed simultaneously. Fifty microliters of a solution containing 75 mM potassium phosphate buffer (pH 7.4), 0.5 mM $NADP^+$, 15 mM $MgCl_2$, 0.5 mM D-glucose, 0.5 mM α-D-glucose-1-P,HK/G6P-DH (2 units ml^{-1}) and UDPG-PP (5 units ml^{-1}) are pipetted into each tube. The samples are incubated for 15–30 min at 30°, followed by a 2-min enzyme deactivation period in a boiling water bath (100°). The actual time required for the complete conversion of ATP and UTP to ADP and UDP-glucose, respectively, should be determined by conducting a time-course experiment using ATP and UTP concentrations similar to those present in the nucleotide extracts. After heat deactivation of the enzymes, the samples are brought to room temperature prior to the actual GTP assay procedure.

Either the initial rise of the light-emission curve, the peak height of luminescence, or some integrated portion of the subsequent reaction kinetics can be used to relate unknown sample extracts to the light emitted from standard GTP solutions. It is recommended that the photometer (or similar light-detection instrument) be interfaced with an analog recorder, in order to follow the reaction kinetics.

CAUTION: All GTP assays should be conducted in reduced illumination, and fluorescence lighting should be avoided because of the chemiluminescent properties of certain glassware and plasticware.

Initial Rise and Peak-Height Measurements. Pipette 1 ml of the firefly luciferase preparation into a small glass vial (18 × 45 mm), and record the background light emission. Using an automatic pipette, withdraw 200 μl of the appropriate sample extract (or GTP standard solution) and position the pipette assembly onto the photometer. Inject the sample into the enzyme preparation and record the photometer digital display; or alternatively, measure the peak height of light emission directly from the strip-chart recorder trace.

NOTE: Reliable initial rise or peak height measurements require specially designed equipment for rapid and reproducible injection velocities to ensure complete mixing of the reagents. To the author's knowledge, only three instruments are commercially available at the present time that have this extended capability. The analytical characteristics and properties of these instruments have recently been evaluated and compared.[8]

Integrated Light-Flux Determinations. Pipette 1 ml of the firefly luciferase preparation into a small glass vial (18 × 45 mm), and record the background light emission. Remove the vial from the photometer, inject 200 μl of the appropriate sample extract (or GTP standard solution) into the enzyme preparation, mix the contents thoroughly for 5 sec, and replace the vial into the photometer. The light flux is integrated, either electronically or manually, over a predetermined portion of the light-emission curve. It is important that integrated light-flux determinations be restricted to the early portion of the reaction (<30 sec) in order to reduce various sources of analytical interference (see Fig. 1 and the section on analytical interferences and troubleshooting, this chapter, pp. 93–94).

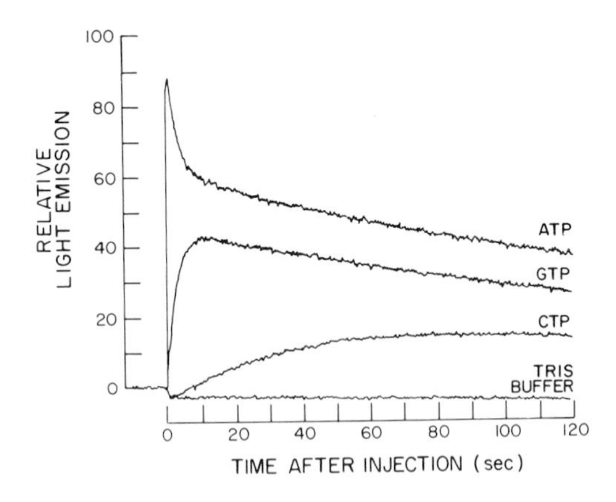

FIG. 1. Light emission kinetics and relative rates of ATP, GTP, and CTP reactivity upon injection of the samples into crude luciferase enzyme preparations. The zero light-emission level represents the enzyme background activity, and values < 0 indicate dilution of this endogenous reactivity. The concentration of all three nucleotides was 20 nM.

GTP, GDP, and GMP Determinations

For each set of guanine nucleotide determinations, 200 μl of the sample are pipetted into a series of 4 disposable culture tubes (12 × 75 mm), labeled (A–D) and placed into a test tube rack. For GTP determinations (tube A), 50 μl of a solution containing 75 mM potassium phosphate buffer (pH 7.4), and 15 mM $MgCl_2$ are added to each tube. For GTP plus GDP determinations (tube B), 50 μl of a solution containing 75 mM potassium phosphate buffer (pH 7.4), 15 mM $MgCl_2$, 0.5 mM PEP, and 20 μg of PK are added to each tube. For GTP plus GDP plus GMP determinations (tube C), 50 μl of a solution containing 75 mM potassium phosphate buffer (pH 7.4), 15 mM $MgCl_2$, 0.5 mM PEP, 10 ng of ATP, 20 μg of PK, and 5 μg of GK are added to each tube. In order to determine the efficiency of coupled reactions (3) and (4), 50 μl of a solution containing 75 mM potassium phosphate buffer (pH 7.4), 15 mM $MgCl_2$, 0.5 mM PEP, 10 ng of ATP, 20 μg of PK, 5 μg of GK, and an internal GMP standard are added to each D tube. The actual concentration of GMP added will depend upon the experimental conditions; however, the internal standard concentration should be kept well within the linear range of the GTP assay and, ideally, should be approximately equal to the concentration of GMP already contained within the sample extract. Standard GTP solutions should also be processed, simultaneously, with each of the four reagents.

In addition to the GTP standard solutions, it is advisable to prepare, and allow to react, several GDP and GMP standards, as well as standard solutions containing known molar ratios of GTP, GDP, and GMP together, in order to monitor the efficiency of reactions (3) and (4). Enzyme blanks (Tris buffer plus the appropriate enzyme solutions) should also be prepared, since certain commercial enzyme preparations contain contaminating nucleotides. All samples are incubated at 30° for 60 min and then place into a boiling water bath (100°, 2 min) in order to denature the enzymes.

NOTE: It is essential that the PK be deactivated since the addition of PEP and active PK to crude firefly luciferase preparations will cause rapid production of ATP from ADP contained within the enzyme mixture. Once the samples have adjusted to room temperature, they are treated as GTP samples, and assayed as described in the preceding section (GTP determinations).

Data Reduction

GTP Data. Determine the net light emission by subtracting the appropriate enzyme blank value from the total light emitted in each of the sample or GTP standard reactions. Construct a standard curve relating net light emission (initial rise, peak or integrated) to GTP concentrations. From this curve, the GTP concentration in each sample extract can be calculated, and, knowing the proportion of the total sample volume actually assayed and the volume (or weight) of material originally extracted, the [GTP] ml^{-1} (or [GTP] $cell^{-1}$, [GTP] g^{-1} of tissue, etc.) of the biological material can be determined.

GTP, GDP, and GMP Data. Determine the net light emission by subtracting the appropriate enzyme blank value from the total light emitted in each of the sample or GTP standard reactions. Standard curves are constructed by plotting net light emission (initial rise, peak, or integrated) versus GTP concentrations. The GTP concentration in each of the A tubes is calculated from the GTP standard curve prepared with A reagents; likewise, the GTP concentration in tubes B, C, and D are calculated from the GTP standard curves prepared with B, C, and D reagents, respectively. GDP concentrations are determined by subtracting tube A determinations from tube B determinations. GMP concentrations are determined by subtracting tube B determinations from tube C determinations, and by correcting for the efficiency of the GK reaction (if necessary), using the internal standard data as follows:

$$\% \text{ recovery} = \frac{[\text{tube D}] - [\text{tube C}]}{[\text{internal standard}]} \times 100$$

$$[\text{GMP}] = \frac{[\text{tube C}] - [\text{tube B}]}{\text{decimal equivalent of } \% \text{ recovery}}$$

Capabilities

The sensitivity of the GTP assay procedure is greatly influenced by the instrument used to detect the light emission. With a commercial ATP photometer (SAI Technology Co., San Diego, California), the procedure as described in this report can detect 1 ng of GTP ml^{-1} of solution ($\sim$2 nM). If additional sensitivity is required, synthetic D-luciferin may be added to saturate the crude enzyme preparation,[21] thereby extending the limit of detection of 1 pg of GTP ml^{-1} ($\sim$2 pM). If measurements are made of the initial rate of light emission, the addition of exogenous NDPK to the crude enzyme preparation will also significantly increase the sensitivity of the GTP assay. The luminescent response is linear to a maximum GTP concentration of 50 ng ml^{-1} of solution (100 nM). For a given operator, variation between replicate standard sample analyses is generally $\pm 3\%$, using the commercial ATP photometer.

[21] D. M. Karl and O. Holm-Hansen, *Anal. Biochem.* **75**, 100 (1976).

Reaction Specificities

As described in this report, the HK/G6P-DH coupled reaction is highly specific for ATP. No hydrolysis of GTP could be detected when a series of standard solutions were treated as experimental samples. Likewise, UDPG-PP showed no reactivity toward GTP at the concentrations and incubation period selected for this reaction. It is essential, however, that the HK/G6P-DH and the UDPG-PP enzyme activities be deactivated, after the appropriate incubation periods, in order to eliminate the possibility of subsequent GTP hydrolysis. GK is specific for GMP as the phosphoryl acceptor, and for ATP as the phosphoryl donor.[22]

PK catalyzes the phosphorylation of both ADP and GDP to ATP and GTP, respectively. This ATP regenerating system ensures maximum conversion of GMP to GTP in coupled reactions (3) and (4). NDPK will catalyze the production of ATP (from ADP) upon the addition of a wide variety of ribose and deoxyribose nucleotide triphosphates. Since ATP and UTP are hydrolyzed prior to the GTP assay, CTP is the only remaining intracellular NTP that may be present at concentrations comparable to GTP. When using initial rise measurements, however, CTP results in $<4\%$ of the light emission resulting from an equimolar concentration of GTP (see Fig. 1).

Analytical Interferences and Troubleshooting

Since the determination of GTP is dependent upon the production of ATP within the crude luciferase preparations [Reaction (5)], followed by an ATP-dependent light-emission reaction [Reaction (6)], the final light emitted from this GTP-dependent coupled light-emission reaction [Reactions (5) and (6)] will be subjected to the same sources of interference as have been described for the quantitative analysis of ATP.[6,23,24] The use of internal standards is strongly recommended in order to correct for all sources of analytical interference. It is important that $MgSO_4$ is not used as the source of Mg^{2+} for enzyme activation during the preliminary steps of this procedure [Reactions (1–4)], since UDPG-PP is strongly inhibited by SO_4^{2-}. Although complete conversion of GMP to GTP may be achieved through the addition of excess ATP, or by increasing the concentration of GK used per reaction tube [Reaction (3)], both of these modifications will result in a significant level of interference on the GTP assay due to the contamination of GK preparations by the enzyme myokinase (MK). Since MK is an

[22] R. P. Miech and R. E. Parks, *J. Biol. Chem.* **240,** 351 (1965).

[23] B. L. Strehler and W. D. McElroy, this series, Volume 3, p. 871.

[24] L. M. Aledort, R. I. Weed, and S. B. Troup, *Anal. Biochem.* **17,** 268 (1966).

extremely heat-stable protein, back production of ATP will occur (2 ADP $\overset{MK}{\rightleftharpoons}$ AMP + ATP) following the heat deactivation of coupled Reactions (3) and (4). By maintaining relatively low concentrations of ATP [i.e., ADP after Reaction (1)] and of GK, the effects of MK contamination are eliminated. If sufficiently high levels of ADP are present within the GTP extracts ($>$200 ng of ADP ml^{-1}), a small amount of light will be produced, after sample injection, owing to MK activity contained within the crude luciferase preparations. The amount of light emitted is less than 1% of the activity resulting from an equimolar concentration of GTP; however, if necessary, this source of interference can be evaluated (and corrected for) by measuring the sum of the concentrations of ATP and ADP within each sample extract[20] and relating these values to the reactivity of standard ADP solutions.

Acknowledgments

The author expresses his appreciation to Dr. O. Holm-Hansen for comments, criticism, and encouragement offered during the course of this research and the preparation of this chapter. Dr. F. Azam and Ms. L. Campbell critically reviewed the original manuscript and offered helpful suggestions for improvement. The methodology described in this report was developed under ERDA contract EY-76-C-03-0010 P.A. 20.

[10] Measurement of the Activity of Cyclic Nucleotide Phosphodiesterases with Firefly Luciferin–Luciferase Coupled Assay Systems

By RICHARD FERTEL and BENJAMIN WEISS

The procedures described below are designed to measure cyclic AMP phosphodiesterase and cyclic GMP phosphodiesterase (EC 3.1.4.17) by reaction sequences using firefly luciferin and luciferase in the final step. These assay systems are simple, sensitive, inexpensive, and reproducible.

ISBN 0-12-181957-4

Intracellular concentrations of adenosine 3′,5′-cyclic monophosphate (cyclic AMP) and guanosine 3′,5′-cyclic monophosphate (cyclic GMP) affect a number of biochemical and physiologic processes in the cell.[1-6] Accordingly, certain cell functions may be controlled by altering the cyclic nucleotide concentrations in that cell. One way to influence the concentration of the cyclic nucleotides is by activating or inhibiting cyclic nucleotide phosphodiesterases, which catalyze the hydrolysis of the cyclic nucleotides to their 5′-monophosphate analogs. For this reason, these enzymes have been the subject of a number of investigations,[6] and a variety of methods have been devised to measure their activity.[7-10] We have developed for these enzymes assay procedures based on the quantitative coupling of the product of the phosphodiesterase reaction [either adenosine 5′-monophosphate (5′-AMP) or guanosine 5′-monophosphate (5′-GMP)] to adenosine-5′-triphosphate [ATP].[11,12] The concentration of ATP in the reaction mixture is then determined by means of the firefly luciferin–luciferase reaction.[13]

Principle of the Assay Systems

Cyclic-AMP Phosphodiesterase

In this reaction sequence, the phosphodiesterase converts cyclic AMP to its degradation product, 5′-AMP. The 5′-AMP, in the presence of a very low concentration of ATP, which serves as a phosphate donor, is converted by the enzyme myokinase (EC 2.7.4.3) to adenosine-5′-diphosphate

[1] G. A. Robison, G. G. Nahas, and L. Triner, eds. *Ann. N. Y. Acad. Sci.* **185** (1971).
[2] P. Greengard, R. Paoletti, and G. A. Robison, eds. "Advances in Cyclic Nucleotide Research," Vol. 1, Raven, New York, 1972.
[3] P. Greengard and E. Costa, eds. *Adv. Biochem. Psychopharmacol.* **3** (1970).
[4] N. D. Goldberg, R. F. O'Dea, and M. K. Haddox, *in* "Advances in Cyclic Nucleotide Research" (P. Greengard and G. A. Robison, eds.), Vol. 2, p. 155. Raven, New York, 1973.
[5] B. Weiss, ed. "Cyclic Nucleotides in Disease." Univ. Park Press, Baltimore, Maryland, 1975.
[6] B. Weiss, and R. Fertel, *Adv. Pharmacol. Chemother.* **14,** 189 (1977).
[7] R. W. Butcher and E. W. Sutherland, *J. Biol. Chem.* **237,** 1244 (1962).
[8] W. J. Thompson and M. M. Appleman, *J. Biochem.* **10,** 311 (1971).
[9] C. R. Filburn and J. Karn, *Anal. Biochem.* **52,** 505 (1973).
[10] W. Y. Cheung, *Anal. Biochem.* **28,** 182 (1969).
[11] B. Weiss, R. Lehne, and S. Strada, *Anal. Biochem.* **45,** 222 (1972).
[12] R. Fertel and B. Weiss, *Anal. Biochem.* **59,** 386 (1974).
[13] B. L. Strehler and J. R. Totter, *Arch. Biochem. Biophys.* **40,** 28 (1952).

(ADP). ADP is then converted to ATP in the presence of phosphoenolpyruvate, which serves as a phosphate donor, and the enzyme pyruvate kinase (EC 2.7.1.40), which catalyzes the reaction.

Under the conditions specified below, the 5′-AMP is rapidly and completely converted to ATP, which is then measured using the firefly luciferin–luciferase reaction. The entire sequence is shown below:

$$\text{Cyclic AMP} \xrightarrow{\text{phosphodiesterase}} 5'\text{-AMP}$$

$$5'\text{-AMP} + \text{ATP} \xrightleftharpoons{\text{myokinase}} 2\ \text{ADP}$$

$$2\ \text{ADP} + 2\ \text{phosphoenolpyruvate} \xrightleftharpoons{\text{pyruvate kinase}} 2\ \text{ATP} + 2\ \text{pyruvate}$$

$$\text{ATP} + \text{luciferin} + O_2 \xrightarrow{\text{luciferase}} \text{AMP} + PP_i + \text{oxyluciferin} + CO_2 + \text{light}$$

Cyclic-GMP Phosphodiesterase

In this reaction sequence, the 5′-GMP formed as the product of the phosphodiesterase reaction alters the equilibrium of the reaction catalyzed by guanosine-5′-monophosphate kinase (EC 2.7.4.8). This leads to a decrease in ATP that is proportional to the concentration of 5′-GMP formed. The ATP concentration is then determined by reaction with firefly luciferin and luciferase. The reaction sequence is as follows:

$$\text{Cyclic GMP} \xrightarrow{\text{phosphodiesterase}} 5'\text{-GMP}$$

$$5'\text{-GMP} + \text{ATP} \xrightleftharpoons[\text{kinase}]{\text{guanosine-5'-monophosphate}} \text{GDP} + \text{ADP}$$

$$\text{ATP} + \text{luciferin} + O_2 \xrightarrow{\text{luciferase}} \text{AMP} + PP_i + \text{oxyluciferin} + CO_2 + \text{light}$$

Reagents Used in the Assay Systems

Reagents for the Cyclic-AMP Phosphodiesterase Assay

Reagent A

Glycylglycine buffer, 150 m*M* pH 8.0
Ammonium acetate, 75 m*M*
Magnesium chloride, 9 m*M*

Calcium chloride, 30 μM
Phosphoenolpyruvate, 0.78 mM
Dithiothreitol, 15 mM
ATP, 3 nM

Reactants in this reagent are all at 3 times their final concentration. The reagent is stable for several months at −4°. On the day of the assay, 3% bovine serum albumin (BSA), myokinase, and pyruvate kinase, which are made fresh or stored for short periods at 4°, are added to make the following reagents:

Reagent B
Reagent A, 1 ml
3% BSA, 10 μl
Myokinase, 1.0 μg
Pyruvate kinase, 0.5 μg

Reagent C
Reagent A, 1 ml
3% BSA, 10 μl
Myokinase, 2.0 μg
Pyruvate kinase, 1.0 μg

Reagents for Cyclic-GMP Phosphodiesterase Assay

Reagent D
Glycylglycine buffer, 150 mM, pH 8.0
Magnesium chloride, 6 mM
Calcium chloride, 30 mM
Dithiothreitol, 15 mM

As described above, the reagent is 3 times the final concentration and is stable for several months if stored at −4°. Guanosine-5′-monophosphate kinase and 3% BSA are either made fresh or stored for short periods at 4°. ATP is either made fresh or stored at −4° for up to 1 year.

On the day of the assay these solutions are added to reagent D to make reagent E as follows:

Reagent E
Reagent D, 1 ml
3% BSA, 10 μl
Guanosine-5′-monophosphate kinase, 20 μg
ATP, 0.1–5 μM, depending on desired assay sensitivity

Reagent for Assay of ATP

Reagent F

Morpholinopropanesulfonic acid (MOPS) buffer, pH 7.8, 10 m*M*
Magnesium sulfate, 10 m*M*
0.5% BSA
15 mg/ml purified luciferin–luciferase

Under certain circumstances it may be desirable to increase the concentration of the luciferin–luciferase mixture, or to add additional luciferase.

Source and Preparation of the Reagents

Myokinase (rabbit muscle), pyruvate kinase (rabbit muscle), and guanosine-5′-monophosphate kinase (hog brain) are obtained from Boehringer Mannheim Biochemicals and stored at 4°. The myokinase and pyruvate kinase are received in a suspension of ammonium sulfate. Before use, the suspensions are centrifuged at 1000 *g* for 10 min, and the resulting precipitates, which contain the enzymes, are dissolved in reagent A to the appropriate concentration. Guanosine-5′-monophosphate kinase, which is in a glycerol solution, is used without further treatment.

A purified luciferin–luciferase mixture is obtained from the instrument division of E.I. DuPont de Nemours & Co. Purified luciferase is obtained from either Sigma Chemical Co. or Calbiochem. All other reagents are obtained from Sigma Chemical Co. or Fisher Scientific.

Purification of Cyclic AMP

Cyclic nucleotides obtained from commercial sources generally contain contaminating nucleotides. Although the concentration of these contaminants is low, they may interfere with the assay. Therefore, the cyclic nucleotides must be purified before use.

Cyclic AMP is purified by precipitating the nucleotide contaminants by the addition of solutions of barium hydroxide and zinc sulfate.[14] To each milliliter of a 100 m*M* solution of cyclic AMP, adjusted to pH 7.5, is added 0.25 ml of 0.25 *M* barium hydroxide and 0.25 ml of 0.25 *M* zinc sulfate. The suspension is mixed and centrifuged, and the resulting supernatant fluid is removed, centrifuged again to remove all traces of the precipitate, and passed through a column (7 × 40 mm) of Dowex 50-X8, 200–400 mesh, hydrogen form. The columns are eluted with water, and 1-ml fractions are

[14] G. Krishna, B. Weiss, and B. B. Brodie, *J. Pharmacol. Exp. Ther.* **163,** 379 (1968).

collected. The eluent is monitored by reading the absorbance at 259 nm. The concentration of cyclic AMP is calculated and adjusted to 3 m*M* on the basis of its molar extinction coefficient ($E = 15{,}400$).

This solution of cyclic AMP is stable for several months when stored at −4°.

Purification of Cyclic GMP

The cyclic GMP is purified by alumina adsorption[15] and ion-exchange column chromatography. One milliliter of a 20 m*M* cyclic GMP solution in 50 m*M* Tris HCl, pH 8.0, is placed on a 7 × 25 mm neutral alumina column. The eluent from this milliliter is discarded, and the column is eluted with 50 m*M* Tris HCl, pH 8.0. The initial milliliter of Tris eluent is discarded. The next 3 ml contain the purified cyclic GMP. One-milliliter portions of these fractions from the alumina column are placed on a 7 × 25 mm column of Dowex 50-X8. The eluent from this milliliter is discarded. The cyclic GMP is then eluted from the column with 2-ml fractions of water. The fractions containing the highest concentration of cyclic GMP, as determined from optical density measurement (252 nm), are combined. The concentration of cyclic GMP is calculated from its molar extinction coefficient ($E = 13{,}700$) and adjusted to 1 mM. Cyclic GMP purified in this manner typically contains less than 0.005% 5′-GMP and is stable for several months if stored at −4°.

Assay Procedure

Cyclic AMP Phosphodiesterase

This assay is performed in 3 steps. In the first step, the following components are added to a 6 × 50 mm tube: 25 μl of tissue sample containing the unknown phosphodiesterase activity, 25 μl of reagent B, and 25 μl of cyclic AMP.

The reaction sequence is initiated with the addition of cyclic AMP (concentrations from 0.5 μM to 1 mM can be used). The samples are incubated for various times at 37°, placed in a boiling water bath for 5 min to stop the reaction, and cooled. In the second step, 25 μl of reagent C are added to each tube, and the samples are incubated at 37° for 1 hr. In the final step, the samples are assayed for ATP by adding 10 μl of reagent F to each sample and recording the emitted light.

[15] A. A. White and T. V. Zenser, *Anal. Biochem.* **41,** 372 (1971).

Cyclic GMP Phosphodiesterase

This procedure is also performed in 3 steps. In the first step, the following components are added to a 6 × 50 mm tube: 25 μl of tissue sample, 25 μl of reagent D, 25 μl of cyclic GMP.

The reaction is initiated with the addition of cyclic GMP (concentrations from 0.1 μM to 1 mM can be used). The samples are incubated for various times at 37°, placed in boiling water bath for 5 min, and cooled. In the second step, 25 μl of reagent E are added to each tube, and the samples are reincubated at 37° for 1 hr. In the final step, ATP concentration is determined by the addition of 10 μl of reagent F as described above.

Measurement of Generated Light

The light emission can be quantitated either by scintillation spectrometer or by instruments designed expressly for this purpose, which are available from E. I. Dupont de Nemours, Inc., Aminco-Bowman, Inc., and SAI Technology, Inc.

Results

Standard Curves Obtained with These Assay Procedures

To determine the activity of an unknown sample of phosphodiesterase, assays are run with a standard curve, which consists of varying concentrations of either 5′-AMP (when cyclic-AMP phosphodiesterase is assayed), or of 5′-GMP (when cyclic-GMP phosphodiesterase is assayed). Standards are assayed under incubation conditions identical to those used for the unknown samples. This procedure is designed to account for the presence of interfering enzymes (e.g., nucleotidases) in the sample. In addition, samples of tissue are tested to determine whether they contain nucleotide contaminants, such as ATP, ADP, or 5′-AMP, which can increase the background of the assay. If the sample does contain high concentrations of such nucleotides, the blank can usually be reduced by preincubating the tissue sample to allow nucleotidases in the tissue to convert these nucleotides to the nucleosides, which do not interfere with the assay.

A typical standard curve for 5′-AMP is linear from 10 to 10,000 pmol and is essentially identical to the curve obtained with ATP alone (Fig. 1). This indicates that under the assay conditions all the 5′-AMP is converted to ATP.

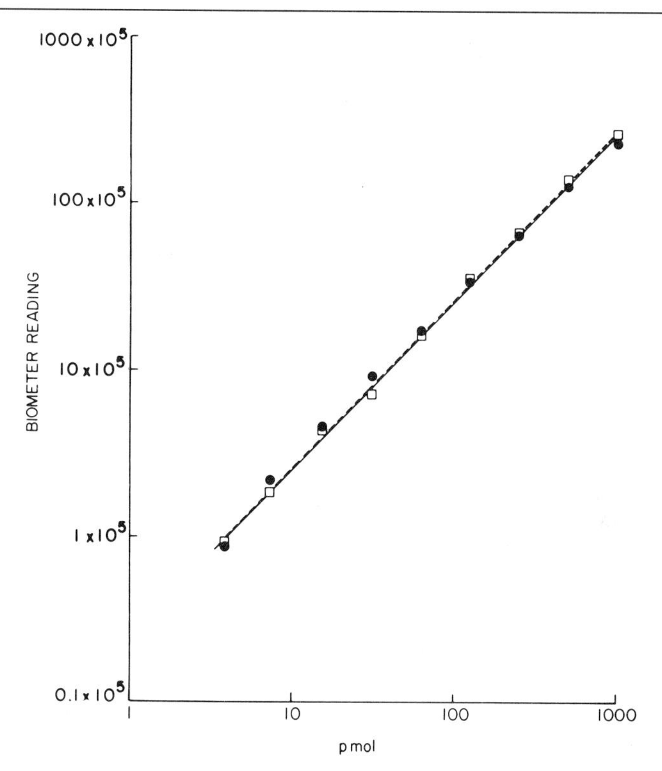

FIG. 1. Conversion of 5′-AMP to ATP. Varying amounts of 5′-AMP (●—●) or ATP (□- - -□) were incubated for 30 min at 37° under conditions described in the test. The light generated by the ATP in the reaction mixture is expressed in arbitrary units. From B. Weiss, R. Lehne, and S. Strada, *Anal. Biochem.* **45,** 222 (1972).

The sensitivity of the assay for 5′-GMP is dependent on the initial concentration of ATP in reagent E (Fig. 2). The lower the initial concentration of ATP, the greater the percentage decrease in ATP with a given concentration of 5′-GMP.

Correlation of Phosphodiesterase Activity with Time of Incubation

The optimal incubation time for determination of both cyclic AMP and cyclic GMP phosphodiesterase in a given tissue may vary, but in general, product formation is directly related to incubation time for at least 60 min (Figs. 3A and 3B).

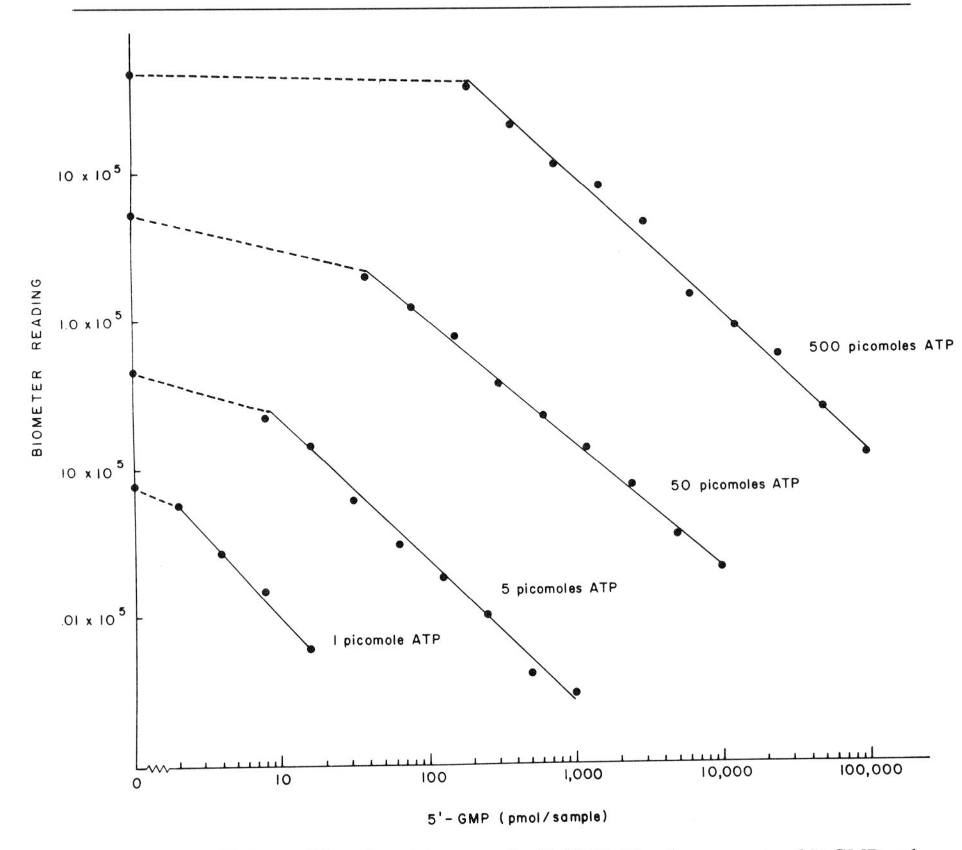

FIG. 2. Sensitivity and linearity of the assay for 5′-GMP. Varying amounts of 5′-GMP and ATP were incubated with reagent E for 60 min at 37°. The amount of ATP that remained is expressed in arbitrary units. Each point represents the mean of five determinations. From R. Fertel and B. Weiss, *Anal. Biochem.* **59,** 386 (1974).

FIG. 3. Hydrolysis of cyclic AMP and cyclic GMP as a function of incubation time. (A) Cyclic AMP phosphodiesterase was measured by incubating a homogenate of rat cerebrum equivalent to 1 μg of tissue with 0.5 mM cyclic AMP at 37° for varying times. Each point represents the mean of 3 separate assays. From B. Weiss, R. Lehne, and S. Strada, *Anal. Biochem.* **28,** 182 (1969). (B) Cyclic GMP phosphodiesterase was measured by incubating a 100,000 g supernatant fraction of rat cerebral homogenate (equivalent to 0.55 μg of protein) with 0.2 mM cyclic GMP at 37° for varying times. Activity was determined as described in the text, using 2 μM ATP in reagent E. Each point represents the mean of 5 determinations. From R. Fertel and B. Weiss, *Anal. Biochem.* **59,** 386 (1974).

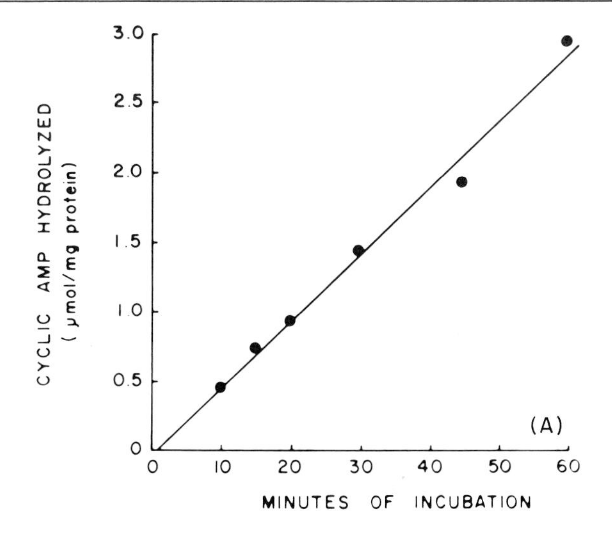

3.0
2.5
2.0
1.5
1.0
0.5
0
CYCLIC AMP HYDROLYZED
(μmol/mg protein)
(A)
0
10
20
30
40
50
60
MINUTES OF INCUBATION

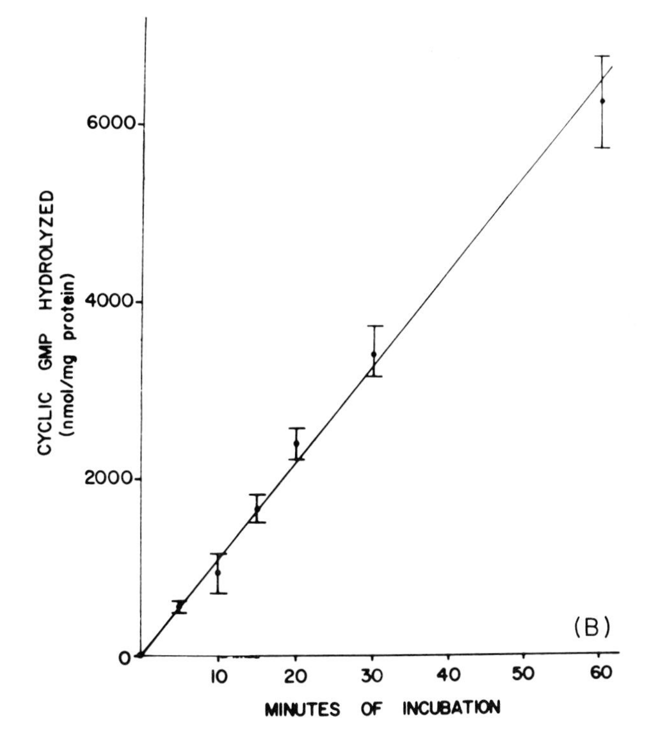

6000
4000
2000
0
CYCLIC GMP HYDROLYZED
(nmol/mg protein)
(B)
10
20
30
40
50
60
MINUTES OF INCUBATION

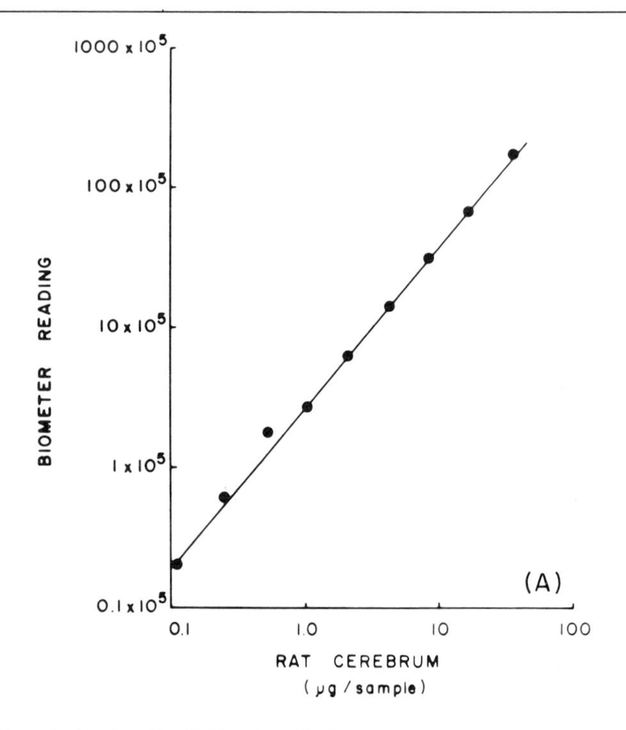

FIG. 4. Hydrolysis of cyclic AMP and cyclic GMP as a function of an increasing concentration of phosphodiesterase. (A) For the cyclic AMP phosphodiesterase, varying concentrations of a homogenate of rat cerebrum were incubated with 0.5 m*M* cyclic AMP for 60 min at 37° as described in the text. From B. Weiss, R. Lehne, and S. Strada, *Anal. Biochem.* **45,** 222 (1972). (B) For cyclic GMP phosphodiesterase, varying concentrations of the 100,000 *g* supernatant fraction of rat cerebral homogenate was incubated with 0.2 m*M* cyclic GMP for 60 min at 37°. Enzymic activity was determined as described in the text, using 2 μM ATP in reagent E. Each point represents the mean of 5 determinations. From R. Fertel and B. Weiss, *Anal Biochem.* **59,** 386 (1974).

Correlation of Phosphodiesterase Activity with the Amount of Tissue Added

Both cyclic AMP and cyclic GMP phosphodiesterase activities increase linearly with increasing amounts of tissue (Fig. 4A, 4B). In both assays, the phosphodiesterase activity of less than 100 ng of cerebral protein can be detected, and there is a linear increase in activity over a wide range of tissue concentrations.

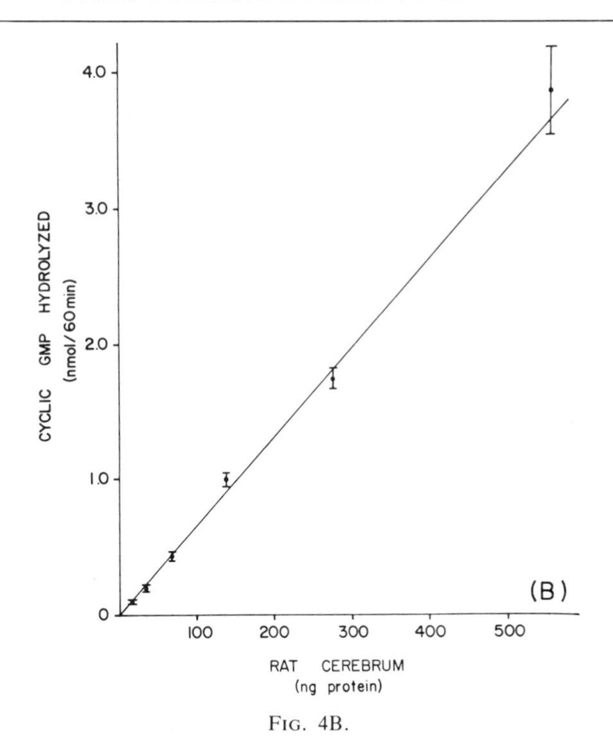

FIG. 4B.

Experimental Results Obtained Using These Assay Procedures

The rapidity and low expense of these procedures, coupled with their high sensitivity, permit the analysis of the large number of samples obtained when the multiple forms of phosphodiesterase are studied. (Several hundred assays per day can easily be performed by one investigator.) For example, analysis of the multiple forms of cyclic-AMP phosphodiesterase of rat cerebrum separated by preparative polyacrylamide gel electrophoresis indicate that there are at least four peaks of cyclic-AMP phosphodiesterase activity (Fig. 5).

These assays also have been used to determine tissue and subcellular distributions of phosphodiesterase isoenzymes, their kinetic parameters, and the response of these enzymes to phosphodiesterase inhibition.[14-20]

[16] R. Fertel and B. Weiss, *Mol. Pharmacol.* **12,** 678 (1976).
[17] B. Weiss, R. Fertel, R. Figlin, and P. Uzunov, *Mol. Pharmacol.* **10,** 615 (1974).
[18] P. Uzunov and B. Weiss, *Biochim. Biophys. Acta* **284,** 220 (1972).
[19] P. Uzunov, H. Shein, and B. Weiss, *Neuropharmacology* **13,** 377 (1974).
[20] P. Uzunov, H. Shein, and B. Weiss, *Science* **180,** 304 (1973).

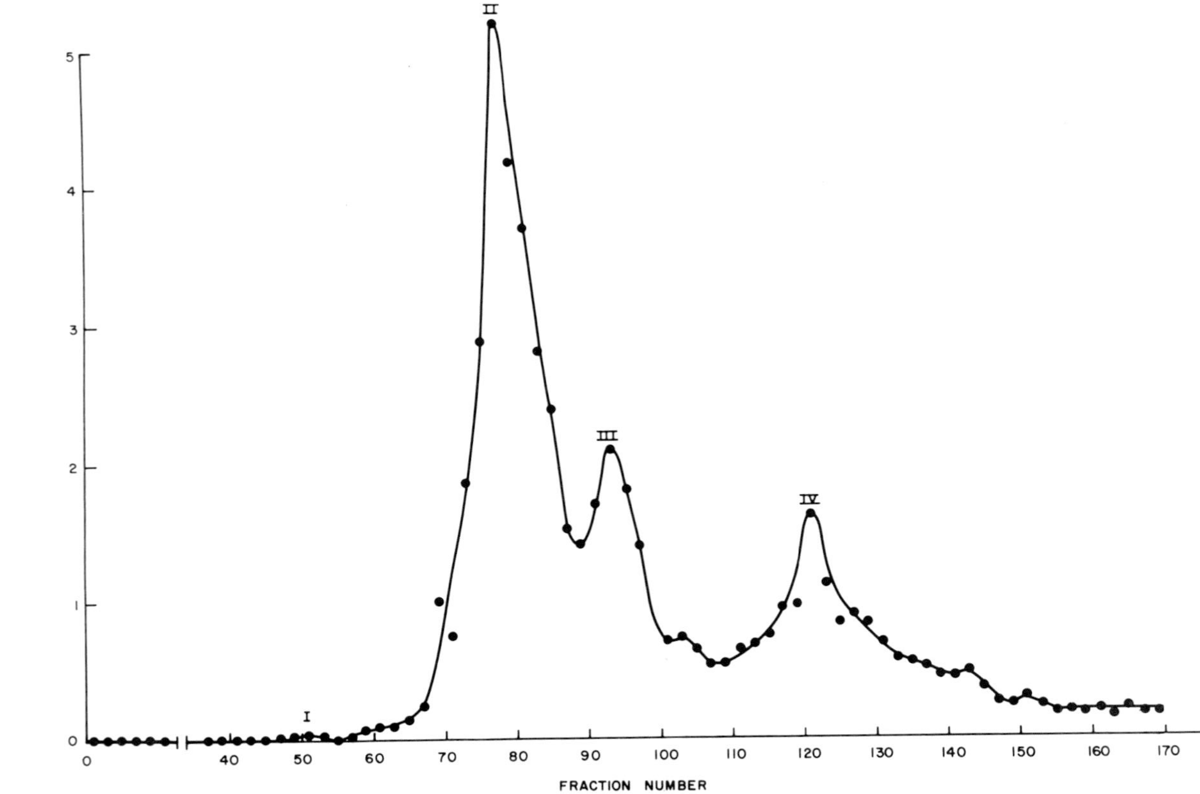

FIG. 5. Rat cerebrum was homogenized in 4 volumes of 0.32 *M* sucrose, sonicated, and centrifuged at 100,000 *g* for 60 min. One milliliter of the supernatant fluid was placed on a preparative polyacrylamide gel electrophoresis apparatus. One-milliliter fractions were eluted from the column, and cyclic AMP phosphodiesterase activity was determined as described in the text, using 0.2 m*M* cyclic AMP as substrate. From B. Weiss, R. Fertel, R. Figlin, and P. Uzunov, *Mol. Pharmacol.* **10,** 615 (1974).

[11] Rapid Microassay for the Calcium-Dependent Protein Modulator of Cyclic Nucleotide Phosphodiesterase[1]

By JOHN C. MATTHEWS and MILTON J. CORMIER

The activities of certain cyclic nucleotide phosphodiesterases and adenylate cyclases from mammalian heart, brain, and other tissues can be stimulated by an endogenous, low molecular weight, calcium-binding protein when calcium is present.[2-9] The functional species is a Ca^{2+}-modulator protein complex that associates with the phosphodiesterase to increase its biological activity over that of the free enzyme.[4-7] The presence of this protein, referred to here as modulator, has been demonstrated in a variety of mammalian tissues and in extracts of many animal species ranging from coelenterates to mammals.[3-14] It is believed that modulator protein functions in the regulation of cellular cyclic nucleotide levels by providing for rapid modulation of cellular cyclic nucleotide phosphodiesterase activity in response to rapid changes in cellular calcium ion concentration.

This chapter describes a procedure for the rapid, sensitive, and quantitative microassay of modulator protein based on its ability to stimulate

[1] This work was supported in part by National Science Foundation grants (BMS 74-06914 and PCM 76-10573). Contribution No. 365 from the University of Georgia Marine Institute, Sapelo Island, Georgia.

[2] C. O. Brostrom, Y. C. Huang, B. M. Breckenridge, and D. J. Wolff, *Proc. Natl. Acad. Sci. U.S.A.* **72,** 64 (1975).

[3] S. Kakiuchi, R. Yamazaki, and H. Nakajima, *Proc. Jpn. Acad.* **46,** 589 (1970).

[4] Y. M. Lin, Y. P. Liu, and W. Y. Cheung, *J. Biol. Chem.* **249,** 4943 (1974).

[5] J. H. Wang, T. S. Teo, H. C. Ho, and F. C. Stevens, *Adv. Cyclic Nucleotide Res.* **5,** 179 (1974).

[6] Y. Teshima and S. Kakiuchi, *Biochem. Biophys. Res. Commun.* **56,** 489 (1974).

[7] T. S. Teo and T. H. Wang, *J. Biol. Chem.* **248,** 588 (1973).

[8] W. Y. Cheung, *Biochem. Biophys. Res. Commun.* **38,** 533 (1970); W. Y. Cheung, *J. Biol. Chem.* **246,** 2859 (1971).

[9] S. Kakiuchi, R. Yamazaki, Y. Teshima, K. Uenishi, and E. Miyamoto, *Biochem. J.* **146,** 109 (1975).

[10] D. M. Watterson, W. G. Harrelson, Jr., P. M. Keller, F. Sharief, and T. C. Vanaman, *J. Biol. Chem.* **251,** 4501 (1976).

[11] F. C. Stevens, M. Walsh, H. C. Ho, T. S. Teo, and J. H. Wang, *J. Biol. Chem.* **251,** 4495 (1976).

[12] J. R. Dedman, J. D. Potter, and A. R. Means, *J. Biol. Chem.* **252,** 2437 (1977).

[13] J. C. Brooks and F. L. Siegel, *J. Biol. Chem.* **248,** 4189 (1973).

[14] D. Waisman, F. C. Stevens, and J. H. Wang, *Biochem. Biophys. Res. Commun.* **65,** 975 (1975).

ISBN 0-12-181957-4

modulator-deficient phosphodiesterase. The adenosine 5′-monophosphate, produced from adenosine 3′,5′-cyclic monophosphate by the phosphodiesterase reaction, is enzymatically converted to adenosine 5′-triphosphate and assayed with the firefly luminescence reaction, using a modified version of the coupled assay described by Weiss *et al.*[15]

Assay Method

The sequence of reactions involved in the assay of modulator protein is given below.[16]

$$\text{Low-activity PDE} \xrightleftharpoons{\text{modulator; Ca}^{2+}} \text{high-activity PDE} \qquad (1)$$

$$\text{cAMP} \xrightarrow[\text{Mg}^{2+}]{\text{PDE}} \text{AMP} \qquad (2)$$

$$\text{AMP} + \text{ATP} \xrightleftharpoons[\text{Mg}^{2+},\ \text{GTP}^{16}]{\text{myokinase}} 2\ \text{ADP} \qquad (3)$$

$$\text{ADP} + \text{PEP} \xrightarrow[\text{Mg}^{2+}]{\text{pyruvate kinase}} \text{ATP} + \text{pyruvate} \qquad (4)$$

$$\text{ATP} + \text{luciferin} + \text{O}_2 \xrightarrow[\text{Mg}^{2+}]{\text{luciferase}} \text{oxyluciferin} + \text{AMP} + \text{PP}_i + \text{CO}_2 + \text{light} \qquad (5)$$

Owing to the requirement for Ca^{2+} in reactions (1) and (2), this portion of the assay must be performed separately since Ca^{2+} inhibits the enzymes in reactions (3)–(5). In addition, since the assay is based on measurement of the initial rate of the luciferase reaction, an incubation period must be allowed for the conversion of AMP to ATP prior to the addition of luciferin and luciferase.

Assay Equipment. A photometer capable of measuring light intensities in the range from 10^7 to 10^{10} photons per second and a strip-chart recorder are required for this assay. Owing to the small volume used in this assay, the photometer must be arranged to collect light from the bottom of small

[15] B. Weiss, R. Lehne, and S. Strada, *Anal. Biochem.* **45,** 222 (1972).

[16] Abbreviations used are GTP, guanosine 5′-triphosphate; ADP, adenosine 5′-diphosphate; PEP, phosphoenolpyruvate; PP_i, pyrophosphate; cAMP, adenosine 3′,5′-cyclic monophosphate; AMP, adenosine 5′-monophosphate; ATP, adenosine 5′-triphosphate; PDE, cyclic nucleotide phosphodiesterase. GTP produces no light with firefly luciferin and luciferase. It is, however, an alternative substrate for myokinase and pyruvate kinase. In this assay, GTP functions as a catalyst in the generation of the initial pool of ATP which is required to perpetuate the cycle.

assay tubes (i.e., 10 × 75 mm).[17] The reader is referred to this volume [41] and [44] for further information concerning suitable photometers and light calibration techniques.

Buffers and Solutions. The following buffers and solutions are required for the assay:

Buffer A: 40 m*M* Tris, 1 m*M* $Mg(Ac)_2$, 1 m*M* cAMP, 0.1 m*M* $CaCl_2$, pH 8.0[18]

Buffer B: 50 m*M* glycylglycine, 15 m*M* DTT, 4 m*M* EGTA, 4 m*M* $Mg(Ac)_2$, 1 m*M* NaN_3, 0.1% w/v BSA, pH 7.8[18]

Solution 1: 10 units of modulator-deficient PDE[19] per milliliter in 40 m*M* Tris, 1 m*M* $Mg(Ac)_2$, pH 8.0

Solution 2: 40 units of myokinase per milliliter, 20 units of pyruvate kinase per milliliter, 10 n*M* GTP, 1 m*M* PEP in buffer B[20]

Solution 3: 0.4 m*M* firefly luciferin plus Sigma firefly lantern extract (8 mg/ml) in buffer B[21]

Preparation of Solution 3. Firefly luciferin is subject to air oxidation so the special precautions outlined here must be taken during its handling to

[17] For unknown reasons this assay is very sensitive to inhibition by contaminants in glassware. The authors recommend that all glassware, especially the assay tubes, be washed with 6 *N* HCl followed by deionized water as the last step.

[18] The characteristics of the assay change markedly as buffer A ages. Therefore, to ensure reproducibility, this buffer should be prepared fresh daily. Buffer B should be prepared fresh each week.

[19] The authors are indebted to Dr. Frank Siegel for kindly providing the purified and lyophilized porcine brain modulator protein and modulator-deficient PDE that were used to develop this assay procedure. Satisfactory procedures for the preparation of modulator-deficient PDE may be found in references cited in footnotes 6–8. Aliquots (125 μl) of solution 1 may be prepared in advance and stored at −80° without significant loss of activity. One unit of PDE is defined as that amount of PDE which is sufficient to produce 1 nmol of AMP per minute when saturated with modulator and Ca^{2+} at pH 8.0 at 25°. One unit of modulator is defined as that amount of modulator required to produce 50% maximal activation of 1 unit of PDE.

[20] Solution 2 must be prepared fresh daily in the amount required, because myokinase and pyruvate kinase are unstable. These enzymes may be purchased as stable suspensions in ammonium sulfate and added to solution 2 directly. One unit of myokinase is defined as that amount of enzyme which is sufficient to produce 1 μmol of ATP per minute at pH 7.5 at 37°. One unit of pyruvate kinase is that amount of enzyme required to produce 1 μmol of pyruvate per minute at pH 7.5 at 37°. A 100 × stock solution of PEP and GTP in buffer B may be prepared weekly and stored at 4°.

[21] Firefly luciferase may be crude or purified without affecting the characteristics of this assay. The pure enzyme was a generous gift from Dr. M. DeLuca. Sigma lyophilized lantern extract is the recommended source of this enzyme owing to its lower cost, more ready availability, and stability in the dry state. The amount of firefly extract employed in solution 3 corresponds roughly to 5×10^{-3} units/ml. One unit of FF luciferase is defined as that amount required to produce 10^{13} protons sec^{-1} at pH 7.8 at 25° in a system containing 0.5 m*M* ATP and 20 μ*M* luciferin.

ensure accuracy. Deoxygenate 5 ml of methanol by bubbling argon through it for at least 5 min. Dissolve approximately 200 μg of D-luciferin in 1 ml of deoxygenated methanol, and maintain this solution under an argon atmosphere. Prepare 1 ml of a 1:10 dilution of the luciferin solution with deoxygenated methanol and place it in a stoppered 1-ml capacity quartz cuvette under argon. Determine the absorption of this solution at 268 and 327 nm. The molar extinction coefficient for luciferin at 327 nm is 18,800, and for luciferin free of oxidation products, the 327 nm/268 nm absorption ratio should be 2.5.[22] However, if this ratio is greater than 1.5 the luciferin is sufficiently pure and the absorption value at 327 nm can be taken as due solely to luciferin. Once the luciferin concentration is accurately known, divide the stock solution into vials so that each vial contains 50 nmol of luciferin. Evaporate the methanol by blowing argon vigorously over the surface of the solution. Store the dried luciferin vials desiccated, in the dark, below 0°. This procedure is sufficient for preparing approximately 14 vials of luciferin. Immediately prior to use, add 1 mg of lyophilized lantern extract and 125 μl of buffer B at 4° to a luciferin vial. This is sufficient luciferin and luciferase for 25 assays. It should be kept at 0°–4° and used within 2 hr of preparation.

Procedure. Step 1: Add 1–5 μl of modulator protein solution (with a chelator concentration < 0.1 mM) to 50 μl of buffer A. Initiate the reaction by adding 5 μl of modulator-deficient PDE (solution 1). Incubate for 20 min at room temperature.

Step 2: Pipette 10 μl of the reaction mixture from step 1 into 100 μl of solution 2 in a 10 $\times$ 75 mm test tube. Incubate for 5 min at room temperature.

Step 3: Pipette 5 μl of firefly luciferin–luciferase (solution 3) into the reaction mixture from step 2. Mix by gently vortexing, place the tube in the photometer sample compartment and record the initial light intensity. This step should be accomplished within 10 sec of the addition of solution 3.

To ensure that modulator is the factor being assayed, control experiments should be performed with no added PDE. In addition, 1 mM EGTA (final concentration) may be added to the assay at step 1, which will prevent the activation of PDE from occurring by removing free Ca^{2+} from the reaction mixture. Interfering amounts of endogenous PDE activity in a modulator sample may be destroyed by heating the samples to 95° for 5 min.

The relationships shown in Figs. 1–3 were determined using the assay method described here. With PDE left out of the complete assay mixture, accurate measurements were obtained for ATP ranging in concentration

[22] E. H. White, F. McCapra, G. F. Field, and W. P. McElroy, *J. Am. Chem. Soc.* **83,** 2402 (1961).

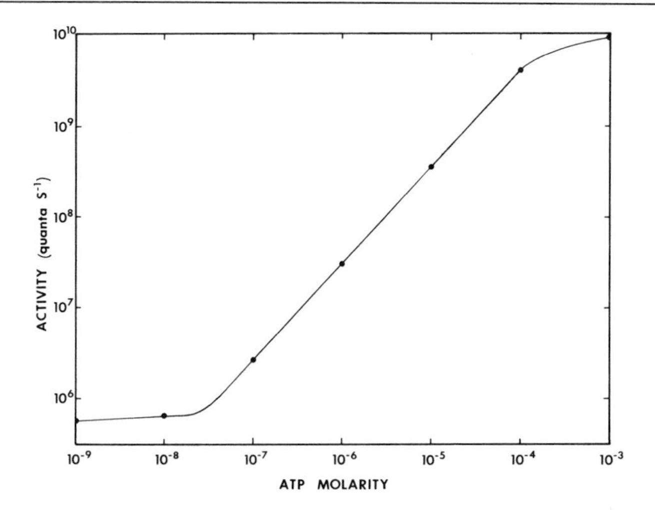

FIG. 1. Dependence of light intensity on ATP concentration in the coupled assay. The data were obtained by adding 10-μl aliquots of known concentrations of ATP in buffer A to step 2 of the assay procedure.

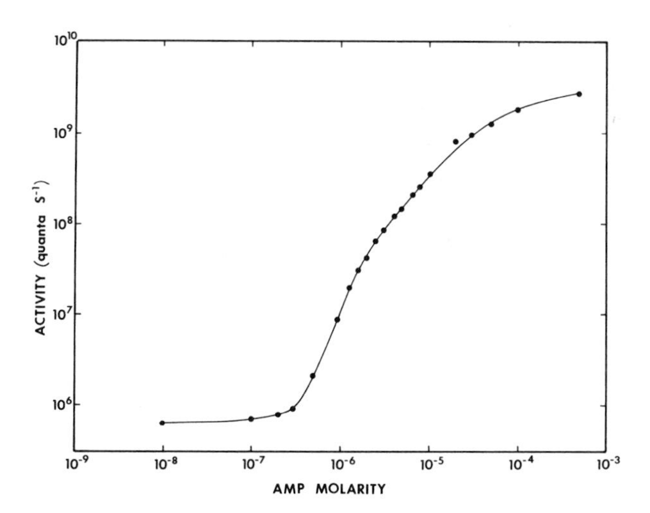

FIG. 2. Dependence of light intensity on AMP concentration in the coupled assay. Data were obtained by adding 10-μl aliquots of known concentrations of AMP in buffer A to step 2 of the assay procedure.

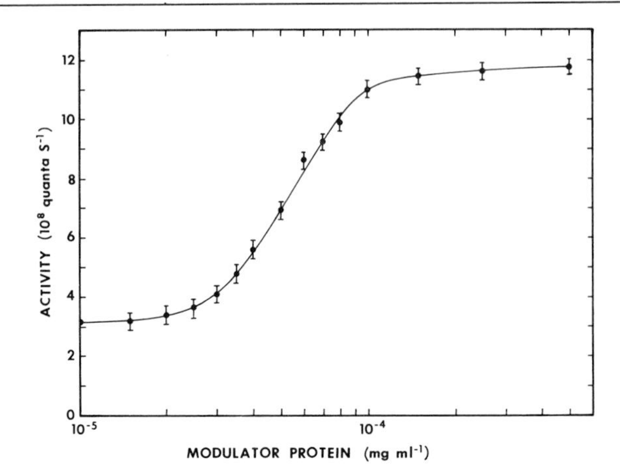

FIG. 3. Dependence of light intensity on modulator protein. Aliquots (5 μl) of known concentrations of porcine brain modulator protein [J. C. Brooks and F. L. Siegel, *J. Biol. Chem.* **248,** 4189 (1973); see also text footnote 19] in buffer A were assayed as described.

from 50 n*M* to 0.5 m*M* (Fig. 1) and for AMP concentrations ranging from 0.5 μM to 0.5 m*M*. (Fig. 2). Modulator concentrations, standardized against porcine brain modulator,[19] were determined over approximately a 10-fold concentration range (Fig. 3). At saturating levels of modulator, a 3- to 4-fold stimulation of the PDE activity was observed.

Advantages of the Luminescence assay. This assay is relatively rapid and, by initiation of steps 1–3 at appropriate time intervals, up to 20 assays can be performed within a 1-hr period. The assay is also sensitive, thus allowing one to conserve material. For example, the amount of PDE used in this assay procedure is approximately 2% of the amount required in more conventional colorimetric assays for PDE.[23] With the exception of modulator-deficient PDE, all reagents used are commercially available.

[23] R. W. Butcher and E. W. Sutherland, *J. Biol. Chem.* **237,** 1244 (1962).

[12] ATP-Labeled Ligands and Firefly Luciferase for Monitoring Specific Protein-Binding Reactions

By ROBERT J. CARRICO, R. DON JOHNSON, and ROBERT C. BOGUSLASKI

Antibodies and many other proteins bind ligands with high affinity and specificity. Binding reactions employing radiolabeled ligands are used extensively for measurement of low levels of ligands and binding proteins in biological materials. In typical competitive binding assays, a ligand, at unknown concentration, and its radiolabeled analog are allowed to compete for a limited number of protein binding sites. Then the protein-bound and free forms of the radiolabeled ligand are separated. The level of radioactivity in either of these fractions can be related to the concentration of competing ligand by means of a standard curve.

Several methods have been devised for replacing radiolabels with enzymes,[1,2] cofactors,[3-5] substrates,[6] fluorescent dyes,[7] and chemiluminescent compounds.[8] These materials avoid the inconvenience associated with radioactivity and usually possess greater stability than their radiolabeled counterparts. Furthermore, in some cases only the protein-bound or the free label is detected by the monitoring reaction, and this simplifies the assay by eliminating the separation step.

Specific binding assays that do not require a separation step can be carried out with ATP as the labeling substance.[5] A ligand is coupled covalently to ATP through a bridging group (see Fig. 1), and low levels of the conjugate can be measured quantitatively with firefly luciferase. When the ligand moiety is complexed with its specific binding protein, the conjugate does not initiate light production. Thus, firefly bioluminescence can be

[1] K. E. Rubenstein, R. S. Schneider, and E. F. Ullman, *Biochem. Biophys. Res. Commun.* **47,** 846 (1972).

[2] E. Engval and P. Perlmann, *Immunochemistry* **8,** 871 (1971).

[3] R. J. Carrico, J. E. Christner, R. C. Boguslaski, and K. K. Yeung, *Anal. Biochem.* **72,** 271 (1976).

[4] H. R. Schroeder, R. J. Carrico, R. C. Boguslaski, and J. E. Christner, *Anal. Biochem.* **72,** 283 (1976).

[5] R. J. Carrico, K. K. Yeung, H. R. Schroeder, R. C. Boguslaski, R. T. Buckler, and J. E. Christner, *Anal. Biochem.* **76,** 95 (1976).

[6] J. F. Burd, R. J. Carrico, M. C. Fetter, R. T. Buckler, R. D. Johnson, R. C. Boguslaski, and J. E. Christner, *Anal. Biochem.* **77,** 56 (1977).

[7] R. A. A. Watson, J. Landon, E. J. Shaw, and D. S. Smith, *Clin. Chim. Acta* **73,** 51 (1976).

[8] H. R. Schroeder, P. O. Vogelhut, R. J. Carrico, R. C. Boguslaski, and R. T. Buckler, *Anal. Chem.* **48,** 1933 (1976).

ISBN 0-12-181957-4

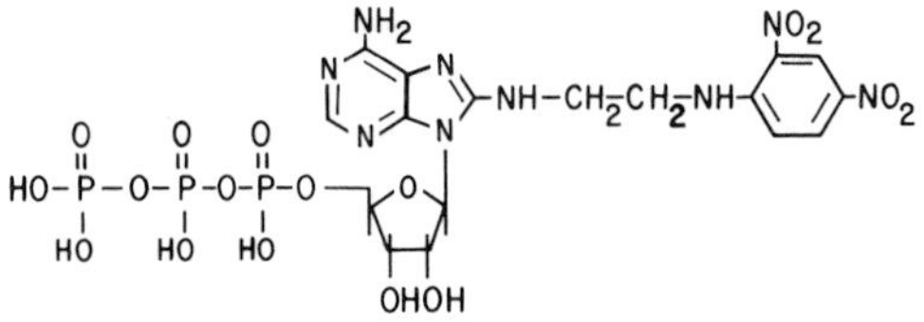

FIG. 1. Structures of ligand–ATP conjugates. (I), C^8-DNP-ATP; (II), N^6-DNP-ATP; (III), P^1-DNP-adenosine 5′-triphosphate ($n = 1$) or P^1-DNP-adenosine 5′-tetraphosphate ($n = 2$).

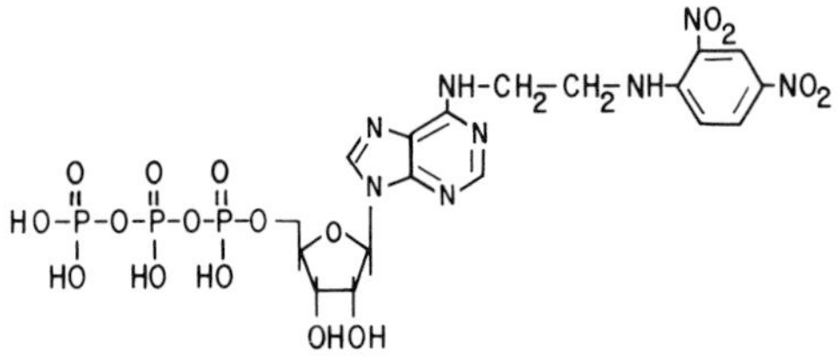

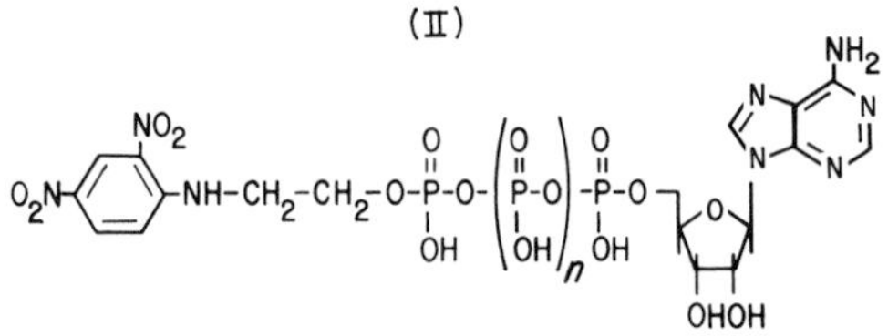

used to measure specific binding proteins or to monitor competitive binding reactions.

Bioluminescent Assay

Luciferase can be isolated from firefly tails by published procedures.[9,10] The methods described here were developed with partially purified luciferase purchased from DuPont (Wilmington, Delaware). In addition to luciferase, the assay medium contains 10 m*M* morpholinopropane sulfonate buffer, pH 7.4, 0.7 m*M* luciferin, and 0.15% bovine serum albumin. One-hundred microliter aliquots of this solution are dispensed into 6 × 50 mm test tubes, which are incubated at 25° for a few minutes. The test tubes are placed in the photometer, and a 10-μl aliquot of the sample to be assayed is injected from a 25-μl Hamilton syringe. The maximum light intensity produced by the reaction is measured.

[9] M. Bény and M. Dolivo, *FEBS Lett.* **70,** 167 (1976).
[10] M. DeLuca and W. D. McElroy, this volume [1].

Light Measurement

Several photometers have been developed for measurement of bioluminescence.[11,12] The work described here was carried out with a DuPont Model 760 luminescence biometer, which automatically records the maximum light intensity in arbitrary units.

Purification of Antibody

Antibodies to specific ligands can be isolated from antisera by the method of Livingstone.[13] This procedure removes enzymes that degrade ATP.

Design of Ligand–ATP Conjugates

The lower limit of sensitivity of a specific protein-binding assay is determined in part by the sensitivity of the bioluminescent assay for the ligand-ATP, which in turn is influenced by the structure of the conjugate. Several conjugates with the ligand coupled through bridging groups at various positions on ATP are shown in Fig. 1. (The synthesis of these is described by Carrico *et al.*[5]) The ligand and cofactor are separated by a bridging arm that has a terminal functional group for covalent attachment of the ligand. The bridging arm also separates the two moieties so that the ligand does not interfere with binding of the cofactor to the enzyme. The optimal length for the bridging arm is unknown; however, it should probably be short so that antibody-bound conjugate will be inactive with enzymes.

The scheme chosen for synthesis of a particular ligand–cofactor conjugate will depend on the chemical and physical properties of the ligand; therefore, detailed methods for synthesis of complete conjugates are not discussed here. Instead, syntheses for adenine nucleoside derivatives which can be coupled to ligands are presented (see Fig. 2).

Synthesis of N^6-(2-Aminoethyl)adenosine 5′-Monophosphate [Fig. 2, (IV)]

6-Chloropurine riboside 5′-phosphate is prepared on a 7.0-mmol scale by the method described by Mosbach[14]; however, this compound is used

[11] G. W. Mitchell and J. W. Hastings, *Anal. Biochem.* **39,** 243 (1971).
[12] H. Rasmussen and R. Nielsen, *Acta Chem. Scand.* **22,** 1745 (1968).
[13] D. M. Livingston, this series, Vol. 34, p. 723.
[14] K. Mosbach, this series, Vol. 34, p. 230.

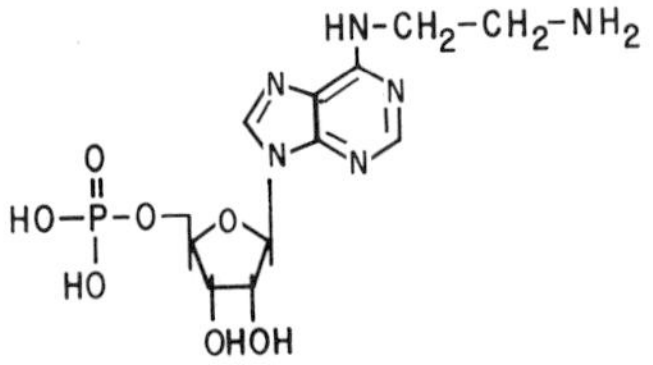

(IV)

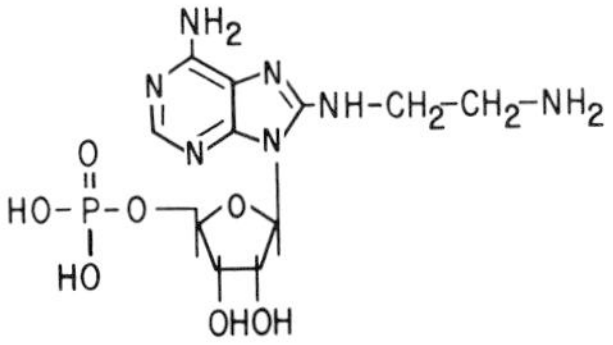

(V)

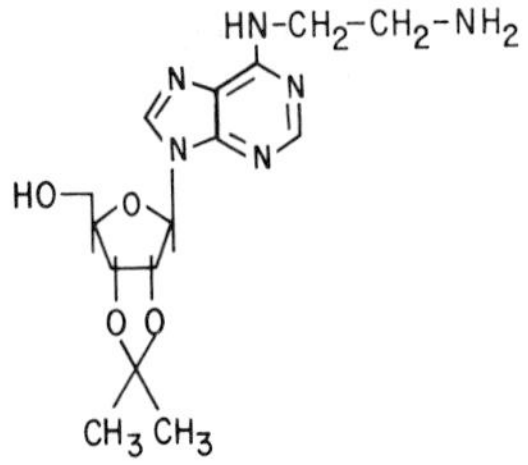

(VI)

FIG. 2. Derivatives of adenosine used for synthesis of ligand–ATP conjugates. (IV), N^6-(2-aminoethyl)adenosine 5′-monophosphate; (V), C^8-(2-aminoethylamino)adenosine 5′-monophosphate; (VI), N^6-(2-aminoethyl)-9-(2′,3′-*O*-isopropylidene-β-D-ribofuranosyl)adenine.

without the purification steps described previously. After filtration to remove the lithium phosphate, the filtrate is mixed with 9.5 ml of ethylenediamine (140 mmol) and allowed to stand at room temperature for 3 hr. The reaction mixture is diluted to 4 liters with water and applied to a 5 × 30 cm column of Dowex 1-X8 resin (acetate form). This column is washed with 3 liters of 10 m*M* NH_4Cl to remove carbonate and then with a linear gradient generated with water (3 liters) and 1 *M* acetic acid (3 liters). Material with an absorption maximum at 264 nm elutes between 1.8 and 2.1 liters of the gradient, and this is concentrated to about 25 ml. While standing at 7°, crystals form to give 1.8 g (65% yield) of N^6-(2-aminoethyl)adenosine 5′-monophosphate of analytical purity. The optical absorption maximum in 0.1 *N* HCl is 264 nm, and the millimolar extinction coefficient is 17.7. During thin-layer chromatography on silica gel, the compound migrates with an R_f of 0.15 in an ethanol:0.5 *M* ammonium acetate (9:4) solvent system.

Synthesis of C^8-(2-Aminoethylamino)adenosine 5′-Monophosphate [Fig. 2, (V)]

8-Bromoadenosine 5′-monophosphate is prepared by a method similar to that described by Lee *et al.*[15] Two grams of AMP (5.8 mmol) are dissolved in 180 ml of 1 *M* sodium acetate–acetic acid, pH 3.9, and 0.44 ml of Br_2 (8.6 mmol) is added and allowed to react at room temperature for 19 hr. The progress of the reaction can be monitored by thin-layer chromatography on silica gel, which separates AMP (R_f 0.34) from 8-bromo AMP (R_f 0.48) when an ethanol: 1 *M* triethylammonium bicarbonate, pH 7.5, (7:3) solvent system is employed. Also, the optical absorption maximum shifts from 257 nm to 262 nm as the reaction progresses. The water in the reaction mixture is removed under vacuum, and the residue is suspended in 50–100 ml ethanol and evaporated to dryness. The treatment with ethanol is repeated twice, and the final residue is dissolved in 450 ml of water and applied to a 2.5 × 55 cm column of Dowex 1-X8 resin (formate form). The chromatogram is developed with a linear gradient generated with 2 liters of water and 2 liters of 0.7 *M* formic acid. Unreacted AMP (λ_{max} = 257 nm) elutes first, followed by 8-bromo-AMP (λ_{max} = 262 nm). The solvent is evaporated under vacuum from the pooled brominated material, and the residue is dried by repeated evaporation from anhydrous ethanol to give 1.9 g (75% yield) of 8-bromo-AMP. The millimolar extinction coefficient for the 8-bromo derivative in dilute acid is 16.4 at 262 nm.

8-Bromo-AMP, 0.9 g (2.2 mmol), and 4 g (60 mmol) ethylenediamine in 25 ml of water are heated at 140° for 2 hr. Then the cooled reaction mixture is chromatographed on a 2.5 × 55 cm column of Dowex 1-X8 resin (bicarbonate form) and the chromatogram is developed with a linear gradient generated with 3 liters of water and 3 liters of 0.5 *M* NH_4HCO_3. A material with an optical absorption maximum of 275 nm elutes between 4.6 and 5.8 liters of the gradient. This purified product is taken to dryness under vacuum, and the residual NH_4HCO_3 is removed by repeated evaporation from water. The residue is dissolved in 20 ml of water by adjusting to pH 8.0 with concentrated NH_4OH, and the solution is filtered and adjusted to pH 5.0 with formic acid. Upon standing at 5°, crystals that form are collected by filtration. A second crystallization gives 2.4 g (27% yield) of analytically

[15] C. Y. Lee, D. A. Lappi, B. Wermuth, J. Everse, and N. O. Kaplan, *Arch. Biochem. Biophys.* **163,** 561 (1974).

pure C^8-(2-aminoethylamino)adenosine 5′-monophosphate. The optical absorption maximum in dilute acid is 275 nm, and the millimolar extinction coefficient is 17.5. The compound has an R_f of 0.23 when examined by thin-layer chromatography on silica gel in an ethanol:0.5 *M* ammonium acetate (9:4) solvent system.

Synthesis of 6-(2-Aminoethyl)-9-(2′,3′-O-Isopropylidene-β-Ribofuranosyl)adenine (Fig. 2, VI)

6-Chloro-9-(2′,3′-*O*-isopropylidene-β-D-ribofuranosyl)purine is synthesized by the method of Hampton and Maguire.[16] Thirty-five grams of 6-chloropurine riboside (0.12 mol) are suspended in 1.5 liters of dry acetone, and 231 g of *p*-toluenesulfonic acid are added. The reaction mixture is allowed to stand at room temperature for 1.5 hr, during which the riboside slowly dissolves. Then the mixture is added slowly to 1.5 liters of stirred 1.0 *M* $NaHCO_3$ in an ice bath. Solid material that forms is removed by filtration, and the filtrate is taken to dryness under vacuum. This residue and the solid from the filtration are combined and added to 1.5 liters of benzene, which is heated to reflux. A Dean–Stark trap is used to remove water from the suspension. The solid material is then removed by filtration of the hot mixture. The filtrate is taken to dryness, and the residue is added in portions to 200 ml of ethylenediamine that is stirred at 5°–10°. This reaction mixture is allowed to stand at room temperature for 2 hr and then the excess ethylenediamine is removed under vacuum. The residue is dissolved in 200 ml of 2-propanol and taken to dryness. The treatment with 2-propanol is repeated. The final residue is dissolved in 200 ml of 0.1 *M* NH_4HCO_3 and adjusted to pH 8.0 with concentrated HCl. Half of this solution was applied to a 5 × 55 cm column of Dowex 50-X2 resin (ammonium form), and the column is washed with 2–3 liters of 0.6 *M* NH_4HCO_3 until the absorbance of the effluent at 265 nm decreases to less than 1.0. The chromatogram is developed with a linear gradient generated with 0.6 *M* NH_4HCO_3 (3 liters) and 1.5 *M* NH_4HCO_3 (3 liters). When the gradient is complete, elution with 1.5 *M* NH_4HCO_3 is continued until the absorbance is less than 10. The UV-absorbing material eluted by the gradient has an absorption maximum at 265 nm and is one component (R_f 0.36) as determined by thin-layer chromatography on silica gel using ethanol:1 *M* triethylammonium bicarbonate, pH 7.5, (7:3) as solvent. The product is taken to dryness under vacuum, and residual NH_4HCO_3 is removed by repeated evaporation (4 or 5 times) from 400–500 ml of water. The yield of analytically pure (VI) is 75%.

[16] H. Hampton and M. H. Maguire, *J. Am. Chem. Soc.* **83,** 150 (1961).

Considerations in Synthesis of Ligand–Adenine Nucleotide Conjugates

The nucleoside derivatives in Fig. 2 are soluble in water, especially at neutral pH. Therefore, they are useful starting materials for coupling to water-soluble ligands by reactions compatible with this solvent. The ligand–AMP conjugate can then be phosphorylated by known procedures to produce the desired ATP conjugate.[17] Unfortunately, (IV) and (V) and their trialkylammonium salts are not soluble in polar organic solvents, such as dimethylsulfoxide, dimethylformamide, and dimethylacetamide. Therefore, synthesis involving water-soluble ligands are carried out with (VI), which is soluble in dimethylacetamide, water, methanol, and chloroform. The protected ligand–adenosine conjugate can be phosphorylated with phosphorus oxychloride in triethylphosphate.[18] The isopropylidene residue is removed from the ligand–adenine nucleotide conjugate by hydrolysis in 90% trifluoroacetic acid at 0°.[19]

Conjugates with the ligand attached to the terminal phosphate of adenosine 5′-tetraphosphate (see Fig. 1) can be synthesized by coupling the ligand to ethanolamine phosphate. When water-insoluble ligands are used, it is necessary to couple them to ethanolamine and then form the phosphate ester at the hydroxyl group. The ligand–ethanolamine phosphate is reacted with diphenylphosphoryl chloride and coupled to ATP by ion exchange.[5,20]

Measurement of Low Levels of Ligand–ATP Conjugates

The bioluminescent assay can detect lower levels of ATP than the ligand–ATP conjugates (see Fig. 3). P^1-DNP–adenosine 5′-tetraphosphate was measured at levels as low as 0.5 n*M* whereas P^1-DNP–adenosine 5′-triphosphate did not initiate light production even at micromolar levels. An enzyme in the firefly luciferase preparation hydrolyzes P^1-DNP–adenosine 5′-tetraphosphate to *N*-DNP–ethanolamine phosphate and ATP. The ATP then serves as substrate for the luciferase.[5] This enzyme might not be present in highly purified luciferase preparations. When P^1-DNP–adenosine 5′-tetraphosphate is added to luciferase, the light intensity increases for about 0.5 min and then decreases with a half-life of 1.2 min. (By comparison, maximum intensity occurs with ATP in less than 1 sec.)

When N^6-DNP–ATP or C^8-DNP–ATP is used, light emission reaches

[17] D. E. Hoard and D. G. Ott, *J. Am. Chem. Soc.* **87,** 1785 (1965).

[18] M. Yoshikawa, T. Kato, and T. Titakenishi, *Bull. Chem. Soc. Jpn.* **2,** 3505 (1969).

[19] J. E. Christensen and L. Goodman, *Carbohydr. Res.* **7,** 510 (1968).

[20] A. M. Michelson, *Biochim. Biophys. Acta* **92,** 1 (1964).

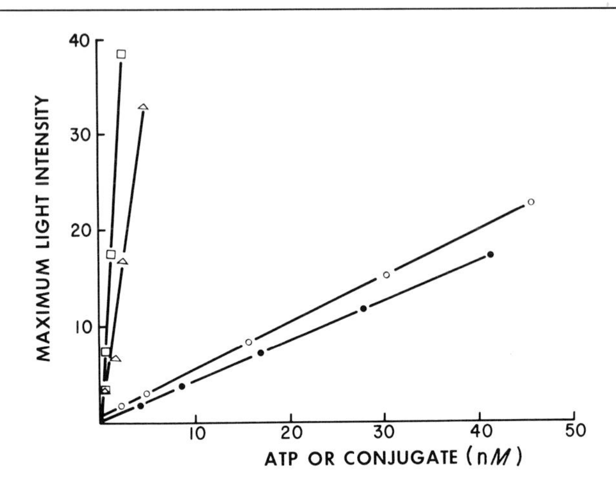

FIG. 3. Maximum light intensities as a function of levels of ATP and ligand–ATP conjugates. The levels of ATP (□), P^1-DNP-adenosine 5′-tetraphosphate (Δ), C^8-DNP-ATP (○), and N^6-DNP-ATP (●) are those in the 10-μl aliquot before addition to the bioluminescent assay.

maximum intensity in 2 min and the decay has a half-life of about 4 min. Maximum light intensities can be used to measure these two conjugates quantitatively at levels as low as 5 n*M* (Fig. 3).

Specific Protein-Binding Reactions

Optimal conditions for binding reactions can vary depending on the binding protein and ligand. The antibody-binding reactions described here were conducted at 25° in 20 m*M* Tris · HCl buffer, pH 7.4, containing 10 m*M* EDTA.

The binding reaction between antibody to the 2,4-dinitrophenyl residue (DNP) and P^1-DNP-adenosine 5′-tetraphosphate is complete within 2 min (see Table I). In control measurements, this antibody does not affect light production by ATP, and nonimmune immunoglobulin does not affect light production by the ligand–ATP conjugate (Table I). Similar results are obtained with the C^8-DNP-ATP and N^6-DNP-ATP. The conjugate can be titrated with antibody to DNP, and, in the presence of excess antibody, the maximum light intensity is about 10% of that observed in the absence of the binding protein (Fig. 4). Thus, the antibody specifically inactivates the ligand–ATP conjugates, and this effect provides a means for measurement of specific binding proteins.

TABLE I

MAXIMUM LIGHT INTENSITIES OBTAINED WITH P^1-DNP–ADENOSINE 5′-TETRAPHOSPHATE AND ATP IN THE PRESENCE OF IMMUNOGLOBULINS[a]

Incubation time (min)	Nonimmune IgG 45 nM P^1-DNP–adenosine 5′-tetraphosphate	Antibody to DNP 45 nM P^1-DNP–adenosine 5′-tetraphosphate	Antibody to DNP 23 nM ATP
0	34	31	63
2	34	4.0	—
30	33	3.5	65

[a] The binding reactions were conducted for the indicated times at 25° in 20 mM Tris · HCl buffer, pH 7.4, containing 10 mM EDTA. Duplicate 10-μl aliquots of the reaction mixtures were assayed.

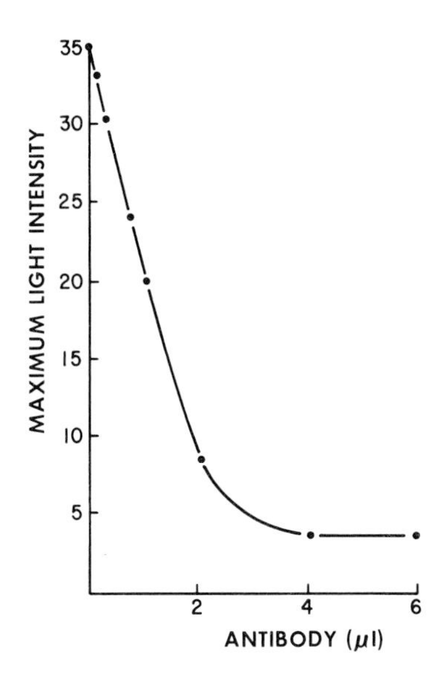

FIG. 4. Effect of antibody on the maximum light intensities produced with P^1-DNP-adenosine 5′-tetraphosphate. Varying amounts of antibody to DNP were incubated for 1.5 hr with 45 nM P^1-DNP-adenosine 5′-tetraphosphate (100 μl total volume). Then 10-μl aliquots of the binding reactions were assayed with the bioluminescent reaction.

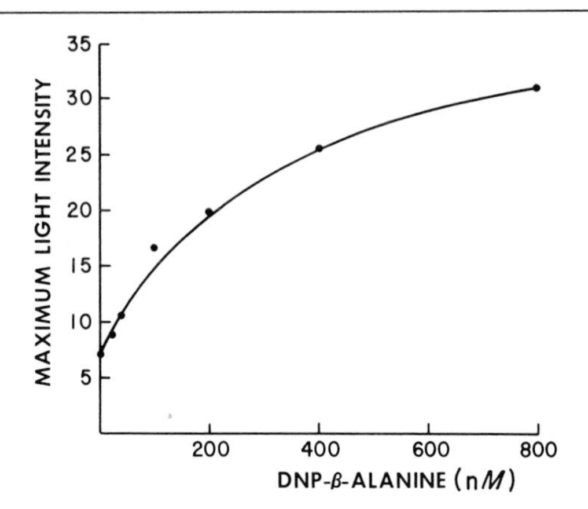

FIG. 5. Results of competitive binding reactions monitored by bioluminescence. Two microliters of antibody to DNP and 45 nM P^1-DNP-adenosine 5′-tetraphosphate were incubated with the indicated levels of DNP–β-alanine for 1 hr at 25° (100 μl total volume). Then duplicate 10-μl aliquots were assayed with the bioluminescent reaction.

The inactivation of the conjugates is relieved by competing, unlabeled ligand (Fig. 5). Competitive binding assays monitored with firefly bioluminescence can measure ligands at levels as low as 10 nM.

Comments

Bioluminescent emission from firefly luciferase can be influenced by materials in the sample being assayed. For example, high levels of salts will decrease production of light.[21,22] Also, protein at levels up to 15 mg/ml increases light production whereas higher levels decrease it. However, when 1.5 mg of bovine serum albumin per milliliter is included with the luciferase, the protein concentration in the sample can vary from 0 to 20 mg/ml without affecting light production significantly.

Many biological fluids, such as serum or plasma, might contain endogenous ATP that can interfere with measurement of ligand–ATP conjugates. However, these fluids often contain enzymes that degrade ATP rapidly,[22] and the endogenous cofactor can be eliminated by incubation of the sample prior to the binding reaction. The degradation of ligand–ATP conjugates by the endogenous enzymes can be inhibited by including EDTA in the buffer for the binding reaction. Ten millimolar EDTA in the sample being assayed does not influence bioluminescence.

[21] L. M. Aledort, R. I. Weed, and S. B. Troup, *Anal. Biochem.* **17**, 268 (1966).
[22] H. Holmsen, I. Holmsen, and A. Bernhardsen, *Anal. Biochem.* **17**, 456 (1966).

Section II

Bacterial Luciferase

[13] Bacterial Bioluminescence: An Overview

By J. W. HASTINGS

Bioluminescent bacteria occur ubiquitously in the ocean and in some nonmarine habitats.[1,2] Bacteria concentrated from a 1-liter sample of seawater, when analyzed by plating on a nutrient medium, will invariably contain luminous bacteria, usually hundreds. These bacteria also occur as symbionts, harbored by their hosts within special (and in different hosts very different) light organs.[2,3] The hosts thus derive their ability to luminesce from the bacteria. Luminous bacteria also occur as intestinal and parasitic symbionts, and although they do *not* contribute to the "phosphorescence" of the ocean (this is due, at least in part, to dinoflagellates), they may be very important in the overall economy of the ocean, especially with regard to the cycling of nutrients and intestinal flora of fish.[4] It is also probable that the bacteria may enjoy more than one "life style": luminous and nonluminous. Mechanisms involved in the control of the luminescent system suggest that its synthesis occurs under certain conditions, presumably where it has functional significance, and that it is repressed by others. Luminous bacteria may thus be able to compete favorably with nonluminous species under conditions where luminescence has no selective advantage.

Luminous bacteria, both those isolated directly from seawater and those from light organs, can generally be grown in the laboratory. Bioluminescent bacteria emit light continuously, but at different intensities, depending upon the specific cellular content of luciferase and substrates driving the reaction. These levels may differ as a consequence of a variety of factors, many being related to the induction and repression of biosynthesis.[4,5] In *Beneckea harveyi* the luminescence of fully induced bacteria may be about 10^3 photons sec^{-1} $cell^{-1}$, which is perhaps 10^4 times greater than that of uninduced cells allowed to "grow out" of their luciferase. Other brighter species may emit at levels up to 10 times higher.

Biochemically, the bacterial bioluminescent system can be viewed (Fig. 1) as a branch of the electron-transport pathway in which electrons from

[1] K. H. Nealson and J. W. Hastings, *in* "The Prokaryotes" (M. Starr, ed.). Springer-Verlag, Berlin and New York, 1978.

[2] E. N. Harvey, "Bioluminescence." Academic Press, New York, 1952.

[3] J. W. Hastings and G. Mitchell, *Biol. Bull.* **141,** 261 (1971).

[4] J. W. Hastings and K. H. Nealson, *Annu. Rev. Microbiol.* **31,** 549 (1977).

[5] K. H. Nealson, *Arch. Microbiol.* **112,** 73 (1977).

ISBN 0-12-181957-4

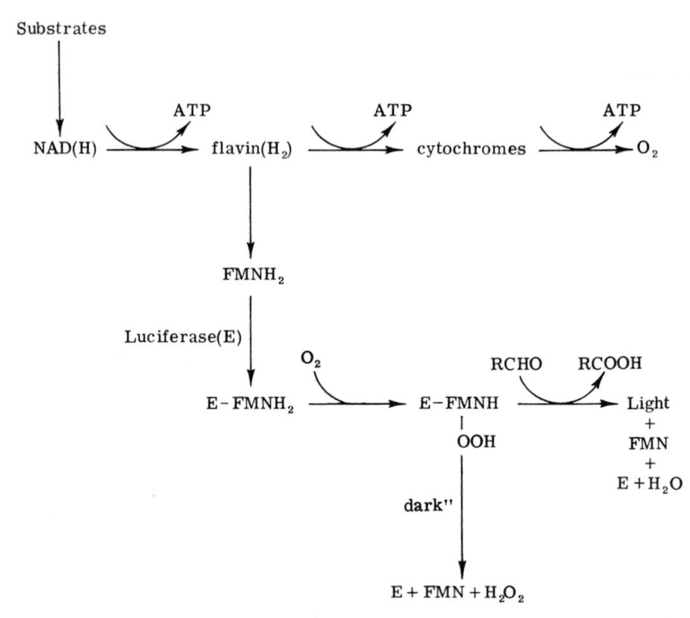

FIG. 1. This scheme depicts the relationship of the luciferase reaction to the electron transport pathway.

reduced substrates are shunted to oxygen at the level of flavin.[2,6] With washed and starved bacteria, luminescence can be stimulated by providing glucose[7]; blocking electron transport at the terminal stage, as for example by cyanide, can similarly stimulate luminescence by increasing the supply of reduced flavin[2,8] and, as we now know, of long-chain aldehyde. Actually, aldehyde may itself be limiting under some conditions, especially during the eclipse period of growth prior to induction.[8] Finally, oxygen may be limiting, but only at extremely low concentrations (below 0.5% O_2). During strict anaerobiosis luminescence ceases, and upon reoxygenation an excess flash of luminescence occurs.[7] This can now be interpreted to be the result of an accumulation of the reduced flavin–luciferase complex.

Biochemically, luciferase may be viewed as a mixed-function oxidase, involving the concomitant oxidation of $FMNH_2$ and RCHO in a reaction that is analogous to reactions with certain flavoprotein hydroxylases.[9] An unusual feature of the reaction is its slowness: at room temperature the time required for a single luciferase cycle to occur is of the order of 10 sec.[10] This

[6] K. L. van Schouwenberg, Doctoral thesis. Delft, The Netherlands, 1938.

[7] F. H. Johnson, K. L. van Schouwenburg, and A. van der Burg, *Enzymologia* **7,** 194 (1939).

[8] K. H. Nealson, T. Platt, and J. W. Hastings, *J. Bacteriol.* **104,** 313 (1970).

[9] J. W. Hastings, and C. Balny, *J. Biol. Chem.* **250,** 7288 (1975).

[10] J. W. Hastings and Q. H. Gibson, *J. Biol. Chem.* **238,** 2537 (1963).

is one of the "slowest" enzymes recorded; a practical consequence of this is that there must be a long-lived luciferase reaction intermediate; indeed, by using low-temperature technology, this has been isolated, purified, and characterized[11,12] as a luciferase–peroxy-reduced flavin.

Bioassays

Although the bacterial luciferase reaction *in vitro* provides many possibilities for measurements of substrates and enzymes coupled to it, living bacteria can themselves provide a unique experimental system for studying various conditions and agents. Determinations *in vivo* can be made instantaneously with ease and accuracy without perturbing the cells, and, when desired, correlated with *in vitro* measurements of luciferase or other enzymes.

The use of luminous bacteria for detecting oxygen in low amounts is classic, having been done by Beijerinck[13] in the detection of oxygen formed by photosynthesis in an extract of crushed clover leaves. Similar and more sophisticated applications have been reported.[14] Luminous bacteria were used for studies of drugs, temperature, and pressure.[15] In recent years extensive studies concerned with anesthetic action in luminous bacteria have been reported.[16]

Relatively few cells are required for *in vivo* measurements. Indeed, when the bacteria are fully induced, the light emission of single cells can be measured using photon counting techniques, and 10^4 cells are adequate for measurement with ordinary photometers. For determinations *in vitro,* aliquots are removed for later analysis; with highly sensitive photometers only about 10^3 cells are required for a single luciferase assay in fully induced cells. With uninduced cells, more are required both *in vivo* and *in vitro* (about a thousandfold).

Transcriptional Control and Autoinduction

Several factors serve to control the synthesis of the bioluminescence system. One of these is an unusual mechanism termed autoinduction. Many

[11] J. W. Hastings, C. Balny, C. LePeuch, and P. Douzou, *Proc. Natl. Acad. Sci. U.S.A.* **70,** 3468 (1973).

[12] C. Balny and J. W. Hastings, *Biochemistry* **14,** 4719 (1975).

[13] M. W. Beijerinck, *Proc. Acad. Sci. Amsterdam* **4,** 45 (1902).

[14] R. Oshino, N. Oshino, M. Tamura, L. Kobilinsky, and B. Chance, *Biochim. Biophys. Acta* **273,** 5 (1972).

[15] F. H. Johnson, H. Eyring, and M. J. Polissar, "The Kinetic Basis of Molecular Biology." Wiley, New York, 1954.

[16] D. C. White, B. Wardley-Smith, and G. Adey, *in* "Progress in Anesthesiology" (B. R. Fink, ed.), Vol. 1, p. 583. Raven, New York, 1975.

species of luminous bacteria excrete into the medium a specific substance, termed autoinducer, which accumulates and, upon reaching a critical concentration, induces the synthesis of the components of the light-emitting system. With dense cultures, autoinducer can be isolated from the medium and purified and then added back at concentrations sufficient to induce cells in culture at low densities.[5,17]

An ecological implication of this phenomenon is that, with bacteria growing free in the seawater, autoinducer could not accumulate, whereas with those growing under confined conditions, for example, symbionts within a light organ, it would. Thus, autoinduction could be viewed as an "environmental sensing" device, restricting the synthesis of the luminescent system to situations where it has functional significance. A practical implication is that the luciferase content of cells may vary by a factor of 10^3 or 10^4 depending upon the status of induction. In growing cells for use in bioassays or for obtaining luciferase it is therefore very important to monitor the specific activity of luminescence.

A second important aspect of the control of the synthesis of the luminescent system, at least in the *Beneckea* species, is that it is subject to catabolite repression; glucose represses luciferase synthesis and cyclic AMP (cAMP) reverses this repression.[4,18] The luciferase system appears to be analogous to catabolite-sensitive systems in other bacteria, and luciferase may be viewed as a "nonessential" enzyme, functioning under certain advantageous conditions while being repressed under others.

Mutants resistant to catabolite repression are readily isolated as those that are brightly luminescent on a medium containing glucose. Although glucose transport mutants might be expected in this class, none have been identified as such,[19] and sugar transport mutants of *B. harveyi* are not bright on a medium with added glucose. Mutants that are nonluminous unless supplied with exogenous cAMP have also been isolated and characterized. The mutant UY-437[20] is pleiotropic, being deficient for a variety of other functions for which cAMP is normally required in enteric bacteria.[21] In the absence of added cAMP, UY-437 is stimulated by low concentrations of cGMP, and inhibited by higher levels.[22] In the presence of added cAMP, cGMP only inhibits.

[17] A. Eberhard, *J. Bacteriol.* **109,** 1101 (1972).

[18] K. Nealson, A. Eberhard, and J. W. Hastings, *Proc. Natl. Acad. Sci. U.S.A.* **69,** 1073 (1972).

[19] K. Nealson and J. W. Hastings, *Bacteriol. Proc.* **72,** 141 (1972).

[20] S. Ulitzur and J. Yashphe, *Biochim. Biophys. Acta* **404,** 321 (1975).

[21] B. Magasanik, *in* "The Lac Operon" (J. R. Beckwith and D. Zipser, eds.), p. 189. Cold Spring Harbor Laboratory, Cold Spring Harbor, New York, 1970.

[22] S. Ulitzur, J. Yashphe, and J. W. Hastings, *Proc. Natl. Acad. Sci. U.S.A.* **73,** 4454 (1976).

There are other compounds that affect the synthesis of luciferase. Exogenous arginine dramatically stimulates the development of luminescence in *B. harveyi* growing in a minimal medium.[23] Arginine is viewed as a coinducer or stimulator, since it has no effect if added without autoinducer.[8] Mutants that have escaped the arginine requirement have been isolated by selecting those that are bright on a minimal medium.[24]

Molecular oxygen is important not only as a substrate for the reaction; in some species it exerts a differential control on growth and the synthesis of the luminous system. In some strains of *P. phosphoreum* and *P. fischeri* the synthesis of luciferase continues at very low oxygen concentrations, so low that growth is limited. The result is that cells are produced with a very high specific luciferase content. In *P. leiognathi* and *B. harveyi*, on the other hand, growth and luciferase synthesis are limited in concert at low oxygen. The practical consequence of this is that with the last-mentioned species maximum luciferase per cell is obtained by growing cells with aeration, whereas in the first two species, both of which are known to occur as symbionts within light organs, maximum luciferase content is achieved in cells grown at low oxygen concentrations where growth is limited. Again, one can perceive an ecological relevance for these latter cases; in such cells, when growing free in seawater, specific luciferase content would be low, whereas within a light organ the oxygen concentration could conceivably be poised at a level appropriate to optimize cellular luciferase content and activity.

Energetics

How much of the cellular energy production is devoted to light emission in the bacteria? Clearly the value has at least two components: the energy required for the biosynthesis of the luminescence system components, e.g., luciferase, and that required for light emission itself. The latter differs depending on the cellular content and activity of the luminous system, which may vary by a factor of a thousand or more. Some years ago Eymers and van Schouwenburg[25,26] deduced from KCN inhibition studies (presumably with fully induced cells) that 20% of the oxygen consumption was utilized in the luminescent system. More recently, Watanabe *et al.*[27] reported that 4.5 photons were emitted per 100 O_2 molecules consumed in *P. phosphoreum* and estimated that 23% of the O_2 uptake passed via

[23] J. J. Coffey, *J. Bacteriol.* **94,** 1638 (1967).

[24] C. A. Waters and J. W. Hastings, *J. Bacteriol.* **131,** 519 (1977).

[25] J. G. Eymers and K. L. van Schouwenburg, *Enzymologia* **1,** 107 (1936).

[26] J. G. Eymers and K. L. van Schouwenburg, *Enzymologia* **1,** 328 (1937).

[27] T. Watanabe, N. Mimara, A. Takimoto, and T. Nakamura, *J. Biochem.* **77,** 1147 (1975).

luciferase, based on a bioluminescence quantum yield of 0.2. There is only a small (perhaps 2-fold), if any, change in oxygen consumption in cells prior to and during induction.[28]

Fully induced bacteria emit somewhere between 10^3 and 10^4 photons per second per cell, depending on the strain.[3,29] Photons at 500 nm are equivalent to about 50 kcal (210 kJ) per einstein, thus being equivalent to about 6 ATP per photon. However, if the quantum yield *in vivo* is only about 0.1, and if *in vivo* a luciferase cycle that fails to emit a photon is equally costly energetically, then the cell foregoes the production of some 60 ATP molecules per photon.

In view of this apparently major energy commitment to luminescence, Ulitzur and Hastings[28] measured ATP levels per cell and found a significant (10-fold) drop prior to and during the period of induction of the luminescence system. However, analysis of ATP levels in different dark mutants allowed them to exclude both the synthesis and the activity of the luminescent system as being directly responsible for the ATP dip. The results with different dark aldehyde mutants suggested that a metabolite belonging to the aldehyde synthesis pathway is involved in the ATP dip.

Mutants

Dark mutants[30] that will emit light upon the addition of specific substances, such as cAMP, fatty acid, or aldehyde, can evidently be used in bioassays for such compounds. Other classes find application as well. Most mutants have been obtained from *B. harveyi* and, with the exception of the two classes mentioned earlier (bright on glucose and on minimal medium), all have been selected originally as dim or dark.[31–34] Actually all "dark" mutants so far reported emit at least some light, but sometimes not much. Photometrically it is possible to detect luminescence levels lower than wild type by a factor as great as 10^{10}.

The general classes of mutants reported include those that involve (1) an alteration in luciferase structure,[35] (2) a block in aldehyde synthesis,[36,37] and (3) a block in the synthesis of the entire luminescent system.[33] The last

[28] S. Ulitzur and J. W. Hastings, *J. Bacteriol.* **133,** 1307 (1978).
[29] J. W. Hastings, *Annu. Rev. Biochem.* **37,** 597. (1968).
[30] T. Cline, this volume [16].
[31] P. Rogers and W. D. McElroy, *Proc. Natl. Acad. Sci. U.S.A.* **41,** 67 (1955).
[32] K. H. Nealson and A. Markovitz, *J. Bacteriol.* **104,** 300 (1970).
[33] T. W. Cline and J. W. Hastings, *Proc. Natl. Acad. Sci. U.S.A.* **68,** 500 (1971).
[34] T. W. Cline and J. W. Hastings, *Biochemistry* **11,** 3359 (1972).
[35] T. W. Cline and J. W. Hastings, *J. Biol. Chem.* **249,** 4668 (1974).
[36] T. W. Cline and J. W. Hastings, *J. Bacteriol.* **118,** 1059 (1974).
[37] S. Ulitzur and J. W. Hastings, *Proc. Natl. Acad. Sci. U.S.A.* **75,** 266 (1978).

category may include those in which there is some alteration in the response to a control element, such as autoinducer, arginine, or some catabolite.[5]

Temperature-sensitive mutants have been reported for all classes, selected initially as darks at 35° which are luminescent at 25°.[33] Luciferase structure mutants that are temperature sensitive were selected as those temperature sensitives whose luminescence is extinguished rapidly upon transfer from a permissive to a nonpermissive temperature. The assignments were confirmed by the property of the isolated luciferase. Such lesions were located in either one of the two subunits. Nonconditional luciferase structure mutants were obtained by screening all dark mutants for alterations in the kinetics of the *in vitro* reaction of the residual luciferase activity. Without exception these lesions were located in the α subunit; all these different lesions have in common the ability to alter the enzymic turnover rate and thus must be at or near the active center.[34]

Aldehyde mutants are selected as those darks that become bright when colonies are exposed to aldehyde vapor.[31,36] Both temperature-conditional and nontemperature-sensitive mutants have been studied. Some mutants have also been found to respond to long-chain fatty acid, but unlike different chain length aldehydes, which are *all* effective over the range of 8–16 carbons, the 14-carbon (myristic) acid is by far the most effective.[37]

These mutants may find wide use in bioassays for a variety of substances, such as fatty acids, aldehydes, lipases, and antilipogenic compounds.[38] Mutants in which very low levels of luciferase are synthesized (but the luciferase has wild-type properties) may be classed as luminescent system (LS) mutants. In one temperature-sensitive LS mutant (TSLS-1), the defect involves also a number of membrane peptides.[39] The mutant UY-437, which requires added cAMP for luminescence, is also of the LS type.[20] Nonconditional LS-type mutants have also been obtained after mutagenesis.[40]

Luciferase

Structure, Subunits

Luciferases have been isolated, purified,[41] and studied[11] from four major groups of luminous bacteria.[4] Although there are significant differences, all bacterial luciferases possess a heterodimeric structure with

[38] S. Ulitzur and J. W. Hastings, this volume [20].

[39] Z. Ne'eman, S. Ulitzur, D. Branton, and J. W. Hastings, *J. Biol. Chem.* **252,** 5150 (1977).

[40] K. H. Nealson, this volume [15].

[41] J. W. Hastings, T. Baldwin, and M. Nicoli, this volume [14].

molecular weights in the range of 76,000 ± 4000. All exhibit specificity for $FMNH_2$, long-chain aliphatic aldehyde (8–16 carbons),[42–44] and slow enzymic turnover. The exact turnover rate is dependent on the chain length of the aldehydes used and may differ significantly and characteristically with different luciferases. Although bacterial luciferases have been thought to be structurally simple, lacking metals, prosthetic groups, and non-amino acid moieties, a glycoprotein with luciferase activity has recently been reported from *P. leiognathi*.[45]

The subunit structure of luciferase, as exemplified by that of *B. harveyi*, is a heterodimer of molecular weight 79,000 whose two nonidentical subunits (α, 42,000; β, 37,000) can be isolated in quantity by DEAE-Sephadex chromatography in 5 *M* urea (see Subunits Isolation[46]). Individual subunits are inactive but can be recombined to yield reconstituted luciferase with activity and properties like those of the native material. From mutant analysis[31] and chemical modification studies[47] it is known that the catalytic properties reside on α. Although the β subunit is required for luciferase activity, its specific function has not been elucidated. The suggestion that it may act by providing conformational stability for α[34] is consistent with the fact that a mutant selected for its lower affinity for substrate ($FMNH_2$) was shown to possess a lesion in the β, not the α, subunit.[48]

Ligand Binding and Chemical Modification of Luciferases

Chemical modification of luciferases and binding by specific probes have been utilized in order to examine substrate binding and the reaction mechanism. The results show that the luciferase dimer possesses a single catalytic site located on the α subunit. Succinic anhydride reacts with lysyl residues to form a succinylated luciferase whose very low but authentic activity exhibits an altered rate of turnover.[49,50] Hybrid luciferase formed with native α and succinylated β has a higher activity with the wild-type turnover rate, whereas that with a native β and succinylated α has the low activity and the altered turnover rate. Luciferase also possesses a particularly reactive sulfhydryl group located on the α subunit in a tryptic peptide

[42] J. W. Hastings, J. A. Spudich, and G. Malnic, *J. Biol. Chem.* **238,** 3100 (1963).

[43] J. W. Hastings, K. Weber, J. Friedland, A. Eberhard, G. W. Mitchell, and A. Gunsalus, *Biochemistry* **8,** 4681 (1969).

[44] T. Watanabe and T. Nakamura, *J. Biochem.* **72,** 647 (1972).

[45] C. V. Balakrishnan and N. Langerman, *Arch. Biochem. Biophys.* **181,** 680 (1977).

[46] S.-C. Tu, this volume [17].

[47] E. A. Meighen, this volume [18].

[48] T. W. Cline, Ph.D. thesis, Harvard University, Cambridge, Massachusetts, 1973.

[49] E. A. Meighen, M. Z. Nicoli, and J. W. Hastings, *Biochemistry* **10,** 4062 (1971).

[50] E. A. Meighen, M. Z. Nicoli, and J. W. Hastings, *Biochemistry* **10,** 4069 (1971).

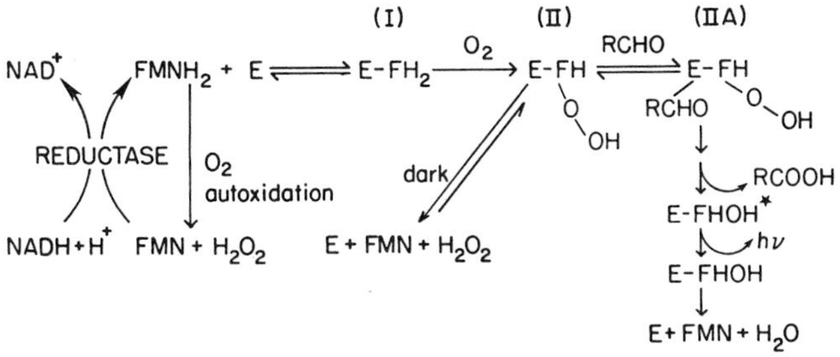

FIG. 2. Reaction sequence and intermediates for the reaction of bacterial luciferase (E) with $FMNH_2$, O_2 and aldehyde (RCHO). An excited intermediate results in light emission.

with the sequence Phe-Gly-Ile-Cys-Arg.[51-53] Its modification, e.g., with maleimides, results in a loss of activity; the addition of FMN, $FMNH_2$, or aldehyde will protect against this loss. A histidyl residue on the α subunit has also been implicated in the luciferase reaction.[54]

Luciferase also has a single binding site for the fluorescent probe 8-anilino-1-naphthalenesulfonate (ANS).[55] ANS acts as a luciferase inhibitor competitive with $FMNH_2$ but does not displace FMN. From energy-transfer studies with luciferases fluorescently labeled at the –SH group, the distance between the SH group and the ANS site was calculated to be in the range of 21–37 Å, and the distance between the ANS and FMN sites 30–58 Å.[56]

Reaction Mechanism of Bacterial Luciferase: The Luciferase Peroxy Flavin Intermediate

In the proposed pathway (Fig. 2) the reaction of reduced flavin with luciferase results in the formation of luciferase-bound $FMNH_2$, designated as intermediate (I). This has been shown to involve a 1:1 reduced flavin–luciferase complex for *B. harveyi*, *P. fischeri*, and *P. phosphoreum* luciferases.[44,57]

[51] J. W. Hastings, A. Eberhard, T. O. Baldwin, M. Z. Nicoli, T. W. Cline, and K. H. Nealson, *in* "Chemiluminescence and Bioluminescence" (M. J. Cormier, D. M. Hercules, and J. Lee, eds.), p. 369. Plenum, New York, 1973.

[52] M. Z. Nicoli, E. A. Meighen, and J. W. Hastings, *J. Biol. Chem.* **249**, 2385 (1974).

[53] M. Z. Nicoli and J. W. Hastings, *J. Biol. Chem.* **249**, 2393 (1974).

[54] J. Cousineau and E. Meighen, *Biochemistry* **15**, 4992 (1976).

[55] S.-C. Tu and J. W. Hastings, *Biochemistry* **14**, 4310 (1975).

[56] S.-C. Tu, C.-W. Wu, and J. W. Hastings, *Biochemistry*. **17**, 987 (1978).

[57] J. E. Becvar and J. W. Hastings, *Proc. Natl. Acad. Sci. U.S.A.* **72**, 3374 (1975).

Reaction of (I) with one molecule of oxygen is extremely rapid,[57a] forming a second intermediate (II), which because of its relatively long lifetime can be purified and studied using low-temperature techniques.[9,11] After isolation and purification by column chromatography at low temperatures ($-20°$ to $-30°$), it can be made to give rise to bioluminescence, with a quantum yield equal to that of the overall reaction, simply by injection into buffer containing aldehyde at $+20°$. Oxygen is not required for this reaction.[58] This oxygen-containing intermediate differs spectrally from both oxidized and reduced flavin; it has absorption in the visible range, peaking at 372 nm, and fluorescence emission centered at about 490 nm. Although the intermediate cannot itself be the emitter, since the aldehyde has not yet reacted, its fluorescence corresponds closely in color to the bioluminescence emission.[12] The intermediate in aqueous solution may also be isolated at 0° to 2°.[58,59]

The intermediate is postulated to be the 4a-peroxy adduct of $FMNH_2$.[11] Spectral properties (absorption and fluorescence) of 4a-substituted flavins resemble those of the luciferase intermediate. Moreover, a synthetic 4a-substituted flavin hydroperoxide does, in fact, emit light when allowed to react with aldehyde.[60] The postulated intermediate (II) suggests that bioluminescence should be obtained from H_2O_2 and FMN in lieu of $FMNH_2$. This has been found to be so.[61,62]

In the absence of aldehyde, intermediate (II) breaks down to yield 1 mol each of H_2O_2 and FMN per mol of intermediate and only a very small amount of bioluminescence (between 10^{-2} and 10^{-4} of that with aldehyde).[63] In the presence of aldehyde, the formation of acid concomitant with light emission has been shown by several workers.[64-67] The other products are FMN and H_2O, not H_2O_2.[9]

[57a] R. P. Presswood and J. W. Hastings, *Fed. Proc., Fed. Am. Soc. Exp. Biol.* **36,** 832 (1977).

[58] J. E. Becvar, S.-C. Tu, and J. W. Hastings, *Biochemistry* **17,** 1807 (1978).

[59] N. Ashizawa, T. Nakamura, and T. Watanabe, *J. Biochem.* **81,** 1057 (1977).

[60] C. Kemal, T. W. Chan, and T. C. Bruice, *Proc. Natl. Acad. Sci. U.S.A.* **74,** 405 (1977).

[61] J. E. Becvar, S.-C. Tu, and J. W. Hastings, *Fed. Proc., Fed. Am. Soc. Exp. Biol.* **35,** 163 (1976).

[62] T. Watanabe and T. Nakamura, *J. Biochem.* **79,** 489 (1976).

[63] J. W. Hastings, Q. H. Gibson, J. Friedland, and J. Spudich, *in* "Bioluminescence in Progress" (F. H. Johnson and Y. Haneda, eds.), p. 151. Princeton Univ. Press, Princeton, New Jersey, 1966.

[64] D. K. Dunn, G. A. Michaliszyn, I. G. Bogacki, and E. A. Meighen, *Biochemistry* **12,** 4911 (1973).

[65] F. McCapra and D. W. Hysert, *Biochem. Biophys. Res. Commun.* **52,** 298 (1973).

[66] O. Shimomura, F. H. Johnson, and Y. Kohama, *Proc. Natl. Acad. Sci. U.S.A.* **69,** 2086 (1972).

[67] A. Vigny and A. M. Michelson, *Biochimie* **56,** 171 (1974).

Although several specific mechanisms have been proposed for the bacterial luciferase reaction,[65,68-70] no one of them is well supported by presently available experimental facts.[70a] It has always seemed that excited FMN could be the emitter in a reaction where oxidized FMN, which is a highly fluorescent molecule, is the product. However, its fluorescence emission is centered at 530 nm whereas the bioluminescence peaks around 490 nm, a difference difficult to attribute solely to environmental effects.[71,72] Moreover, luciferase-bound FMN is actually nonfluorescent.[73,74] The fact that the isolated luciferase intermediate (II) exhibits a fluorescence emission attributable to its flavin moiety, which exactly matches the bioluminescence emission,[12] suggests that the emitting species in bioluminescence is a similar substituted flavin such as (E-FHOH)* (Fig. 2).

[68] J. N. Lowe, L. L. Ingraham, J. Alspach, and R. Rasmussen, *Biochem. Biophys. Res. Commun.* **73**, 465 (1976).

[69] A. Eberhard and J. W. Hastings, *Biochem. Biophys. Res. Commun.* **47**, 348 (1972).

[70] C. Kemal and T. C. Bruice, *J. Am. Chem. Soc.* **99**, 7064 (1977).

[70a] R. P. Presswood, P. Shannon, R. Spencer, J. E. Becvar, J. W. Hastings, and C. Walsh, *Fed. Proc., Fed. Am. Soc. Exp. Biol.* **37**, 1719 (1978).

[71] M. J. Cormier, J. Lee, and J. E. Wampler, *Annu. Rev. Biochem.* **44**, 255 (1975).

[72] G. Mitchell and J. W. Hastings, *J. Biol. Chem.* **244**, 2572 (1969).

[73] T. O. Baldwin, *Biochem. Biophys. Res. Commun.* **57**, 1000 (1974).

[74] T. O. Baldwin, M. Z. Nicoli, J. E. Becvar, and J. W. Hastings, *J. Biol. Chem.* **250**, 2763 (1975).

[14] Bacterial Luciferase: Assay, Purification, and Properties

By J. W. Hastings, Thomas O. Baldwin, and Miriam Z. Nicoli

$$FMNH_2 + RCHO + O_2 \rightarrow FMN + RCOOH + H_2O + h\nu$$

Principles

The activity of bacterial luciferase is measured as the initial maximum light intensity (which is a measure of initial velocity) upon mixing the enzyme, aldehyde, and oxygen with reduced flavin mononucleotide ($FMNH_2$). Luciferase activity may also be measured by means of the total light produced. Although luciferases from different bacterial species differ in several ways, including physical, kinetic, and purification properties, all catalyze this same reaction.

It must be emphasized that one of the principles involved in the assay of bacterial luciferase differs completely from that for most enzymes, includ-

ISBN 0-12-181957-4

ing many other luciferases: no catalytic turnover is involved.[1] This is because the time required for a catalytic cycle is long compared to the "lifetime" of the substrate, $FMNH_2$, which is autoxidized rapidly, within 1 sec at 20°.[2]

$$FMNH_2 + O_2 \rightarrow FMN + H_2O_2$$

Although the reduced-flavin substrate is available for less than 1 sec, light emission nevertheless continues for 10 or 20 sec or longer at 20°, depending upon the specific luciferase and conditions. The activity measured is thus the result of intermediates formed in the luciferase-reduced flavin reaction during the first fraction of a second. No enzymic turnover is involved; the lifetime of the luminescence is then a direct measure of the time required for a catalytic cycle, i.e., the enzyme turnover number.

A second important feature of the reaction is that luciferase can exhibit both oxidase and monooxygenase activities. In the absence of aldehyde an enzyme intermediate is formed that decomposes to give H_2O_2 and FMN without light emission. However, unlike the nonenzymic autoxidation, this oxidation is slow, involving an intermediate with a lifetime measured in seconds.

In lieu of utilizing $FMNH_2$ directly, one may use a coupled assay and reduce the flavin *in situ*. A suitable soluble flavin reductase may be isolated and purified from extracts of luminous bacteria.[3]

$$NADH + H^+ + FMN \rightarrow NAD^+ + FMNH_2$$

With this system the reaction continues because $FMNH_2$ is continuously replenished; the initial light intensity is a function of the product of the concentrations of two enzymes, luciferase and reductase.[4]

Reagents, Materials, and Instrumentation for Assay[5]

*Buffer
*FMN (Sigma), monosodium salt
*Decanal, dodecanal, or tetradecanal (Aldrich)
Bovine serum albumin (BSA) (Sigma, fraction V powder)
β-NADH (Sigma)
H_2 gas

[1] J. W. Hastings and Q. H. Gibson, *J. Biol. Chem.* **238,** 2537 (1963).
[2] Q. H. Gibson and J. W. Hastings, *Biochem. J.* **83,** 368 (1962).
[3] Q. H. Gibson, J. W. Hastings, G. Weber, W. Duane, and J. Massa, *in* "Flavins and Flavoproteins" (T. P. Singer, ed.), p. 341. Elsevier, Amsterdam, 1966.
[4] J. W. Hastings, W. H. Riley, and J. Massa, *J. Biol. Chem.* **240,** 1473 (1965).
[5] Items marked with an asterisk (*) are required for all assay methods.

Flask, 100-ml, round-bottom, with 14/20 joint and rubber serum stopper to fit snugly (16 mm, hooded)

Sodium dithionite (sodium hydrosulfite, $Na_2S_2O_4$)

Platinized asbestos, fibrous, 5% (Matheson Coleman & Bell)

*Photometer apparatus

*Plastic syringes and hypodermic needles to fit vial compartment (18 G, 19 G, or 20 G; $2\frac{1}{2}$- or 3-inch)

*Scintillation vials for assay mixture, 20-ml, glass or plastic

*Sonicator or Triton X-100

The FMN may be made up as a 5 m*M* stock solution (0.239 g/100 ml), stored at $-20°$ in 1-ml aliquots, and thawed and diluted to the desired concentration for use on a given day. Flavin solutions should be protected from light during storage and use. The concentration of an FMN solution may be checked by its visible absorbance ($\epsilon_{450\ nm} = 12{,}200\ M^{-1}\ cm^{-1}$).[6] Aldehyde stock "solutions" (emulsions) should be prepared fresh about every 8 hr. BSA solutions may be made up in a phosphate buffer as a 20 × concentrated stock [4% BSA (w/v), i.e., 20 g in 500 ml of 1.0 *M* phosphate, pH 7.0] and stored in 10-ml aliquots at $-20°$. Dilution (1:20) produces the buffer solution (0.2% BSA, 50 m*M* phosphate, pH 7.0) used in the assays.

A stable, inexpensive solid-state photomultiplier photometer suitable for measurements of bioluminescence has been described.[7] A photometer housing with a compartment holding a scintillation vial[8] is convenient for the bacterial luciferase reaction. The vial is placed so that the photomultiplier is directly beneath the vial. The light-tight compartment is drilled with holes of a diameter to exactly fit the syringe needle (Nos. 18 G, 19 G, or 20 G is suitable) without a light leak, and reagents to initiate the reaction are added rapidly by injection. A scintillation counter may also be employed, but (1) it must be operated without coincidence, (2) it must be used only at the appropriate low light intensities with solutions which are emitting usually not above 10^8 or 10^9 quanta per second, (3) it must be adapted by one means or another to accommodate for the time interval between initiation of the reaction and the beginning of counting, and (4) counts are usually integrated on such a machine. A simple fluorometer or a modified spectrophotometer may also be used. Finally it should be mentioned that the stopped-flow apparatus[1,2] is well suited to assay and study of bacterial luciferase.

[6] L. G. Whitby, *Biochem. J.* **54,** 437 (1953).

[7] G. W. Mitchell and J. W. Hastings, *Anal. Biochem.* **39,** 243 (1971).

[8] J. W. Hastings and G. Weber, *J. Opt. Soc. Am.* **53,** 1410 (1963).

Assay Methods

Luciferase

The first three methods are nonturnover assays utilizing $FMNH_2$ without cycling. In the fourth a coupled assay is used, with the flavin reduced repeatedly by NADH. In the nonturnover assays the reaction system is separated into two components, one aerobic (containing O_2 and aldehyde) and the other anaerobic (containing $FMNH_2$). The enzyme may be in either component. The reaction is initiated by rapid mixing of one solution with the other. The three methods described differ in the procedure by which the flavin is reduced.

The conditions for optimal activity vary somewhat with different luciferases; in general, suitable conditions include a temperature between 20° and 25°, 50 m*M* phosphate buffer, pH 7, 0.2% BSA, and an aldehyde at a concentration of about 0.1 m*M*, depending somewhat on the chain length of the aldehyde and the luciferase being used.

Catalytically Reduced Flavin (the Standard Assay). The FMN (50 μ*M*, in H_2O) is placed in a small flask (plugged with a rubber serum stopper) with a small amount (~40 mg) of platinized asbestos and bubbled with H_2 using hypodermic needles for entry and exit ports and metal or silicone tubing (which is far less gas-permeable than Tygon or rubber) for connections. The H_2 exit port should lead to a fume hood or outside the building. The progress of the reduction (which takes about 20 min) may be followed by the loss of yellow color and green fluorescence of the FMN. Dry platinized asbestos should not be added to a flask containing residual H_2; an explosion may result. Caution in the use of H_2 should be exercised.

The $FMNH_2$ for each assay (1.0 ml) is withdrawn in a syringe without exposure to O_2, and the luciferase reaction is initiated by vigorous injection of the $FMNH_2$ directly into a vial containing the other components of the reaction mixture, in the housing in front of the photomultiplier tube, so that the reaction rate may be followed continuously. The turbulence caused by the injection of $FMNH_2$ provides for mixing, and the vigor of the injection is vital for reproducible assays. The mixture in the vial (1.0 ml) contains 0.2% BSA (w/v), 50 m*M* phosphate (pH 7.0), luciferase (1 μ*M* to 1 n*M*), and aldehyde (e.g., decanal, dodecanal, or tetradecanal) at a concentration optimal for luminescence activity. A water emulsion of the aldehyde (0.1% v/v) for this assay may be prepared by sonication or by suspension in a 0.1% Triton X-100 solution. The optimal aldehyde concentration depends upon the type of luciferase (bacterial strain) and on the aldehyde chain length. With decanal or dodecanal, and *Beneckea harveyi* luciferase, 10–15 μl of a 0.1% sonicated emulsion is optimal; higher concentrations may be inhibitory in this assay method. This inhibition does not occur if aldehyde is added after $FMNH_2$ (two injections are required). In the standard assay aldehyde inhibition is less severe with the added 0.2% BSA, which for unknown reasons also results in a higher quantum yield.[9] The luciferases from *Photobacterium fischeri* and *P. phosphoreum* require substantially higher amounts of aldehyde (100–200 μl) for optimal activity. The *P. fischeri* enzyme is not inhibited by higher aldehyde concentrations. Care should be taken to add the aldehyde to the solution before adding the luciferase, to avoid denaturation due to locally high aldehyde concentrations; however, the aldehyde should not be diluted into the buffer more than about 5 min prior to the assay. BSA is not required in the assay for *Photobacterium phosphoreum luciferase.*[10]

Photochemically Reduced Flavin. A deoxygenated solution of FMN (50 μM) containing 1 mM EDTA as an electron donor[11,12] may be readily reduced photochemically. The FMN solution (1.0 ml) is reduced by irradiation with a "black" (near u.v.) light in a stoppered container or in a syringe immediately prior to injection into the rest of the reaction mixture, as described for the catalytically reduced flavin assay. The photochemical method is particularly convenient for work with a stopped-flow apparatus.[1,2]

Chemically Reduced Flavin (the Dithionite Assay). This method involves initiation of the reaction by injection of 1.0 ml of air-equilibrated buffer containing aldehyde emulsion (0.01%, v/v) into a vial containing an anaerobic solution of luciferase and $FMNH_2$, which forms the enzyme–substrate complex, E:$FMNH_2$. The luciferase is added to 1.0 ml of FMN (50 μM) in 50 mM phosphate (or Bis-Tris), pH 7.0, and just prior to the assay sodium dithionite is added either as a powder (a few grains)[13,14] or, more quantitatively, as a solution.[15] Such a solution may be prepared by dissolving 150 mg of $Na_2S_2O_4$ in 10 ml of O_2-free water and kept anaerobic by flushing with argon or N_2. (Commercial N_2 should be "scrubbed" by bubbling through another dithionite solution before the working dithionite solution.) The volume of dithionite solution to be added to the FMN–luciferase mixture is determined by trial and error as the minimum required (generally ~10–20 μl) to remove the dissolved O_2, fully reduce the flavin, and still give optimum luminescence upon injection of the aldehyde emulsion. A fresh dithionite solution should be prepared at least every 4 hr.

[9] A. L. Baumstark, T. W. Cline, and J. W. Hastings, *Arch. Biochem. Biophys.* in press (1978).
[10] T. Nakamura and K. Matsuda, *J. Biochem,* **70,** 35 (1971).
[11] W. J. Nickerson and G. Strauss, *J. Am. Chem.* **82,** 5007 (1960).
[12] G. Strauss and W. J. Nickerson, *J. Am. Chem. Soc.* **83,** 3187 (1961).
[13] E. A. Meighen and J. W. Hastings, *J. Biol. Chem.* **246,** 7666 (1971).
[14] E. A. Meighen and R. E. MacKenzie, *Biochemistry* **12,** 1482 (1973).
[15] S.-C. Tu and J. W. Hastings, *Biochemistry* **14,** 4310 (1975).

This assay is especially useful when low $FMNH_2$ concentrations (<10 μM) are to be used, e.g., in K_m determinations, or for mutant luciferases with lower aldehyde binding affinities,[16] since high aldehyde concentrations are required to saturate, and these would be inhibitory in the other assay methods. Aldehyde is not inhibitory in this assay even at high concentrations; this is presumably due to the fact that aldehyde is added after formation of the luciferase–$FMNH_2$ complex.

The NADH–FMN Coupled Assay. In this assay the $FMNH_2$ is generated from NADH + FMN by a flavin reductase. The light emission from luciferase is thus a function of the product of the concentrations of the two enzymes.[4] In crude extracts of luminous bacteria they occur together, but it is desirable to be able to vary the concentrations independently, which requires the use of purified enzymes. The purification of the reductase has been described elsewhere,[17-21] and the luciferase purification is described below.

The assay is initiated by the addition of 0.1 ml of 0.1 m*M* NADH to a vial containing 1 ml of luciferase, reductase, FMN (4 μM), BSA (0.2%, w/v), and decanal (0.001%, w/v) in 50 m*M* phosphate buffer, pH 7.0. The light intensity rises rapidly to a maximum and continues at a relatively constant level until the NADH is exhausted, with the luciferase turning over continuously. With the *P. fischeri* enzymes the reaction is inhibited if NADH is added to the enzyme mixture first and the reaction is initiated by addition of FMN. The reason for this is not known; it does not occur with the *B. harveyi* enzymes.

Photoexcitable Luciferase

Some light emission can be obtained by flash irradiation of luciferase preparations. FMN is not required; O_2 and aldehyde are needed for emission, but need not be present during irradiation.[22-24] This activity has been purified and shown to be a consequence of the association of a flavinlike chromophore with luciferase to form a species termed photoexcitable luciferase, which is inactive in the chemically ($FMNH_2$) initiated reac-

[16] T. W. Cline and J. W. Hastings, *Biochemistry* **11,** 3359 (1972).
[17] K. Puget and A. M. Michelson, *Biochimie* **54,** 1197 (1972).
[18] W. Duane and J. W. Hastings, *Mol. Cell. Biochem.* **6,** 53 (1975).
[19] E. Gerlo and J. Charlier, *Eur. J. Biochem.* **57,** 461 (1975).
[20] E. Jablonski and M. DeLuca, *Biochemistry* **16,** 2932 (1977).
[21] G. A. Michaliszyn, S. S. Wing, and E. A. Meighen, *J. Biol. Chem.* **252,** 7495 (1977).
[22] Q. H. Gibson, J. W. Hastings, and C. Greenwood, *Proc. Natl. Acad. Sci. U.S.A.* **53,** 187 (1965).
[23] J. W. Hastings and Q. H. Gibson, *J. Biol. Chem.* **242,** 720 (1967).
[24] G. W. Mitchell and J. W. Hastings, *Biochemistry* **9,** 2699 (1970).

tion.[25,26] This chromophore absorbs maximally at 375 nm and when luciferase-bound is fluorescent, with an excitation maximum at 375 nm and the emission maximum at 495 nm; the chromophore is postulated to be a 4a-substituted flavin, which is a false, or dead-end intermediate in the bioluminescence reaction. For assay, the photoexcitable luciferase is placed in a Vycor or quartz syringe, irradiated, and then promptly transferred to the vial containing aldehyde and buffer in front of the phototube.[27]

Growth and Harvesting of Cells

Luciferases have been purified from four different species of luminous bacteria, *Beneckea harveyi, Photobacterium fischeri, P. phosphoreum,* and *P. leiognathi*. Although there are differences, similar procedures appear to be involved; purification procedures for only the first two species have been reported in detail.

For the inoculum care should be taken to start with a fresh single bright colony, since dim variants low in luciferase occur readily in older cultures.[28] The liquid medium may be seawater complete (SWC) or NaCl complete. One liter of SWC contains 800 ml of seawater + 200 ml of distilled water (or 1000 ml of distilled water + 32 g of Marine Mix, Utility Chemical Co., Paterson, New Jersey), 5 g of tryptone (Difco or Gibco), 3 g of yeast extract (Difco or Gibco), and 2 ml of glycerol. NaCl complete contains (in 1 liter of distilled water) 7 g of $Na_2HPO_4 \cdot 7\,H_2O$, 1 g of KH_2PO_4, 0.5 g of $(NH_4)_2HPO_4$, 0.1 g of $MgSO_4$, 30 g of NaCl, and tryptone, yeast extract, and glycerol as in SWC. For solid media, 12 g of agar per liter are added.

In most cases the system is "autoinducible," so the specific luciferase content of the cells may be very different at different times.[29] Thus the luminescence *in vivo* should be monitored in order to assure that harvesting (by centrifugation) is done at the time of maximum emission, and promptly so, since light emission and luciferase content (activity) decline quickly after the peak. Cells are harvested by centrifugation and stored in the deep freeze at −20° for later extraction.

Beneckea harveyi

The following detailed procedures are for the growth and harvesting of cells in 175 liters of medium in a New Brunswick fermentor.

[25] S.-C. Tu, C. A. Waters, and J. W. Hastings, *Biochemistry* **14,** 1970 (1975).
[26] S.-C. Tu and J. W. Hastings, *Biochemistry* **14,** 1975 (1975).
[27] J. W. Hastings, Q. H. Gibson, and C. Greenwood, *Photochem. Photobiol.* **4,** 1227 (1965).
[28] J. W. Hastings and K. H. Nealson, *Annu. Rev. Microbiol.* **31,** 549 (1977).
[29] K. Nealson, T. Platt, and J. W. Hastings, *J. Bacteriol.* **104,** 313 (1970).

A tube containing 10 ml of liquid medium is inoculated from a slant and allowed to grow overnight at 30° to 34° with vigorous aeration. Cells are then streaked on plates and incubated at 30° to 34° for 24 hr for selection and isolation of a single bright colony. This is used to inoculate a tube containing 10 ml of liquid medium at noon on the day prior to the fermentation and incubated with shaking at 34°. At midnight, two 2-liter flasks, each containing 800 ml of medium, are inoculated with 2 ml of the culture from the shake tube. These shake flasks are incubated at 34° with vigorous aeration.

At 7 AM the contents of the two flasks are uniformly distributed into 8 flasks, each containing 800 ml of medium. These flasks are allowed to grow until 10 AM at 34° with vigorous aeration and then used to inoculate the 175 liters of seawater or NaCl complete medium contained within the fermentor. The optical density (660 nm) and the luminescence intensity of the culture are monitored, and the cells are harvested when the luminescence intensity stops increasing, which occurs just as the culture enters stationary phase ($\sim 4 \times 10^9$ cells/ml). The yield from such a fermentation is generally between 1500 and 2000 g of packed wet cells.

The harvesting of the cells deserves special attention. In stationary-phase cells, the luciferase is degraded with a half-time of about 30 min. It is therefore important for maximal luciferase content to harvest the cells rapidly using a chilled continuous-flow centrifuge. Cooling of the cells in the fermentor is also advantageous; if rapid-chilling equipment is not available, the contents of the fermentor can be cooled by addition of 20 liters of frozen seawater or 3% NaCl (water without salt may cause some cell lysis).

The cells are removed from the centrifuge and transferred into large plastic bags for storage. For easy lysis later, it is important that the cells be frozen at −20° (rather than −70° or lower) and that the cells be frozen in a thin sheet about 2 cm thick, rather than in a more cubic block.

Other Species

The growth and harvesting of *Photobacterium fischeri* is similar to that described for *B. harveyi* except that NaCl complete medium is always used (not SWC), and the growth temperature of choice is about 26° (not to exceed about 28°), with some resultant increase in the growth times recommended above. Furthermore, for some members of this species, a higher luciferase content per cell is obtained when aeration is less vigorous so that the oxygen concentration is growth-limiting.

Photobacterium phosphoreum is a low-temperature species that should also be grown on NaCl complete medium, but at even lower temperatures, preferably not above 18°. Like *P. fischeri*, some strains of *P. phosphoreum* produce more luciferase per cell when the O_2 concentration is growth-limiting. *Photobacterium leiognathi* is less well studied than the other three, but like *B. harveyi*, it requires vigorous aeration.

Purification

The scheme described here is applicable to the different luciferases with only minor modifications. It involves osmotic lysis, DEAE-cellulose batch adsorption, and DEAE-Sephadex column chromatography.[30,31] A detailed protocol for purification of the luciferase from *B. harveyi* is given below and outlined in Table I. Attempts to scale up the procedure, for example to 1 kg of cells, have met with difficulties in the first step.

Reagents and Materials

1. *Lysis*

Frozen luminous bacterial cells, 500 g (wet weight)
3% NaCl, 100 ml
Cold distilled water, 5 liters

2. *DEAE-Cellulose Batch Adsorption*

Büchner funnel (preferably 10-inch diameter)
Miracloth (Calbiochem or local restaurant supplier)
DEAE-cellulose: 200 g of Whatman microgranular DE-52, preswollen and regenerated as described below
NaOH, 0.25 *M*, 4 liters
H_3PO_4, 0.2 *M*, 2 liters
Phosphate, 1 *M*, pH 7.0, 200 ml
Tris base
Phosphate, 0.15 *M*, pH 7.0 (cold), 400 ml
Phosphate, 0.35 *M*, pH 7.0 (cold), 1.5 liters
Ammonium sulfate
Dialysis buffer (0.25 *M* phosphate, pH 7.0, 1 m*M* EDTA, 0.5 m*M* DTT), cold, 8 liters

3. *DEAE-Sephadex Column Chromatography*

Chromatography column, 5 × 60 cm
DEAE-Sephadex A-50 (Pharmacia)
Phosphate, 0.35 *M*, pH 7.0, 1 m*M* EDTA, 0.5 m*M* DTT (~6–8 liters)

4. *Preparation of pH 7.0 Phosphate Buffers*. Phosphate buffers may conveniently be prepared by mixing appropriate volumes of 2 *M* NaH_2PO_4

[30] A. Gunsalus-Miguel, E. A. Meighen, M. Z. Nicoli, K. H. Nealson, and J. W. Hastings, *J. Biol. Chem.* **247**, 398 (1972).

[31] T. O. Baldwin, M. Z. Nicoli, J. E. Becvar, and J. W. Hastings, *J. Biol. Chem.* **250**, 2763 (1975).

TABLE I
PURIFICATION OF *Beneckea harveyi* LUCIFERASE[a]

Step and fraction	Volume (ml)	A_{280}	Total A_{280} $\times 10^{-3}$	Activity[b] (q sec^{-1} ml^{-1} $\times 10^{-14}$)	Total activity[b] (q $sec^{-1} \times 10^{-17}$)	Specific activity[b] (q sec^{-1} $A_{280}^{-1} \times 10^{-14}$)
Step 1						
Lysate[c]	5700	27.1	154	0.56	3.2	0.020
Step 2						
Supernatant after DEAE-cellulose adsorption	5900	20.0	118	0.045	0.27	0.002
Step 3						
a. 0.15 *M* phosphate eluate	295	30.7	9.1	0.40	0.12	0.013
b. 0.35 *M* phosphate eluate	1330	13.2	15.3	2.09	2.8	0.16
c. 40% ammonium sulfate supernatant	1420	8.2	11.6	1.69	2.4	0.21
d. 75% ammonium sulfate supernatant	1500	2.0	3.0	0.017	0.025	0.008
e. 75% ammonium sulfate precipitate after dialysis	121	72.0	8.7	15.0	1.8	0.21
Step 4 DEAE-Sephadex column (conservative pool, about 1/3 of activity, precipitated with 75% ammonium sulfate and dialyzed)	18	21.0	0.38	31.1	0.56	1.5

[a] From 500 g of wet cell paste.
[b] Activity was determined in the standard assay with decanal [15 μl of a 0.1% (v/v) emulsion], and is expressed in quanta sec^{-1} according to the light standard of J. W. Hastings and G. Weber [*J. Opt. Soc. Am.* **53,** 1410 (1963)].
[c] Absorbance at 280 nm is that of an aliquot after centrifugation to remove cell debris.

and 2 *M* K_2HPO_4 and diluting, or (for a large volume of a particular buffer concentration) by dissolving the appropriate weights of the mono- and dibasic salts. If care is taken with the proportions of the mono- and dibasic salts, no further pH adjustment is usually required.

a. Stock solutions, 2*M*
   NaH_2PO_3: 276.0 g of $NaH_2PO_4 \cdot H_2O$ per liter
   K_2HPO_4: 348.4 g of anhydrous K_2HPO_4 per liter
b. Phosphate buffers, pH 7.0
   NaH_2PO_4, 2*M*, 195 ml
   K_2HPO_4, 2*M*, 305 ml
   Dilute to 1 liter for 1 *M* phosphate, etc.
c. Phosphate, 0.35 *M*, pH 7.0 (20 liters)
   $NaH_2PO_4 \cdot H_2O$, 376.8 g
   K_2HPO_4 (anhydrous), 743.8 g
   Dilute to 20 liters

Beneckea harveyi

Step 1. Osmotic Lysis. A batch of 500 g of frozen cells (freezing and thawing aids in cell lysis) is allowed to thaw overnight at 4° and thoroughly mixed with about 100 ml of cold 3% NaCl such that a thin cell suspension results. This suspension is then poured slowly (5–10 min) into 5 liters of cold distilled water with continuous stirring to rapidly disperse the cells, this being important to achieve good lysis. The mixture should be kept at 4°. Complete lysis and solubilization requires about 30 min, and can be monitored by assaying a small aliquot of the lysate for luciferase activity (standard assay) before and after centrifugation. If an appreciable percentage of the activity is pelleted, sonication may be used to complete the lysis.

Step 2. DEAE-Cellulose Batch Adsorption. Regeneration of the DEAE-cellulose prior to use in this step is important for good binding and recovery. Preswollen Whatman microgranular DE-52 (200 g) is washed extensively on a Büchner funnel using Miracloth as a support, first with water (~8 liters), then with 4 liters of 0.25 *M* NaOH followed by another 8 liters of water. The DEAE-cellulose is converted to the phosphate form by washing on the filter with 2 liters of dilute (about 0.2 *M*) H_3PO_4, then rinsing with 8 liters of water, followed by 200 ml of 1 *M* phosphate, pH 7.0, and finally about 2 liters of water to decrease the ionic strength. The pH of the last wash should be checked; it should be about 7.0. The regenerated cellulose may be stored as a thick slurry in water at 4°. For storage for more than a day or two, it should be layered with a few drops of toluene (which should be washed out before use) to retard microbial growth.

All the steps below involving batch adsorption and elution can be carried out in the laboratory at room temperature, but all solutions used for washing and elution are kept cold, as are the eluates. Solid Tris base (about 1 g) is added to the lysate to raise the pH (from about 6.2 to between 6.5 and 6.8); if the pH is adjusted too high, the viscosity of the lysate will increase and subsequent filtration steps will be difficult. About 200 ml (settled volume) of the regenerated DEAE-cellulose are then added to the lysate with gentle and continuous stirring. The pH of the supernatant should be checked and, if necessary, adjusted to about 6.8 by addition of a small amount (1–2 g) of solid Tris base. After about 15 min, a small volume of the suspension is centrifuged to remove the cellulose, and the activity should be about 10% of that of the lysate prior to addition of the cellulose. If more activity remains unbound, an additional 50 ml of cellulose is added, and the supernatant activity is checked again. Cellulose additions should be continued until 90% of the luciferase activity is bound to the cellulose. Extreme care should be taken not to add too much cellulose, as elution of the luciferase will then require an excessively large volume of buffer.

After adsorption of the luciferase activity, the cellulose is filtered and rinsed with five 100-ml aliquots of distilled water by vacuum filtration using a Büchner funnel and Miracloth. Excess liquid can be removed from the cake by pressing it with the bottom of a beaker.

Step 3. Elution. The vacuum is released, and the cake is moistened with 200 ml of 0.15 *M* phosphate, pH 7.0. After 5 min the liquid is removed by filtration and another wash with 0.15 *M* phosphate is carried out. This procedure elutes the flavin reductase[30] and other weakly binding proteins but should release no more than 20% of the total luciferase activity. This should be verified before proceeding.

The luciferase is then eluted from the DEAE-cellulose by repeated (5–10) washings with 200-ml aliquots of 0.35 *M* phosphate, pH 7.0. Five minutes should be allowed after each addition before application of the vacuum. Each 200-ml eluate should be assayed, and elution should be continued until between 80% and 90% of the luciferase activity has been obtained. The luciferase is then concentrated and collected as the ammonium sulfate precipitate between 40% and 75% of saturation, resuspended in a minimum volume of 0.25 *M* phosphate, pH 7.0, 1 m*M* EDTA, 0.5 m*M* DTT, and dialyzed overnight at 4° against two 4-liter volumes of the same buffer.

Step 4. DEAE-Sephadex Column Chromatography. A column (5 × 50 cm) of DEAE-Sephadex A-50 is equilibrated overnight at 4° with 0.35 *M* phosphate buffer, pH 7.0, 1 m*M* EDTA, 0.5 m*M* DTT (40 ml/hr). Since no gradient is employed, this same column can be used several times without regeneration so long as bacterial growth on the column does not occur. The dialyzed luciferase sample is applied. Luciferase from *Beneckea harveyi* is eluted with the equilibration buffer (0.35 *M* phosphate, pH 7.0, 1 m*M* EDTA, 0.5 m*M* DTT), no gradient being required, at a flow rate of about 40 ml/hr. Fractions (10 ml) are collected and the absorbance at 280 nm read and luciferase activity assayed (standard assay, above). The fractions of highest specific activity are pooled, and may be concentrated by precipitation with ammonium sulfate at 75% of saturation. The purity at this stage is about 80–90%, depending upon the choice of fractions pooled from the column. The ammonium sulfate paste may be stored frozen or it may be dialyzed against 0.25 *M* phosphate as described above and stored frozen in this buffer. The photoexcitable luciferase (luciferase with tightly bound flavin chromophore) of *B. harveyi* is partially resolved from the luciferase on the column described.[30] However, with *P. fischeri* preparations, it cochromatographs with the luciferase.

Step 5. Further Purification. For further purification, the enzyme from step 4 may be rechromatographed on DEAE-Sephadex or on Sephadex G-200. Another alternative that seems to be very successful in removing the final 5–10% impurities is chromatography on aminohexyl-Sepharose-4B[32] or aminohexyl-Sepharose-6B. The method of preparation of the column material is that described by March *et al.*[33]; the ligand employed in the reaction with CNBr-activated Sepharose is 1,6-diaminohexane. During the CNBr activation the material held in a sintered-glass funnel is agitated gently by bubbling N_2 through the fritted glass disk. The aminohexyl-Sepharose is packed into a column (1.5 × 30 cm) at 4° and equilibrated with 50 m*M* phosphate, pH 7.0, 1 m*M* EDTA, 0.5 m*M* DTT. The protein sample (about 100 mg in the same buffer, i.e., only about 10% of the pool from the DEAE-Sephadex column) is applied to the column, and the column is eluted at 40 ml/hr with a linear gradient of 250 ml of the equilibration buffer and 250 ml of the same buffer 0.25 *M* in phosphate. Fractions (7 ml) are collected; the luciferase elutes in tubes 20 to 30. The relative purity of a typical preparation of *B. harveyi* luciferase following DEAE-Sephadex column chromatography (step 4) and following the additional aminohexyl–Sepharose-6B column is demonstrated by SDS polyacrylamide gel electrophoresis (Fig. 1).

Other Species

Photobacterium fischeri luciferase is slightly more acidic than the *B. harveyi* enzyme, and requires 0.45 *M* phosphate, pH 7.0, for elution from the DEAE-cellulose. The 40–75% ammonium sulfate precipitate is

[32] J. Cousineau and E. A. Meighen, *Biochemistry* **15,** 4992 (1976).
[33] S. C. March, I. Parikh, and P. Cuatrecasas, *Anal. Biochem.* **60,** 149 (1974).

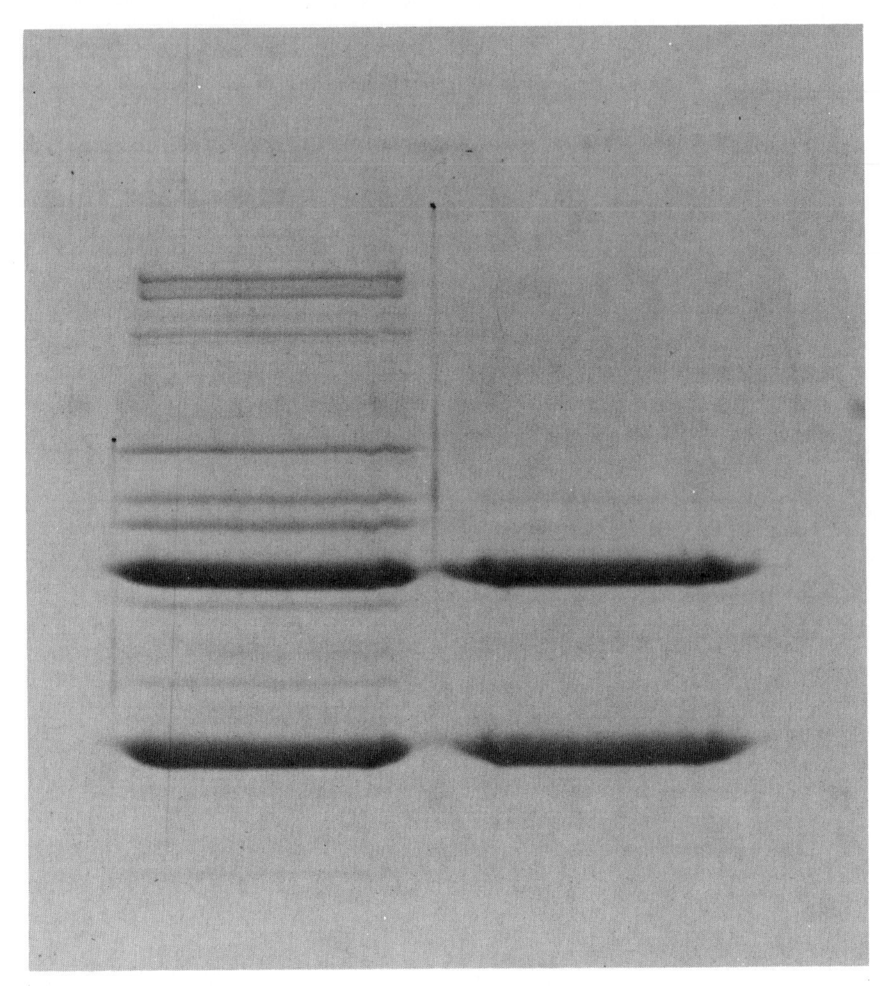

FIG. 1. *Beneckea harveyi* luciferase after sodium dodecyl sulfate polyacrylamide gel electrophoresis. *Left:* Pooled material from DEAE-Sephadex column (step 4). *Right:* Same preparation after chromatography on aminohexyl–Sepharose 6B. Slab gel electrophoresis of 5-μg samples of protein was carried out by the procedure of U. K. Laemmli, *Nature* **227,** 680 (1970).

dialyzed against *0.35 M* phosphate, pH 7.0, 1 m*M* EDTA, 0.5 m*M* DTT for loading on the DEAE-Sephadex column. The column is equilibrated with this same buffer; after application of the *P. fischeri* enzyme to the column, the column is first washed with about 600 ml of the equilibration buffer and then the luciferase activity is eluted with a linear gradient of 600 ml of equilibration buffer and 600 ml of 0.5 *M* phosphate, 1 m*M* EDTA, 0.5 m*M* DTT.

Photobacterium phosphoreum luciferase has been purified by Nakamura and his co-workers from a strain isolated from a cuttlefish.[10,34] Meighen and his co-workers[35] have purified the enzyme from strain number 844 of the National Collection of Marine Bacteria of Scotland, and have also separated the subunits of the enzyme. Procedures similar to those used for *B. harveyi* are applicable.

The luciferase from *Photobacterium leiognathi* has been less extensively studied than the others, but its isolation from one strain was reported some years ago.[36,37] A luciferase isolated from another strain of *P. leiognathi,* S-1, has been recently reported to be a glycoprotein.[38]

Properties

All the bacterial luciferases so far studied appear to have molecular weights in the range of 77,000 ± 5000 with an $\alpha\beta$ heterodimeric structure.[39] Luciferase concentration can be roughly estimated from the absorbance at 280 nm using an extinction coefficient of about 1.0 (0.1%, 1 cm). However, since the enzymes from different species do differ somewhat in physical and chemical properties, this section will focus on generalities, with reference to specifics only for the *P. fischeri* and *B. harveyi* enzymes.

Stability

The luciferase from *B. harveyi* is stable under the conditions of purification; the *P. fischeri* enzyme is only slightly less so. Conditions leading to loss of activity include extremes of pH <6.0 or >8.5 (*P. fischeri*) or 9.5 (*B. harveyi*), exposure to reagents that react with sulfhydryl, imidazole, or amino groups, temperatures above 35°, and bacterial contamination. Bacterial luciferases are exceedingly labile to proteases,[40] and often, even slightly bacterial contamination can lead to rapid loss of activity due to extracellular proteases.[41] Multivalent anions (phosphate, sulfate, pyrophosphate, citrate, etc.) at concentrations of 0.1 *M* or higher significantly stabilize bacterial luciferase from inactivation due to heat, urea, and proteases.[42] The enzyme is stable to repeated freezing and thawing, but

[34] T. Watanabe and T. Nakamura, *J. Biochem.* **79,** 489 (1976).
[35] E. A. Meighen, personal communication.
[36] J. P. Henry and A. M. Michelson, *C.R. Acad. Sci. Ser.* D **270,** 1947 (1970).
[37] F. Lavelle, J. P. Henry, and A. M. Michelson, *C.R. Acad. Sci. Ser.* D **270,** 2126 (1970).
[38] C. V. Balakrishnan and N. Langerman, *Arch. Biochem. Biophys.* **181,** 680 (1977).
[39] S.-C. Tu, this volume [17].
[40] T. O. Baldwin, *Fed. Proc., Fed. Am. Soc. Exp. Biol.* **33,** 1441 (1974).
[41] T. O. Baldwin, this volume [22].
[42] P. L. Riley, Senior Thesis, University of Illinois at Urbana, 1977.

does not survive lyophilization. Luciferases should be stored frozen, in ammonium sulfate or in phosphate buffer, pH 7.0, with 0.5 m*M* DTT to maintain a reducing environment and 1 m*M* EDTA to retard microbial growth.

Activity Units

Activity measurements are usually recorded as the initial maximum light intensity upon mixing in one of the first three (nonturnover) assays described above, under specified conditions of temperature, pH, buffer, and aldehyde, in units of quanta sec^{-1}. Calibration of the particular instrument requires a standard; the two in common use differ by a factor of about three.[43,43a] The activity thus reported is a rate and could be converted to the more usual units of molecules sec^{-1} if one knew the bioluminescent quantum yield. For several reasons it is more desirable to retain the use of the units quanta sec^{-1} for rate and to quote quantum yields independently. Specific activity may be expressed as the rate per milligram of protein, i.e., quanta sec^{-1} mg^{-1}. Some of the differences in the literature may be attributable to differences in conditions (e.g., temperature or pH), which are not always specified, or to differences in the light standard used. It is also possible that differences in quantum yield (and thus apparent activity) occur owing to factors or differences in luciferase as yet undetermined.[44] It is thus advisable to estimate enzyme purity by polyacrylamide gel electrophoresis rather than by specific activity comparisons with literature values. However, with the light standard of Hastings and Weber,[7] the approximate specific activities of the luciferases from *B. harveyi, P. fischeri,* and *P. phosphoreum*[35] (22°, pH 7.0, low ionic strength buffer, optimal concentration of dodecanal) are 2×10^{13} quanta sec^{-1} mg^{-1}, 4×10^{14} quanta sec^{-1} mg^{-1}, and 4×10^{13} quanta sec^{-1} mg^{-1}, respectively. The *B. harveyi* enzyme specific activity with decanal is about 1.6×10^{14} quanta sec^{-1} mg^{-1}.

Since the standard assay does not involve turnover, the total light emitted in such a reaction represents the total product from a single enzymic cycle. Thus, unlike other enzymes where the total product obtained is related to the amount of substrate used, not to the quantity of enzyme, the photon yield is directly proportional to the amount of luciferase. Factors that affect the *activity* of luciferase (*rate* at which light is produced), such as pH and temperature, have little or no effect on the

[43] J. W. Hastings and G. Reynolds, *in* "Bioluminescence in Progress" (F. H. Johnson and Y. Haneda, eds.), p. 45. Princeton Univ. Press, Princeton, New Jersey, 1966.

[43a] P. R. Michael and L. R. Faulkner, *Anal. Chem.* **48,** 1188 (1976).

[44] C. A. Reeve, M. Z. Nicoli, and T. O. Baldwin, *Fed. Proc., Fed. Am. Soc. Exp. Biol.* **36,** 917 (1977).

total photon yield. Total light is thus a valid measure of the amount of enzyme. Using a simple photomultiplier, it is quite easy to assay as little as 10 pg ($\sim 10^{-16}$ mol) of *B. harveyi* luciferase; using more sophisticated photomultiplier tubes and single-photon counting electronics, it is possible to detect the activity of 100–1000 molecules.

Quantum Yield

As just mentioned, activity is measured by the light intensity in photons per second. This is proportional to molecules per second, the two being related by the quantum yield. The overall bioluminescence quantum yield (Q_B) is an expression of the number of quanta emitted per molecule of substrate utilized, enzyme turned over, or product formed. The yield can be expressed as a function that takes into account side reactions, efficiency of populating the excited state, and efficiency of emission from the excited state, which may be designated as Q_C, Q_{EX}, and Q_F, respectively.[45] Thus, the product of these terms gives the overall bioluminescence quantum yield $Q_B = Q_C \times Q_{EX} \times Q_F$. For the bacterial luciferase reaction under optimum conditions, the bioluminescence quantum yield per catalytic cycle is estimated to be, within a factor or two, 0.1.[28]

Substrate Specificity

A considerable number of different flavin isomers and analogs have been tested for light production with luciferase.[46] All were found to be active with the enzyme to a greater or lesser extent, but FMN has by far the highest quantum yield. Meighen and MacKenzie[14] have found that the negative charge supplied by the phosphate is required both for good binding and for a high quantum yield. Riboflavin thus functions poorly in the light-producing reaction, but is more active with added inorganic phosphate or sulfate.

Aldehydes of carbon chain length 8 and greater serve well in the bioluminescence reaction.[47-49] Alcohols and other long-chain compounds are competitive inhibitors.[50] Care should be taken in supplying aldehyde to

[45] M. J. Cormier, J. Lee, and J. E. Wampler, *Annu. Rev. Biochem.* **44,** 255 (1975).

[46] G. W. Mitchell and J. W. Hastings, *J. Biol. Chem.* **244,** 2572 (1969).

[47] J. W. Hastings, J. A. Spudich, and G. Malnic, *J. Biol. Chem.* **238,** 3100 (1963).

[48] J. W. Hastings, K. Weber, J. Friedland, A. Eberhard, G. W. Mitchell, and A. Gunsalus, *Biochemistry* **8,** 4681 (1969).

[49] T. Watanabe and T. Nakamura, *J. Biochem.* **72,** 647 (1972).

[50] J. W. Hastings, Q. H. Gibson, J. Friedland, and J. Spudich, *in* "Bioluminescence in Progress" (F. H. Johnson and Y. Haneda, eds.), p. 151. Princeton Univ. Press, Princeton, New Jersey, 1966.

the enzyme since high aldehyde levels can inhibit some luciferases, particularly the *B. harveyi* enzyme.

Stoichiometry and Products of the Reaction

The stoichiometry of the bacterial bioluminescence reaction has been the subject of some controversy in recent years.[45] However, the experiments of Becvar and Hastings[51] demonstrated quite unambiguously that the stoichiometry of the light-producing reaction is one $FMNH_2$ per luciferase molecule ($\alpha\beta$). The luciferase peroxyflavin intermediate contains one flavin chromophore per luciferase.[52] The stoichiometry with respect to O_2 and aldehyde is thought to be one of each per enzyme (or per $FMNH_2$), although the data are not so compelling as for the $FMNH_2$ stoichiometry.[49]

Inhibitors

Several inhibitors of the luciferases from *B. harveyi* and *P. fischeri* have been described.[14,50,53] Although the studies with inhibitors such as 2-diethylaminoethyl-2,2-diphenylvalerate (SKF), *N*,*N*-diethyl-2,4-dichloro(6-phenylphenoxy)ethylamine (DPDA), 2,3-dichloro(6-phenylphenoxy)ethylamine (DPEA), and 8-anilino-1-naphthalenesulfonate (ANS) have not yielded structural information regarding the active center of the enzyme, these studies do describe the types of compounds that bind to one or more sites on the enzyme. Information of this sort may prove to be useful in developing specific affinity chromatography systems for use in the purification of the luciferase. Recently, Holzman and Baldwin have utilized an analog of SKF bound to Sepharose-4B for the rapid purification of the luciferase from *B. harveyi* to about 80% homogeneity.[54]

Acknowledgments

The studies on which these procedures are based have included important contributions from many individuals including Ann Gunsalus-Miguel, Pattie Dobson, and other former associates. The work was supported in part by grants from the National Science Foundation, BMS 74-23651 (J.W.H) and PCM 76-00452 (T.O.B.), and a United States Public Health Service Postdoctoral Fellowship (M.Z.N.).

[51] J. E. Becvar and J. W. Hastings, *Proc. Natl. Acad. Sci. U.S.A.* **72,** 3374 (1975).
[52] J. W. Hastings and C. Balny, *J. Biol. Chem.* **250,** 7288 (1975).
[53] K. H. Nealson and J. W. Hastings, *J. Biol. Chem.* **247,** 888 (1972).
[54] T. F. Holzman and T. O. Baldwin, unpublished observations.

[15] Isolation, Identification, and Manipulation of Luminous Bacteria

By KENNETH H. NEALSON

The luminous bacteria are characterized by their ability to emit visible light, the product of an enzyme (bacterial luciferase)-catalyzed reaction. As far as is known, all luminous bacteria utilize the same mechanism of light emission and have similar luciferases. As a group they represent a rather homogeneous array of marine species, which have in the past been placed into many different bacterial genera. For a discussion of this rather confusing nomenclatural history, the reader is referred to two recent review articles.[1,2] Presently there is a general agreement that the light-emitting bacteria should be placed into two genera, each with two or more species, although some differences exist as to the recommended names for these groups (see Table II). In this report, the generic names *Beneckea* and *Photobacterium,* as recommended by Reichelt and Baumann,[1,3,4] are used.

The genus *Beneckea (Lucibacterium)* is characterized by the presence of a single, sheathed polar flagellum, inducible (on solid medium) peritrichous flagella, and the ability to grow on a wide variety of carbon compounds as the sole source of carbon and energy. The genus is widely distributed in warm marine environments and contains many nonluminous species. The various species are quite similar to one another, but can be distinguished by their nutritional properties as well as by *in vitro* DNA/DNA homologies.[1,3–5]

The second genus, *Photobacterium (Vibrio),* is composed of three distinct species, all of which contain luminous members, which can be rather easily separated on the basis of nutritional versatility, flagellation, and temperature sensitivity. All three species of this group have been found in symbiotic association with luminous fish.[2]

Both genera can be isolated from a variety of marine habitats including free living (from seawater), saprophytic, detrital, parasitic, and commensal (as gut-tract bacteria of marine fishes). Light organ symbiosis, on the other hand, appears to be restricted to the genus *Photobacterium;* no members of the genus *Beneckea* have been reported as light-organ symbionts.

[1] P. Baumann and L. Baumann, *Annu. Rev. Microbiol.* **31,** 39 (1977).
[2] J. W. Hastings and K. H. Nealson, *Annu. Rev. Microbiol.* **31,** 549 (1977).
[3] J. L. Reichelt and P. Baumann, *Arch. Microbiol.* **94,** 283 (1973).
[4] L. Baumann and P. Baumann, *Microbios. Lett.* **3,** 11 (1976).
[5] J. L. Reichelt, P. Baumann and L. Baumann, *Arch. Microbiol.* **110,** 101 (1976).

ISBN 0-12-181957-4

TABLE I
MEDIA FOR LUMINOUS BACTERIA

	SWC	SWMH	SSWM[d]	SSWC[e]	BGM[f]	LM[f]	LYO
Seawater	750 ml	750 ml					
Dist. water	250 ml	250 ml	1000 ml	1000 ml	444 ml	444 ml	750 ml
ASW[a]					500 ml	500 ml	250 ml
NaCl			30 g	30 g			
NaH_2PO_4			5.3 g	5.3 g			
K_2HPO_4			2.1 g	2.1 g			
$(NH_4)_2HPO_4$			0.5 g	0.5 g			
$MgSO_4$			0.1 g	0.1 g			
$CaCO_3$						1.0 g	
Tris[b]					50 ml	50 ml	
$FeSO_4 \cdot 7H_2O$					0.028 g	0.028 g	
$K_2HPO_4 \cdot 3H_2O$					0.075 g	0.075 g	
NH_4Cl					1.0 g	1.0 g	
HEPES[c]		50 ml					
Glycerol	3 ml	3 ml	3 ml	3 ml	3 ml	3 ml	
Peptone	5.0 g			5.0 g		5.0 g	5.0 g
Yeast extract	0.5 g			0.5 g		5.0 g	5.0 g

[a] ASW contains per liter: 31.1 g of NaCl; 1.5 g of KCl; 24.7 g of $MgSO_4 \cdot 7H_2O$, and 2.9 g of $CaCl_2 \cdot 2H_2O$.
[b] Tris buffer stock is 1 *M* (121.1 g of Tris per liter) adjusted to pH 7.5.
[c] HEPES buffer stock is 1 *M* at pH 7.5.
[d] From A. H. Farghaley, *J. Cell. Comp. Physiol.* **36,** 165 (1950).
[e] From K. H. Nealson and A. Markovitz, *J. Bacteriol.* **104,** 300 (1970).
[f] From J. L. Reichelt and P. Baumann, *Arch. Mikrobiol.* **94,** 283 (1973).

Media for Growth and Storage

While a variety of media for the growth of luminous bacteria have been published, and several are presented here (Table I), a simple seawater nutrient medium is usually sufficient for the isolation and study of luminous bacteria. The seawater may be natural, dried (as from aquarium suppliers), or a defined artificial seawater (ASW). Nearly all luminous species exhibit maximum growth and luminescence from 300 to 500 m*M* Na^+[6]; natural seawater is thus routinely diluted to 75% with distilled water, eliminating the precipitates occasionally encountered when full strength seawater is autoclaved.

Artificial Seawater (ASW)

ASW is prepared according to the formula of MacLeod.[7] The salts are dissolved separately, then combined to give a stock solution with the

[6] J. L. Reichelt and P. Baumann, *Arch. Microbiol.* **97,** 329 (1974).
[7] R. A. MacLeod, *Microbiol. Sea* **1,** 95 (1968).

following concentrations: NaCl, 400 mM; $MgSO_4 \cdot 7H_2O$, 100 mM; KCl, 20 mM; and $CaCl_2 \cdot 2H_2O$, 20 mM.

Seawater Complete (SWC)

SWC is easy to prepare and is a widely used medium that supports the growth of all luminous species. In any complete medium, most luminous bacteria produce and excrete organic acids, even when grown aerobically,[8] so for long-term experiments, or for storage, care should be taken to add buffer. Either tris(hydroxymethyl)aminomethane (Tris) or N-2-hydroxyethylpiperazone-N'-2-ethanesulfonic acid (HEPES) at 50 mM and pH 7.5 are adequate; solid calcium carbonate (1 g/liter) will also suffice.

Seawater Minimal (SWMH)

This minimal medium, while not completely defined, is acceptable for the study of carbon metabolism in a minimal medium. Acid production may occur in minimal media, depending upon the carbon source used, so a HEPES buffer is routinely included.

Farghaley's Minimal Medium (SSWM and SSWC)

SSWM is a minimal synthetic seawater medium first described by Farghaley.[9] This medium does not support the growth of *Photobacterium* species well, but is adequate for species of *Beneckea*. However, even *Beneckea* species display severe limitation of both growth and bioluminescence when grown in SSWM. In this medium a strong positive response of luminescence to arginine is seen for some strains. When either SWMH or BGM media are used, the cells are less limited for luminescence. SSWC is simply SSWM plus peptone and yeast extract. It is an adequate buffered complex medium for all luminous species.

Artificial Seawater Media (BGM and LM)

These media support the growth of all species, and both minimal and complete media are recommended for physiological studies. If desired, Tris buffer can be replaced with HEPES buffer. BM is the same as BGM, but without glycerol.

Lyophilization Medium (LYO)

For lyophilization of the luminous bacteria, it has been found to be advantageous to suspend the cells in a medium of low ionic strength before

[8] E. G. Ruby and K. H. Nealson, *Appl. Environ. Microbiol.* **34,** 164 (1977).
[9] A. H. Farghaley, *J. Cell. Comp. Physiol.* **36,** 165 (1950).

freezing and lyophilization. LYO broth is an excellent medium for this purpose.

Growth of the Bacteria on Solid Medium

Growth on solid medium (1–2% agar) is similar for all species with the exception that many *Beneckea* strains induce peritrichous flagella and grow as spreading colonies. SWC plates support the growth of all species. On such an unbuffered complex medium, however, luminescence usually ceases in a few days, because the cells excrete organic acids into the medium, lowering the pH to a level that is suboptimal for light emission and growth. Several general (though not necessarily diagnostic) features can be seen on SWC plates: (1) colonies that produce a yellow cell-associated pigment after several days of growth are *P. fischeri;* (2) those colonies which, after several days of growth at 18°, are small and white are *P. phosphoreum;* (3) spreading colonies (indicative of peritrichous flagella) are *Beneckea* species.

Storage of Luminous Bacteria

Storage of the luminous bacteria on plates or agar slants is often difficult, as the various species respond to temperature quite differently (Table II). In general, *Beneckea* species are killed by refrigerator storage, and *P. phosphoreum* and *P. fischeri* survive poorly at higher temperatures (above 20°). All species store poorly on minimal media. Furthermore, all species tend to form dark variants and lose luminosity during storage.

We have found that SWC soft agar (with 1 g of calcium carbonate and 7–8 g of agar per liter) stab cultures (2–3 ml of medium in 1-dram vials) stored at 18° will survive for a year or more. These vials are stabbed with inoculum to the bottom, then sealed with melted paraffin. Using this method, our stock cultures need to be transferred only once a year.

Because of the tendency of luminous bacteria to form dark variants, one must always check for luminosity when transferring cultures. Failure to do this may lead to loss of luminescence, as the dark forms often "take over."[2]

For long-term storage, both lyophilization and cold storage have been successful. Lyophilization is done with LYO broth. Cells from several overnight SWC slants are suspended in LYO broth, quick-frozen in a Dry Ice acetone bath, lyophilized, sealed under vacuum, and stored at room temperature. Cold storage is accomplished simply by mixing 1 ml of a log-phase culture (about 5×10^8 cells/ml) with 1 ml of glycerol and freezing in liquid nitrogen. Cultures are stored at $-85°$ until needed.

TABLE II

NOMENCLATURE, TEMPERATURE PROPERTIES, AND NATURAL LOCATIONS OF MARINE LUMINOUS BACTERIA

Nomenclatural designations[a]			Temperature		Natural habitats[g]				
Reichelt and Baumann[b]	Hendrie *et al.*[c]	Chumakova *et al.*[d]	Growth range[e] (°C)	Storage temp.[f] (°C)	Light-organ symbionts	Gut symbionts	Free living	Saprophytes	Parasites[h]
B. harveyi *B. splendida*	*L. harveyi*	*L. harveyi* (*P. belozerskii*)	10–37	18	–	+	+	+	+
P. fischeri	*V. fischeri*	Group I (*P. fischeri*)	4–25	18	+	+	+	+	+
P. leiognathi	*P. mandapamensis*	Group II	4–35	18	+	+	+	+	–
P. phosphoreum	*P. phosphoreum*		0–25	4	+	+	+	+	+

[a] Generic abbreviations: *B.* = *Beneckea*, *L.* = *Lucibacterium*, *P.* = *Photobacterium*, *V.* = *Vibrio*.

[b] See J. L. Reichelt and P. Baumann, *Int. J. Syst. Bacteriol.* **25,** 208 (1975); and *Arch. Microbiol.* **94,** 283 (1973).

[c] M. S. Hendrie, W. Hodgkiss, and J. Shewan, *J. Gen. Microbiol.* **64,** 151 (1970); this nomenclature is currently adopted by Bergey's Manual of Determinative Bacteriology.

[d] R. I. Chumakova, B. F. Vanyushin, N. A. Kokurina, B. F. Yoros'eva, and S. E. Medvedeva, *Microbiology (USSR)* **41,** 613 (1972). These authors originally designated *P. belozerskii* as a new species, but it appears to be a *Beneckea*. On the basis of their data, these authors could not distinguish their group II from *P. leiognathi* or *P. phosphoreum*.

[e] Numbers stated represent a combination of those in the literature and those observed in the author's laboratory.

[f] Recommended storage temperatures are based on personal experience of the author and P. Baumann.

[g] These data are mainly from the author's laboratory.

[h] Literature concerning parasitic isolates is sparse, and no parasitic isolates of *P. leiognathi* have yet been reported.

Isolation and Enrichment Culture Techniques

No specific advantage for luminescence per se in culture is known, but it is possible to isolate luminous bacteria from a variety of "natural enrichments" discussed below. Similarly, no good methods are available to enrich for any of the species of luminous bacteria in preference to any other. To this end, it would seem that temperature could be used to some degree to favor or exclude certain species. In general, enrichment at low temperature, 0°–5°, will yield only *P. phosphoreum;* 5°–10° will yield *P. phosphoreum* and *P. fischeri;* higher temperatures favor *P. leiognathi* and *Beneckea*. From the properties of the bacteria listed in Table II, it would seem possible to use temperature to enrich directly for each species. No reports of such enrichment techniques have been made. Recently our laboratory has received some strains of psychrophilic luminous bacteria which, if grown at temperatures above 18°, become permanently dark. Thus, if one wants to isolate luminous bacteria from the deep sea, or other low-temperature environments, plating must be done at appropriately low temperatures, or they may not be recognized as luminous at all.

Direct Isolation from Seawater

With very few exceptions luminous bacteria are found only in the marine environment.[2] This is apparently related to their universal requirement for the sodium ion.[6] They are prevalent in neaı shore waters, averaging 1–5 bacteria per milliliter,[10,11] and can easily be isolated simply by plating seawater onto seawater nutrient plates. For studies in which many isolates are needed, the soft agar isolation method can be used: (a) Sterile SWC soft agar (0.75%) is melted, and 4-ml aliquots are dispensed into small tubes and held at 45°; (b) seawater, 1 or 2 ml, is added to each tube; the contents are mixed, and immediately poured onto the surface of a cool SWC plate; (c) plates are incubated at 18° for 2 days; (d) all bright spots are picked using sterile toothpicks, and isolated as single colonies on SWC agar plates at 18°–20°.

Since some species of luminous bacteria are killed by high temperature, it is important to pour the samples as soon as possible onto cool plates (within seconds). We have found that this method is acceptable and accurate for all four species. Some *P. fisceri* and most *P. phosphoreum* strains are inhibited in growth above 20°, so one must take care to do isolations at 20° or less.

[10] E. G. Ruby and K. H. Nealson, *Limnol. Oceanogr.* **23,** 530 (1978).
[11] E. G. Ruby, Ph.D. thesis, Scripps Institution of Oceanography, La Jolla, California.

Isolation of Attached Forms

The usual ("classic") enrichment for luminous bacteria takes advantage of the fact that these bacteria commonly attach to the surfaces of marine animal material. One has merely to take any marine fish or shrimp (which has not been washed in fresh water) and incubate it for a day or two at 20°. During this incubation the fish should be kept damp by partial immersion in seawater. Such efforts nearly always reward the viewer with spots of luminous bacteria on the surfaces of the incubated organism, which can then be picked and isolated as single colonies.

Other excellent sources of luminous bacteria are the gut tracts of marine fishes and squids. Direct plating of the gut contents of many marine fish often yields high (10^6/ml or greater) numbers of luminous bacteria.[11] Whether or not some species of fish yield higher or more consistent numbers of luminous bacteria is not presently known. Nor is it known whether or not such associations are transient or permanent. Although published data are not abundant, it is apparent from work in our laboratory that fish fecal matter is another excellent source of luminous bacteria.

Isolation of Parasitic Forms

It is often possible to collect crustaceans that are glowing because they are infected with luminous bacteria. Such organisms, when their body fluids are examined, invariably yield homogeneous cultures of luminous bacteria.[12,13] It is not clear whether the luminous bacteria are actual pathogens or merely opportunists.

Isolation of Symbiotic Forms

The luminous bacteria from a variety of symbiotic associations have been cultured.[2,14-19] When the surface of the light organ is sterilized with ethanol, the organ is opened using aseptic procedures, and the organ fluid is plated, 10^9–10^{10} luminous bacteria per milliliter of fluid are commonly

[12] K. H. Nealson, unpublished results.
[13] E. N. Harvey, "Bioluminescence." Academic Press, New York, 1952.
[14] E. G. Ruby and K. H. Nealson, *Biol. Bull.* **151,** 574 (1977).
[15] J. Reichelt, K. H. Nealson, and J. W. Hastings, *Arch. Mikrobiol.* **112,** 157 (1976).
[16] E. N. Harvey, "Bioluminescence." Academic Press, New York, 1952.
[17] E. G. Ruby and J. G. Morin, *Deep Sea Res.* **25,** 161 (1978).
[18] J. M. Fitzgerald, *Arch. Microbiol.* **112,** 153 (1977).
[19] J. W. Hastings and G. W. Mitchell, *Biol. Bull.* **141,** 261 (1971).

observed.[14,17-19] As far as is known, all such symbioses are species specific.[14,15,17,18]

It is clear from examination of Table II that the luminous species are widely distributed in the marine environment. All species appear to have the capacity to occupy any of the niches, with the exception that no *Beneckea* species has yet been identified as a light organ symbiont, and no parasitic *P. leiognathi* have been observed. Also the growth ranges of the *P. phosphoreum* apparently restrict it to deep or cold waters. Such environments also exclude the other species.

Location and Isolation of Luminous Bacteria

The location and isolation of luminous bacteria is greatly aided by a rheostat-controlled red light and sterile toothpicks. The red light is attenuated to the point at which the blue (glowing) colonies are visible above it, and sterile toothpicks, which obviate the need for a burner in the dark room, can be used to pick the bright spots onto SWC plates for single-colony isolation. This approach is also valuable for mutant isolation (see this volume [16]).

Identification of the Species

Based on the work of Reichelt and Baumann, it is now possible to routinely identify the luminous bacteria.[3] A slight modification of the features used by these authors is shown in Table III. In the top part of the first column are the 19 characters that we routinely use because they are easily tested by replica plating. The procedure is described below.

Nutritional Versatility

Media: (A) 556 ml of BM [with ASW (undiluted), Tris, $FeSO_4 \cdot 7H_2O$, $K_2HPO_4 \cdot 3H_2O$, and NH_4Cl]
(B) 444 ml of distilled water + 12 g of Noble agar

Carbon sources are filter sterilized; A and B are autoclaved separately; then all components are mixed and cooled to 45°, and plates are poured.

Plates are made for each of the carbon sources to be tested. Plates without a carbon source (zero carbon) are also prepared. When the plates are dry, master plates are made by patching (5 × 5 mm patches) onto LM (or SWC) 9 strains per plate. Master plates are grown overnight at 18°, then replica plated, using sterile velveteen pads, to the appropriate test plates. One master plate can be used for eight tests. A zero carbon plate is done first to test for agar digesters, then the seven test plates, the last of which is

TABLE III

DIAGNOSTIC TAXONOMIC CHARACTERS OF THE LUMINOUS BACTERIA[a]

Characters	*B. harveyi*	*P. fischeri*	*P. leiognathi*	*P. phosphoreum*	Amount of carbon source used	Time of scoring (days)
Growth at 4°	−	(−)[b]	−	+	−	
Growth at 35°	+	(+)[b]	(+)[b]	−	−	1
Amylase	+	(−)[b]	−	−	−	4
Lipase	+	−	−	−	−	2,4,6
Gelatinase	+	−	−	−	−	4
Growth on:						
Maltose	+	+	−	+	0.2%	3
Cellobiose	+	+	−	−	0.2%	3
Gluconate	+	(−)[b]	+	+	0.1%	3
Glucuronate	+	−	−	(+)[b]	0.1%	3
Mannitol	+	+	−	−	0.1%	5
Proline	+	+	+	+	0.1%	3
Lactate	+	−	+	(−)[b]	0.2%	5
Pyruvate	+	−	+	−	0.1%	
Acetate	+	−	+	−	0.05%	10
Propionate	+	−	−	−	0.05%	10
Heptanoate	+	−	−	−	0.05%	10
D-α-Alanine	+	(−)[b]	−	−	0.1%	10
L-Tyrosine	+	−	−	−	0.04%	10
α-Ketoglutarate	+	−	−	−		10
Moles % GC	46.5 ± 1.3	39.8 ± 1.1	41.5 ± 0.7	42.9 ± 0.5		
Flagellation[c]	SP → Pr	SP	P	P		
Polar flagella	1	2–8	1–3	1–3		
PHB[d] accumulation	−	−	+	+		
Gas from glucose	−	−	−	+		
Luciferase kinetics:	Slow	Fast	Fast	Fast		

[a] Adapted from original system of J. L. Reichelt and P. Baumann, *Arch. Mikrobiol.* **94,** 283 (1973). *B.* = *Beneckea; P.* = *Photobacterium*.
[b] Parentheses ( ) indicate that the trait exhibits a small degree of variability.
[c] S, sheathed flagella; P, polar flagella; Pr, peritrichous flagella; SP → Pr, indicates a conversion from one mode of flagellation to the other.
[d] Poly-β-hydroxybutyrate.

usually a test for one of the extracellular enzymes using a complex medium and serves as a positive control.

Replica plates are incubated at 18°. Because growth occurs at different rates on the various media, the plates are examined after different times (see Table III).

Extracellular Enzymes

The plates for extracellular enzyme determinations are replica plated last, as positive controls. The media for all these tests are similar. They have a BM base with 5 g of yeast extract and 15 g of agar.

Gelatinase: Add 50 g of gelatin slowly and boil to dissolve.

Amylase: Add 2 g of soluble starch.

Lipase: Add 10 ml of Tween 80 (polyethylene sorbitan monocleate).

Gelatinase. After 4 days' growth, acidic mercuric chloride ($HgCl_2$, 15 g; H_2O, 100 ml; conc. HCl, 20 ml) is poured onto the surface of each plate. A positive gelatinase is indicated by a clear zone.

Amylase. After 4 days' growth, Gram's iodine solution is poured over the surface of each plate. A positive amylase is indicated by a clear (uncolored) zone.

Lipase. These plates must be streaked separately, five per plate, as colonies tend to spread. Examine colonies for insoluble Ca^{2+} salts of fatty acids released by lipase activity. Check after 2, 4, and 6 days of growth.

For other tests and specific methods, see references cited in footnotes 2 and 3.

Taxonomic Analysis by Luciferase Assay

An interesting feature of the bacterial luminous system relates to differences in the enzyme luciferase isolated from different bacterial species. Data from several years of analysis of natural isolates (including over 3000 strains of all four species) have shown that an unambiguous separation of the two genera can be made on the basis of luciferase kinetics. When the flavin-induced luciferase assay is used,[20] utilizing dodecanal, *Beneckea* exhibits a slow decay, while luciferases from any of the *Photobacterium* species exhibit a fast decay. The difference is quite distinct, as shown in Fig. 1.

Thus, after pure colonies are isolated, a cell lysate can be made, the luciferase can be assayed, and the *Beneckea* species can be identified by their slow kinetics. The *Beneckea*-specific tests can then be eliminated (see Table I), and the taxonomic work is greatly aided. At present, no other

[20] J. W. Hastings, this volume [13].

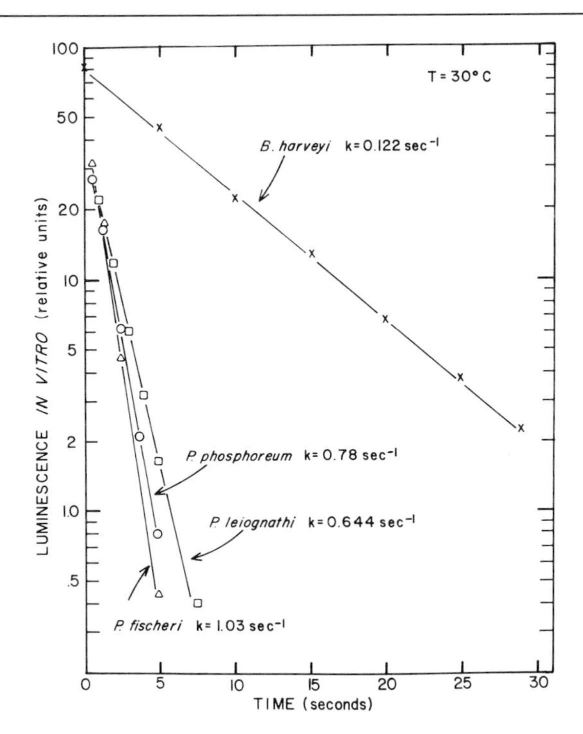

FIG. 1. Luminous bacteria that have grown overnight on seawater complete (SWC) plates are removed by scraping with a clean spatula and placed into 5 ml of cold lysis buffer (10 m*M* EDTA; 1 m*M* DTT) in a small tube. After lysis for 20 min, the extracts are assayed by the reduced flavin assay. Twenty microliters of lysate are added to a scintillation vial containing 10 μl of aldehyde suspension (0.1 ml of dodecanal in 10 ml of water) in 1 ml of 10 m*M* phosphate buffer (pH 7.1) in a light-tight chamber, and 1 ml of $FNMH_2$ is injected. The resulting flash of light is recorded on a strip chart recorder and plotted on semilog paper to determine rate constants. When dodecanal is used, luciferase from *Beneckea* species displays slow-decay kinetics (X——X), while others (*Photobacterium* species) show a fast decay (○, △, □). Since the decay is very temperature dependent, controls of known species must be done. Typically, a 6- to 10-fold difference in decay rates is seen, so that the species can be easily distinguished even without replotting the data. For further details of the luciferase assay, see this volume [14].

specific methods are available using luciferase to identify the bacterial species. A brief description of the technique is included in the legend to Fig. 1.

Physiological Considerations

The physiological properties of the luminous bacteria require that one exercise particular care in liquid growth experiments. Although all species of luminous bacteria lend themselves quite well to experiments in broth cultures, they display marked species-specific differences in their characteristics when studied in liquid media. The development of the luminous system is dependent upon the stage and conditions of growth; thus the amount of luciferase obtained and its activity *in vivo* are subject to experimental control. The major factors to consider are discussed below.

Growth in Complete Media

Autoinduction and Inhibitors. Nearly all luminous bacteria display a lag in the development of luminescence during growth; the pattern has been called autoinduction and is due to at least two separate factors.

First, in a complete medium there appears to be an inhibitor (or inhibitors), at present unidentified, which specifically delays the synthesis of the luminous system.[21,22] The inhibitor(s) can be removed by growing bacteria until induction occurs; a variety of luminous and nonluminous species can be used to remove the inhibitory effect. This inhibition is not seen in mineral media[23] and is thought to be associated with the yeast extract.[21]

Second, the bacteria produce a small molecule (autoinducer) which accumulates in the medium until it reaches a concentration at which it induces the synthesis of the luminous system.[23] The autoinducers are species specific (i.e., each bacterial species produces an autoinducer that is active only for that species). These two effects combine to make the time of growth at which maximal luminescence occurs, and the final amount of luminescence, unique to each strain and medium combination—critical information to have if one wishes to harvest luciferase or to study light emission *in vivo*.

Catabolite Repression. Nutrients have pronounced effects on the level of bioluminescence, and these effects vary between the species. *Beneckea harveyi* strains are sensitive to catabolite repression; glucose severely represses luciferase synthesis, and cyclic AMP reverses this repression.[24] Other sugars also repress, but to a lesser degree. Glycerol shows no repression.

For the *Photobacterium* species, catabolite repression has not been examined carefully. Sugars exert some effects, but generally not permanent repression.[2] Clearly this feature must be understood before the cells

[21] E. S. Kempner and F. E. Hanson, *J. Bacteriol.* **95,** 975 (1968).
[22] A. Eberhard, *J. Bacteriol.* **109,** 1101 (1971).
[23] K. H. Nealson, T. Platt, and J. W. Hastings, *J. Bacteriol.* **104,** 313 (1970).
[24] K. Nealson, A. Eberhard, and J. W. Hastings, *Proc. Natl. Acad. Sci. U.S.A.* **69,** 1037 (1972).

are used for study of enzyme production or of luminescence activity *in vivo*.

Oxygen. Although oxygen is a substrate for luciferase, its presence in excess is not always beneficial for the development of the luminous system. For *P. phosphoreum* and *P. fischeri*, more luciferase is synthesized and the cells become brighter at lower oxygen tension.[25] For *P. leiognathi* and *Beneckea harveyi*, lower oxygen results in less luciferase synthesis and activity. Thus, growth at the wrong oxygen tension (either too high or too low) can lead to low yields of luciferase, and it is wise to consider this characteristic when doing experiments.

In fact it should be mentioned that some species of *Photobacterium* actually are sensitive to oxygen. *Photobacterium phosphoreum* isolates often form clumps when grown at low cell densities with rapid shaking. This clumping can be alleviated by slowing the shaking rate, or by moving the cells away from visible light (Nealson, unpublished observation).

Acid Production. When grown anaerobically on glucose, all luminous bacteria are fermenters of the mixed acid type; they do not ferment amino acids or glycerol. When grown aerobically they also excrete acids; growth on amino acids results in accumulation of acetate in the medium. When grown aerobically on sugars a species-specific (and sugar specific) excretion of pyruvic acid into the medium occurs.[8] Clearly, it is important to have this information. For some species, acid production is so great as to decrease the pH of the medium and inhibit luciferase synthesis and expression while still at a low cell density.

Although all species have not yet been characterized with regard to all the variables discussed here, it should be clear that one must know what organism he has. It is a wise precaution to check the parameters discussed above. Taken together, inhibitors, autoinduction, catabolite repression, oxygen, and acid production represent a complex of effectors that can be manipulated to greatly increase or decrease luminescence.

Growth in a Minimal Medium

When some species are grown in certain liquid minimal media, growth is much slower and marked effects on luminescence are seen. The medium of Reichelt and Baumann (BGM) supports the growth of all species, while that of Farghaley supports the growth of *Beneckea* but not *Photobacterium*. We have used a seawater minimal medium (SWMH) with good success also. Insofar as is known, all the properties described in complete media hold

[25] K. H. Nealson and J. W. Hastings, *Arch. Microbiol.* **112,** 9 (1977).

also for minimal medium, with the exceptions discussed below, which have been studied primarily in SSWM medium.

Arginine and Salt Effects. When grown in SSWM, some strains of *Beneckea harveyi* are almost completely dark; autoinduction operates, and some luciferase, about 1% of that synthesized in complete medium, is made. Two separate changes in the growth medium can lead to brighter cells: the addition of arginine and the lowering of the salt (NaCl) concentration.[25]

The addition of arginine specifically increases both the synthesis and the expression of the luminous system with no effect on growth rate. Millimolar amounts of arginine are required for maximum expression, and the mechanism is as yet unknown. Cells grown in SSWM minimal medium are enhanced in growth and luminescence by the addition of glucose. Conversely, when arginine is added, glucose acts to repress bioluminescence.[24]

If *B. harveyi,* strain B392, is grown in medium containing low salt (1–2% NaCl), much more luminescence is obtained than with the usual 3%. This effect is apparently at the level of luciferase synthesis, and the mechanism, like that of the arginine effect, is not known.

The different species of luminous bacteria have not yet been systematically examined in minimal media. However, the situation for *B. harveyi* suggests that the situation may be extremely complicated, and affected by many variables. It is suggested that the study of a new strain begin with a survey of the above variables, and their effects on the synthesis and activity of the luminous system.

[16] Isolation and Characterization of Luminescence System Mutants in Bacteria

By THOMAS W. CLINE

Mutants altered with regard to the light-emitting system have been classed in three major groups: (1) those with structural alterations in the luciferase, (2) those with defects in the aldehyde system, and (3) those in which the synthesis or control of the luminescence system is somehow altered.[1] This paper will focus on the identification and characterization of mutants with altered luciferases. With the exception of the aldehyde class,

[1] J. W. Hastings and K. H. Nealson, *Annu. Rev. Microbiol.* **31,** 549 (1977).

ISBN 0-12-181957-4

all other mutants are placed in the third group without specific identification. Temperature-conditional mutants have been isolated in all groups.

With a few exceptions mutants are picked by virtue of their failure to emit light at the wild-type level. Although there are no techniques currently available by which one may enrich cultures for nonluminescent variants, the isolation and characterization of a wide variety of dark mutant types is readily accomplished with a minimum of effort. The human eye is extremely efficient in detecting variations in colony luminescence, and the *in vitro* luciferase assay is sufficiently rapid and sensitive to allow for its use as an effective mutant screening tool. Reversions and suppressor mutations can be identified based on their ability to restore luminescence capacity to dim or dark mutant strains. The following technique has been used to isolate mutants of *Beneckea harveyi* induced by *N*-methyl-*N*-nitro-*N'*-nitrosoguanidine (MNNG).[2-5]

Mutagenesis

Late log or early stationary phase cells grown in a seawater complete medium[6] are pelleted by centrifugation, then resuspended in 1–2 ml of fresh medium to a concentration of about 10^{11} cells/ml. Crystals of MNNG are added to the suspension, and the culture is vigorously agitated. Aliquots are taken at increasing intervals over a period of 30 min, diluted 10^{-5} in fresh medium to decrease mutagen concentration, incubated at 25° for 60 min, and then plated at 36° for viable count determinations. These cultures are stored at 4° during the time required for viable count determination (about 12 hr). The aliquot in which there was a survival of 5–10% is then plated (on several dozen plates) so as to yield a maximum of 1200 colonies per standard petri plate, and these are used for mutant selection. It has recently been found[7] that luminescence declines in proportion to viable count following treatment of cultures with MNNG in several luminescent species. This drop in luminescence can reportedly be used as a rough indication of a particular strain's sensitivity to the mutagen. After the appropriate standardization, it can even be used as an immediate measure of culture viable count whenever it may be desirable to avoid the 4° storage step used above.

[2] T. W. Cline and J. W. Hastings, *Proc. Natl. Acad. Sci. U.S.A.* **68,** 500 (1971).
[3] T. W. Cline and J. W. Hastings, *Biochemistry* **11,** 3359 (1972).
[4] T. W. Cline and J. W. Hastings, *J. Bacteriol.* **118,** 1059 (1974).
[5] T. W. Cline and J. W. Hastings, *J. Biol. Chem.* **249,** 4668 (1974).
[6] K. H. Nealson, this volume [15].
[7] K. H. Nealson, unpublished observation.

Mutant Detection and Isolation

Identification of dim and dark colonies is greatly facilitated by the use of a spectral-quality red lamp whose intensity is controlled by a variable transformer. Plates are illuminated from the side at an appropriate intensity. The wild-type colonies emit blue light and reflect the red, and look blue-white, while the dark mutant colonies only reflect light and thus stand out red. As colonies age they darken at their centers. Generally plates should be screened before colonies begin to dim. (However, some colonies have been observed that darken more slowly than the wild type.[8] These are often "minimal bright" mutants.[9]) The desired colonies are removed with sterile toothpicks and are immediately replated at 36° for single-colony isolation. Approximately 0.5% of the colonies are dim or dark following the MNNG procedure described. When the single colonies reach peak luminescence, a visual estimate is made of their luminescence, and each is placed in one of four to six intensity categories, ranging from wild-type luminescence level to nonluminescent to the well dark-adapted eye. A streak from a single colony of each variant is made on two plates of fresh medium. It is helpful to plate together those mutants which have similar levels of luminescence at 36°. One set of plates is incubated at 22°, another at 36°. Once again the mutants are graded for intensity of luminescence. Temperature-sensitive luminescence mutants are thus identified (they constitute 10–20% of 36° darks and dims). About 10–30% of the darks and dims become brightly luminescent upon exposure to aldehyde. These are referred to as aldehyde mutants. The culture plates at 36° are inverted over a petri dish containing decanal, and the luminescence is determined after about 5 sec. Among the many (175) aldehyde mutants examined, none have been found to also possess alterations in luciferase structure.

To distinguish those temperature-sensitive mutants in which the luciferase itself is thermally unstable, plates grown at 22° are transferred to a 36° warm room. After 60 min those that have a temperature-sensitive luciferase are dark owing to the denaturation of luciferase. With other temperature-sensitive classes (aldehyde, luminescence system synthesis), the loss of luminescence following transfer to the nonpermissive temperature takes much longer. For this test it is important that the cells not be past their peak luminescence, and not respond to exogenous aldehyde. About 1% of the strains initially selected as dark or dim at 36° possess temperature-sensitive luciferase by this *in vivo* test.

[8] T. W. Cline, Mutational alteration of the bacterial bioluminescence system. Ph.D. thesis. Harvard University, Cambridge, Massachusetts, 1973.

[9] C. A. Waters and J. W. Hastings, *J. Bacteriol.* **131,** 519 (1977).

Luciferase Structural Mutants

The luminescence system is unique in that there are phenotypic tests that permit us to rapidly and unambiguously identify mutants possessing structural alterations in the luciferase. These phenotypes include thermal stability, altered turnover rate, and substrate ($FMNH_2$) binding affinity, all of which are easily and rapidly measured *in vitro*. As many as one hundred dark or dim mutant strains can be screened for by one worker in the course of one day. This is done by preparing crude extracts[10] of each strain and testing the luciferase in each. Even though the mutant is dim or dark and luciferase activity is low, compared to wild type, there is almost always sufficient activity for purposes of screening.

Strains to be screened are streaked onto seawater complete solid medium[6] (four lawns per 100 × 15 mm petri plate) and grown at 22° for 24 hr. Cells are removed from the plates with a damp Q-tip and suspended in 4 ml of distilled water at 4°. The cells are allowed to lyse overnight. If lysis is not complete, the suspensions are sonicated. The lysates are then centrifuged to remove cell debris (this is essential). Extracts are brought to 50 m*M* phosphate, pH 7, by the addition of the appropriate amount of concentrated buffer.

Thermal Stability

Temperature-sensitive mutants in which the luciferase is thermally unstable may be selected as described above as those strains which lose luminescence more rapidly upon transfer from permissive to nonpermissive temperatures. However, there are many mutants in which there are smaller changes in the thermal stability, not readily detected by the above *in vivo* screening. Such changes, while small, may signal important alterations in the luciferase structure. Mutant luciferases that are less stable than the wild type are easily identified by the following *in vitro* test. One-half milliliter of cell extract is added to a test tube at 4°, one extract per tube. The tubes are then transferred to a water bath at 44.5°, incubated for 10 min, then returned to 4°. Each heated extract is assayed for luciferase activity, along with an assay of an unheated sample of the same extract. With *B. harveyi* wild-type extract, the activity is reduced to 25% of its initial value. Reduction of activity to less than 10% of its initial level indicates a mutant luciferase. The choice of incubation time and temperature depends on the thermal stability of the native luciferase and the degree of stringency

[10] J. W. Hastings, T. O. Baldwin, and M. Z. Nicoli, this volume [14].

desired for the test. Lesions causing the thermal instability phenotype occur in both α and β subunits[11] in approximately equal numbers.[3]

Altered Kinetics

Luciferase has a very low turnover number; a single catalytic cycle lasts several (5–50) seconds at room temperature.[12] This value is directly measured in the standard assay as the rate of decay of luminescence.[12] Luciferase structural mutants in which the modification is at or near the catalytic center may be expected to exhibit an alteration in this decay rate. This can be screened for by simply observing the rate of decay of luminescence in the standard assay. Mutant luciferases selected in this fashion may exhibit alterations in many other parameters, including substrate binding and thermal stability. Using about 50 μl of extract, the luciferase is assayed by injection of $FMNH_2$[10] with decanal (an aldehyde with a rapid turnover rate and high activity). Samples of wild type are interspersed every ten mutant assays to control for variations in assay conditions. Rate constants are determined for the decay of luminescence. Variations of more than 15% from the wild-type value reliably indicate an altered luciferase. Observed at 36°, about 3% of dark and very dim mutant strains (those with less than 5% of the wild-type luminescence) and about 2% of moderately dim mutants (those with 5–25% wild-type luminescence) possessed luciferases altered in this fashion. All luciferases so selected have been found to have defects in the α subunit.[3]

$FMNH_2$ Affinity

Although many mutant luciferases that exhibit altered decay rates also have a decreased affinity for $FMNH_2$, structural mutations that affect the reduced flavin binding site without necessarily altering turnover rate can be screened for by an *in vitro* scheme based directly on measurement of the K_m for $FMNH_2$ by the dithionite assay.[10] Assays are performed using about 25 μl of extract at two concentrations of flavin (2.5 μM and 25 μM), chosen so that the wild-type *B. harveyi* luciferase will exhibit 20% more activity at the 10× higher flavin concentration. Luciferase with a 2-fold lower affinity for $FMNH_2$ will show a 60% increase in activity at the higher flavin concentration. The variability in this assay is somewhat greater than in the standard $FMNH_2$ assay; however, extracts in which a substantial alteration in flavin affinity is indicated can be quickly retested for confirmation. From extracts

[11] S.-C. Tu, this volume [17].

[12] J. W. Hastings and Q. H. Gibson, *J. Biol. Chem.* **238,** 2537 (1963).

of 202 mutants with a wide range of *in vivo* luminescence, two strains were identified which produced luciferase with substantially reduced affinity for $FMNH_2$.[8] One had an alteration in the α subunit, the other in the β subunit. These mutants had such minor changes in turnover rate that they would not have been identified by the kinetic *in vitro* screen described above. Both of these mutant luciferases do, however, have a substantially decreased thermal stability; although they would probably not be considered temperature-sensitive mutants by the *in vivo* criterion (see above), they would be selected for in the *in vitro* test for thermally unstable luciferases. The strain with the β subunit-defective luciferase was treated a second time with MNNG, followed by a search for colonies with increased *in vivo* luminescence. This search yielded a luciferase with a mutation that restored wild-type $FMNH_2$ binding affinity but did not restore wild-type thermal stability.

[17] Preparation of the Subunits of Bacterial Luciferase

By SHIAO-CHUN TU

Bacterial luciferase is a heterodimer, designated $\alpha\beta$. Results of luciferase chemical[1] and mutational[2] modifications show that α is the catalytic subunit. α and β subunits are not active individually; the specific function of β, however, is not clear. The interaction affinity between the two subunits is apparently high, so the isolation of individual luciferase subunits by ion-exchange column chromatography requires the presence of a denaturant, such as urea.[3-5] The procedures described here are for *Beneckea harveyi* luciferase, in which the α and β subunits have molecular weights of about 42,000 and 37,000, respectively.[4,6]

[1] E. A. Meighen, M. Z. Nicoli, and J. W. Hastings, *Biochemistry* **10,** 4069 (1971).
[2] T. W. Cline and J. W. Hastings, *Biochemistry* **11,** 3359 (1972).
[3] J. Friedland and J. W. Hastings, *Proc. Natl. Acad. Sci. U.S.A.* **58,** 2336 (1967).
[4] J. W. Hastings, K. Weber, J. Friedland, A. Eberhard, G. W. Mitchell, and A. Gunsalus, *Biochemistry* **8,** 4681 (1969).
[5] A. Gunsalus-Miguel, E. A. Meighen, M. Z. Nicoli, K. H. Nealson, and J. W. Hastings, *J. Biol. Chem.* **247,** 398 (1972).
[6] E. A. Meighen, L. B. Smillie, and J. W. Hastings, *Biochemistry* **9,** 4949 (1970).

ISBN 0-12-181957-4

General Methods

Reagents

Na/K phosphate buffer, 1 *M*, pH 7.0: dissolve 106.25 g of K_2HPO_4 and 53.82 g of $NaH_2PO_4 \cdot H_2O$ in water to a final volume of 1 liter; pH is checked after a 10-fold dilution and readjusted if necessary

EDTA, 0.1 *M*, neutral

Urea, 8 *M*. Urea (Merck) solutions are deionized by stirring with a mixed-bed resin (Amberlite MB-1, Mallinckrodt), and used immediately. Freshly prepared solutions using the Ultrapure urea (Schwarz/Mann) need not be deionized.

Luciferase Preparation and Activity Measurements. Procedures for the purification of luciferase from cells of *B. harveyi* and the standard assay method (at 23° and using decanal as a substrate) for the measurement of luciferase activity are both described in this volume [chapter 14] pp. 135–152. The initial maximal intensity (I_0) of *in vitro* assays is expressed in quanta sec^{-1} using a liquid scintillation sample as a standard.[7] Luciferase concentrations are determined using an absorbance coefficient (1 mg ml^{-1} cm^{-1}) of 1.2 at 280 nm.[8] The purified luciferase usually has a specific activity of 2×10^{14} quanta sec^{-1} mg^{-1} as determined by the standard assay.

Preparation of Luciferase Subunits

All phosphate buffer solutions are prepared using the above-listed reagents, with appropriate dilutions and additions of other chemicals as indicated. All operations are conducted at $5 \pm 1°$.

To 170 mg of *B. harveyi* luciferase in 4 ml of 40 m*M* phosphate, pH 7, 1 m*M* EDTA, 1 m*M* dithiothreitol, 1.6 g of urea (Ultrapure grade from Schwarz/Mann) are added and dissolved. The sample is immediately applied to a DEAE-Sephadex column (2.8×35 cm) preequilibrated with 40 m*M* phosphate, pH 7, containing 1 m*M* EDTA, 1 m*M* dithiothreitol, and 5 *M* urea. The column is eluted with a linear gradient using 400 ml of the equilibration buffer and 400 ml of 0.12 *M* phosphate, pH 7, 1 m*M* EDTA, 1 m*M* dithiothreitol, 5 *M* urea. The flow rate is set at 10 ml/hr, and 3.5-ml fractions are collected. After such a single step of column chromatography, α and β subunits can be well separated (Fig. 1) with good reproducibilities. For more diluted luciferase samples, the use of up to 15 ml of sample

[7] J. W. Hastings and G. Weber, *J. Opt. Soc. Am.* **53,** 1410 (1963).

[8] S.-C. Tu, T. O. Baldwin, J. E. Becvar, and J. W. Hastings, *Arch. Biochem. Biophys.* **179,** 342 (1977).

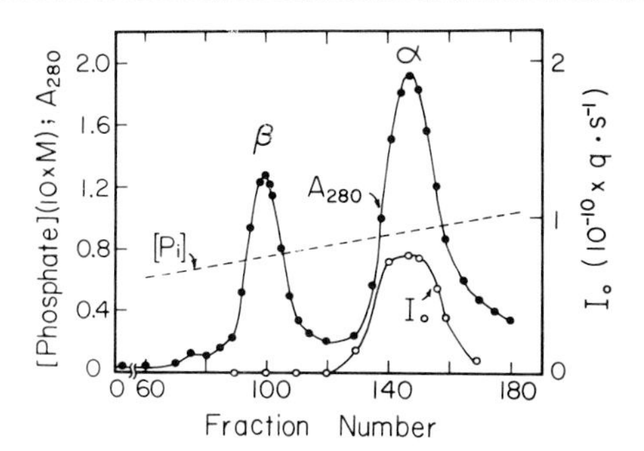

FIG. 1. Chromatography of *Beneckea harveyi* luciferase subunits on DEAE-Sephadex column. The protein elution pattern is recorded as sample absorbances at 280 nm (●). Residual luciferase activities (○) in 10-μl aliquots of selected fractions are also measured. Details are described in the text. Reproduced from S.-C. Tu, J. C. Makemson, J. E. Becvar, and J. W. Hastings, *J. Biol. Chem.* **252,** 803 (1977).

solution for loading the column does not impair the degree of subunit resolution.

Subunits of *Photobacterium fischeri* luciferase can also be resolved and isolated by the same procedures described above, except that the α subunit elutes from the column prior to the β subunit.[3]

Properties

For measurements of residual luciferase activities in individual fractions, 10-μl aliquots are each diluted 100-fold into a recovery solution (0.1 *M* phosphate, pH 7, 1 m*M* EDTA, 10 m*M* dithiothreitol, 0.2% bovine serum albumin), and the total activities are measured after standing at 5 ± 1° for approximately 60 hr (Fig. 1). No residual activities are found in β fractions whereas some α fractions exhibit low levels of luciferase background activity, due to contamination of trace amounts of β in these fractions, which are usually $<1\%$ of the corresponding maximal reconstitutable activities (Fig. 2).

Concentrations of α and β subunits can be calculated based on absorbance coefficients (1 mg ml^{-1} cm^{-1}) at 280 nm of 1.2 and 0.65, respectively.[5] Both α and β samples are active in reconstituting luciferase activity when renatured in the presence of their respective complementary subunits (Fig. 2). It is estimated, based on maximal reconstituted activities, that 10% and 30% of the α and β molecules, respectively, are active materials. Somewhat different results have been obtained with different preparations. The poor yields of α and β subunits active in reconstitution are attributed to the

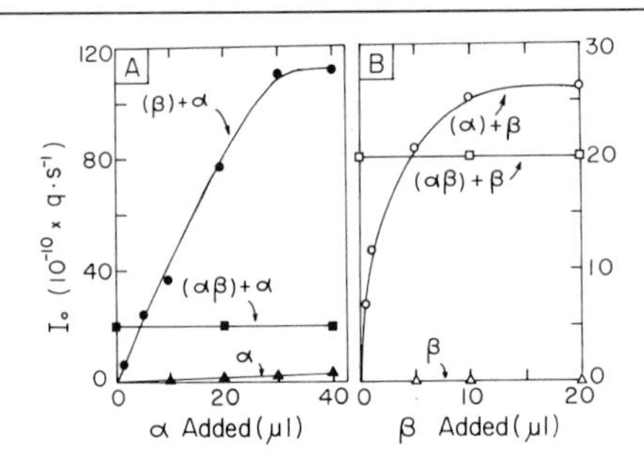

FIG. 2. Reconstitution of luciferase activity from isolated subunits. The α and β subunits used are from tubes No. 150 (1.5 μg/μl) and No. 100 (1.9 μg/μl), respectively, shown in Fig. 1. (A) To 2-ml recovery solutions, each containing no (▲) or 5 μl (●) of the β samples, various amounts of the α sample are added and total activities are determined after the samples have stood at 5 ± 1° for approximately 60 hr. The background luciferase activities in α samples (▲) are subtracted from the corresponding β-containing samples for presentations of the reconstituted activities (●). Similarly, 1 μg of native luciferase is also titrated with α samples, and final activities (minus α background activities) are shown (■). (B) Similar to (A) except that 5 μl of the α sample (○) and 1 μg of luciferase (□) are titrated with β samples, which by themselves do not contain any background luciferase activity (△). The background luciferase activity in the 5 μl of α sample (4×10^9 quanta sec^{-1}) is subtracted from samples containing this subunit (○) for presentations. Activities of 280 and 400×10^{10} quanta sec^{-1} are expected for 5 μl of α and β samples, respectively, if complete reconstitution had been achieved upon titration with the complementary subunit of each. Reproduced from S.-C. Tu, J. C. Makemson, J. E. Becvar, and J. W. Hastings, *J. Biol. Chem.* **252,** 803 (1977).

prolonged exposure to urea during the process of chromatography. Prompt renaturation of luciferase subunits after treatment with 5 *M* urea often results in 80–100% activity recoveries.[9] The inactive protein species in the isolated α and β samples do not seem to perturb the activity of a diluted native luciferase sample (Fig. 2).

[9] J. M. Friedland and J. W. Hastings, *Biochemistry* **6,** 2893 (1967).

[18] Preparation of Luciferases Containing Chemically Modified Subunits

By EDWARD A. MEIGHEN

Chemical modification of the heteropolymeric bacterial luciferases ($\alpha\beta$) may be used to define the location and type of amino acid residues involved in the bioluminescent reaction and the effect of environment on their

ISBN 0-12-181957-4

reactivity. Investigations on bacterial luciferases are greatly facilitated by the slow turnover of the enzyme permitting the elucidation of the effect of chemical modification not only on activity, but on specific steps in the catalytic mechanism. Since neither subunit (α or β) has yet been assigned a functional activity by itself, chemical reagents must be used that alter the catalytic properties without dissociation of the two nonidentical subunits. The modified luciferase ($\alpha_m\beta_m$) can then be complemented with native α or β subunits to form hybrid luciferases containing one native and one modified subunit ($\alpha\beta_m$ or $\alpha_m\beta$). Ideal reagents for such studies are ethoxyformic anhydride, a histidine-specific reagent, and sulfhydryl reagents, such as *N*-ethylmaleimide, since both a histidyl residue[1] and a cysteinyl residue[2] have been implicated in the functional activity of *Beneckea harveyi* luciferase, and the activity of luciferase can be altered after limited reaction with these reagents. Reagents that alter the catalytic properties of luciferase after more extensive modification, such as succinic anhydride,[3] can also be used providing that the subunit interactions are not extensively disrupted. These studies permit comparison of the properties of subunits of different bacterial luciferases and the elucidation of the role in the bioluminescent reaction of the nonidentical subunits within the intact dimeric structure.

Reagents

Bacterial luciferase, stored as an ammonium sulfate suspension. Enzyme activity is measured by the standard or the dithionite assay.[4]

Luciferase subunits, isolated by DEAE-Sephadex chromatography in urea.[5]

Ethoxyformic anhydride (Eastman). Stock solutions are prepared in ethanol and stored at −20°. Radioactive ethoxyformic anhydride, 1-ethyl-^{14}C-labeled, can be synthesized by the method of Melchior and Fahrney[6] or obtained from New England Nuclear. Molar concentrations are determined by the increase in absorbance at 240 nm after reaction of the reagent (~ 0.1 m*M*) with 10 m*M* *N*-acetylhistidine in 0.1 *M* phosphate buffer, pH 6.1, based on an extinction coefficient of 3600 M^{-1} cm^{-1} at 240 nm.[7]

[1] J. Cousineau and E. Meighen, *Biochemistry* **15,** 4992 (1976).
[2] M. Z. Nicoli, E. A. Meighen, and J. W. Hastings, *J. Biol. Chem.* **249,** 2385 (1974).
[3] E. A. Meighen, M. Z. Nicoli, and J. W. Hastings, *Biochemistry* **10,** 4062 (1971).
[4] The procedures for purification and enzyme assay are described in this volume by Hastings [13].
[5] See S.-C. Tu, this volume [17].
[6] W. B. Melchior and D. Fahrney, *Biochemistry* **9,** 251 (1970).
[7] J. J. Holbrook and V. A. Ingram, *Biochem. J.* **131,** 729 (1973).

N-Ethylmaleimide (Eastman) stock solutions (10 mM) are prepared fresh daily. *N*-[1-^{14}C]Ethylmaleimide is obtained from New England Nuclear.

Succinic anhydride (Eastman).

Phosphate buffers, prepared by mixing appropriate amounts of NaH_2PO_4 and K_2HPO_4

Modification of Histidyl Residues. Reaction with Ethoxyformic Anhydride

Luciferase is transferred by gel filtration on BioGel P-10 into 0.1 M phosphate, pH 6.1, just prior to chemical modification. One milliliter of the luciferase solution (0.1 to 2.0 mg/ml) at 0° is mixed with 10 μl of a solution of ethoxyformic anhydride, and the loss of activity is followed with time. For kinetic studies of the rate of inactivation, sufficient reagent should be added so that 50% inactivation occurs in less than 20 min since the reagent is slowly hydrolyzed under these solvent conditions. The reaction is stopped with about 10% activity remaining by addition of 50% by volume of 0.3 M histidine, 0.5 M phosphate, pH 7.0. Although the reaction can be conducted at a different pH, bacterial luciferase is relatively unstable at lower pH whereas at higher pH, particularly above pH 7, the specificity of the reagent for histidyl residues is decreased. Regeneration of activity after incubation of an aliquot in 14 mM mercaptoethanol, 0.6 M NH_2OH, 0.1 M phosphate, pH 7.4, at 22° provides indirect support that inactivation arises from histidyl modification. In this case, activity is measured by the dithionite assay, since the NH_2OH in the sample interferes in the standard assay.

The extent of modification of histidyl residues is determined by following the increase in absorbance at 240 nm during the course of the reaction using an extinction coefficient of 3600 M^{-1} cm^{-1} for the formation of ethoxyformyl histidyl derivatives.[7] Alternatively, the enzyme is modified with ^{14}C-labeled ethoxyformic anhydride. The modified enzyme is separated from radioactive by-products by dialysis or gel filtration, and the amount of radioactivity incorporated into the enzyme is determined. The degree of incorporation of [^{14}C]ethoxyformyl groups into each of the subunits (α and β) can be determined after separation of the subunits by DEAE-Sephadex chromatography in urea, or sodium dodecyl sulfate (SDS) gel electrophoresis.[1] In the latter case, the polypeptide bands, stained with Coomassie brilliant blue in 25% isopropanol:10% acetic acid, are excised, dissolved overnight at 50° in 30% H_2O_2 in sealed glass scintillation vials, and the radioactivity is determined. Between 80 and 90% of the applied radioactivity is recovered in this procedure.

The specificity of the reagent for histidyl residues is determined by comparing the number of radioactive ethoxyformyl residues incorporated and the number of ethoxyformylhistidyl derivatives measured spectrophotometrically. Any effects on the quaternary structure are determined by sedimentation velocity or gel filtration experiments.

Modification of Cysteinyl Residues. Reaction with N-Ethylmaleimide

Luciferase is dialyzed against 20 m*M* phosphate, pH 7.0, to remove reducing agents, just prior to modification. The reaction is initiated by addition of the appropriate volume of 10 m*M* *N*-ethylmaleimide in 20 m*M* phosphate, pH 7.0, to the enzyme solution at 25°, and the loss of activity is determined with time. The reaction is terminated with 1–10% activity remaining by addition of mercaptoethanol to a final concentration of 30 m*M*, and the modified enzyme is dialyzed against the desired buffer.[2] If the luciferase is unstable under the above conditions, the reaction should be conducted at 0°.

The extent of modification with time may be determined by the decrease in *N*-ethylmaleimide absorbance at 305 nm upon reaction with luciferase using an extinction coefficient of 620 M^{-1} cm^{-1}.[8] Since high protein concentrations are necessary to measure the absorbance change, it is generally preferable, however, to use ^{14}C-labeled *N*-ethylmaleimide to measure the stoichiometry of the reaction. The extent of incorporation of radioactivity into each subunit is measured after separation of the subunits by DEAE-Sephadex chromatography in urea or SDS gel electrophoresis.[2]

The reagent is highly specific for cysteinyl residues.[9] Hydrolysis of the modified enzyme and determination of the amount of *S*-succinylcysteine and ethylamine can be used to check the specificity.[10]

Modification of Lysyl Residues. Reaction with Succinic Anhydride

Succinylated luciferase is prepared by the addition of solid succinic anhydride in small increments to a 1% solution of luciferase in 50 m*M* Tris-chloride, pH 8.0, at room temperature. A 3- to 5-fold molar excess of succinic anhydride per lysyl residue (1–2 mg/ml) is added in order to inactivate luciferase without causing extensive dissociation of the nonidentical subunits.[3] The pH is maintained at 8.0 by the addition of 1.0 *N* NaOH until the completion of the reaction (~ 20 min). The modified enzyme is then dialyzed against the desired buffer. The temperature of the reaction should be decreased if the luciferase is unstable under the above conditions.

[8] J. D. Gregory, *J. Am. Chem. Soc.* **77**, 3922 (1955).
[9] J. F. Riordan and B. L. Vallee, this series, Vol. 25, p. 449.
[10] D. G. Smyth, O. O. Blumenfeld, and W. Konigsberg, *Biochem. J.* **91**, 589 (1964).

The number of free amino groups in the modified luciferase is determined by the procedure described by Fraenkel-Conrat.[11] Gel filtration or sedimentation velocity experiments are conducted to determine that a substantial proportion of the modified enzyme maintains a dimeric structure. If necessary, the modified dimer is purified prior to investigation of its properties and complementation with native subunits.

Preparation of Luciferases Containing Native and Chemically Modified Subunits

The preparation of hybrid luciferases containing native and modified subunits is most simply accomplished by complementation of the modified enzyme ($\alpha_m\beta_m$) with an excess of either the native α or β subunit to form the hybrids $\alpha\beta_m$ and $\alpha_m\beta$, respectively. The modified luciferase is dissociated by incubation for 30 min in 5 *M* urea, 10 m*M* DTT, 0.1 *M* phosphate, pH 7, at 0°. The dissociated modified subunits (α_m and β_m) and a molar excess of either the unmodified α or β subunit, also in 5 *M* urea, are diluted into a renaturation buffer of 0.2% bovine serum albumin, 10 m*M* DTT, 1 m*M* EDTA, 0.1 *M* phosphate, pH 7.0, so that the final protein concentration is in the range of 1 to 100 μg/ml in order to obtain maximum renaturation and complementation. The final urea concentration should be less than 0.2 *M*, and if necessary the sample can be dialyzed against the renaturation buffer to lower the urea concentration. The sample is incubated at 4° to 6° until maximum activity is obtained (24–72 hr). The specific activity of the hybrid is calculated based on the maximum amount of hybrid that could be formed.

The major difficulty in this approach is that low specific activities may reflect a decreased capability of the modified subunits to complement with the native subunits as well as a decreased activity of the hybrid. Depending on the properties of the modified subunits, a number of different approaches may be used to investigate this problem. (1) A fixed amount of either the unmodified α or β subunit is complemented with a molar excess of the modified luciferase ($\alpha_m\beta_m$). The activity, after correction for the background activity of the added $\alpha_m\beta_m$, can be compared directly to that obtained if the same amount of native subunit is mixed with an excess of the complementary native subunit. If necessary the background activity of $\alpha_m\beta_m$ can be eliminated by separation of the modified subunits (α_m and β_m) by procedures similar to that for the native subunits,[5] prior to addition of a molar excess of one of the modified subunits to a limited amount of the complementary native subunit. (2) A second approach requires purification

[11] G. Fraenkel-Conrat, this series, Vol. 4, p. 247.

of the hybrid species and determination of the specific activity. The difficulty in this approach is that renaturation must be conducted in the absence of bovine serum albumin, resulting in a lower extent of complementation, and that further manipulation of the hybrid could lead to some inactivation.

Applications

Investigation of the Role of the Nonidentical Subunits in Luciferase

Hybrid luciferases have been prepared by complementation of the unmodified subunits of *B. harveyi* luciferase with subunits modified with ethoxyformic anhydride,[1] *N*-ethylmaleimide,[2] or succinic anhydride.[3] Hybrids containing β subunits with about 7 modified lysyl or 3 modified histidyl residues ($\alpha\beta_m$) had 60–80% the specific activity of the native enzyme. In contrast, hybrids containing α subunits with about 7 modified lysyl, 1.0 modified cysteinyl, or 1.5 modified histidyl residues ($\alpha_m\beta$) had only 2–15% the specific activity of the native enzyme and the same activity as luciferase with both subunits modified ($\alpha_m\beta_m$). An increase in the K_m for $FMNH_2$ (7-fold) and the decay rate of luminescence (1.5-fold) was also observed for the hybrid containing an α subunit with modified lysyl residues.[12] Such studies can be used to investigate the role of the nonidentical subunits of luciferase in the bioluminescent reaction and illustrate for *B. harveyi* luciferase that changes in the catalytic parameters arise from modification of the α subunit, but not the β subunit.

Investigation of the Active Site of Luciferase

Comparison of the active sites of luciferases isolated from different strains would indicate which amino acid residues have been conserved and are essential for the catalytic function of luciferase in the bioluminescent reaction. The use of reagents that inactivate luciferases after only limited modification of the enzyme makes isolation and comparison of the peptides containing the modified residues a feasible approach. Two reagents may meet the requirements for such experiments: *N*-ethylmaleimide, which has been shown to inactivate luciferases isolated from three different strains,[2,13,14] and ethoxyformic anhydride, which has been shown to inacti-

[12] E. A. Meighen, M. Z. Nicoli, and J. W. Hastings, *Biochemistry* **10,** 4069 (1971).

[13] T. O. Baldwin, M. S. Carrie, M. Z. Nicoli, and T. W. Cline, *Fed. Proc., Fed. Am. Soc. Exp. Biol.* **34,** 681 (1975).

[14] Luciferase, purified from *Photobacterium phosphoreum* (NCMB 844), is rapidly inactivated with *N*-ethylmaleimide or ethoxyformic anhydride (E. Meighen and I. Bogacki, unpublished experiments).

vate luciferases isolated from two different strains.[1,14] Good evidence is available for *B. harveyi* luciferase to indicate that inactivation arises from modification of a single cysteinyl residue or histidyl residue, respectively, on the α subunit.[1,2] In this regard, the tryptic peptide containing the modified cysteinyl residue has been isolated and characterized.[15]

The use of other chemical reagents can also provide information about the environment of the active site. Nicoli and Hastings[16] have shown that long-chain alkylmaleimides inactivate *B. harveyi* luciferase more rapidly than *N*-ethylmaleimide, providing evidence for a hydrophobic pocket at or near the active site. Moreover, luciferase inactivated with *N*-octylmaleimide could still bind FMN,[17] suggesting that modification of the cysteinyl residue may not prevent the binding of substrates. This conclusion is supported by studies showing that substrates could protect luciferase against inactivation with a histidyl reagent even if the cysteinyl residues had been previously modified with sodium tetrathionate.[18] In contrast, substrates did not protect luciferase against inactivation with a sulfhydryl reagent if the histidyl residues had been first modified with ethoxyformic anhydride.[18]

The chemical modification of luciferase with reagents of different specificity should provide information about the role of other amino acid side chains in the bioluminescent reaction. However, such studies must be approached with some degree of caution since many reagents, even with a relatively high degree of reactivity for a specific amino acid side chain, may still react with sulfhydryl groups.[19] If such reagents are used, it may be necessary to reversibly protect any reactive cysteinyl residues in luciferase. The reaction of sodium tetrathionate (10 m*M*) with *B. harveyi* luciferase in 0.1 *M* phosphate, pH 8.0, at 0° protects this enzyme against irreversible inactivation with *N*-ethylmaleimide.[18] The enzyme, modified with tetrathionate, can be reactivated by incubation in 40 m*M* mercaptoethanol, 80 m*M* cysteine, 0.1 *M* phosphate, pH 8.0, at 0°. Studies using tetranitromethane, *N*-bromosuccinimide, and trinitrobenzene sulfonic acid, reagents with specificity for tyrosyl, tryptophanyl, and lysyl residues, respectively, have shown that prior modification of the cysteinyl residues of luciferase with sodium tetrathionate protected the enzyme against inac-

[15] M. Z. Nicoli, Ph.D. thesis, Harvard University, Cambridge, Massachusetts, 1972.

[16] M. Z. Nicoli and J. W. Hastings, *J. Biol. Chem.* **249,** 2393 (1974).

[17] M. Z. Nicoli, T. O. Baldwin, J. E. Becvar, and J. W. Hastings, *in* "Flavins and Flavoproteins" (T. P. Singer, ed.), p. 87. Elsevier, Amsterdam, 1976.

[18] J. Cousineau and E. Meighen, *Can. J. Biochem.* **55,** 433 (1977).

[19] L. A. Cohen, *in* "The Enzymes" (P. D. Boyer, ed.), 3rd ed., Vol. 1, p. 147. Academic Press. New York, 1970.

tivation by these reagents.[20] Consequently, chemical modification studies to investigate the role of other amino acid residues in luciferase are clearly limited to the use of reagents and conditions for which there is very limited reactivity of the cysteinyl residues.

[20] J. Cousineau, Ph.D. thesis, McGill University, Montreal, Canada, 1976.

[19] Quantitation of Malate, Oxaloacetate, and Malate Dehydrogenase

By PHILIP E. STANLEY

L-Malate and oxaloacetate are estimated using malate dehydrogenase and monitoring the change in level of NADH with dehydrogenase–luciferase complex derived from *Photobacterium fischeri*. For both substrates the minimum measurable level is a few picomoles but is dependent on the sensitivity and background of the equipment used as well as the activity and background of the dehydrogenase–luciferase complex. A liquid scintillation spectrometer is used as the light-detection device.

Malate dehydrogenase may be measured using the same approach, but with saturating levels of malate and excess NAD.

Quantitation is effected by ensuring that the component to be measured is the only limiting item in the enzymic reaction. Initial reaction rates are used as a measure of the component under investigation. Recent reviews concerned with analytical bioluminescence are to be found in references cited in footnotes 1–3.

Assay Principle

$$\text{L-Malate} + \text{NAD} \leftrightarrow \text{oxaloacetate} + \text{NADH}$$

This reaction is mediated by malate dehydrogenase (L-malate:NAD oxidoreductase, EC 1.1.1.37). At pH 7.2 and 22° the equilibrium constant for the porcine heart enzyme is around 2×10^{-12},[4] thus favoring heavily the

[1] E. Schram, *in* "Liquid Scintillation Counting: Recent Developments" (P. E. Stanley and B. A. Scoggins, eds.), p. 383 Academic Press, New York, 1974.

[2] P. E. Stanley, *in* "Liquid Scintillation Counting" (M. A. Crook and P. Johnson, eds.), Vol. 3, p. 253. Heyden, London, 1974.

[3] P. E. Stanley, *in* "Liquid Scintillation. Science and Technology" (A. A. Noujaim, C. Ediss, and L. J. Wiebe, eds.), p. 209. Academic Press, New York, 1976.

[4] J. R. Stern, S. Ochoa, and F. Lynen, *J. Biol. Chem.* **221,** 61 (1956).

ISBN 0-12-181957-4

formation of malate. The conditions for the reverse reaction, necessary to measure malate, can be made more favorable by increasing the pH to 9.5 and using a large excess of NAD or by using acetyl pyridine NAD.

The basis of the reactions used to measure NADH is as follows:

$$\text{NADH} + \text{FMN} \xrightarrow{\text{dehydrogenase}} \text{NAD} + \text{FMNH}$$

$$\text{FMNH} \xrightarrow[\text{long-chain aldehyde}]{\text{luciferase}} \text{light (max. 490 nm)}$$

With FMN, long-chain aldehyde dehydrogenase and luciferase in an appropriate excess, the amount of light produced will be proportional to the amount of NADH present at that instant. This relationship holds for levels of NADH over several orders of magnitude.

There appears to be some variation in the stability and activity of commercially available preparations of the enzyme, and this may be due in part to the bacterial species from which it has been derived. Thus the amount of enzyme protein mentioned in the assay protocol serves only as a guide and may need adjustment.

Equipment, Materials, and Reagents

Equipment. A sensitive light-detecting unit is required and a liquid scintillation counter is a suitable instrument. It is necessary to switch off the coincidence circuit between the two photomultipliers to count the single photons produced in this reaction.[5,6] Pulses from either or both photomultipliers may be counted with a pulse-height analyzer (channel) set for a tritium spectrum. The background count may be as high as 25,000 cpm and acceptable; however, the optimum signal-to-noise ratio can be achieved by adjusting the level of the lower discriminator setting[5,6] on the pulse-height analyzer. The pulse-height analyzer should not be subjected to a counting rate in excess of 10^6 cpm; otherwise spectral distortion will occur.[6] The sample chamber should be maintained at room temperature, 20° being quite acceptable. The equipment should be operated in a room lit only by tungsten light. Sunlight and fluorescent lighting should be strenuously avoided, as they can cause the spurious production of light and thus adversely affect the assay.

The reaction rate may be followed by plotting the counts obtained from sequential printouts, but a continuous assessment of rate is best achieved

[5] P. E. Stanley and S. G. Williams, *Anal. Biochem.* **29,** 381 (1969).

[6] P. E. Stanley, *Anal. Biochem.* **39,** 441 (1971).

by connecting a rate meter to the pulse-height analyzer output and plotting the rate, in analog form, on a chart recorder.[6] Alternatively the pulse-height analyzer output may be connected to a multichannel analyzer operated in multiscale mode.[6] Thus for a 256-channel unit, each channel would be opened sequentially for, say, 1 sec, and the counts produced in that time would be stored in that channel. At the end of the 256-sec sequence, the results can be displayed on a video screen, an *X-Y* plotter, or readout in digital form for subsequent analysis.

As an alternative to the liquid scintillation counter, there are now several commercially available units; some operate as pulse or quantum counters, and others operate as current-measuring devices or photometers.[2]

Vials. Ordinary glass scintillation vials are suitable. They should be kept free of dust and away from sunlight or fluorescent lighting. Failure to do the latter will result in a high and variable level of phosphorescence, which in turn means a very variable blank.

Polythene vials may be used, but these are generally subject to phosphorescence, even more so than glass, and further, light transmission through their walls is more variable.[7]

The use of small volumes, e.g., 2 ml, of liquid in scintillation vials can give rise to problems concerned with the optical collection of light. This depends on the design of the detector assembly of the instrument used, but more so on the variability of the shape of the base of vial and thus the height of the column of liquid reagents. As an alternative and to avoid this problem, a small vial to hold the reagents may be placed inside the scintillation vial and held snugly in its neck. A glass vial may be used or alternatively a polystyrene autoanalyzer cup capable of holding about 2 ml. The scintillation vial may be filled with water to act as a constant-temperature bath for the inner vial.

Reagents. The chemicals used are all commercially available. Dehydrogenase/luciferase is obtainable as bacterial luciferase from the Sigma Chemical Co., St. Louis, Missouri; malate dehydrogenase is obtainable from Boehringer GmbH., Mannheim, as porcine heart enzyme suspended in ammonium sulfate. A minimum specific activity of 1000 units/mg is suitable. Decanal, the long-chain aldehyde which is a cofactor for the luciferase, is obtainable from Aldrich Chemical Co., Milwaukee, Wisconsin. Dithiothreitol is obtainable from Calbiochem, La Jolla, California.

Before they are used, all reagents should be centrifuged or filtered (Millipore, 1.2 μm) to remove microorganisms and particulate material, since these can produce spurious results. Reagents can be kept at room

[7] J. E. Corredor, D. G. Capone, and K. E. Cooksey, *Anal. Biochem.* **70,** 624 (1976).

temperature while in use with the exception of the luciferase and malate dehydrogenase.

FMN should be prepared weekly and stored in the dark at +4°, and NAD, NADH, luciferase as well as internal standards of oxaloacetate and L-malate should be prepared daily. The long-chain aldehyde, decanal should be prepared daily as a solution in ethanol or phosphate buffer.

General Method

Adjust one pulse-height analyzer on the liquid scintillation spectrometer to count tritium (or optimize for signal-to-noise ratio), having previously switched off the coincidence circuit and selected one or both photomultipliers. For temperature equilibration the sample chamber should have been maintained at room temperature for at least 48 hr. Select repeat sample count mode and a preset counting time of 0.04 or 0.1, and preset counts on all channels to maximum.

Bring all reagents to the temperature of the sample changer with the exception of the enzymes, which should be kept on crushed ice.

Place in a suitable vial the reagents listed in the assay protocol except for NAD(H), and the component to be measured. Mix them well by vortexing after the addition of each component, and be sure to maintain the same sequence and time interval between each addition. Immediately place the vial in the counter (or other detector) and print the results for at least 3 × 0.1 or preferably 6 × 0.04 min counting sequences (counters equipped with shorter counting times and/or concurrent counting and printing capability are best suited for this type of work) over a period of 0.5 to 1 min. This provides data for part of the reagent background and mainly serves to alert the worker to the presence of an unexpected high background from, for example, phosphorescence. Then retrieve the sample vial and add the NAD(H) using the same preset counting times over a period of about 1 to 1.5 min. These data provide the true background from which the results are to be calculated. Again unload the sample vial and initiate the assay reaction by adding the unknown and then recount for a period of 2 min (at least 8 counting sequences). Plot the counts against elapsed time or printing sequence. Obtain by hand or by a suitable computer curve-fitting procedure the rate of reaction at zero time, i.e., when the component to be assayed was added, and read this result from a standard curve. Typical results are shown in Fig. 1 for oxaloacetate and Fig. 2 for malate; standard curves are presented in Figs. 3 and 4. Other means of calculating results are mentioned in the Results section.

As the procedure is not an absolute one, calibration curves are mandatory. Further, as this is an enzymic procedure and is likely to be influenced

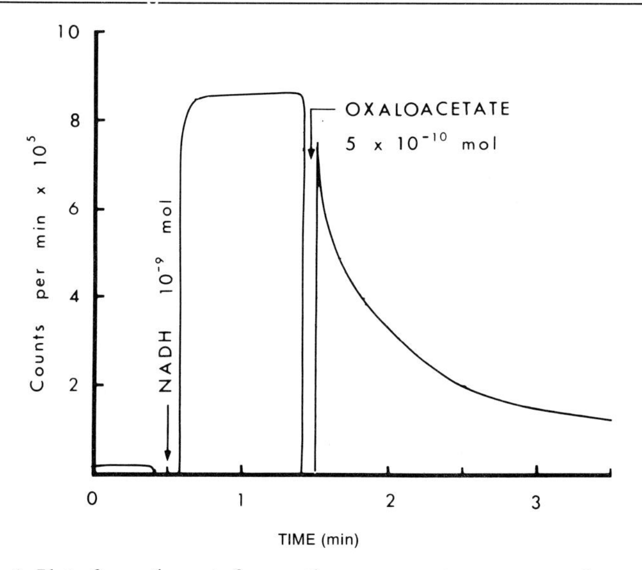

FIG. 1. Plot of counting rate for reaction sequence to measure oxaloacetate.

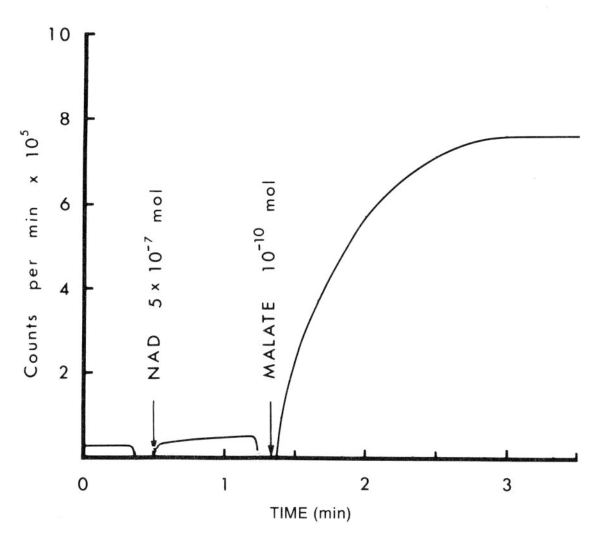

FIG. 2. Plot of counting rate for reaction sequence to measure malate.

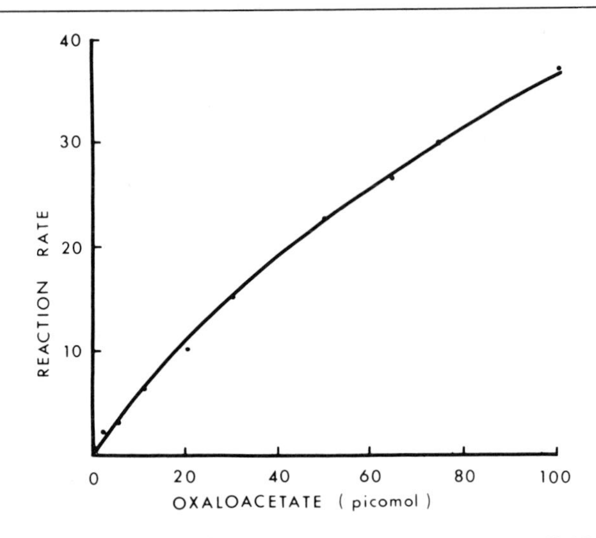

FIG. 3. Calibration curve for oxaloacetate. Vertical axis is a measure of initial reaction rate and is expressed in arbitrary units.

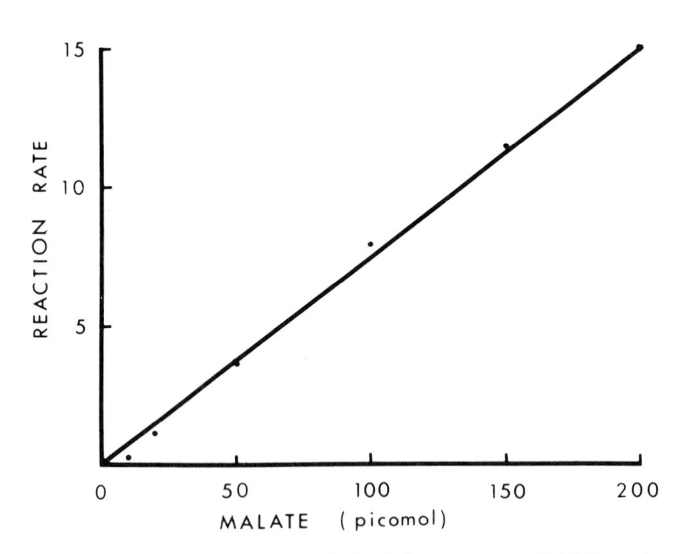

FIG. 4. Calibration curve for malate. Vertical axis is a measure of initial reaction rate and is expressed in arbitrary units.

by inhibitors in the sample, it is necessary to assay the unknown a second time, but on this occasion after the addition of a known amount of the component. This amount is usually about one and a half times the apparent value of the unknown. The recovery factor can then be included in the calculation of the result. In addition, colored matter in the assay solution can absorb some of the light before it reaches the photomultiplies. Thus the internal standard serves to correct for both inhibition and light absorption.

Assay Protocols

Assay Protocol for Malate

Use the procedure as described in General Method.

Reagents should be added as follows[8]: Tris · HCl,[9] 50 mM, pH 9.5, 2 ml; β-mercaptoethanol, 2 μl; decanal, 10 μl/ml in ethanol, 20 μl; flavin mononucleotide, 0.2 mM, 10 μl; luciferase,[10] 1 mg/ml (phosphate, 50 mM, pH 7.0), 20 μl; malate dehydrogenase (0.5 mg/ml, ammonium sulfate suspension), 10 μl; NAD, 50 mM, 10 μl; L-malate to be assayed, 10–200 pmol in approximately 50 μl.

Assay Protocol for Oxaloacetate

Proceed as described in General Method.

Reagents should be added as follows: Phosphate (potassium/sodium), 50 mM, pH 7.0, 2 ml; β-mercaptoethanol, 2 μl; decanal, 10 μl/ml in ethanol, 20 μl; flavin mononucleotide, 0.2 mM, 10 μl; luciferase,[10] 1 mg/ml (phosphate, 50 mM, pH 7.0), 20 μl; malate dehydrogenase (0.5 mg/ml, ammonium sulfate suspension), 10 μl; NADH, 10–1000 pmol (depends on level of oxaloacetate to be assayed); oxaloacetate, 10–1000 pmol in approximately 50 μl.

Assay Protocol for Malate Dehydrogenase

The procedure is a modified version of that adopted for the determination of malate. Ten nanomoles of malate are used, and the light-producing

[8] When carrying out the assay in an autoanalyzer cup it is necessary to use a total volume of around 1 ml. Thus the mentioned volume should be halved.

[9] Some batches of luciferase are inhibited by Tris · HCl, and a glycine buffer, pH 10, may be substituted.

[10] As mentioned previously, luciferase stability can vary from batch to batch. Inconsistent results frequently point to this as the problem, and it can be often solved by preparing the luciferase in phosphate buffer containing bovine serum albumin (10 mg/ml) and dithiothreitol (1 mg/ml). As the last-mentioned protects sulfhydryl groups, β-mercaptoethanol need not be included.

reaction is initiated by addition of the malate dehydrogenase sample. If the pH of 9.5 is not suitable for the dehydrogenase in question, then it may be necessary to employ a pH of 7 and to follow the reverse reaction used for the assay of oxaloacetate. The concentration of oxaloacetate and NADH will need to be selected according to the sensitivity required. It is, of course, necessary to employ an internal standard of malate dehydrogenase to correct for inhibition and colored compounds just as for the assay of malate and oxaloacetate.

Results

Results are obtained by reference to a standard curve and corrected by a factor determined using an internal standard. The present author has found that the reaction rate at zero time is a suitable measure when a multichannel analyzer or rate-meter recording is available. However, if only the counts from the spectrometer are available then the equilibrium value of NADH level is a more convenient measure. As a further alternative, the count rate at a fixed time after adding the unknown may be used.

It will be apparent that, although the forward reaction is less favored from the equilibrium viewpoint, it makes for greater convenience for the assay of malate since the worker has the relatively easy task of measuring an increase of light over and above a relatively small background level. This is in direct comparison with the assay of oxaloacetate, where the worker is confronted with the task of measuring a decrease, perhaps a small one, from a high level of light output. In the latter technique it is necessary to appropriately reduce the amount of added NADH to a level some 20–50% above that required for the reduction of all the oxaloacetate likely to be encountered. The range of the assay can thus be altered by adjusting the level of NADH; new calibration curves are required for each level of NADH. The effect of using only this slight excess of NADH can be observed in Fig. 3, where a true curve, rather than a straight line is seen. Here about 120 μmol of NADH were used; above a level of 50 μmol of oxaloacetate, a significant portion of the NADH is used rather quickly, and despite the favorable equilibrium the reaction rate apparently slows appreciably.

The coefficient of variation is around 7% for levels of malate and oxaloacetate about 50 pmol and 15% at lower values. It appears that the repeatability for pipetting of reagents and their mixing is critically important.

[20] Bioassay for Myristic Acid and Long-Chain Aldehydes

By S. ULITZUR and J. W. HASTINGS

Principle

The light-emitting reaction in bacteria involves a mixed-function oxidation of $FMNH_2$ and long-chain fatty aldehyde catalyzed by bacterial luciferase.

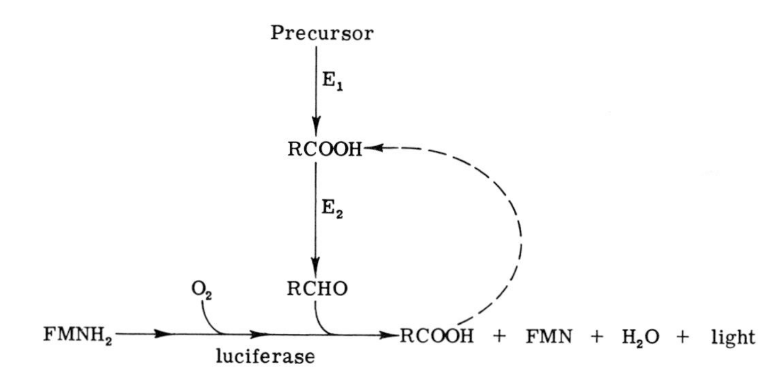

A dim aldehyde mutant[1] of the luminous bacterium *Beneckea harveyi* emits light upon the exogenous addition of both long-chain aldehydes and fatty acids.[2] The bioassay with the mutant is very sensitive, so that picomoles of aldehyde or fatty acid can be detected in a procedure that requires less than a minute.

Bacterial Strains

The aldehyde mutant M17 was obtained from the wild type of a marine luminous bacterium, *B. harveyi* MAV #392,[3] by mutagenesis with nitrosoguanidine.[1] This mutant, unlike certain other aldehyde mutants,[2,4] is stimulated to emit light by both aldehyde and fatty acid.[2] Similar mutants have also been obtained from *Photobacterium fischeri*.[2]

[1] T. W. Cline, Mutational alteration of the bacterial bioluminescence system of the bacterial bioluminescence system. Ph.D. thesis. Harvard University, Cambridge, Massachusetts, 1973.

[2] S. Ulitzur and J. W. Hastings, *Proc. Natl. Acad. Sci. U.S.A.* **75,** 266 (1978).

[3] J. L. Reichelt and P. Baumann, *Arch. Mikrobiol.* **94,** 283 (1973).

[4] T. W. Cline and J. W. Hastings, *Proc. Natl. Acad. Sci. U.S.A.* **68,** 500 (1971).

ISBN 0-12-181957-4

Bacterial Growth

Bacteria are grown in a complex medium consisting of a buffered artificial seawater (ASW) containing 20 m*M* morpholinopropanesulfonic acid (MOPS), pH 7.3, and 5 g of Casamino amino acids (Difco) per liter. ASW contains (grams/liter):NaCl, 17.55; KCl, 0.75; $MgSO_4 \cdot 7\ H_2O$, 12.3; $CaCl_2 \cdot 2\ H_2O$, 1.45; $K_2HPO_4 \cdot 3\ H_2O$, 0.075.

Cells to be used for the assay are grown by inoculating a loop or colonial isolate into 50 ml of medium in a 250-ml Erlenmeyer flask. The culture is shaken vigorously at a constant temperature between 25° and 30° until a cell density of about 10^9 cells ml^{-1} is reached (about 200 Klett Units, filter No. 66, or 1.5 OD_{660} units in the Coleman Jr. spectrophotometer).

The time of harvesting is of great importance because the response per cell to both aldehyde and fatty acid is maximal at the end of the logarithmic phase of growth, and the latter declines rapidly, indicating that the intracellular activity of the enzyme(s) converting fatty acid to aldehyde does not remain high for a long period. In experiments where a very high sensitivity to myristic acid is required, it is recommended to monitor the response per cell during the later stages of growth and to harvest cells just at the time of maximal response. Cells are harvested by centrifugation for 10 min at 10,000 *g*, washed twice, and resuspended in the cold in 50% of the original volume using ASW with MOPS buffer, pH 6.5. These cells can be utilized over a period of some 2–4 hr.

Light Determinations

Any simple photomultiplier photometer[5] or a fluorometer or light-scattering instrument capable of detecting an intensity of 10^7 to 10^8 quanta sec^{-1} may be used. Geometrical considerations, such as the housing and the location of the sample, can be important in detection capability. More-sensitive devices, such as scintillation counters (operating without coincidence on the ^3H setting at room temperature), the ATP analyzer, or single-photon detectors, may also be used but without any great advantage in this case, since the detection sensitivity is limited by the background emission of the mutant cells. If the more-sensitive devices are used, the quantity of cells used in the assay must be scaled down.

The Assay

The fatty acids and the aldehydes to be tested may be dissolved either in ethanol or in nonionic detergents such as Triton X-100. A final concentra-

[5] G. W. Mitchell and J. W. Hastings, *Anal. Biochem.* **39,** 243 (1971).

tion of 1% ethanol or 0.05% Triton X-100 may be used with no adverse effects on the results. All measurements should be carried out at some fixed temperature, preferably between 20° and 30°. One milliliter of ASW–MOPS-buffer, pH 6.5, is placed in a carefully cleaned scintillation vial, and 0.1 ml of the cold M17 culture is added. The background light emission, which should normally not exceed 1×10^8 quanta sec^{-1}, is monitored until it is stabilized (1–2 min). Myristic acid standard (in 1–10 μl volume) is added, mixed, and the resulting increase in luminescence is recorded. The maximum response should be 500–1000 times background with optimal amounts of myristic acid ($\sim 5\ \mu M$) and should require between 10 sec and 1 min to reach the peak level. Over a range of 0.01 to 2 μM the response is a linear function of the fatty acid concentration (Fig. 1). Similar standard curves may be established for aldehydes.

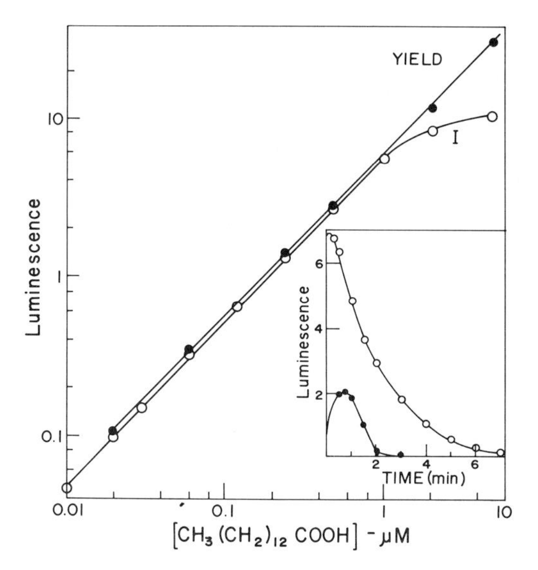

FIG. 1. Effect of concentration of tetradecanoic acid on bioluminescence. *Beneckea harveyi* M17 cells were grown at 25° in complex medium to a density of 10^9 cells ml^{-1}. The culture was then diluted 1:20 in artificial seawater–morpholinopropanesulfonic acid (ASW-MOPS) buffer at pH 6.5, and luminescence of 1-ml aliquots was determined in the presence of different amounts of tetradecanoic acid. Luminescence intensity without additions, 1×10^7 quanta sec^{-1}. Total quanta (yield) were obtained by integration of the full time course of the reaction, which with small amounts of acid lasted only a few minutes (see inset), but with large amounts took hours. Intensity units in quanta $sec^{-1}\ ml^{-1} \times 10^9$; total light yield in quanta $ml^{-1} \times 5 \times 10^{12}$.

Inset. An aliquot (0.1 ml) of *B. harveyi* (M17) cells was diluted into 0.9 ml of ASW-MOPS buffer at pH 6.5. Tetradecanoic acid (○) and tetradecanal (●) in ethanol (0.5 nmol of each) were added, and the luminescence response with time was recorded. The yields were 2.2×10^{13} and 4.4×10^{12} quanta with fatty acid and aldehyde, respectively.

Factors Affecting the Bioassay

Specificity: Chain Length and Inhibitors. With fatty acids the system is quite specific for myristic acid (see table). While longer and shorter chain lengths are far less effective, they have no inhibitory effect on the response to myristic acid. On the other hand, certain unsaturated fatty acids, such as palmitoleic, oleic, or linoleic acids, are potent competitive inhibitors.[2]

With aldehydes the system is far less specific; chain lengths ranging from 8 to 14 carbons result in responses that are comparable within an order of magnitude (see table).

Nonspecific Inhibitors. Generally any chemical or solvent that may affect the bacterial metabolism, especially the respiration or cell integrity, would be expected to compromise the bioassay. More specific problems

RESPONSE OF M17 CELLS TO DIFFERENT FATTY ACIDS, FATTY ACID DERIVATIVES, AND ALDEHYDES[a]

Lipid	Relative maximum luminescence intensity	Total quanta/1 μmol
Ethanol	<0.0001	$< 10^{9}$
Decanoic acid	0.3	1×10^{11}
Dodecanoic acid	2.0	4×10^{11}
Tridecanoic acid	56	6×10^{12}
Tetradecanoic acid	100	1.5×10^{13}
Pentadecanoic acid	45	5×10^{12}
Hexadecanoic acid	5.5	1.5×10^{11}
Octadecanoic acid	0.1	$< 10^{9}$
Myristoleic (*cis*-9,10) acid (14:1)	8.5	2×10^{12}
αHydroxymyristic acid	12.1	
Myristyl alcohol	<0.0001	$< 10^{9}$
Methyl myristate ester	< 0.0001	$< 10^{9}$
Palmitoleic (*cis*-9,10) acid (16:1)	2.5	4.5×10^{11}
Linoleic (*cis*-9, *cis*-12) acid (18:2)	0.01	$< 10^{9}$
Octanal	2.7	3.8×10^{10}
Nonanal	8	6×10^{11}
Decanal	10	7×10^{11}
Undecanal	1.5	5×10^{10}
Dodecanal	4	1.2×10^{11}
Tetradecanal	12	2×10^{11}
Hexadecanal	4	2.5×10^{11}

[a] M17 cells were grown in a complex medium to a cell density of about 10^{9} cells ml^{-1}. The culture was diluted 1:20 in artificial seawater (ASW) buffer at pH 6.5 in a final volume of 1 ml, and the luminescence in the presence of 1 μmol of different fatty acids and aldehydes (all dissolved in ethanol) was determined. The activity with tetradecanoic (myristic) acid was defined as 100.

are encountered with compounds that form complexes with fatty acids or aldehydes. For example, bovine serum albumin may drastically alter the response of the M17 cells to the lipids.

pH. The pH of the cell suspension greatly affects the sensitivity of the system to fatty acids, but not to aldehydes. M17 cells show a maximal response to fatty acids below about pH 6.5 (Fig. 2).

Other Applications

In addition to the direct determination of fatty acids (especially myristic) and aldehydes, the bioassay may be easily linked to any enzymic or hydrolytic system where fatty acid or aldehyde are liberated.

Examples include the determination of lipase and phospholipase activity. Two commercially available (Sigma) substrates may be used to determine lipase and phospholipase activity: trimyristin and L-α-phosphatidylcholine dimyristoyl. It is recommended to purify these substrates before use to assure that no residual fatty acids are present. Since some of the suggested enzymic assays for lipase and phospholipase contain organic solvents, care should also be taken to remove these before use in the assay with luminous bacteria.

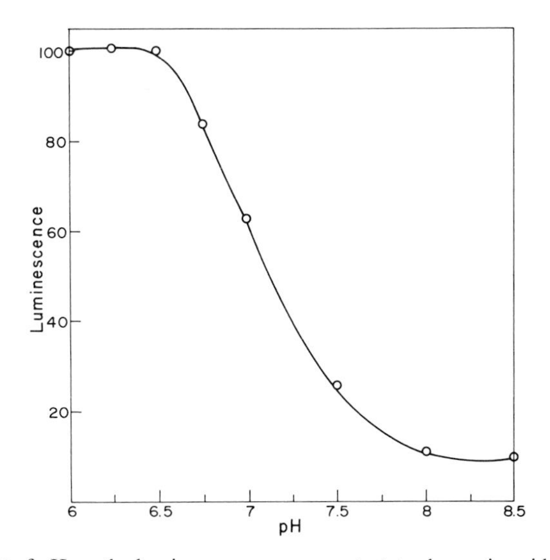

FIG. 2. Effect of pH on the luminescence response to tetradecanoic acid. M17 cells were grown in complex medium to a density of 10^9 cells ml^{-1}. The culture was centrifuged in the cold and the cells were resuspended in artificial seawater containing imidazole · HCl (50 m*M*) buffer at the different pH values. The maximal luminescence in the presence of 1 μM tetradecanoic acid was recorded. Ordinate, luminescence in relative units; abscissa, pH.

To test lipase or phospholipase activity, the enzymes are incubated with the proper substrate under appropriate conditions; aliquots are taken with time and boiled to inactivate the enzyme, and the liberated myristic acid is assayed. Alternatively, the substrate and the lipase or the phospholipase can be incubated together with the M17 cells. Since as little as 10 pmol of myristic acid can be detected, low enzymic activity can be readily followed within short time periods. Moreover, it is possible to use substrates where myristic acid is linked only to one of the glycerol positions (e.g., the β position), while nonactive fatty acids, such as stearic acid, are linked to the α and α' positions. Using such substrates, one can determine the specificity of the lipase for the α' or β' positions.

When testing lipase or phospholipase activity, one must be alert to the possibility that there may be inhibitory unsaturated fatty acids coming from contaminating triglycerides. One way to control for this is to incubate the lipase or phospholipase without the addition of the myristic acid-containing substrate. Samples are then taken with time from the assay mixture and tested for their effect on the response of M17 cells to known concentrations of myristic acid. When no unsaturated fatty acids (or other inhibitors) are present, the response of the M17 cells will be the same in the presence and in the absence of the tested sample.

A bioassay for antilipogenic compounds can also be carried out with this system. Antilipogenic compounds, such as cerulenin[6] and C-55,[7] inhibit bioluminescence,[8] providing the basis for the assay.

[6] A. Omura, *Bacteriol. Rev.* **40,** 681 (1976).

[7] T. Ohno, J. Awaya, T. Kesado, S. Nomura, and S. Omura. *Antimicrob. Agents Chemother.* **6,** 387 (1974).

[8] S. Ulitzur and I. Goldberg. *Antimicrob. Agents Chemother.* **12,** 308 (1977).

[21] The Luciferase Reduced Flavin Hydroperoxide Intermediate

By J. W. HASTINGS and JAMES E. BECVAR

$$FMNH_2 + E + O_2 \longrightarrow E\text{-}FMNH\text{-}OOH$$

$$E\text{-}FMNH\text{-}OOH + RCHO \longrightarrow cq\,(h\nu) + E + FMN + RCOOH + H_2O$$

$$E\text{-}FMNH\text{-}OOH \xrightarrow{\text{spontaneous}} E + FMN + H_2O_2$$

where E = bacterial luciferase; $FMNH_2$ = reduced riboflavin 5′-phosphate; E-FMNH-OOH = the luciferase-bound 4a-peroxy-reduced flavin; cq = bioluminescent quantum yield.

ISBN 0-12-181957-4

Principles

First, the turnover number for luciferase in the presence of saturating amounts of all substances is very low; i.e., the time required for a single catalytic cycle is long,[1,2] namely, 2–20 sec at 20° depending on the luciferase and the chain length of aldehyde used.[3,4] Because of a sufficiently positive enthalpy of activation for the rate-limiting step preceding light emission, turnover (emission) can be effectively prevented by lowering the temperature.[5]

Second, the reaction of oxygen with luciferase-bound reduced flavin occurs in the absence of aldehyde to produce the same intermediate as is involved in the light reaction.[1,6] This bound peroxy reduced flavin intermediate decomposes spontaneously at a relatively slow rate, having a half-life (*Beneckea harveyi*) of about 10 sec at 20°. The high enthalpy of activation for the spontaneous decomposition reaction has the import that, by lowering the temperature (if only to 0°), the intermediate is sufficiently stabilized to allow its chromatographic isolation. In this way, one can obtain an essentially homogeneous solution of enzyme molecules, each of which is poised at the same state in catalysis, a true chemical intermediate in the overall reaction. During the slow spontaneous decomposition, the amount of intermediate remaining at any time can be assayed by mixing an aliquot of the sample with aldehyde and simultaneously increasing temperature.[2]

Materials

Luciferase purified from *Beneckea harveyi*, *Photobacterium fischeri*, or *P. phosphoreum*[7,8]
Sephadex G-25 column (1 × 18 cm)
NaCl, 0.5 *M*, in 0.3 *M* phosphate buffer, pH 7.0 (elution buffer)
Phosphate buffer, 0.1 *M*, pH 7.0 (assay buffer)
Long-chain aldehyde; decanal, dodecanal, or tetradecanal are satis-

[1] J. W. Hastings and Q. H. Gibson, *J. Biol. Chem.* **238,** 2537 (1963).
[2] J. W. Hastings, C. Balny, C. LePeuch, and P. Douzou, *Proc. Natl. Acad. Sci. U.S.A.* **70,** 3468 (1973).
[3] J. W. Hastings, J. A. Spudich, and G. Malnic, *J. Biol. Chem.* **238,** 3100 (1963).
[4] J. W. Hastings, K. Weber, J. Friedland, A. Eberhard, G. W. Mitchell, and A. Gunsalus, *Biochemistry* **8,** 4681 (1969).
[5] J. W. Hastings, Q. H. Gibson, and C. Greenwood, *Proc. Natl. Acad. Sci. U.S.A.* **52,** 1529 (1964).
[6] J. W. Hastings, *Annu. Rev. Biochem.* **37,** 597 (1968).
[7] T. O. Baldwin, M. Z. Nicoli, J. E. Becvar, and J. W. Hastings, *J. Biol. Chem.* **250,** 2763 (1975).
[8] J. W. Hastings, T. O. Baldwin, and M. Z. Nicoli, this volume [14].

factory (10 ml of aldehyde in 10 ml of water dispersed by sonication)
Sodium hydrosulfite (solid)
Flavin mononucleotide, 5 m*M*

Procedure[9]

Isolation procedures can be carried out in a walk-in cold room with cold baths and jackets to maintain temperatures at 0°. A small (1 × 18 cm) Sephadex G-25 column is preequilibrated at 0° with buffer containing 0.5 *M* NaCl in 0.3 *M* phosphate buffer, pH 7.0. A flow rate of between 1 and 2 ml/min is used. To 1 ml of buffer containing about 1 mg of luciferase, an excess (about 2-fold) of FMN is added and then just enough solid sodium hydrosulfite to bleach the solution. This anaerobic mixture is promptly applied to the column and chromatographed, whereupon the enzyme-containing fraction migrates ahead of the hydrosulfite and reaction with oxygen occurs. The intermediate elutes with the void volume, well separated from free low-molecular-weight species such as free oxidized flavin, H_2O_2, and dithionite oxidation products. Larger quantities of enzyme (e.g., 10 mg) can be used.

Chromatography may also be carried out at lower temperatures, down to about −30°, by using a buffer composed of 50:50 ethylene glycol–phosphate buffer.[2]

Assay

The activity is measured by rapidly mixing small aliquots of the intermediate with long-chain aldehyde in buffer at 25°. A sample (0.1 ml) is taken up in a cooled syringe and injected into buffer (1.8 ml, 0.1 *M* phosphate, pH 7) containing aldehyde appropriate in amount and chain length for the luciferase species used (e.g., for *P. fischeri* intermediate, 0.1 ml of 0.1% sonicate of decanal, for final assay concentration of 0.25 m*M*).

Properties

Composition, Stability, and Reactivity. The intermediate as isolated contains about 1 mol of bound flavin chromophore per mole of enzyme.[2] In the absence of aldehyde, the intermediate decomposes to form equimolar amounts of FMN, H_2O_2, and free luciferase at a rate that depends not only on the temperature, but also on the buffer, ionic strength, and type of

[9] J. E. Becvar, S.-C. Tu, and J. W. Hastings, *Biochemistry* **17,** 1807 (1978).

luciferase (species of bacteria).[9,10] At 0°, the half-life at high ionic strength can be longer than 30–40 min for the intermediate prepared from *B. harveyi* luciferase; for the *P. fischeri* intermediate, the half-life can be more than an hour. The half-life is shorter at lower ionic strength. For later use, samples may be stored at lower temperatures, either by freezing, or in the liquid state by mixing with ethylene glycol.

In the presence of aldehyde, the intermediate is less stable at lower temperatures and will emit light even below 0°. The products of the reaction in the presence of aldehyde are FMN, RCOOH, H_2O, and light. An important point is that dissolved oxygen is not required for, nor does it affect the rate of, reaction of the peroxy intermediate either with or without aldehyde.[9]

Absorption Spectra. With all luciferases, the spectrum for the isolated intermediate at wavelengths longer than 300 nm exhibits a single absorbance peak in the 370 nm region ($\epsilon \sim 1 \times 10^4 M^{-1} cm^{-1}$). Incubation in the absence of aldehyde results in the slow loss of potential for luminescence and at the same time a dramatic absorbance increase in the 450 nm region, producing a spectrum resembling that of oxidized flavin. Three clear isosbestic points occur during this reaction (approximately 329, 368, and 401 nm for the *B. harveyi* intermediate and 333, 382, and 404 nm for *P. fischeri*).[11] The product has been identified as FMN.[13]

Decomposition Kinetics and Light Yield. The loss in luminescence potential and increase in 450 nm absorbance are processes obeying first-order kinetics. Although the rate constant describing both processes is dependent on the temperature, buffer, ionic strength, and type of luciferase used, both kinetic processes occur in parallel at the same rate under all conditions so far examined.

By comparing the bioluminescence capacity remaining in the sample with the amount of flavin chromophore still remaining in the intermediate form, quantum yields of 0.19 and 0.27 per flavin have been reported for *B. harveyi* and *P. fischeri* intermediates, respectively.[9]

Fluorescence Spectra. The intermediate of *B. harveyi* exhibits fluorescence emission in the 500 nm region which increases dramatically (about 5-fold) upon irradiation, with an ultimate fluorescence quantum yield equal at least to that of free FMN.[13] Similar fluorescence properties were not evident with the intermediate of *P. phosphoreum*.[14]

[10] J. W. Hastings and C. Balny, *J. Biol. Chem.* **250,** 7288 (1975).

[11] The exact values for the isosbestic points with *B. harveyi* reported in this publication[9] are not in agreement with other observations.[2,12] The reason for this discrepancy is not clear.

[12] J. W. Hastings, *Energy Transform. Biol. Systems, Ciba Found. Symp. 1975.*

[13] C. Balny and J. W. Hastings, *Biochemistry* **14,** 4719 (1975).

[14] N. Ashizawa, T. Nakamura, and T. Watanabe, *J. Biochem.* **81,** 1057 (1977).

[22] Bacterial Luciferase as a Generalized Substrate for the Assay of Proteases

By THOMAS O. BALDWIN

In the past few years, the importance and diversity of protease-mediated cellular control processes have become commonly recognized. As a result, interest in a rapid, sensitive, and generally applicable method for the detection and assay of protease activities has risen steadily. Bacterial luciferase, under nondenaturing conditions, is highly susceptible to inactivation by a wide diversity of protease activities.[1] This protease sensitivity has been exploited as a method for detection of protease activities.[2]

It is well known that most peptide bonds in denatured proteins are hydrolyzed by proteases at a much faster rate than the same bonds in the conformationally more defined native protein. This fact is the basic principle underlying most methods commonly used for the assay of proteolytic activities. Certain proteases, however, serve such highly specific functions that the only way known to detect the activity is to supply the natural substrate. An elegant example, described by P. Setlow, is the high-molecular-weight protease within spores of *Bacillus megaterium* which, during outgrowth, hydrolyzes a peptide bond defined by a five-residue sequence in storage polypeptides.[3,4] The specificity of this protease is so great that its activity can be monitored only by supplying the natural substrates, the storage polypeptides. While the realization that proteases of such high specificity do exist will virtually certainly lead to the discovery of more such enzymes, it is doubtful that such rigid specificity requirements will prove to be commonplace. Rather, it appears likely that many, if not most, proteolytic enzymes will hydrolyze a variety of peptide bonds and that a generalized assay method that furnishes a variety of susceptible bonds may be utilized to detect these enzymes. However, failure of a generalized assay to detect proteolytic activity in a given situation cannot be taken as a demonstration of the absence of proteases. Furthermore, detection of a proteolytic activity by a generalized assay cannot be taken as proof that the activity is responsible for any particular biological phenomenon. Such a correlation, rather, requires support from studies such as mutant analysis and/or chemical modification.

[1] T. O. Baldwin, *Fed. Proc., Fed. Am. Soc. Exp. Biol.* **33,** 1441 (1974).
[2] D. Njus, T. O. Baldwin, and J. W. Hastings, *Anal. Biochem.* **61,** 280 (1974).
[3] P. Setlow, personal communication (1977).
[4] P. Setlow, *J. Biol. Chem.* **251,** 7853 (1976).

ISBN 0-12-181957-4

Commonly Used Protease Assays

A wide variety of protease assays are currently available, each with particular advantages and shortcomings. Most of these methods employ a denatured protein as substrate. Denatured proteins exist in many conformations and are therefore subject to proteolytic attack in a vast number of positions not accessible in the native conformation. Because proteolysis of a denatured substrate involves cleavage at many different sites, protease assays utilizing denatured substrates should be little affected by substrate conformational changes resulting from changes in pH, ionic environment, etc.

General protease assays using protein substrates customarily measure an increase in trichloroacetic acid-soluble material[5,6] or the solubilization of dyes attached to insoluble proteins.[7] All these methods suffer from the inconvenience of a centrifugation step required to resolve the substrate from the products of proteolysis. Furthermore, detection of solubilized material by spectroscopic methods (including ninhydrin, fluorescamine, and ultraviolet absorbance) may be hampered by contaminating substances in crude extracts that either themselves absorb or fluoresce, or react with the chromogenic reagent to yield a false positive reaction.

Bacterial Luciferase as a Protease Substrate

The use of bacterial luciferase to assay for the action of proteases has several advantages. The activity of the luciferase may easily be assayed at concentrations in the picomolar range. The assay is not hindered by contaminating materials in crude extracts, unless of course the absorbance at 490 nm in the luciferase assay mixture is above $\sim 2\ cm^{-1}$. (The luciferase assay involves detection of light emitted around 490 nm; see this volume [14]). The method does not entail the use of radioisotopes with accompanying problems. And, above all else, the assay is rapid and simple.

The use of the luciferase method is limited to assay conditions under which the luciferase activity is stable, i.e., that the luciferase is not inactivated in the absence of protease. Furthermore, conditions that protect the luciferase from proteolytic attack must be avoided. These include in particular the use of high concentrations of phosphate, pyrophosphate, sulfate, citrate, arsenate, chloride, and perhaps a number of other anions

[5] M. L. Anson, *J. Gen. Physiol.* **22,** 79 (1938).
[6] E. H. Reimerdes and H. Klostermeyer, this series, Vol. 45, p. 26.
[7] G. L. Moore, *Anal. Biochem.* **32,** 122 (1969).

known to bind to the luciferase and stabilize it to such agents as heat, urea, and proteases.[8,9]

Methods

Assay of Bulk Solution

Reagents

Bacterial luciferase (*Beneckea harveyi*): 1 mg/ml in 50 m*M* phosphate, 0.2 *M* NaCl, 0.5 m*M* DTT, pH 7.0

Assay buffer: 20m*M* phosphate, 0.2% BSA, pH 7.0

$FMNH_2$: Prepare by bubbling H_2 through a solution of FMN (50 μM) containing a small amount of platinized asbestos in a round-bottom flask plugged with a serum stopper.

n-Decyl aldehyde: Prepare by dispersing 10 μl of *n*-decyl aldehyde in 10 ml of deionized water with a probe sonicator.

To 100 μl, or some other convenient volume of the solution suspected of containing protease activity, add 10 μl of the luciferase solution and incubate at 25° for 10 min. Withdraw 5 μl of the solution, dilute into 1 ml of the assay buffer in a scintillation vial, add 15 μl of the decanal suspension, place the vial in the photomultiplier housing, and inject 1 ml of the $FMNH_2$ solution through a 19-gauge hypodermic needle with sufficient force to give thorough mixing. Compare the initial maximum light level with two control samples. The first control consists of luciferase added to a boiled aliquot of the test sample, and the second control consists of luciferase diluted into buffer without test sample, both allowed to incubate for the same period of time. Since the luciferase assay is reproducible within $\pm 10\%$, several assays should be performed within a time period short compared with the time of incubation.

Assay of Column Fractions

Column fractions may be assayed either by withdrawing aliquots from individual tubes and assaying as above, or by diluting 10 μl of luciferase directly into the tubes to be assayed (containing as much as 20 ml of solution in each tube). After a designated period of incubation, 10 μl is withdrawn from each tube and assayed for luciferase activity. Each tube assayed should yield the same level of luciferase activity, except the tubes containing the protease activity. In the tubes containing protease, the remaining luciferase is inversely proportional to the concentration of protease added. For most purposes, the amount of luciferase added (10 μg in 20

[8] P. L. Riley, Senior Thesis, Department of Biochemistry, University of Illinois, Urbana, 1977.

[9] P. L. Riley and T. O. Baldwin, manuscript in preparation (1978).

ml in this example) is negligible and should not interfere with further experiments.

Biochemistry of the Assay

Bacterial luciferase from *B. harveyi* is a heterodimer ($\alpha\beta$) in which the catalytic center is confined to the α subunit.[10,11] With all proteases utilized in studies of the structure of the luciferase,[2] the fate of the enzyme has been the same. The pseudo-first-order rate of the inactivation is directly proportional to the concentration of protease at constant initial luciferase concentration. Gel electrophoresis in sodium dodecyl sulfate (SDS) of the products of digestion with trypsin, chymotrypsin, and subtilisin shows that only bonds of the α subunit are hydrolyzed during the time of these experiments; the rate of loss of α subunit is the same as the rate of loss of light-producing activity. The action of the proteases produces a band with mobility indicative of a molecular weight of about 28,000 and smaller fragments. High speed sedimentation equilibrium analysis in the Beckman Model E ultracentrifuge shows, however, that, in nondenaturing buffers, luciferase inactivated to less than 1% of its initial activity with either trypsin or chymotrypsin has essentially the same molecular weight as the native enzyme. These findings indicate the existence of a flexible region(s) on the α subunit which contains bonds that satisfy the specificity of a wide variety of proteolytic activities. Hydrolysis of bonds within this flexible stretch of the α polypeptide results in loss of light-producing capacity, but does not cause a general disintegration of the tertiary and quaternary structure of the enzyme. It has therefore been proposed that the susceptible region(s) lies close to the catalytic center of the enzyme.[8,9,12,13]

The molecular events that accompany proteolytic inactivation of the luciferase from *Photobacterium fischeri* are qualitatively the same as described above for the *B. harveyi* enzyme.[14] This finding suggests that the protease-labile region is of some importance, since it has been retained through a substantial evolutionary divergence.[15] A definitive demonstration of this proposal must await more experiments and perhaps the determination of the crystal structures of the luciferases.

Acknowledgment

This work was supported in part by the National Science Foundation (grant PCM 76-00452).

[10] E. A. Meighen, M. Z. Nicoli, and J. W. Hastings, *Biochemistry* **10,** 4069 (1971).
[11] T. W. Cline and J. W. Hastings, *Biochemistry* **11,** 3359 (1972).
[12] T. O. Baldwin, J. W. Hastings, and P. L. Riley, *J. Biol. Chem.*, in press (1978).
[13] E. A. Lerner, M. P. Frosch, M. V. Merritt, and T. O. Baldwin, *Biol. Bull.* **153,** 435 (1977).
[14] R. Pruzan, Senior thesis, Department of Biochemistry, University of Illinois, Urbana, 1977.
[15] J. L. Reichelt, P. Baumann, and L. Baumann, *Arch. Microbiol.* **110,** 101 (1976).

[23] Immobilization of Bacterial Luciferase and Oxidoreductase and Assays Using Immobilized Enzymes

By EDWARD JABLONSKI and MARLENE DELUCA

The immobilization of an enzyme generally results in increased stability and the additional property of reusability, while retaining specificity for substrates and effectors. The immobilized enzyme can be as useful an analytical tool as the soluble counterpart, with the added advantages of easier handling and decreased cost. The extreme sensitivity and specificity of the luminescent enzymes from firefly and marine bacteria for the determination of ATP, NADH, and NADPH and for monitoring reactions producing these compounds has been shown in preceding chapters of this volume.

This article describes the immobilization of the NADH and NADPH:FMN oxidoreductases and luciferase from *Beneckea harveyi* onto arylamine glass beads. These immobilized enzymes are individually active and also function to produce light via a coupled reaction utilizing NADH or NADPH.[1]

$$NAD(P)H + H^+ + FMN \rightarrow NAD(P)^+ + FMNH_2 \quad (1)$$

$$FMNH_2 + O_2 + RCHO \rightarrow FMN + RCOOH + H_2O + h\nu \quad (2)$$

The availability of highly purified oxidoreductases specific for NADH or NADPH has led to the development of an immobilized enzyme system capable of continuous monitoring of reactions producing NADH or NADPH.[2]

Preparation and Immobilization of Enzymes

Bacteria

Beneckea harveyi strain No. 392 (formerly MAV) was obtained from Dr. K. H. Nealson of the Scripps Institute of Oceanography. The cells are grown at 25° in a medium containing, per liter, the following components: 3 ml of glycerol, 30 g of NaCl, 0.1 g of $MgSO_4$, 3.7 g of Na_2HPO_4, 1.0 g of KH_2PO_4, 0.5 g of $(NH_4)_2HPO_4$, 5 g of bacto-peptone (Difco), and 3 g of

[1] E. Jablonski and M. DeLuca, *Proc. Natl. Acad. Sci. U.S.A.* **73,** 3848 (1976).
[2] E. Jablonski and M. DeLuca, *Anal. Biochem.,* in press.

ISBN 0-12-181957-4

yeast extract (Difco). The cells are harvested by centrifugation at maximum *in vivo* luminescence and may be stored as a frozen cell paste.

Enzyme Preparation

Luciferase is purified according to the method of Gunsalus-Miguel *et al.*,[3] as modified by Baldwin *et al.*[4] The oxidoreductases are separated from the luciferase during chromatography on a DEAE-Sephadex A-50 column, the last step in the purification of luciferase.[5] For the preparation of immobilized enzyme rods that are useful in determining the total amount of NADH or NADPH in a sample, the partially purified NADH and NADPH oxidoreductases obtained from the Sephadex A-50 column can be used. These fractions from the DEAE-Sephadex A-50 column are combined and concentrated by precipitation between 35% and 75% ammonium sulfate. The precipitate is then dissolved in 0.1 *M* potassium phosphate, pH 7.0, containing 1 m*M* EDTA and 0.5 m*M* DTT and dialyzed against this solution. The preparation is then stored frozen at $-20°$. This oxidoreductase preparation has a specific activity of 0.15–0.20 μmol of NADH oxidized per minute per milligram of protein at pH 7.0. These partially purified enzymes are not useful for the preparation of immobilized enzymes capable of continuously monitoring reactions producing NADH or NADPH. This is due to the presence of trace amounts of various pyridine-linked dehydrogenase activities, which can be immobilized along with luciferase and oxidoreductase and thus interfere in the detection of the soluble forms of these dehydrogenases or their substrates.

NADH and NADPH:FMN oxidoreductases may be obtained in a highly purified form by first separating the NADH and NADPH:FMN oxidoreductase activities as they elute from the DEAE-Sephadex A-50 column. The individual enzymes are then concentrated by ammonium sulfate precipitation and chromatographed on a Sephadex G-100 column. The final step is passage through specific affinity columns as described by Jablonski and DeLuca.[6] Luciferase from the DEAE-Sephadex A-50 column may also be further purified by chromatography on the Sephadex G-100 column followed by passage through both affinity columns to remove any various NAD(P)H-linked dehydrogenase activities. These highly purified enzymes are used to prepare immobilized enzyme rods that are spe-

[3] A. Gunsalus-Miguel, E. A. Meighen, M. Z. Nicoli, K. H. Nealson, and J. W. Hastings, *J. Biol. Chem.* **247,** 398 (1972).

[4] T. O. Baldwin, M. Z. Nicoli, J. E. Becvar, and J. W. Hastings, *J. Biol. Chem.* **250,** 2763 (1975).

[5] E. Gerlo and J. Charlier, *Eur. J. Biochem.* **57,** 461 (1975).

[6] E. Jablonski and M. DeLuca, *Biochemistry* **16,** 2932 (1977).

cific for either NADH or NADPH and are useful for continuously monitoring reactions producing NADH or NADPH.

Immobilization of Bacterial Luciferase and Oxidoreductase

Immobilization Procedure. P-arylamine/CPG-550A was manufactured by Corning Biological Products, Medfield, Massachusetts and purchased from Pierce Chemical Company. The arylamine glass beads are cemented to a plain glass capillary tube 1.7 mm in diameter and 4.0 cm long, which has been sealed at both ends. The capillary tube is first coated with a thin layer of Duro-E-PoxE5 cement and gently rolled in a weighed amount of glass beads. A 1/4-inch length on one end of the capillary tube should be left free of glue or glass beads to serve as a handle. The capillary tube with the glass beads is placed upright and allowed to dry overnight. The enzyme can then be coupled to the beads by the diazotization procedure of Weetall and Filbert.[7] The diazotization process is carried out at 0°–4°. Eight to ten rods are placed in a glass vial and covered with 10 ml of 2 *N* HCl; then 250 mg of solid $NaNO_2$ are added. This reaction mixture is then placed in a desiccator and evacuated for 20 min to remove trapped gas bubbles. After the reaction is completed, the activated rods are rinsed with 500 ml of cold distilled water containing 1% w/v sulfamic acid and immersed separately in 0.5 ml of an enzyme solution for 6 hr.

All enzymes to be immobilized should be dialyzed for 16 hr against 0.1 *M* potassium phosphate, pH 7.0, in order to remove DTT, which is a necessary component for maintaining enzyme activity during purification and storage. Failure to remove DTT will reduce the enzyme activity recovered upon immobilization. The less-pure luciferase and oxidoreductases are mixed in a ratio of 2 mg of protein of luciferase preparation to 3 mg of protein of oxidoreductase preparation in 1 ml of 0.1 *M* potassium phosphate, pH 7.0.[1] The highly purified enzymes are combined in a ratio of 1 mg/ml luciferase to 0.03–0.05 mg/ml of either NADH:FMN oxidoreductase or NADPH:FMN oxidoreductase, and brought to a final concentration of 4 mg of protein per milliliter by the addition of bovine serum albumin (BSA).[2] This enzyme solution is also buffered by 0.1 *M* potassium phosphate, pH 7.0.

After coupling, the bound enzyme rods are rinsed with cold 1 *M* NaCl followed by cold 0.1 *M* potassium phosphate, pH 7.0, to remove any noncovalently bound enzyme. The immobilized rods are then incubated in 0.1 *M* potassium phosphate, pH 7.0, containing 1 m*M* DTT and 1% w/v BSA for 24 hr at 0°–4°. The rods may then be stored in 0.1 *M* potassium

[7] H. H. Weetall and A. M. Filbert, this series, Vol. 34, p. 59.

phosphate pH 7.0, containing 0.5 m*M* DTT and 15% v/v glycerol at −20°. The activity of the immobilized enzymes increases with storage at −20°, reaching maximal activity after about a week. The immobilized enzymes are then stable for several months.

Enzyme Assays. All assays are performed at 23°. Soluble luciferase is assayed by the injection of 0.1 ml of 1.5×10^{-4} M $FMNH_2$, photoreduced in the presence of 5 m*M* EDTA in 0.1 *M* potassium phosphate, pH 7.0,[8] into 0.5 ml of a solution containing luciferase, 0.0005% v/v decanal, and 0.1 *M* potassium phosphate buffer, pH 7.0. The initial peak light intensity is linear with respect to added luciferase. Immobilized luciferase is assayed with the same concentration of substrates. The rod with the immobilized enzymes is placed in the buffer solution containing decanal and $FMNH_2$ is injected. All light-emission measurements are obtained with an Aminco Chem-Glow photometer and an Aminco recorder.

The soluble oxidoreductases are assayed by monitoring the initial rate of oxidation of NADH or NADPH by the loss in absorbance at 340 nm using a Cary Model 14 recording spectrophotometer. The assay reaction is initiated by adding 0.1 ml of 2.0 m*M* NADH or NADPH in 0.1 *M* potassium phosphate, pH 7.0, to 0.9 ml of a solution containing oxidoreductase, 0.13 m*M* FMN and 0.1 *M* potassium phosphate, pH 7.0. When the immobilized oxidoreductases are assayed, the rod with the bound enzymes is dipped into a cuvette and mixed for 1-min intervals, then removed; the absorbance at 340 nm is measured. The NADH:FMN oxidoreductase and the NADPH:FMN oxidoreductase exhibit maximal activity at pH 8.5 and pH 5.6, respectively.[9] Assays are performed at pH 7.0 in order to maintain the stability of the immobilized enzymes. The soluble oxidoreductases are assayed at pH 7.0 to compare the activities of the soluble and immobilized forms.

The soluble oxidoreductases or luciferase may be assayed in a coupled reaction. The maximum initial light intensity is measured upon the addition of 0.1 ml of NADH or NADPH into 0.5 ml of 0.1 *M* potassium phosphate, pH 7.0, containing 2.3 μ*M* FMN, 0.0005% v/v decanal, luciferase, and oxidoreductase. When NADH or NADPH is added in saturating amounts of 0.2 m*M* initial concentration, light intensity is proportional to the product of the concentrations of the two enzymes.[10] Otherwise light intensity is linear with reduced pyridine nucleotide concentration. When the immobilized enzymes are being used to assay NADH or NADPH, the rod is immersed into the solution containing FMN and aldehyde, and NADH or

[8] J. Lee, *Biochemistry* **11,** 3350 (1972).

[9] E. Jablonski and M. DeLuca, *Biochemistry,* **17,** 672 (1978).

[10] J. W. Hastings, W. H. Riley, and J. Massa, *J. Biol. Chem.* **240,** 1473 (1965).

NADPH is injected directly into the solution. Activity is expressed in arbitrary light units.

The continuous monitoring of a reaction producing NADH or NADPH using the immobilized luminescent enzyme rods is performed on an Aminco Chem-Glow photometer with an Aminco integrator-timer. A given enzyme or substrate is assayed by monitoring the rate of increasing light intensity produced by the immobilized luminescent enzymes. This is achieved by either calculating a slope of light intensity vs time or determining the total units of light produced in a constant time period by integration. In practice the latter method has been found to be faster and more accurate, eliminating the error involved in tracing out a slope produced by a recorder. For a typical assay, light production is initiated by immersing the immobilized enzyme rod into 0.5 ml of a mixture containing all the necessary components at optimal concentrations except for the enzyme or substrate to be assayed. Light output is integrated for a set period of 60 sec. All assays of this type are performed at pH 7.0, near the optimal pH for the coupled reaction of oxidoreductase and luciferase. The rod is removed, rinsed several times with 0.1 *M* potassium phosphate, pH 7.0, containing 0.5 m*M* DTT at room temperature, and reused.

Results

Assay of NADH and NADPH using Partially Purified Immobilized Enzymes

Table I gives typical results for the immobilization of oxidoreductase and luciferase obtained from the DEAE-Sephadex column without further purification.[1] About 1.3 mg of total protein are bound per rod, to which 10–15 mg of arylamine glass beads had been cemented. It appears that luciferase is largely inactivated by immobilization, retaining only 0.05% of its activity. Oxidoreductase, as measured by NADH oxidation, retains 10% of its activity upon immobilization, while 3% of the light-emitting activity is retained for the coupled assay, using saturating concentrations of all substrates. In all three cases, greater than 50% of the activity was removed from the original mixture.

A single rod of this type has been used for over 100 consecutive assays of either NADH or NADPH without any loss of activity. After the completion of an assay, the immobilized enzyme rod is removed from the assay solution and rinsed gently with 25 ml of 0.1 *M* potassium phosphate, pH 7.0, with 0.5 m*M* DTT at room temperature. The rod is then left in this rinse buffer until the next assay is ready to be performed. We have kept immobilized enzyme rods for several months, stored in phosphate buffer with

TABLE I
BINDING OF LUCIFERASE AND OXIDOREDUCTASE TO GLASS RODS[a]

Enzyme sample assayed	Luciferase (relative light units/ml)	Oxidoreductase (μmol of NADH oxidized per ml/min)	Coupled assay (relative light units/ml)	Protein (mg)
Original mixture	7.0×10^6	0.29	4.2×10^5	2.56
Supernatant	2.0×10^6	0.10	2.0×10^4	1.25
Rods[b]	2.5×10^3 (0.05)	0.02 (10.3)	1.2×10^4 (3.0)	1.31 (51)

[a] Enzymic activities of a mixture of soluble luciferase and oxidoreductase prior to coupling to the beads (original mixture). After the coupling procedure, the mixture was again assayed (supernatant). The amount of activity associated with the rods was also assayed. Luciferase was assayed by injection of $FMNH_2$ into the reaction mixture. Oxidoreductase was assayed by disappearance of absorbance at 340 nm, and the coupled assay is the light intensity obtained upon injection of NADH into the reaction.

[b] The percentage of activity recovered on the rods was calculated by dividing the activity found on the rods by the activity lost from the original mixture. The results are given in parentheses.

DTT and 15% glycerol at −20° with no loss of activity, indicating that the enzymes do not leach off of the arylamine glass beads and are covalently bound.

The immobilized luminescent enzymes are sensitive to NADH or NADPH concentration. With a single rod and repetitive injections of NADH or NADPH, the peak light intensity is reproducible within a variation of ±3%. Because no precautions are taken in placing the rod in the test tube in any fixed geometry, apparently the photon detection is not affected by small changes in the position of the rod in front of the phototube.

Figure 1 shows the initial peak light intensity obtained in response to varying concentrations of NADH and NADPH using the immobilized enzyme rod.[1] The intensity of light emission is linear with NADH concentration from 10 n*M* up to 0.5 m*M*, after which the light intensity begins to reach a constant level owing to saturation with NADH. These concentrations of NADH correspond to between 1 pmol and 50 nmol of NADH in the assay solution. The corresponding curve for NADPH is 10-fold higher in this case. The range of linearity is 0.1 μM to 2 m*M*. Since there are two separate oxidoreductases in *B. harveyi*, the limit of detection of NADPH may be lowered by increasing the ratio of the NADPH-specific enzyme to the NADH-specific enzyme. This may be achieved by utilizing fractions containing predominantly the NADPH:FMN oxidoreductase eluted from the DEAE-Sephadex A-50 column.

The immobilized enzymes exhibit properties similar to the corresponding soluble forms in pH and substrate concentration optima. The coupled

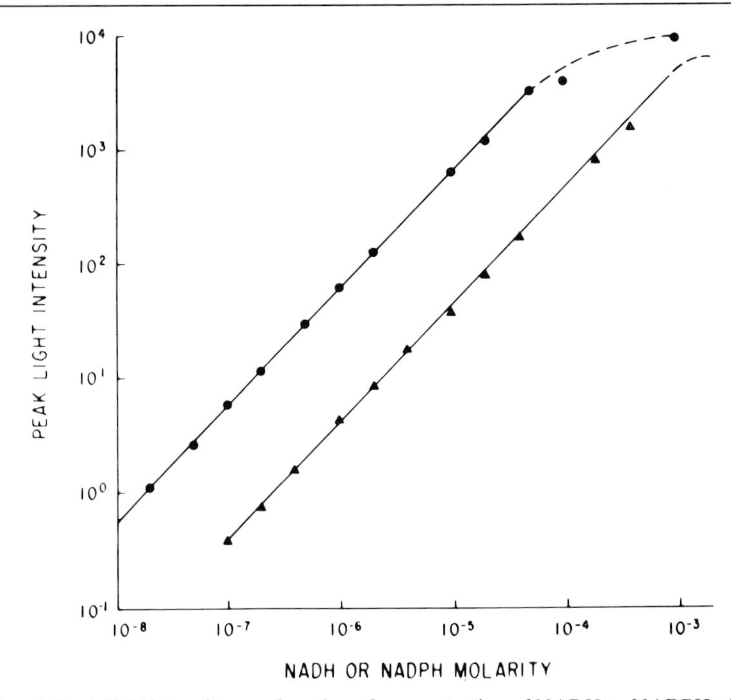

FIG. 1. Peak light intensity as a function of concentration of NADH or NADPH. A single rod with immobilized enzymes was used to test each substrate. ●——●, NADH; ▲——▲, NADPH.

reaction exhibits optimal activity at pH 6.5 with a shoulder at pH 7.5 when using NADH as a substrate. We routinely assay at pH 7.0 because the immobilized enzymes appear to be most stable at this pH.

The immobilized enzymes exhibit a property uncharacteristic of the soluble enzymes, which appears to be a diffusion-limited process for light emission in the coupled reaction. In a typical assay for NADH or NADPH using soluble luciferase and oxidoreductase, if one injects saturating NADH or NADPH, the light intensity rises to a peak level and remains constant until NADH is oxidized to a less than saturating concentration. The intensity then starts to decrease owing to a limiting amount of NADH or NADPH. Figure 2 shows the time course of light emission from the rods after the injection of 1 μM NADH.[1] The light intensity rises to a peak, then decays to a base line level. If the tube containing the rod and substrate is mixed by inverting it once, there is another rise in light intensity followed again by decay to base line. This mixing can be repeated many times until the NADH has been completely utilized. If the chamber containing the test tube and phototube is mechanically vibrated after the first addition of

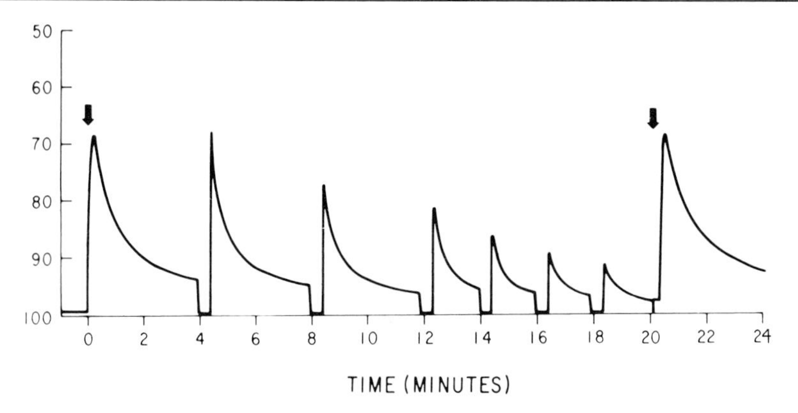

FIG. 2. Time course of light production from immobilized enzymes. At the first arrow, 1 μM NADH was injected into assay solution. Each successive peak occurred after the assay tube had been removed from the photometer, mixed by inverting it once, and replaced in the photometer. At the second arrow, another aliquot of 1 μM NADH was injected into the assay solution.

NADH, the decay in light intensity is similar to that of the soluble forms of the enzymes. The most logical explanation is that the diffusion of some or all components of the reaction is the rate-limiting process for the bound enzymes. Similar results have been obtained with other immobilized enzymes.[11-13]

It has been determined that exposure of the immobilized enzymes to concentrations of NADH or NADPH greater than 0.1 mM results in the emission of light from the rods for a long period of time after the removal of NADH or NADPH. This is due to the retention of these substrates on or within the porous glass supports. The rods must be washed extensively and incubated in buffer and DTT for up to an hour before residual NADH or NADPH is utilized, and the background light emission is reduced to an acceptable level for another assay to be performed.

Continuous Monitoring of Reactions Producing NADH or NADPH Using Purified Enzymes

Highly purified luciferase and oxidoreductases are necessary for the preparation of immobilized enzyme rods that are useful in assaying enzymes or their substrates that are involved in producing reduced pyridine nucleotides. These purified immobilized enzyme rods emit light at a low

[11] A. K. Sharp, G. Kay, and M. D. Lilly, *Biotechnol. Bioeng.* **11,** 363 (1969).
[12] R. Goldman, O. Kedem, and E. Katchalski, *Biochemistry* **10,** 165 (1971).
[13] G. Brown, D. Thomas, and C. Guillon, *J. Cell. Comp. Physiol.* **36,** 105 (1969).

intensity in the absence of the soluble pyridine-linked dehydrogenase to be assayed. This is due to trace amounts of pyridine-linked dehydrogenases present in the luciferase or oxidoreductase preparations. These dehydrogenases have been immobilized onto the arylamine glass supports along with luciferase and oxidoreductase. Plots of total light vs enzyme or substrate concentration do not go through the origin because of this low-intensity background emission. The retention of soluble substrates or enzymes from the previous assay also contributes to the background emission, but this can be reduced to a constant level by washing in buffer and DTT. Each of these immobilized enzyme rods has retained constant activity for a minimum of 50 assays over a week's time.

Enzyme Assays Using Immobilized Enzyme Rods

Lactate Dehydrogenase. LDH produces NADH according to the following reaction:

$$\text{L-Lactate} + \text{NAD}^+ \rightleftharpoons \text{pyruvate} + \text{NADH} + \text{H}^+ \tag{3}$$

Figure 3 gives the total initial light output for 60 sec vs picomoles of soluble LDH using an immobilized NADH:FMN oxidoreductase-luciferase rod.[2] The assay for LDH is linear with light output in the range of 0.003 to 0.70 pmol. The molecular weight of rabbit muscle LDH is 140,000. The LDH used in this study had an activity of 8.4×10^2 IU/mg.

Glucose-6-phosphate Dehydrogenase. GPDH produces NADPH according to the following reaction:

$$\text{D-Glucose 6-phosphate} + \text{NADP} \rightleftharpoons \text{D-6-phosphogluconolactone} + \text{NADPH} + \text{H}^+ \tag{4}$$

Figure 4 gives the total initial light output for 60 sec vs picomoles of soluble GPDH using immobilized NADPH:FMN oxidoreductase and luciferase.[2] The assay is linear with light output for soluble GPDH in the range of 0.0015 to 0.10 pmol. The molecular weight of GPDH is 102,000. The activity of the GPDH used in this study was 2.6×10^2 UI/mg.

In addition, low levels of soluble malate dehydrogenase, alcohol dehydrogenase, and hexokinase have been determined using these immobilized enzyme rods.[2]

Substrate Assays with Immobilized Enzyme Rods

Glucose. D-Glucose leads to the formation of NADPH through the reaction hexokinase coupled to GPDH. Figure 5 shows the total light emitted over 60-sec intervals from 0 to 5 min as a function of picomoles of glucose using immobilized luciferase and NADPH:FMN oxidoreductase

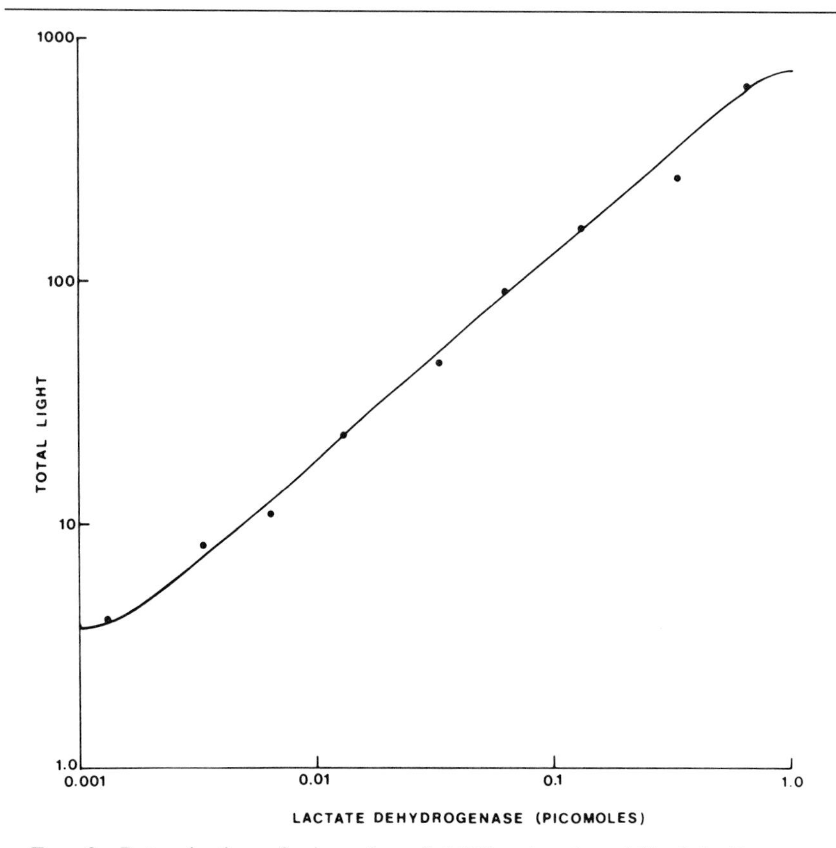

FIG. 3. Determination of picomoles of LDH using immobilized luciferase and NADH:FMN oxidoreductase. Final concentrations of the soluble components were: 20 μM decanal, 2.3 μM FMN, 16 mM L-lactate, 0.31 mM NAD^+, and soluble LDH, all in 0.1 M potassium phosphate, pH 7.0.

and soluble hexokinase and GPDH.[2] It is apparent that an incubation period is not necessary to achieve maximum sensitivity. The high enzyme concentrations eliminate any lag in NADPH production from glucose. The assay for glucose is linear with light output from 150 to 15,000 pmol.

Alcohol. Ethanol leads to the production of NADH through the reaction of alcohol dehydrogenase.

$$\text{Ethanol} + NAD^+ \rightleftharpoons \text{acetaldehyde} + \text{NADH} + H^+ \quad (5)$$

In order to obtain an immobilized enzyme rod that would emit light in response to ethanol without the addition of soluble ADH, ADH was immobilized along with luciferase and NADH:FMN oxidoreductase on the arylamine glass supports. The immobilized enzymes were prepared as

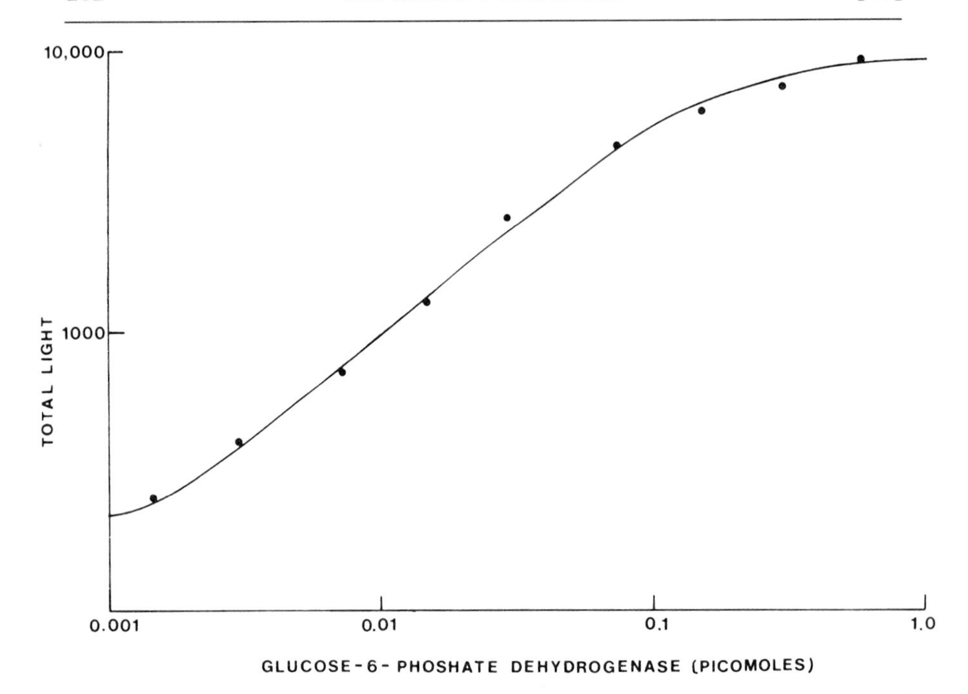

FIG. 4. Determination of picomoles of GPDH using immobilized luciferase and NADPH:FMN oxidoreductase. Final concentrations of the soluble components were: 20 μM decanal, 12 μM FMN, 3.3 mM D-glucose 6-phosphate, 0.2 mM $NADP^+$, and soluble GPDH, all in 0.1 M potassium phosphate, pH 7.0.

described in the Preparation section above with the addition of ADH to 6 μg/ml in the coupling mixture.

Table II shows the percentage recovery of activity for three enzymes upon immobilization.[2] The individual activities of the immobilized enzymes were determined as described earlier. Comparisons of the soluble and immobilized ADH in conjunction with the immobilized NADH:FMN oxidoreductase and luciferase indicated that the immobilized ADH was more stable and enabled faster and more accurate determination of ethanol over a wider range of concentrations. Figure 6 shows the total initial light output for 60 sec as a function of percentage of ethanol using the immobilized luciferase–NADH:FMN oxidoreductase–ADH rods.[2] The assay is linear with light output for ethanol concentrations of 0.0004% to 0.015%.

Comments

With use the immobilized enzymes will slowly lose activity, dependent upon the system and conditions of the assays. Washing the rods exten-

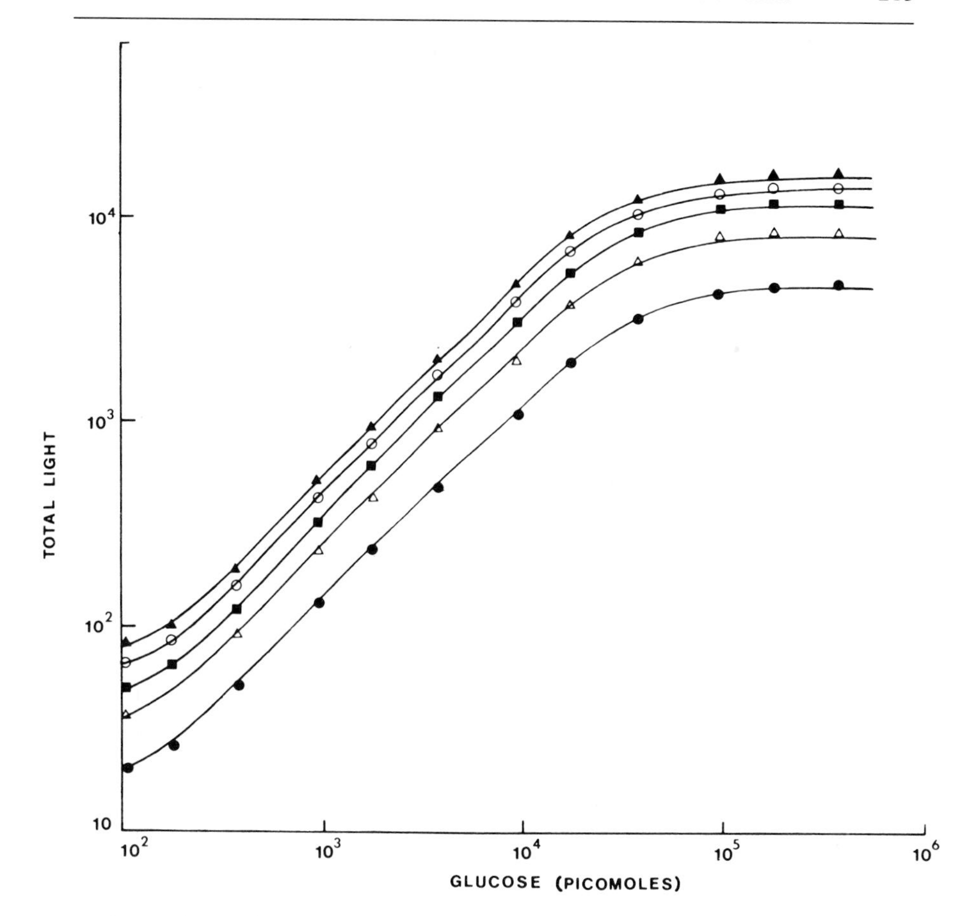

FIG. 5. Determination of picomoles of glucose using immobilized luciferase and NADPH:FMN oxidoreductase and soluble hexokinase and GPDH. Final concentrations of the soluble components were: 20 μM decanal, 12 μM FMN, 0.58 mM ATP, 12 mM $MgCl_2$, 0.22 mM $NADP^+$, 1.4 IU/ml hexokinase, and 1.4 IU/ml GPDH, all in 0.1 M potassium phosphate, pH 7.0. Total light output for 60 sec was measured after various periods of incubation: ●——●, 0 min; △——△, 1 min; ■——■, 2 min; ○——○, 3 min; ▲——▲, 4 min.

sively to remove background light emission caused by exposure to high concentrations of NADH or NADPH may shorten the lifetime of a given rod. Exposure to acidic or basic pH or high temperature for an extended period of time can also adversely affect the lifetime of the rods. Activity will also be lost through the actual removal of arylamine glass beads from rods by handling.

At this writing it is no longer possible to purchase arylamine glass beads from the Pierce Chemical Company. Underivatized glass beads may be obtained from Electro-Nucleonics Inc., Fairfield, New Jersey, and the arylamine derivative may be prepared by the method of Weetall and Filbert.[7]

TABLE II
PERCENT RECOVERY OF ACTIVITY OF PURIFIED ENZYMES UPON IMMOBILIZATION

Enzyme	Percent activity lost from coupling mixture	Percent activity found on rods	Percent activity recovered on rods[a]
NADH:FMN oxidoreductase	68	3.3	4.9
Luciferase	53	0.11	0.21
Alcohol dehydrogenase[b]	10	2.1	21.3

[a] Percent activity recovered on rods is the percent activity found on the rods divided by the percent activity lost from the coupling mixture. This is a measure of the activity an enzyme retains upon immobilization.

[b] Soluble ADH was assayed by following the production of NADH with 8 m*M* NAD and 0.35 *M* ethanol in 0.1 *M* phosphate, pH 7.0. When immobilized ADH was assayed, the rod was dipped into the assay solution in a cuvette which was mixed for 1-min intervals, then removed, and the absorbance at 340 nm was measured.

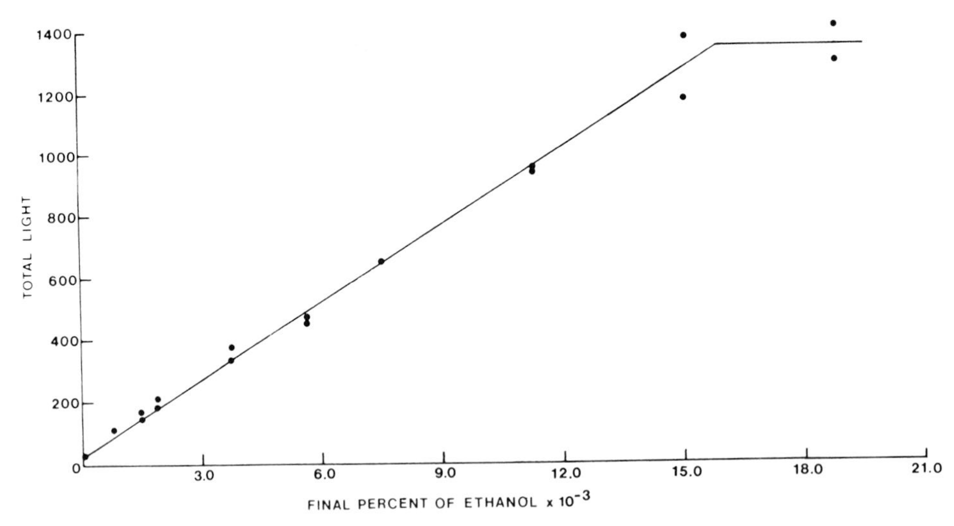

FIG. 6. Determination of final percentage of ethanol using immobilized luciferase and NADH:FMN oxidoreductase and ADH. Final concentrations of the soluble components were: 20 μM decanal, 2.3 m*M* FMN, 30 μM NAD^+, and ethanol, all in 0.1 *M* potassium phosphate, pH 7.0.

[24] Quantitation of Picomole Amounts of NADH, NADPH, and FMN Using Bacterial Luciferase

By PHILIP E. STANLEY

This procedure is of value for measuring nanomole to picomole amounts of NADH, NADPH, and FMN, and thus is suitable for the assay of amounts below those generally accessible by spectrophotometry and spectrophotofluorometry.

Purified bacterial luciferase is used for the procedure, and this is available commercially. Most preparations contain the dehydrogenase, which promotes light production some twenty times more efficiently with NADH than with NADPH. Thus the sensitivity of the assay of NADPH is only 5% of that for NADH. The ultimate sensitivity of the assays for these compounds depends on the enzyme blank or background production of light from the enzyme preparation as well as the background of the light-measuring device. A liquid scintillation spectrometer is a suitable unit for measuring the number of photons produced, as are a number of sensitive photometers marketed to measure chemiluminescence and bioluminescence.

Quantitation is effected by ensuring that the component to be measured is the only rate-limiting item of the coupled enzyme reaction.

Recent reviews and symposia concerned with analytical bioluminescence are to be found in references cited in footnotes 1–4.

Assay Principle

Many preparations of bacterial luciferase, including those commercially available, contain the two enzymes necessary for the assays. The following two equations tersely describe the reactions involved:

$$\text{NADH} + \text{FMN} \xrightarrow{\text{dehydrogenase}} \text{FMNH} + \text{NAD} \qquad (1)$$

[1] P. E. Stanley, *in* "Liquid Scintillation Counting" (M. A. Crook and P. Johnson, eds.), Vol. 3, p. 253. Heyden, London, 1974.

[2] E. Schram, *in* "Liquid Scintillation Counting: Recent Developments" (P. E. Stanley and B. A. Scoggins, eds.), p. 383. Academic Press, New York, 1974.

[3] P. E. Stanley, *in* "Liquid Scintillation. Science and Technology" (A. A. Noujaim, C. Ediss, and L. J. Weibe, eds.), p. 209. Academic Press, New York, 1976.

[4] "Proceedings of the Second Biannual ATP Methodology Symposium" (G. A. Borun, ed.). SAI Technology Company, San Diego, California, 1977.

ISBN 0-12-181957-4

$$\text{FMNH} + O_2 \xrightarrow[\text{long-chain aldehyde}]{\text{luciferase}} \text{light (max. 490 nm)} \qquad (2)$$

Under suitable conditions the rate of production of photons is proportional to the concentration of NADH, NADPH, and FMN. These enzymes may be coupled to other reactions in which NADH is the indicator species. See, for example, the chapter concerning the estimation of malate, oxalocetate, and malate dehydrogenase in this volume [19].

The procedure used measures the light produced over a specific time interval following initiation of the reaction. It differs therefore from the other approach often adopted, where flash height is measured.[5-8] This latter method uses larger amounts of enzyme and requires that the reaction be initiated in the detector chamber, since the flash occurs within a few seconds or so. This approach cannot readily be achieved in the scintillation spectrometer without the use of a special reaction vessel and a multichannel analyzer.[9,10] In both procedures reproducible and rapid mixing on initiation of the reaction is of paramount importance to produce reliable results.

Equipment and Reagents

Equipment. The liquid scintillation spectrometer, found in most laboratories, is an efficient means for counting photons provided it is set up in a proper fashion. To use it to count the single photons produced in the bioluminescence reaction, it is necessary to switch off the coincidence circuitry. One or both photomultipliers may be employed for counting, but it is usual to select that photomultiplier with the lowest background. In this way the signal-to-noise ratio is improved, but the ultimate sensitivity of the assay also depends on the background of light produced by the enzyme. The background of the photomultiplier may be as high as 25,000 cpm in the tritium channel normally used, but this is quite acceptable. However the signal-to-noise ratio may be optimized by adjusting the discriminators, particularly the lower one, instead of using the standard pulse-height settings for tritium.[1,10] The pulse rate from the photomultipliers to the analyzer

[5] S. E. Brolin and S. Hjertén, *Mol. Cell. Biochem.* **17,** 61 (1977).

[6] A. Ågren, S. E. Brolin, and S. Hjertén, *Biochim. Biophys. Acta* **500,** 103 (1977).

[7] S. E. Brolin, *Bioelectrochem. Bioenerg.* **4,** 257 (1977).

[8] J. Lee, C. L. Murphy, G. J. Faini, and T. L. Baucom, *in* "Liquid Scintillation Counting: Recent Developments" (P. E. Stanley and B. A. Scoggins, eds.), p. 403. Academic Press, New York, 1974.

[9] P. E. Stanley, *in* "Liquid Scintillation Counting: Recent Developments" (P. E. Stanley and B. A. Scoggins, eds.), p. 421. Academic Press, New York, 1974.

[10] P. E. Stanley, *Anal. Biochem.* **39,** 441 (1971).

should not exceed 10^6 cpm, as above this value there may be distortion of the pulse-height spectrum.[1,10] Some spectrometers are fitted with live timing to permit higher counting speeds; however, this mode should not be used, since the set counting time may be substantially exceeded, and, as the enzymic production of photons is time dependent, this method will give an incorrect appraisal of the reaction rate. When high count rates are encountered, either the volume used should be scaled down or the luciferase concentration reduced.

As the assay depends on a count of individual photons, other sources of light will interfere. Thus to avoid high and sometimes variable backgrounds, and consequently spurious results, these sources should be avoided. They include phosphorescence of glass and plastic sample tubes caused by exposing them to sunlight or fluorescent lights. Thus, it is best done by conducting the assay and operating the spectrometer or photometer in a room lit only by tungsten lights.

Temperature is an important parameter to control in the assay, and the massive detector assembly of the spectrometer is a useful heat sink provided the unit has been operating at room temperature (20° + 5°) for at least 72 hr, with the temperature control device working if it is fitted. A convenient means of storing reagents at a constant temperature during the assay is thus in or on the sample changer mechanism. Ideally the temperature range during analysis should be no greater than 0.5°. The optimum temperature for one luciferase is around 28°,[10] and this is shown in Fig. 1.

A number of photometers suitable for measurement of bioluminescence are available commercially. These include JRB Model 3000 ATP photome-

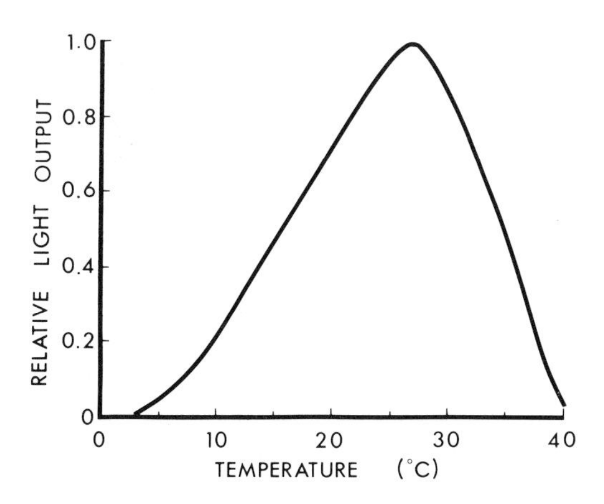

FIG. 1. Effect of temperature on light output of luciferase used in the assay of NADH.

ter, SAI Technology Co., San Diego, California; Chem-Glow photometer, American Instrument Co., Silver Spring, Maryland; Celltester Model 1030, Lumac Systems AG, Basel, Switzerland.

Vials. With the liquid scintillation spectrometer, the assay may be carried out in glass, polystyrene, polyethylene, or polypropylene tubes (approximately 50 mm long by 15 mm in diameter) or polystyrene autoanalyzer cups (long form, 40 mm long by 15 mm in diameter). They should be clean, dry, free from dust, and stored away from direct sunlight and fluorescent lighting. The tubes or cups should fit snugly into the neck of a standard liquid scintillation vial constructed of glass or polypropylene. To obtain consistent results, the same scintillation vial should be used throughout the assay run, as considerable variation has been observed from vial to vial in this type of work.[11] For constant-temperature work, the scintillation vial can be filled with distilled water and equilibrated in a water bath prior to use. Thus the reaction vial and its contents can be maintained for several minutes close to the desired temperature. The scintillation and reaction vial may remain uncapped if evaporation has been shown not to be a problem.

Special reaction vials are often available with commercially available photometers.

Reagents. Bacterial luciferase derived from *Photobacterium fischeri*, bovine serum albumin, and flavin monucleotide (FMN) is obtainable from Sigma Chemical Co., St. Louis, Missouri; dithiothreitol from Calbiochem, La Jolla, California; and decanal from Aldrich Chemical Co. Inc., Milwaukee, Wisconsin.

Potassium phosphate, 100 mM, pH 7.0, is prepared and filtered through a 0.2-μm filter to remove extraneous material, which can give rise to spurious results. This buffer is used to prepare 200 μM FMN and a standard solution of 100 μM NADH as well as luciferase at 1 mg/ml containing also 10 mg of bovine serum albumin per milliliter and 6 mM dithiothreitol. The FMN solution when kept at 4° is usable for up to a week, but the NADH and luciferase should be prepared as required and kept for no more than 8 hr. They should be stored in ice.

A decanal suspension is prepared daily using 100 μl of aldehyde in 100 ml of phosphate buffer contained in a stoppered tube. It should be mixed occasionally by swirling or placing in an ultrasonic bath. The aldehyde solution oxidizes in the presence of air, and an excess of decanal should always be present to obtain reliable results.

There appears to be some batch-to-batch diversity in activity of commercially available preparations of bacterial luciferase, and indeed the

[11] J. E. Corredor, D. G. Capone, and K. E. Cooksey, *Anal. Biochem.* **70,** 624 (1976).

stability is also variable. This may be due in part to the bacterial species from which it has been derived. Small amounts of enzyme, e.g., 1 mg, which have been preweighed into a tube and stored for a few days at $-20°$ sometimes lose activity, presumably owing to oxidation. Enzyme should thus be removed from bulk supply only when required. It will be realized from the foregoing that the protocol to be described serves only as a guide to the amount of enzyme required, and the added bovine serum albumin and dithiothreitol may not be required at the level stated.

At the time of printing, two kits had just been released commercially for the measurement of either NADH or NADPH by Lumac Systems AG, Basel, Switzerland. Two bacterial enzyme preparations. Lumase® (NADH) and Lumase-P® (NADPH) are employed. The author has had an opportunity to make a short preliminary approaisal of Lumase® using the Celltester 1030 photometer and found it to be apparently quite satisfactory for measurements of the nucleotide down to nanogram amounts.

General Method

Adjust one pulse-height analyzer on the liquid scintillation spectrometer to count tritium (or optimize for signal/noise rates), having previously switched off the coincidence circuit and selected one or both photomultipliers to count the sample. As the assay is very dependent on a constant temperature, the sample changer of the detector should have been maintained at room temperature for a minimum of 72 hr prior to the assay to allow equilibration. The instrument should be in an area free from drafts and in a room lit by tungsten lights. The high voltage to the photomultipliers should have been set for normal scintillation spectrometry and stabilized for at least 72 hr.

Select the repeat single sample count mode and a preset counting time of 0.04 or 0.1 min, and preset counts on all channels to maximum.

Bring all reagents to the temperature of the sample changer by storing them there for at least an hour prior to use. Bacterial luciferase solution should be kept on crushed ice. Reaction vials should also be stored in the sampler changer to reach reaction temperature.

Place in the reaction vial the reagents listed in the assay protocol except for the standard sample of NAD(P)H. Mix them well by vortexing *briefly* after the addition of each component. Be sure to maintain the same sequence and time interval for each addition. Immediately place the vial in the spectrometer (or other detector) and print the counts resulting for at least 1 min. The last printed value gives the background and serves to alert the user to an unexpectedly high value. Retrieve the reagent vial and add the NAD(P)H standard or sample and then recount for a period of 1–2 min.

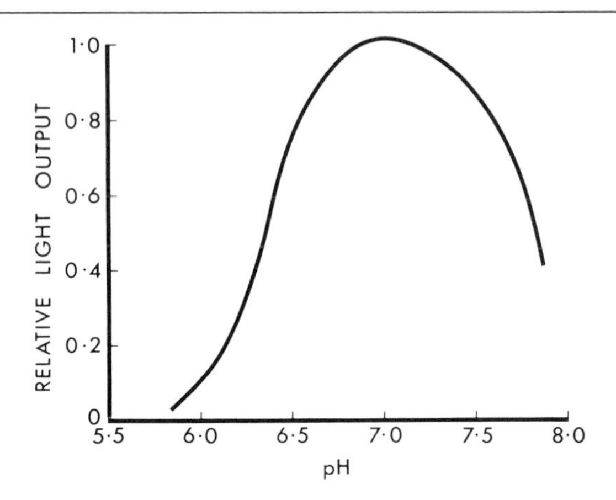

FIG. 2. Effect of pH on light output of luciferase used in the assay of NADH.

The final count gives a value for either producing a standard curve or evaluating a sample. Normally the count rate will increase for about 0.5–1 min after addition of NAD(P)H, after which time it will remain more or less constant for 1–2 min. Select one count, at a fixed time on the plateau, and use this to prepare a standard curve to encompass a 100-fold range below 10^{-9} mol. The samples should be of the same volume as the standards and should not be grossly colored or effect a substantial change in the pH of the assay mixture, as this will affect the light output (Fig. 2).

The procedure is not an absolute one, and calibration curves should be prepared daily. As this is an enzymic procedure, it is likely to be influenced by inhibitors in the sample, and further, as light is produced any colored matter may absorb the photons produced and thus prevent them from reaching the detector. Consequently, it is necessary to assay the sample a second time after the addition of a known amount of NAD(P)H. This amount, in a small volume relative to the sample, should be about one and a half times the apparent NAD(P)H level of the sample. The recovery factor can then be included in the calculation of the result. This approach of using an internal standard thus allows for a correction of enzymic inhibition as well as light absorption by the standard.

Assay Protocols

Assay Protocol for NAD(P)H

The procedure is described under General Method.

Pipetting of small volumes should be carried out with units capable of providing a coefficient of variation of less than 0.2%. SMI pipettes (Emeryville, California) are entirely suitable. [An alternative and more convenient means is to employ a pipettor dilutor (Hamilton, Cavro, Micromedic, for example), in which case the phosphate buffer is used as the diluent following each dispensing. The initial volume of buffer, 2 ml, is therefore not used, but is supplied during the pipetting procedure.]

Reagents should be added as follows:

- Potassium phosphate, 100 m*M*, pH 7.0, 2 ml
- FMN, 200 μ*M*, 10 μl
- Decanal, saturated solution, 50 μl
- Luciferase, 1 mg/ml in phosphate buffer containing 10 mg of bovine serum albumin per milliliter and 6 m*M* dithiothreitol, 10 μl
- NAD(P)H, 50 μl to start the reaction (1 nmole or less)

The amount of luciferase and need for bovine serum albumin will vary from batch to batch of luciferase. Dithiothreitol protects the sulfhydryl groups of luciferase necessary for enzymic activity.

A typical calibration curve is presented in Fig. 3. For some reaction vessels, e.g., autoanalyzer cups, it may be necessary to limit the total volume of reagents. In this case, the above volumes should be scaled down accordingly.

Assay Protocol for FMN

FMN may be assayed using a modification of the above protocol. FMN should be used to start the reaction, and the mixture should contain 10 nmole

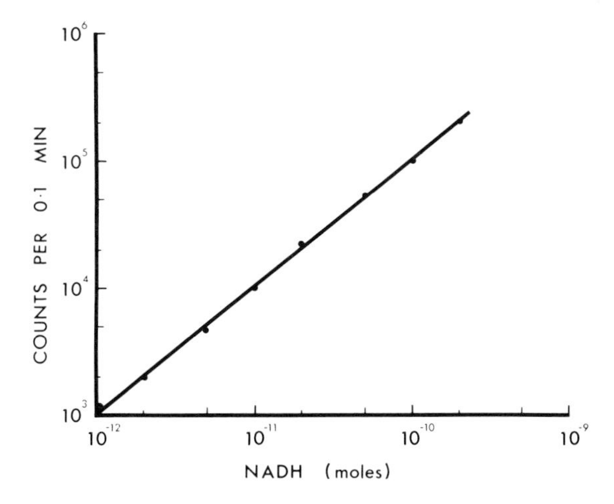

FIG. 3. A calibration curve for NADH using the liquid scintillation spectrometer.

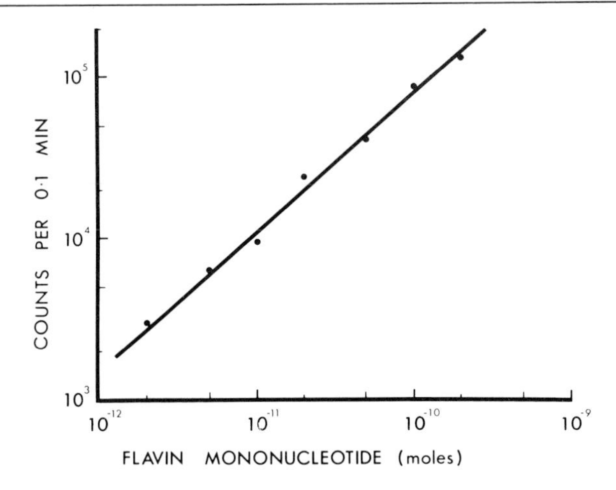

FIG. 4. A calibration curve for FMN using the liquid scintillation spectrometer.

of NADH. The useful range for FMN is from a nanomole down to a few picomoles. A typical calibration curve is presented in Fig. 4.

Discussion

The assay described is quite reproducible, and most problems can be traced to one of the following: dust or microorganisms in the reagents on the surface of the glassware; poor pipetting of reagents; oxidation of the decanal solution; variable backgrounds associated with phosphorescence of reaction vessels or "light activation" of luciferase due to fluorescent lighting of the laboratory; particulate matter in the luciferase solution.

The importance of the internal standard cannot be overemphasized, since it serves to check for inhibition as well as light absorption in each assay.

It will be noticed that the slope of the line illustrated for the FMN standard curve is less than one. This is attributed to photon absorption by the FMN solution.

Since currently commercially available luciferase preparations are only partially purified, the interference in assays by contaminating enzymes should be assessed using appropriate controls. The same word of caution should also be made when assays are to be conducted on samples containing enzymes or substrates that unexpectedly produce NAD(P)H or FMN when mixed with the luciferase system.

[25] Luminous Bacteria as an Oxygen Indicator

By B. CHANCE and R. OSHINO

The need to determine precisely the oxygen affinity of a variety of oxidases at the concentration 10^{-7} to 10^{-10} M has necessitated a variety of special techniques, on which the work of Schindler stands out as being one of the most detailed.[1,2] With the recognition of the limitations of manometric and polarographic techniques, the need for technology between 10 and 100 times more sensitive became apparent, particularly as theoretical calculations based upon the velocity of the reaction of oxygen with cytochrome oxidase together with the first measurements of the turnover indicated the affinities to be of the order of 10^{-8} M.[3] A second and important consideration was the need to measure rates of respiration at very low oxygen concentrations, since one of the more useful equations for study of the oxygen reaction involves both the rate of oxygen utilization and the oxygen concentration.[1] These parameters, together with the relative insensitivity of bacterial luminescence to inhibitors of most heme-enzyme reactions, make the bacteria themselves ideal oxygen indicators of mitochondrial oxygen requirements.

It should be pointed out in the beginning that the high oxygen affinity of the luminous system of the bacteria does not reside in any special speed of reaction of the luciferase system with oxygen over and above that of cytochrome oxidase; it relies specifically upon the low turnover number of the system as indicated by the simple equation. The oxygen affinity (K_M) is approximated by the ratio of the turnover number (TN in electrons per second) to the velocity constant for the oxygen reaction (k_1 in M^{-1} $\sec^{-1}$).[1,2]

$$K_M = (\mathrm{TN}/4)\,(1/k_1) \qquad (1)$$

The cytochrome oxidase system turns over under physiological conditions about 10 times per second, whereas the rate for the luciferase system of bacteria may be one-tenth to one-hundredth of this.

One possible interference in the bacterial system with some of the oxidases is the requirement for salt. We have found, however, that between 1% and 3% NaCl or 0.5 M mannitol are satisfactory for the bacteria and,

[1] F. J. Schindler, Oxygen Kinetics in the Cytochrome Oxidase-Oxygen Reaction, Ph.D. thesis, University of Pennsylvania, Philadelphia, 1964.

[2] B. Chance, *J. Gen. Physiol.* **49,** 163 (1965).

[3] B. Chance, R. Oshino, and N. Oshino, this series, Vol. 54, p. 499, in press.

ISBN 0-12-181957-4

under the conditions of our experiment, for mitochondria as well. Submitochondrial particles are less sensitive to the osmotic conditions.

Culture Conditions

The method of culturing the bacteria has been described recently in *Methods in Enzymology*[3] and, for convenience, is repeated here:

Photobacterium phosphoreum, P. fischeri, Beneckea harveyi (MAV) are used mostly for biochemical research. They are grown in McElroy's medium[4]: NaCl, 3 g; $Na_2HPO_4 \cdot 7H_2O$, 7 g; KH_2PO_4, 1 g; $(NH_4)_2HPO_4$, 0.5 g; $MgCl_2$, 0.1 g; glycerol, 3 ml; bactopeptone, 5 g; and yeast extract, 3 g in 1 liter of distilled water. The pH is adjusted to 7.0–7.2 with NaCl. Bacteria are cultured in 2-liter flasks containing 100 ml of the culture medium at 15° to 25° with vigorous shaking for 15–25 hr. As biosynthesis of luciferase depends on cell growth, luminescence increases steeply during the late log phase of a growth curve of the bacteria. In the case of *P. phosphoreum*, luminescence reaches maximal intensity at 10^8 to 10^9 cells per milliliter of culture medium. Cells are harvested by centrifugation at 6000 *g* for 10 min, then washed with 3% NaCl–0.1 *M* phosphate buffer, pH 7.0, and suspended in the same medium at concentrations of about 500 mg, wet weight, of cells per milliliter. This bacterial preparation can be kept at 0° to 5° for a few days without significant loss of activity. Occasionally, during repeated culture, the brightness of the cells decreases. In this case it becomes difficult to get sufficient luminescent intensity with a minimum amount of bacteria. It is recommended to restart the whole culture process by selecting brilliant colonies on agar slants.

Methods for Preparation

Extremely simple optical systems can be employed in the preparation of bacterial luminescence, and one of the simplest is shown in Fig. 4 of Schindler,[1] which has, in addition to simplicity, a number of fundamental advantages. In the flask, its "fluid inlet" and magnetic stirrer are enclosed in a light-tight box. Removal of a shutter allows a photomultiplier to view the flask. Reaction volumes of 5 liters are convenient for calibration of luminescence as indicated below. As described recently,[3] 5-ml reaction vessels can be used; however, it is important to minimize the surface to volume ratio and the liquid to gas interface, for very small oxygen concentrations are to be measured.

First, the use of a large volume allows calibration with aliquots of air-saturated buffer to give a precisely known amount of oxygen without

[4] A. A. Green and W. D. McElroy, this series, Vol. 2, p. 847.

difficulties in the measurement of the dilution factor (0.1 ml of air-saturated buffer gives 0.5 μM oxygen in the reaction volume, for example, 5 liters). Second, the use of a larger vessel where the material under observation is well separated from the inlet connections ensures that small oxygen leaks are not effective in altering the oxygen concentration of the material under oxidation. Third, the use of a large vessel together with an end-on photomultiplier ensures sufficient light-gathering by the photomultiplier.

Ancillary Measurements

At the same time as oxygen tension is measured by bacterial suspension, it is usually desirable to measure other components. For example, cytochrome *c* may be measured by time-sharing; cytochrome *c* is measured spectrophotometrically during one interval and luminescence is measured in another interval. This apparatus is described by Green and McElroy.[4]

Performance

The calibration curve taken from Schindler[1] has as its lowest measurement point 20 nM, and oxygen concentrations can be detected down to 0.1 nM. It is appropriate to note that previous estimates of the "oxygen concentration corresponding to just-perceptible light" were not made with the same instrumental refinement as that used here, and Schindler's value may be taken as a reliable one at least for the particular components and configuration.

Chance and Oshino[3] gave a calibration curve that usefully covers the range from approximately 10^{-8} M and to 10^{-6} M, which is reproduced in their Fig. 3.

Detailed Experimental Conditions

Chance and Oshino[3] gave details on the reaction mixture mentioned above, 3% NaCl, 0.1 M phosphate buffer pH 6–8, 10 mM glucose, and 1 nM catalase. An ingenious method of depleting the reaction volume of oxygen is to add a dialysis bag full of yeast cells while the reaction mixture is bubbled with highly purified argon. This causes relatively rapid disappearance of luminescence, which might otherwise take as long as 20 min, and the process might result in some inactivation of the accompanying mitochondria. The luminescence is usually measured as the peak of the "flash" on adding oxygen, and conditions should be arranged so that the peak intensity is appropriately measured. The intensity is allowed to subside to a minimal value between additions of oxygen. When the system is sufficiently dilute, serial additions of oxygen without waiting for subsidence of luminescence are possible.[1] These two conditions are identified for the non-steady-state system and for steady-state titration.[4]

Applications

Since not many oxidases other than cytochrome oxidase have oxygen affinities in the region below 1 μM, the main application of this method has been in detailed studies of the oxygen requirement for cytochrome function in mitochondria. A number of papers describing such studies have been published.[1,2,5] Obvious applications to other types of cytochrome oxidase and cytochrome $a + a_3$ can be visualized, but as far as is known, little use has been made of the method.

Applications in measuring oxygen gradients in Krogh models have been proposed recently by Quistorff *et al.*,[5] where the luminous bacteria are immobilized in a gel through which a very small hole is made, usually by withdrawing a thread embedded that had been in the agar during solidification. Gases are allowed to flow through this hole, and the graph of the oxygen gradient around the Krogh capillary is subjected to mathematical analysis according to the diffusion equations.[5]

[5] B. Quistorff, B. Chance, and A. Hunding, *in* "An Experimental Model of the Krogh Tissue Cylinder: Two-Dimensional Quantitation of the Oxygen Gradient" (I. A. Silver and M. Erecińska, eds.), p. 127. (3rd Symp. Int. Soc. Oxygen Transport to Tissue (ISOTT). Academic Press, New York, 1977.

[26] Purification of a Blue-Fluorescent Protein from the Bioluminescent Bacterium *Photobacterium phosphoreum*

By JOHN LEE and PRASAD KOKA

A new protein has recently been isolated from extracts of bioluminescent bacteria,[1] and its properties suggest that it has a function in the bioluminescent reaction.[2] It is a blue-fluorescent protein (BFP) with a fluorescence spectral maximum at 476 nm and absorption maxima at 418 nm (ϵ 4000 M^{-1} cm^{-1}) and 274 nm (ϵ 15,000 M^{-1} cm^{-1}).[2] The fluorescence

[1] R. Gast and J. Lee, *Int. Congr. Photobiol.*, Abstr., p. 324 (1976).
[2] R. Gast and J. Lee, *Proc. Natl. Acad. Sci. U.S.A.* **75,** 833 (1978).

ISBN 0-12-181957-4

spectrum of BFP purified from *Photobacterium phosphoreum* is identical, both in spectral maximum and shape, to the bioluminescence from these bacteria. The reaction of the purified luciferase with $FMNH_2$, oxygen, and aldehyde produces an *in vitro* bioluminescence with a maximum shifted to about 490 nm. However, inclusion of BFP in this reaction mixture shifts back the *in vitro* emission to make it identical to that *in vivo*. This is accompanied by an increase in the light yield and a change in the kinetics of light emission. To explain these effects, Gast and Lee[1,2] proposed that BFP functions as a sensitizer of the bioluminescence.

Further support for this proposal is the fact that BFP is found associated with bacterial luciferase, more strongly in the case of *Photobacterium fischeri* than of *P. phosphoreum*.[3] In the first steps of purification of *P. phosphoreum* luciferase, a salt precipitate of the crude material is subject to gel chromatography (Sephadex G-75), and somewhat less than one-half the BFP coelutes with luciferase activity near the column front.[3] The "free BFP" is retarded on this column, since it has a low molecular weight (22,000).[2]

The separation of the "luciferase-associated" BFP and its partial purification have been described already.[3] This work will describe the processing of the free BFP from *P. phosphoreum* and the use of affinity chromatography to purify it to homogeneity.

Experimental

Photobacterium phosphoreum strains "A13" (J. Fitzgerald, Monash University) and ATCC 11040, were maintained at room temperature on a solid agar nutrient medium containing in 100 ml of solution at pH 7.4: sodium chloride, 3.4 g; dibasic sodium phosphate, 0.25 g; monobasic potassium phosphate, 0.21 g; dibasic ammonium phosphate, 0.05 g; magnesium sulfate, 0.01 g; bacto-peptone (Difco Laboratories, Detroit, Michigan), 1 g; glycerol, 0.3 ml; bacto-agar (Difco), 1.5 g. Liquid cultures had the same composition except for the omission of bacto-agar. Unless otherwise indicated, the preparative procedures were carried out at below 5° in a standard buffer containing: sodium phosphate, 50 m*M*; EDTA, 0.3 m*M*; 2-mercaptoethanol, 5 m*M*, pH 7.0. All chemicals were of the best commercial grades. Cibacron Blue F-3-GA conjugated to Sepharose was a gift of J. Travis.[4] Fluorescence assays were made with an Aminco-Bowman spectrofluorometer with the following medium-resolution slit arrangement: (mm) excitation monochromator → 2,3,3 (sample), 2,2, emission mono-

[3] R. Gast, I. Neering, and J. Lee, *Biochem. Biophys. Res. Commun.* **80,** 14 (1978).
[4] J. Travis, J. Bowen, D. Tewksburg, D. Johnson, and R. Pannell, *Biochem. J.* **157,** 301 (1976).

chromator, 1, photomultiplier. Wavelengths are instrumental values, and spectra are uncorrected. Absorbance at the excitation wavelength was adjusted to below 0.1 to avoid intensity distortion due to geometrical factors and self-absorption. Units of fluorescence intensity are arbitrary, but constant day-to-day sensitivity of the instrument was maintained by reference to a solution of fluorescein in 0.1 *N* NaOH. Absorption measurements were made on a Cary 14 spectrophotometer. Photometric measurements were made with reference to the luminol chemiluminescence quantum yield standard.[5]

Growth of Bacteria

Liquid cultures of composition described above were made in 400-liter batches. Inoculations were of about 1 liter of cell density 100 Klett units (Klett-Summerson colorimeter), and growth was at room temperature with vigorous stirring (250 rpm) and aeration (6 ft^3/min). Antifoam (Dow-Corning 30% silicone) was added in minimum amounts. A maximum light output of about 5000 photons sec^{-1} $cell^{-1}$ (strain A13) was reached after about 20 hr, and the cells were harvested (Sharples centrifuge, 50,000 rpm, 5°, 4 hr) to yield about 1 kg of wet cell cake.

If the growth was made at 12°, a maximum light output of 20,000 photons sec^{-1} $cell^{-1}$ was obtained with a corresponding increase in the yield of BFP. However, with this same inoculum about 3 days of growth were required before harvesting.

Extraction and Initial Steps

About 500 g of wet cell cake were suspended in buffer (1 liter) containing phenylmethylsulfonylfluoride (1 μM) to retard proteolysis. The bacteria were disrupted by passing this suspension twice through a French press (9000 psi), and cell walls were removed by centrifugation (60 min, 25,000 *g*, Sorvall RC-5, GSA Head). The centrifugate was decanted and made 30% saturation concentration in ammonium sulfate by slowly adding solid with stirring over a period of 30 min. Without delay the solution was centrifuged (30 min, 25,000 *g*). The pellet was discarded, and the concentration of ammonium sulfate was then raised to 80% saturation; after 2 or more hours, the solution was again centrifuged (30 min, 25,000 *g*) to collect the precipitated proteins.

[5] J. Lee, A. S. Wesley, J. F. Ferguson, and H. H. Seliger, *in* "Bioluminescence in Progress" (F. H. Johnson and Y. Haneda, eds.), p. 35. Princeton Univ. Press, Princeton, New Jersey, 1966.

The relative concentration of BFP was monitored by the fluorescence at 470 nm with excitation at 420 nm. The total amount of BFP then is proportional to the total fluorescence (Table I), the 470 nm fluorescence intensity times the total volume of sample. Similarly, the protein concentration is proportional to the absorbance at 280 nm and total protein to the total absorbance. The relative purity of BFP is given by its specific fluorescence, that is, total fluorescence divided by total absorbance.

The steps in the purification of BFP from *P. phosphoreum*, strain A13, are listed in Table I. Lower yields of BFP were found with the type strain. Both luciferase and BFP were precipitated by 80% ammonium sulfate, but up to and including this stage, the high concentration of protein with its resultant light scattering did not allow a reliable estimate of the BFP concentration to be made by fluorescence. However, a separate slow gel chromatography previously made of a sample of the 30% supernatant[3] showed about the same amount of separated BFP as the first G-75 Sephadex column in Table I. Therefore we have set the 80% ammonium sulfate fraction at 100 arbitrary units of BFP (470 nm) fluorescence.

The 80% pellet was redissolved in about 200 ml of buffer and applied to a column of Sephadex G-75 (10 × 60 cm, 2 ml/min). Two fractions of BFP are eluted, one along with the luciferase activity near the column front and

TABLE I
PURIFICATION OF BLUE-FLUORESCENT PROTEIN FROM *Photobacterium phosphoreum*

Step and fraction	Total fluorescence[a]			Total absorbance at 280 nm	Fluorescence/absorbance[b]
	420 nm	470 nm	520 nm		
1. Cell lysate	—	—	—	47,000	—
2. Ammonium sulfate					
30% Supernatant	—	—	—	60,000	—
80% Pellet	860	100	260	47,000	0.0021
3. Sephadex G-75					
Luciferase fraction[c]	(5)	(40)	(36)	(23,600)	(0.0017)
Free BFP fraction	4	60	30	3,300	0.018
4. DEAE-Sephadex A-50					
BFP fraction	6	40	24	180	0.22
5. Sephadex G-75 Superfine					
BFP fraction	3	20	4	24	0.833
6. Sepharose					
BFP fraction	2	18	1	8	2.25

[a] The fluorescence intensity was observed for 370 → 420, 420 → 470, 470 → 520 nm.

[b] Total fluorescence (470 nm)/total absorbance (280 nm).

[c] Not further processed; see R. Gast, I. Neering, and J. Lee., *Biochem. Biophys. Res. Commun.* **80,** 14 (1978).

the other retarded, the free BFP.[3] Table I describes the purification of the free fraction.

Ion Exchange

The free BFP fraction was loaded onto a column of DEAE-Sephadex A-50 (2 × 20 cm, buffer) and then washed with buffer containing phosphate (0.1 *M*, 500 ml) and with phosphate (0.15 *M*, 500 ml) and then eluted with phosphate (0.2 *M*, 1 liter).

Figure 1 shows the elution profile of this column assayed by fluorescence with the fractions started collecting after application of the 0.2 *M* buffer. Besides BFP, in the 600–850 ml fraction there are two other fluorescent materials. One has a maximum at 420 nm when excited at 370 nm and appears to be an irreversible denaturation product of BFP, since its fluorescence intensity increases at the expense of a decrease in 470 nm fluorescence as a solution of BFP ages. This happens particularly in the absence of 2-mercaptoethanol, the BFP decreasing to one-half its fluorescence in 1.5 days at 5°; 2-Mercaptoethanol increases this to about 7 days. The other has a maximum fluorescence at 520 nm when excited at 470 nm, suggestive of a flavoprotein. Another protein with a flavinlike fluorescence eluted at around 500 ml.

The BFP fraction was collected from 600 to 850 ml. Although a slight

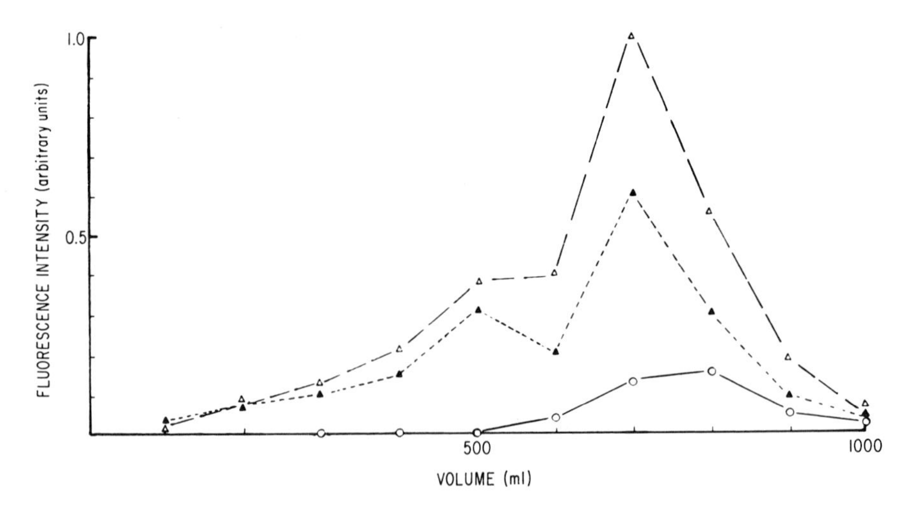

FIG. 1. Elution of blue-fluorescent protein (BFP) from DEAE-Sephadex A50 with phosphate (0.2 *M*). The fluorescence 420 → 470 nm (△——△) is a measure of BFP concentration. Two fluorescent contaminants are 370 → 420 nm (○——○) and 470 → 520 nm (▲——▲).

separation of the 420 nm fluorescence was observed, it was not sufficient to provide a useful separation.

The small amount of luciferase carrying over from step 3 (Table I),[3] did not bind to the DEAE and eluted with the first wash.

Gel Filtration

The BFP fraction was concentrated to about 20 ml by "Diaflo" membrane ultrafiltration (YM-10 membrane; Amicon Corporation, Lexington, Massachusetts) and applied to a column of Sephadex G-75, Superfine grade (4.5 × 100 cm). Three yellow bands eluted from this column. The first, corresponding to the absorbance maximum at the volume of 250 ml in Fig. 2, was a nonfluorescent flavoprotein, judging from its flavinlike absorption spectrum, and is probably the same protein as that found after slow molecular sieving of the luciferase-associated BFP ("FP" in Fig. 2 of Gast *et al.*[3]).

The next band was BFP (500 ml) accompanied by somewhat less of the 520-nm fluorescence than in the step before (Fig. 1), but with the same proportion of 420 nm fluorescence. The crosses in Fig. 2 were obtained by dividing the fluorescence at 470 nm by absorbance at 275 nm, near the BFP maximum.[2] This ratio is roughly constant in the volume range 500–600 ml, indicating relative homogeneity of BFP with respect to these parameters.

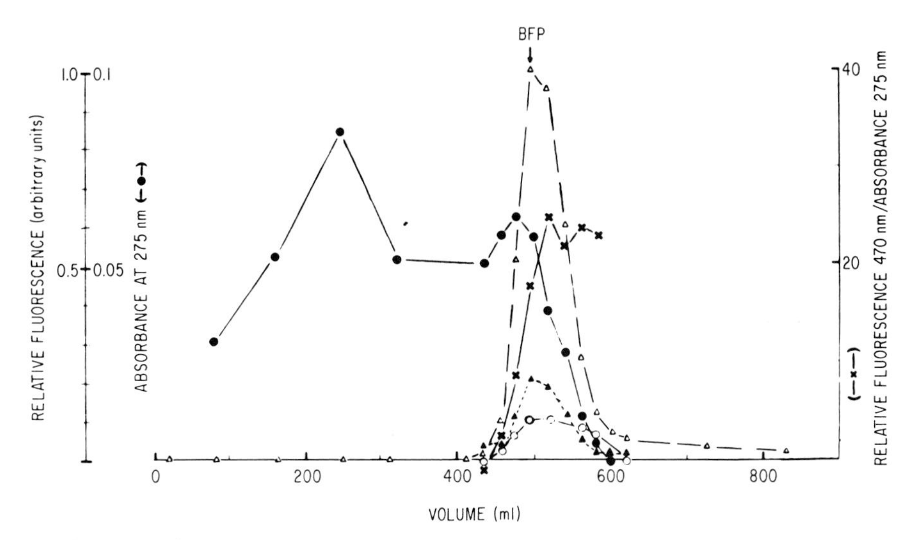

FIG. 2. Elution of blue-fluorescent protein (BFP) (498 ml) on Sephadex G-75 superfine (10 ml/hr). The column front is around 100 ml and the symbols are the same as Fig. 1. The fluorescence/absorbance (X——X) is an indication of homogeneity of the BFP fraction.

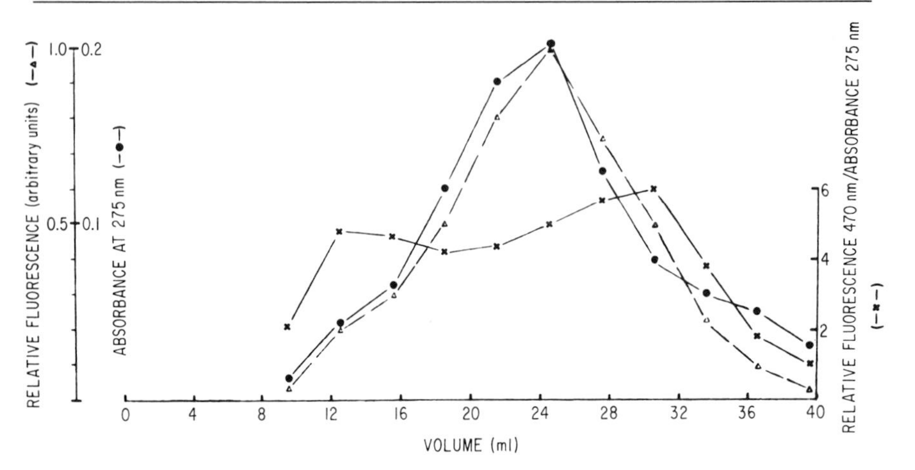

FIG. 3. Elution of blue-fluorescent protein by buffer at pH.6.5, from a Blue-Sepharose affinity column. Fractions are started immediately on application of the sample to the column.

The last band eluted at a much higher volume, off the range of Fig. 2. It was identified as FMN by its absorption spectrum, fluorescence spectrum, quantum yield, and activity in the luciferase bioluminescence reaction.

Affinity Chromatography

The BFP fraction was reconcentrated by the above Diaflo method to 2 ml and applied to a Cibacron Blue-Sepharose 4B column (6.5 $\times$ 2.0 cm) equilibrated with buffer at pH 6.5. It was eluted with this same buffer, and the results are seen in Fig. 3. The BFP was recovered in 90% yield and showed no content of the 420 nm or 520 nm fluorescent materials. This method is much more convenient and efficient than the repeated superfine chromatography first tried.[3] The flavoprotein eluted at a higher salt concentration, and the 420 nm was found in the fractions in front of BFP. The crosses again indicate that the fluorescence of BFP is homogeneous in the fractions from 12 to 32 ml.

Disc Gel Electrophoresis in SDS

Two of the fractions from Fig. 3 were analyzed by gel electrophoresis using the method of O'Farrell.[6] The fractions at 24.5 and 27.5 ml were lyophilized, redissolved to a concentration of 8 mg/ml, denatured by boil-

[6] P. H. O'Farrell, *J. Biol. Chem.* **250,** 4007 (1975).

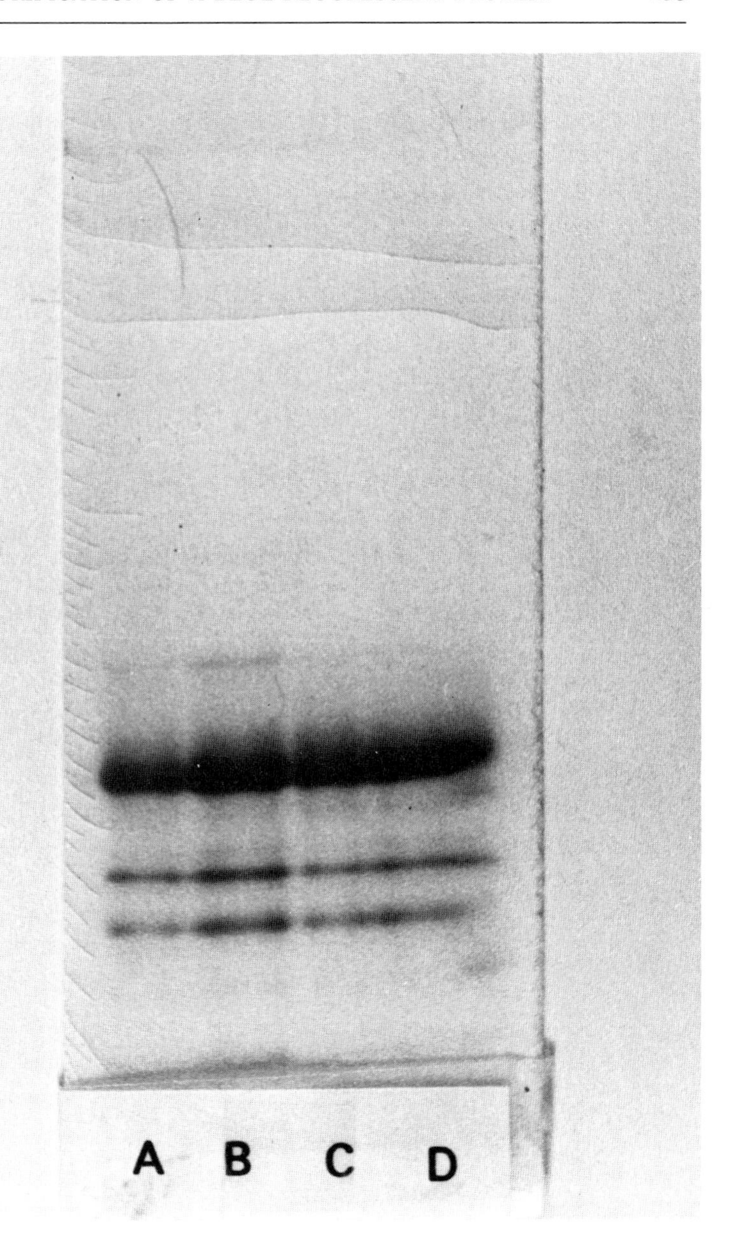

FIG. 4. Disc gel electrophoresis in sodium dodecyl sulfate by the method of P. H. O'Farrell [*J. Biol. Chem.* **250,** 4007 (1975)]. Acrylamide concentration gradient is from 10% at the top to 17% at the bottom. The blue-fluorescent protein samples are from the affinity step in Fig. 3. Fraction at 24.5 ml: A, 150 μg; B, 300 μg; fraction at 27.5 ml: C, 150 μg; D, 300 μg.

ing (5 min) in SDS (2%) with 2-mercaptoethanol (2%), and applied to the gel. Electrophoresis was for 7 hr at 150 V, and then the gels were stained with Coomassie Brilliant Blue R250, and fixed with aqueous methanol (50%) containing acetic acid (7.5%).

Figure 4 shows one major-staining and heavily overloaded protein band, migrating to a distance corresponding to a molecular weight of around 20,000. There are only two significant contaminants, and they are of lower molecular weight and are at levels, for instance in sample A, of about 5 μg (total), although the nonlinearity and high sensitivity of the Coomassie Blue stain do not allow these levels to be estimated quantitatively. However, it can be concluded that the BFP has a very satisfactory purity at this stage.

Acknowledgments

We thank G. T. Herrington and W. J. Brown, III, for valuable technical assistance, and J. Linn and Dr. R. Makula for growing the bacteria. We also thank Dr. L. Dure for running the gel electrophoresis and Dr. J. Travis for the gift of Blue Sepharose.

This work was supported by National Science Foundation grant BMS 74-19890.

Section III

Renilla reniformis Luciferase

[27] Applications of *Renilla* Bioluminescence: An Introduction[1]

By MILTON J. CORMIER

Renilla bioluminescence can be used as a versatile tool for the rapid, sensitive, and specific assay of important biochemicals. It can also provide a unique approach to some problems of interest to neurobiologists and those interested in the molecular aspects of biological energy transfer.

This introductory chapter is intended to provide the reader with some general information on the features of the *Renilla* bioluminescence system including the characteristics of the proteins and substrates involved in the light-emitting process. The molecular basis for *Renilla* bioluminescence, as well as certain features of its nerve-linked control, is now understood.[2-5] What follows is a description of the characteristics of these substrates and proteins and how they may be utilized for analytical purposes. Although the work described here deals specifically with the anthozoan coelenterate *Renilla reniformis*, all bioluminescent anthozoans so far examined provide a source of similar proteins and substrates to those found in *Renilla*.[2]

Requirements for PAP-Dependent Light Emission. The requirement for 3′,5′-diphosphoadenosine (PAP) in *Renilla* bioluminescence was described 15 years ago.[6] It is involved as a cofactor in the conversion of luciferyl sulfate [(I), Fig. 1] to coelenterate-type luciferin [(II), Fig. 1] by the enzyme luciferin sulfokinase.[5] The reaction, illustrated below, is specific for PAP.

$$\text{Luciferyl sulfate} + \text{PAP} \xrightarrow{\text{luciferin sulfokinase}} \text{luciferin} + [\text{PAPS}] \quad (1)$$

There is no evidence that 3′-phosphoadenosine-5′-phosphosulfate (PAPS) is a product, and this is indicated by the brackets in Eq. 1. However, the reaction is reversible, since luciferyl sulfate is readily produced from luciferin and PAPS in the presence of luciferin sulfokinase.

[1] This work was supported in part by the National Science Foundation (BMS 74-06914) and ERDA (AT-38-1-635). Contribution No. 366 from the University of Georgia Marine Institute, Sapelo Island, Georgia.

[2] M. J. Cormier, K. Hori, Y. D. Karkhanis, J. M. Anderson, J. E. Wampler, J. G. Morin, and J. W. Hastings, *J. Cell. Physiol.* **81,** 291 (1973).

[3] M. J. Cormier, K. Hori, and J. M. Anderson, *Biochim. Biophys. Acta* **346,** 137 (1974).

[4] M. J. Cormier, J. Lee, and J. E. Wampler, *Annu. Rev. Biochem.* **44,** 255 (1975).

[5] M. J. Cormier, *in* "Bioluminescence in Action" (P. Herring, ed.). Academic Press, New York, in press.

[6] M. J. Cormier, *J. Biol. Chem.* **237,** 2032 (1962).

ISBN 0-12-181957-4

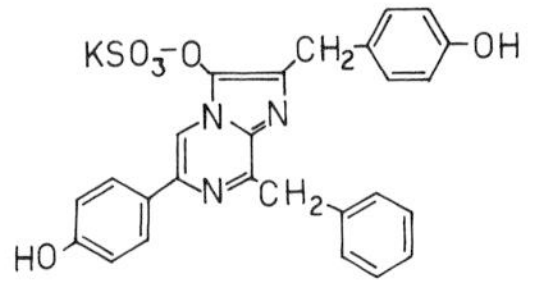

Luciferyl sulfate (I) 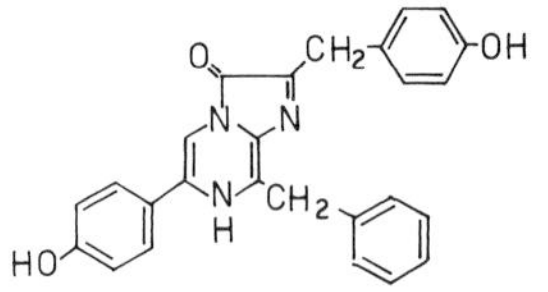 Coelenterate-type luciferin (II)

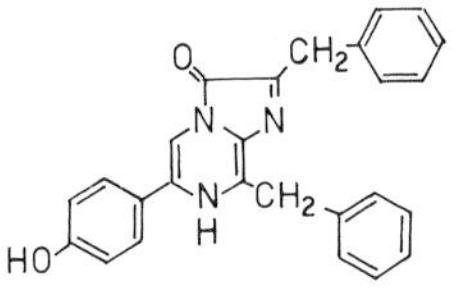

Benzyl luciferin (III) 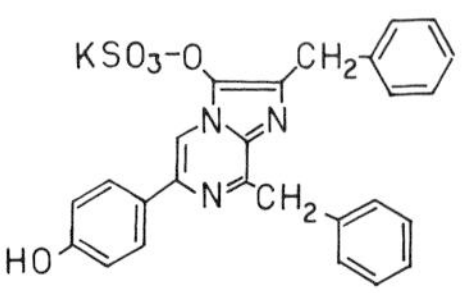 Benzyl luciferyl sulfate (IV)

FIG. 1. Structures of fully active *Renilla* luciferins and luciferyl sulfates.

In order to produce light as an end product, the luciferin sulfokinase reaction [Eq. (1)] must be coupled to a second reaction, catalyzed by luciferase, as illustrated by Eq. (2).

$$\text{Luciferin} + O_2 \xrightarrow{\text{luciferase}} \text{oxyluciferin} + CO_2 + \text{light} \qquad (2)$$

Thus the assay for PAP involves a coupled enzyme luminescence assay utilizing luciferyl sulfate as the substrate. As outlined in this volume [28], the two enzymes, luciferin sulfokinase and luciferase, fractionate together during the first few purification steps, providing a convenient source of these two proteins for assay purposes. It is this partially purified material that is used for routine PAP assays.

Although luciferin sulfokinase has not been purified and characterized, *Renilla* luciferase has been recently purified to homogeneity from crude extracts with an overall recovery of 24%.[7] Luciferase is active as a single polypeptide chain monomer of molecular weight 35,000. It is a glycoprotein containing 3% carbohydrate and is considerably hydrophobic. Its content of hydrophobic and aromatic amino acid residues is high resulting in an $\epsilon_{280\text{ nm}}^{0.1\%}$ of 2.1 and an average hydrophobiscity value of 1200 cal/residue. Luciferase also contains 3 free SH groups but no disulfide linkages. One of these SH groups is located in or near the active center of the enzyme.[5] The purified enzyme has a specific activity of $1.8 = 10^{15}$ *hv* sec^{-1} mg^{-1} and a turnover number of 111 μmol min^{-1} μmol^{-1} enzyme.

Luciferase catalyzes the reaction shown in Fig. 2.[5] Luciferin binds to a

[7] J. C. Matthews, K. Hori, and M. J. Cormier, *Biochemistry* **16,** 85 (1977).

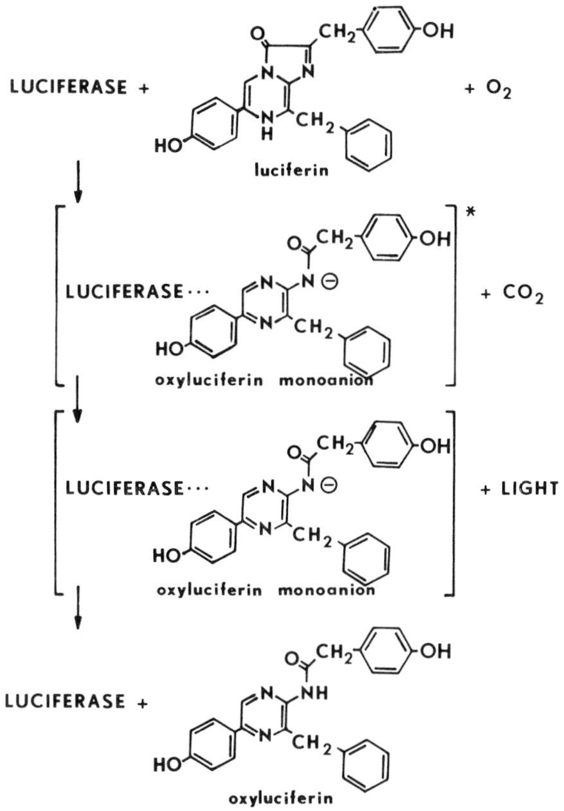

FIG. 2. Chemistry of the *Renilla* luciferase-catalyzed oxidation of coelenterate-type luciferin.

single binding site on the enzyme with a dissociation constant of 3×10^{-8} *M*. *Renilla* luciferase can be classified as an oxygenase since O_2 is apparently incorporated into one of the products, oxyluciferin, while a second product, CO_2, is derived from carbon 3 of luciferin. The initial product of the reaction is an electronic excited state of luciferase–oxyluciferin monoanion complex which relaxes to its ground state with the production of blue light (λ_{max} = 480 nm). The quantum yield for this reaction is 5.5%, but the excitation yield is high, i.e., in the range 21–100%.[5,8]

In the past, the assay for PAP depended on the availability of luciferyl sulfate [(1), Fig. 1] which had to be isolated from *Renilla*. This process was tedious, and the amounts of luciferyl sulfate obtained were small. In 1973

[8] W. W. Ward and M. J. Cormier, *Photochem. Photobiol.*, **27**, 389 (1978).

the structural requirements for a fully active luciferin [(III), Fig. 1] were elucidated, and its structure was confirmed by synthesis.[9] This compound, termed benzyl luciferin, serves as well as the native compound [(II), Fig. 1] in the luciferase-catalyzed luminescence reaction. For example, the quantum yields, kinetics, and emission spectra are indistinguishable when either (II) or (III) is used as substrate. More recently it has been found that benzyl luciferyl sulfate [(IV), Fig. 1] can replace native luciferyl sulfate [(I), Fig. 1] in the coupled enzyme assay for PAP described above. Since the synthesis of (IV) presents fewer problems than the synthesis of (I), the former compound is routinely used as a source of synthetic substrate in the assay for PAP.

Requirements for in Vitro Nonradiative Energy Transfer. As outlined above the luciferase-catalyzed oxidation of luciferin produces a blue emission (λ_{max} = 480 nm) which has a broad wavelength distribution. On the other hand, a careful examination of light emission from *Renilla in vivo* reveals a green emission (λ_{max} = 509 nm) whose wavelength distribution is characteristically narrow and structured.[10,11] This green emission is due to the fact that, prior to excited-state formation, another protein, the *Renilla* green-fluorescent protein (GFP) can participate in the bioluminescence reaction by acting as an energy-transfer acceptor. *Renilla* GFP has been isolated, purified to homogeneity, and characterized.[8,12] It exists as a dimer of identical subunits (molecular weight = 27,000), each containing at least one covalently linked chromophore of unknown structure. The chromophore is responsible for the characteristic visible absorption and intense fluorescence of GFP (Q_F = 80%).

The fluorescence emission of *Renilla* GFP is identical to the green *in vivo* luminescence of *Renilla*, suggesting that it is the *in vivo* emitter.[3,11] In addition, the green *in vivo* emission can be duplicated *in vitro* by the addition of GFP to a luciferase–luciferin reaction mixture. The effect is an immediate shift in spectral distribution from blue to green as illustrated in Fig. 3. Furthermore, the bioluminescence quantum yield for luciferin oxidation increases more than 3-fold in the presence of GFP.[8,12] These effects occur with protein concentrations as low as 1 μM GFP and 1 nM luciferase, thus precluding trivial energy transfer as a mechanism. We have recently shown that a rapid equilibrium exists between luciferase and GFP resulting in the formation of a functional energy-transfer complex prior to excited-state formation.[8] This complex exists as one luciferase molecule associated with one GFP dimer molecule, the interaction between the two proteins

[9] K. Hori, J. E. Wampler, J. C. Matthews, and M. J. Cormier, *Biochemistry* **12,** 4463 (1973).
[10] J. G. Morin and J. W. Hastings, *J. Cell. Physiol.* **77,** 313 (1971).
[11] J. E. Wampler, K. Hori, J. W. Lee, and M. J. Cormier, *Biochemistry* **10,** 2903 (1971).
[12] W. W. Ward and M. J. Cormier, *J. Phys. Chem.* **80,** 2289 (1976).

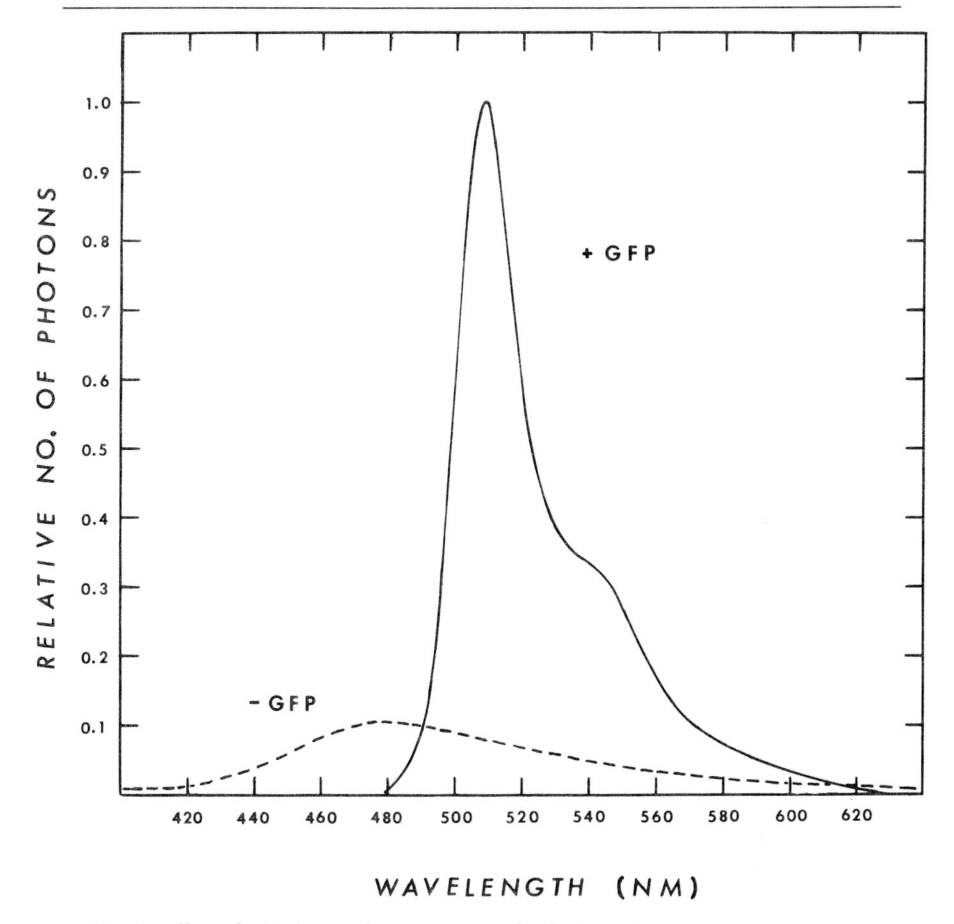

FIG. 3. Effect of added green fluorescent protein (GFP) on the wavelength distribution of *Renilla in vitro* luminescence. – – –, Luminescence of a luciferase–luciferin reaction; ——, effect of added GFP.

being highly specific. The energy-transfer complex thus allows the calculated critical distance (~30 Å) to be achieved between the luciferase-bound oxyluciferin monoanion excited state and the chromophore of GFP.[11,12] Excitation energy of the enzyme-bound oxyluciferin monoanion is thus transferred to the chromophore on GFP by a nonradiative process resulting in the emission of green light. A schematic representation of these events is shown in Fig. 4.

During energy transfer the number of blue and green photons emitted can be determined independently and simultaneously from a single reaction mixture, thus providing a quantitative measure of energy transfer.[8,12] In

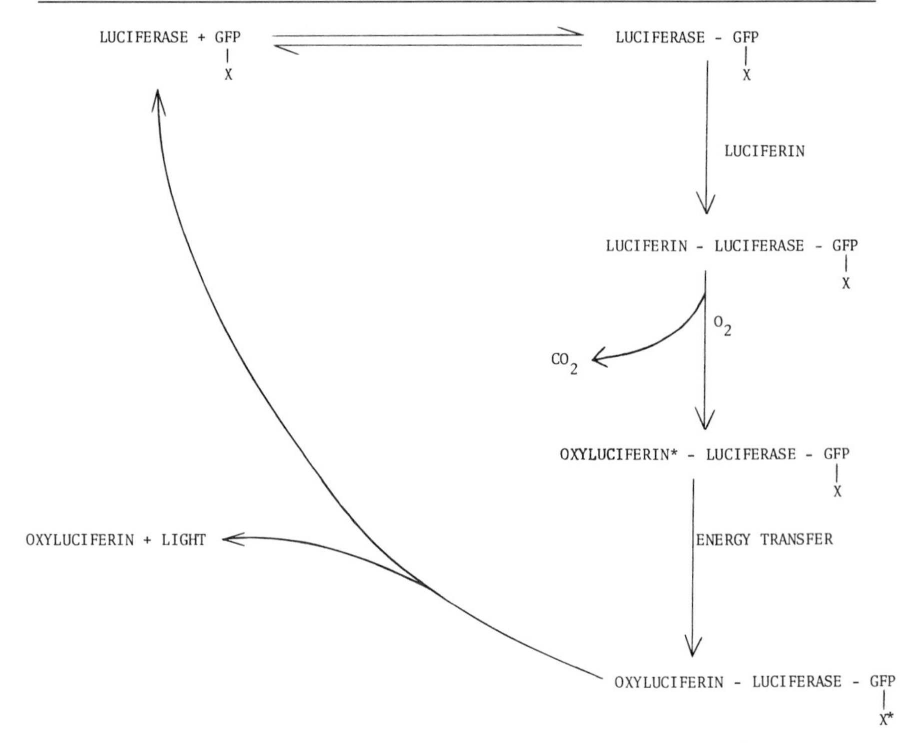

FIG. 4. Protein–protein interaction model for nonradiative energy transfer in *Renilla* bioluminescence. GFP, green-fluorescent protein; *X*, the unknown chromophore on GFP.

addition, such measurements represent a sensitive and rapid quantitative approach to the study of protein–protein interaction.

Requirements for Calcium-Triggered Bioluminescence. In addition to luciferin sulfokinase, luciferase, and GFP discussed above, a fourth protein, termed a calcium-triggered luciferin-binding protein ($BP-LH_2$), was recently discovered in *Renilla;* it can link the nerve net to bioluminescence.[13] $BP-LH_2$ has since been purified 6000-fold to homogeneity from *Renilla* crude extracts and characterized.[14] It is a single polypeptide chain protein of molecular weight 18,500 having an isoelectric point of 4.3 and containing one free SH group, one disulfide linkage, and no tryptophan. The protein contains two high-affinity Ca^{2+} binding sites ($K_d = 1.4 \times 10^{-7}$ *M*) and, when isolated in the absence of Ca^{2+}, it also contains 1 mol of coelenterate-type luciferin [(II), Fig. 1] noncovalently bound to the protein.

[13] J. M. Anderson, H. Charbonneau, and M. J. Cormier, *Biochemistry* **13,** 1195 (1974).

[14] M. J. Cormier and H. Charbonneau, *in* "Calcium Binding Proteins and Calcium Function" (R. H. Wasserman *et al.*, eds.), p. 481. Elsevier, North-Holland, Amsterdam, (1977).

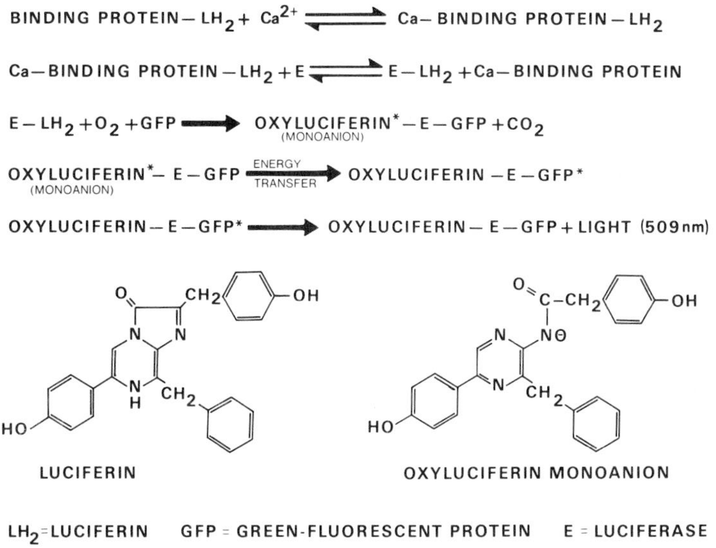

BINDING PROTEIN—LH_2 + Ca^{2+} ⇌ Ca—BINDING PROTEIN—LH_2

Ca—BINDING PROTEIN—LH_2 + E ⇌ E—LH_2 + Ca—BINDING PROTEIN

E—LH_2 + O_2 + GFP → OXYLUCIFERIN*—E—GFP + CO_2
(MONOANION)

OXYLUCIFERIN*—E—GFP (MONOANION) —ENERGY TRANSFER→ OXYLUCIFERIN—E—GFP*

OXYLUCIFERIN—E—GFP* → OXYLUCIFERIN—E—GFP + LIGHT (509 nm)

LUCIFERIN

OXYLUCIFERIN MONOANION

LH_2 = LUCIFERIN GFP = GREEN-FLUORESCENT PROTEIN E = LUCIFERASE

FIG. 5. Pathway to calcium-triggered green luminescence *in vitro*.

The probable *in vivo* function of BP-LH_2 is understood from an examination of its interaction with the isolated proteins involved in *Renilla* bioluminescence as illustrated in Fig. 5. When Ca^{2+} is absent, BP-LH_2 will not produce light with luciferase because its bound luciferin is unavailable. As seen in Fig. 5 the binding of Ca^{2+} to BP-LH_2 allows the transfer of luciferin to luciferase, apparently via a conformational change in BP-LH_2, resulting in a bioluminescent oxidation.[14] An *in vitro* mixture of the three proteins seen in Fig. 5 will, upon the introduction of Ca^{2+} and O_2, produce a flash of green light identical to that observed *in vivo*.

BP-LH_2, luciferase, and GFP are apparently sequestered within a membrane *in vivo* and are thus protected from Ca^{2+} in the resting state. In fact, these three proteins can be isolated from *Renilla* as a prepackaged unit within a membrane bounded vesicle termed lumisome.[4,15] The origin of this vesicular membrane, i.e., whether it is derived from the plasma membrane or some other, has not been established with certainty. Lumisomes have also been isolated in highly purified form and found to contain BP-LH_2, luciferase, and GFP.[15]

It does appear that most, if not all, of the luciferase, BP-LH_2, and GFP in *Renilla* are somehow prepackaged and ready for triggering by the nerve net.[5] In addition to their being isolatable in the form of lumisomes, the

[15] J. M. Anderson and M. J. Cormier, *J. Biol. Chem.* **248,** 2937 (1973).

concentrations of these three proteins in *Renilla* have been found to be approximately equal.[5] Furthermore, the total amount of light that can be produced from a single *Renilla* as the result of electrical or chemical stimulation can be accounted for by the amount of BP-LH_2 in a single animal.

Lumisomes will produce a flash of green light in the presence of Ca^{2+} and dissolved oxygen under appropriate conditions. One way in which this Ca^{2+}-triggered luminescence is accomplished may be related to the *in vivo* control of bioluminescence. For example, by establishing an Na^+ gradient across the lumisomal membrane, i.e., with Na^+ isotonic on the inside and low in concentration on the outside, a flash of green light is observed upon external Ca^{2+} addition.[16,17] Whereas any monovalent cation will suffice to establish isotonicity on the outside of the lumisomal membrane, rapid Ca^{2+} influx, as measured by bioluminescence, is specifically dependent upon internal Na^+.[5]

Thus BP-LH_2 serves as the terminal link between nerve excitation and bioluminescence, with Ca^{2+} acting as a second messenger.[14] Changes in membrane polarization could result in a rapid influx of Ca^{2+} across the lumisomal membrane, turning the bioluminescence system on. A rapid resequestering of Ca^{2+} could turn the system off. Just how all of this occurs at the molecular level is of interest to the neurobiologist. In this regard, it is clear that the lumisomal system provides a unique tool for studying the mechanisms of rapid Ca^{2+} transients across excitable membranes.

[16] J.-P. Henry, *Biochem. Biophys. Res. Commun.* **62,** 253 (1975).
[17] J. M. Anderson and M. J. Cormier, *Biochem. Biophys. Res. Commun.* **68,** 1234 (1976).

[28] A Bioluminescence Assay for PAP (3′,5′-Diphosphoadenosine) and PAPS (3′-Phosphoadenylyl Sulfate)[1]

By JAMES MICHAEL ANDERSON, KAZUO HORI, and MILTON J. CORMIER

Sulfate incorporation into biological molecules is dependent upon preactivation of the sulfate by attachment to a nucleotide.[2] In most systems

[1] This work was supported in part by the National Science Foundation (BMS 74-06914 and PCM 76-10573) and ERDA (AT-38-1-635). Contribution No. 367 from the University of Georgia, Marine Institute, Sapelo Island, Georgia.
[2] H. D. Peck, Jr., *in* "The Enzymes" (P. D. Boyer, ed.), 3rd ed., Vol. 10, p. 651. Academic Press, New York, 1974.

ISBN 0-12-181957-4

that have been studied, the activated form of sulfate used is 3′-phosphoadenosine-5′-phosphosulfate (PAPS), which is formed by two separate reactions, as follows:

$$\text{ATP} + \text{SO}_4^{2-} \xrightarrow[\text{Mg}^{2+}]{\text{ATP-sulfurylase}} \text{APS} + \text{PP}_i$$

$$\text{ATP} + \text{APS} \xrightarrow[\text{Mg}^{2+}]{\text{APS-kinase}} \text{PAPS} + \text{ADP}$$

The sulfate group of PAPS in turn can be transferred to any of several acceptors by means of a sulfotransferase, as indicated below:

$$\text{R-OH} + \text{PAPS} \xrightarrow{\text{sulfotransferase}} \text{R-SO}_4 + \text{PAP}$$

Sulfotransferases will transfer sulfate to many different compounds, such as steroids,[3] phenolic compounds,[4] aromatic amines,[5] and cerebrosides.[6] Several of the resultant compounds have tremendous importance in mammalian physiology, such as the requirement for sulfatides in the formation of the myelin sheath of nerve cells.[6]

Most of the chemistry of PAPS production and utilization has been studied by use of radioactive sulfur. The radioactive techniques require that the final sulfur-containing compound be separated from contaminating labeled sulfur compounds. The assay techniques can be very sensitive, but also very tedious. Another approach has been to use a sulfotransferase with the reaction run in the reverse direction of that shown above. The assay then becomes dependent on 3′,5′-diphosphoadenosine (PAP). The reaction used is the PAP-dependent transfer of sulfate from *p*-nitrophenyl sulfate to phenol, which is followed spectrophotometrically.[4] An alternative, and potentially much more sensitive, assay for PAP and PAPS became available with the discovery of the PAP requirement for bioluminescence in the sea pansy, *Renilla reniformis*.[7] Using materials supplied by us, Stanley *et al.*[8] used such an assay to examine the synthesis of PAPS by plant tissue extracts. This luminescence assay involves the transfer of sulfate from luciferyl sulfate [(VI), Fig. 1] to PAP and is catalyzed by the enzyme luciferin sulfokinase.[9] The resulting product is luciferin [(II), Fig. 1], the bioluminescence substrate. A second enzyme, luciferase, catalyzes the

[3] S. S. Singer and S. Sylvester, *Endocrinology* **99,** 1346 (1976).

[4] J. D. Gregory and F. Lipmann, *J. Biol. Chem.* **229,** 1081 (1957).

[5] A. B. Roy, *Biochem. J.* **74,** 49 (1960).

[6] A. A. Farooqui, G. Rabel, and P. Mandel, *Life Sci.* **20,** 569 (1977).

[7] M. J. Cormier, *J. Biol. Chem.* **237,** 2032 (1962).

[8] P. E. Stanley, B. C. Kelley, O. H. Tuovinen, and D. J. D. Nicholas, *Anal. Biochem.* **67,** 540 (1975).

[9] M. J. Cormier, K. Hori, and Y. D. Karkhanis, *Biochemistry* **9,** 1184 (1970).

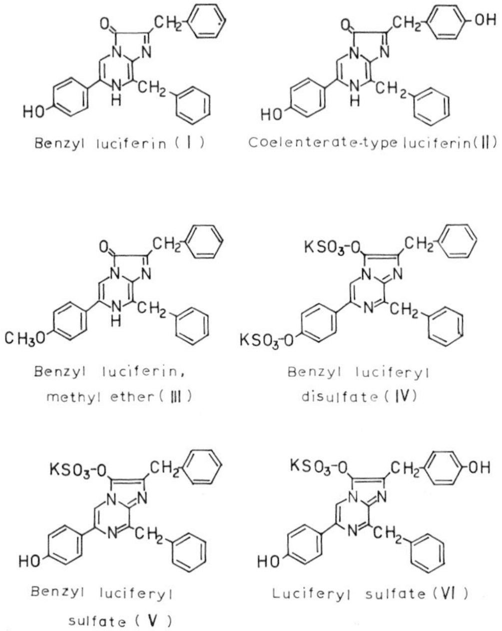

Benzyl luciferin (I) Coelenterate-type luciferin(II)

Benzyl luciferin, methyl ether (III) Benzyl luciferyl disulfate (IV)

Benzyl luciferyl sulfate (V) Luciferyl sulfate (VI)

FIG. 1. Structure of *Renilla* luciferins, luciferin derivatives, and luciferyl sulfates.

oxidation of luciferin by O_2, resulting in the production of light. This coupled enzyme assay is illustrated below:

$$\text{PAP} + \text{luciferyl sulfate} \xrightarrow{\text{luciferin sulfokinase}} \text{luciferin} + [\text{PAPS}]$$

$$\text{Luciferin} + O_2 \xrightarrow{\text{luciferase}} \text{oxyluciferin} + CO_2 + \text{light}$$

Since the reaction produces light, the assay can be made extremely sensitive to low levels of PAP. The chemistry of the bioluminescence reactions on which the PAP-dependent luminescent assay is based is reviewed in this volume [27].

The main impediment to the further development of this bioluminescence assay for PAP was the absence of a readily available source of luciferyl sulfate. This problem has been solved with the synthesis of a fully active analog of *Renilla* luciferyl sulfate. A secondary problem was the low activity of the enzyme preparation from *Renilla reniformis*. This problem has also been solved by using modified enzyme isolation techniques. The assay technique presented here is sensitive to 0.1 pmol of PAP in the assay, but this should not be taken as the limit of sensitivity, as further purification of the enzymes to remove contaminating PAP could increase the sensitivity by 2 additional orders of magnitude. It is the intent of this chapter to describe the procedures used for the bioluminescence assay for PAP, the chemical synthesis of luciferyl sulfate, and the preparation of the enzymes required.

Preparation of Luciferase and Luciferin Sulfokinase

Preliminary Considerations. The procedure given here for the isolation of luciferin sulfokinase and luciferase is not intended to produce pure enzymes, but rather to produce a suitable enzyme mixture for the assay of PAP. The method does provide certain important features, such as concentrating the enzymes and removing contaminants that affect the sensitivity of the PAP assay. The more important contaminants are PAP, luciferin, and inhibitors of luciferase. The procedure takes into account the fact that both luciferase and luciferin sulfokinase are very unstable during the initial stages of the isolation procedure and the fact that *Renilla* homogenates contain a slimy material that tends to clog chromatographic columns. Some of this slime can be removed with adsorption to alumina gel, and most of the rest by gel filtration. The enzyme mixture can then be concentrated on DEAE-cellulose and most of the PAP and luciferase inhibitors can also be removed by further gel filtration. The final contaminant, luciferin, is removed in the last step, which involves the use of DEAE-Agarose. This step also concentrates the enzymes.

Luciferin sulfokinase activity is labile but can be stabilized by the continuous presence of β-mercaptoethanol. Therefore, 50 μl of β-mercaptoethanol are added to each liter of all the buffers used throughout the preparation and storage of the enzymes. It may also be necessary to add additional β-mercaptoethanol to enzyme solutions that have been frozen and thawed. All steps of the isolation procedure through the second gel filtration step must be carried out in a continuous fashion because of enzyme instability. Storage of the enzymes following the DEAE-cellulose step should especially be avoided. All procedures should be carried out at 4° using deionized water and glassware free of any trace of dichromate ions. Solutions used for column equilibration and elution are made up in standard buffer (Tris · HCl, pH 7.5 at 25°, 1.0 m*M*; EDTA, neutral Tris salt, 0.2 m*M*) unless otherwise stated.

Obtaining Live Animals. Although *Renilla reniformis* was used for all the procedures developed for this chapter, the biochemistry of the bioluminescent reactions has been found to be the same for all the alcyonarian coelenterates which have been tested (i.e., those from *Renilla, Cavernularia, Ptilosarcus, Acanthoptilum,* and *Stylatula*).[10] Presumably any bioluminescent alcyonarian coelenterate could be used as a source of luciferase and luciferin sulfokinase. However, the chromatographic properties of these proteins show species differences, and so, use of a species other than *Renilla reniformis* could require modification of the isolation procedure.

Renilla reniformis can be collected in large numbers off the coast of Georgia. *R. mulleri* can be obtained from the north Gulf coast of Florida or can be purchased from the Gulf Specimen Co., Inc., P. O. Box 237, Panacea, Florida 32346. *R. kollikeri, Ptilosarcus guerneyi, Stylatula elongata,* and *Acanthoptilum gracile* can be collected from the coast of California or purchased from Pacific Bio-Marine Laboratories, Inc., P. O. Box 536, Venice, California 90291.

Pretreatment of Live Animals. Live animals can be used fresh, or they can be relaxed and frozen for indefinite storage. The relaxing procedure was developed originally to preserve the activity of the calcium-triggered luciferin binding protein discussed in this volume [27], but the procedure increases the yields of the other bioluminescent proteins as well. Relaxed and frozen animals must be stored at $-80°$ in order to retain maximum enzyme activities. The enzyme isolation procedures described here utilize relaxed and frozen animals. The use of live animals would require an increase in the ionic strength of the homogenizing medium to prevent enzyme losses during the alumina adsorption step. Live animals are relaxed by placing 25 g (wet weight) of animals into 1 liter of relaxing medium: $Mg(OAC)_2$, 300 m*M;* EGTA, 20 m*M;* Tris base, 50 m*M;* solution pH 7.8 at 25°. A large tray should be used so as to position the animals in a shallow layer. The animals are exposed to relaxing medium for a total of 45 min, a change to fresh relaxing medium occurring every 15 min. Experience has shown that each relaxing medium bath can be used up to 5 times with fresh animals prior to discarding it. The animals are then quick-frozen by dropping them into liquid N_2; they must be maintained at $-60°$ to $-80°$ or lower for storage purposes.

Relaxing medium is made by dissolving 144.6 g of magnesium acetate in 1 liter of deionized water and dissolving 3.8 g of NaOH, 13.6 g of Tris base, and 17.2 g of EGTA in a second liter of water. The two solutions are mixed by slowly adding the Tris-EGTA solution to the magnesium acetate solution with stirring, and the mixture is brought to an overall volume of 2.5

[10] M. J. Cormier, K. Hori, Y. D. Karkhanis, J. M. Anderson, J. E. Wampler, J. G. Morin, and J. W. Hastings, *J. Cell. Physiol.* **81,** 291 (1973).

liters. The pH of the final solution should be 7.8 and there should be no precipitate.

Isolation of Enzymes

Relaxed and frozen *Renilla* (150 g) are homogenized in 600 ml of homogenizing buffer (Tris · HCl, pH 7.5 at 25°, 1.0 m*M*; EDTA, neutral Tris salt, 1.0 m*M*) to which is added 2.5 ml of β-mercaptoethanol. Homogenization is accomplished with a Willems Polytron PT 20st, set at full speed, for 30 sec. The resulting homogenate is then centrifuged at 14,000 *g* for 10 min. To the resultant supernatant, alumina gel (Amphojel, Wyeth Laboratories Inc.) is added (3.3 ml/100 ml) with continuous stirring. The alumina gel suspension should be uniformly mixed prior to use. The alumina gel is then removed by centrifugation at 10,000 *g* for 5 min.

The alumina gel-treated supernatant is then passed through a BioGel P-100 column (Bio-Rad, 12 × 12 cm bed volume) in order to remove contaminating salt and slime; 125-ml fractions are collected, and the most active enzyme fractions are saved using care to avoid contamination with the salt volume.

The desalted homogenate is then loaded onto a DEAE-cellulose column (Whatman DE-22, 3 × 9 cm) followed by a 50-ml wash of 80 m*M* NaCl. The enzyme mixture is then pulsed off the column with 600 m*M* NaCl and 48 m*M* sodium sulfanilate. The resultant peak enzyme fractions are immediately passed through a Sephadex G-100 column (Pharmacia, 6 × 54 cm). At this stage the resulting enzyme mixture is fairly stable and may be stored overnight at 4°.

The Sephadex-treated enzyme mixture is then loaded onto DEAE-Agarose (DEAE-BioGel A, 2 × 13 cm). Luciferase and luciferin sulfokinase are eluted from this column by using a 500-ml linear gradient starting with solution A and ending with solution B.

Solution A: Tris · HCl, 1.0 m*M*, pH 7.5 at 25°; EDTA, neutral Tris salt, 0.1 m*M*; $CaCl_2$, 0.5 m*M*

Solution B: NaCl, 400 m*M*; HEPES, 5.0 m*M*, pH 7.6 at 25°; EDTA, neutral Tris salt, 1.0 m*M*

The calcium chloride discharges luciferin from the luciferin-binding protein, thus removing all contaminating luciferin from the enzyme mixture. The luciferase elutes, with peak activity, about 16 ml in front of the luciferin sulfokinase, so that extra fractions should be taken on the front side of the sulfokinase peak. The overall procedure should yield about 70 ml of a highly active mixture of luciferin sulfokinase and luciferase with an overall yield of 35%. The enzyme mixture is only partially stable at 4°, and some loss will occur in only 24 hr. Therefore, the enzyme mixture should be divided into amounts sufficient for a day's use and should be frozen at −80°, where it can be stored for at least several months.

One unit of enzyme activity is defined as that amount of protein which generates 5.0×10^{14} photons per second at peak intensity at 25°. Using a 5% quantum yield for bioluminescence and the luminol light standard of Lee and Seliger,[11] one can convert photons per second to conventional enzyme units. Thus one can calculate that 1 enzyme unit will convert 1 μmol of luciferyl sulfate per minute to oxidized products and light. Using this definition, 150 g of relaxed-frozen *Renilla* should yield 4.6×10^{-3} units of enzyme activity. It should be emphasized that the unit refers to the combined activities of luciferase and luciferin sulfokinase.

Synthesis of Benzyl Luciferyl Sulfate

Preliminary Considerations. Although it is possible to isolate approximately 5 mg of luciferyl sulfate from 40,000 *Renilla*,[12] the process is tedious and costly. The synthetic route, although complex, is a preferable source of material. Benzyl luciferin [(I), Fig. 1] is a fully active analog of coelenterate-type luciferin [(II), Fig. 1]. Both compounds are oxidized by *Renilla* luciferase at equal rates and produce the same quantum yields and emission spectra. Furthermore, the sulfonated forms of (I) and (II) [i.e., (V) and (VI), Fig. 1] are equally effective as substrates for the PAP-dependent luciferin sulfokinase. Since the synthesis of benzyl luciferyl sulfate [(V), Fig. 1] presents fewer problems than the synthesis of luciferyl sulfate [(VI), Fig. 1], details for the synthesis of the former compound are presented here. The procedure involves the synthesis of the methyl ether of benzylluciferin [(III), Fig. 1], which is then demethylated to yield benzylluciferin. Benzylluciferin is then sulfonated to yield benzyl luciferyl disulfate [(IV), Fig. 1]. This is then converted to benzyl luciferyl sulfate [(V), Fig. 1] by selective removal of one sulfate group with aryl sulfatase. These procedures are outlined below.

The synthesis of benzyl luciferyl sulfate via benzylluciferin is described below, and the procedures outlined are straightforward. However, both benzylluciferin, and all the intermediates to benzyl luciferyl sulfate, are potentially labile molecules, and so certain precautions are necessary for their safe storage and handling. Benzylluciferin (I) and its derivative (III)

[11] J. Lee, A. S. Wesley, J. F. Ferguson, and H. H. Seliger, *in* "Bioluminescence in Progress" (F. H. Johnson and Y. Haneda, eds.), p. 35. Princeton Univ. Press, Princeton, New Jersey, 1966.

[12] K. Hori, Y. Nakano, and M. J. Cormier, *Biochim. Biophys. Acta.* **256,** 638 (1972).

are stable as dry powders kept in a desiccator. Solutions of benzylluciferin and (III) are reasonably stable if kept anaerobic and neutral or, preferably, acidic. Basic solutions should be avoided. In contrast, benzyl luciferyl sulfate and the disulfate are not autoxidizable. Furthermore, they are not stable in acidic solutions since the enol-sulfate linkage is labile under these conditions. Benzyl luciferyl sulfate is stable in solutions maintained near pH 7.5 and may be stored at $-20°$ for months in this way.

Synthesis Procedure. Compound (III) is prepared by the condensation of the methyl ether of 2-amino-3-benzyl-5(*p*-methoxyphenyl)pyrazine (etioluciferin) with the diethyl acetal of benzyl glyoxal. The synthesis of these starting materials has been previously described.[13,14] The diethyl acetal of benzylglyoxal (68 mg) is added to the methyl ether of etioluciferin (97 mg) dissolved in 5 ml of ethanol. The mixture is cooled in an ice bath, and 5 ml of 60% ethanol–6 *N* HCl (1 : 1) are added. The test tube containing the reaction mixture is placed in a Dry Ice acetone bath, sealed under an argon atmosphere, and heated for 2 hr at 110°. The sealed tube is cooled and opened, and the solvent is removed at 50° under a stream of argon gas. The residue is dissolved in 10 ml of 80% methanol–0.01 *N* HCl[15] and filtered. The brownish filtrate is chromatographed on a 4 × 20 cm column of LH-20 (Sigma) that has previously been equilibrated with 80% methanol–0.01 *N* HCl. This same solvent is used to elute the product which appears as a yellow-fluorescent band on the column. The yellow-fluorescent fraction, approximately 15 ml, is collected and concentrated to approximately 0.5 ml with argon. This is done in order to precipitate the product. The resulting yellow precipitate is separated by centrifugation and dried *in vacuo* over KOH pellets. The yield of (III) is 55–65 mg (40–45%).

Under an argon atmosphere, anhydrous pyridine hydrochloride (1 g) is added to (III) (40 mg) in a thick (1.5 mm) walled Pyrex tube. The tube containing the reaction mixture is placed in a Dry Ice acetone bath and sealed under argon. The sealed tube is first heated in an oil bath at 160° until the solid is melted, and then heated at 210° for 2 hr. The reaction mixture is cooled in a Dry Ice acetone bath and the seal is broken. When gas evolution has ceased, 5 ml of 80% methanol–0.1 *N* HCl[16] is added and the solvent is then removed at 50° under a stream of argon gas. The residue is dissolved in 10 ml of 80% methanol–0.01 *N* HCl, and filtered. The brownish filtrate is chromatographed on LH-20 as described for the preparation of (III). The product is also collected, concentrated, and dried as described for the synthesis of (III). The yield of (I) is 30 mg (80%).

[13] H. D. Dakin and H. W. Dudley, *J. Biol. Chem.* **18,** 29 (1914).
[14] Y. Kishi, H. Tanino, and T. Goto, *Tetrahedron Lett.* **27,** 2747 (1972).
[15] Methanol, 80%, made 0.01 *N* with 12 *N* HCl.
[16] Methanol, 80%, made 0.1 *N* with 12 *N* HCl.

The procedure used for the synthesis of (IV) is a modification of those reported for the synthesis of sulfate esters of alcohols and phenols.[17,18] Although the procedure given is for small-scale preparations, it should be possible to operate on a larger scale. In scaling up, precautions must be taken to maintain the pH of luciferyl sulfate solutions near neutrality since the enol-sulfate linkage is very acid labile. For small-scale preparations anhydrous pyridine (5 ml) is added to sulfamic acid (3 mg) in a tube. The tube is then placed in a Dry Ice acetone bath and sealed under argon. The sealed tube is heated at 110° for 1 hr and allowed to cool. Then, the resulting suspension (containing undissolved sulfamic acid) is added to (I) (1 mg) in a tube. This tube is placed in a Dry Ice acetone bath, sealed under argon, and heated at 110° for 30 min. The tube is allowed to cool, the seal is broken, and the pyridine is removed at 90° under a stream of argon gas. One milliliter of 1 *M* K_2HPO_4 is added to the residue, and the mixture is heated at 80° for 5 min. The solution is cooled and filtered. Four milliliters of methanol are added dropwise to the almost colorless filtrate, and the precipitate is removed by centrifugation. The supernatant is chromatographed on a 4 × 8 cm LH-20 column that has been equilibrated with 80% methanol–10 m*M* phosphate buffer, pH 7.5, and eluted in the same solvent. Fifteen-milliliter fractions are collected at the rate of 60 ml/hr and the fractions are monitored by measuring the absorbance at 266 nm, the absorption maximum of (IV). Most of the desired product is located in fractions 44–47.

Selective removal of the sulfate group located at the *p*-hydroxyphenyl position of (IV) is achieved by treatment with aryl sulfatase (EC 3.1.6.1). The four major fractions of (IV) are combined and concentrated to 4 ml by flushing with argon at 45°. Four milliliters of methanol are added dropwise, the resulting precipitate is removed by centrifugation, and the solution is concentrated to 2 ml as described above. Two milliliters of 10 m*M* phosphate buffer (pH 7.5) are added to the solution of (IV) followed by 0.1 mg of aryl sulfatase. The mixture is allowed to stand at 25° for 24 hr in the dark. Sixteen milliliters of methanol are then added, and the resulting precipitate is removed by centrifugation. The product is purified by chromatography on LH-20 as outlined above for (IV). The 15-ml fractions are monitored by measuring the absorbance at 272 nm, the absorption maximum for (V) in 80% methanol.[12] The Σ_{mM} for (V) in methanol–phosphate buffer is 40 at 272 nm. Most of the desired product is located in fractions 34–36. The yield of (V) starting from compound (I) is 0.85–0.9 mg (70–75%). Although this represents a relatively small amount of material, 1 mg of benzyl luciferin sulfate is sufficient for over 16,000 assays.

[17] S. Yamaguchi, *Nippon Kagaku Zasshi* **80,** 171 (1959).
[18] S. Yamaguchi, *Nippon Kagaku Zasshi* **80,** 204 (1959).

The three major fractions of (V) are combined and concentrated at 50° to 1 ml under a stream of argon gas. Phosphate buffer (1.0 mM, pH 7.5) is added to dilute the sample to the desired concentration for assay purposes. The Σ_{mM} for (V) in phosphate buffer, pH 7.5 is 40 at 268 nm, the absorption maximum for (V) in aqueous solution at pH 7.5.

In our experience, the benzyl luciferyl sulfate associated with the major chromatographic fractions is >90% pure. The absorption characteristics of these fractions should be similar to that reported earlier for luciferyl sulfate.[12] If they are similar, this generally indicates a high degree of purity and is usually sufficient for assay purposes. A direct determination of purity of a given column fraction is also possible, providing the photometer being used is calibrated with a light standard. All the work reported here relied on the luminol light standard of Lee *et al.*[11] Although several different types of standards are available, the user should be aware that they may result in calibration differences of approximately 3-fold (see also this volume [44]). Using an instrument calibrated with the luminol light standard, a column fraction representing highly purified benzyl luciferyl sulfate should yield the following relationship between absorbance and activity. Assume that a column fraction has an absorbancy of 0.4 at 272 nm. If this fraction is diluted 1:100 into 10 mM phosphate buffer (pH 7.5) and 10 μl of this dilution is used in the assay procedure described below with saturating amounts of PAP, then the total number of photons generated should be 3.3×10^{10}.

Determination of PAP

Assay Procedure. The assay for PAP is done in test tubes containing 0.8 ml of 20 mM HEPES, pH 7.6 at 25°, 100 mM NaCl, and 1.0 mM $CaCl_2$. To this assay mixture is added 0.1 ml of the enzyme mixture and 25 μl of a 5.0 μM solution of luciferyl sulfate. The reaction is initiated by addition of 0.1 ml of a PAP solution ranging in concentration from 10^{-9} to $10^{-7} M$. The final concentration of all the assay medium constituents is as follows: HEPES, pH 7.6, 16 mM; NaCl, 80 mM; $CaCl_2$, 0.8 mM; EDTA, 0.1 mM; Luciferyl sulfate, 125 nM; PAP, 0.1–10 nM; enzyme, 3×10^{-6} to 6×10^{-6} units.

Light emission is followed using a strip-chart recorder until the maximum intensity is reached (Fig. 2). This value is used for all PAP determinations, since it shows the smallest variance in multiple determinations. All light measurements are done using a photometer similar to the one described by Anderson *et al.* in this volume [41].

Sensitivity of the PAP Assay. All preparations of luciferin sulfokinase contain a background level of PAP, which is not easily removed. The multistep procedure for preparing the enzyme mixture, described above,

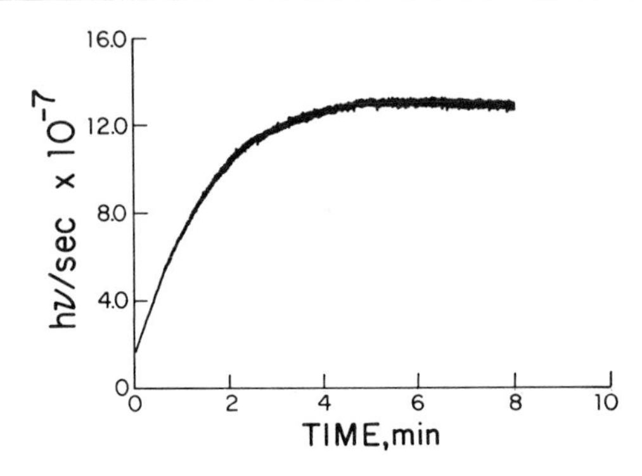

FIG. 2. Time course of 3,5′-diphosphoadenosine (PAP)-dependent bioluminescence. Reaction was initiated with addition of luciferyl sulfate. The PAP in the assay was 5.8 pmol. See text for details.

removes much of the PAP, but the remaining PAP contamination sets the lower limit to the sensitivity of the assay for added PAP. In this procedure the detectable limit for PAP in a test solution is 1 pmol and in the assay is 0.1 pmol. If a method could be devised for complete removal of this contaminating PAP, greater sensitivity of at least two orders of magnitude ought to be realized. The basal level of light intensity, produced by the contaminating PAP, is determined for each enzyme preparation, and this value is subtracted from all subsequent measurements.

Increases in the sensitivity should be realized by using saturating levels of luciferyl sulfate in the assay. However, the use of increased levels of luciferyl sulfate is not possible since it is a potent competitive inhibitor of luciferase.[19] The maximum usable concentration of luciferyl sulfate is 125 n*M*. Sensitivity has been increased 10-fold over that obtained by Stanley *et al.*[8] in a similar assay technique. This is largely because the present enzyme isolation procedure was not available and because we were able to furnish only limiting quantities of native luciferyl sulfate at the time. The sensitivity also represents a 50-fold increase over that obtained by Gregory and Lipmann[4] using a colorimetry-based assay for PAP that utilizes *p*-nitrophenylsulfate and phenol sulfokinase.

Calibration Curve for PAP. The relationship of light output to PAP concentration is linear up to 0.15 μM, at which point saturation of the enzyme occurs (Fig. 3A). Thus generation of a calibration curve is a simple procedure requiring only a few experimental points to determine the relationship between light output and PAP concentration. This linear function

[19] J. C. Matthews, K. Hori, and M. J. Cormier, *Biochemistry* **16,** 5217 (1977).

must be determined for every enzyme mixture, since the ratio of luciferase to luciferin sulfokinase, and the contaminating levels of PAP, vary from preparation to preparation. For concentrations of added PAP below 10 nM, a separate calibration curve is required (Fig. 3B), which uses twice the normal enzyme concentration in the assay. The increased enzyme level increases the signal-to-noise ratio for low PAP concentrations.

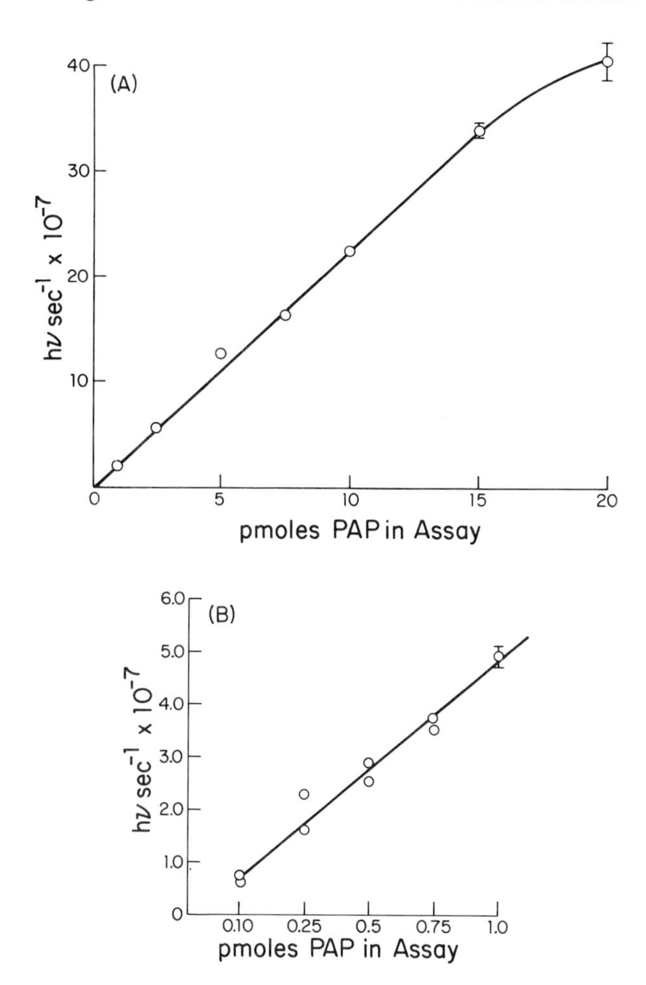

FIG. 3. Relationship between luminescence intensity and 3,5′-diphosphoadenosine (PAP) concentration. Reactions were initiated by addition of luciferyl sulfate. See text for details. (A) For PAP concentrations above 10^{-8} M, the enzyme concentration was 3×10^{-6} unit. Where error bars are not shown, they are within the encircled experimental points. (B) For PAP concentrations below 10^{-8} M, the enzyme concentration was 6×10^{-6} unit. Either error bars or all points taken are shown.

Utility of the Assay

In order to demonstrate the applicability of the assay technique, the PAP and PAPS levels in several tissues were tested using the bioluminescent technique described above. Table I shows that PAP was found in every tissue tested, and the levels were easily detected by the bioluminescent assay. The lowest level of PAP we measured was found in spinach leaves, and this level was still two orders of magnitude above the lower limit of sensitivity of the assay. Since the extraction method used hydrolyzed PAPS to PAP, the data in Table I show values for the level of PAP plus PAPS. If PAPS is protected from hydrolysis, the independent levels of both PAP and PAPS can be determined. Subsequent acidification of the sample at 37° for 30 min will quantitatively convert the PAPS to PAP. Stanley *et al.*[8] have shown that the bioluminescent assay for PAP can be used successfully for following the enzymic formation of PAPS in tissue extracts of yeast and green plants. The increase in PAPS is easily followed by hydrolyzing

TABLE I
PAP LEVELS IN TISSUE[a]

Tissue source	PAP (nmol/g fresh wt.) Bioluminescence assay	PAP (nmol/g fresh wt.) Phenol sulfokinase assay[b]	Percent loss of PAP in control
Rabbit			
Liver	17.0	18.0	0
Lung	5.5	3.5	0
Heart	5.4	—	6.3
Kidney	14.0	3.0	8.4
Bacteria			
Photobacterium fischeri	2.4	—	13
Green plants			
Potato tuber	0.68	—	45
Spinach leaves	0.25	—	75

[a] Five grams of tissue were homogenized in 50 ml of 0.1 *N* HCl using the Willems Polytron. After removal of debris by centrifugation, the homogenate was incubated at 37° for 30 min to convert PAPS to PAP. The resulting solution was neutralized with NaOH and buffered with HEPES to pH 7.6. Two controls were run for each tissue tested. In the first, a known amount of PAP was added to the assay medium to determine any inhibition of the assay by added sample. The second was an internal standard (5.4 *M* synthetic PAP) added to a second set of tissue samples prior to homogenization to determine the amount of PAP lost during the extraction process. All values in the table were corrected for the loss found in the controls. The values obtained from the bioluminescence assay are compared with those available from the phenol sulfokinase assay.

[b] From J. D. Gregory and F. Lipmann, *J. Biol. Chem.* **229,** 1081 (1957).

aliquots of the reaction mixture and assaying the resultant PAP by the bioluminescent technique.

Tissue levels of PAP have also been investigated by use of a phenol sulfokinase coupled assay technique.[4] It is interesting to note that the levels of PAP found in rabbit liver and lung by the phenol sulfokinase method agree favorably with those determined by the bioluminescent technique (Table I). However, the phenol sulfokinase assay for PAP is quite laborious whereas the bioluminescence assay is simpler and provides a more rapid and sensitive assay for both PAP and PAPS.

Bioluminescence Assay for Aryl Sulfatase. Aryl sulfatase will selectively remove the sulfate group located at the *p*-hydroxyphenyl position of luciferyl disulfate [(IV), Fig. 1] as discussed earlier. This reaction provides a basis for a bioluminescence assay for aryl sulfatase. This could be accomplished by coupling aryl sulfatase with the luciferin sulfokinase–luciferase reaction as illustrated below:

$$\text{Benzyl luciferyl disulfate} \xrightarrow{\text{aryl sulfatase}} \text{benzyl luciferyl sulfate} + \text{sulfate}$$

$$\text{Benzyl luciferyl sulfate} + \text{PAP} \xrightarrow{\text{luciferin sulfokinase}} \text{benzylluciferin} + [\text{PAPS}]$$

$$\text{Benzylluciferin} + O_2 \xrightarrow{\text{luciferase}} \text{oxyluciferin} + CO_2 + \text{light}$$

[29] Protein–Protein Interactions as Measured by Bioluminescence Energy Transfer in *Renilla*[1]

By WILLIAM W. WARD and MILTON J. CORMIER

Protein–protein interactions (interactions between distinct and separate polypeptide chains) occur in a variety of biochemical systems. Classical examples include (1) proteins containing multiple subunits, such as hemoglobin and aspartate transcarbamylase; (2) self-associating oligomeric proteins, such as glycogen phosphorylase and glutamate dehydrogenase, where association–dissociation behavior may play a role in the *in vivo* control of activity; (3) protein self-assembly

[1] This work was supported in part through grants to M.J.C. (NSF, BMS 74-06914 and ERDA, AT-38-1-635) and to W.W.W. (NIH, 1 F32 EY 05104-01, and Rutgers Res. Council, 07-2312).

ISBN 0-12-181957-4

systems, such as those of the muscle proteins myosin and actin, the microtubule protein tubulin, and the various viral coat proteins; (4) antigen–antibody systems; and (5) multienzyme complexes, such as the pyruvate dehydrogenase complex, where several oligomeric enzymes combine to produce a functional particle.

Several bioluminescence systems are also known to involve protein–protein interactions. Firefly luciferase, for example, is active as a 50,000 molecular weight monomer at low protein concentrations, but at higher protein concentrations, in low ionic strength buffer, the monomers associate to form fully active dimers.[1a,2] The active form of bacterial luciferase is a dimer of nonidentical subunits.[3] The separated α and β subunits, each with molecular weights of about 40,000, are inactive, but they can be reassociated with restoration of catalytic activity.[4] Luciferases isolated from the crustacean *Cypridina*,[5] the earthworm *Diplocardia*,[6] and the dinoflagellate *Gonyaulax*[7] also appear to contain multiple subunits. Furthermore, there are at least three bioluminescence systems in which a protein-bound luciferin (luciferin-binding protein) is thought to interact *in vivo* with a luciferase via protein–protein interaction. Such systems have been described in *Gonyaulax*,[8] in the mollusk *Pholas*,[9] and in an anthozoan coelenterate, the sea pansy *Renilla reniformis*.[10,11] In the sea pansy, and in several other anthozoan coelenterates, luciferase[12] can also be shown to associate with an accessory protein known as the green-fluorescent protein (GFP).[13] This latter association functions in the sea pansy to alter the color of bioluminescence. During the lifetime of the transient association between luciferase and GFP, excitation energy from enzyme-bound oxyluciferin is transferred to an unknown chromophore on the GFP, resulting in green light emission from the complex (see this volume [27]).

In this chapter we will present experimental evidence for protein–protein interaction between luciferase and GFP in the *Renilla* system and

[1a] J. L. Denburg and W. D. McElroy, *Biochemistry* **9**, 4619 (1970).
[2] M. DeLuca, *Adv. Enzymol.* **44**, 37 (1976).
[3] J. Friedland and J. W. Hastings, *Proc. Natl. Acad. Sci.* U.S.A. **58**, 2336 (1967).
[4] J. W. Hastings, K. Weber, J. Friedland, A. Eberhard, G. W. Mitchell, and A. Gunsalus, *Biochemistry* **8**, 4681 (1969).
[5] F. I. Tsuji, R. V. Lynch, III, and C. L. Stevens, *Biochemistry* **13**, 5204 (1974).
[6] R. Bellisario, T. E. Spencer, and M. J. Cormier, *Biochemistry* **11**, 2256 (1972).
[7] N. Krieger, D. Njus, and J. W. Hastings, *Biochemistry* **13**, 2871 (1974).
[8] M. Fogel and J. W. Hastings, *Arch. Biochem. Biophys.* **142**, 310 (1971).
[9] J.-P. Henry, M. F. Isambert, and A. M. Michelson, *Biochim. Biophys. Acta* **205**, 437 (1970).
[10] J. M. Anderson, H. Charbonneau, and M. J. Cormier, *Biochemistry* **13**, 1195 (1974).
[11] M. J. Cormier and H. Charbonneau, *in* "Calcium Binding Proteins and Calcium Function" (R. H. Wasserman *et al.*, eds.), p. 481. Elsevier, Amsterdam, 1977.
[12] J. C. Matthews, K. Hori, and M. J. Cormier, *Biochemistry* **16**, 85 (1977).
[13] W. W. Ward and M. J. Cormier, *Photochem. Photobiol.* **27**, 389 (1978).

describe how this interaction can be measured by the assay of *in vitro* energy transfer. We hope, that in describing this bioluminescence energy-transfer system, we may also be providing a model to aid in the understanding of more complex interacting systems, such as radiationless energy transfer in the photosynthetic systems.

Energy Transfer in Coelenterates

In many bioluminescent coelenterates the color in the *in vivo* light emission is green whereas their soluble *in vitro* systems produce blue light. This color difference results from the presence *in vivo* of a fluorescent accessory protein, GFP, GFPs were first implicated in coelenterate bioluminescence in the hydrozoan jellyfish *Aequorea*,[14] where an absorption and reemission mechanism for energy transfer was suggested. Later it was suggested that similar proteins also exist in the hydrozoan *Obelia*[15] and the anthozoan *Renilla*,[16] where they may serve as acceptors of radiationless energy transfer from excited-state oxyluciferin molecules.

The first of these proteins to be purified and to be assayed for energy-transfer ability was the *Aequorea* GFP.[17] In agreement with the original authors,[14] we interpret the results of their *in vitro* studies with this protein[17] to mean that, in *Aequorea*, energy transfer is not a radiationless process but proceeds by the trivial route of absorption and reemission. Subsequent attempts to demonstrate radiationless energy transfer with *Aequorea* GFP have also failed,[13] and the system in *Obelia* has not been pursued. We have, however, shown that the *in vitro* energy transfer process in *Renilla* is clearly radiationless.[13,18] This conclusion is based primarily on the fact that the bioluminescence quantum yield for luciferin oxidation increases by a factor of three with prior addition of GFP to a luciferin–luciferase reaction (see Fig. 3 of M. J. Cormier, this volume [27]). Several other lines of evidence in support of this conclusion will follow.

In Vitro Assay

Dual-Phototube Photometer

Assays for energy transfer in the soluble *Renilla* system were performed with a dual-phototube photometer similar to that described by

[14] F. H Johnson, O. Shimomura, Y. Saiga, L. C. Gershman, G. T. Reynolds, and J. R. Waters, *J. Cell. Comp. Physiol.* **60,** 85 (1962).
[15] J. G. Morin and J. W. Hastings, *J. Cell. Physiol.* **77,** 313 (1971).
[16] J. E. Wampler, K. Hori, J. W. Lee, and M. J. Cormier, *Biochemistry* **10,** 2903 (1971).
[17] H. Morise, O. Shimomura, F. H. Johnson, and J. Winant, *Biochemistry* **13,** 2656 (1974).
[18] W. W. Ward and M. J. Cormier, *J. Phys. Chem.* **80,** 2289 (1976).

Wampler.[19] Light from the luminescent source, collimated by passage through an f/1 planoconvex lens, was split into two perpendicular beams by a prismatic beam splitter. Each emerging beam was filtered through a narrow-bandpass interference filter before striking the photocathode of an "end-on" photomultiplier tube. The filters (Ditric Optics, Marlboro, Massachusetts) were selected for maximum transmission at 480 nm and 510 nm in order to correspond with the emission peaks of the blue reaction (luciferin and luciferase) and the green reaction (luciferin, luciferase, and GFP), respectively. Phototube signals were separately displayed or electrically divided and displayed as the ratio of phototube outputs. Algebraic manipulations of the data, described earlier,[13,18] were used to generate a numerical value for the "percent green pathway" (number of luciferin molecules in a reaction whose oxidation ultimately leads to green light emission divided by the number of luciferin molecules that generate blue light).

Protein Isolation

Renilla luciferase was isolated from the sea pansy *R. reniformis* and was purified to homogeneity as a monomeric protein with a molecular weight of 35,000 as previously described.[12] Using procedures described elsewhere,[19a] the *Renilla* GFP was isolated from the same species and purified 12,000-fold to homogeneity with 26% overall recovery of fluorescence units. The purified *Renilla* GFP exists as a stable dimer of identical subunits, each with a molecular weight of 27,000. A covalently bound chromophore, whose chemical structure is presently unknown, is responsible for the protein's intense visible absorption band (λ_{max} = 498 nm, ϵ^{M}_{498} = 266,000) and fluorescence emission band (λ_{max} = 509 nm, ϕ_{fl} = 80%).[13]

Assay Conditions and Sensitivity

Assays of *in vitro* energy transfer were performed at room temperature in 10 × 75 mm test tubes containing 200-μl final volumes of assay buffer (1 m*M* Tris · HCl, 0.1 m*M* EDTA, pH 8.0). In addition to buffer, the final assay solution included luciferase (1–100 n*M*), luciferin (0.1–10 μM), and green-fluorescent protein (0.01–1 μM). Although the luminescent reaction can be initiated with either luciferase or luciferin, for convenience we routinely initiated the reaction by adding 2 μl of an anaerobic methanolic solution of luciferin from a push-button Hamilton syringe. The source of luciferin for

[19] J. E. Wampler, *in* "Analytical Application of Bioluminescence and Chemiluminescence" (E. W. Chappelle and G. L. Picciolo, eds.), p. 105 (NASA Publ. No. SP-388) National Aeronautics and Space Administration, Washington, D.C., 1975.

[19a] W. W. Ward and M. J. Cormier, *J. Biol. Chem.*, manuscript submitted (1978).

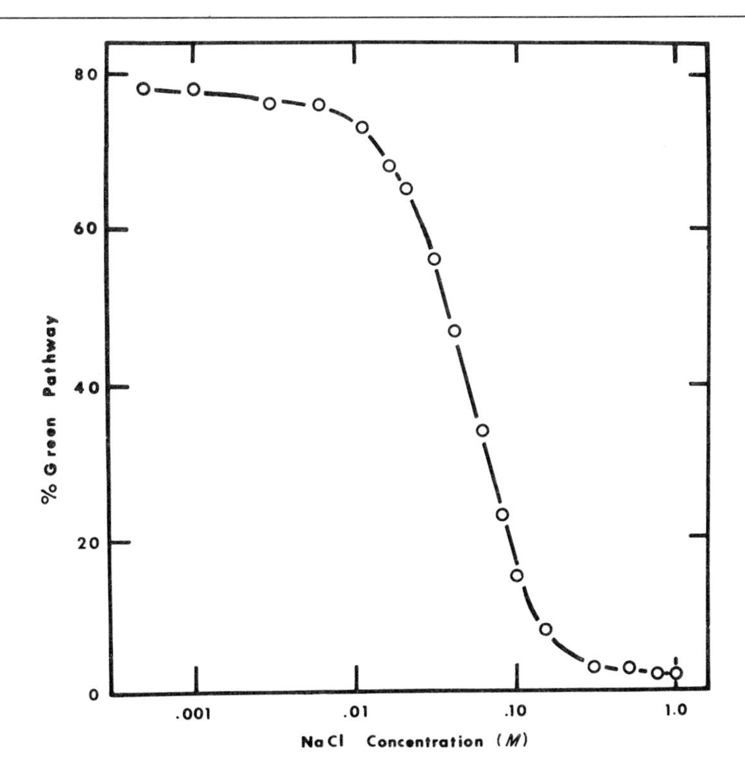

FIG. 1. The effect of ionic strength on the percentage of luciferin molecules that emit via the green bioluminescence pathway (% green pathway). The concentrations of green-fluorescent protein (1.0 μM) and luciferase (~10 LD nM) were held constant. The pH 8.0 buffer contained 1 mM Tris · HCl, 0.1 mM EDTA, and varying amounts of NaCl to achieve the desired ionic strength. Reproduced, with permission, from W. W. Ward and M. J. Cormier, *Photochem. Photobiol.*, **27,** 389 (1978). Pergamon Press, Ltd.

these assays was the synthetic benzyl analog of luciferin, which was prepared by Dr. K. Hori[20] or Dr. R. Hart.

The energy-transfer assay (percent green pathway) was found to be relatively independent of temperature (0°–35°) and pH (6–9). However, as shown in Fig. 1, the assay is highly dependent upon the ionic strength of the assay buffer, especially in the 0.01 to 0.10 M range (when NaCl is varied). From the data shown in Fig. 2, the assay sensitivity is seen to range from 0.01 to 1 μM GFP. Thus, under these assay conditions, the limit of detection of energy transfer function approaches 2 pmol of GFP. For reasons that we do not presently understand, the percent green pathway does not

[20] K. Hori, J. E. Wampler, J. C. Matthews, and M. J. Cormier, *Biochemistry* **12,** 4463 (1973).

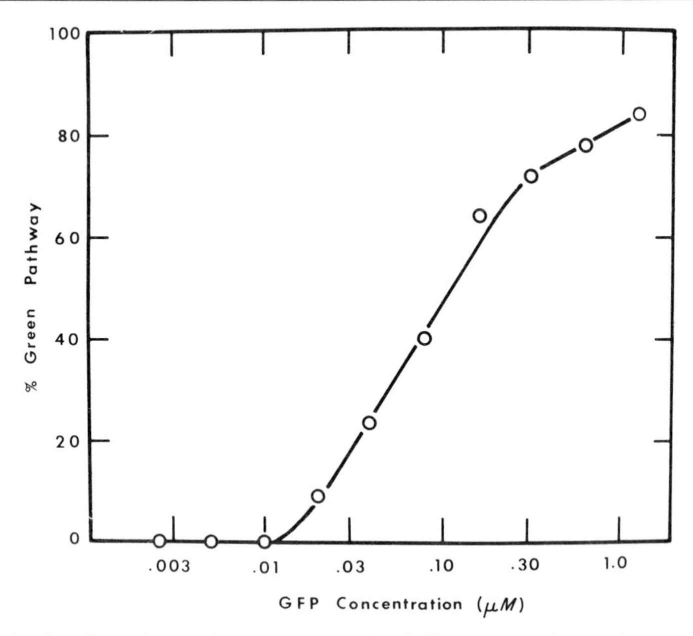

FIG. 2. The effect of green-fluorescent protein (GFP) concentration on the percentage of luciferin molecules that emit via the green bioluminescence pathway (% green pathway). Luciferin and luciferase concentrations were both 30p*M* and the pH 8.0 buffer contained 5 m*M* Tris·HCl and 0.1 m*M* EDTA. Reproduced, with permission, from W. W. Ward and M. J. Cormier, *Photochem. Photobiol.*, **27**, 389 (1978) Pergamon Press, Ltd.

vary with luciferase concentration, even when the enzyme concentration is varied over six orders of magnitude.

Energy-Transfer Mechanism

Electronic energy transfer between a donor molecule, D, and an acceptor molecule, A, may occur by one of the following two mechanisms. In the first case, an excited-state donor, D*, emits light, which is then absorbed by molecule A and reemitted with the fluorescence characteristics of A. The maximum quantum yield attainable by this trivial mechanism is the product of the fluorescence quantum yields of donor and acceptor molecules $(\phi_{fl,D}) \cdot (\phi_{fl,A})$. Thus addition of an acceptor molecule at any concentration cannot increase the overall fluorescence quantum yield of a system operating by trivial transfer. In the second mechanism, known as radiationless (or resonance) energy transfer, D* transfers its excitation energy directly to A, a molecule with which it is in resonance. Because the efficiency of excited-state production (D→D*) may be much greater than the fluorescence quantum yield of D, it is possible to increase the overall fluorescence quantum yield of a resonance energy transfer system by the selection of a suitable acceptor molecule, A.

The fact that sensitization with *Renilla* GFP triples the bioluminescence quantum yield of luciferin oxidation[13] strongly argues in favor of the second mechanism—radiationless energy transfer. Furthermore, because the efficiency of resonance energy transfer is inversely proportional to the 6th power of the distance between donor and acceptor chromophores,[21] and because, in *Renilla*, energy transfer is relatively efficient, it is necessary that the donor, oxyluciferin, and the acceptor, GFP, be in close proximity. In homogeneous solution, such proximity can be achieved at high concentrations of donor and acceptor (approaching millimolar). But, as previously shown (Fig. 2), maximum energy transfer efficiency is observed at micromolar concentrations in the *Renilla* system. Thus a mechanism must exist to increase the proximity of these chromophores in dilute solutions. Evidence indicates that excited-state oxyluciferin is firmly bound to luciferase following the enzymic oxidation of luciferin,[12,22] and that the GFP chromophore is covalently bound to the apoprotein.[13] Thus it is apparent that the two proteins, luciferase and GFP, must interact in dilute solution and, in doing so, must also orient the enzyme-bound oxyluciferin and the GFP chromophore to facilitate efficient energy transfer. Following this model, the data of Fig. 1 suggest that moderately high salt concentrations disrupt this protein–protein interaction, possibly by masking electrostatic bonds.

Protein–Protein Interaction

The data of Figs. 1 and 2 are suggestive of a mechanism involving protein–protein interaction, but alone they do not constitute sufficient evidence. Therefore, to establish this mechanism more clearly, the following three additional experiments were performed.

Protein Species Specificity

The mechanism we have postulated for *Renilla* energy transfer would be expected to involve fairly specific chemical interactions between luciferase and GFP. One way to test this supposed specificity in an energy-transfer assay would be to substitute for *R. reniformis* luciferase (or for GFP) the homologous protein from another species of bioluminescent

[21] T. Förster, *Discuss. Faraday Soc.* **27,** 7 (1959).

[22] J. C. Matthews, K. Hori, and M. J. Cormier, *Biochemistry* **16,** 5217 (1977).

TABLE I
SPECIES SPECIFICITY IN ANTHOZOAN ENERGY TRANSFER[a]

Green-fluorescent protein	Luciferases				
	R. reniformis	*R. kollikeri*	*R. mülleri*	*S. elongata*	*A. gracile*
Renilla reniformis	+	+	+	0	+
Renilla kollikeri	+	+	+	0	+
Stylatula elongata	0	?	?	+	0
Acanthoptilum gracile	+	+	+	+	+

[a] The symbol + indicates full energy transfer, 0 indicates no energy transfer, and ? indicates insufficient material for analysis. Tabular data have been abbreviated from W. W. Ward and M. J. Cormier, *Photochem. Photobiol.*, **27,** 389 (1978). Pergamon Press, Ltd.

coelenterate. Efficient cross-reactions would be expected between closely related species, but not between distantly related ones. Such an experiment, involving 31 distinct cross-reactions with highly purified extracts, has been performed,[13] and a portion of the data (the anthozoan cross-reactions) is shown in Table I. All the sea pansies (*Renilla* sp.) cross-react within that genus; however, the sea pens, *Stylatula* and *Acanthoptilum*, show some interesting differences. For example, *Stylatula* luciferase transfers energy to *Stylatula* GFP and to *Acanthoptilum* GFP, but all other interspecies cross-reactions involving *Stylatula* were unsuccessful. On the other hand, the interspecies cross-reactions involving *Acanthoptilum* and other anthozoans, with one exception, were all successful. However, interclass cross-reactions were unsuccessful between members of the class Anthozoa (shown in Table I) and two other coelenterate classes (which are not shown), the hydrozoans and the ctenophorans.[13] Since the luciferins,[23,24] and the GFP chromophores[25] are almost certainly the same in all these species, the differences in cross-reactivities appear to involve differences in protein–protein interaction.

Chemical Modification of GFP

If the interactions between two proteins are relatively specific, it should be possible to disrupt such interactions by chemically modifying amino acid side chains on one or both of the proteins. Ideally, the modification

[23] K. Hori, H. Charbonneau, R. C. Hart, and M. J. Cormier, *Proc. Natl. Acad. Sci. U.S.A.* **74,** 4285 (1977).

[24] M. J. Cormier, K. Hori, Y. D. Karkhanis, J. M. Anderson, J. E. Wampler, J. G. Morin, and J. W. Hastings, *J. Cell. Physiol.* **81,** 291 (1973).

[25] J. E. Wampler, Y. D. Karkhanis, J. G. Morin, and M. J. Cormier, *Biochim. Biophys. Acta* **314,** 104 (1973).

should not drastically affect the functional properties of the protein, and it should be reversible. From preliminary screening, it was learned that modification of some or all of the twenty amino groups of GFP effectively disrupts energy transfer without altering the spectral properties of the chromophore. Therefore, a study of amino group modification reactions was undertaken.[13] Representative results of that study are shown in Table II. The modification of 10 amino groups by citraconic anhydride almost completely destroys the energy-transfer function of GFP, but it leaves the fluorescence intensity of the protein unchanged. With the milder reagent, dimethylmaleic anhydride, similar results were found; however, this reaction could be reversed with full restoration of energy-transfer function. Both of these reagents leave negative charges in place of the positively charged amino groups. A similar blockage of energy transfer was found with reagents that retain or abolish the positive charge.[13] The fact that the chromophore retains its spectral properties during these reactions indicates that amino group modification interferes, not with the gross properties of the chromophore, but, more likely with the ability of GFP to recognize and/or bind to luciferase.

Complex Formation

Green-fluorescent protein and luciferase can be extracted from *Renilla reniformis* in approximately equal amounts. Because of their similar size and charge, these proteins are difficult to separate by conventional column chromatography techniques, especially in the early stages of purification. In later stages, however, the two proteins are completely resolved. Thus

TABLE II
CHEMICAL MODIFICATION OF GREEN-FLUORESCENT PROTEIN (GFP) AMINO GROUPS[a]

Modifying reagent	Relative fluorescence	Percent green pathway	Modified amino groups	Free amino groups
1. Control	100	72	0	20
2. Citraconic anhydride	95	4	10	10
3. Dimethylmaleic anhydride	103	1	—	—
4. Reversal of No. 3	87	74	—	—

[a] Control and experimental samples were treated 20 hr at 0° in 6 *M* guanidine · HCl. The concentration of modifying reagent was 5 m*M*. The reversal of dimethylmaleic anhydride inhibition of energy transfer was achieved by exhaustive dialysis. The numbers of free amino groups were determined colorimetrically with trinitrobenzene sulfonic acid. Tabular data have been abbreviated from W. W. Ward and M. J. Cormier, *Photochem. Photobiol.* **27**, 389 (1978). Pergamon Press, Ltd.

there is little indication during purification that luciferase and GFP may exist as a stable complex. However, a transient complex between these proteins would not necessarily be detected in the course of purification. We have looked for such a complex by chromatographing a narrow band of concentrated luciferase through a Sephadex G-100 column preequilibrated with GFP at 1 μM concentration.[13] The results of this experiment, shown in Fig. 3, indicate that, under conditions of energy transfer, luciferase and GFP can form a complex whose apparent molecular weight by gel filtration

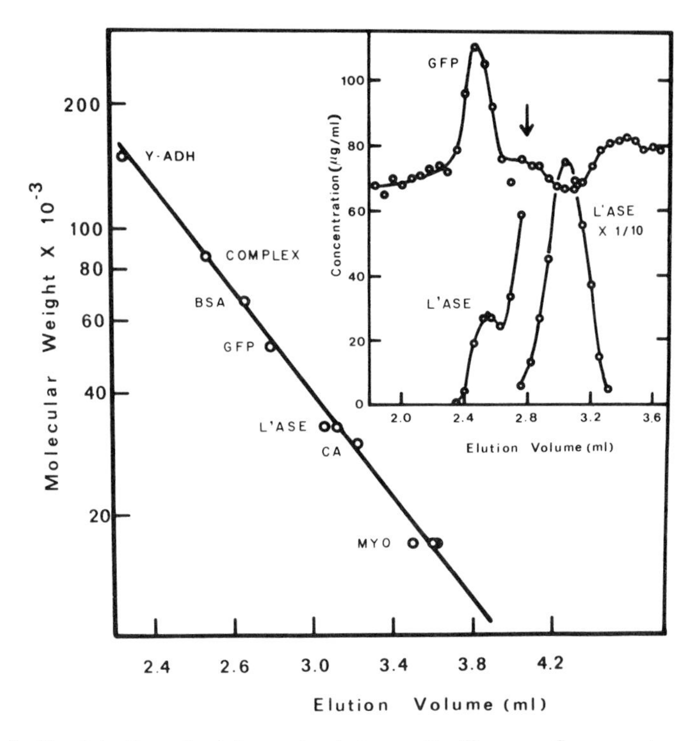

FIG. 3. The detection of a 1:1 complex between *Renilla* green-fluorescent protein and luciferase by the gel-filtration method of J. P. Hummel and W. J. Dreyer [*Biochim. Biophys. Acta* **63,** 530 (1962)]. The G-100 superfine Sephadex column was calibrated with yeast alcohol dehydrogenase (Y-ADH), bovine serum albumin (BSA) green-fluorescent protein (GFP), luciferase (L'ASE), carbonic anhydrase (CA), and myoglobin (MYO). Catalase and sodium nitrite were included as void volume and salt volume markers, respectively. Two other standards, ovalbumin and horseradish peroxidase, which gave highly aberrant elution profiles, have been excluded from the calibration curve. The inset shows the elution profile resulting from the chromatographing of 50 μl of 0.13 mM luciferase through a column equilibrated with 1.4 μM GFP. The arrow indicates the position where GFP elutes when not complexed with luciferase. Reproduced, with permission, from W. W. Ward and M. J. Cormier, *Photochem. Photobiol.*, **27,** 389 (1978). Pergamon Press, Ltd.

(86,000) is the sum of the molecular weights of luciferase (35,000) and GFP dimer (54,000). Furthermore, integration of the chromotography peak gave a GFP/luciferase ratio of 1.3, a number that we feel is within experimental error of unity.

As a model for studying protein–protein interaction and/or biological energy transfer, the *Renilla* GFP system may have certain advantages over existing systems. As compared with photosynthesis,[26] which involves numerous interacting proteins and chromophores, the *Renilla* system, with two proteins and two chromophores, is exceedingly simple. Furthermore, the components of the *Renilla* system can be isolated, purified, and then recombined *in vitro* with recovery of energy-transfer function. The energy-transfer assay, which is, in fact, a measure of protein–protein interaction, is sensitive and highly specific.

The system has obvious limitations as well. The simplicity of the system, in a sense, is one of its drawbacks, since it limits the number of parameters that can be studied. We have not, for example, been able to detect fluorescence energy transfer from enzyme-bound oxyluciferin to GFP. Foremost among the limitations, however, are the difficulties associated with collecting and processing hundreds of pounds of sea pansies to obtain 100-mg quantities of pure proteins—problems not uncommon to other bioluminescence systems. This relative scarcity of material thus limits studies of primary and tertiary structure, necessary for sophisticated work on protein–protein interaction. However, with the application of modern genetic technology, perhaps even these problems can be overcome.

[26] E. Gantt, C. A. Lipschultz, and B. Zilinskas, *Brookhaven Symp. Biol.* **28,** 347 (1977).

Section IV

Aequorin

[30] Introduction to the Bioluminescence of Medusae, with Special Reference to the Photoprotein Aequorin

By FRANK H. JOHNSON and OSAMU SHIMOMURA

Observations on luminescent medusae have a long and distinguished background, beginning as early as the first century, when Caius Plinius Secundus ("Pliny the Elder," 23–79 A.D.) wrote an account of the "Pulmo marinus," describing the brightly luminous slime that could be rubbed from its surface onto various objects, making them glow as though on fire, including walking sticks, which he alleged could then be used like a torch to light the way in the dark.[1] Pliny was evidently referring to the organism we now call *Pelagia noctiluca,* a scyphozoan jellyfish common in the Bay of Naples, not far from where Pliny was living in the year 79, unfortunately so close to Vesuvius that his complex of careers as explorer, soldier, lawyer, politician, naturalist, and writer in all fields of knowledge was brought to an abrupt termination. Pliny was familiar not only with luminescent jellyfish, but also with the edible clam, which has been referred to by common names such as "dactylus" and "piddock." It too is well known in the Mediterranean, as well as in parts of the southern coast of France, where it has now become something of a rarity owing to the ravages of gourmet-inclined members of the populace. It is noteworthy that the luminescent system of this clam, bearing the scientific name *Pholas dactylus,* is the one to which the terms "luciferin" and "luciferase" were first applied by Dubois[2] in the belief that the emission of its light resulted from the activity of a specific substrate–enzyme system, i.e., luciferin–luciferase system, analogous to, but chemically different from, that of the elaterid beetle *Pyrophorus,* with which he had earlier demonstrated the existence of two such components.[3] Recent research on the *Pholas* system by Michelson and his colleagues[4–11]

[1] E. N. Harvey, "A History of Luminescence from Earliest Times until 1900." *Philos. Soc.*, Philadelphia, Pennsylvania, 1957.
[2] R. Dubois, *C. R. Soc. Biol. Ses. Fil.* **39,** 564 (1887).
[3] R. Dubois, *C. R. Soc. Biol. Ses. Fil.* **37,** 559 (1885).
[4] J.-P. Henry, M. F. Isambert, and A. M. Michelson, *Biochim. Biophys. Acta* **205,** 437 (1970).
[5] J.-P. Henry, M. F. Isambert, and A. M. Michelson, *Biochimie* **55,** 83 (1973).
[6] J.-P. Henry and A. M. Michelson, *Biochim. Biophys. Acta* **205,** 451 (1970).
[7] J.-P. Henry and A. M. Michelson, *Biochimie* **55,** 75 (1973).
[8] J.-P. Henry, C. Monny, and A. M. Michelson, *Biochemistry* **14,** 3458 (1975).
[9] A. M. Michelson, *Biochimie* **55,** 465 (1973).
[10] A. M. Michelson, *Biochimie* **55,** 925 (1973).
[11] A. M. Michelson and M. F. Isambert, *Biochimie* **55,** 619 (1973).

ISBN 0-12-181957-4

has confirmed some of the observations and conclusions of Dubois, while greatly extending basic knowledge concerning the properties of the components and biochemistry of the system. As it turns out, the *Pholas* system has certain points in common with the *Aequorea* system under consideration in the present chapter, and these points tend to place both systems among the "photoprotein" types,[12] which comprise a somewhat different category from that of a substrate–enzyme system in the usual sense of the term.

Nomenclature

Although the present authors have consistently adhered to *Aequorea aequorea* as the name of the jellyfish to which most of their research on luminescent medusae has been devoted (Fig. 1), some authors have preferred to use a different designation of the same species, namely, *Aequorea forskalea*.[13-19] According to the priorities generally observed in biological classification, however, the species name *aequorea* should have precedence over *forskalea*. The circumstances have been clearly and concisely summarized by M. E. Johnson and Snook[20] as follows (under "Family—AEQUORIDAE," p. 69; quotation from page 71): "Until recently this medusa has been known as *Aequorea forskalea* Peron and Lesueur but H. B. Bigelow has pointed out that by right of priority the name given by Forskal thirty-four years earlier should be used. As Forskal called it *Medusa aequorea*, and *Medusa* is no longer applicable while the genus *Aequorea* has long been recognized, the name now becomes *Aequorea aequorea* (Forskal)." Thus, there seems to be no real argument.

The Nonrequirement of Molecular Oxygen

The requirement of molecular oxygen has long remained one of the seemingly most universal characteristics of bioluminescence systems.

[12] O. Shimomura and F. H. Johnson, "Bioluminescence in Progress" (F. H. Johnson and Y. Haneda, eds.), p. 495. Princeton Univ. Press, Princeton, New Jersey, 1966.
[13] E. N. Harvey, *Biol. Bull. Woods Hole* **41,** 280 (1921).
[14] J. W. Hastings, G. Mitchell, P. H. Mattingly, J. R. Blinks, and M. van Leeuwen, *Nature (London)* **222,** 1047 (1969).
[15] C. C. Ashley, *Endeavour* **30,** 18 (1971).
[16] R. Llinás, J. R. Blinks, and C. Nicholson, *Science* **176,** 1127 (1972).
[17] J. E. Brown and J. R. Blinks, *J. Gen. Physiol.* **64,** 643 (1974).
[18] S. R. Taylor, R. Rüdel, and J. R. Blinks, *Fed. Proc., Fed. Am. Soc. Exp. Biol.* **34,** 1379 (1975).
[19] J. R. Blinks, F. G. Prendergast, and D. G. Allen, *Pharm. Rev.* **28,** 1 (1976).
[20] M. E. Johnson and H. J. Snook, "Seashore Animals of the Pacific Coast." Macmillan, New York, 1927.

FIG. 1. Negative print from color transparency taken by O. Shimomura of living *Aequorea aequorea* in a seawater holding tank. The dark spots around the edge of the umbrella are contracted tentacles. One photogenic organ, not clearly visible in this photograph, is located on both sides of each tentacle, but nowhere else in the organism.

Modern experiments have shown, however, that under very special conditions, not normally encountered in nature, a bioluminescent reaction can occur without the immediate participation of oxygen. For example, luminescence on the (anaerobic) addition of a suitable aldehyde to an aqueous solution of luciferase extracted from luminous bacteria and containing a "long-lived intermediate" resulting from the prior addition of reduced flavin mononucleotide plus oxygen.[21]

A natural exception to the oxygen requirement does occur among a few primitive species, namely, radiolarian protozoa, medusae, and ctenophores. In fact, in experiments with a "hemispherical species of medusa," possibly identical with the classical "Pulmo marinus" of Pliny, Spallanzani, and others, Macartney demonstrated, more than a century and a half ago, that bioluminescence can take place in a vacuum, presumably in the absence of a significant quantity of oxygen. In his words[22]: "Some of the scintillating and hemispherical species of medusa, contained in a small glass jar, were introduced into the receiver of an air pump, and the air being exhausted, they shone as usual when shaken; if any difference could be perceived, the light was more easily excited, and continued longer in a vacuum."

Whether or not a nearly perfect vacuum was achieved by Macartney with the apparatus available in 1810, later experiments by Harvey and Korr[23,24] left scarcely any room for doubt that the light-emitting reaction not only of medusae, but of radiolarians and ctenophores as well, could take place in practically the complete absence of molecular oxygen. A reasonable hypothesis to account for such a circumstance could hardly be advanced without some substantial knowledge of the chemical nature of the system, but as pointed out some time ago[25] (p. 307), the hydromedusan *Aequorea* is not an anaerobic organism, and " . . . the gist of the problem is: How far antecedent to the actual light-emitting reaction is a step requiring molecular oxygen no longer needed?" This question calls for elucidation of the mechanism of the light-emitting reaction, an objective that as yet has been only partially achieved, as indicated in the discussions that follow.

[21] J. W. Hastings, Q. H. Gibson, J. Friedland, and J. Spudich, *in* "Bioluminescence in Progress" (F. H. Johnson and Y. Haneda, eds.), p. 151. Princeton Univ. Press, Princeton, New Jersey, 1966.

[22] J. Macartney, *Philos. Trans. R. Soc. London* **100,** 258 (1810).

[23] E. N. Harvey and I. M. Korr, *J. Cell. Comp. Physiol.* **12,** 319 (1938).

[24] E. N. Harvey, "Bioluminescence." Academic Press, New York, 1952.

[25] F. H. Johnson and O. Shimomura, *in* "Photophysiology" (A. C. Giese, ed.), Vol. 7, p. 275. Academic Press, New York, 1972.

Stability of the Medusa System

Despite the repeated failure of attempts by Harvey[24] and other investigators to find evidence of a luciferin–luciferase system in medusae, the system has long been known to remain potentially active after the death and even after partial liquefaction of a specimen, e.g., the "Pulmo marinus" (*Pelagia noctiluca*). Thus Spallanzani[26,27] found that a bright luminescence resulted when fresh water was poured onto such a partially autolyzed, dead specimen. In studying the hydromedusae *Aequorea, Halistaura,* and *Phialidium* at Friday Harbor, Harvey[13] noted that not only could they be dried over $CaCl_2$ and luminesce again on moistening with water, but that these organisms, or the roughly excised photogenic areas of the largest of the three (*Aequorea*), could be squeezed through several layers of cheese cloth to yield a brei that remained luminescent for as much as 9 hr at room temperature. After the light had disappeared on standing, it could be briefly restored by adding fresh water, or saponin or certain other glycolytic agents, but a second addition of such agents would not elicit further luminescence.[13] Consideration of these and other properties led Harvey to conclude that the luminescence of coelenterates was fundamentally extracellular, associated with the breakdown of cells and dissolution of particles.

The seeming stability of components of the *Aequorea* system against drying, autolysis, etc., offered sufficient encouragement in regard to the possibility of obtaining cell-free luminescent extracts and their active components that one of the present authors (F. H. J.) began research on this problem in 1959. Although endeavors to obtain a luciferin and luciferase from this species by various procedures were unsuccessful, it was found possible to reversibly inhibit the light emission of a luminescent brei by adding suitable concentrations of urea, urethane, or ammonium sulfate.[28] The brei was prepared essentially according to Harvey's method, involving excising the circumoral ring of tissues containing the photogenic organs of *Aequorea* and squeezing a number of such rings in a mass through several layers of cheesecloth, or preferably a linen handkerchief, yielding a preparation that has since been referred to as a "squeezate."[29] The active

[26] L. Spallanzani, *Mem. Mat. Fis. Soc. Ital. Sci. Modena Verone* **7**, 271 (1794).

[27] L. Spallanzani, "Travels in the two Sicilies and some parts of the Apennines" (translated from Italian, 4 vol.), London, 1793. See Chapter 27, on phosphorescent medusae.

[28] F. H. Johnson, H.-C. Sie, and Y. Haneda, *in* "Light and Life" (W. D. McElroy and B. Glass, eds.), p. 206. Johns Hopkins Press, Baltimore, Maryland, 1961.

[29] O. Shimomura, F. H. Johnson, and Y. Saiga, *J. Cell. Comp. Physiol.* **59,** 227 (1962); **62,** 1 (1963).

material in a squeezate could be mostly precipitated by ammonium sulfate, and the concentrate thus obtained gave a brilliant greenish-blue light merely upon decanting the supernatant and adding several volumes of distilled water.

Cell-Free Extracts and Cation Requirements

In regard to obtaining a cell-free extract of *Aequorea* capable of a bioluminescent reaction in aqueous solution, the first indications of success were obtained by Shimomura *et al.*[29] with extracts made with diluted acetic acid, then neutralized with bicarbonate. The clear, colorless solution emitted measurable light on addition of a small volume of boiled photogenic tissues, thus giving the impression that luciferase and luciferin components had finally been separated from this type of organism. In due course, however, it was found that instead of boiled extracts of *Aequorea*, addition of ordinary tap water, or better, seawater, would bring on the light. Moreover, tap water or seawater brought on the light much faster than distilled water. The possibility naturally suggested itself that the active substance in tap water or seawater was some inorganic ion the same as, or analogous to the Mg^{2+} required in the luminescence of firefly extracts. Different salts available in the local stockroom were then tested for activity like that of tap water. Only two were really effective, namely Ca^{2+} and Sr^{2+}, the Sr^{2+} proving to have only about one one-hundredth the potency of the former. Limited activity of a few others was attributed to contamination with Ca^{2+}.

The cation requirement, together with the accumulated indirect evidence that some sort of specific enzyme, or at least a protein, was an essential component, pointed the way toward developing a procedure for purification, simply by chelating the Ca^{2+} at the start, then assaying each step by adding Ca^{2+} in a suitable manner. A chelator that proved to be satisfactory for this purpose was EDTA (ethylenediaminetetraacetic acid). The basic procedure that yielded a small amount of very active, evidently very pure, material was modified with various refinements and improvements, resulting in greater convenience and greater yields of highly purified material as described below. It is convenient to state at once, however, that the active material from *Aequorea*, first obtained in a convincingly pure state, turned out to be a specific, conjugated protein, which was given the name "aequorin",[29] and was later considered to belong in a category of bioluminescent systems for which the term "photoproteins" was introduced.[12] It is perhaps well to emphasize here the fundamental distinction between the photoprotein type and the luciferin–luciferase type of system, both of which involve a specific protein. In the photoprotein type, however, light emission does not normally depend directly on catalytic activity of an enzyme, and the total light is proportional to the amount of the specific protein. In the luciferin–luciferase type, on the other hand, light is produced as a result of the aerobic oxidation of the substrate luciferin, catalyzed by the enzyme luciferase, and the total light is normally proportional to the amount of luciferin.

Harvesting and Initial Processing of the Photoprotein Aequorin

The initial "fractionation" was accomplished with a pair of scissors, an operation that has been appropriately referred to as "circumcision,"[19] consisting of cutting off a thin ring of tissue from the outer margin of the umbrella. The photogenic organs (or organelles) of *Aequorea* are all located in that region, occurring as small, yellowish masses, one on each side of the base of each tentacle adjacent to the circumoral canal. The excised ring was dropped into cold seawater in a beaker in an ice bath. Specimens that occur along the coast of Washington in midsummer average about 3 inches in diameter, rarely attaining a maximum of 6 inches, with an average volume of 50 ml. A skillfully excised ring would be expected to have a volume of not more than 0.5 ml, thus representing less than 2% of the total volume of the specimen. Rings that are cut too thick are apt to include a lot of mucus and other impurities that may seriously interfere with the purification process; if the rings are cut too thin, photogenic raw material is apt to be lost. With efficient teamwork, aided by a couple of practiced, skilled technicians operating a specially designed cutting machine (Fig. 2),[30] it has been possible to collect and initially process, in one long, hard-working day, more than 5000 individual specimens of *Aequorea* at the Friday Harbor Laboratories of the University of Washington. With the facilities that have ordinarily been at the author's disposal, however, it has been advisable not to undertake to process so large a number in a single day; everything considered, a couple of thousand specimens per day is perhaps optimum, although this depends upon the rather erratic, unpredictable abundance of collectable specimens. During periods when they are difficult to find, it is unfortunately impractical to delay processing until a favorable number are at hand—for example, in a holding tank supplied with adequately running seawater. Under such conditions, although a significant loss of active photogenic material is not likely to occur within 2 or 3 days, the individual specimens gradually shrink, and the margin of the umbrella tends to fold inward, making the ring difficult to cut off.

[30] F. H. Johnson, *Naval Res. Rev.*, p. 16, February (1970).

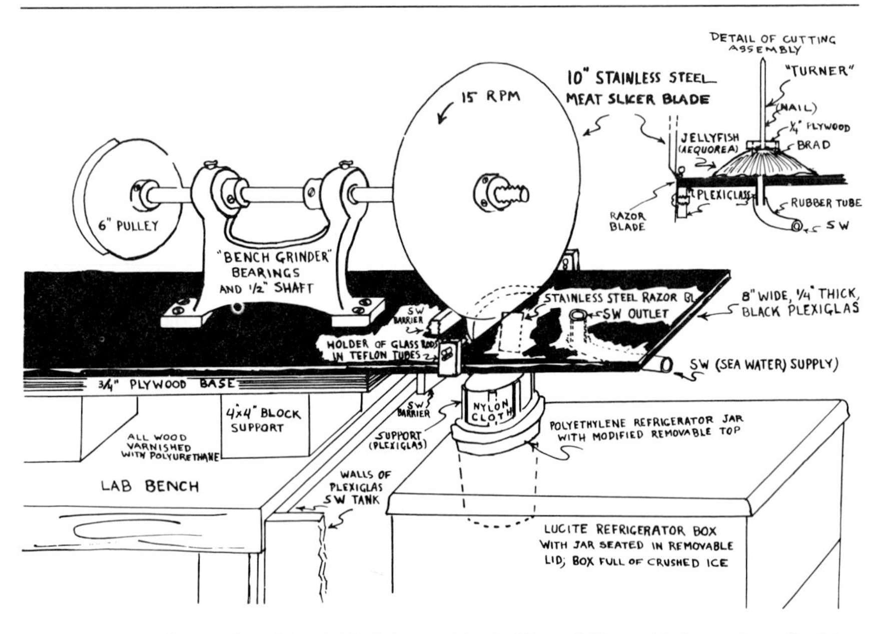

FIG. 2. Machine designed by F. H. Johnson (April–May, 1969) to aid the cutting off of thin rings of tissue containing the photogenic organs of *Aequorea* (F. H. Johnson, *Naval Res. Rev.*, p. 16, February, 1970). Specimens are first positioned, oral side down, over inflowing seawater on a platform of black plexiglass, thus inflating the body and filling the internal canals, including the circumoral canal, with seawater. The specimen is then rotated, with the aid of a "turner" (upper right-hand corner of the figure) against cutting edges consisting of a slowly rotating meat blade and a fixed razor blade. A Teflon-covered rod (not shown in the figure) is so placed that the periphery of the specimen passes under it while the main body is kept away from the cutting edges. The excised ring drops into a ring-container beneath the blades, in a polyethylene box containing crushed ice. Excess seawater flows through a nylon barrier at the top of the ring-container, then off the top of the refrigerator box into a plexiglass seawater tank with drain to the outside.

Although Blinks *et al.*[19] have described a different method, the present authors have found the following procedures most generally satisfactory. Five hundred rings from specimens averaging 3–4 inches in diameter are accumulated in a beaker containing 800 ml of cold seawater, kept chilled in an ice bath for a period up to a few hours when necessary. The seawater is drained from 500 such rings through a nylon cloth, and the rings are transferred to a flask of appropriate size containing 1 liter of cold, saturated ammonium sulfate (ACS grade) containing 50 m*M* EDTA–2Na, adjusted to a pH between 6.0 and 6.5 by NaOH. The flask is closed with Parafilm, and the active material is dislodged by shaking the flask violently for about a minute. It is advisable to assay a 0.1-ml portion of the fluid by adding 5 ml of 10 m*M* calcium acetate and recording the total light quickly emitted in a

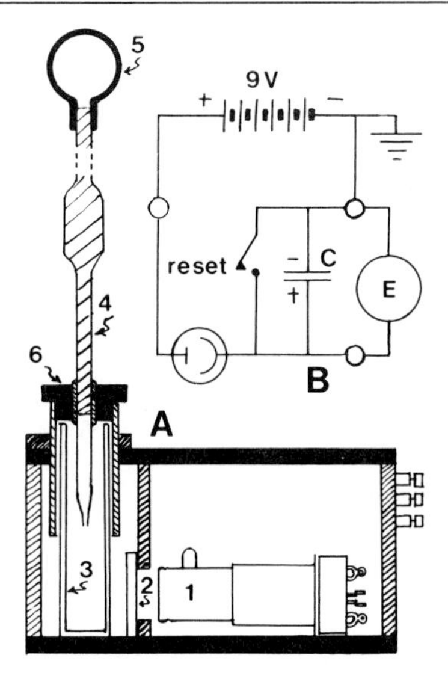

FIG. 3. Diagram of a simple, inexpensive photometer and phototube circuit. [From O. Shimomura and F. H. Johnson, *Nature (London), New Biol.* **78,** 287 (1972).] (A) Light-proof housing of the R-330 phototube (1), separated by a glass window (2) from a flat-bottomed tube (3) 93 mm deep by 16 mm in diameter, to which 0.1 ml of the aequorin material is added, followed by 5 ml of 10 m*M* calcium acetate from the pipette (4), which is wound with black tape extending from the rubber bulb (5) to the cover (6). Luminescence is immediately elicited on addition of the calcium acetate solution. The cathode current of the phototube is fed into a capacitor on which the charge, proportional to total light emitted, is measured by a suitable voltmeter with a high input resistance of at least 10^{12} ohms. (B) Diagram of the circuit we have used, with a very low leakage capacitor (C) of 0.005 or 0.01 microfarad, a Keithley Model 621 electrometer (E), and a 9 V transistor radio battery. Most modern pH meters can be adapted for use in place of the more versatile Keithley.

suitable photometer, e.g., as used by Johnson and Shimomura[31] (Fig. 3). The amount of aequorin present can be estimated on the basis that 1 mg of pure aequorin emits, on addition of an excess of calcium acetate, 4.5×10^{15} ($\pm$ 10%) photons at 25°, λ_{max} = 469 nm.

After the flask has been shaken violently and amount of aequorin assayed, a wad of rings is lifted out on a large plastic fork; the wad is cut into smaller chunks and replaced into the flask, followed by vigorous shaking and assaying a second time. If only slight increase in activity is noted, further shaking is unnecessary.

A convenient and sometimes more effective procedure, alternative to shaking as described above, is to transfer the mass of rings, after draining off the seawater, to a suitable volume of cold ammonium sulfate–EDTA

[31] F. H. Johnson and O. Shimomura, *Nature (London), New Biol.*, **78,** 287 (1972).

solution in the large mixing bowl of a heavy-duty electric cake mixer, and subject the rings to slowly rotating blades for about a minute; one then proceeds as above with assaying, cutting the wad of rings into smaller masses, and operating the mixer again. With either alternative, the remaining steps are the same, as follows. Care must be taken throughout to maintain the temperature of the material within the range of an ordinary refrigerator, i.e., 0 to 4°.

At this stage, most of the photogenic material has been dislodged from the tissues and is present in the form of small particles that are easily squeezed through 50- to 100-mesh Dacron curtain cloth, which retains the bulk of the ring tissue. The average activity to be expected at this stage should be on the order of 10^{12} to 10^{13} photons per 0.1 ml. After filtering through Dacron cloth, 80 ml (dry) of analytical grade Celite is added to the fluid and filtration carried out on Whatman No. 3 paper on a 15-cm Büchner funnel, previously coated with a little of the same Celite poured onto the filter paper from a suspension in EDTA solution. The particles of active material stay in the filter cake which can now be removed, broken into smaller pieces, and transferred to a flask containing 400 ml of cold 50 m*M* EDTA-2Na, pH 6.5. The pieces of filter cake are easily dispersed by shaking the flask, and the active material goes into homogeneous solution as the particulate material undergoes lysis. The Celite and tissue debris are removed by filtering on Whatman No. 4 paper layered on No. 3 paper on an 18.5-cm Büchner funnel previously coated with Celite as described above. The filter cake is washed with 80 ml of EDTA solution. The filtrate in both cases is collected in a chilled 2-liter suction flask containing at the start 280 g of $(NH_4)_2SO_4$. The flask is then placed in a refrigerator with occasional stirring and kept at least overnight in order for precipitation to go to completion. The precipitate can be quantitatively recovered either by centrifuging 15 min at 15,000 rpm in a refrigerated centrifuge at 0°, or by filtering on paper with the aid of Celite. It is important to use only analytical grade Celite throughout, because lesser grades are apt to contain enough Ca^{2+} to fire off the luminescent reaction of all the aequorin present, thereby reducing the yield to zero. Material that has been brought to this stage in the above manner undergoes practically no loss during storage for several days at 0°, or in a few months with Dry Ice in a freezer. The same does not hold for material prior to precipitation with ammonium sulfate–EDTA solution, inasmuch as that material undergoes rapid loss of luminescence potency within a very few days, even at 0°.

Purification

Aequorin can be purified satisfactorily by successive column chromatographies at 0–4°, first on Sephadex G-75 or G-100 (Pharmacia) in a 5 cm in diameter 80 cm long column with 10 m*M* Tris·HCl buffer, pH 7–8, containing 10 m*M* EDTA and 1 *M* $(NH_4)_2SO_4$. A column of this size is suitable for material from 1–4 batches of 500 rings each. Under these conditions aequorin tends to undergo considerable aggregation with the result that it is eluted near the void volume. The same column is used for the second chromatography, but with the $(NH)_4SO_4$ replaced with 0.2 *M* NaCl in the Tris·HCl buffer. A reversal of the aggregation now occurs and aequorin is eluted mostly as a monomer. The third chromatography is performed on a 2.5 cm in diameter, 10 cm long column of DEAE-cellulose prepared with 10 m*M* Tris·HCl buffer containing 10 m*M* EDTA. Elution is by linear gradient concentration of NaCl from 0.05 to 0.35 *M*. The final two, i.e., the fourth and fifth, chromatographies are the same as the first and second, except that a smaller column is used; it is convenient at this stage to combine the material from several of the first three steps, aggregating up to 10,000 rings. The overall yield in this purification should be somewhat more than 30%, of which about half is practically pure and the remainder is of less purity.

Storage and Retrieval of Purified Aequorin

Purified aequorin can be stored in the same manner as the extract prior to chromatography, in saturated $(NH_4)_2SO_4$ containing some EDTA, at 0°, or better in a freezer. For use, the material can be centrifuged after thawing, and the precipitate taken up in the desired solvent, usually a buffer at pH higher than pH 7.5 in order to avoid aggregation. If the desired concentration of aequorin for use is so great that a volume of solvent cannot be added in an amount to sufficiently dilute the $(NH_4)_2SO_4$ and EDTA present in the precipitate, then the solvent can be quickly and easily exchanged with a more satisfactory one by filtration through a small column of Sephadex G-25 or BioGel P-6 (Bio-Rad) suitably prepared with the desired buffer. In the event that a very small volume of concentrated aequorin in a desired buffer is wanted, one drop of a concentrated solution of aequorin can be filtered through a microcolumn of 2.5–3 mm internal diameter and 4–5 cm gel height, equilibrated with that buffer. Such a column can be conveniently made from plastic tubing with a disk of 400-mesh nylon cloth (Pharmacia) across the opening at the base, held in place by a cylindrical plastic plug of the same optical density as the column, firmly juxtaposed with the aid of a short length of Tygon tubing surrounding the connection. An outlet for the eluate can be provided by a capillary-sized bore in the plastic plug. Small droplets of eluate can be assured by sharpening the exit area of the plug. In addition to the obvious importance of safeguarding against contamination with Ca^{2+} from all possible sources, we recommend including at least 1–10 μM EDTA in any aequorin solution as the minimum necessary protection.

TABLE I
CONCENTRATION OF Ca^{2+} IN VARIOUS KINDS OF TEST TUBES

Kind of test tubes[b]	Concentration of Ca^{2+} (M)[a]		
	At 1 min after adding solution to dry tubes[c]	At 1 min after adding solution to rinsed tubes	At 24 hr after adding solution to rinsed tubes
Polycarbonate (Kimble)	$< 10^{-8}$	$< 10^{-8}$	$< 10^{-8}$
Polypropylene (Nalgene)	$< 10^{-8}$	$< 10^{-8}$	$< 10^{-8}$
Vycor (silica glass, Corning)	2.6×10^{-8}	2.0×10^{-8}	2.0×10^{-8}
Kimax (borosilicate glass, Kimble)	3.3×10^{-8}	3.0×10^{-8}	1.3×10^{-7}
Pyrex (borosilicate glass, Corning)	4.0×10^{-8}	3.6×10^{-8}	1.5×10^{-7}
Boron-free glass (Corning)	5×10^{-8}	$< 10^{-8}$	3.0×10^{-8}
Soda-lime glass (A. H. Thomas)	2×10^{-7}	5.6×10^{-8}	1×10^{-6}

[a] Concentration found in 15 ml of 2 mM sodium acetate solution, which was placed in test tubes having inside diameters of 1.6–1.8 cm, at 25° [O. Shimomura and F. H. Johnson, *Symp. Soc. Exp. Biol.* **30,** 41 (1976)].

[b] Brand new test tubes were washed with detergent, rinsed carefully with deionized distilled water, then oven dried, except the plastics, which were air dried. The cleaned test tubes were used in the test within 3 days.

[c] Test tubes left 1 month after cleaning showed highly irregular results, with the average concentration of Ca^{2+} about 10 times over the present results.

Although a method of protection has been suggested,[17] intended to provide "EDTA-free" aequorin solution by adding Chelex 100 (Bio-Rad) beads, we have found that an easily detectable amount of a calcium-chelating substance gradually but constantly leaches out of these beads into the solution, making the designation "EDTA-free" utterly meaningless. In this same context, it should be noted that the aequorin test for Ca^{2+} has made it possible to demonstrate the leaching of calcium from various kinds of glass (Table I).[32]

Properties of Electrophoretically Pure Aequorin: Molecular Weight

Although data concerning the molecular weight of aequorin has been sought by a number of investigators as well as by different methods, some wide discrepancies have resulted, owing mostly, we believe, to differences in the conditions and concentrations of aequorin employed, but to some extent also to unsuspected differences in behavior of aequorin in the

[32] O. Shimomura and F. H. Johnson, *Symp. Soc. Exp. Biol.* **30,** 41 (1976).

process of gel filtration. For example, using a trace amount of aequorin and a Sephadex column, the molecular size appeared to correspond to that of a globular protein having a molecular weight of 21,000–23,000.[14,33,34] A similar molecular size was indicated by sodium dodecyl sulfate gel electrophoresis.[35] On the other hand, sedimentation analysis indicated a molecular weight of 30,000,[34,35] a value in conformance with data on the content of amino acids and functional chromophore. Our most recent data[35a] in studies of the phenomenon of aggregation have indicated that, even though aggregation was undetected earlier at a pH higher than 8,[34,35] strong aggregation at pH 8 occurs when there is a high concentration of $(NH)_4SO_4$ in the buffer. Thus, when a sample containing more than 0.5 mg ml^{-1} of aequorin was chromatographed on a column of Sephadex G-75, using 10 m*M* Tris·HCl buffer, pH 7.8, containing 10 m*M* EDTA and 1 *M* $(NH)_4SO_4$, the protein was eluted at close to the void volume, indicating an apparent molecular weight of more than 100,000, a misleading value clearly due to aggregation. The extent of aggregation varied with the preparation and in any case was completely reversible, however, by removing the $(NH)_4SO_4$. The purification procedure described above was designed with this phenomenon in mind. The correct molecular weight of the monomer, according to recent data[35a], now appears to be approximately 21,000, close to the values proposed by Blinks *et al.*[19]

Other Physical Properties

Aqueous solutions of aequorin are practically nonfluorescent. Concentrated solutions have a straw-yellow color due to a weak absorption peak at 460–465 nm. The value of $E_{1cm}^{1\%}$ at 280 nm was found to be 27.0.[34] The spectrum of the functional chromophore, as indicated by the difference spectrum between aequorin and the protein residue of aequorin, shows a strong absorption peak at 300 nm, in addition to the peak at 460 nm.[36] An isoelectric point has been found at between pH 4.2 and 4.9.[37] Aequorin is rather highly soluble in aqueous buffer solutions, and complete precipitation requires not less than saturated $(NH)_4SO_4$. According to our most recent data on luminescence potency and molecular weight, 1 mg of pure

[33] J. R. Blinks, P. H. Mattingly, B. R. Jewell, and M. van Leeuwen, *Fed. Proc., Fed. Am. Soc. Exp. Biol.* **28,** 781 (1969).

[34] O. Shimomura and F. H. Johnson, *Biochemistry* **8,** 3991 (1969).

[35] Y. Kohama, O. Shimomura, and F. H. Johnson, *Biochemistry* **10,** 4149 (1971).

[35a] O. Shimomura, and F. H. Johnson, "Detection and Measurement of Free Calcium Ions in Cells" (C. C. Ashley and A. K. Campbell eds.), North Holland, Amsterdam (1978).

[36] O. Shimomura, F. H. Johnson, and H. Morise, *Biochemistry* **13,** 3278 (1974).

[36a] O. Shimomura, and F. H. Johnson, *Proc. Natl. Acad. Sci. U.S.A.* **75,** 2611 (1978).

[37] J. R. Blinks, *Int. Congr. Physiol. Sci. (Proc.) 25th,* Vol. 9, p. 68 (1971).

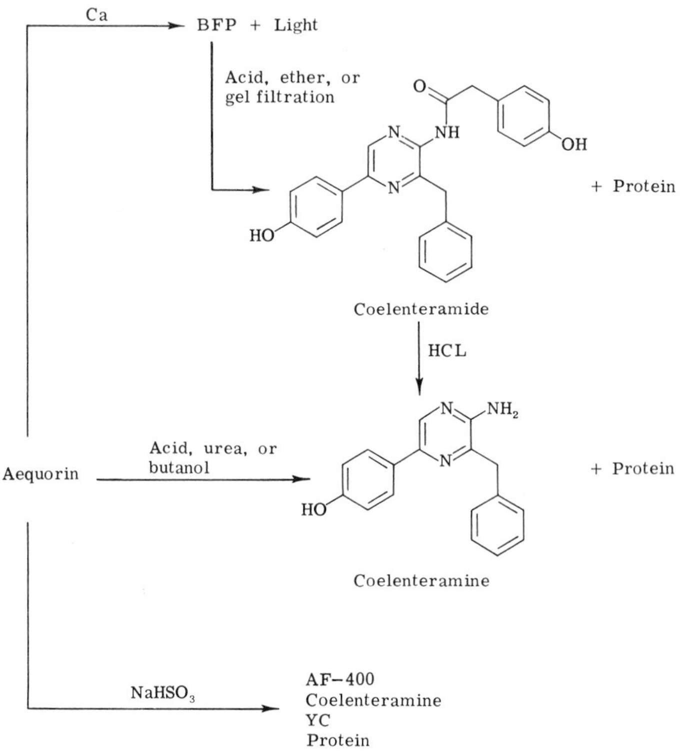

FIG. 4. Chief reactions of aequorin [O. Shimomura and F. H. Johnson, *Symp. Soc. Exp. Biol.* **30** (1976)]. BFP, blue-fluorescent protein product of the light-emitting reaction triggered by calcium; AF-400, blue-fluorescent compound having an absorption maximum at 400 nm; YC, a yellow compound responsible for the yellow color of native aequorin. The recently established structure of YC is indicated in the text (p. 285).

aequorin emits 4.3×10^{15} photons at the peak wavelength of 470 nm when triggered by an excess of Ca^{2+} at 25° and a pH between 5.5 and 8.5, with a quantum yield of 0.15.

Chief Reactions of Aequorin and Structure of the Chromophore

Figure 4[32] diagrams some of the chief reactions of aequorin and shows the structures of the two product compounds, which we have named "coelenteramine" and "coelenteramide" because of their chemical nature as well as widespread distribution among bioluminescent systems of both the photoprotein and luciferin–luciferase type among coelenterates. Determination of these structures[38,39] has proved to be a major breakthrough

[38] O. Shimomura and F. H. Johnson, *Biochemistry* **11,** 1602 (1972).
[39] O. Shimomura and F. H. Johnson, *Tetrahedron Lett.* **31,** 2963 (1973).

not only in regard to chemical aspects of the bioluminescent reaction of aequorin, but also to arriving at the structure of the luciferin of various other types of luminescent organisms (cf. Johnson and Shimomura, 1976).[40]

Referring to Fig. 4, the Ca^{2+}-triggered luminescent reaction of aequorin results in a blue-fluorescent product, a protein that we have named BFP, (blue-fluorescent protein). Various treatments of BFP, e.g., with acid, ether, or gel filtration, splits it into two products, a protein moiety plus a substance, which we have named coelenteramide, having the structure shown in Fig. 4. Neither of these two products of splitting BFP is fluorescent. However, on simple mixing together in aqueous solution containing calcium, coelenteramide recombines with the protein to form BFP identical to the original with its characteristic fluorescence.[39] The binding constant has been found to be $2 \times 10^5 \ M^{-1}$ (Morise *et al.*[41] When treated with $NaHSO_3$, aequorin yields 4 products: coelenteramine, AF-400 (a blue-fluorescent compound having an absorption maximum at 400 nm), YC (a yellow compound presumably responsible for the yellow color of native aequorin) and the residual part of the protein. The amount of AF-400 increases with time, whereas the amount of coelenteramine decreases with time.[36]

The structure of YC is as follows (I) according to evidence awaiting publication.[36a] Structure (II) or structure (III) must be the precursor of YC that is present in aequorin as judged by the method of producing YC through $NaHSO_3$ treatment of aequorin.

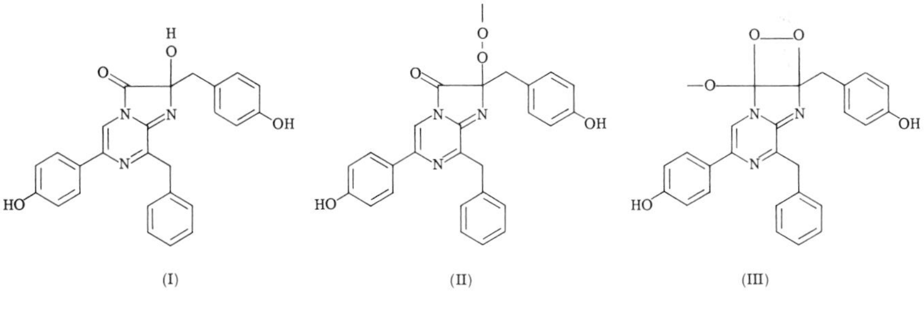

(I) (II) (III)

The Light-Emitter in Aequorin Luminescence

The coinciding of the bioluminescence emission spectrum of aequorin with the fluorescence spectrum of BFP·2 Ca (Fig. 5) points to excited BFP, or more precisely the excited state of coelenteramide bound in BFP, as the

[40] F. H. Johnson and O. Shimomura, *Trends Biochem. Sci.* **1,** 250 (1976).

[41] H. Morise, O. Shimomura, F. H. Johnson, and J. Winant, *Biochemistry* **13,** 2656 (1974).

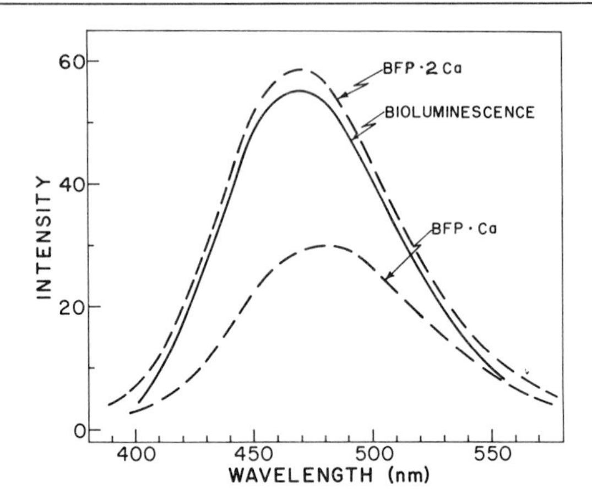

FIG. 5. Emission spectrum of the bioluminescent reaction of aequorin triggered by calcium, and fluorescence emission spectra of BFP·2Ca and of BFP·Ca on excitation at 340 nm [O. Shimomura and F. H. Johnson, *Nature (London)* **227,** 1356 (1970)].

light-emitter of this system.[42] Note that the product coelenteramide can be converted back to coelenteramine by acid hydrolysis (Fig. 4).

The Light-Emitter *in Vivo*: GFP, the Green-Fluorescent Protein

The luminescence emitted by the photogenic organs of a living specimen of *Aequorea* is greenish in quality, with maximum intensity at 509 nm, indistinguishable to the eye from the fluorescence emitted by the same organs when illuminated by a "black light" from a mercury vapor lamp with a filter eliminating most of the visible radiation. The luminescence of aequorin solutions, on the other hand, is decidedly bluish, with a peak emission at 470 nm. The difference is due to the fact that a green-fluorescent protein (GFP) is present in the photogenic organs, closely associated with aequorin. Absorption- and fluorescence-emission spectra were measured on partially purified preparations some time ago,[43] and several authors have suggested that the *in vivo* luminescence of *Aequorea* involves an energy transfer from the light-emitter of aequorin (excited BFP) to the GFP.[43-48]

[42] O. Shimomura and F. H. Johnson, *Nature (London)* **227,** 1356 (1970).

[43] F. H. Johnson, O. Shimomura, Y. Saiga, L. C. Gershman, G. T. Reynolds, and J. R. Waters, *J. Cell. Comp. Physiol.* **60,** 85 (1962).

[44] F. H. Johnson, *in* "Comprehensive Biochemistry" (M. Florkin and E. H. Stotz, eds.), Vol. 27, p. 79. Elsevier, Amsterdam, 79. 1967.

[45] J. G. Morin and J. W. Hastings, *J. Cell. Comp. Physiol.* **77,** 313 (1971).

[46] J. E. Wampler, K. Hori, J. W. Lee, and M. J. Cormier, *Biochemistry* **10,** 2903 (1971).

[47] M. J. Cormier, K. Hori, and J. M. Anderson, *Biochim. Biophys. Acta* **346,** 137 (1974).

[48] M. J. Cormier, J. Lee, and J. E. Wampler, *Annu. Rev. Biochem.* **44,** 255 (1975).

More recently, GFP has been well purified and crystallized.[41] When sufficiently concentrated in solution with aequorin, the excitation energy arising from addition of calcium to the mixture is transferred by the Förster mechanism to GFP, and light is emitted with the same spectral distribution of intensity as that of GFP when excited by ultraviolet. Such transfer of energy is greatly facilitated by first coadsorbing the two compounds on a suitable adsorbant, such as DEAE-cellulose or DEAE-Sephadex. When GFP and aequorin are treated in this manner, the addition of Ca^{2+} results in a greenish luminescence that is essentially identical to that of an intact, living specimen of *Aequorea* (Fig. 6).[41] By a similar procedure, the addition of Ca^{2+} to aequorin coadsorbed with flavin mononucleotide (FMN) leads to the transfer of the energy of excitation to FMN, resulting in the orange-yellow luminescence characteristic of that substance (Fig. 6).

Green-fluorescent proteins occur in association with the bioluminescent systems of species other than *Aequorea*, including the luciferin–luciferase system of the coelenterate *Renilla*.[46] Although it is reasonable to believe that GFPs of various species are fundamentally alike in chemical nature and function, some differences in detailed properties are also to be expected. The GFP of *Aequorea* in particular has been found to occur in several isomeric forms, which have been designated GFPa, GFPb, and GFPc, as the most abundant among at least six green-fluorescent components found in analytical polyacrylamide gel electrophoresis of samples of GFP that had been purified through six steps of column chromatographies.[41] The molecular weights of GFPa, GFPb, and GFPc, estimated by SDS polyacrylamide gel electrophoresis, were 69,000, 53,000, and 34,000, respectively. The amino acid composition and other properties are described in detail by Morise *et al.*[41]

The Bioluminescence Reaction and Regeneration of Aequorin

Figure 7 diagrams a postulated mechanism of the Ca-triggered luminescent reaction of aequorin, and regeneration of the active photoprotein from synthetic coelenterazine and the product BFP.[47] The luminescent reaction involves the binding of two Ca^{2+} to one molecule of aequorin, based on the recently revised molecular weight mentioned above. The intimate mechanism of the intramolecular reaction involved in the bioluminescence of aequorin has been discussed at some length[36] prior to a knowledge of the structure of the yellow compound YC. This structure is now known and the most probable mechanism of bioluminescence of aequorin, according to the evidence at hand now, can be represented as follows.

Figure 7 indicates that the blue-fluorescent protein product (BFP) of the calcium-triggered luminescent reaction is split into two nonfluorescent compounds, i.e., a protein moiety and the chromophore coelenteramide,

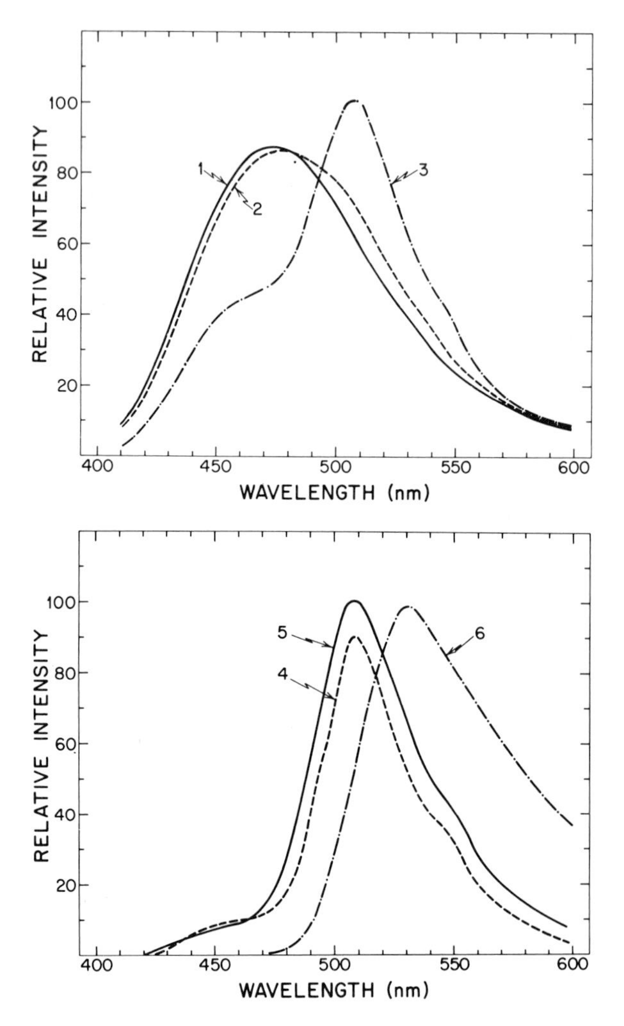

FIG. 6. Emission spectrum of the calcium-triggered bioluminescence of aequorin, and spectral changes due to the addition of GFP (green-fluorescent protein) in solution or as a coadsorbant on DEAE-cellulose or DEAE-Sephadex; the influence of added flavin mononucleotide (FMN) is similarly illustrated [H. Morise, O. Shimomura, F. H. Johnson, and J. Winant, *Biochemistry* **13,** 2656 (1974)]. Each luminescent reaction, except (5), was triggered by addition of 30 μl of 10 m*M* calcium acetate to the following mixtures: (1) 270 μl of buffer solution containing 11 μg of aequorin, (2) 270 μl of buffer solution containing 11 μg of aequorin plus 46 μg of GFP, (3) 270 μl of buffer solution containing 11 μg of aequorin plus 46 μg of GFP plus 100 μl of finely ground DEAE-cellulose or DEAE-Sephadex suspension (see text), (4) 250 μl of buffer solution containing 11 μg of aequorin plus 460 μg of GFP plus 100 μl of the DEAE-cellulose or DEAE-Sephadex suspension, which was centrifuged and the precipitate then resuspended in buffer solution making the total to 270 μl, (5) luminescence of photogenic organs of *Aequorea* (6) the same as (4) except that GFP was replaced with 50 μg of FMN. The buffer solution used was 10 m*M* sodium phosphate containing 1 m*M* EDTA (pH 7.0). All measurements were made with a 10-mm cell, with slit width of 2 mm, at 25°. Reprinted with permission from *Biochemistry* **13,** 2656–2662 (1974). Copyright by the American Chemical Society.

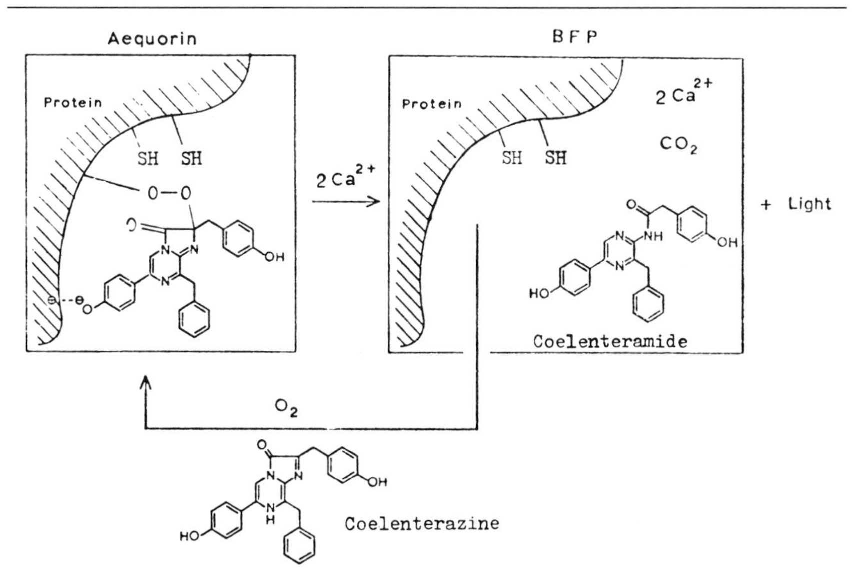

FIG. 7. Postulated mechanism of luminescence and regeneration of aequorin [from O. Shimomura and F. H. Johnson, *Nature* (*London*) **256**, 236 (1975)]. Calcium-triggered luminescence of aequorin produces the blue-fluorescent protein (BFP), which contains dissociable coelenteramide. The protein part of BFP regenerates aequorin on aerobic incubation with coelenterazine.

by means of acid, ether, or gel filtration, and the latter of these components is converted to coelenteramine by acid hydrolysis. Referring to Fig. 7, regeneration of active, calcium-triggerable aequorin takes place when the protein moiety of spent aequorin is allowed to stand at 5° in the presence of an excess of synthetic coelenterazine plus 2 m*M* 2-mercaptoethanol, 5 m*M* EDTA, and oxygen. If an excess of Ca^{2+} is also present during this process, luminescence occurs as fast as the active aequorin is formed, but this is vastly slower than the bioluminescent reaction of native, untreated aequorin, i.e., of aequorin with Ca^{2+} when not limited by the rate of formation. Under the above conditions, the yield at 5° of active aequorin amounted to 50% in 25–30 min, and almost 90% of the initial amount in 3 hr. On removal of the reagents by passing the product mixture through a column of Sephadex G-25 after 20 hr, the material so obtained was identical to the original aequorin in absorption spectrum (λ_{max} 281 and 460 nm, $A_{460}/A_{281} = 0.030$), luminescence spectrum (λ_{max} 470 nm), luminescence activity (1.55×10^{15} photons ml^{-1} at 25° when A_{280} is 1.0 cm^{-1}), kinetics of luminescence in the presence and in the absence of air (Fig. 8),[49] and

[49] O. Shimomura and F. H. Johnson, *Nature (London)* **256,** 236 (1975).

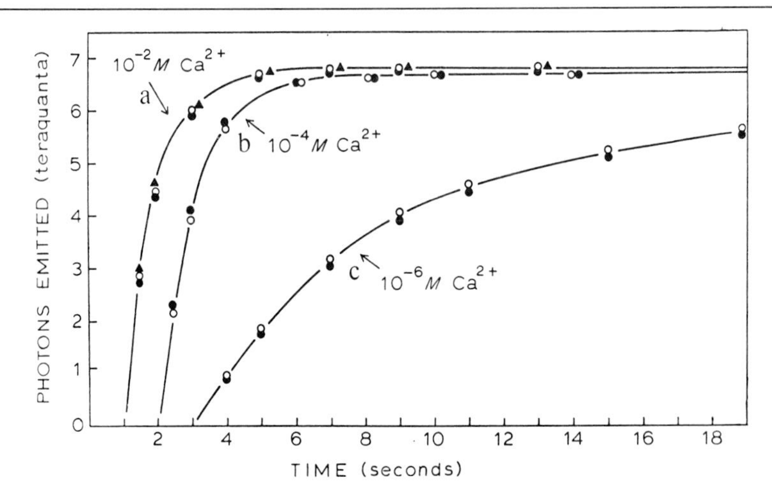

FIG. 8. Ca^{2+}-triggered luminescence of native aequorin (○) and regenerated aequorin (●). Curves for 10^{-2} (a), 10^{-4} (b), and 10^{-6} *M* Ca^{2+} (c) are offset at 1, 2, and 3 sec, respectively, for 0 time. Luminescence was initiated by quick addition of 3 ml of 10 m*M* Tris · HCl buffer (pH 7.2) containing calcium acetate to aequorin solutions consisting of 3 μl of aequorin stock solution containing 0.2 m*M* EDTA (A_{280} 1.44 cm^{-1} for both native and regenerated aequorin) plus 0.3 ml of water. To avoid contamination of the more dilute (10^{-4}, 10^{-6} *M*) Ca^{2+} solutions, plastic rather than glass containers and pipettes were used throughout, except for a Hamilton syringe to measure the aequorin stock solution. The absence of any effect due to the presence of oxygen is shown by data (▲) on luminescence of regenerated aequorin with 10^{-2} *M* Ca^{2+} in a vessel thoroughly evacuated by a vacuum pump. Temperature: 24°. From O. Shimomura and F. H. Johnson, *Nature (London)* **256,** 236 (1975).

chromatographic behavior on DEAE-cellulose and Sephadex G-100. Regeneration is thus confirmed. Regeneration was not influenced by various concentrations of EDTA or by the addition of H_2O_2, but it did not take place in an evacuated reaction vessel without oxygen (cf. Fig. 9).[49]

In the regeneration process described above, the protein moiety of spent aequorin can be logically regarded as an enzyme catalyzing a luminescent oxidation of coelenterazine. In this view, aequorin itself must be considered to represent an enzyme intermediate, as hypothetically suggested by several investigators.[45-48,50] By its known properties, however, this is an indissociable intermediate.[29,34] Moreover, the slow production of dim light when Ca^{2+} is present during the regeneration must be regarded as a low-grade bioluminescence, not in the same class as the brilliant flash resulting from the reaction of aequorin and Ca^{2+}. Thus, the biological significance of the light, if any, as far as the organism is con-

[50] J. W. Hastings and J. G. Morin, *Biochem. Biophys. Res. Commun.* **37,** 493 (1969).

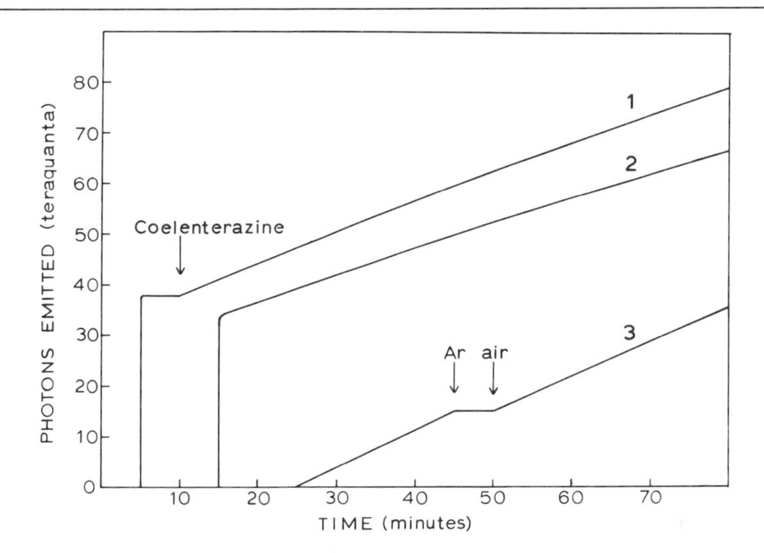

FIG. 9. Ca^{2+}-triggered luminescence of aequorin and regenerated aequorin. Curve 1: 4 ml of 10 m*M* calcium acetate was added to 8.7 μg of aequorin in 1 ml of buffer at 5 min, followed by addition of 6 μg of coelenterazine in 20 μl of methanol at 10 min; curve 2: 8.7 μg of protein moiety of BFP and 6 μg of coelenterazine were added to 1 ml of buffer; the mixture was kept at 5° for 3 hr, then placed in the light-measuring apparatus at 0 min, and 4 ml of 10 m*M* calcium acetate were added to this mixture at 15 min; curve 3: 4 ml of 10 m*M* calcium acetate containing 6 μg of coelenterazine were added to 1 ml of buffer containing 8.7 μg of protein moiety of BFP at 25 min; bubbling the mixture with Ar gas started at 45 min; Ar was switched to air at 50 min, then bubbling was stopped after 15 sec. The buffer (pH 7.5) in all cases consisted of 10 m*M* Tris·HCl, 5 m*M* EDTA, and 2 m*M* 2-mercaptoethanol. The amount of aequorin was estimated from A_{280} (E 1%–1 cm = 27.0) [O. Shimomura and F. H. Johnson, *Biochemistry* **8**, 3991 (1969)]. The amount of protein moiety of BFP was calculated assuming that aequorin yielded the same amount of the protein moiety. All luminescence measurements were carried out at 20°. The quick flash at 5 min is caused by native aequorin; all other light emissions are caused by regenerated aequorin. From O. Shimomura and F. H. Johnson, *Nature (London)* **256**, 236 (1975).

cerned would seem a priori to reside only in the Ca^{2+}-triggered reaction. In addition, the catalytic activity of the protein moiety engaged in the regeneration of aequorin must be considered to be very low grade, at least with respect to rate of reaction, inasmuch as it has a turnover number of 1 or 2 per hour. In this respect again, the system can scarcely be considered on par with a luciferin–luciferase system in the usual sense of the terms.

[31] Practical Aspects of the Use of Aequorin as a Calcium Indicator: Assay, Preparation, Microinjection, and Interpretation of Signals

By JOHN R. BLINKS, PATRICK H. MATTINGLY, BRIAN R. JEWELL, MENNO VAN LEEUWEN, GARY C. HARRER, and DAVID G. ALLEN

The photoprotein aequorin has been employed increasingly as a biological calcium indicator since its use for the purpose was first proposed in 1963.[1] The properties that make aequorin (and probably other calcium-sensitive photoproteins) particularly suitable for this are: (1) ease of signal detection, (2) high sensitivity to Ca^{2+}, (3) relative specificity for Ca^{2+}, and (4) lack of toxicity. Disadvantages of the photoproteins are: (1) their scarcity, (2) their large molecular size, (3) the fact that they are consumed in the luminescent reaction, (4) the nonlinear relation between [Ca^{2+}] and light intensity, (5) the influence of the chemical environment on the sensitivity to [Ca^{2+}], and (6) the limited speed with which the luminescent reaction can follow very rapid changes in [Ca^{2+}]. These properties of the photoproteins and specific applications to which aequorin has been put have been discussed extensively elsewhere[2-4] and will not be enlarged upon here. The prime application of aequorin is as a Ca^{2+}-indicator inside living cells, but it is also useful in certain situations *in vitro* where conventional methods (atomic absorption spectroscopy and ion-sensitive electrodes) are unsatisfactory because of an unfavorable combination of the following factors: (1) a need to determine ionized rather than total calcium concentration, (2) small sample size, (3) very low [Ca^{2+}], and (4) a need to follow fairly rapid changes in [Ca^{2+}].

The unique position of aequorin as an intracellular calcium indicator is now being challenged by the new generation of calcium-sensitive microelectrodes[5] and the metallochromic dyes, such as Arsenazo III.[6] Nevertheless, aequorin seems to have clear advantages over the other methods for use in muscle cells or other preparations subject to movement, and

[1] O. Shimomura, F. H. Johnson, and Y. Saiga, *Science* **140,** 1339 (1963).

[2] J. R. Blinks, F. G. Prendergast, and D. G. Allen, *Pharmacol. Rev.* **28,** 1 (1976).

[3] J. R. Blinks, *Photochem. Photobiol.* **27,** 423 (1978).

[4] J. R. Blinks, *Ann. N. Y. Acad. Sci.*, **307,** 71 (1978).

[5] W. Simon, D. Ammann, M. Oehme, and W. E. Morf, *Ann. N. Y. Acad. Sci.* **307,** 52 (1978).

[6] A. Scarpa, F. J. Brinley, T. Tiffert, and G. R. Dubyak, *Ann. N. Y. Acad. Sci.* **307,** 86 (1978).

ISBN 0-12-181957-4

under circumstances in which image intensification is used to determine the distribution of Ca^{2+}. Aequorin is sufficiently difficult to obtain and to work with that its use should not be undertaken casually. This chapter is intended to provide practical details important in the use of aequorin as a biological Ca^{2+}-indicator.

Assay of Aequorin

General Considerations. A fast and reliable method for measuring the concentration of active aequorin in solutions is essential in any laboratory where the photoprotein is used. Repeated measurements of luminescent activity are needed to guide the course of extraction and purification of the protein. It is also important to make frequent checks of the activity of aequorin solutions as they are used experimentally, and quantitative interpretation of experimental results may depend on knowledge of the exact concentration of active aequorin in a particular sample. Measurements of protein concentration do not suffice, even when the aequorin sample is known to be pure, because they give no indication of the extent to which the aequorin has been discharged or inactivated since the purification was carried out.

The basic principle in the assay of aequorin activity is obvious: one measures the total amount of light emitted when all the aequorin in a test sample is discharged with Ca^{2+}. Various types of photometers have been designed for work with bioluminescent systems (see this volume [30, 41, 43]), and many of these might be suitable for the assay of the photoprotein. Essential features are (1) ease and reproducibility of mixing, (2) a wide range of sensitivities, and (3) outputs proportional to both instantaneous light intensity and the total (integrated) light emitted.

Apparatus. Figure 1 shows the field version of an assay apparatus designed specifically for use with aequorin. Several copies of this apparatus have been in use in our laboratory for nearly 10 years and have proved to be reliable and trouble-free. The apparatus consists of a photometer unit that houses the reaction cuvette and photomultiplier, and a control unit containing all the associated electronics, including the high-voltage power supply for the photomultiplier. These two units, plus a cuvette, a light standard, and an injection syringe, are all that is needed to assay aequorin, and all of this equipment will fit in a large shoebox. Purchased components cost less than $300. The electronics are easily assembled by an amateur, but the construction of the housing for the cuvette and photomultiplier requires a skilled machinist.

Figure 2 shows the construction of the photometer unit. The cuvette (A) is cylindrical, with a flat bottom (ordinary 12 × 35 mm shell vials are satisfactory for most purposes). It is inserted from below into a slightly

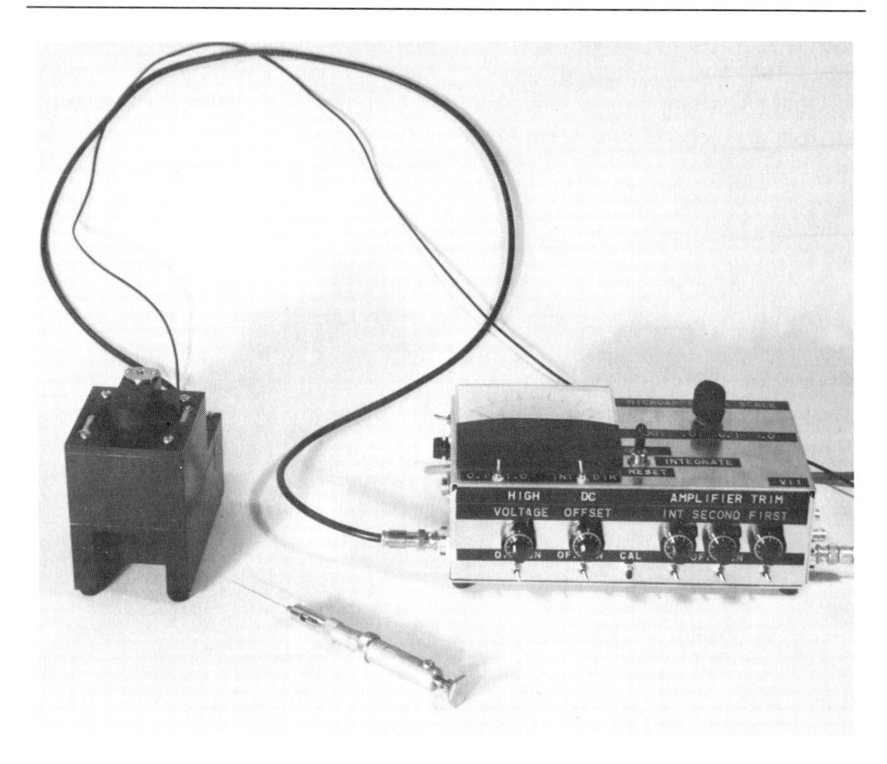

FIG. 1. Apparatus for the assay of aequorin. The photometer unit housing the cuvette and photomultiplier is at the left; all the associated electronics are in the control unit at the right. A Hamilton CR 700-20 automatic syringe is in the foreground.

larger cylindrical reaction chamber, which is located eccentrically in a rotor (B). When the rotor is turned through 180°, the cuvette is directly over the photomultiplier. A lightproof rubber septum (C) is held in place at the top of the reaction chamber by a long stainless steel needle guide (D), which assures that needles inserted through the septum will be centered over the reaction cuvette. The rotor housing (E) has a cap (F) which contains a detent mechanism that centers the reaction chamber precisely over either the cuvette insertion port or the photomultiplier port. The rotor rests on a black plastic bearing plate (G) that is penetrated by these two ports. The cuvette insertion port (toward the right in Fig. 2) is slightly larger in diameter than the reaction cuvette. The photomultiplier port is slightly smaller than the reaction cuvette and serves as a window to the photomultiplier without letting the cuvette drop or catch as it is rotated past. These ports are straddled by a pair of grooves that are concentric with the axis of

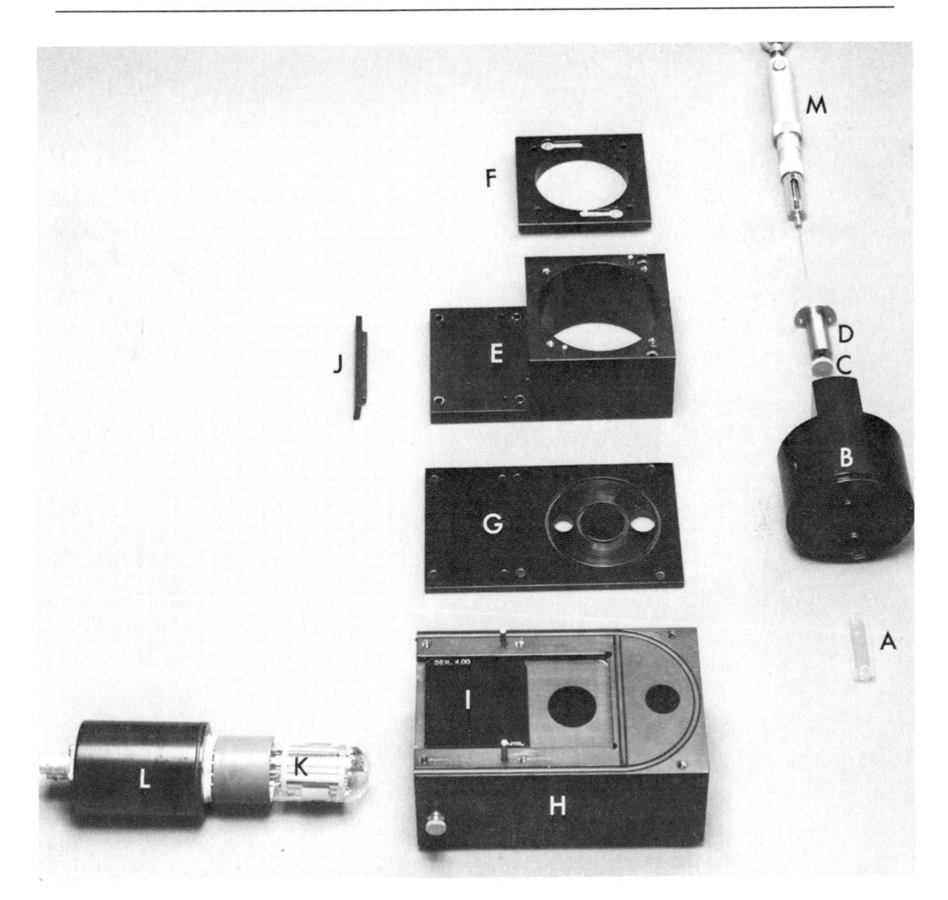

FIG. 2. Construction of the photometer unit. A, Cuvette; B, rotor containing reaction chamber; C, rubber septum; D, needle guide; E, rotor housing; F, rotor-housing cap; G, plastic bearing plate; H, photomultiplier housing; I, neutral-density filter; J, filter-channel cover; K, photomultiplier tube (EMI 9781B); L, photomultiplier base; M, spring-loaded automatic syringe. All black parts except (G) are anodized aluminum.

the rotor. The grooves accept corresponding projections on the bottom of the rotor that serve both as bearings for the rotor and as light seals which prevent any light entering from the cuvette insertion port or the top of the rotor from reaching the photomultiplier. The block (H) that houses the photomultiplier has a channel milled in its upper surface to accommodate a neutral-density filter (I). The filter can be slid into or out of the light path between the cuvette and the photomultiplier by tilting the photometer unit appropriately. Filters can be changed by unscrewing a filter channel cover (J). (Lines visible near the edge of the block are rubber light seals.) The dynode chain for the photomultiplier (K) is inside the socket holder (L), which slides into the block (H) with an O-ring light seal and is fixed in position with a knurled screw.

The control unit (schematic diagram in Fig. 3) provides a continuously variable high-voltage power source for the photomultiplier, converts the anode current of the tube to a voltage signal, amplifies that signal, and integrates it as a function of time. The operational amplifiers that perform the latter three functions and most of the other small components in the circuit are mounted on a printed circuit board (made for us by the Mayo Foundation Section of Engineering, and available at cost). The amplified voltage signal or its integral can be fed to a meter on the control unit and

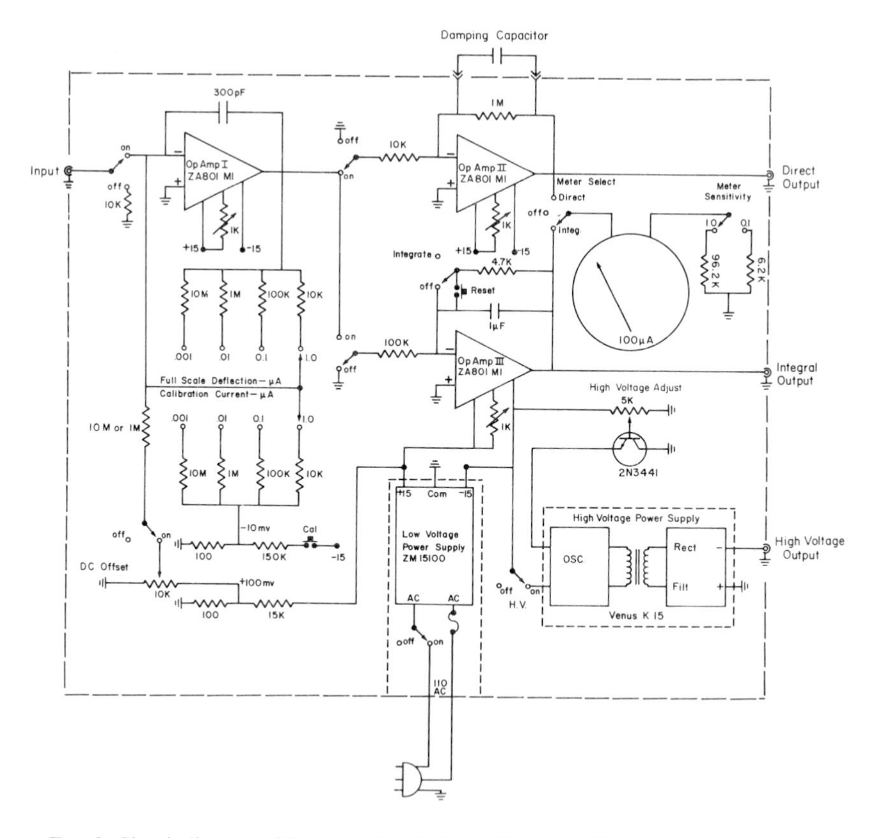

FIG. 3. Circuit diagram of the control unit. Operational amplifiers and low-voltage power supply were obtained from Zeltex Inc. (940 Detroit Ave., Concord, California 94518), the high-voltage power supply from Venus Scientific, Inc. (Farmingdale, New York 11735). All other components are available from standard electronics suppliers.

read off directly or fed through output jacks to an external recording device (e.g., pen recorder). Photomultiplier shot noise can be damped to any extent desired by plugging external capacitors into jacks provided. The gain of the amplifier is variable over four orders of magnitude; the sensitivity of the system can be reduced as necessary by means of neutral-density filters. (These should be calibrated individually with aequorin as the light source.)

Procedure. A standard volume (usually 1 ml) of a pH-buffered solution containing a high concentration of Ca^{2+} (e.g., 20 m*M*) is put into a cuvette, which is then inserted into the reaction chamber and rotated into position over the photomultiplier. A standard amount of the solution to be assayed (usually 10 μl) is drawn into a spring-loaded automatic syringe (Hamilton CR 700-20, labeled M in Fig. 2). The needle of the syringe is then pushed into the needle guide to the hilt, piercing the rubber septum. The dimensions of the needle guide are such that when this is done the tip of the needle will be centered in the cuvette and about 5 mm below its top (well above the fluid level in the cuvette). The integrator is reset, and the injection is made. Because of the high ejection velocity of the spring-loaded syringe, mixing is rapid and fairly complete. (The flat bottom of the cuvette causes the injected stream to break up with a great deal of turbulence; tubes with rounded bottoms tend to reflect the jet of injected fluid upward, with the result that fluid may be thrown out of the top of the cuvette.) Light intensity rises very rapidly to a peak, then decays approximately exponentially (Fig. 4). The peak light intensity is influenced somewhat by variations in mixing speed. Therefore, the total light (integrated signal) is the more suitable measure of aequorin content. Repeated measurements can be made without replacing the solution in the cuvette provided that the solution is well

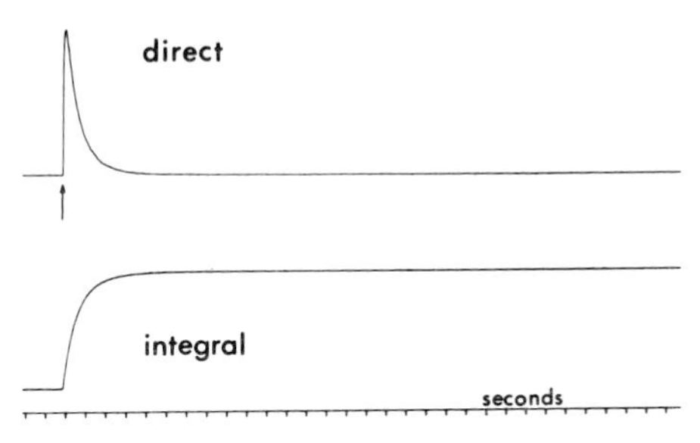

FIG. 4. Oscillograph record of an aequorin assay. The arrow marks the time of injection of aequorin into the cuvette housed in the photometer unit; the direct signal has been damped considerably (time constant 50 msec) by plugging an external capacitor (0.05 μF) into the jacks provided for this purpose on the control unit. The aequorin concentration in the sample is directly proportional to the final amplitude of the integrated signal.

buffered and contains much more Ca^{2+} than can be bound by any chelator introduced with the samples.

Influence of Reaction Conditions. The light yield is essentially independent of the pH of the solution in the cuvette between pH 6.0 and 8.5, but drops off steeply on either side of this range. It is essential that the Ca^{2+} solution be pH-buffered if the aequorin solution to be assayed contains chelators, because these substances release protons when they combine with Ca^{2+}.

The Ca^{2+} concentration of the solution in the cuvette is not critical, but should be high enough (>0.1 m*M*) to give a maximal reaction rate. Usually it is desirable to use a [Ca^{2+}] much higher than that so that multiple samples of solutions containing calcium binding substances can be assayed without changing the solution in the cuvette, but solutions of very high salt concentration (e.g., 1 *M* $CaCl_2$) reduce the light yield and should be avoided. We routinely use a solution containing 20 m*M* $CaCl_2$ and 25 m*M* piperazine-*N*,*N'*-bis(2-ethanesulfonic acid) (PIPES) buffer, pH 7.0.

The salt concentration of the aequorin solution usually has no influence on the assay. However, when concentrations of ammonium sulfate in the sample are high enough to salt out the aequorin, the light yield is reduced somewhat, even when the samples are well stirred. The reduction increases progressively as the ammonium sulfate concentration in the aequorin sample is increased above 1.5 *M*, and amounts to about 20% in a saturated solution (4 *M*). The decrease cannot be attributed to the binding of Ca^{2+} by SO_4^{2-} because the free [Ca^{2+}] remains high enough to discharge the aequorin at the maximal rate even after multiple shots of a saturated ammonium sulfate solution have been introduced into the cuvette.

The quantum yield of the luminescent reaction of aequorin is influenced by temperature, dropping by some 25% as the temperature of the reaction cuvette is increased from 0° to 20°. It is necessary, therefore, to perform aequorin assays at a standard temperature if the results are to be comparable.

Reproducibility and Linearity. When successive 10-μl aliquots of a given aequorin solution are assayed as outlined above, the coefficient of variation is about 2% if the integrated light signal is used as the index of aequorin concentration. The chief source of variation appears to be the size of a small air bubble that forms in the needle of the Hamilton syringe in the wake of each forceful ejection. When the syringe is refilled, the bubble is drawn up into the barrel. If pains are taken to dislodge it and to expel it from the needle, the coefficient of variation can be reduced below 2%. On the other hand, if multiple bubbles are allowed to accumulate, or if the technique of filling the syringe is not uniform, the coefficient of variation may rise to 5% or more.

The light yield of the aequorin assay is directly proportional to the concentration of aequorin in the sample over as wide a range as we have examined. Figure 5 shows the results of an experiment in which a concentrated solution of aequorin was diluted serially over eight orders of magnitude. The line has a slope of 1.0 and was drawn to fit the points in the upper half of the graph. At very low aequorin concentrations the points tend to fall slightly but consistently below the line. This reflects the instability of aequorin in extremely dilute solutions rather than a lack of linearity of the assay: the deviation of these points from the line increases rapidly with time after dilution. In the experiment of Fig. 5, pains were taken to measure the activity of each solution as soon as possible after the dilution.

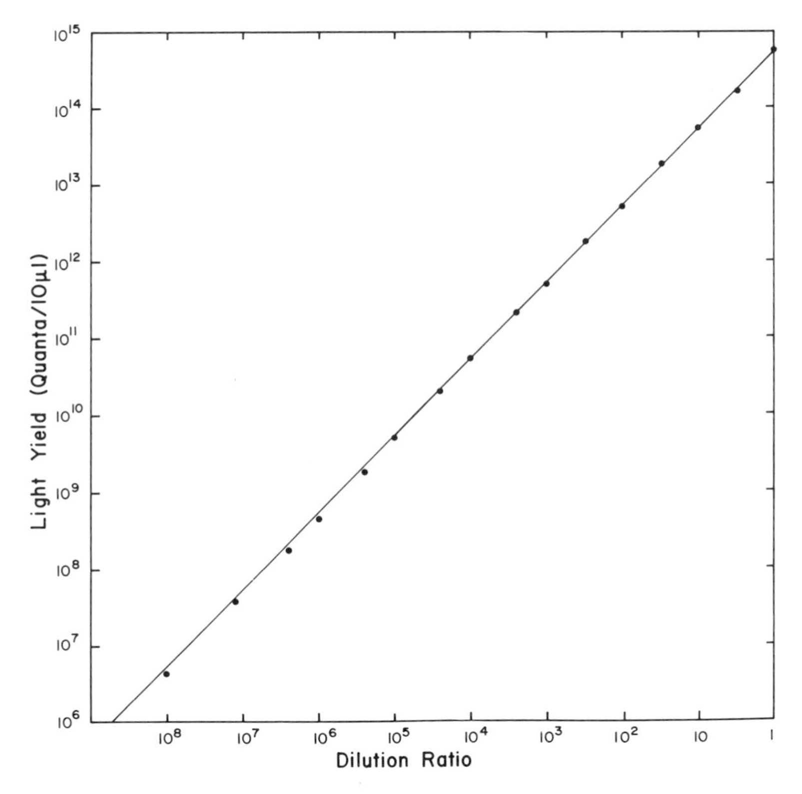

FIG. 5. Linearity of aequorin assay. Log–log plot of the results of an experiment in which assays were made as a concentrated solution of aequorin was diluted serially in a solution of 10 m*M* EDTA, pH 7.0. The initial aequorin concentration was about 6 mg/ml. The photometer was calibrated against the radioactive light standard of J. W. Hastings and G. Weber [*J. Opt. Soc. Am.* **53,** 1410 (1963)]. The line has a slope of 1.0.

Calibration. The problem of calibration can be addressed at two levels—relative and absolute. For many purposes a relative calibration will suffice. An arbitrary scale of "light units" can be established for use within a given laboratory, and this is usually all that is needed except for the determination of quantum yields or for the comparison of light measurements from one laboratory to another. A stable light source that can be put into the photometer in place of the cuvette is necessary for making day-to-day checks of the sensitivity of the photometer (see this volume [44]). Radioactive light sources are ideal for this purpose. We have used two sorts—a solid one containing radium and a liquid one containing [^{14}C]hexadecane[7]—and have checked their stability by comparing them against each other from time to time using the same photomultiplier tube. Scrapings from old-fashioned radium watch dials make very useful light standards provided the luminescent paint does not also have a phosphorescent "memory" of exposure to light. A layer of particles on the bottom of a cuvette can be sealed in place with a transparent silicone potting compound, such as Sylgard (Dow Corning Corp., Midland, Michigan 48640). It should be noted that two photometers that have been adjusted to give identical readings with a particular light standard may not give equal readings when a given sample of aequorin is assayed in the two units. This is because the emission spectrum of the standard is unlikely to be the same as that of aequorin, and the photomultipliers may have different spectral sensitivities. The best way to compare the sensitivities of two photometers is with the same sample of aequorin itself.

Absolute calibrations are much more difficult and are best made with a liquid light standard that emits a known number of photons per unit of volume and time (see this volume [44]). If a volume of the standard equal to that of the assay reaction mixture is placed in a vessel having the same dimensions as the reaction cuvette, the calibration can be made without the need for any geometrical corrections. Differences in the emission spectra of the light standard and of aequorin still must be taken into account, however, and the best way to do this is with a quantum counter. (A quantum counter is a chamber containing a fluorescent dye that absorbs light of various wavelengths and reemits a constant fraction of the quanta with a fixed emission spectrum.[8]) The portable assay unit described here has no provision for the insertion of a quantum counter; it is best calibrated with an aequorin solution that has just been assayed against a known light standard in a more elaborate unit equipped with a quantum counter. Once this has been done, a watch-dial light source can be used to maintain

[7] J. W. Hastings and G. Weber, *J. Opt. Soc. Am.* **53,** 1410 (1963).

[8] J. N. Demas and G. A. Crosby, *J. Phys. Chem.* **75,** 991 (1971).

constant sensitivity provided the same light source and photomultiplier are always used together and the spectral sensitivity of the photomultiplier does not change. Obviously, the unit must be recalibrated whenever the photomultiplier tube is replaced, and ideally the calibration should be checked at intervals even if no components are changed.

Preparation and Handling of Aequorin

Control of Ca^{2+} Contamination. Although the regeneration of spent aequorin has been accomplished *in vitro*,[9] the procedure is not practical for routine use, and for most purposes aequorin must be regarded as a reagent that is consumed in the course of the luminescent reaction. This means that the photoprotein must be rigorously protected from exposure to Ca^{2+} during all stages of its preparation, storage, and handling. During the various steps of extraction and purification (see below), this is best accomplished by the presence of a soluble chelator, such as ethylenediaminetetraacetic acid (EDTA). Once the photoprotein has been purified, it usually must be freed of chelators before it can be used experimentally. At this point it becomes critical to take stringent precautions to avoid calcium contamination, since free calcium concentrations as low as 0.1 μM can lead to a fairly rapid loss of luminescent activity, especially when the ionic strength of the solution is low. Calcium leaches out of ordinary glassware no matter how carefully the glass has been cleaned, so it is usually best to use scrupulously cleaned plastic vessels to handle unprotected photoproteins. Water must be distilled at least twice, preferably in quartz, and stored in plastic bottles. Alternatively, ordinary distilled water can be run through a column of chelating resin (Chelex 100) before it is used. If unprotected aequorin is to be dissolved in solutions other than distilled water, those solutions also must first be passed through Chelex columns, because even reagents of the highest grade contain enough calcium to discharge aequorin. Because the use of Chelex columns is so crucial to work with photoproteins, and these columns are not widely used in other work, we offer the following suggestions for their use.

1. Chelex 100 chelating resin is manufactured by Bio-Rad Laboratories, Richmond, California 94804. It is available as spherical beads of several sizes. For most purposes the coarsest beads (50–100 mesh) seem to be satisfactory, and these are the easiest to use.
2. An all-plastic column should be used, with a nylon bed support net fine enough to retain beads of the size chosen even when they are shrunk by

[9] O. Shimomura and F. H. Johnson, *Nature (London)* **256,** 236 (1975).

exposure to acid. A 0.9 × 30 cm column, such as the Pharmacia K 9/30, is satisfactory.

3. Wash the beads in several changes of 0.1 *N* HCl and then in distilled water. Use a plastic vessel; change solutions by decantation.

4. Fill the column no more than half full with acid-washed Chelex beads. (The bed will approximately double its volume when made alkaline.)

5. Pass several liters of 0.1 *N* KOH (or other base, depending on the cation of the solution to be decalcified) through the column and collect it in a clean plastic vessel.

6. Remove the Chelex from the column and acid-wash it again. Rinse with distilled water in a plastic vessel. Neutralize with the base prepared in step 5. Stir continuously except when decanting samples of supernatant to check the pH. Never put the pH electrode into the main body of the solution. It takes many minutes for the pH to equilibrate after each addition of base.

7. Replace the Chelex in the column, wash with distilled water, run the solution to be treated through the column. Discard the first few column volumes, or more if the pH of a buffered solution is changed by passage through the column. (Large volumes of fluid may be lost in this way if the pH of the Chelex slurry in step 6 is not brought very close to that of the solution to be passed through it.)

The affinity of Chelex 100 for Ca^{2+} is not particularly high (binding constant approximately $4.6 \times 10^3\ M^{-1}$ at pH 7.0),[2] and therefore batch treatment of solutions with Chelex is relatively unsatisfactory as a means of removing contaminating Ca^{2+}. For the same reason, it is important not to stir or repour Chelex columns during use. Chelex that has bound Ca^{2+} must be kept away from the outflow end of the column, and the beads must be rewashed in acid if it is necessary to repour the column. Chelex can be acid-washed and reused more or less indefinitely. EDTA should not be used to remove calcium from Chelex columns because it binds to the Chelex and leaches off slowly thereafter. Obviously, Chelex cannot be used to decalcify solutions that contain Mg^{2+} or other cations with a high affinity for the resin.

Sources of Aequorin. One of the major disadvantages presently associated with the use of photoproteins is the difficulty of supply. Aequorin is the only photoprotein so far isolated in quantities sufficient to make its widespread use as a calcium indicator a practical matter. It became available commercially in 1975 (through Sigma Chemical Co., St. Louis, Missouri 63178), but the commercial preparation is very expensive and of disappointing quality. We have tested two samples (one purchased in 1975 and one in 1976, but marked with the same batch number) and found them unsuitable for use in cells without further purification. Most investigators with a serious interest in the use of photoproteins have collected and purified the materials themselves. The favorite spot for collecting *Aequorea* is Friday Harbor, Washington, where the University of Washington maintains a well-equipped marine biological station and large numbers of *Aequorea* are present regularly in the late summer and early fall.

Collection and Handling of Aequorea. The photoprotein of *Aequorea* is contained in specialized cells (photocytes) that occur in small bilobular clumps (photogenic masses) between the tentacular bulbs along the free margin of the jellyfish. The photogenic masses are most easily identified by the intense green fluorescence that they exhibit when illuminated with ultraviolet light. This fluorescence is due to the "green-fluorescent protein" (GFP), which occurs in the photocytes in association with aequorin and is responsible for the fact that bioluminescence of *Aequorea* is green, whereas the light emitted by aequorin is blue (see preceding chapter). Luminescence occurs intracellularly in response to chemical, electrical, or gentle mechanical stimulation. However, the photocytes are readily dislodged and easily disrupted, with the result that showers of luminous particles (and probably some free aequorin as well) are liberated into the seawater when the medusae are handled roughly. The lost photocytes are not regenerated during any period over which it is feasible to store the medusae, so it is essential that the animals be collected and handled as gently as possible. Thus, although it would often be possible to collect *Aequorea* quickly and in large numbers by means of drag nets or seines, it is not worthwhile to do so. The jellyfish must be collected individually and transferred to tanks of running seawater as soon as possible. (The plastic mesh "skimmers" sold for removing leaves and insects from swimming pools make suitable collecting nets.) The amount of active aequorin extractable from roughly handled specimens does increase modestly if the medusae are stored in uncrowded tanks for several hours. This may reflect the regeneration of aequorin that was discharged by luminescence but not actually lost during the collection process.

Extraction of Aequorin. In any biochemical purification process, it is important to start with material that is not only as rich as possible in the compound of interest, but also as free as possible of substances that are likely to interfere with subsequent steps of the procedure. In the case of coelenterates, the slimy constituents of the mesoglea are particularly troublesome, since they tend to clog chromatography columns. Thus the extraction procedure should be designed not only to recover as much of the aequorin in the jellyfish as possible, but also to minimize the mass of mesoglea that is disrupted in the process. *Aequorea* have two particularly favorable characteristics from this standpoint: (1) The anatomy of the jellyfish is such that it is relatively easy to cut off a thin circumferential ring of tissue containing all of the photogenic masses, reducing the amount of tissue to be extracted by a factor of nearly 100. (2) The photocytes are easily dislodged, and the vast majority of them can be removed from the circumferential rings simply by shaking them briskly. This can be done with minimal disruption of the mesoglea, producing a suspension of photocytes, nematocysts, and certain other cells that is relatively free of slime. These cells are then harvested, washed free of Ca^{2+} and as much slime as possible, and then lysed to release the aequorin into free solution.

The device that we use for cutting the margins from jellyfish is shown in Fig. 6. This apparatus evolved by a process of progressive simplification from earlier designs shown to us by Johnson and Shimomura (see preceding chapter) and by Ridgway. It performs as well as more complex machines, and has the advantages of economy, portability, and simplicity. "Cutting boards" of this type are conveniently mounted on trays in or over the tanks used to store the *Aequorea;* the receptacle for the cut margins is kept cold by the seawater running through it. The cutting boards can easily be moved from tank to tank to avoid the unnecessary handling involved in transferring the jellyfish to a stationary cutting machine. As many boards can be used simultaneously as manpower permits, with the result that the cutting process need not serve as a bottleneck when collecting is good. With five people participating, we have found it possible to collect and process as many as 10,000 specimens per day.

During the cutting process the jellyfish and the rings cut from them should be handled as gently as possible to minimize the number of photocytes dislodged. The receptacles for cut rings are emptied frequently into a plastic bucket of seawater that is stored in a refrigerator. They can be left there for many hours without apparent loss of luminescent activity, and therefore it is our practice to process the whole day's harvest of rings together at the end of the day.

The extraction procedure that we use differs from that of Johnson and Shimomura (see preceding chapter) primarily in that we dislodge the photocytes from the marginal rings by shaking the rings in seawater rather than in a solution of 50 m*M* EDTA in saturated ammonium sulfate. In side-by-side comparisons of the two methods, we have found that the seawater method gives extracts that are very much cleaner, with aequorin recoveries only marginally lower than the EDTA method. On three occasions we have compared the two methods directly by using them to extract equal volumes of rings from the same batch. The yields of active aequorin were not significantly different (seawater method 96% of ammonium sulfate–EDTA method), but the seawater extracts were always strikingly cleaner, as evidenced by lower turbidity, viscosity, and OD_{280} per unit of aequorin activity. The reasons for this are evidently (1) that the photocytes

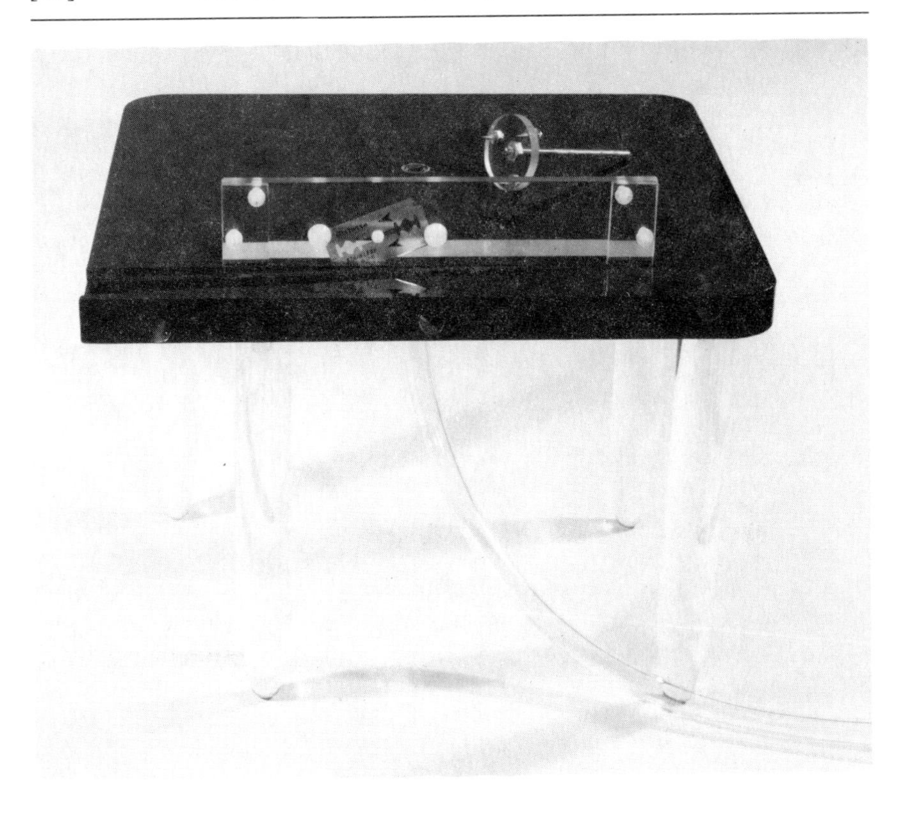

FIG. 6. Simple apparatus for cutting the margins from specimens of *Aequorea*. The top of the table tilts downward toward the lower left corner. During use, a jet of seawater emerges from the hole in the middle of the table and flows toward the gutter on the near side. The jellyfish is placed on the table over the jet in order to spread its free margin out evenly. The specimen is then speared with the three-pronged rotor (shown resting on top of the table) and slid over against the clear plastic fence. The height of the fence and the angle of the razor blade are adjustable; when the fence is adjusted properly, about 2 mm of the free margin of the umbrella projects underneath. The jellyfish is then advanced to the razor blade and rotated through 360° by turning the handle of the rotor between the fingers. The severed free margin floats down the gutter into a receptacle (not shown) consisting of a plastic sieve in a small plastic bucket. All parts are made of plastic except the razor blade and the tines and handle of the rotor, which are stainless steel.

are less easily dislodged in the hypertonic solution and therefore more shaking is required than in seawater, and (2) that the mesoglea tends to disintegrate more in the Ca^{2+}-free solution. The biggest deterrent to the use of the seawater method is a psychological one. It is undeniably distressing to witness the brilliant luminescence that occurs during the shaking process, and to think of the aequorin that is being discharged in this way. Nonetheless the amount lost is a small percentage of the total, and a price well worth paying for a much cleaner extract.

We have found it convenient to shake volumes of 500–1000 ml of cut rings at a time, in a 4-liter Erlenmeyer flask about half full of seawater and rings. The water that the rings have been stored in should not be discarded, as it usually contains a substantial number of dislodged photocytes. To minimize the amount of shaking to which photocytes are subjected after they have been dislodged, we recommend that the seawater be changed repeatedly during the shaking process. We do this by pouring the contents of the flask into a 50-mesh nylon net stretched over a plastic bucket. The rings are then put back into the flask with a liter of fresh seawater and shaking is resumed. One can readily follow the progress of the shaking procedure by illuminating the flask with a long-wavelength ultraviolet lamp; shaking should be continued until almost all of the green fluorescence has disappeared from the rings, but it is important not to disrupt the rings any more than is necessary. We find that four sequences of about 25 strokes each are usually sufficient. About three-fourths of the total yield is obtained in the first 25 shakes, and a similar fraction of the remainder in each successive sequence. It may sometimes be desirable to process the product of the first sequence separately, as its specific activity is by far the highest. The marginal rings are not greatly altered, either in volume or in gross appearance, by being shaken in seawater. In addition to other cells and debris, the product contains green fluorescent particles ranging in size from several hundred micrometers down to less than 0.5 μm in diameter. The smaller particles are evidently fragments of photocytes that have rounded up and sealed themselves off in such a way as to protect the aequorin within from the Ca^{2+} of the seawater in which they are suspended. Of course, any aequorin that has been released into free solution will have been discharged immediately.

The particles containing the aequorin are harvested in a cake of silica analytical filtering aid, such as Celite (Johns-Manville), in a large Büchner funnel. The filtrate will contain a small amount of luminous activity because some of the smallest photocyte fragments pass through the filter. No more than 2% of the total luminescent activity should be lost in this way, however. The filter cake (which should not have been allowed to crack) is then washed with a hypertonic solution containing EDTA. This removes the Ca^{2+} of the seawater, as well as a certain amount of soluble matter, without lysing the photocytes. We use a solution of 50 m*M* EDTA, pH 8.0, 75% saturated with ammonium sulfate for this purpose because it is available as a by-product of a later stage in the process. The aequorin is then extracted from the photocytes by passing a hypotonic solution of EDTA through the filter cake. For this purpose we use about 2000 ml of 50 m*M* EDTA, pH 8. The filtrate (henceforth referred to as the first extract) is a clear, greenish fluid that contains about 90% of the aequorin originally in the filter cake. Most of the remainder can be extracted from the cake by removing it from the Büchner funnel, stirring it vigorously in 50 m*M* EDTA, pouring the slurry back into the Büchner funnel, and washing the cake again. However, this time the filtrate (the "second extract") is turbid, and has a much lower specific activity than the first extract. The first and second extracts are salted down separately by stirring them overnight with an excess of ammonium sulfate. When the solutions are fully saturated they are decanted from the solid ammonium sulfate and mixed with one-third volume of 50 m*M* EDTA to bring the final ammonium sulfate concentration to 75% of saturation. (This is the concentration at which the aequorin is just fully salted out.) The precipitated protein is then harvested by filtration and stored on Dry Ice for shipment to the home laboratory. The filtrate, which consists of 50 m*M* EDTA, 75% saturated with ammonium sulfate, is saved for use in washing the initial filter cake (see above) in subsequent extractions.

Purification of Aequorin

OVERALL SEQUENCE. The purification sequence that we currently employ includes the following steps: (1) ammonium sulfate fractionation of the crude extract (25–75% of saturation cut); (2) gel filtration on Sephadex G-50 (fine); (3) ion-exchange chromatography on QAE-Sephadex A-50 with pH step and salt gradient elution; (4) gel filtration on Sephadex G-50 (fine); (5) ion-exchange chromatography on DEAE-Sephadex A-50 with pH step and salt gradient elution.

All these steps are carried out at low temperature (0°–5°) and in the presence of EDTA to minimize the discharge of aequorin. The sequence was adopted after a great deal of experimentation in which the theoretical desirability of combining procedures that separate molecules on the basis of different physicochemical properties was balanced against practical considerations of the effectiveness, yield, and the compatibility of successive steps.

RATIONALE. The ammonium sulfate fractionation is the least effective of the steps in terms of purification, but is included because it can be incorporated with very little extra effort and virtually no loss of aequorin. With increasing concentration of $(NH_4)_2SO_4$, aequorin begins to precipitate at 25% of saturation, and is completely salted out at 75%. Therefore, we make the practice of harvesting our aequorin from a solution that is 75% saturated in ammonium sulfate, and later redissolving it for the first gel-filtration step in a solution that is 25% saturated in the salt. Unfortunately, most of the proteins in the crude extract are retained along with aequorin in this 25–75% cut. We have not found zone precipitation to be enough more effective in this respect to be worth the additional effort.

Gel filtration and ion-exchange chromatography are both very effective in separating aequorin from other proteins and are well suited to processing large amounts of crude extracts. The two techniques complement each other well, since the extract must be loaded onto the ion-exchange column in a solution of low ionic strength, and gel filtration can be used to reduce the salt concentration at the same time that it performs its primary separatory function. Two stages of gel filtration are necessary to achieve the maximum benefit from that procedure. Both gel-filtration runs are carried out on G-50 Sephadex because this is the most highly cross-linked gel that includes aequorin within its fractionation range, and therefore the gel-filtration medium that gives the best separation.[10] The Sephadex is equilibrated and eluted with 10 m*M* EDTA, pH 5.5. The concentration of EDTA was chosen to be high enough to give good protection against Ca^{2+} contamination and low enough not to interfere with the binding of aequorin to the ion-exchange media. The solution is made to pH 5.5 because this is the lowest pH at which both aequorin and the green-fluorescent protein (GFP) are charged enough to bind firmly to anion exchangers. EDTA alone serves as an adequate pH buffer.

QAE- and DEAE-Sephadex are used in preference to other anion exchangers such as DEAE-cellulose, because of their higher capacity and ability to bind aequorin in the presence of higher salt concentrations. They do have the modest disadvantage that their bed volumes change as a function of pH and ionic strength. The properties of QAE- and DEAE-Sephadex with respect to aequorin are not obviously different. Since both exchangers work well, we use them both in the hope that we may thereby take advantage of any subtle differences that may exist in their separatory properties. The pH step in the elution pattern makes use of the fact that aequorin and the GFP have different isoelectric points.[2]

Practical details. In our laboratory the initial gel-filtration run is carried out on a 10 × 100 cm column; the first extracts from as many as 5000 jellyfish can be loaded onto such a column at once in volumes of up to 300 ml. There are three protein peaks (as judged by optical density at 280 nm) in the elution profile. The initial one is in the void volume; the second contains

[10] M. K. Joustra, in *Protides Biol. Fluids, Proc. Colloq.* **14,** 533 (1967).

the GFP (K_{av} = 0.183) and aequorin (K_{av} = 0.201); the third comes off with the ammonium sulfate that was contained in the load. The relative magnitudes of the three peaks depend greatly on the extraction technique used earlier. With very clean extracts the second peak can be considerably higher than the others. With dirty extracts it may be no more than a shoulder on the void-volume peak.

The GFP always elutes slightly ahead of the aequorin, and although the peaks are sharp, they are too close together to allow significant separation of the two proteins. For practical purposes, one can select the active fractions almost as well by monitoring for green fluorescence as by measuring aequorin luminescence, and we have constructed an automatic device for doing just that. The most active fractions (containing in the aggregate about 60% of the luminescent activity loaded onto the Sephadex G-50 column) are then pooled and percolated through the QAE-Sephadex column. Tails with lower luminescent activity are salted down for subsequent reprocessing.

For the first stage of ion-exchange chromatography we use a column 5 cm in diameter containing 4 g (dry weight) of QAE-Sephadex (A-50) that has been equilibrated with 10 m*M* EDTA at pH 5.5. The initial bed depth is about 5 cm. When the product of the initial gel filtration run is passed through this column, virtually all of the aequorin and GFP bind to the top few millimeters of the bed. The GFP is then removed selectively by eluting the column with 5 m*M* EDTA containing 5 m*M* sodium acetate at pH 4.75. When a large amount of aequorin is bound to the column, it then becomes visible as an orange band near the top of the column. The aequorin itself is eluted with a linear sodium chloride salt gradient (500 ml, 0–1 *M* NaCl, in 10 m*M* EDTA, pH 5.5) and frozen at −70° for subsequent processing. The fractions pooled for further processing contain 80–85% of the luminescent activity of the initial load applied to this column.

The products of 5 to 10 QAE-Sephadex runs are pooled for a second stage of gel filtration. They are concentrated initially to a volume of no more than 75 ml by ultrafiltration over an Amicon PM-10 membrane, and then loaded onto a 5 × 100 cm column of Sephadex G-50 (fine) that has been equilibrated with 10 m*M* EDTA at pH 5.5. The column is eluted with the same solution. This column yields three protein peaks, which correspond to those obtained in the first gel filtration run, except that this time the aequorin peak is by far the largest and is well separated from the others. The ratio of aequorin activity to OD_{280} is determined for each active fraction, and those fractions having ratios within 10% of the maximum are pooled for application to a DEAE-Sephadex column. This pool usually includes about 85% of the luminescent activity applied to the gel filtration column.

We generally make the second ion-exchange column about twice as long as the first, using 10 g (dry weight) of DEAE-Sephadex (A-50) in a column of 5 cm diameter. This time there should be little or no GFP present; the aequorin binds to the top few millimeters of the column as a dense apricot-orange band. The same elution pattern is used as the QAE-Sephadex column in order to remove any residual GFP with a pH step. The aequorin is then eluted as before with a salt gradient, and the most active fractions are frozen or lyophilized for future use. The recovery of aequorin activity is usually around 85%. More than 95% of the protein in these fractions runs in a single band on heavily loaded SDS gels. Ordinary polyacrylamide gels show multiple luminescent bands, but no bands of nonluminescent protein. The multiple luminescent components (isoaequorins) can be shown by isoelectric focusing to have isoelectric points between pH 4.2 and pH 4.7. The relative prominence of the various isoaequorins appears to vary from batch to batch.

YIELD. Table I summarizes the various steps in the extraction and purification of aequorin and the approximate recoveries to be expected in each. It should be noted that the overall recovery is somewhat greater than the 20% indicated if the aequorin recovered in second extracts and recycled chromatography tails is taken into consideration. The greatest single loss of luminescent activity in the purification process occurs during the initial gel filtration run. (The luminescent activity recovered in all fractions seldom exceeds 70% of that loaded onto the column.) Over the years we have tried a great many modifications of technique in efforts to reduce this loss, but have not found that any made a significant difference. We have used gel filtration media of three types (Sephadex, BioGel P-30, and Ultrogel

TABLE I
CUMULATIVE RECOVERIES OF ACTIVE AEQUORIN THROUGHOUT THE PREPARATION PROCESS

Stage of preparation	Luminescent activity remaining (%)
Undisturbed *Aequorea* in ocean	100
Aequorea in holding tank	81
Rings after cutting	70
First extract	57
Ammonium sulfate cut	55
First gel filtration	33
QAE-Sephadex	27
Second gel filtration	23
DEAE-Sephadex	20

AcA54) and various pore sizes (Sephadex G-50, G-75, and G-100). We have run the columns with various pH levels between 5.5 and 9.0, various EDTA concentrations, various aequorin concentrations, and various salt concentrations including 1 *M* ammonium sulfate. We have added glycerol, sucrose, cysteine, KCN, and various protease inhibitors to the eluent, all to no avail. The mechanism of the loss remains a mystery, but it seems not to be specific to gel filtration. Similar losses result if the crude extract is dialyzed against 10 m*M* EDTA. The losses are much less severe in all gel filtration runs after the first. For further discussion see Blinks, Prendergast, and Allen.[2]

Desalting, Lyophilization, and Storage. As it comes from the purification process, aequorin usually contains about 10 m*M* EDTA and a substantial concentration of a salt that has been used to elute the protein from an ion-exchange column. These substances must be removed before the photoprotein can be used for most purposes. Furthermore the protein is usually stored for some time, and in some cases shipped before it is used. Aequorin in free solution loses its luminescent activity over a period of weeks, even in the presence of EDTA, and this deterioration is much faster at room temperature than in the cold. These problems have been addressed in two rather different ways. Johnson and Shimomura have followed the practice of salting out the aequorin with ammonium sulfate as soon as it is purified and storing it in a freezer as the ammonium sulfate precipitate. The precipitate, which is sufficiently stable at room temperature to be mailed without great loss of activity, is redissolved and desalted by gel fitration immediately before use. In this case, the gel-filtration column (usually Sephadex G-25) is equilibrated and eluted with a Chelex-treated solution of composition dictated by the nature of the experiment in which the aequorin is to be used. This method cannot be used when the experiment calls for aequorin dissolved in distilled water, because EDTA is not readily separated from aequorin when gel-filtration columns are eluted with distilled water.[2] Additional disadvantages are that (1) it is inconvenient to perform a desalting run in conjunction with each experiment, (2) the volume of solution prepared is often much greater than is needed for a particular purpose, and (3) it is not feasible to prepare highly concentrated solutions of aequorin by this method.

We have used a different approach to get around these problems. Large amounts of aequorin are freed of EDTA at one time by gel filtration on a 2.5 × 100 cm column of Sephadex G-25 equilibrated and eluted with a Chelex-treated solution of 10 m*M* ammonium acetate. (Ammonium acetate is used because some salt must be present in the eluent to prevent EDTA from being excluded from the Sephadex on the basis of charge; the presence of even a modest concentration of salt also reduces the sensitivity of aequorin to contaminating Ca^{2+} (see below). The ammonium acetate sublimes away during lyophilization, leaving the aequorin virtually salt-free.) The aequorin-containing fractions are collected in plastic tubes containing beads of a calcium-binding resin. Aliquots containing the resin beads are then quickly frozen and lyophilized in acid-washed glass ampules, which are sealed under vacuum.

Because nothing better was available, we have used Chelex-100 in our lyophilization ampules in the past, despite the resin's relatively low affinity for Ca^{2+} ($K \simeq 4.6 \times 10^3 M^{-1}$). Recently a new calcium-binding resin with a very much higher affinity for calcium ($K \simeq 10^9 M^{-1}$) has become available.[11] This material, which consists of parvalbumin immobilized on polyacrylamide beads, is known as Calex, and is available from Sigma Chemical Co. We are currently using Calex to protect aequorin during lyophilization, but have not had enough experience with it yet to say whether unexpected problems will be encountered.

A loss of aequorin activity of some 10% may be expected during the gel-filtration carried out to remove EDTA from the aequorin solution. Losses during lyophilization and storage range from 0 to 100%, depending on the care that is taken to avoid calcium contamination. The product, which consists of dry aequorin adhering to beads of resin, is stable indefinitely in an ultracold (−75°C) freezer, for weeks (at least) at refrigerator temperatures, and for several days (at least) at room temperature. The aequorin can be dissolved just before use in whatever solution and at whatever concentration is appropriate to the needs of the particular experiment. No more protein need be dissolved than will be required, and if small volumes of fluid are added to an ampule, solutions of extraordinarily high protein concentration can easily be prepared. The beads of resin tend to crack when lypohilized, but any fragments produced can be filtered out readily with a fine ($< 0.1\ \mu m$) Nuclepore or Millipore filter in a miniature holder made for the purpose (see Fig. 7). This filter makes it possible to work with samples of only a few microliters.

Introduction into Cells

The most difficult problem associated with the use of aequorin as an intracellular calcium indicator is getting the protein into the cell. To date the only way in which aequorin has been introduced into intact cells has been by injection. Some giant cells can be cannulated or fitted with axial perfusion capillaries, and once this has been achieved the introduction of the photoprotein is relatively straightforward. Smaller cells can be injected with glass micropipettes, though the difficulty of injection increases rapidly with decreasing cell size, and only a few types of cells under 50 μm in

[11] P. Lehky, M. Comte, E. H. Fischer, and E. A. Stein, *Anal. Biochem.* **82,** 158 (1977).

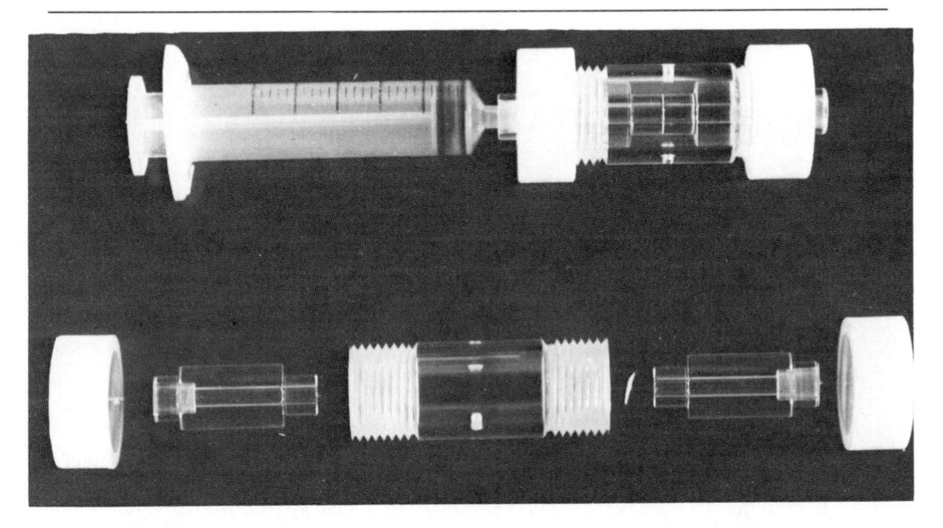

FIG. 7. Filter holder for use with microliter volumes. A disk of fine (<0.1 μm) filter material cut with an ordinary paper punch is soaked in EDTA, and then clamped between the two plastic cylinders. The holes through these cylinders have Luer tapers at their outer ends for the insertion of plastic plugs (not shown) or a syringe. The holes in the middle of the threaded case allow the unit to be washed thoroughly after it is assembled. A drop of aequorin solution is placed on one side of the filter disk and then forced through it by air pressure from the syringe. The filtered solution is removed from the other side with a fine polyethylene tube.

diameter have so far been injected successfully. The approach that has been used most widely in work with aequorin is to apply pressure from cylinders of compressed gas to fine-tipped pipettes containing a small amount of fluid at the tip. Although this technique, which is generally credited to Strumwasser,[12] is fairly widely used, no detailed description of it has been published. We offer the following observations on its application to the microinjection of aequorin.

In our experience the most satisfactory pipettes for use in small cells have been drawn from capillary tubing with a glass filament fused to the inside surface [such as Omega dot (Frederick Haer, Inc., P. O. Box 337, Brunswick, Maine 04011) or "Kwik-fil" (W-P Instruments, Inc., P. O. Box 3110, New Haven, Connecticut 06515)]. With pipettes made from glass of this sort, it is feasible to eject aequorin solutions through tips small enough to have resistances greater than 10 megohms in 3 *M* KCl. Pipettes pulled with free strands of fiberglass inside, after the method of Tasaki *et al.*,[13] have been unsatisfactory in our hands because of the difficulty of

[12] F. Strumwasser, *in* "Circadian Clocks" (M. Aschoff, ed.), p. 442. North-Holland Publ., Amsterdam, 1965.

[13] K. Tasaki, Y. Tsukahara, S. Ito, M. J. Wayner, and W. Y. Yu, *Physiol. Behav.* **3,** 1009 (1968).

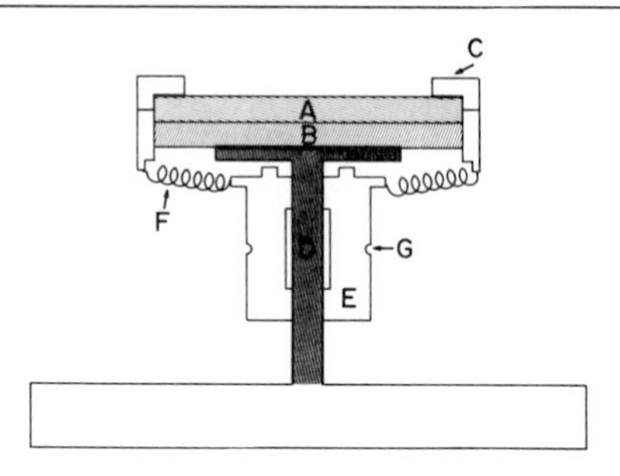

FIG. 8. Micropipette beveler. A modification of the design of B. R. Kripke and T. E. Ogden [*Electroencephalogr. Clin. Neurophysiol.* **36,** 323 (1974)] that eliminates the need for ball bearings by removing the thrust of the driving belt from the rotating plate. Vertical section through the rotor. A and B, Disks of optically flat glass (stock No. 2155, Edmund Scientific Co., Great Barrington, New Jersey 08007); C, anodized aluminum ring cemented to upper glass disk; D, stationary steel post; E, brass pulley; F, springs connecting pulley to rotating ring (there are four); G, groove for driving belt. Plate A, coated on top with a fine abrasive such as 0.05 μm alumina, rotates on plate B, which is stationary. Plates A and B are separated by a film of oil; heavy lines indicate cement. Ring C is in two parts to facilitate the separation of plates A and B during disassembly.

controlling Ca^{2+} contamination. The glass used to draw the micropipettes must, of course, be carefully cleaned (e.g., by soaking it in concentrated H_2SO_4 and hot HCl) and free of particles of any kind. It is helpful to fire-polish the ends of the glass before washing it in order to prevent the generation of glass chips and to avoid shaving particles from the Teflon packing of the electrode holder as pipettes are inserted. Some workers have found it helpful to bevel the tips of pipettes to be used for injection. Several techniques for doing this have been described.[14-17] Figure 8 illustrates a modification of the bearing arrangement of Kripke and Ogden[15] that was constructed in our laboratory. Beveling systems are available commercially from W-P Instruments, Inc.; from David Kopf Instruments, 7324 Elmo Street, Tujunga, California 91042; and from the Narishige Scientific Instrument Laboratory, Tokyo, Japan. Silver wires should not be used inside pipettes to monitor the resistance during beveling, because traces of Ag^+ rapidly inactivate aequorin. Platinum wires can be used in conjunction with monitoring systems that use alternating current. The tips of pipettes made from filament-containing capillaries usually fill satisfactorily if a few

[14] J. Barrett and D. G. Whitlock, *in* "Intracellular Staining in Neurobiology" (S. B. Kater and C. Nicholson, eds.), p. 297. Springer-Verlag, Berlin and New York, 1973.

[15] B. R. Kripke and T. E. Ogden, *Electroencephalogr. Clin. Neurophysiol.* **36,** 323 (1974).

[16] K. T. Brown and D. G. Flaming, *Science* **185,** 693 (1974).

[17] J. J. Chang, *Comp. Biochem. Physiol.* **52A,** 567 (1975).

microliters of aequorin solution are placed in the butt end of the pipette. Bubbles in the tip usually do not interfere with the use of the pipette because the crevices on either side of the filament fill by capillarity and form a conductive channel past the bubble.

The concentration of aequorin in the solution for injection is usually arrived at empirically by a series of compromises. It is obviously desirable not to have to inject large volumes of solution, particularly into small cells. On the other hand, the difficulty of injection increases with increasing protein concentration because of increasing viscosity and a tendency for pipettes to become blocked. For work with vertebrate muscle, we have usually used aequorin concentrations in the range 1–5 mg/ml. Because it has been reported[18] that aequorin is particularly prone to aggregation in acid solutions, it is probably wise to buffer injection solutions at slightly alkaline pH levels. In most situations it is also desirable to increase the conductivity of the solution by including an electrolyte: the micropipette can then be used as a microelectrode, and it is possible to determine when cells have been penetrated by recording the potential difference between the electrode and the bathing medium. We normally dissolve aequorin in an isotonic solution of KCl, buffered to pH 8.0 with 1 m*M* *N*-2-hydroxyethylpiperazine-*N*′-2-ethanesulfonic acid (HEPES) and freed of calcium by passage through a Chelex column. The solution is filtered through a small disk of 0.08 μm Nuclepore filter material in the holder shown in Fig. 7 immediately before it is used to fill the pipette.

Pipette holders of the general type first described by Grundfest *et al.*[19] permit simultaneous electrical recording from and the application of gas pressure to the pipettes. Figure 9 shows the modification of that design used in our laboratory. One end of the holder is equipped with a threaded fitting that can be tightened to compress a Teflon collar around the pipette. The opposite end contains an insulated pin with an attached wire that is threaded into the pipette as the pipette is inserted into the holder. The customary chlorided silver wire has been replaced by platinum because silver rapidly inactivates aequorin. Gas pressure (instead of oil as in the original design) is applied to the interior of the holder (and pipette) through a sidearm. The Teflon collar is capable of holding pipettes securely in the face of applied pressures at least as high as 10 atm.

Another essential piece of equipment for pressure injection is a convenient system to control pressure applied to the pipette. The operator should be able to turn the pressure on or off at the flick of a switch with decompression of the system when the pressure is turned off, and he should be able to

[18] Y. Kohama, O. Shimomura, and F. H. Johnson, *Biochemistry* **10**, 4149 (1971).
[19] H. Grundfest, C. Y. Kao, and M. Altamirano, *J. Gen. Physiol.* **38**, 245 (1954).

FIG. 9. Pipette holder for microinjection. (A) Intact assembly.

adjust the pressure applied to the pipette while seated at the microscope. Figure 10 is a diagram of a simple and convenient pressure system first shown to us by Dr. J. E. Brown and illustrated here with his permission. Purely mechanical systems of this sort can be constructed conveniently from miniature components (MTV-3 toggle valves and MAR-1 pressure regulator) available from Clippard Instrument Laboratory, Inc., 7390 Colerain Road, Cincinnati, Ohio 45239. Electrical control may be desirable for applying rapid pressure pulses.[20,21] In this case it is convenient to use solenoid valves with power requirements low enough so that they can be operated directly from transistorized circuits. Valves of this sort are avail-

[20] B. Rose and W. R. Loewenstein, *J. Membr. Biol.* **28,** 87 (1976).

[21] R. Llinás and C. Nicholson, *Proc. Natl. Acad. Sci. U.S.A.* **72,** 187 (1975).

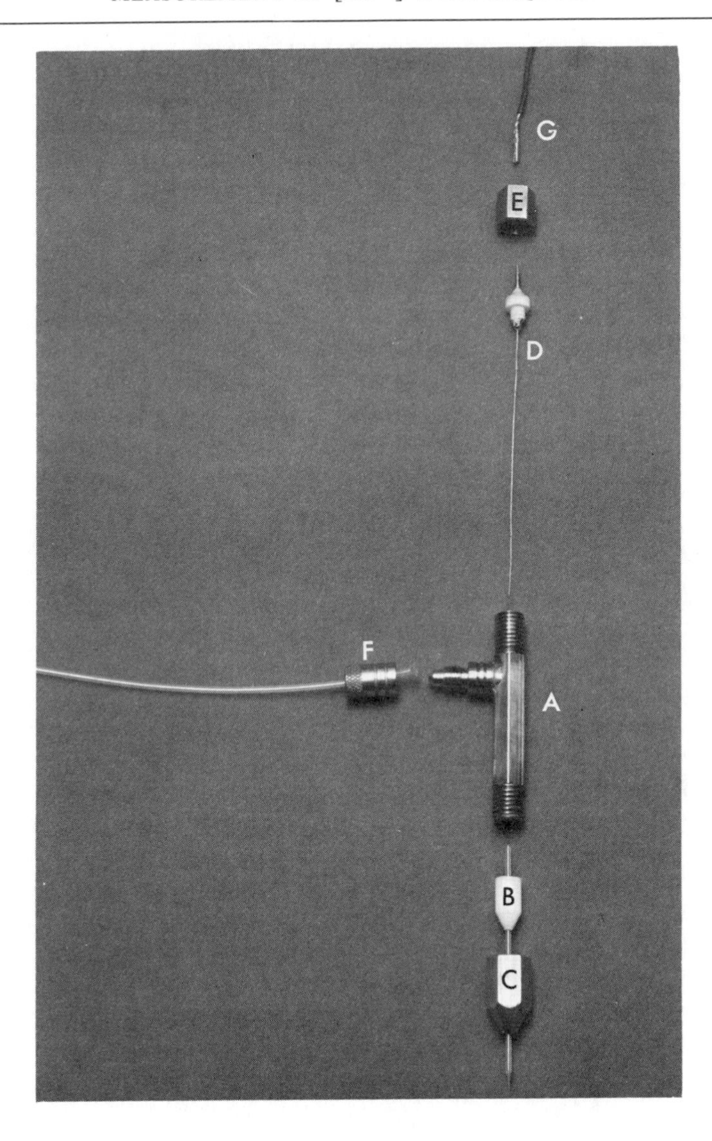

FIG. 9. Pipette holder for microinjection. (B) Component parts of the holder: A, body with pressure sidearm; B, Teflon collar; C, threaded compression fitting with tapered cavity; D, platinum wire attached to insulated pin; E, threaded cap; F, flared polyethylene tubing with pressure fitting; G, wire to preamplifier. The pressure sidearm is a Clay Adams Model A-1025 plastic tubing adapter, size B (7541).

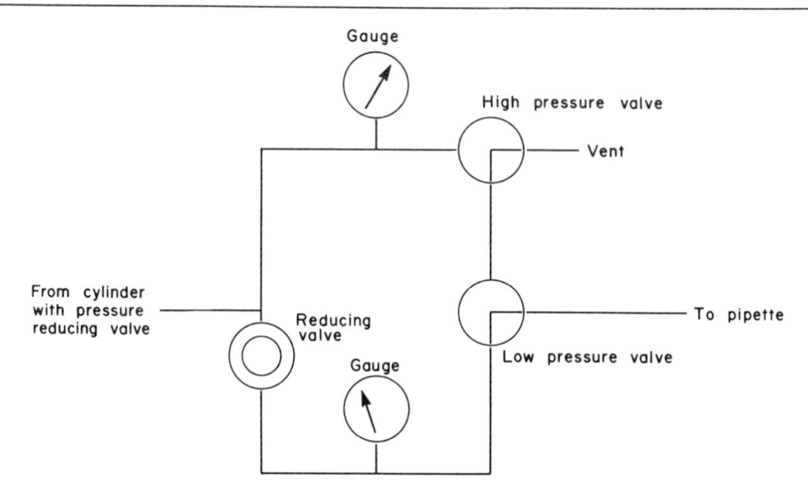

FIG. 10. Pressure system for microinjection. Three-way pressure valves can be operated either mechanically or electrically (see text).

able from Northeast Fluidics, Inc., Amity Road, Bethany, Connecticut 06525.

Because microinjection is almost prohibitively difficult in small cells, an alternative technique for getting the protein into cells is sorely needed. The large molecular size-to-charge ratio of the photoproteins is an obstacle to the use of iontophoresis. Liposome fusion presents intriguing possibilities, but attempts to use this approach have so far been unsuccessful. It may be possible to introduce photoproteins into some cells by temporarily rendering the cell membrane permeable to large molecules. Campbell and Dormer[22] have succeeded in getting obelin into erythrocyte ghosts which could be resealed subsequently with the active photoprotein inside. It remains to be seen whether this approach can be used in other types of cells without greatly interfering with their function.

Light Recording

Aequorin is most useful, both *in vitro* and *in vivo,* under conditions in which its reaction rate is low. It is particularly well suited to the measurement of [Ca^{2+}] in small volumes, and because it binds calcium, it should be used in low concentrations whenever possible. For all these reasons it is often important to be able to detect very low light levels in work with aequorin. Fortunately, photomultipliers and image intensifiers capable of

[22] A. K. Campbell and R. L. Dormer, *Biochem. J.* **152,** 255 (1975).

detecting small numbers of photons are readily available. Many aspects of the use of photomultipliers are discussed in Section IX of this volume and need not be repeated here. A few special points specific to their use with photoproteins do deserve mention, however. For optimum sensitivity the photocathode should have a high quantum efficiency in the blue (λmax for aequorin = 469 nm), and for low, dark current it should have low sensitivity in the red and infrared. Bialkali cathodes meet these requirements reasonably well.

In setting up a photomultiplier for work on aequorin-injected cells one is almost inevitably faced with the need to compromise between sensitivity and convenience. For optimum sensitivity the photomultiplier should have a large photocathode, be mounted as close to the preparation as possible, and be separated from it by as few reflecting interfaces as possible. In the extreme, the front face of the photomultiplier tube can be coupled optically (e.g., with Q2-3067 optical couplant, Dow-Corning Corp., Midland, Michigan 48640) to the wall of the chamber containing the preparation.[23] Although optically efficient, this arrangement has the serious disadvantage that the photomultiplier cannot be equipped with a shutter. In the absence of a shutter, it is impossible to examine the preparation in situ without exposing the photocathode to bright light, which greatly increases the dark current for many hours thereafter. For these reasons some workers have separated the photomultiplier from the tissue bath by a distance at least great enough to install a slide-type shutter, and accepted the attendant decrease in sensitivity. Still others have made use of microscope lenses[20,24] or fiber-optic probes[25,26] to allow the photomultiplier to be placed a substantial distance away from the preparation. In such circumstances electronic shutters (such as those available from Vincent Associates, 1255 University Avenue, Rochester, New York 14607) are very convenient. Water-immersion microscope objectives of numerical aperture as high as 0.60 have been used to record light from aequorin-injected cells[20], and salt water immersion lenses of N.A. >1.0 are available (from E. Leitz, Inc.). The light-gathering ability of such lenses is higher than that of glass fiber-optic probes, which usually have a numerical aperture of about 0.5 and accept only about 70% of the incident light within the cone defined by that aperture. There is an additional transmission loss in conventional glass fiber-optic probes of about 0.3% per centimeter. The losses in microscope lens systems are variable, but increase with the number of optical elements in

[23] E. B. Ridgway, J. C. Gilkey, and L. F. Jaffe, *Proc. Natl. Acad. Sci. U.S.A.* **74,** 623 (1977).
[24] J. E. Brown and J. R. Blinks, *J. Gen. Physiol.* **64,** 643 (1974).
[25] R. Llinás, J. R. Blinks, and C. Nicholson, *Science* **176,** 1127 (1972).
[26] R. Eckert, D. Tillotson, and E. B. Ridgway, *Proc. Natl. Acad. Sci. U.S.A.* **74,** 1748 (1977).

the light path. An advantage of lens systems is that the gathered light can be focused to a small spot. This makes it possible to use photomultipliers with small photocathodes and correspondingly low dark currents. Although they do not provide the flexibility of fiber-optic probes, excellent light guides can be made out of polished rods of acrylic plastic.[2,20,32] In our experience their efficiency is considerably higher than that of fiber-optic guides.

One of the important advantages of the photoproteins as Ca^{2+} indicators is the readiness with which they lend themselves to the localization of Ca^{2+} by means of image intensification. Although microscopic image intensification has been used successfully for more than a decade in studies of bioluminescence, it is just beginning to see application in situations where photoproteins have been introduced as calcium indicators.[20,27,28] The application of image intensification to microscopy was pioneered by G. T. Reynolds, who has written two extensive reviews[29,30] on the subject. In one of those papers[30] Reynolds describes his apparatus incorporating a four-stage magnetically focused image tube and television camera, and the interested reader is referred there for details of his experimental setup. It should be emphasized, however, that this is a field of rapidly advancing technology, and an investigator who plans to set up an image intensification system would be wise to determine the best way of meeting his particular needs with currently available equipment, rather than to copy a particular setup described in the literature. For example, it would be pointless now to duplicate Reynolds' optical system for numbering individual television frames as they are recorded. Inexpensive units are now available commercially (e.g., from Thalner Electronic Laboratories, Inc. 7235 Jackson Road, Ann Arbor, Michigan 48103) to insert this information into the television picture electronically. Similarly, investigators setting up high-performance image intensification systems in the future may wish to consider using television cameras equipped with silicon target image tubes, silicon intensified target (SIT) tubes, or image isocons, rather than the plumbicon used in Reynolds' apparatus.

Some television cameras now available "off the shelf" incorporate various sorts of image intensification systems as integral parts of the camera, or even of the camera tube. Integrated systems of this sort can take advantage of efficiencies in the coupling of successive stages that are not possible when the image intensifier is separate from the television camera. They also have the advantages of being self-contained, relatively inexpensive, and simple to use. Commercial television systems intended for night-

[27] D. L. Taylor, G. T. Reynolds, and R. D. Allen, *Biol. Bull.* **149,** 448 (1975).

[28] J. C. Gilkey, E. B. Ridgway, L. F. Jaffe, and G. T. Reynolds, *Biophys. J.* **17,** 277a (1977).

[29] G. T. Reynolds, *Adv. Opt. Electron Microsc.* **2,** 1 (1968).

[30] G. T. Reynolds, *Quart. Rev. Biophys.* **5,** 295 (1972).

time surveillance have been used successfully in some types of work with aequorin,[20] but we have found them insufficiently sensitive for use with aequorin-injected muscle fibers.

Magnetically focused image intensifiers are probably the only ones presently capable of providing the combination of high gain and low noise required for the localization of natural concentrations of Ca^{2+} in most types of cells. Unfortunately, they cannot be coupled directly to television camera tubes because the magnetic field interferes with the function of the camera. Image intensifiers of very high gain will also be necessary whenever the need for very high framing rates requires that photographic recording be used. (Ordinary television systems operate at 60 frames per second.) Under such circumstances the intensifier tube may also have to be equipped with special high-speed phosphors. This can be done only at a sacrifice of gain, and that makes the use of the most powerful type of tube all the more important.

In the course of this discussion it should be noted that there is an element of ambiguity in the terms "gain" and "sensitivity" as applied to image intensifiers. The gain of an image intensifier reflects both the quantum efficiency of the initial photocathode and the amplification of the photoelectrons once they are displaced from it. Increased amplification is desirable only up to the point at which individual photoelectrons become clearly detectable as spots on the film or television screen. Beyond that point, increases in gain are useful only to the extent that they result from increased quantum efficiency at the initial photocathode.

Interpretation of Light Measurements

Aequorin has a number of properties that complicate the interpretation of light signals obtained from systems containing the photoprotein. The following characteristics must be taken into consideration whenever plans are made to use aequorin as a Ca^{2+} indicator:

1. Aequorin is consumed in the luminescent reaction. Corrections must be made for this if the [Ca^{2+}] in the compartment under study rises high enough or for long enough to discharge an appreciable fraction of the aequorin present.

2. The speed with which light intensity follows sudden changes in [Ca^{2+}] is limited. When aequorin is mixed very rapidly with a solution of saturating [Ca^{2+}], light intensity rises with a halftime of about 6 msec at 20°.[31] There is some uncertainty[31,32] about whether the rise time is influ-

[31] J. W. Hastings, G. Mitchell, P. H. Mattingly, J. R. Blinks, and M. van Leeuwen, *Nature (London)* **222,** 1047 (1969).

[32] G. Loschen and B. Chance, *Nature (London), New Biol.* **233,** 273 (1971).

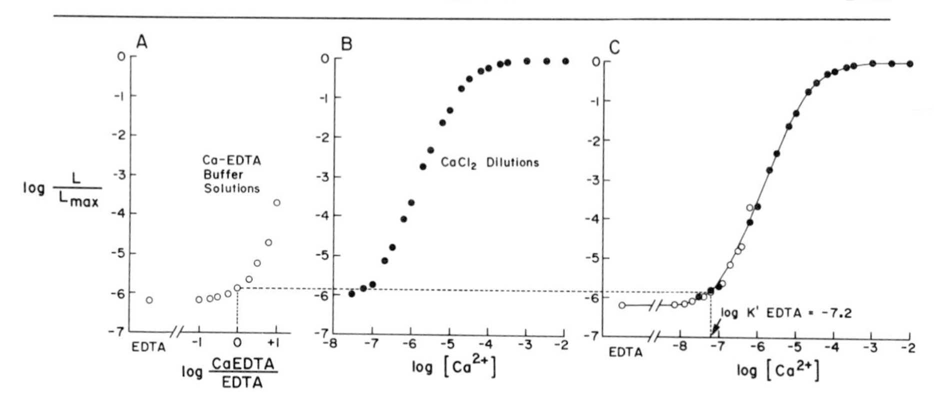

FIG. 11. Determination of a calcium concentration-effect curve for aequorin. Double logarithmic plots; L/L_{max} indicates light intensity as a fraction of that achieved in saturating $[Ca^{2+}]$. Panel A shows experimental points determined with a series of Ca-EDTA (ethylenediaminetetraacetic acid) buffers (○) and with simple dilutions of $CaCl_2$ (●). The point marked EDTA was determined in the EDTA solution used for the calcium buffers with no added Ca^{2+}. Panel B contains points determined with simple dilutions of $CaCl_2$. In panel C, the Ca^{2+}-buffer points have been shifted as a group to give the best fit to the points determined with $CaCl_2$ dilutions. This has the effect of establishing the Ca-EDTA binding constant appropriate to the particular conditions of the experiment, and greatly increases the confidence that can be placed in the lower part of the concentration-effect curve. All solutions in this experiment contained 150 m*M* KCl, 5 m*M* piperazine-*N,N'*-bis(2-ethanesulfonic acid) (PIPES), pH 7.0. The [EDTA] was 1 m*M* in the calcium-buffered solutions. All determinations were made at 21°.

enced by the range of $[Ca^{2+}]$ over which the change occurs (see Blinks *et al.*[2] for discussion), but the rise is certainly never faster than in saturating $[Ca^{2+}]$. Even then, it is slow enough to significantly distort the time course of signals arising from rapid Ca^{2+} transients such as those occurring in vertebrate skeletal muscle.[33]

3. The rate of the luminescent reaction is independent of $[Ca^{2+}]$ both at very high and at very low calcium concentrations (see Figs. 11 and 12). The fact that the reaction saturates at high $[Ca^{2+}]$ comes as no surprise, but it has only recently been appreciated[34] that the calcium concentration-effect curve for aequorin also flattens out at very low $[Ca^{2+}]$. This imposes a lower limit on the range of $[Ca^{2+}]$ detectable with aequorin that is independent of the sensitivity of the light-measuring apparatus or the amount of aequorin present.

4. The relation between $[Ca^{2+}]$ and light intensity is nonlinear. The calcium concentration-effect curve is sigmoid on a log–log plot, and has a

[33] J. R. Blinks, R. Rüdel, and S. R. Taylor, *J. Physiol. (London)* **277,** 291 (1978).
[34] D. G. Allen, J. R. Blinks, and F. G. Prendergast, *Science* **196,** 996 (1977).

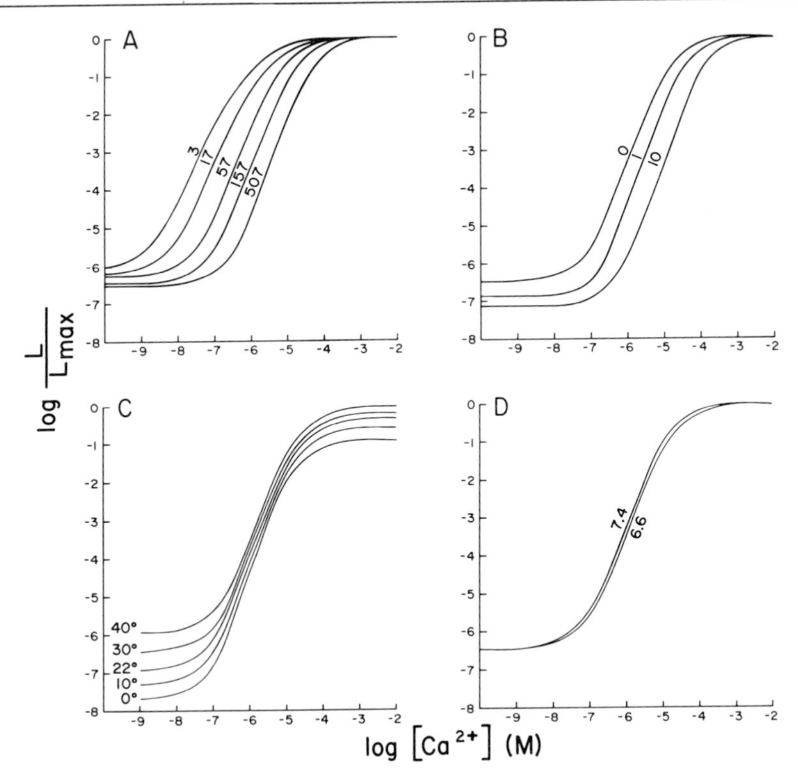

FIG. 12. Influence of reaction conditions on sensitivity of aequorin to Ca^{2+}. Curves were drawn from results of D. G. Allen and J. R. Blinks, to be published.)

(A) Influence of monovalent salt concentration. Concentrations of K^+ as indicated (m*M*). Curves were determined as in Fig. 11 except that both 1 m*M* EGTA and 1 m*M* (1,2-cyclohexylenedinitrilo)tetraacetic acid (CDTA) were used as calcium buffers. For the curve in 3 m*M* K^+, 1 m*M* PIPES was used as the pH buffer, in all other curves the concentration of PIPES was 5 m*M*. The balance of the salt concentration was made up with KCl. Temperature 21°, pH 7.0. (B) Influence of [Mg^{2+}]. Concentrations of $MgCl_2$ as indicated (m*M*). Curves were determined as in Fig. 11 with 1 m*M* EGTA as the only calcium buffer. All solutions contained 150 m*M* KCl, 5 m*M* PIPES, at pH 7.0. Temperature 21°. (C) Influence of temperature. All curves determined as in Fig. 11 in solutions containing 150 m*M* KCl, 5 m*M* PIPES, pH 7.0. The pH of the solutions was adjusted to 7.0 at each of the temperatures studied; 1 m*M* EGTA was the only calcium buffer used. (D) Influence of pH. Curves determined as in Fig. 11 in solutions containing 150 m*M* KCl, 5 m*M* PIPES, at pH 6.6 and 7.4. Both EGTA (1 m*M*) and CDTA (1 m*M*) were used as buffers.

maximum slope of about 2.5 under physiological conditions (Figs. 11 and 12). The steepness of this curve means that changes in light intensity tend to give an exaggerated impression of the changes in [Ca^{2+}] responsible for them. An additional consequence is that local regions of high [Ca^{2+}] tend to dominate patterns of light emission. In a system where the aequorin concentration is uniform, an amount of Ca^{2+} added at one point will give a much brighter signal when its concentration is locally high than after it has diffused to reach a uniform concentration. This behavior complicates the quantitative interpretation of photomultiplier records, but may be helpful in work with image intensification, since it tends to increase contrast.

5. The sensitivity of aequorin to [Ca^{2+}] is influenced profoundly by a number of experimental factors of physiological importance (Fig. 12). Temperature, ionic strength, and [Mg^{2+}] have marked effects on the relation of luminescence to [Ca^{2+}]; although less important, pH also influences the sensitivity of aequorin somewhat. One must also be alert to the possibility that experimental solutions might contain as yet unrecognized substances capable of altering the sensitivity of aequorin to Ca^{2+}.

Several approaches have been used for the translation of measurements of light intensity into absolute Ca^{2+} concentrations. If the [Ca^{2+}] to be measured is constant and high enough to give luminescence above the calcium-independent level, it may be estimated as follows: a photoprotein is added to the test system, and small amounts of a series of Ca^{2+}–EGTA buffers are then added to find the mixture (of known [Ca^{2+}]) that just does not change the light emission from the system.[35,36] [EGTA, ethyleneglycolbis (β-aminoethylether)-N,N'-tetraacetic acid, is also described as ethylenebis(oxyethylenenitrilo)tetraacetic acid. The use of EGTA and other chelators as calcium buffers has been reviewed by Caldwell.[37]] This null method for determining [Ca^{2+}] has the advantages that it is relatively simple and that it is not necessary to know either the composition of the test solution or the transfer function of the measuring system (including the photoprotein). However, it has the disadvantages that without additional information it is not possible to relate light intensity to [Ca^{2+}] except at the single [Ca^{2+}] where the null measurement was made, the repeated injection of Ca^{2+}–EGTA buffers into the same region of the same cell is not always technically feasible, and in many cells the [Ca^{2+}] under resting conditions is likely to be low enough so that aequorin luminescence is nearly independent of [Ca^{2+}].

[35] P. F. Baker, A. L. Hodgkin, and E. B. Ridgway, *J. Physiol. (London)* **218,** 709 (1971).

[36] R. Di Polo, J. Requena, F. J. Brinley, Jr., L. J. Mullins, A. Scarpa, and T. Tiffert, *J. Gen. Physiol.* **67,** 433 (1976).

[37] P. C. Caldwell, *in* "Calcium and Cellular Function" (A. W. Cuthbert, ed.), p. 10. Macmillan, New York, 1970.

Other approaches to the "calibration" of light measurements depend on knowledge of the calcium concentration-effect curve for the photoprotein under conditions comparable to the experimental ones, and the definition of at least one point on that curve in the experimental system for use as a scaling factor. It is particularly important that the reference concentration-effect curve be determined under the same conditions of temperature, ionic strength, and $[Mg^{2+}]$ to be used in the experiment (Fig. 12). Ideally, the calibration curve should be determined with photoprotein from the same sample used in the experiment, since minor variations in the calcium-independent luminescence seem to exist in aequorin from different batches.

For the determination of calcium concentration-effect curves we have usually used an apparatus essentially like that illustrated in Figs. 1 and 2, except that it is equipped with a series of neutral density filters mounted in a wheel, and a temperature-controlled cuvette holder. Ten-microliter aliquots of a dilute (<1 μM) solution of EDTA-free aequorin are injected into 1-ml volumes of the test solutions. Calibration curves are generally determined with the combined use of simple dilutions of $CaCl_2$ and calcium buffers.[34] This permits the construction of a calibration curve covering a wider range of $[Ca^{2+}]$ than would be possible with dilutions or one Ca^{2+}-buffer alone. Furthermore, if the segments of the curve determined with $CaCl_2$ and with a buffer can be made to overlap, they provide a precise measure of the binding constant of the chelator under the exact conditions of the experiment (Fig. 11).

The determination of full calcium concentration-effect curves requires meticulous attention to the avoidance of contamination, not only by Ca^{2+}, but also by Ag^+.[34] (Very low concentrations of Ag^+ greatly increase the calcium-independent luminescence of aequorin.) Cuvettes must be scrupulously clean, and all solutions used must be Chelex-treated before being used to make $CaCl_2$ dilutions. Even before Chelex treatment, pH electrodes must not be put into the main body of the solution because the reference electrodes usually contain silver, and Chelex has a low affinity for Ag^+. Silver contamination can be checked for by adding CN^-, which binds Ag^+ very tightly and eliminates Ag^+-induced luminescence. (The cyanide solution must be Chelex-treated, however, or added in the presence of a chelator such as EGTA or EDTA). When low calcium concentrations are used, the aequorin must be added to the test solution before the cuvette is put into the photometer. This avoids contamination that may be introduced into the tip of the needle when it is passed through the rubber septum. Fortunately, this contamination is not a problem when the $[Ca^{2+}]$ is so high that the light intensity decays appreciably in the time required to insert the cuvette into the photometer.

The main obstacle to calibrating measurements of luminescence made *in vitro* is that one needs to know the ionic strength and [Mg^{2+}] of the test solution. It may sometimes be possible to get around this need by adding large and equal concentrations of, say, KCl and Mg^{2+} to both the test and calibration solutions. However, the possibility that doing so might alter the binding of Ca^{2+} to constituents of the solution must not be lost sight of. If equal volumes of the test and calibration solutions are studied under identical conditions in the same photometer with the same sample of photoprotein, the need for a scaling factor is eliminated.

When photoproteins are used as intracellular calcium indicators uncertainties about the composition of the medium are likely to be a more serious problem. Furthermore, a method for establishing a scaling factor is indispensable for quantitative interpretation. Two different approaches to this have been proposed. One is to determine the resting intracellular [Ca^{2+}], if it is detectable, by the null method of finding a Ca^{2+}-EGTA buffer mixture that does not change the light output when it is injected into the cell.[35] (In doing this, one must not fail to determine that the injection of a buffer mixture of lower [Ca^{2+}] reduces the resting glow; otherwise there is no assurance that the resting glow was not due to calcium-independent luminescence.) The other approach is to lyse the cell while recording light emission under optical conditions identical to those that were used to make a series of physiological observations immediately beforehand.[38,39] The necessary scaling factor can then be derived from the integrated light signal recorded during the discharge of all the photoprotein in the cell.

The Binding of Calcium to Photoproteins

Most indicators bind the substance to which they are sensitive, and thus to some extent must alter the concentration that they are supposed to measure. Photoproteins are no exception in this respect, and it will be important to define those conditions under which the binding of calcium by the proteins might seriously interfere with their use as calcium indicators. To do this properly, one must know the number and the affinities of all calcium-binding sites on the photoprotein molecule both before and after the luminescent reaction. (There is at present no information on whether the affinities change as a result of the reaction.) To assess the impact of Ca^{2+}-binding by the indicator on the system being studied, one must also know something about the calcium-buffering capacity of that system. Our information on all these points is fragmentary at best. Conventional ap-

[38] D. G. Allen and J. R. Blinks, *in* "Detection and Measurement of Free Calcium Ions in Cells" (C. C. Ashley and A. K. Campbell, eds.). Elsevier, Amsterdam, in press, 1978.

[39] D. G. Allen and J. R. Blinks, *Nature (London)* **273**, 509 (1978).

proaches such as equilibrium dialysis, Chelex partitioning, gel filtration, or ultrafiltration can give information about the binding of calcium to the spent protein only. Although there would seem to be no reason why such studies could not be performed on spent photoproteins, the only published information on Ca^{2+}-binding comes from a single report[40] of equilibrium dialysis measurements on aequorin in which the observations are not extensive enough to be very useful.

Estimates of the affinity of the reactive sites of unspent aequorin for calcium can and have been made from measurements of the relation between calcium concentration and luminescence, but different estimates can be derived from the same data depending on the reaction model that one adopts.[4] However, for purposes of assessing the potential influence of aequorin as a "calcium sink" we shall follow the worst-case principle and assume that each aequorin molecule has three Ca^{2+}-binding sites, each with the affinity constant $K = 6.2 \times 10^4\ M^{-1}$.[34] These values were used to calculate the series of curves in Fig. 13, which show how, in the absence of any other calcium-binding substances, the fraction of the total calcium that is bound to aequorin might be expected to vary as a function of the Ca^{2+} and aequorin concentrations. It is self-evident that the fraction of calcium bound will increase with increasing aequorin concentration and decrease with increasing calcium concentration. When the aequorin concentration is as high as 1 μM, the fraction of calcium bound is unacceptably high (about 16%) over most of the range of [Ca^{2+}] in which aequorin is a useful Ca^{2+} indicator. As the aequorin concentration is reduced the situation improves, and in the presence of 0.1 μM aequorin less than 2% of the calcium is bound at any [Ca^{2+}]. This would seem to be an acceptable level for most applications, and as a rule of thumb, we would suggest that whenever aequorin is used as a Ca^{2+} indicator in systems where the [Ca^{2+}] is unbuffered, the photoprotein concentration be kept below 0.1 μM. This may seem like a rather low concentration, but in an efficient photometer 10^{-14} mol of aequorin is enough to allow the detection of light levels down to and including the calcium-independent luminescence. A tenth of a microliter of a 0.1 μM aequorin solution contains this amount of photoprotein, and thus it should be possible to determine full calcium concentration-effect curves on samples as small as 0.1 μl without significant distortion resulting from the binding of calcium to aequorin. Still smaller samples could be used if the [Ca^{2+}] to be measured was relatively high, if the binding of more than 2% of the calcium to the indicator could be tolerated, or if calcium-buffering substances were present in the sample. The last point is extremely important in connection with the use of photoproteins in biological systems. Figure 13 and the preceding discussion are predicated on the assumption

[40] O. Shimomura and F. H. Johnson, *Nature (London)* **227,** 1356 (1970).

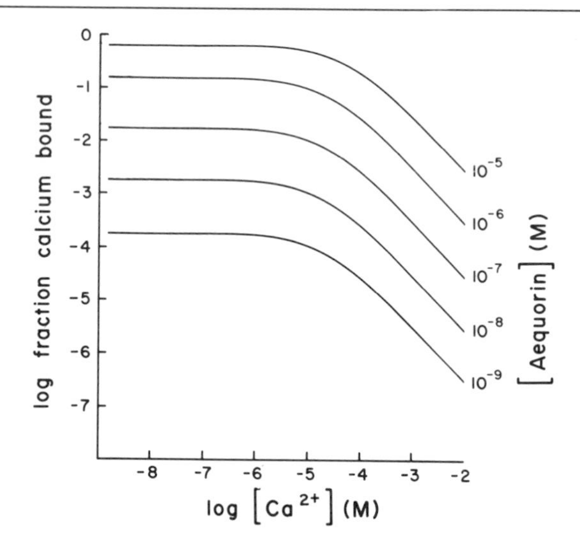

FIG. 13. The binding of Ca^{2+} to aequorin. Each curve indicates the fraction of calcium in a solution that would be bound to a particular concentration of aequorin at various free calcium concentrations. The curves are calculated on the basis of the assumptions that aequorin molecules have three calcium binding sites, each with an affinity constant $K = 6.2 \times 10^4$ M^{-1}, and that no other calcium-binding substances are present. It should be emphasized that these estimates are subject to revision and are used only for purposes of illustration (see text). From J. R. Blinks, *Ann. N. Y. Acad. Sci.* **307,** 71 (1978), with permission of the publisher.

that the photoprotein is to be used in a system containing no other calcium-binding substances. In most biological systems this is not the case; calcium concentrations are often well buffered, and it may be possible to introduce relatively high concentrations of photoproteins into cells without binding an appreciable fraction of the calcium in the system or greatly altering the [Ca^{2+}] to be measured. Such cases have to be judged individually on the basis of information about the calcium-binding properties of the system under study.

Acknowledgments

Original work described in this paper was supported by USPHS grants HE 03738 and HL 12186. Parts of the work were done during the tenure of an Established Investigatorship (to J. R. B.) and a Visiting Scientist Award (to B. R. J.) of the American Heart Association, and a British–American Research Fellowship (to D. G. A.) from the American Heart Association and the British Heart Foundation. Use of the facilities of the Friday Harbor Laboratories, University of Washington, is gratefully acknowledged. The apparatus shown in Figs. 2 and 6–10 was constructed in the department workshop by Mr. Merlin Neher.

Section V

Cypridina

[32] Introduction to the *Cypridina* System

By FRANK H. JOHNSON and OSAMU SHIMOMURA

The luminescence system of *Cypridina* has been studied from various points of view, including not only biochemical and physical chemical, but also anatomical, embryological, histological and others.[1] The literature bearing on any of these aspects frequently refers to the organisms as "ostracod crustaceans," without further remarks concerning the meaning of this designation. Thus, despite the allergic reaction of some biochemists to taxonomic details, and despite some differences among specialists in regard to the best schemes of classification, it seems appropriate to consider briefly here the nature of ostracods and their relation to other kinds of luminescent organisms, especially among crustacea.

The Ostracods

Crustacea in general are usually considered to constitute one of six classes under the great phylum Arthropoda, namely (with popularly known examples): Crustacea (see below), Merostomata (horseshoe crabs), Arachnoidea (spiders), Pantopoda (sea spiders), Myriapoda (millipeds), and Hexapoda, or Insecta (insects). Following van Morkhoven,[2] it is convenient to recognize two subclasses of crustacea: the Malacostraca (lobsters, crabs, shrimps, etc.) and the Entomostraca, the latter comprising five natural groups or orders: Trilobita (trilobites, all extinct), Branchiopoda (water fleas), Ostracoda (seed shrimps, including *Cypridina, Pyrocypris*), Copepoda (*Metridia, Oncaea*) and Cirrepedia (barnacles). The property of luminescence in entomostracan crustaceans is found among ostracods and copepods. In melacostracan crustaceans it occurs among mysids as well as decapod and schizopod, or euphausiid shrimps. The euphausiid shrimp group is noteworthy in that (1) luminescent organs are possessed by all but one (*Bentheuphausia*) of the 11 genera in the single family (Euphausiidae) of the order Euphausiacea or Schizopoda, and (2) the eyelike photogenic organs are among the most elaborate known, complete with lens, reflectors and other histological features including internal,

[1] E. N. Harvey, "Bioluminescence." Academic Press, New York, 1952.

[2] F. P. C. M. van Morkhoven, "Post-Paleozoic Ostracoda. Their Morphology, Taxonomy, and Economic Use." Elsevier, Amsterdam, 1962.

ISBN 0-12-181957-4

precisely arranged rods that suggest the possibility of bioluminescent light being emitted in the form of a laser.[2a]

Ostracod crustacea are characterized by small bodies, enclosed within hinged, bivalve shells of carbonaceous material, ranging in size from 0.5 mm to 4 or 5 mm long. Dried specimens often resemble sesame seeds in size and in superficial appearance. The abdomen bears two pairs of legs, but locomotion is mainly accomplished by a means that is unique in the animal kingdom, namely by motion of one or both of two pairs of antennae in the head region. A deep-sea form, *Gigantocypris,* is said to attain a length of 1 inch,[1] and although it is thought to be luminous, this possibility apparently remains to be confirmed by direct observation.

The distribution of ostracods in nature is practically ubiquitous[3] wherever there is water: in the sea, in brackish bays, in freshwater ponds, in marshes, swamps, lakes, rivers, in leaf litter of forest floors, and occasionally in specialized habitats, such as that of the genus *Entocythere,* which dwells among the gills of crayfishes. The property of bioluminescence is confined to certain marine species. The total number of living species of ostracods is much fewer than the several thousand species known from fossil forms. Fossilized shells have been found in the upper Cambrian, of about 500 million years ago. Fossil species are distinguished primarily on the basis of structure of the carapace, which may be smooth or characterized by specific patterns of various kinds of markings.[4] The fossil forms have been found of much practical importance to petrologists, especially for correlating stratigraphic formations in drilling wells for petroleum.

In addition to their significance to petrologists, and their occasional significance to investigators interested in bioluminescence, ostracods have one other claim to fame, namely, by far the largest sperm cells in the whole animal kingdom are produced by members of this group of organisms, attaining in fact a length several times that of the entire animal, and kept folded between the two leaves of the carapace.[5] By an odd quirk of nature, among some of the freshwater species, males are never found, e.g. *Candonna* and *Cypris*.[6] The females of these examples, as well as of all parthenogenetic ostracods, retain unimpaired as if through a sort of evolutionary nostalgia, the receptaculum seminis, used normally for storing the spermatozoa derived from the male.

[2a] J.-M. Bassot, *in* "Bioluminescence in Progress" (F. H. Johnson and Y. Haneda, eds.), p. 605. Princeton Univ. Press, Princeton, New Jersey, 1966.

[3] H. S. Puri, *Proc. Symp. Crustaceans,* Part I, p. 457, Marine Biological Association of India, Mandapam Camp., 1966.

[4] C. L. Fenton and M. A. Fenton, "The Fossil Book." Doubleday, Garden City, New York, 1958.

[5] B. Grzimek, ed., "Grzimek's Animal Life Encyclopedia," Vol. I, Lower Aniamls, Ostracoda, p. 450. Van Nostrand-Reinhold, Princeton, New Jersey, 1974.

[6] G. Smith, *in* "Cambridge Natural History" (S. F. Harmer and A. E. Shipley, eds.), Vol. 4, p. 108. Macmillan, New York, 1909.

The Genus *Cypridina*

It is interesting to note that one of the first of the recorded instances in which luminescence of the sea was clearly and definitely attributed to the prevalence of large numbers of brightly luminescent, tiny living organisms, evidently involved ostracod crustacea, as reported (1760) by Commander Godeheu de Riville[7]:

> The 14th of July, 1754, about 9 at night, in 8 degrees 47 minutes north and 73 degrees longitude east of the meridian of Paris, the sea appeared all on fire; I viewed it from the ship's gallery with the utmost surprise. The sea, whose surface was but slightly agitated, was covered over with small stars; every wave which broke about us dispersed a most vivid light, in complexion like that of a silver tissue electrified in the dark—the more distant waves, which appeared as confounded together, formed at the horizon the appearance of a plain covered with snow; and the wake of our ship, whose brightness lasted a long while, was of a lovely and luminous white, interspersed with brillant and azure points . . . . I was struck with the light issuing from certain small bodies, which frequently kept fixed to the rudder when the sea left it. I caused some of the water to be taken up, and strained into a vessel through a handkerchief; after which operation, I observed that the filtered water was no longer luminous, but that the handkerchief was covered with brilliant points, which were fixed to it. Some of these I took upon the end of my finger, when they insensibly lost their light; and as they resembled the eggs of fishes in form and size, I had the curiosity to examine one of them by day-light, with a strong magnifying glass, and was astonished to perceive a sensible motion within it; and being somewhat in doubt as to what I saw, I turned it about several ways to be satisfied, by placing it on my nail in a drop of water. But how greatly was my surprize increased when on surveying it attentively, I perceived it to be surrounded with a brilliant liquor, which every one in the room saw as well as myself: Another of these small insects, by the help of a pair of corn tongs, I fixed to the side of a glass cup, but this pressure, though very slight, was evidently too powerful for so delicate an animal; for notwithstanding the light of two candles, we saw issuing from its body a blueish and luminous liquor, the glare of which extended to the distance of two or three lines into the water. I took it upon the pinch of the tongs, and it was no sooner placed under the microscope than it yielded again a considerable quantity of the same azure liquor; notwithstanding I had the satisfaction of seeing it still full of life, and moving with much vivacity.

The remarks in the foregoing quotation,[7] together with original drawings reproduced here in part (Figs. 1 and 2), leave no real doubt that a species of ostracods, possibly *Cypridina,* was largely responsible for the particular example of luminescence in the Indian Ocean in the area of the Maldive Islands that was described by Commander Godeheu de Riville about the middle of the eighteenth century. The possibility that the ostracod was *Cypridina hilgendorfii* (Fig. 3), the same as that studied so intensively by Harvey (see below), can neither be ruled out nor affirmed from the information available at this date.

[7] Godeheu de Riville, *Mem. Math. Phys. Acad. R. Sci. Paris,* **3,** 269 (1760); English translation in *Gentleman's Mag.* **38,** 408 (1768).

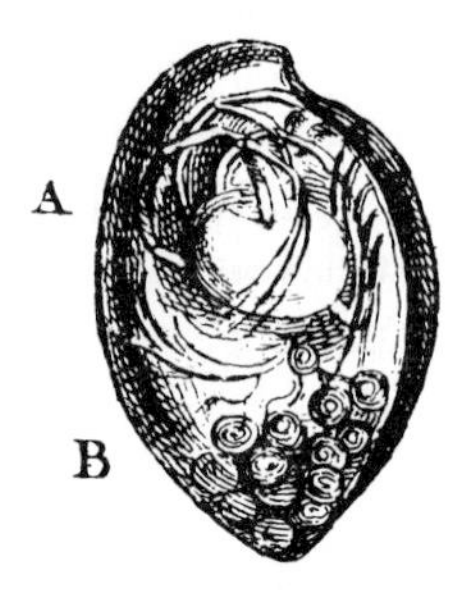

FIG. 1. Drawing by Godeheu de Riville (1760, Plate X, p. 276) of one of the small animals, possibly a species of *Cypridina,* responsible for widespread luminescence of the sea during the summer of 1754 in the Indian Ocean. The drawing was made through a strong magnifying glass, in daylight. The whole body of the animal within the shell is represented at A. At B were "several globules, which seem attached to the body by fine threads." The explanations of this figure and also of Fig. 2 are from the English translation (Godeheu de Riville, 1768, p. 408) of the original 1760 article in French; see text footnote 7).

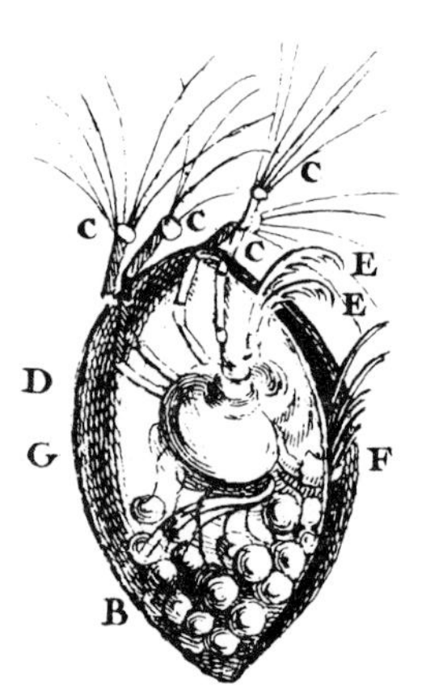

FIG. 2. Drawing from the same source as Fig. 1, of an animal "in agitation," showing two pairs of antennae, the second of which pair is used for swimming, protruding from the body through spaces between the valves. These structures (CCCC) were referred to as "four movable horns fixed round the head; formed of several articulations. D, the head. E E, two feet armed with claws [identifiable as 1 pair of mandibles], F, a large foot armed with an indented claw [identifiable as the caudal furca], G, the body of the animal. B, globules which contain the azure liquor." The globules indicated at B are probably eggs, though they may have been contaminated with some of the "azure liquor" which results from mixed secretions of specialized glands that presumably represent evolutionarily modified salivary glands, now producing the specific substrate luciferin, and specific enzyme luciferase of the light-emitting system of *Cypridina*. These two components complete the requirements for a bioluminescent reaction on mixing together after secretion into the seawater habitat. The authors are indebted to Dr. Y. Haneda of the Yokosuka City Museum, Dr. Louis S. Kornicker of the Smithsonian Institution, and Professor Ida Thompson of the Princeton Geology Department for helpful information concerning various aspects of the ostracod crustacea.

FIG. 3. Photograph of living specimens of *Cypridina hilgendorfii*, greatly enlarged beyond their usual length of 3–5 mm. By courtesy of Dr. Y. Haneda.

Species of Ostracod Crustacea, of *Cypridina* in Particular

Problems and individual judgments involved in the classification of ostracods, as in many other groups of organisms, have led to changes and revisions by various authorities. It seems unfortunate in the present context that the property of bioluminescence has generally been viewed as having little or no taxonomic significance. As a result, a firm list of luminescent species is seldom encountered, though a summary of species of planktonic oceanic ostracods of the genus *Conchoecia* that have been reported to be luminescent has been recently compiled.[8] In a comprehensive, detailed treatment of Ostracoda, G. W. Müller[9] includes 1719 living species, mostly of marine habitat, 921 of which he regards as "sichere," but the remaining 798 as "unsichere," and with little information concerning luminescence, even pertaining to the 14 species he includes in the genus *Cypridina*. This genus itself has been divided into 5 subgenera,[10] the latter

[8] M. V. Angel, *Proc. R. Soc. Edinburgh Sect. B* **73,** 213 (1972).
[9] G. W. Müller, "Das Tierreich," *K. Preuss. Acad. Wiss. Berlin*, 31 Liefering, p. 7 ff, 1912.
[10] T. Skogsberg, *Zoologiska Bidrag Fran Uppsala Suppl. Bd. 1*, pp. 1–784 (1920).

including *Cypridina* and *Vargula*. A key to the species of *Vargula*, notably including *V. hilgendorfii, V. harveyi*, and the new, luminescent species *V. tsujii*, is given by Kornicker and Baker.[11] The present paper attempts to avoid confusion that might arise through the changes and complications in the literature on classification by adhering to the nomenclature adopted by the author(s) of each specific reference cited.

The most important species in the present context is the one that has received such intensive investigation by Harvey and his followers, usually referred to as *Cypridina hilgendorfii*, which until recently has been abundant in the coastal waters of Japan, where the numbers have now been reduced, apparently as a result of increase in pollution. A related, luminescent species, *Cypridina noctiluca*, has been common near Hachijo Island of Japan and in the tropical waters of Palao, Java, and Malay.[12]

No important biochemical differences have been reported among the luminescent species of *Cypridina*, and a thorough study of the system of *Cypridina serrata*[13] has indicated that it is essentially identical to that of *C. hilgendorfii*, as would be expected, especially in view of the previously demonstrated interchangeability of luciferin and luciferase extracts from different genera, *Cypridina* and *Pyrocypris*, to give a luminescent reaction.[14] A nonluminous species, *C. bairdii*, is interesting in that it apparently produces luciferin but not luciferase; the addition of *C. hilgendorfii* luciferase preparations to crude extracts presumably containing luciferin from *C. bairdii* results in light emission.[15] These experiments would be more convincing, of course, if conducted with highly purified components from both species.

Cypridina hilgendorfii

The late Professor E. Newton Harvey, whose broadly encompassing, scholarly studies of bioluminescent organisms and their light-emitting systems earned him, among his peers and students, the honorific title "Dean of Bioluminescence," began his research in this area as a young instructor in the Biology Department at Princeton University. His first paper on bioluminescence dealt with the temperature limits for light production in a population of luminous bacteria.[16] Shortly thereafter he experimented with

[11] L. S. Kornicker and J. H. Baker, *Proc. Biol. Soc. Wash.* **90,** 218 (1977).
[12] Y. Haneda, *in* "The Luminescence of Biological Systems" (F. H. Johnson and Y. Haneda, eds.), p. 335. Am. Assoc. Adv. Sci., Washington, D.C., 1955.
[13] F. I. Tsuji, R. V. Lynch, III, and Y. Haneda, *Biol. Bull.* **139,** 386 (1970).
[14] E. N. Harvey, "Living Light." Princeton Univ. Press, Princeton, New Jersey, 1940.
[15] Y. Haneda, *Rec. Oceanogr. Works Jpn.* **12**(2), N.S. **1**(1), 103 (1953).
[16] E. N. Harvey, *Biochem. Bull.* **2**(7), 456 (1913).

firefly luminescence. In 1916, in Japan at the Misaki Marine Station, he first became acquainted with the organism that the Japanese call "umi botaru," meaning "sea firefly," and the species involved, i.e., *C. hilgendorfii*, remained of topmost interest in his luminescence research throughout his life.[17] His objectives, in part, were to isolate the luciferin component and determine its chemical structure, as well as to purify the luciferase component and determine its major properties. On a number of occasions Harvey expressed a view to the effect that: "For chemical investigation ostracods are by far the most favorable animals which exist. The large luminous glands, the abundant formation of secretion, and the fact that, when quickly dried, the ability to luminesce on moistening is retained indefinitely make these organisms unique".[18] At the time when he wrote that statement, he had at hand some potentially active *Cypridina* specimens that had been kept in a dried state for 30 years. The unique advantages that so impressed Harvey, however, might now be considered more than offset by certain innate disadvantages, especially the extraordinary instability of the luciferin to oxygen in aqueous solutions, and the misleading impression that could arise from the brightness of the luminescence elicited on moistening small amounts of powdered, dried specimens as to the quantity of luciferin present. In point of fact, Harvey realized that the amount of luciferin in his dried specimens was indeed small. For example, he found that sufficient luminescence to be detected by the dark-adapted eye was produced by crude extracts of luciferin diluted to equal one part of the dried material in 400,000,000 parts of water. On the estimate that the animal contained 1% of luciferin by weight, the concentration of luciferin in the above dilution would be 1 in 40,000,000,000 comparable to the concentration needed to detect the fluorescence of fluorescein, or chemiluminescence of Luminol, i.e., 1 in 10,000,000,000.[19] In similar experiments with very pure luciferin and luciferase, it was found that visible light occurred with a dilution of 1 part of luciferin in 100,000,000,000 of buffer (20 pM luciferin).[20] The potency of luciferase was tested at the same time, and taking the molecular weight as 50,000, visible light with luciferin was found at a concentration of less than 10^{-7} μg/ml ($= < 1$ fM). A reasonably likely estimate of the concentration of active components contained in one individual, living *Cypridina* of wet weight equal to 4 mg, was made that the amount averages 1 μg each of luciferin and luciferase, i.e., a ratio of approximately 100 molecules of substrate to 1 molecule of enzyme.

[17] F. H. Johnson, *Biogr. Mem. Natl. Acad. Sci. U.S.A.* **39,** 193 (1967).
[18] E. N. Harvey, "Bioluminescence," p. 305, Academic Press, New York, 1952.
[19] E. N. Harvey, *Science* **57,** 501 (1923).
[20] F. H. Johnson, O. Shimomura, and Y. Saiga, *Science* **134,** 1755 (1961).

Cypridina Luciferin

Experiments on purification made it evident that the dried material contained very large amounts of impurities which, though inactive in the luminescent reaction, had chemical properties otherwise sufficiently similar to those of luciferin that separation of the latter from the former, by any biochemical procedures available prior to about the early 1950s, was an enormously difficult problem. Success in isolating this luciferin was finally achieved by Shimomura *et al.* in Japan,[21] with the aid of column chromatography, with rigorous exclusion of molecular oxygen throughout, and with the use, at the start, of approximately 10 times the amount of dried material that apparently was ever used at one time in Harvey's laboratory. Thus, it seems that Harvey seldom if ever risked the investment of more than 50 g of dried *Cypridina* in any one series of experiments. The maximum purification was thought to be accomplished by Anderson's benzoylation procedure,[22] involving reversibly converting the luciferin in a crude extract obtained from defatted, dried material to an inactive, fat-soluble product by adding benzoyl chloride and later hydrolyzing this product back to the original luciferin by HCl in absence of oxygen. It was estimated that by repeating the procedure, giving two cycles of purification, the increase in potency amounted to about 2000 times as compared with the starting material. Shimomura *et al.*[21] began their procedure with some 500 g of dried specimens and obtained a yield of a few milligrams of a crystalline product having an estimated potency some 40,000 times that of the starting material. Since no data are available that would permit a quantitative comparison of the potencies of the starting material in these two instances, and since the initial potencies are subject to wide variation due to various factors, including the season of collecting, the methods of drying and of keeping the specimens, etc., the increase in potencies can be considered to provide only a very rough index to the amount of purification. There is no doubt, however, that the crystalline product just referred to was of vastly greater purity than any achieved previously, and its properties were largely confirmed by later investigations when improved procedures, such as extraction of luciferin from live organisms placed in methyl alcohol precooled to very low temperatures with Dry Ice,[23] eventually resulted in much greater yields of the crystalline product. Some misleading or equivocal conclusions are rather to be expected when based on incompletely purified components of an enzyme–substrate system. To a certain

[21] O. Shimomura, T. Goto, and Y. Hirata, *Bull. Chem. Soc. Jpn.* **30,** 929 (1957).

[22] R. S. Anderson, *J. Gen. Physiol.* **19,** 301 (1935).

[23] Y. Haneda, F. H. Johnson, Y. Masuda, Y. Saiga, O. Shimomura, H.-C. Sie, N. Sugiyama, and I. Takatsuki, *J. Cell. Comp. Physiol.* **57,** 55 (1961).

extent this expectation has indeed been realized in regard to the luminescence system of *Cypridina*. The remarkable fact is that much of the early research on this system, prior to its full purification, has resulted in useful data, very nearly correct as borne out subsequently through studies with undoubtedly very highly purified preparations. For present purposes there seems to be no real point in reviewing the earlier work in detail, except to mention a few aspects of special interest, as follows.

First, a conclusion based on experiments with partially purified luciferin, to the effect that cyanide combined irreversibly with luciferin, giving a product that would not lead to a light-emitting reaction with luciferase[24] was widely held for a long time and influenced speculations concerning the chemistry of luciferin. Reinvestigation of this problem with essentially fully purified luciferin and very highly purified luciferase showed that no loss in luminescence activity occurred in crystalline luciferin left standing for 2 days in liquid HCN at room temperature before evaporating off the HCN and testing the luminescence potency of the residue.[25] On the other hand, when cyanide was added to extremely dilute luciferin in aqueous solution, the light-emitting activity with luciferase was quickly abolished, though no such effect was observed with the same concentration of cyanide and a larger concentration of luciferin. For example, 1 m*M* KCN at room temperature and neutral pH, in 2 min abolished the luminescence activity of 10 n*M* luciferin, whereas the same concentration of KCN, under similar conditions, had essentially no effect on the luminescence activity of 1 μM luciferin. Quantitative data on the influence of such factors as purity of the distilled water solvent, pretreatment of the solvent with ultraviolet light, effects of reducing agents, effects of low concentrations of heavy-metal ions, or charcoal, etc. indicated that cyanide acts in conjunction with heavy-metal impurities to reduce, by a process of catalytic destruction, the light-emitting potency of very dilute concentrations of luciferin.[25] On the basis of the evidence at hand at the time (1940), it was natural to conclude that cyanide combined irreversibly with *Cypridina* luciferin, all the more so because azide also reduced the light-emitting capacity of corresponding preparations of luciferin, but in a manner reversible by dilution. The remarkable fact is that, despite the undoubted presence of an unknown but very likely considerable amount of impurities, and a considerable element of uncertainty in what appeared to be the most logical assumptions, some substantially correct conclusions regarding the properties of the luciferin substrate as well as other aspects of the luminescence system were arrived at, in some instances by altogether ingenious methods. For example, the

[24] A. C. Giese and A. M. Chase, *J. Cell. Comp. Physiol.* **16,** 237 (1940).
[25] F. H. Johnson, O. Shimomura, and Y. Saiga, *J. Cell. Comp. Physiol.* **59,** 265 (1962).

molecular weight of luciferin was estimated as between 250 and 570 on the basis of rapid oxidation by ferricyanide,[26] or between 800 and 2400 on the basis of the presumed combination with cyanide[24]; the molecular weight of crystallized luciferin, according to several methods, was better estimated to be 470.[27] Final elucidation of the chemical structure required the combined evidence of numerous approaches—IR, NMR, mass spectra, elementary analysis, visible and UV absorption, etc.—contributed through the cooperative efforts of investigators in this country and Japan.[28] Figure 4 shows the structure of this complicated, unstable molecule, as deduced by Kishi *et al.*[28] (1966) and confirmed by total synthesis shortly thereafter.[29] It is worthy of note that the intact molecule is not the unalterable requirement for a light-emitting reaction with *Cypridina* luciferase, although it results in the fastest reaction of any of the derivatives tested. The desguanyl derivative reacts at a rate of 0.002 to 0.0001 as fast,[30-32] but ultimately yields almost the same total amount of light as a corresponding number of molecules of native luciferin.

Biological Distribution of *Cypridina* Luciferin

The occurrence in nature of a substance with the structure of *Cypridina* luciferin (Fig. 4) outside of a luminescent system is unknown. For many years it was thought that this substance was unique even among luminescent systems, as judged by numerous attempts to obtain a light-emitting "cross-reaction" between *Cypridina* luciferin and extracts of other types of luminescent organisms prepared in a manner that they potentially contained the luciferase component of the other type.[1] The only exceptions to this "rule" were found among luminescent organisms that were fairly closely related biologically, e.g., the two genera of ostracods, *Cypridina* and *Pyrocypris*, as mentioned above,[1,14] or different types of fireflies.[33,34] Ultimately, however, a reciprocal cross-reaction was discovered between partially purified extracts of internal photogenic organs of the self-luminous

[26] A. M. Chase, *J. Cell. Comp. Physiol.* **33,** 113 (1949).

[27] Y. Hirata, O. Shimomura, and S. Eguchi, *Tetrahedron Lett.* **5,** 4 (1959).

[28] Y. Kishi, T. Goto, Y. Hirata, O. Shimomura, and F. H. Johnson, *in* "Bioluminescence in Progress" (F. H. Johnson and Y. Haneda, eds.), p. 89. Princeton Univ. Press, Princeton, New Jersey, 1966.

[29] Y. Kishi, T. Goto, S. Inoue, S. Sugiura, and H. Kishimoto, *Tetrahedron Lett.* **29,** 3445 (1966).

[30] O. Shimomura, unpublished observations, 1964, 1971.

[31] Y. Hirata, *Kagaku No Ryoiki* **80,** 91 (1967). In Japanese.

[32] T. Goto, *Pure Appl. Chem.* **17,** 421 (1968).

[33] E. N. Harvey, *Science* **46,** 241 (1917).

[34] W. D. McElroy and E. N. Harvey, *J. Cell. Comp. Physiol.* **37,** 83 (1951).

PHOTOGEN	PHOTAGOGIKON
Firefly luciferin	λ_{max} = 552–582
Cypridina and fish luciferin	λ_{max} = 462
Coelenterazine	Coelenteramide λ_{max} = 460 (*Oplophorus*), 470 (*Aequorea*), 490 (*Renilla*)
Bacterial luciferin	Oxidized form of FMNH λ_{max} = 470–490
Latia luciferin	Flavin derivative (?) λ_{max} = 535
Earthworm luciferin	(?) λ_{max} = 490

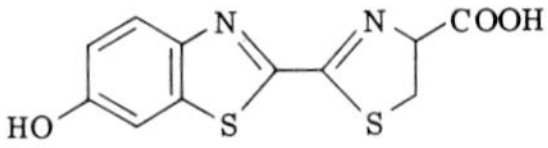

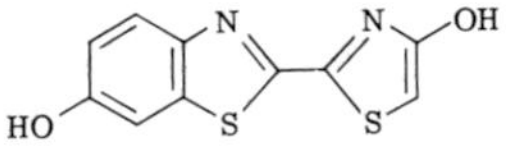

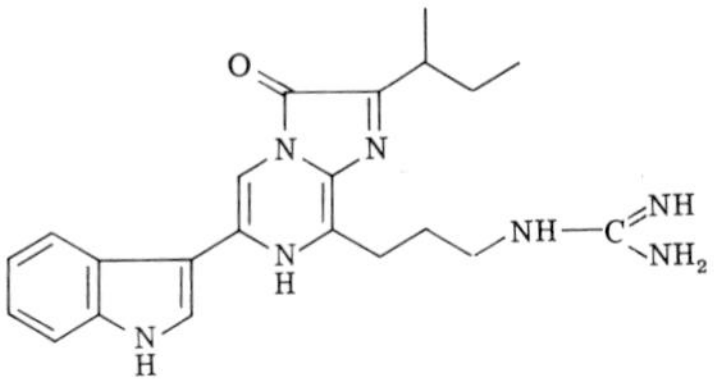

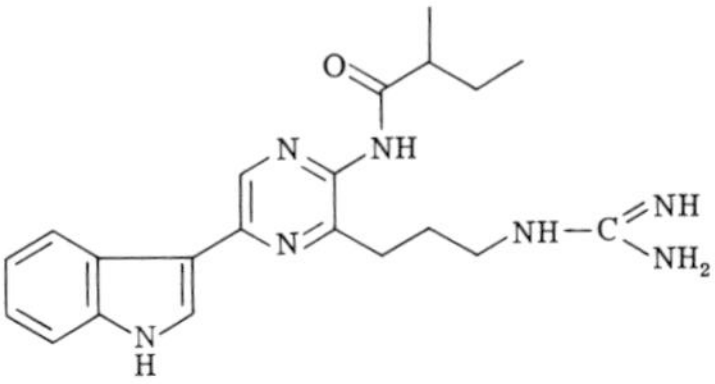

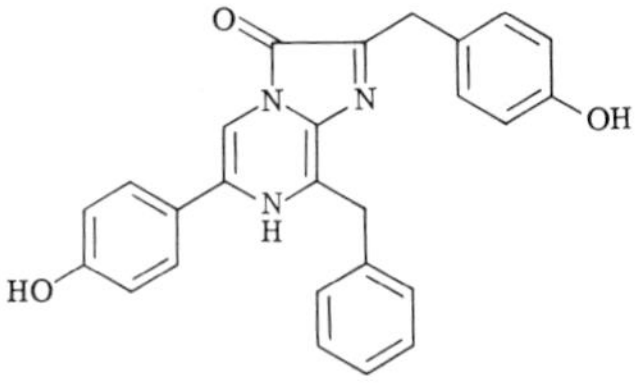

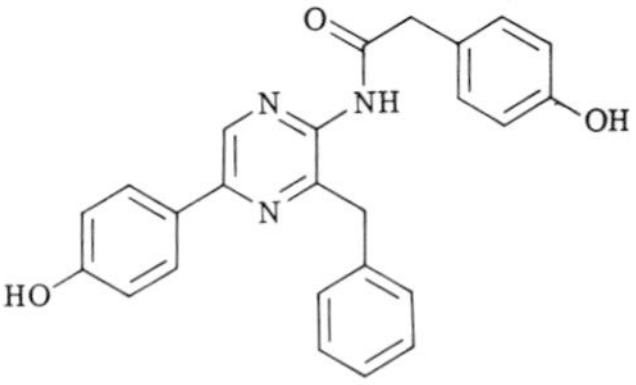

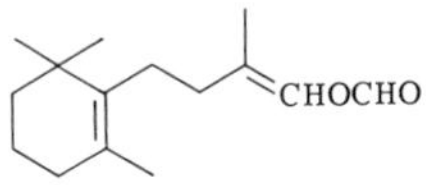

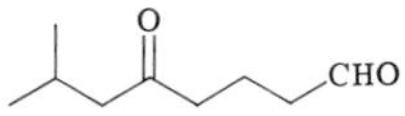

FIG. 4. Chemical structures of photogens and photagogika [corresponding to Fig. 1, p. 251, of F. H. Johnson and O. Shimomura, *Trends Biochem. Sci.* **1,** 250 (1976)].

fishes *Parapriacanthus beryciformes* and *Apogon ellioti* and similar extracts of *Cypridina*.[35,36] Partially purified luciferin from *Apogon* that revealed certain properties similar to those of *Cypridina* luciferin was obtained from photogenic organisms of another self-luminous fish, *P. beryciformes,* and found to have major chemical properties identical to those of *Cypridina* luciferin.[37] At the present time, the source of components of the luminescence systems in these fishes has not been established beyond doubt, and it remains possible that they are derived by harvesting in some manner these components from specimens of *Cypridina* ingested as part of the diet of the fishes.[38]

Until lately, not only *Cypridina* luciferin was generally thought to be unique, i.e., to occur only in the luminescent system of this particular type of organism, but likewise practically all other luciferins from different types of luminescent organisms,[38a] unless the types were biologically rather closely related as indicated above. The evidence for this widely held belief was based in part on the failure of many attempts to demonstrate "cross-reactions" in the components extracted from distantly related species, but in part also on considerations of additional indirect evidence. As late as 1961, such additional evidence was summarized by Glass[39] as follows: "Johnson and his coworkers do not regard either the differences in fluorescent and chemiluminescent spectra of the varieties of luciferin found in different organisms, or the disjunct phylogenetic occurrence of bioluminescence, or the failure to obtain cross-reactions between different 'luciferins' and 'luciferases', or the specific requirements for different cofactors as sufficient grounds for supposing the several bioluminescent systems to be entirely different in chemical nature. These workers are strengthened in this view by their recent finding of a luminescent cross-reaction between the luciferin and luciferase of a fish, *Apogon,* and the crustacean *Cypridina*. . . . Other contributors to the symposium on the subject of bioluminescence do not seem to share this view, or at least are noncommittal while emphasizing the distinctiveness of the systems they

[35] F. H. Johnson, Y. Haneda, and H.-C. Sie, *Science* **132,** 422 (1960).

[36] H.-C. Sie, W. D. McElroy, F. H. Johnson, and Y. Haneda, *Arch. Biochem. Biophys.* **93,** 286 (1961).

[37] F. H. Johnson, N. Sugiyama, O. Shimomura, Y. Saiga, and Y. Haneda, *Proc. Natl. Acad. Sci. U.S.A.* **47,** 486 (1961).

[38] Y. Haneda, F. H. Johnson, and O. Shimomura, *in* "Bioluminescence in Progress" (F. H. Johnson and Y. Haneda, eds.), p. 533. Princeton Univ. Press, Princeton, New Jersey, 1966.

[38a] T. Goto and Y. Kishi, *Angew. Chem. Int. Ed. Engl.* **7,** 407 (1968).

[39] W. D. McElroy and B. Glass (eds.), "Light and Life," p. 849. Johns Hopkins Press, Baltimore, Maryland, 1961.

have been investigating." It should be noted that these remarks were made at a time when the structure of not a single type of luciferin had been established, and only two examples of fully purified luciferins had been obtained, namely those of *Cypridina*[21] and the firefly.[40] All evidence concerning the properties of all other luciferins, therefore, was based on studies of impure preparations. In the light of present knowledge, the arguments summarized in the above quotation are intrinsically shaky, and the view they were taken to support was definitely as unjustified as the attempt to dismiss with obvious disdain any alternative.

The discovery of further examples of cross-reactions would seem to add support to the view of nonuniqueness just mentioned, but the reliability of such evidence is uncertain when the reactions are observed with incompletely purified components. The luciferin of the fish *Parapriacanthus*, however, was purified and crystallized, and found to be identical to that of *Cypridina* in its major chemical properties.[37] Another example of a cross-reaction between the luciferin of *Cypridina* and of a very different type of luminescent fish, the "midshipman," *Porichthys porosissimus,* was discovered later,[41] though without demonstrating similarity of chemical properties in pure preparations of the cross-reacting components. Additional examples of cross-reactions are now known.[42–44]

Progress in identifying the chemical structure of key reactants among different luminescent systems has by now revealed some surprising relationships. Figure 4 provides a convenient summary, showing all the structures thus far established for the luciferins of different types of luminescent organisms, referred to in this figure as "photogens" as a term intended to include not only luciferin in the usual sense, i.e., the specific substrate of an enzyme which catalyzes the light-emitting reaction, but also the chromophore or active group of a "photoprotein" of a nonenzymic type of luminescence system such as that of *Aequorea* (see this volume, p. 272). The term "photagogikon" is intended to refer to the "light-emitter", i.e., the molecule or complex that is brought into the excited state and then emits.[45] Fig. 4 illustrates the fact that the luciferin of *Cypridina* and certain fish has a similar structural nucleus to that of coelenterazine, a substance that is not

[40] B. Bitler and W. D. McElroy, *Arch. Biochem. Biophys.* **72,** 358 (1957).

[41] M. J. Cormier, J. M. Crane, Jr., and Y. Nakano, *Biochem. Biophys. Res. Commun.* **29,** 747 (1967).

[42] M. J. Cormier, D. M. Hercules, and J. Lee (eds.), "Chemiluminescence and Bioluminescence." Plenum, New York, 1973.

[43] M. J. Cormier, K. Hori, Y. D. Karkhanis, J. M. Anderson, J. E. Wampler, J. G. Morin, and J. W. Hastings, *J. Cell. Physiol.* **81,** 291 (1973).

[44] M. J. Cormier, K. Hori, and J. M. Anderson, *Biochim. Biophys. Acta* **346,** 137 (1974).

[45] F. H. Johnson and O. Shimomura, *Trends Biochem. Sci.* **1,** 250 (1976).

only identical with the natural luciferin of *Renilla*[45a] and certain other coelenterates, but is also the photogen in the photoprotein aequorin from the jellyfish *Aequorea*. The remaining photogens, though distinctly different in total structure, are alike in being types of aldehydes, i.e., in luminous bacteria, the gastropod *Latia*, and luminous earthworms. No less impressive is the identical product of the luminescent reaction of coelenterazine, namely coelenteramide, throughout the coelenterates that have been examined in this regard, including the jellyfish *Aequorea*, the sea pansy *Renilla*, the sea pen *Leioptilus*, the sea cactus *Cavernularia*, and the decapod shrimp *Hoplophorus* (=*Oplophorus*) and *Heterocarpus*.[46-48a] Coelenterazine occurs also in a sulfated form in the luminescence system of the "firefly squid" *Watasenia*, a cephalopod mollusk.[49,50] It should be emphasized that the terms "*Renilla* luciferin" and "*Oplophorus* luciferin" are functionally defined and do not refer to a single compound of definite structure. The term "*Oplophorus* luciferin" was introduced at practically the same time as the term "coelenterazine," which refers to a specific compound. "*Renilla* luciferin" has been used to designate two different synthetic compounds, as well as a third compound, i.e., native *Renilla* luciferin, different from the other two.[51,52] In view of these facts, and considering the widespread significance as well as chemical nature implied by the term "coelenterazine," we believe its use is justified to the exclusion of the other two.

Cypridina Luciferase

In sharp contrast to the instability of *Cypridina* luciferin in aqueous solution containing oxygen, the luciferase of this system was early discovered to be remarkably stable, with or without oxygen. In fact, as Harvey noted,[53] a crude luciferase solution, prepared by making cold, aqueous extracts of dried specimens, can be left standing in a tall tube until partially putrified by the growth of bacteria and will emit light when the tube is shaken to dissolve O_2. This observation, together with many experiments on the effects of reducing agents on spent luciferin, led to the conclusion

[45a] S. Inoue, H. Kakoi, M. Murata, T. Goto, and O. Shimomura, *Tetrahedron Lett.*, 2685 (1977).

[46] O. Shimomura and F. H. Johnson, *Proc. Natl. Acad. Sci. U.S.A.* **72,** 1546 (1975).

[47] O. Shimomura and F. H. Johnson, *Nature (London)* **256,** 236 (1975).

[48] S. Inoue and H. Kakoi, *Chem. Commun.* **1976,** 1056 (1976).

[48a] O. Shimomura, S. Inoue and T. Goto, *Chem. Lett.* **1975,** 247 (1975).

[49] S. Inoue, H. Kakoi, and T. Goto, *Tetrahedron Lett.* **34,** 2971 (1976).

[50] S. Inoue, H. Taguchi, M. Murata, H. Kakoi, and T. Goto, *Chem. Lett.* **1977,** 259 (1977).

[51] K. Hori and M. J. Cormier, *Proc. Natl. Acad. Sci. U.S.A.* **70,** 120 (1973).

[52] K. Hori, J. E. Wampler, J. C. Matthews, and M. J. Cormier, *Biochemistry* **12,** 4463 (1973).

[53] E. N. Harvey, *J. Gen. Physiol.* **1,** 133 (1918).

that the luminescent oxidation of luciferin with luciferase was reversible by various agents, possibly involving a simple dehydrogenation–hydrogenation, as in the oxidation and reduction of methylene blue. Moreover, one of Harvey's associates, R. S. Anderson,[54] found evidence that the luciferin of *Cypridina* underwent two types of oxidation: (1) spontaneous with atmospheric O_2, or under the influence of agents such as ferricyanide, in absence of luciferase, emitting no observable light and reversible by such agents as $Na_2S_2O_4$, and (2) an irreversible (by any known means) light-emitting oxidation, in the presence of *Cypridina* luciferase and molecular oxygen. The hypothesis of this dual oxidation served to explain some otherwise anomalous kinetics of the light-emitting reaction, differing according to the relative amounts of fully reduced luciferin at the start, in mixtures of fully reduced plus reversibly oxidized luciferin, the latter portion being subject to progressive reduction during the course of the luminescent reaction, presumably by impurities that were undoubtedly present. Simultaneous oxidation of the freshly reduced form, along with the initially present reduced form, contributed to the total amount of light produced, as well as to the complicated rate of production.[54]

The relation of the data obtained early with crude and only partially purified components of the *Cypridina* system, to the properties of the essentially fully purified system, is now understandable.[32] The early idea of an oxidation involving a simple dehydrogenation, i.e., loss of two hydrogens,[1] cannot hold for the luminescent oxidation with luciferase, whereby CO_2 is produced.[55] The evidence available to date is that the oxidation of *Cypridina* luciferin in the presence of luciferase includes at least the reactions schemed in Fig. 5.

Reaction Mechanism

The intimate mechanism of light emission in both bioluminescent and purely chemiluminescent reactions is a problem of much interest. Two major alternative hypotheses have been proposed with reference to three different bioluminescence systems, namely, the firefly, *Cypridina,* and the sea pansy *Renilla,* as diagrammed in Fig. 6. These systems are similar in that all require molecular oxygen and produce CO_2 as a product.[55-58] A

[54] R. S. Anderson, *J. Cell. Comp. Physiol.* **8,** 261 (1936).

[55] H. Stone, *Biochem. Biophys. Res. Commun.* **31,** 386 (1968).

[56] O. Shimomura and F. H. Johnson, *in* "Chemiluminescence and Bioluminescence" (M. J. Cormier, D. M. Hercules, and J. Lee, eds.), p. 337. Plenum, New York, 1973.

[57] P. J. Plant, E. H. White, and W. D. McElroy, *Biochem. Biophys. Res. Commun.* **31,** 98 (1968).

[58] M. DeLuca, M. E. Dempsey, K. Hori, J. E. Wampler, and M. J. Cormier, *Proc. Natl. Acad. Sci. U.S.A.* **68,** 1658 (1971).

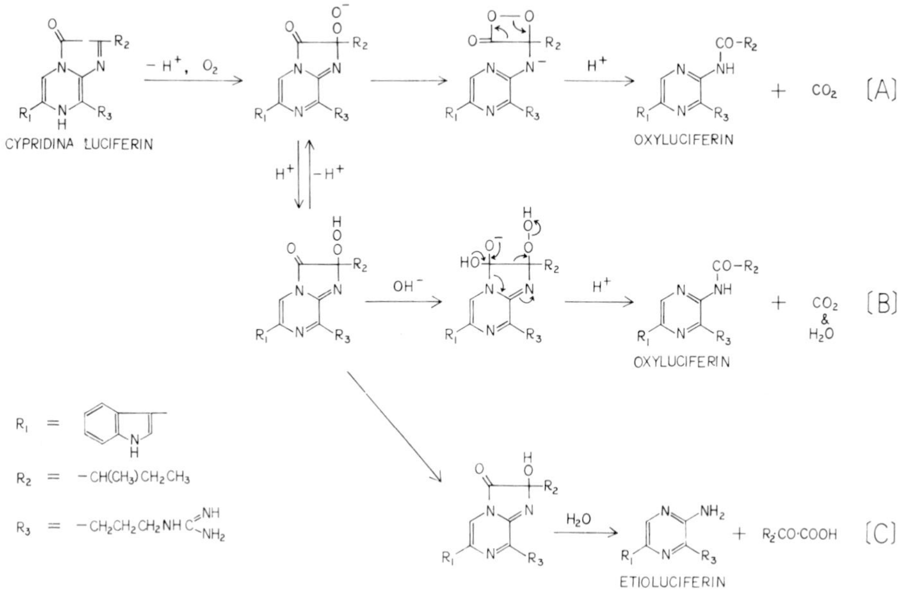

FIG. 5. Oxidative reactions of *Cypridina* luciferin according to various lines of evidence, as diagrammed in Fig. 2, p. 342, of O. Shimomura and F. H. Johnson[56] (superscript numbers refer to text footnotes). Pathway (A) was first suggested by McCapra and Chang[62] in reference to the chemiluminescence of a synthetic analog; pathway (B) is the same as the mechanism proposed by DeLuca and Dempsey[63]; pathway (C) was proposed[56] to explain the direct formation of etioluciferin and of the α-keto acid. In pathway (C), an intermediary tertiary alcohol, formed probably by a base-catalyzed oxygen elimination or by a reaction of hydroperoxide with unreacted luciferin, is hydrolyzed. The available data indicate that, in the luciferase-catalyzed reaction, pathway (A) predominates, and the ratio of the amount of luciferin oxidized in pathways (A), (B), and (C) might be 7:1:1. Our recent data,[70] however, indicate that the contribution of (B) and (C) would be less than this and could be zero. In chemiluminescence without luciferase, pathway (C) seems to be significantly involved. Since the maximum quantum yield of the luciferase-catalyzed luminescence is approximately 0.3,[113] pathway (A) must be the light-emitting reaction. No evidence is available at present as to whether pathways (B) and (C) give rise to luminescence. In regard to the source of oxygen in the CO_2 produced in the three different systems of *Cypridina*, the firefly, and sea pansy *Renilla*, however, substantial evidence is available that in all three systems one of the oxygens in the CO_2 liberated comes from molecular oxygen,[71] and the other from the luciferin involved, as indicated in Fig. 6.

mechanism that involves a dioxetane intermediate, scheme A, has been proposed for the luminescence of both firefly and *Cypridina* luciferin.[59-62]

[59] T. A. Hopkins, H. H. Seliger, E. H. White, and M. W. Cass, *J. Am. Chem. Soc.* **89,** 7148 (1967).

[60] F. McCapra, Y. C. Chang, and V. P. Francois, *Chem. Commun.* **1968,** 22 (1968).

[61] E. H. White, E. Rapaport, H. H. Seliger, and T. A. Hopkins, *Bioorg. Chem.* **1,** 92 (1971).

[62] F. McCapra and Y. C. Chang, *Chem. Commun.* **1967,** 1011 (1967).

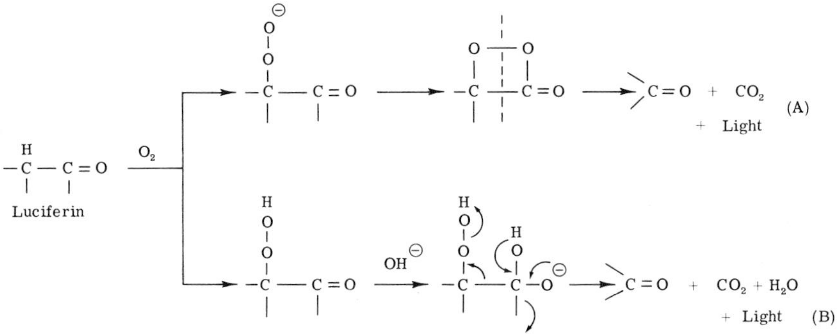

Luciferin $\xrightarrow{O_2}$

(A) $\rightarrow\ >C{=}O + CO_2$ + Light

(B) $\rightarrow\ >C{=}O + CO_2 + H_2O$ + Light

FIG. 6. Proposed oxidative mechanisms in pathways (A) and (B) of Fig. 5, based on evidence from various sources as discussed by O. Shimomura, T. Goto, and F. H. Johnson.[71]

In (A), one O of the product CO_2 originates from molecular oxygen. To test this scheme, DeLuca and Dempsey[63,64] studied the labeling of the product CO_2 with ^{18}O in the bioluminescence reaction of firefly luciferin. Their data indicated that one O of the CO_2 was labeled when the reaction was carried out in $H_2{}^{18}O$ medium with $^{16}O_2$ gas, but not in $H_2{}^{16}O$ medium with $^{18}O_2$ gas. They proposed reaction scheme B instead of A. In chemiluminescence of firefly luciferin, the data reported by White *et al.*[65] were consistent with scheme A but did not completely rule out scheme B,[66] in contrast to the data of DeLuca *et al.*,[67] which supported scheme B but failed to rule out scheme A. Studies on the bioluminescence of *Cypridina* luciferin by labeling product CO_2 with ^{18}O fully supported scheme A.[68-70] In regard to *Renilla* luciferin, however, one O of the product CO_2 was reported to arise from solvent H_2O in both bioluminescence[58] and chemiluminescence,[67] which would seem odd in view of the structural similarity between *Renilla* luciferin and *Cypridina* luciferin.

The most recent study unambiguously supports scheme A but not scheme B for the bioluminescence of the luciferin in the firefly[71,71a] and

[63] M. DeLuca and M. E. Dempsey, *Biochem. Biophys. Res. Commun.* **40,** 117 (1970).

[64] M. DeLuca and M. E. Dempsey, *in* "Chemiluminescence and Bioluminescence" (M. J. Cormier, D. M. Hercules, and J. Lee, eds.), p. 345. Plenum, New York, 1973.

[65] E. H. White, J. D. Miano, and M. Umbreit, *J. Am. Chem. Soc.* **97,** 198 (1975).

[66] J. W. Hastings and T. Wilson, *Photochem. Photobiol.* **23,** 461 (1976).

[67] M. DeLuca, M. E. Dempsey, K. Hori, and M. J. Cormier, *Biochem. Biophys. Res. Commun.* **69,** 262 (1976).

[68] O. Shimomura and F. H. Johnson, *Biochem. Biophys. Res. Commun.* **44,** 340 (1971).

[69] O. Shimomura and F. H. Johnson, *Biochem. Biophys. Res. Commun.* **51,** 558 (1973).

[70] O. Shimomura and F. H. Johnson, *Anal. Biochem.* **64,** 601 (1975).

[71] O. Shimomura, T. Goto, and F. H. Johnson, *Proc. Natl. Acad. Sci. U.S.A.* **74,** 2799 (1977).

[71a] J. Wannlund, M. DeLuca, K. E. Stempel, and P. D. Boyer, *Biochem. Biophys. Res. Commun.* **81,** 987 (1978).

Renilla,[71b] and also possibly the deep-sea shrimp *Oplophorus*.[72] These studies tend to disprove the validity of other very recent data.[72a]

Minimum Components

Despite the uncertainties that are inevitably involved in interpretations based on data obtained with impure preparations of specific substances, and the occasional mistakes, eventually shown by study of pure preparations, to have been made, evidence was accumulated through the dedicated investigations by Harvey and his associates that the minimum components needed for natural luminescence of the *Cypridina* system were simply luciferin, luciferase, water, and molecular oxygen. The qualification of "natural" is used here to exclude the spontaneous chemiluminescence of luciferin under certain conditions in aprotic solvents, such as dimethyl sulfoxide (DMSO). Evidence stemming from the properties of the first crystalline luciferin[21] indicating that it is an indole derivative, together with the observations of Cormier and Eckroade[73] and of Totter[74] that many indole derivatives exhibit easily visible chemiluminescence in DMSO in the presence of either H_2O_2 or sodium persulfate, as well as Philbrook *et al.*'s observation of the brilliant chemiluminescence of skatole (3-methylindole) in alkaline DMSO,[75] suggested that *Cypridina* luciferin might exhibit the same phenomena.

A number of indole derivatives, some of which were synthesized for possible bearing on the structure of *Cypridina* luciferin (Stachel *et al.*, 1966) and others that were already well known, have been tested for chemiluminescence in DMSO. A visible or measurable chemiluminescence has been found in an impressively large number of them.[76,77] The luciferin

[71b] R. C. Hart, K. E. Stempel, P. D. Boyer, and M. J. Cormier, *Biochem. Biophys. Res. Commun.* **81,** 980 (1978).

[72] O. Shimomura and F. H. Johnson, *Proc. Natl. Acad. Sci.* U.S.A. **75,** 2611 (1978).

[72a] F. I. Tsuji, M. DeLuca, P. D. Boyer, S. Endo, and M. Akutagawa, *Biochem. Biophys. Res. Commun.* **74,** 606 (1977).

[73] M. J. Cormier and C. B. Eckroade, *Biochim. Biophys. Acta* **88,** 99 (1964).

[74] J. R. Totter, personal communication, 1963.

[75] G. E. Philbrook, J. B. Ayers, J. F. Garst, and J. R. Totter, *Photochem. Photobiol.* **4,** 869 (1965).

[76] F. H. Johnson, H.-D. Stachel, E. C. Taylor, and O. Shimomura, *in* "Bioluminescence in Progress" (F. H. Johnson and Y. Haneda, eds.), p. 67. Princeton Univ. Press, Princeton, New Jersey, 1966.

[77] N. Sugiyama, M. Akutagawa, T. Gasha, and Y. Saiga, *in* "Bioluminescence in Progress" (F. H. Johnson and Y. Haneda, eds.), p. 83. Princeton Univ. Press, Princeton, New Jersey, 1966.

of *Cypridina*, however, exhibits chemiluminescence due more to the pyrazine skeleton than to the indole component.[78]

The important point in the present context, however, is that the evidence has long indicated that *Cypridina* luminescence results from the activity of a simple enzyme–substrate system that requires no specific cofactors, either organic or inorganic, beyond water and dissolved oxygen, despite the results of recent studies[79] that have led to an interpretation that *Cypridina* "luciferase is a metalloenzyme containing calcium." This interpretation was derived from experiments with purified luciferase, but incompletely purified luciferin, with special reference to the effects of various inorganic ions and chelating agents on the activity of the enzyme. Lynch *et al.* observed that EDTA and EGTA were effective inhibitors at a concentration of 0.1 m*M* whereas Ca–EDTA and Ca–EGTA did not inhibit. Their results were interpreted in line with the fact that EDTA and EGTA "inhibit metalloenzymes by forming a coordination complex with the constituent metal ion, resulting in either an *in situ* block of the apoenzyme or removal of the metal ion." Their data, however, do not exclude the alternative possibility that inhibitory effects of EDTA and EGTA by themselves occur by an action directly on the protein. Moreover, some evidence was already available supporting the view that *Cypridina* luciferase contains no functional metallic component.[80] Second, it has been demonstrated that various substances influence the luminescence of the *Cypridina* system very differently according to the amount and nature of impurities present, as well as the actual concentration of the luciferin.[25] Third, because complicated effects on this system, even when both luciferin and luciferase are used in highly purified preparations, are brought about by chelating agents alone and in combination with various cations in a concentration equal to 1 m*M*.[80] For example, 1 m*M* KCN caused no inhibition when added alone, but this same concentration of KCN when added together with 1 m*M* $CuSO_4$ caused an inhibition of 100%; the addition of 1 m*M* $CuSO_4$ without KCN caused an inhibition of 93%. In regard to the action of EDTA, it was shown that, with both luciferin and luciferase very highly purified, 0.5 m*M* EDTA caused an inhibition of 90%, but the inhibition was reduced by simultaneous addition of various cations each at 1 m*M*, namely (with, in parentheses, the percent inhibition caused in the presence of 0.5 m*M* EDTA): Cu^{2+} (80), Zn^{2+} (0), Mn^{2+} (5), Co^{2+} (5), Mg^{2+} (70), Ba^{2+} (0), and Ca^{2+} (0). Addition of these ions at 1 m*M* without EDTA resulted in

[78] T. Goto, S. Inoue, and S. Sugiura, *Tetrahedron Lett.* **36,** 3873 (1968).

[79] R. V. Lynch, F. I. Tsuji, and D. H. Donald, *Biochem. Biophys. Res. Commun.* **46,** 1544 (1972).

[80] O. Shimomura, F. H. Johnson, and Y. Saiga, *J. Cell. Comp. Physiol.* **58,** 113 (1961).

the following percent inhibitions (in parentheses after each ion): Cu^{2+} (93), Zn^{2+} (15), Mn^{2+} (10), Co^{2+} (5), Mg^{2+} (0), Ba^{2+} (0), and Ca^{2+} (−10). Further evidence is desirable regarding the action of EDTA and Ca^{2+} on this system.

Purification of *Cypridina* Luciferase

The first notable success toward purification of *Cypridina* luciferase was that of McElroy and Chase,[81] who obtained a preparation with an estimated increase of some 150 times the activity of the starting material. Their procedure involved fractional precipitation by means of $(NH_4)_2SO_4$ and acetone, and adsorption onto and elution from calcium phosphate gel. As judged by the product obtained subsequently by other investigators,[80] with an estimated increase in potency of some 4000 times that of fresh specimens, the material of McElroy and Chase was nevertheless remarkably good, and they stated that their "results indicate that only two components, as extracted from dried *Cypridina*, luciferase and luciferin, are necessary for light emission."[81]

In the experience of the authors of the present article, however, any procedure whatever that involves precipitation of this luciferase is likely to result in a marked loss of activity. For this reason, the very convenient, and in fact rather effective, method they found of purification by precipitation of luciferase with rivanol (=Acrinol=Ethodin=6,9-diamino-2-ethyoxyacridine lactate) was abandoned in favor of repeated column chromatography on DEAE-cellulose, giving the purest preparations in the best yields.[80]

The results of investigations by these authors were in fair agreement, but there were some discrepancies. For example, Shimomura *et al.*[80] found the molecular weight to be between 48,500 and 53,000, whereas Tsuji and Sowinski[82] found it to be about 80,000, although this value was revised to 70,000 in a later study by Lynch, Tsuji, and Donald.[79] Another study by Shimomura *et al.* of the properties of *Cypridina* luciferase has indicated a best value of 53,000 for the molecular weight.[83]

There is evidently an optimum concentration of salts, at ca. 0.05 to 0.1 *M* for maximum activity of *Cypridina* luciferase. Figure 7 shows the effects of adding various salts in concentrations up to 1 *M*, on the activity of a highly purified preparation of luciferase, which was then dialyzed against 1 m*M* phosphate buffer, pH 7, and finally diluted to 1 part in 100,000 parts of

[81] W. D. McElroy and A. M. Chase, *J. Cell. Comp. Physiol.* **38,** 401 (1951).

[82] F. I. Tsuji and R. Sowinski, *J. Cell. Comp. Physiol.* **58,** 125 (1961).

[83] O. Shimomura, F. H. Johnson, and T. Masugi, *Science* **164,** 1299 (1969).

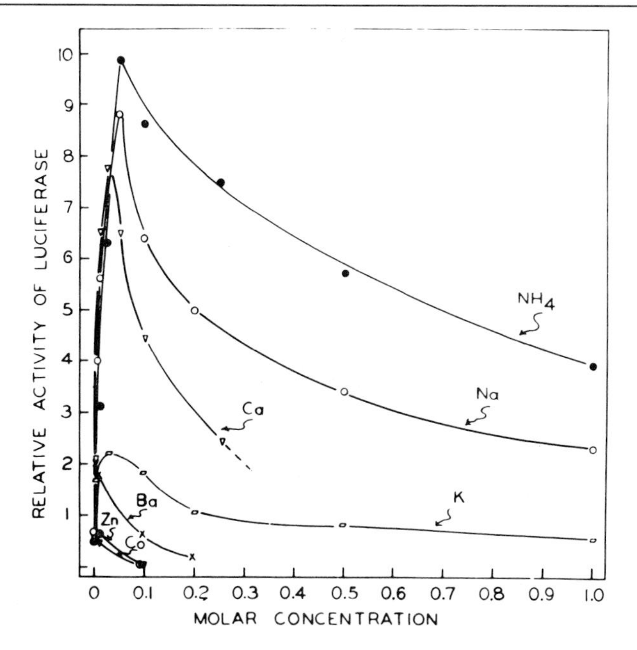

FIG. 7. Activity of luciferase at 25° as a function of the concentration of various cations. The pH of the solutions, just after measuring enzyme activity, ranged between 5.9 and 6.8. In only two instances, namely, at 0.1 *M* $CoCl_2$ and 0.1 *M* $ZnCl_2$, was an initial adjustment of pH made, in each instance by adding a trace of NaOH. From Fig. 7, p. 120 of O. Shimomura, F. H. Johnson, and Y. Saiga, *J. Cell. Comp. Physiol.* **58,** 113 (1961).

specially purified, deionized distilled water.[80] It will be noted that, under the conditions involved, as much as a 20-fold increase in activity results from adding small amounts of NH_4, and almost as much by similar amounts of Na or Ca. Other cations were less effective toward increasing the activity, and heavy-metal ions proved inhibitory, including Hg and Ag, as expected. Even with ions that were most favorable to activity at concentrations of 0.1 *M* or less, the increase in activity rapidly became less at higher concentrations. This dual action has been accounted for as effects of the ions on conformation of the enzyme in a manner that favors formation of the activated complex, according to the theory set forth by Neville and Eyring[84] and applied to the activity of hen egg lysozyme, pig heart fumarate hydratase, and ox liver arylsulfatase B. Application by Neville of the same theory to data on the effects of NaCl as that shown in the present Fig. 7 has resulted in a calculated, smooth curve that practically coincides with the

[84] W. M. Neville and H. Eyring, *Proc. Natl. Acad. Sci. U.S.A.* **69,** 2417 (1972).

one in Fig. 7 (cf. Johnson, Eyring, and Stover's[85] Fig. 2.8, p. 74; details of the treatment are given on pp. 74–76).

Additional Properties of *Cypridina* Luciferase

The following points of interest regarding *Cypridina* luciferase are from the investigation by Shimomura *et al.*,[80] except as noted otherwise. Their usual purification procedure involved 5 successive column chromatographies on DEAE-cellulose, saving only those fractions that had the highest ratios of luciferase activity to absorption at 278 nm, aggregating up to 75% of the activity placed on the column. The product had an absorption spectrum typical of simple proteins, with a sharp maximum at 277 nm, which is within the range of 275–280 nm reported for the fractions with highest activity obtained by McElroy and Chase[81] as well as within the 275–278 nm reported by Osborne and Chase.[86] Data from the Spinco analytical centrifuge indicated a sedimentation constant $s = 3.93$ Svedberg units. The diffusion constant measured in a Model 38 Perkin-Elmer boundary electrophoresis apparatus at 20° gave an averaged value of 7.65×10^{-7} using the zero time method of Lundgren and Ward,[87] calculated according to area, or 7.625×10^{-7} according to inflection. The average of these two values when corrected to the viscosity of water at 20° is 7.88, or approximately 7.9×10^{-7}. The isoelectric point as determined by electrophoresis with the Perkin-Elmer apparatus was found to be about 4.35 by extrapolation of the curve pertaining to the amount of migration at higher pH; below pH 4.5 the enzyme was rapidly inactivated, even at 0°. Paper electrophoresis indicated an isoelectric point of 4.7 in 40 m*M* sodium phosphate or potassium biphthalate buffer, each containing 0.2 *M* NaCl. The pH-activity curve indicated a maximum at 7.5 at room temperature in a solvent having a favorable salt composition of 50 m*M* NaCl and 5 m*M* sodium phosphate buffer. The stability of luciferase was essentially unaffected over a broad range between about pH 5 and pH 10, in several different buffers. The Michaelis–Menten constant was computed to be 0.52 μM at 25°, assuming 470 as the molecular weight of luciferin. The actual molecular weight of luciferin (free form) is 423, and that of luciferin dihydrochloride is 496.[28] The effects of various substances on catalytic activity showed that very slight or no effect was caused by 1 m*M* sodium diethyldithiocarbamate, 8-hydroxyquinoline, or *o*-phenanthroline. Similar results for these substances were reported by Lynch *et al.*[79] Shimomura *et al.*[80] also found

[85] F. H. Johnson, H. Eyring, and B. J. Stover, "The Theory of Rate Processes in Biology and Medicine" Wiley (Interscience), New York, 1974.

[86] A. W. Osborne and A. M. Chase, *J. Cell. Comp. Physiol.* **44**, 49 (1954).

[87] H. P. Lundgren and W. W. Ward, *in* "Amino Acids and Proteins" (D. M. Greenberg, ed.), p. 312. Thomas, Springfield, Illinois, 1951.

little or no effect of 1 m*M* KCN, or half-saturated *p*-chloromercuribenzoic acid. Strong inhibition was caused by 0.5 m*M* EDTA in a manner that was reversible by dialysis or by addition of 1 m*M* Zn, Mn, Ca, or Co, but not by Cu, which is strongly inhibitory of itself, and not by the weakly chelated cations Mg or Ba. The effects of these and additional metal ions, including mono, di-, and trivalent examples, were studied by Lynch *et al.*[79] Their results in some instances were complex and difficult to interpret, and not all were consistent with those of Shimomura *et al.*[80] It seems possible that the discrepancies might be due to the use of a more highly purified preparation of luciferin by Shimomura *et al.* and possibly also to seemingly anomalous effects sometimes encountered in the combined action of metal ions and chelating agents on this system. For example, at concentrations of 1 m*M*, 8-hydroxyquinoline causes no inhibition, and Zn^{2+} only a 15% inhibition, but when added together they cause a 100% inhibition.[80] Moreover, some of the results obtained by Lynch *et al.*[79] could be due to effects of such substances as $K_3Fe(CN)_6$ on the luciferin rather than the luciferase.

It is noteworthy that in a remarkable number of instances the numerical values given above are close to, or not very different from, those arrived at previously through ingenious experiments with incompletely purified material. Thus, Kauzmann, Chase, and Brigham[88] computed a value of 1.24 μmol/liter for the Michaelis–Menten constant, using 300 as a reasonable value of the then unknown molecular weight of luciferin. The maximum activity in 10 m*M* sodium phosphate buffer was found by Chase [89] to be at pH 7.2. Values for the diffusion constant, sedimentation constant, and molecular weight prior to the availability of quite pure luciferase, were also in good agreement with the values obtained later with considerably more highly purified material, as determined by Chase and collaborators,[90-92] as follows: $s_{20} = (5.3 \pm 0.2) \times 10^{-13}$ sec; $D = 7.4 \times 10^{-7}$; and MW $= (7.0 \pm 0.5) \times 10^4$. The molecular weight reported by Tsuji and Sowinski[82] was 79,650, and their value for the sedimentation constant was $s_{20} = 4.58$. A later paper by Tsuji *et al.*[92a] gives 68,000 as the molecular weight.

"Simplicity" of the *Cypridina* System

Everyone who has devoted serious study to chemical aspects of bioluminescence has soon become acquainted with the terms "luciferin"

[88] W. J. Kauzmann, A. M. Chase, and E. H. Brigham, *Arch. Biochem.* **24,** 281 (1949).
[89] A. M. Chase, *J. Cell. Comp. Physiol.* **31,** 175 (1948).
[90] A. M. Chase, *J. Cell. Comp. Physiol.* **45,** 13 (1955).
[91] G. A. Fedden and A. M. Chase, *Biochim. Biophys. Acta* **32,** 176 (1959).
[92] A. M. Chase and R. Langridge, *Arch. Biochem. Biophys.* **88,** 294 (1960).
[92a] F. I. Tsuji, R. V. Lynch III, and C. L. Stevens, *Biochemistry* **13,** 5204 (1974).

and "luciferase," as well as with the origin of these terms in connection with the experiments of Raphael Dubois.[93-95] Dubois was apparently the first to obtain a luminescent reaction *in vitro* by mixing at room temperature two extracts prepared from photogenic material, one in hot water, sufficiently hot to extinguish the glow of bioluminescence practically at once, and the other in cold water, wherein the glow of bioluminescence gradually diminished and finally disappeared. His first experiments were with the West Indian elaterid beetle *Pyrophorus*,[93] and later[95] with the somewhat more readily accessible, boring clam *Pholas dactylus* of the Mediterranean and southern coasts of France. On the basis of the evidence he assembled, he concluded that emission of light *in vitro* on mixing two such extracts had characteristics of the oxidation of a relatively heat-stable, specific substance, to which he gave the name "luciférine," catalyzed by a relatively heat-labile specific enzyme for which he introduced the term "luciférase." Specificity was indicated by the absence of a light-emitting reaction, subsequently known as a "luciferin–luciferase reaction" on mixing *Pholas* luciferin solution with a *Pyrophorus* luciferase solution or vice versa, i.e., there was no "cross-reaction." Dubois carried out experiments on bioluminescent systems for many years and wrote a book on the general subject of life and light, published in 1914.[96] His views concerning the chemical nature and properties of luciferin and luciferase were sometimes seriously misled—for some time he thought luciferase was a diastase—owing to the fact that the active components were never separated from numerous impurities in his extracts. It seemed that *Pyrophorus* and *Pholas* systems were fundamentally alike, except for certain specific differences; e.g., the luciferin of *Pholas* was more easily destroyed by heat than was that of *Pyrophorus,* and light emission with *Pholas* luciferin was not strictly dependent on *Pholas* luciferase, but could be elicited by nonspecific oxidizing agents such as $KMnO_4$. Recent studies with the aid of modern methods have been carried out by Michelson and associates.[97-102] These have indicated that the luciferin of *Pholas* is a protein, as in fact Dubois had thought.

[93] R. Dubois, *C. R. Soc. Biol.* **37,** 559 (1885).
[94] R. Dubois, *Bull. Soc. Zool. Fr.* **11,** 1 (1886).
[95] R. Dubois, *C. R. Soc. Biol.* **39,** 564 (1887).
[96] R. Dubois, "La vie et la lumière." Alcan, Paris, 1914.
[97] J.-P. Henry and A. M. Michelson, *Biochim. Biophys. Acta* **205,** 451 (1970).
[98] J.-P. Henry, M. F. Isambert, and A. M. Michelson, *Biochim. Biophys. Acta* **205,** 437 (1970).
[99] J.-P. Henry, C. Monny, and A. M. Michelson, *Biochemistry* **14,** 3458 (1975).
[100] A. M. Michelson, *Biochimie* **55,** 465 (1973).
[101] A. M. Michelson, *Biochimie* **55,** 925 (1973).
[102] A. M. Michelson and M. F. Isambert, *Biochimie* **55,** 619 (1973).

The luminescence brought on by various agents and combinations thereof evoke the question whether or not the *Pholas* system should more properly be considered an example of the "photoprotein" type, along with those of *Aequorea*, *Chaetopterus*, and *Meganyctiphanes*,[103] rather than the luciferin–luciferase type in Dubois' sense of the terms.

As it has turned out, the *Cypridina* system, which apparently never became available to Dubois to investigate, comes perhaps closest of all known systems to his concept of the mechanism of bioluminescence as involving simply an enzyme-catalyzed oxidation of a specific, organic substrate. The weight of present evidence is heavily in favor of the view, held early by Harvey and by the majority of investigators ever since, that the only components required for bioluminescence of *Cypridina* are its luciferin and luciferase, in addition, of course, to water and molecular oxygen. Dubois lived until 1929 and was still writing on the subject of biophotogenesis the year before.[104] It was not until much later that the requirement of a specific cofactor in any luminescence system was discovered, namely that of ATP for luminescence of firefly extracts.[105] It is now evident that the same requirement is true of all, and possibly only, fireflies, including *Pyrophorus*, studied by Dubois in his investigations which from the biochemical point of view were frontier-breaking. Various other cofactors have by now been found to be required in a number of other systems, photoprotein as well as luciferin–luciferase types.[103]

Kinetics of *Cypridina* Bioluminescence: Rate of Decay

An early quantitative study of kinetics of the *Cypridina* reaction was made by Amberson,[106] who used what would now be considered a primitive method of measuring the intensity of luminescence, namely, a moving photographic film exposed through a narrow slit to the light of a reaction mixture of crude luciferin and luciferase prepared from dried *Cypridina*. By careful calibration of density of the image as a function of distance from the origin, i.e., in terms of time after mixing, through knowing the lateral speed of the film on a kymograph, Amberson showed that the intensity of luminescence diminished with the kinetics of a reaction that was first order with respect to concentration of luciferin, much as Dubois would very likely have supposed. Amberson's method, however, was limited to reaction mixtures wherein the intensity of luminescence decayed rather slowly,

[103] F. H. Johnson and O. Shimomura, *in* "Photophysiology" (A. C. Giese, ed.), Vol. 7, p. 275. Academic Press, New York, 1972.

[104] R. Dubois, *in* "Dictionnaire de Physiologie" (C. Richet, ed.), tome X, 277 (1928).

[105] W. D. McElroy, *Proc. Natl. Acad. Sci. U.S.A.* **33**, 342 (1947).

[106] W. R. Amberson, *J. Gen. Physiol.* **4**, 517 (1922).

frequently measured for as long as 2 hr. Stevens[107] used a device for visual photometry and studied faster reactions, easily effected by regulating the relative concentrations of luciferase in the reaction mixture with luciferin, having a half-life of some 17 sec. Still later, Harvey and Snell,[108] using high concentrations of luciferase and recording luminescence with the aid of a string galvanometer and photoelectric cell, measured short flashes of luminescence, half complete in 0.5–1 sec. The decay rate was exponential throughout the major portion of the time course according to each of the three methods and periods of reaction time involved. These results were essentially all confirmed later with preparations of much greater purity in extensive research by Chase and his collaborators.[109,110] Their data make it clear that the luciferase-catalyzed, light-emitting oxidation of luciferin follows the kinetics characteristic of a reaction that is first order with respect to luciferin concentration and with a specific rate constant proportional to the amount of active luciferase, as illustrated in Fig. 8. This figure illustrates also the wide range of luciferin concentrations over which the final yield of light, or "total light" is the same and thus proportional to the initial amount of luciferin, even though specific rate, i.e., first-order rate constant, differed widely.

In the experiments by Kauzmann, Chase, and Brigham,[88] the Michaelis–Menten constant was computed for two temperatures. On the basis of a few assumptions, including a value of 300 for the molecular weight of luciferin, values of $1.24 \times 10^{-6} M$ at 22°, and $0.88 \times 10^{-6} M$ at 15°, respectively, were thus arrived at. Later, using the molecular weight of 470 for luciferin, and carrying out the experiments with very highly purified luciferase as well as luciferin, Shimomura *et al.*[80] computed the constant to be $0.52 \times 10^{-6} M$ at 25°. These numerical values are remarkably close. The significance of the Michaelis constant, however, and the changes in it at different temperatures, are open to alternative interpretations, as Kauzmann *et al.*[88] pointed out; i.e., it could represent either a difference in rate constants of successive reactions or a true dissociation constant of the luciferin–luciferase complex.

Kinetics of *Cypridina* Bioluminescence: Rate of Rise

The rate at which luminescence appears and reaches a maximum on mixing the reactants *in vitro* has been studied by the flow method of

[107] K. P. Stevens, *J. Gen. Physiol.* **10,** 859 (1927).
[108] E. N. Harvey and P. A. Snell, *J. Gen. Physiol.* **14,** 529 (1931).
[109] A. M. Chase, *Arch. Biochem.* **23,** 385 (1949).
[110] A. M. Chase, *Methods Biochem. Anal.* **8,** 61 (1960).

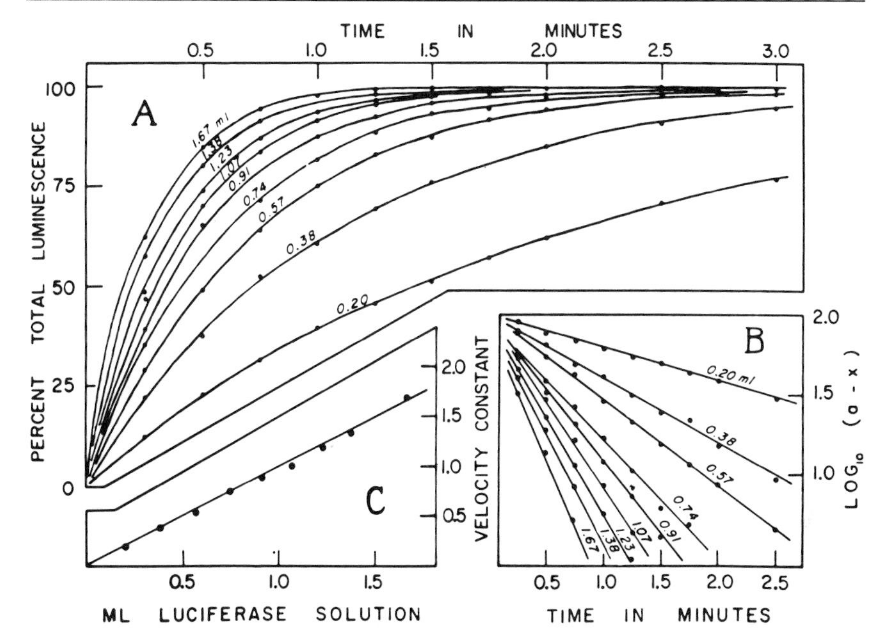

FIG. 8. Kinetics of the luminescent reaction of partially purified luciferin and luciferase of *Cypridina* (From Chase[110] p. 99, Fig. 15). (A) Integrated light produced as a function of time, with various amounts of luciferase (in terms of milliliters of an enzyme preparation added per 20 ml of reaction mixture), with the same initial concentration of luciferin. (B) The data of (A) plotted in a manner of a first-order reaction. (C) Relation between the first-order velocity constants, taken as positive rather than negative numbers, according to the slopes of the lines in (B), as a function of the concentration of luciferase. The data in this figure demonstrate that (1) the first-order velocity constant, representing in fact the specific reaction rate constant, is proportional to the concentration of luciferase, and (2) with a given concentration of luciferase the reaction velocity is proportional to the amount of luciferin.

measuring rapid reactions, using crude aqueous extracts of *Cypridina*.[111] Unfortunately, similar experiments remain to be done with highly purified extracts. The data are illustrated in Fig. 9, which includes for comparison the rate of development of luminescence on adding a salt solution containing dissolved oxygen to a suspension of luminous bacteria that had become dark due to anaerobic conditions.

Figure 9 shows that when a solution of luciferin containing oxygen is mixed with a solution of luciferase containing oxygen, luminescence comes on less rapidly (time to reach 1/2 maximum intensity, or 1/2 rise time =

[111] B. Chance, E. N. Harvey, F. H. Johnson, and G. Millikan, *J. Cell. Comp. Physiol.* **15,** 195 (1940).

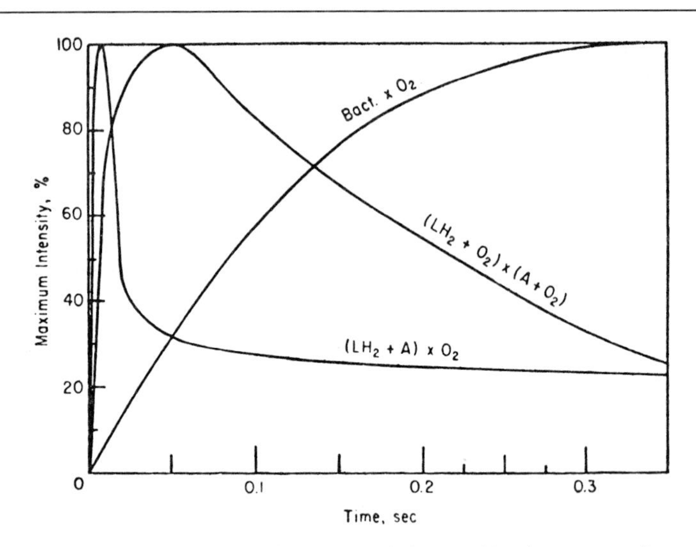

FIG. 9. Curves showing the development and decay of luminescence when oxygen is admitted to luciferin (LH_2) and luciferase (A) mixed in the absence of oxygen [(LH_2 + A) × O_2]; when luciferin and luciferase both containing oxygen are mixed [(LH_2 + O_2) × (A + O_2)]; and when oxygen is admitted to luminous bacteria without oxygen (Bact. × O_2). From Fig. 13, p. 207, of B. Chance, E. N. Harvey, F. H. Johnson, and G. Millikan, *J. Cell. Comp. Physiol.* **15,** 195 (1940).

0.006 sec) than when the luciferin and luciferase are first mixed together in absence of oxygen and are then mixed with a solution containing oxygen (1/2 rise time = 0.002 sec). In both these instances, however, the 1/2 rise time is notably shorter than when luminous bacterial cells that have become dark through oxygen lack are suddenly mixed with a salt solution containing oxygen, whereupon the 1/2 rise time amounts to 0.08 sec. All experiments were done at room temperature.

Efficiency and Quantum Yield

The term "efficiency" is subject to some differences in meaning, according to points of view and circumstances. "Quantum yield" ("q.y.," for which the symbol ϕ may be used) is perhaps a somewhat more precise term and may be defined for present purposes as the ratio between the number of photons produced and number of luciferin molecules reacted. Obviously, any critical values of quantum yields depend on measurements, in absolute units, of light produced by known quantities of luciferin in an adequately purified reaction mixture. With the aid of thoroughly purified luciferin and a portion of the luciferase purified by Shimomura *et al.*,[80] quantum yield of *Cypridina* luminescence was critically measured by Johnson *et al.*[112] Taking into account various sources of errors, a best value for quantum yield with respect to luciferin at 4° amounted to 0.28 ± 15%, with an estimated maximum value of 0.29 ± 17%. The quantum yield with respect to oxygen consumed was found to be the same as for luciferin oxidized.

The above data were obtained under conditions of concentration of reactants, salt composition of buffered solvents, temperature, etc., intended to be optimal. Further study of factors influencing the quantum yield, in effort to contribute to an understanding of the reaction mechanisms, was undertaken by Shimomura and Johnson[113] with special reference to the influence of such factors as temperature, pH, salts, and relative concentrations of luciferin or luciferase, etc., many of which factors had been found to affect the luminescence of preparations of various degrees of purity.[22,54,89,109,114-119]

The data of Shimomura and Johnson,[113] illustrated in Fig. 10, show the quantitative variation in fluorescence as well as bioluminescence intensity under the influence of different factors. These results, together with data on fluorescence of the light-emitting oxyluciferin–luciferase complex (see below), provide suggestive evidence in explanation of the observed variations in quantum yield, including a very low value of 0.03 for chemiluminescence of luciferin in organic solvents. Analysis of the degradation products indicates a different pathway of luciferin oxidation having the low quantum yield, evidently due to a reduced population of the excited state, rather than to such mechanisms as internal and external quenching that result in radiationless transitions of the excited state (cf. Johnson *et al.*[85]). The highest quantum yield found for bioluminescence of the *Cypridina* system was 0.305 (ca. 0.31) with 5 μg of luciferin in 10 m*M* sodium phosphate buffer, pH 6.1, at 7°, containing 0.5 mg of luciferase, in 5 ml of reaction mixture. In 99% D_2O instead of H_2O, but under otherwise similar conditions, the quantum yield was slightly higher, amounting to 0.33 (with 5 μg of luciferin).

[112] F. H. Johnson, O. Shimomura, Y. Saiga, L. C. Gershman, G. T. Reynolds, and J. R. Waters, *J. Cell. Comp. Physiol.* **60,** 85 (1962).

[113] O. Shimomura and F. H. Johnson, *Photochem. Photobiol.* **12,** 291 (1970).

[114] R. S. Anderson, *J. Cell. Comp. Physiol,* **4,** 517 (1933).

[115] R. S. Anderson, *J. Am. Chem. Soc.* **59,** 2115 (1937).

[116] A. M. Chase, *Arch. Biochem.* **23,** 385 (1949).

[117] A. M. Chase, *in* "Bioluminescence in Progress" (F. H. Johnson and Y. Haneda, eds.), p. 115. Princeton Univ. Press, Princeton, New Jersey, 1966.

[118] A. M. Chase and P. B. Lorenz, *J. Cell. Comp. Physiol.* **25,** 53 (1945).

[119] A. M. Chase, F. S. Hurst, and H. J. Zeft, *J. Cell. Comp. Physiol.* **54,** 115 (1959).

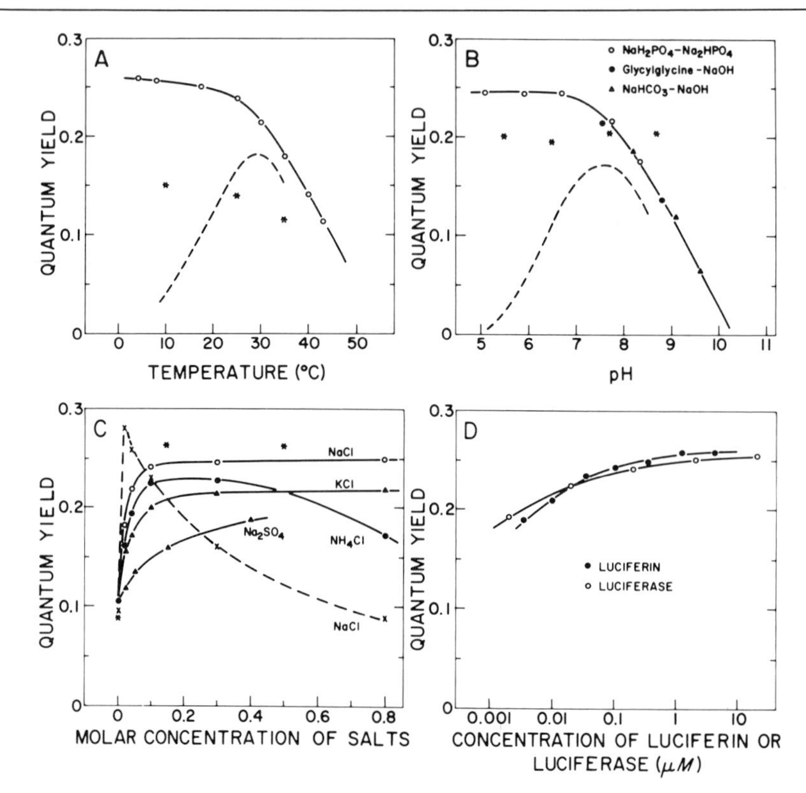

FIG. 10. Quantum yield of the *Cypridina* bioluminescent reaction (—), maximum light intensity (– – –) and fluorescence intensity of the oxyluciferin–luciferase light-emitting complex (arbitrary units, single stars), as influenced by (A) temperature, (B) pH, (C) concentration of salts, and (D) concentration of luciferin and luciferase. Reaction mixtures, each in a final volume of 5 ml: sodium phosphate buffer (50 m*M* for A, 10 m*M* for C and D) or 50 m*M* buffer as specified in B, containing 0.1 *M* NaCl (for A, B, and D), and for quantum yields 1 μg of luciferin plus 20 μg of luciferase. For intensity curves, 10 μg of luciferin and 0.05 μg of luciferase were used; in (D) 100 μg of luciferase was used for the luciferin curve and 1 μg of luciferin for the luciferase curve. Except in (A) the temperature was 25°, and except in (B) the pH was 6–8 throughout. Fluorescence intensities were measured with an Aminco-Bowman spectrofluorometer at 462 nm, excited at 360 nm, using 1 ml of a solution containing 2 μg of oxyluciferin and 100 μg of luciferase. From Fig. 1, p. 293, of O. Shimomura and F. H. Johnson, *Photochem. Photobiol.* **12,** 291 (1970).

The Light-Emitting Molecular Complex

On first consideration, it would appear likely that the light-emitter in the bioluminescent oxidation of a specific substrate molecule would be a product of this oxidation rather than the initial reactant, at least whenever the oxidative reaction furnishes the energy for excitation. In the *Cypridina* system convincing evidence has been found that the light emitter consists of the oxyluciferin–luciferase complex (Fig. 11).[83] Addition of oxyluciferin to luciferase solutions resulted in a great increase in fluorescence above a previously negligibly low level, thus making it possible to do fluorometric titrations of luciferase with this substance. The resulting data indicated that 1 mol of oxyluciferin binds hydrophobically with 1 mol of luciferase. The dissociation constant, computed on the assumption that the increase in fluorescence is proportional to the amount of oxyluciferin–luciferase so formed, turned out to be 0.3 μM at about 0°. The conclusion that the equimolar complex of oxyluciferin with luciferase in an excited state is the light-emitter in *Cypridina* luminescence, and that there is only one hydrophobic binding center for oxyluciferin on the luciferase molecule, suggests that the catalytic center for the light-emitting reaction is also only one, and that these centers, if not identical, must be very close together.

Luminescence as a "Reaction Rate Tool"

One of the most important results of the investigations of *Cypridina* luminescence from the point of view of kinetics was the evidence that the brightness of the light provides an index to the instantaneous reaction velocity of the system. The potential usefulness of this circumstance, however, was a long time in becoming realized. In fact, the idea that the intensity of chemiluminescence is determined by the rate of the underlying reaction goes back at least to Trautz,[120] who came to this conclusion in connection with his comprehensive studies of chemiluminescence of phenol and aldehyde compounds. Trautz was evidently carried away with this idea when he concluded that any reaction would produce luminescence if the reaction velocity were sufficiently increased. Moreover, as Harvey[14] (p. 155) has emphasized, "It is a *particular* reaction occurring at an optimum velocity that results in light." He pointed out that *Cypridina* luciferin can be oxidized rapidly at high temperatures or by ferricyanide without the production of light, whereas in the presence of luciferase it can be oxidized much less rapidly but with the production of bright luminescence.

The important point in the present context is that bioluminescence, to the extent that its brightness can be taken as a visible indicator of the reaction velocity of the system, offers a unique "tool" with virtually unparalleled advantages for investigating the kinetics of biological reactions *in vivo* as well as *in vitro*. Thus, each instantaneous measurement of light intensity is in effect equivalent to the slope of a line relating rate of reaction to time in an ordinary enzyme-catalyzed process. In instances where bioluminescence is normally an intracellular process and in a steady state—for example, in a population of luminous bacteria under favorable

[120] M. Trautz, *Z. Phys. Chem.* **53,** 1 (1905).

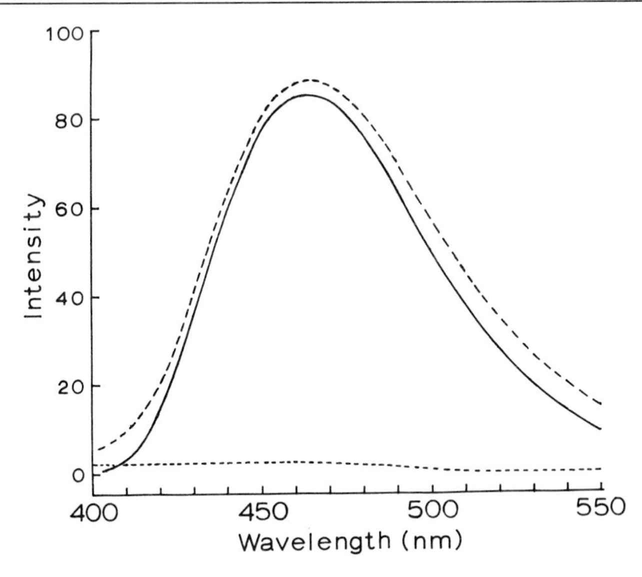

FIG. 11. The light-emitter in the bioluminescent reaction of the *Cypridina* system, i.e., the oxyluciferin–luciferase complex, as demonstrated by spectrographic evidence. Luminescence spectrum of luciferin plus luciferase (—) and fluorescence spectrum for the equimolar (3 μM) mixture of oxyluciferin and luciferase on excitation at 360 nm (– – –) in a mixture of 50 mM sodium phosphate and 0.1 M NaCl, pH 7.4, at about 0° within 15 sec of the mixing. In reference to the latter spectrum, fluorescence without oxyluciferin (----) and without luciferase (not shown) were both negligibly low. From O. Shimomura, F. H. Johnson, and T. Masugi, *Science* **164,** 1299 (1969). Copyright 1969 by the American Association for the Advancement of Science.

physiological conditions—the influence of various factors, such as changes in temperature and pressure, can be analyzed without in any way injuring or disrupting the cells in order to obtain accurate, quantitative data concerning the immediate effect of the factor involved. Moreover, when the factor itself is known to, or is apt to, cause damage to the cells, especially on long exposure, measurements of luminescence intensity can be made so quickly that a truly minimum time of exposure is required. Thus, the phenomenon of bioluminescence comes very close to fulfilling the ideals of measuring an intracellular reaction in a way that does not harm the structure or integrity of the cell and at the same time yielding an instantaneous, accurate value for reaction velocity of the limiting reaction.

With these extraordinary advantages, made convincingly evident through studies of *Cypridina* luminescence, investigations of the steady-state luminescence of living bacterial cells provided a key to a rational understanding of the biological action of temperature, increased hydrostatic pressure, and various chemical agents including narcotics and anesthetics such as urethane, alcohol, ether, and others, as well as interrelations in the combined action of these factors which had previously been thought to act largely independently of each other. The initial observations were due to collaborative research by Brown, Johnson, and Marsland,[121–123] and the theoretical basis for the quantitative interpretation of the data was provided by the Erying theory[124,125] of absolute reaction rates. The details, which are much beyond the scope of the present article, are available in a monograph[85] and to some extent in a very concise, recent review by the same authors.[126] It seems appropriate, however, to list, without further discussion here, the major points of general interest achieved through application of the bioluminescence "tool," as follows.

The temperature-activity curve, from temperatures well below to temperatures well above the so-called optimum, was accounted for primarily on the basis of the well-known increase, with rise in temperature, in specific reaction rate constant of the limiting enzyme reaction, characterized by ordinary values for activation energy, together with a shift in a reversible denaturation equilibrium of the enzyme toward an inactive state, characterized by a large heat and entropy of reaction. Both the rate and equilibrium reaction evidently involved volume increases of activation and of reaction, respectively, as shown by counteracting effects of moderately increased hydrostatic pressures of a few hundred atmospheres. Because the equilibrium reaction involved much larger volume increases than the rate process, the net effect of pressure depended on which of these reactions was quantitatively the more important in the control of the overall process under the conditions involved, and since this varied with temperature, the net effects of pressure likewise varied with temperature. In most of the large amount of previous research on the biological effects of pressure, the temperature was not taken into account, or even mentioned, much less realized to be a significant factor influencing the results obtained.[127] The research with bacterial luminescence showed that increased pressure caused a marked inhibition at low temperatures, little effect near the optimum of this process in the particular species, and a marked increase at temperatures well above the optimum, where the intensity at normal pressure was already reduced by the reversible thermal denaturation of the

[121] D. E. Brown, F. H. Johnson, and D. A. Marsland, *J. Cell. Comp. Physiol.* **20,** 151 (1942).
[122] F. H. Johnson, D. E. Brown, and D. A. Marsland, *Science* **95,** 200 (1942).
[123] F. H. Johnson, D. E. Brown, and D. A. Marsland, *J. Cell. Comp. Physiol.* **20,** 269 (1942).
[124] H. Eyring, *J. Chem. Phys.* **3,** 107 (1935).
[125] S. Glasstone, K. J. Laidler and H. Eyring, "The Theory of Rate Processes." McGraw-Hill, New York, 1941.
[126] F. H. Johnson, H. Eyring, and B. J. Stover, *Annu. Rev. Biophys. Bioeng.* **6,** 111 (1977).
[127] M. Cattell, *Biol. Rev. Cambridge Philos. Soc.* **11,** 441 (1936).

catalyst. This denaturation reaction was found to be intimately involved in the inhibitory action of certain lipid-soluble narcotics—alcohol, ether, chloroform, and others—through a mechanism that more satisfactorily accounted than previously for the temperature relations of the action of these drugs. Convincing additional evidence derived from the totally new relation unexpectedly discovered between the potency of effects of these drugs on bioluminescence and hydrostatic pressure; increased hydrostatic pressures of a couple of hundred atmospheres practically abolished the effects of a moderately inhibitory concentration of 2–3% ethanol on bacterial luminescence.[123] Similar results were obtained later in regard to abolishing the alcohol narcosis of tadpoles,[128,129] as well as in regard to the alcohol reduction of the action potential of single axons.[130] More recently, the successful use of increased helium pressure to reverse anesthesia in mice under the influence of various types of anesthetics has been reported by Lever *et al.*[131] These results of research on various types of biological phenomena provide noteworthy justification of fundamental research on bioluminescence and the more general implications thereof. In sum, basic research on bioluminescence has served (1) to identify reversible protein denaturation as a controlling factor in a normal physiological process; (2) to bring together in one simplified (no doubt somewhat oversimplified), rational and quantitative theoretical basis, three previously almost wholly separate fields of endeavor, i.e., the biological effects of temperature, pressure, and narcotics; (3) to clarify the important relation between temperature and both the magnitude and direction, i.e., inhibitory or stimulatory, of increased pressure on biological processes; and (4) discover a new phenomenon of some general importance, namely, the pressure reversal of narcosis or anesthesia.

[128] F. H. Johnson and E. Flagler, *Science* **112,** 91 (1951).
[129] F. H. Johnson and E. Flagler, *J. Cell. Comp. Physiol.* **37,** 15 (1951).
[130] C. S. Spyropoulos, *J. Gen. Physiol.* **40,** 849 (1957).
[131] M. J. Lever, K. W. Miller, W. D. M. Paton, and E. B. Smith, *Nature (London)* **231,** 368 (1971).

[33] *Cypridina* Luciferin and Luciferase

By FREDERICK I. TSUJI

Cypridina hilgendorfii is a small (~3 mm long) marine ostracod crustacean commonly found in the waters along the south coast of Japan. When disturbed, it produces a blue luminescence by ejecting luciferin and luciferase from separate glands into seawater. A luminous cloud is formed,

METHODS IN ENZYMOLOGY, VOL. LVII

ISBN 0-12-181957-4

which is readily visible to the dark-adapted eye. The light results from the oxidation of luciferin by molecular oxygen, catalyzed by luciferase.[1] The products of the reaction are light, oxyluciferin, and CO_2.[2,3] The mechanism of this reaction has been studied extensively, as mentioned in this volume [32], but there is still controversy regarding the source of oxygen in the CO_2.[4–9] Whether the oxygens originate from molecular oxygen, water medium, or both has been a matter of much debate and investigation.

Collection of *Cypridina*

Cypridina is a nocturnal animal, which lives in sand during the day and comes out to feed at night. It is best collected on moonless nights in fall or winter. A piece of bait (e.g., a fish head) is attached to a long string and lowered to the sandy bottom near shore. After 15–30 min, the line is drawn in. If the animals are present, they will be found attached to the bait, usually in large numbers. They can be easily dislodged by swishing the bait in a bucket of seawater. After the collection is completed, the organisms are strained by pouring the contents of the bucket through a fine mesh screen just small enough to retain the organisms, but large enough to allow other smaller organisms and particles to pass through. The organisms are washed several times with fresh seawater and then freeze-dried from the frozen state. The desiccated organisms may be stored with a desiccant in a sealed container for many years without loss of activity.

Other species of luminescent *Cypridina* are found in many parts of the world, especially in the Indo-Pacific region, but they are often free-swimming and difficult to collect in large numbers.[1,10–12] Furthermore, most species are usually too small for use in preparative work.

[1] E. N. Harvey, "Bioluminescence." Academic Press, New York, 1952.

[2] H. Stone, *Biochem. Biophys. Res. Commun.* **31,** 386 (1968).

[3] Y. Kishi, T. Goto, Y. Hirata, O. Shimomura, and F. H. Johnson, *in* "Bioluminescence in Progress" (F. H. Johnson and Y. Haneda, eds.), p. 89. Princeton Univ. Press, Princeton, New Jersey, 1966.

[4] F. McCapra and Y. C. Chang, *J. Chem. Soc., Chem. Commun.* **1967,** 1011 (1967).

[5] O. Shimomura and F. H. Johnson, *Biochem. Biophys. Res. Commun.* **44,** 340 (1971).

[6] O. Shimomura and F. H. Johnson, *Biochem. Biophys. Res. Commun.* **51,** 558 (1973).

[7] O. Shimomura and F. H. Johnson, *in* "Chemiluminescence and Bioluminescence" (M. J. Cormier, D. M. Hercules, and J. Lee, eds.), p. 337. Plenum, New York, 1973.

[8] O. Shimomura and F. H. Johnson, *Anal. Biochem.* **64,** 601 (1975).

[9] F. I. Tsuji, M. DeLuca, P. D. Boyer, S. Endo, and M. Akutagawa, *Biochem. Biophys. Res. Commun.* **74,** 606 (1977).

[10] Y. Haneda, *in* "The Luminescence of Biological Systems" (F. H. Johnson, ed.), p. 335. Am. Assoc. Adv. Sci., Washington, D.C., 1955.

[11] F. I. Tsuji, R. V. Lynch III, and Y. Haneda, *Biol. Bull.* **139,** 386 (1970).

[12] L. S. Kornicker and J. H. Baker, *Proc. Biol. Soc. Wash.* **90,** 218 (1977).

Luciferin

Natural Luciferin

The dried organisms are ground to a fine powder with a mortar and pestle or in a mill. They are ground rapidly so as to minimize exposure to moisture, which accelerates luciferin oxidation. Luciferin is hygroscopic and readily autoxidizes. The powder is placed in an asbestos thimble with a layer of glass wool at the top and is defatted with cold benzene (15 hr), followed by extraction with ethyl ether (4 hr), using an apparatus specially designed for this purpose (Fig. 1). The apparatus allows continuous extraction with cold solvent without heating the powder. The extraction should be carried out in a well-ventilated hood and extreme caution should be used in handling benzene because of its leukemogenic properties. The solvent is drained from the thimble, and the powder is evacuated overnight to remove the last traces of solvent.

Subsequent steps in the purification are carried out in the presence of 99.995% argon, as follows.

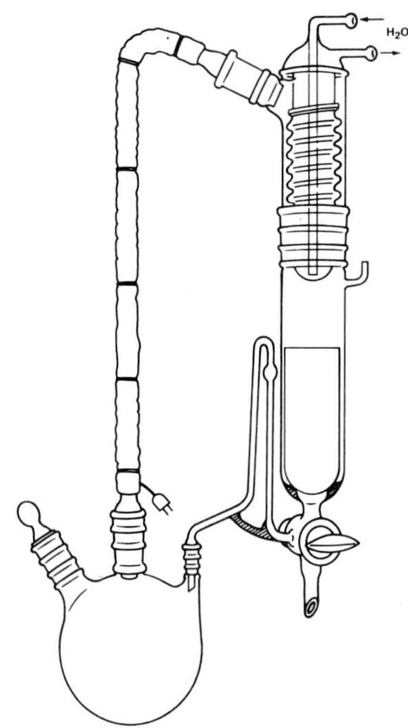

FIG. 1. Cold-solvent extraction apparatus. A heating mantle is used to heat the solvent in the flask; the solvent vapor travels up the left column (wrapped and heated with a heating tape) and enters a modified Soxhlet apparatus; the vapor condenses, and the solvent drips onto the thimble, accumulates, and siphons back into the flask; stopcock at bottom is for draining thimble at end of experiment; small opening at the side is used for equalizing pressure.

Step 1. Place 100 g of powder in a sealable glass vessel with inlet and outlet ports, then add 500 ml of absolute methyl alcohol and a bar magnet. Bubble the mixture with argon for 3 hr while stirring with a magnetic stirrer, then turn off the argon, close off the vessel, and continue the stirring for another 24 hr at 4°.

Step 2. Centrifuge mixture at 10,000 *g* for 20 min at 4°, decant, and save supernatant. Wash residue once with absolute methyl alcohol, centrifuge, decant, and combine supernatants. Dry residue in air for use in preparing luciferase.

Step 3. Evacuate the combined supernatants to about 50 ml with a flash evaporator, add 100 ml of *n*-butyl alcohol, continue evaporating to about 125 ml, and chill in an ice bath. Centrifuge the mixture at 5000 *g* for 20 min at 4°, decant supernatant, and discard residue.

Step 4. Evacuate the supernatant to complete dryness using a vacuum pump and an ethyl alcohol–CO_2 trap. The last traces of *n*-butyl alcohol should be removed (which may require 2 days of pumping) before proceeding to the next step.

Step 5. Dissolve the residue in 150 ml of ice-cold 0.5 *N* HCl and transfer to a separatory funnel.

Step 6. Extract with 40-ml portions of freshly distilled ethyl ether, discarding the upper ether layer each time and continue extracting until the ether extract becomes clear.

Step 7. Add 20 ml of ice-cold *n*-propyl alcohol and extract successively with 40, 30, 30, and 30 ml of ice-cold *n*-butyl alcohol. Combine the extracts and bubble with argon for 1 hr in a vessel with inlet and outlet ports. Vessel should be immersed in an ice bath during the bubbling.

Step 8. Divide the *n*-butyl alcohol extract into small aliquots (about 10 ml) and evacuate to complete dryness using a vacuum pump and an ethyl alcohol–CO_2 trap. Continue the evacuation for 2 days to remove the last traces of *n*-butyl alcohol.

Step 9. Dissolve the residue in a small volume (about 15 ml) of a solvent mixture consisting of ethyl acetate, ethyl alcohol, and water (5:2:3 by volume, upper layer) and place on a powdered cellulose column, 7.5 × 10 cm, previously equilibrated with the same solvent mixture,[13,14] saturated with argon. Elute with the argon-saturated solvent and collect 10-ml fractions. Luciferin moves as a yellow band, followed by another, much slower-moving, yellow band, inactive with luciferase.

Step 10. Remove 0.01–0.05-ml aliquots from each fraction, evacuate to dryness with a vacuum pump (a few minutes), and assay with *Cypridina* luciferase (see Assay Procedure).

[13] F. I. Tsuji, A. M. Chase, and E. N. Harvey, *in* "The Luminescence of Biological Systems" (F. H. Johnson, ed.), p. 127. Am. Assoc. Adv. Sci., Washington, D.C., 1955.

[14] F. I. Tsuji, *Arch. Biochem. Biophys.* **59**, 452 (1955).

Step 11. Combine the fractions showing the highest activity and evacuate to dryness with a vacuum pump using an ethyl alcohol–CO_2 trap. Continue the evacuation for 2 days in order to remove the last traces of solvent.

The luciferin at this stage can be stored for many months in a vacuum desiccator with a desiccant, e.g., silica gel or Drierite.

Further purification is carried out by rechromatographing on a powdered cellulose column or by means of thick-layer chromatography.[9] The latter procedure involves the following steps.

Step 1. Dissolve a portion of the luciferin in a few drops of absolute methyl alcohol, apply horizontally to a preparative cellulose glass plate (cellulose, MN 300, 0.5 mm) held in a box flushed with argon, and chromatograph with the previously described solvent mixture saturated with argon at 4°. The position of luciferin is readily identified by its characteristic yellow color, yellow fluorescence in long ultraviolet light, and activity with luciferase.

Step 2. Scrape off the cellulose with the luciferin from the plate, stir the scrapings in spectral grade absolute methyl alcohol, centrifuge for 5 min, and quickly filter the supernatant through an ultrafine sintered-glass filter under negative pressure.

Step 3. Transfer aliquots of the yellow luciferin solution to small, preweighed screw-cap vials and evacuate for 2 days using a vacuum pump and an ethyl alcohol–CO_2 trap.

Step 4. Weigh the vials, replace the caps, and seal in glass tubes with silica gel and argon.

Other methods for the purification and crystallization of luciferin have been described.[15,16]

The structure of luciferin has been determined.[17-19] Luciferin is able to form a free base (molecular weight, 470) and also the dihydrochloride and dibromide salts. The absorption spectrum of luciferin shows peaks at 265 and 430–435 nm, and a shoulder at 310 nm; when oxidized with luciferase, the 430–435 nm peak disappears and a new peak is formed at 365 nm.[13,14] For luciferin in methyl alcohol, $\epsilon_{265} = 21{,}300\,M^{-1}\,cm^{-1}$ and $\epsilon_{433} = 8700\,M^{-1}\,cm^{-1}$.[16] Other spectral data have also been published.[15,17,19]

[15] O. Shimomura, T. Goto, and Y. Hirata, *Bull. Chem. Soc. Jpn.* **30,** 929 (1957).

[16] Y. Haneda, F. H. Johnson, Y. Masuda, Y. Saiga, O. Shimomura, H.-C. Sie, N. Sugiyama, and I. Takatsuki, *J. Cell. Comp. Physiol.* **57,** 55 (1961).

[17] Y. Kishi, T. Goto, Y. Hirata, O. Shimomura, and F. H. Johnson, *Tetrahedron Lett.* **1966,** 3427 (1966).

[18] Y. Kishi, T. Goto, S. Eguchi, Y. Hirata, E. Watanabe, and T. Aoyama, *Tetrahedron Lett.* **1966,** 3437 (1966).

[19] Y. Kishi, T. Goto, Y. Hirata, O. Shimomura, and F. H. Johnson, *in* "Bioluminescence in Progress" (F. H. Johnson and Y. Haneda, eds.), p. 89. Princeton Univ. Press, Princeton, New Jersey, 1966.

There is strong evidence that some of the fish luciferins may be similar, if not identical, to *Cypridina* luciferin.[20-23]

Synthetic Luciferin

The total synthesis of *Cypridina* luciferin has been accomplished by using acrolein and indole to synthesize 2-amino-5-benzamidovaleramidine dihydrobromide and 3-indolylglyoxal, respectively.[24-27] The two are then condensed to obtain 2-amino-3-(3-benzamidopropyl)-5-(3-indolyl)pyrazine. Hydrolysis of the product in methanolic KOH yields etioluciferamine, which when condensed with *S*-ethylisothiourea gives etioluciferin. Etioluciferin is then condensed with D-2-oxo-3-methylvaleric acid to form an intermediate, which upon reduction and dehydration yields luciferin. An improved route for the synthesis of *Cypridina* luciferin from etioluciferin has also been published.[28,29]

Luciferase

Luciferase is prepared either from the residue recovered from the purification of luciferin (step 2) or from freshly ground powder. For critical studies, the luciferin is sacrificed and the ground powder is used directly without defatting. There is no evidence, however, that luciferase prepared from the recovered residue differs in any way from native luciferase.

In the procedure described below,[30] the initial steps in the purification involving fractional precipitation with acetone and ammonium sulfate are essentially the same as those used by McElroy and Chase.[31]

Step 1. Extract 100 g of residue or powder with 800 ml of distilled water while stirring continuously for 18 hr at 4°.

[20] E. H.-C. Sie, W. D. McElroy, F. H. Johnson, and Y. Haneda, *Arch. Biochem. Biophys.* **93,** 286 (1961).

[21] F. H. Johnson, N. Sugiyama, O. Shimomura, Y. Saiga, and Y. Haneda, *Proc. Natl. Acad. Sci. U.S.A.* **47,** 486 (1961).

[22] F. I. Tsuji, Y. Haneda, R. V. Lynch III, and N. Sugiyama, *Comp. Biochem. Physiol.* **40A,** 163 (1971).

[23] F. I. Tsuji, B. G. Nafpaktitis, T. Goto, M. J. Cormier, J. E. Wampler, and J. M. Anderson, *Mol. Cell. Biochem.* **9,** 3 (1975).

[24] Y. Kishi, T. Goto, S. Inoue, S. Sugiura, and H. Kishimoto, *Tetrahedron Lett.* **1966,** 3445 (1966).

[25] S. Sugiura, S. Inoue, Y. Kishi, and T. Goto, *Yakugaku Zasshi* **89,** 1616 (1969).

[26] S. Sugiura, S. Inoue, Y. Kishi, and T. Goto, *Yakugaku Zasshi* **89,** 1652 (1969).

[27] Y. Kishi, S. Sugiura, S. Inoue, and T. Goto, *Yakugaku Zasshi* **89,** 1657 (1969).

[28] S. Inoue, S. Sugiura, H. Kakoi, and T. Goto, *Tetrahedron Lett.* **1969,** 1609 (1969).

[29] S. Sugiura, S. Inoue, and T. Goto, *Yakugaku Zasshi* **90,** 707 (1970).

[30] F. I. Tsuji and R. Sowinski, *J. Cell. Comp. Physiol.* **58,** 125 (1961).

[31] W. D. McElroy and A. M. Chase, *J. Cell. Comp. Physiol.* **38,** 401 (1951).

Step 2. Centrifuge at 750 *g* for 20 min at 4° and dialyze supernatant against many changes of distilled water for 18 hr at 4°.

Step 3. Centrifuge at 20,000 *g* for 20 min at 4°, discard precipitate, and freeze-dry supernatant.

Step 4. Slowly add ice-cold acetone while stirring to a 1.5% water solution of powder (w/v) chilled in an ice bath until the final concentration is 30% (v/v). Allow to stand overnight at 2°.

Step 5. Centrifuge at 750 *g* for 30 min at 4° and discard precipitate.

Step 6. Add cold 0.1 *M* sodium phosphate buffer, pH 5.5, to supernatant in the ratio of 8 ml of buffer to 70 ml of supernatant.

Step 7. Add ice-cold acetone to raise concentration to 60% (v/v) and allow to stand for 1 hr in an ice bath, then centrifuge at 750 *g* for 30 min at 4° and discard supernatant.

Step 8. Dissolve precipitate in 100 ml of cold 0.1 *M* sodium phosphate buffer, pH 6.8, and add solid ammonium sulfate to 40% saturation while keeping flask in ice bath. Allow to stand 30 min.

Step 9. Centrifuge at 750 *g* for 30 min at 4°, discard precipitate, and add solid ammonium sulfate to 60% saturation while still keeping flask in ice bath. Allow to stand 60 min.

Step 10. Centrifuge at 750 *g* for 30 min at 4°, discard supernatant, and dissolve precipitate in a small volume of 0.2 *M* sodium phosphate buffer, pH 6.8.

Step 11. Dialyze against many changes of distilled water at 4° and freeze-dry.

For DEAE-cellulose chromatography,[30] carry out the following steps.

Step 12. Dissolve 100 mg of the above powder in 10 ml of 70 m*M* sodium phosphate buffer, pH 6.8, and dialyze overnight against the buffer.

Step 13. Centrifuge at 750 *g* for 20 min at 4°, discard precipitate, and place supernatant on a powdered DEAE-cellulose column, 1.5 × 37.5 cm, previously equilibrated with 70 m*M* sodium phosphate buffer, pH 6.8.

Step 14. Elute column using a linear gradient of sodium chloride in 70 m*M* sodium phosphate buffer, pH 6.8 (luciferase elutes at a concentration close to 0.35 *M* sodium chloride); collect fractions.

Step 15. Assay fractions for activity (see Assay Procedure), combine active fractions, dialyze at 4° against many changes of distilled water, and freeze-dry. Store powder until needed at −15°.

Further purification of luciferase is achieved by means of preparative acrylamide gel electrophoresis.[32]

Step 1. Dissolve 100 mg of column-purified luciferase in 5 ml of upper buffer (Tris-glycine) containing 10% sucrose and layer over upper gel of preparative acrylamide gel electrophoresis apparatus.[33]

[32] F. I. Tsuji, R. V. Lynch III, and C. L. Stevens, *Biochemistry* **13,** 5204 (1974).
[33] T. Jovin, A. Chrambach, and M. A. Naughton, *Anal. Biochem.* **9,** 351 (1964).

Step 2. Run electrophoresis for approximately 18 hr at 4° and a constant current of 27 mA; collect 7-ml fractions.

Step 3. Assay for activity (see Assay Procedure), combine active fractions, dialyze against many changes of distilled water at 4°, and freeze-dry. Store powder at −15° until needed.

The specific activities of the luciferase purified by DEAE-cellulose chromatography and acrylamide gel electrophoresis are approximately 460 and 500–800 times that of the starting *Cypridina* powder, respectively. The molecular weight is about 68,000, whereas the subunit molecular weight is approximately 12,000. There is also evidence that luciferase requires a divalent metal ion, possibly calcium, for activity.[34] The enzyme is rapidly inactivated by EDTA and EGTA, but not by certain analogs, and spectrographic analysis shows the presence of about five calcium atoms in each molecule. Amino acid analysis reveals a high content of aspartic and glutamic acids. Free sulfhydryl groups do not appear to be present.

A method for purifying *Cypridina* luciferase has been independently developed by Shimomura *et al.*[35,36] Stone has also described a method for purifying luciferase.[2]

There are some similarities between *Cypridina* luciferase and some of the fish luciferases, but differences have also been found in their physical-chemical properties, viz., *Cypridina* luciferase cross-reacts with *Apogon* (fish) luciferin, but differs from *Apogon* luciferase in immunological and chromatographic properties.[37]

Assay Procedure

The optimum pH of the luciferin–luciferase reaction in phosphate buffer is 7.1.[32] The assay is usually carried out in 0.1–0.2 *M* sodium phosphate buffer or 0.1–0.2 *M* Tris · HCl buffer, pH 7.2. Luciferin is dissolved in 0.1 or 0.3 *N* HCl and kept as a working solution in an ice bath. In acidic solution and at low temperature, luciferin keeps its activity reasonably well for 1–2 hr. An even better method is to keep the luciferin solution in a vessel equipped with a stoppered opening, and inlet and outlet ports for flushing with argon. The luciferin is pipetted through the opening. Just prior to initiating the reaction 0.1–0.3 ml of the working solution of luciferin is diluted with buffer. The buffer concentration is made high enough to neutralize the HCl introduced with the luciferin. Luciferase is also dis-

[34] R. V. Lynch III, F. I. Tsuji, and D. H. Donald, *Biochem. Biophys. Res. Commun.* **46,** 1544 (1972).

[35] O. Shimomura, F. H. Johnson, and Y. Saiga, *J. Cell. Comp. Physiol.* **58,** 113 (1961).

[36] O. Shimomura, F. H. Johnson, and T. Masugi, *Science* **164,** 1299 (1969).

[37] F. I. Tsuji and Y. Haneda, *in* "Bioluminescence in Progress" (F. H. Johnson and Y. Haneda, eds.), p. 137. Princeton Univ. Press, Princeton, New Jersey, 1966.

solved in the same buffer. If purified luciferase is used, a stock solution is first prepared in distilled water. This solution is then used to prepare a working solution by diluting with either distilled water or buffer. The stock solution is stored at $-15°$ and is usable for many months. The working solution is also stored at $-15°$, but its activity may decrease significantly. An aliquot of the working solution, usually 0.05–0.1 ml, is diluted with buffer and pipetted into the reaction cell, which is placed in a holder before the photomultiplier within the light-tight compartment of the photometer. The lid of the compartment is closed, and the shutter occluding the photomultiplier is opened. The luciferin solution, previously diluted with buffer, is then withdrawn into a syringe and injected into the luciferase solution through a small hole in the lid.

The concentrations of luciferin and luciferase of the working solutions are determined by the final reaction volume (as determined by the capacity of the cell and the geometry around the photomultiplier) and by the sensitivity of the photometer. To assay for luciferase a saturating concentration of luciferin is injected. The measured light intensity is directly proportional to the luciferase concentration. To assay for luciferin, the emitted light is integrated. The total light is directly proportional to the luciferin concentration. For fractions possessing exceptionally high luciferin or luciferase activity, aliquots are diluted with distilled water; for fractions with low activity, aliquots are assayed directly without dilution. When the luciferin or luciferase concentration is unknown, a few trial runs are usually necessary to estimate the appropriate conditions. It is important that solvents and other dissolved compounds that are potentially inhibitory be removed by evacuation or dialysis, or by diluting with excess buffer, before carrying out the assay. For examples of some assays, see the publications dealing with purification of luciferin and luciferase.[22,32] For an example involving high concentrations of luciferin and luciferase, see a recent publication.[9]

A paper discussing various aspects of luciferin and luciferase assays, and light measurement, has been previously published.[38] This volume also contains an article on light measurement.[39]

[38] A. M. Chase, *Methods Biochem. Anal.* **8**, 61 (1960).
[39] H. H. Seliger, this volume [44].

Section VI

Earthworm Bioluminescence

[34] Assaying Hydrogen Peroxide Using the Earthworm Bioluminescence System[1]

By MICHAEL G. MULKERRIN and JOHN E. WAMPLER

Diplocardia bioluminescence involves a two-step reaction where hydrogen peroxide and *Diplocardia* luciferin (*N*-isovaleryl-3-aminopropanal) react to form a fairly stable intermediate, which subsequently reacts with the protein *Diplocardia* luciferase to produce light and unknown products.[2-4] Since luciferin can be easily synthesized and the preparation of luciferase is straightforward, an assay for hydrogen peroxide using the *Diplocardia* system has been developed and tested. This assay can also be coupled to peroxide producing oxidases to give a convenient system for monitoring their action at low substrate or enzyme concentrations. This bioluminescence assay for hydrogen peroxide is sensitive in the range of nanomolar concentrations of peroxide and has advantages over polarographic methods,[5,6] direct spectrophotometric analysis,[7] and chemical methods[8,9] in its broad range of linearity, tolerance of turbidity, specificity, and ability to measure the rate of peroxide production in coupled assays.

Preparation of *Diplocardia* Luciferase[4]

The bioluminescent earthworm *Diplocardia longa* is found in the southern part of the state of Georgia and can be collected in large numbers during the months from spring to fall. Coelomic cells containing the bioluminescent system[10] are collected by shocking the live worms, using a magneto generator. The worms are suspended in 0.1 *M* potassium phosphate buffer, pH 7.5, for shocking. The coelomic cells are collected by allowing them to settle at 4° in a graduated cylinder. The cells are then

[1] Studies were supported by the National Science Foundation, Grants PCM-74-08663 and PCM-76-15842.

[2] R. Bellisario, T. E. Spencer, and M. J. Cormier, *Biochemistry* **11,** 2256 (1972).

[3] H. Ohtsuka, N. G. Rudie, and J. E. Wampler, *Biochemistry* **15,** 1001 (1976).

[4] N. G. Rudie, Studies of the Physiology and Chemistry of the Bioluminescent Earthworm, *Diplocardia longa,* Ph.D. dissertation, University of Georgia, Athens, 1977.

[5] R. K. Bonnischen, B. Chance, and H. Theorell, *Acta Chem. Scand.* **1,** 685 (1947).

[6] K. G. Schick, V. G. Magearu, N. L. Field, and C. O. Huber, *Anal. Chem.* **48,** 2186 (1976).

[7] R. F. Beers and J. W. Sizer, *J. Biol. Chem.* **195,** 133 (1952).

[8] W. A. Patrick and H. B. Wagner, *Anal. Chem.* **21,** 1279 (1949).

[9] M. L. Issacs, *J. Am. Chem. Soc.* **44,** 1662 (1922).

[10] N. G. Rudie and J. E. Wampler, *Comp. Biochem. Physiol.* **59A,** 1 (1978).

ISBN 0-12-181957-4

washed twice with the same buffer and stored by freezing the combined slurry of cells from 100 worms. Once frozen, the coelomic cells can be stored at −20° for a year with approximately 50% loss in activity.

Fresh or frozen coelomic cells from 100 earthworms are mixed with 1 liter of 0.1 *M* borate buffer ($NaBO_3$, pH 7.5) containing per liter 33 mg of catalase (Sigma), 0.125 g of sodium azide, and 75 mg of dithiothreitol (DTT). The mixture is homogenized with a Waring blender for 3 min at 4°. The blender is used to achieve good mixing, since lysis of the cells occurs during the buffer change. The homogenization time should be kept short to prevent foaming. All subsequent steps are carried out at 4°.

A second batch of cells (100 worms) is then homogenized with a second liter of the homogenization buffer. The combined homogenates are centrifuged at 27,300 *g* for 30 min. The pellets from this centrifugation are combined with a third liter of borate buffer, rehomogenized, and centrifuged. The resulting pellets are discarded, and the combined homogenates are applied to a DEAE-cellulose column (35 × 4 cm) previously equilibrated with the borate buffer (without catalase). DTT is very effective in preventing the formation of slime, which normally occurs upon lysis of the coelomocytes, and thus makes the purification much more tractable. Catalase in the homogenizing buffer prevents bioluminescence as well as the irreversible inactivation of luciferase due to reaction with peroxide.[2,4]

The slime components and luciferin are washed through the DEAE column with 2 liters of borate buffer at a flow rate of 60 ml/hr. A linear elution gradient is then run using the 0.1 *M* borate buffer (1 liter) and 1 liter of the same buffer containing 0.2 *M* sodium chloride. Elution is completed using an additional liter of this latter buffer. The optimum elution rate is about 60 ml/hr. Luciferase fractions that trail throughout the last tubes of this elution are pooled and dialyzed versus three 10-volume changes of 10 m*M* phosphate buffer (potassium phosphate, pH 6.8, with 0.0125 g of sodium azide per liter) with 4 hr of dialysis per volume change. This dialyzate is applied to a 30 × 3.5-cm CM-cellulose column previously equilibrated with the 10 m*M* phosphate buffer and eluted with a linear gradient using 1 liter of this buffer and 1 liter of the same buffer containing 0.2 *M* sodium chloride. The pooled active fractions are dialyzed as described above and then applied to a hydroxyapatite column (20 × 2.5 cm). The column is washed with 1 liter of 10 m*M* phosphate buffer, and the activity is eluted with 0.2 *M* phosphate, pH 7.5, at 20 ml/hr maximum rate.

Diplocardia luciferase prepared by this procedure is dialyzed versus 0.1 *M* phosphate buffer, pH 7.5, with 0.125 g of sodium azide per liter. It can be kept dialyzing in this buffer for several weeks with slow loss of activity. A unit of luciferase activity[2] has been defined as the amount of enzyme needed to produce 10^{11} photons per second in a standard assay involving 0.02 ml of luciferin (0.25 mg/ml in methanol), luciferase, and 0.1 *M* phosphate buffer (pH 7.5) with a total volume of 0.9 ml and initiating the reaction by injection of 0.2 ml of 22 m*M* H_2O_2.

Preparation of *Diplocardia* Luciferin[3]

Diplocardia luciferin can be readily prepared from commercial isovaleryl chloride (Aldrich) and 3-aminopropanol (Aldrich). Isovaleryl chloride, 1.2 g dissolved in 5 ml of benzene, is added dropwise to a mixture of 2.7 g of 3-aminopropanol and 0.7 g of potassium carbonate in 45 ml of acetonitrile at $-20°$. The reaction mixture is stirred for 30 min at $-20°$ and for 30 min longer at room temperature. It is then filtered and evaporated to dryness under an exhaust hood. The remaining oily residue is dissolved in chloroform and washed with 1 *N* HCl in water, 10% potassium carbonate solution, and pure water. The remaining chloroform solution is dried for 3 hr over sodium sulfate, filtered, and evaporated to dryness with 95% yield of the luciferin alcohol (1.51 g of *N*-isovaleryl-3-aminopropanol).

Luciferin alcohol from several preparations (7 g) is dissolved in 44 ml of methylene chloride and oxidized to the aldehyde by rapidly adding a solution of 13.5 g of pyridinium chlorochromate (98%, Aldrich) in 132 ml of methylene chloride. This mixture is stirred at room temperature for 2 hr. The solution is then diluted to 1 liter with anhydrous ether and filtered through silica gel (6 × 14 cm). The residue from the reaction vessel is taken up in 30% ethanol in hexane and put through this same column. The column is washed with 4 liters of 30% ethanol in hexane and the total eluate is flash-evaporated to yield a brown, viscous oil. This oil is further purified by filtering it through a second silica gel column (35 × 4 cm) and washing with 2 liters of ethyl acetate. The total eluate is again flash-evaporated to yield a yellow oil. Two additional passes through silica gel (2 × 50 cm) eluting with 2.5% methanol in chloroform give luciferin free of colored contaminants, and flash evaporation yields a colorless oil (60% yield).

Assays for luciferin activity during the preparation are tedious, involving drying of a portion of each fraction before assaying. Once specific columns have been used, however, repetitive purification can be carried out with only a few spot checks to ensure that luciferin is eluting in the proper fractions.

Assay

Procedure

Assays are carried out using a photometer of our own design.[11] The cell compartment includes a masked cell holder for 18 × 150 mm disposable

[11] J. E. Wampler, *in* "Analytical Applications of Bioluminescence and Chemiluminescence" (E. W. Chappelle and G. L. Picciolo, eds.), p. 105 (NASA Publ. No. SP-388). National Aeronautics and Space Administration, Washington, D.C., 1975.

test tubes. All solutions are made up and stored at 4°. For each day's assays, fresh luciferin solutions are made up using 0.25 mg/ml in 0.1 *M* potassium phosphate buffer (pH 7.5) and a peroxide standard curve is obtained appropriate for the concentration range to be measured and the luciferase concentration to be used. Peroxide solutions for the standard curve are prepared by volumetric dilution of stock peroxide (30%, Fischer) where the actual titer of hydrogen peroxide in the stock has been determined by volumetric titration.[12]

The assay procedure is as follows.

1. One hundred microliters of the sample solution in 0.1 *M* phosphate buffer are placed in an assay tube along with 20 μl of luciferin solution. This mixture is incubated at ice temperatures for at least 30 sec prior to placing it in the photometer.

2. The tube is placed in the photometer, closing the cell compartment and opening the shutter, and then a base line signal is recorded.

3. Fifty microliters of luciferase solution (minimum concentration of 1 unit/ml) is injected using an automatic syringe (Hamilton Co., Model CR-700-200).

The light signals are recorded on a strip-chart recorder (Gulton, Model TR-711) and integrated using digital techniques.[11] Since absolute light yields vary with the specific activity of the luciferase preparation, all assays in a given series should be performed with the same enzyme preparation.

Linearity and Sensitivity

Under the standard conditions described above the reaction has nearly ideal kinetics for analytical purposes with a slow rise to peak light intensity followed by a long emission plateau lasting for minutes. Both the peak intensity and the steady-state intensity can be used for assay (Fig. 1). Assay linearity is maintained over many orders of magnitude (Fig. 2).

Preincubation of the sample and luciferin is required, but the incubation can be short (30 sec) and the assay is insensitive to variable incubation times from 30 sec to several minutes.

Effects of Extraneous Materials

A variety of organic and inorganic solutes have been added to the assay reaction to test its sensitivity to extraneous solutes. Table I shows the effects of these reagents on the assay. Of the substances tested, only two inhibiting conditions were found, inhibition by ferric ions and inhibition in

[12] W. W. Scott, *in* "Standard Methods of Chemical Analysis," p. 1539. Van Nostrand-Reinhold, Princeton, New Jersey, 1922.

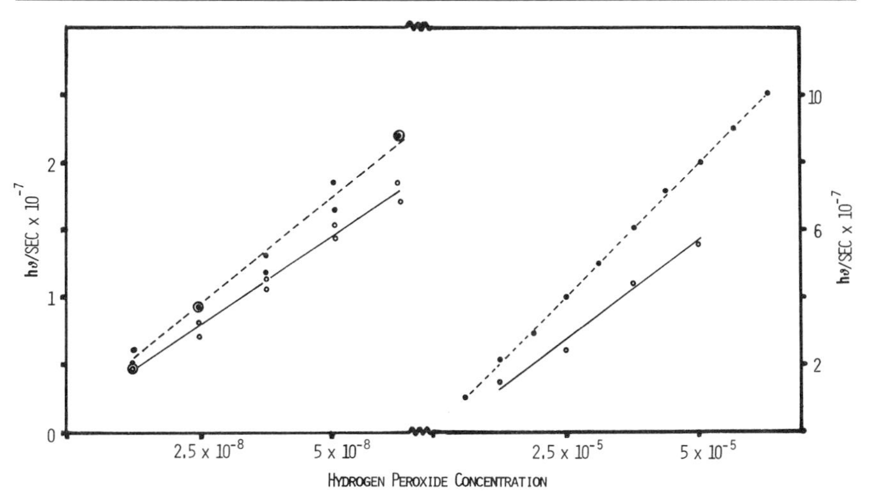

FIG. 1. Linearity of the hydrogen peroxide assay using the *Diplocardia* bioluminescence system. Dashed lines are for peak height measurements (●), and solid lines are for the plateau region of the emission profile (○). Reactions are carried out at 4° in 0.1 *M* potassium phosphate buffer, pH 7.5, using preincubated 0.12-ml samples containing the peroxide and luciferin (0.005 mg). Reactions were initiated by injection of 0.05 ml of luciferase solution. For the data at the left, the concentration of luciferase used was 12 units/ml; for the data at the right, 0.4 unit/ml was used. Concentrations of peroxide on the graph are the actual concentration in the assay.

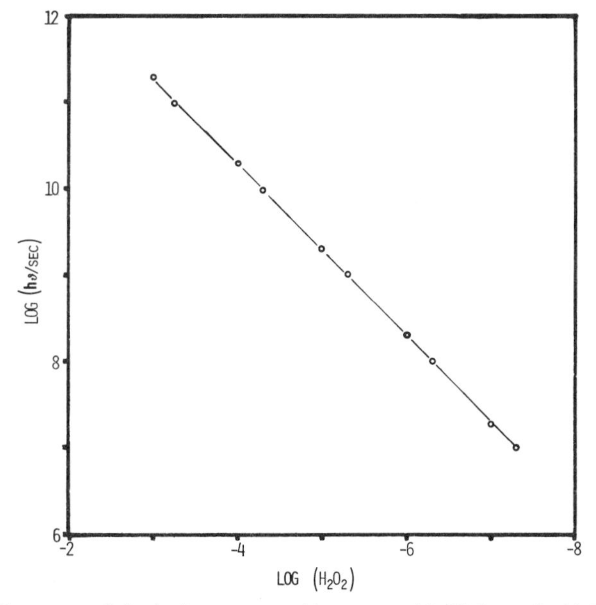

FIG. 2. The range of the hydrogen peroxide assay with *Diplocardia* bioluminescence. Averages of five assays each with the experimental conditions given in Fig. 1, using a luciferase solution containing 60 units/ml.

TABLE I
EFFECTS OF ADDED SOLUTES ON EARTHWORM BIOLUMINESCENCE ASSAY FOR HYDROGEN PEROXIDE[a]

Conditions	Range of triplicate assays (% of standard assay)
Variations of buffer (0.1 *M*, pH 7.5)	
Potassium phosphate (standard assay)	100
+ 0.05 *M* NaCl	120–136
+ 0.2 *M* NaCl	120–160
+ 0.5 *M* NaCl	240–256
Tris·HCl	12
+ 0.5 *M* NaCl	56–64
HEPES	75–77
Phosphate + 0.2 *M* $(NH_4)_2SO_4$	123–125
+ 1 *M* $(NH_4)_2SO_4$	156–172
With added metal ions (saturating)	
Fe II	98–103
Fe III	18
Mg	100–105
Mn	115
Zn	105
Ca	98–113
Cu I Cu II	250–300
With other additions	
BSA (6 mg/ml)	93–102
Casein (0.03 %)	117–118
Yeast extract (0.03%)	123–128
Sucrose (3%)	105–109

[a] Standard assay using 0.12 ml of 0.1 *M* phosphate buffer, pH 7.5, containing 0.03 mg of luciferin per milliliter and 60 μM H_2O_2 with the reaction initiated by injection of 0.05 ml of luciferase solution.

organic buffers. Both of these inhibitions are explained by the properties of the assay. Ferric ion inhibition should be expected owing to its action to degrade hydrogen peroxide. The buffer inhibition appears to be due to an ionic strength effect that can be demonstrated both in phosphate to a small extent and in Tris buffers (see Table I). Other inhibitors have been tested[2,4] with strong inhibition by metal chelators explained by the copper content of luciferase.

Since phosphate buffer limits the solubility of extraneous metals and shows no pronounced salt effect, 0.1 *M* phosphate, pH 7.5, was chosen as the standard assay buffer. Under these conditions saturating concentrations of the metal ions tested (Table I) had no effect with two exceptions. When ferric ions are present, the reaction is inhibited as pointed out above, and solutions of copper made up by dissolving copper(I) chloride [copper(I) is not stable in water, but dismutates to form copper(II)] stimulated the reaction several fold.

Extraneous protein and organic substances have been tested (Table I) with stimulation by casein and yeast extracts. With yeast extracts there is a high background indicating the presense or formation of peroxide by these solutions. Even with these materials present, however, assays can be performed over several orders of magnitude of peroxide concentration with good linearity. Taken together, the data of Table I indicate that the bioluminescent assay for hydrogen peroxide is quite insensitive to extraneous solutes.

Coupled Assays

Galactose oxidase and putrescine oxidase have both been coupled to the assay using very low concentrations of these enzymes (1.8×10^{-5} unit/ml for galactose oxidase and 5×10^{-4} unit/ml for putrescine oxidase).[13] The lack of sensitivity to extraneous protein and the tolerance of moderate turbidity inherent in a bioluminescence assay favor its use over those involving a color reaction or nonspecific potentiometric method. A variety of other oxidases should also be assayable with *Diplocardia* bioluminescence.

[13] One unit of galactose oxidase is the amount of enzyme needed to cause a change in absorbance at 425 nm of one per minute at 25° in the coupled assay using peroxidase and *o*-toluidine (Sigma). A unit of putrescine oxidase is defined as the amount of enzyme required for conversion of 1 n*M* of oxygen per second.

Section VII

Pholas dactylus

[35] Purification and Properties of *Pholas dactylus* Luciferin and Luciferase

By A. M. MICHELSON

The luminescent mollusk *Pholas dactylus* was once prevalent in the entire Mediterranean Basin and on the Atlantic coast of Europe. Until lately, the species could be found in a few isolated areas, particularly on the coast of Brittany (France) but is now in the process of total extermination. This ecological and scientific disaster is due to several reasons, including consumption of the clams as a gastronomic delicacy and their use as a luminescent bait by fishermen, but perhaps the most fatal aspect has been pollution of the coastal regions, particularly by oil tankers. Thus in November 1976 no collections could be made since all the animals were dead in the interior of their holes bored in the rocks. Since we have not as yet been able to find new colonies, all work on this system (which with respect to bioluminescence presents a variety of interesting, perhaps unique, details) has been stopped. This report must thus be mainly of theoretical interest, and is probably an epitaph.

Pholas dactylus luminescence was repeatedly examined by R. Dubois in the period 1885–1925, but apart from a few notes on λ_{max} of emission and the effects of reducing agents on crude extracts,[1-3] in recent times biochemical studies have been pursued exclusively in Paris.

Bioluminescence is due to mucal excretions from specialized cells that are localized in specific organs as described by Harvey;[4] details of ultrastructure and morphology of these organs have been presented by Bassot.[5] Stimulation of the animal thus results in production of a protective cloud of light.

Assay Techniques

Luciferin. Purification of luciferin can be followed by oxidation with a standard solution of luciferase and measurement of light output (I_{max} and total light emitted in a given time) the reaction being initiated by injection of

[1] W. D. McElroy and H. H. Seliger, *Adv. Enzymol.* **25,** 119 (1963).

[2] H. H. Seliger and W. D. McElroy, "Light: Physical and Biological Action," p. 168. Academic Press, New York, 1965.

[3] P. E. Plesner, *Publ. Staz. Zool. Napoli* **31,** 43 (1959).

[4] E. N. Harvey, "Bioluminescence," p. 310. Academic Press, New York, 1952.

[5] J. M. Bassot, *in* "Bioluminescence in Progress" (F. H. Johnson and Y. Haneda, eds.), p. 561. Princeton Univ. Press, Princeton, New Jersey, 1966.

ISBN 0-12-181957-4

the luciferase.[6] A more convenient routine assay is to measure total light emitted on addition of ferrous ions to solutions of the luciferin in phosphate buffer[7] using oxidation by the superoxide radicals produced. The cuvette contains the aliquot of luciferin in 2 ml of 0.15 *M* phosphate buffer, pH 7.0, containing 0.75 *M* NaCl. Reaction is initiated by injection of 1 ml of a degassed aqueous solution of 0.3 m*M* ferrous sulfate. The reaction is complete in 2 or 3 min; total light emitted is proportional to the luciferin present in the cuvette over a very wide range. With a suitable light-measuring apparatus, femtomole quantities of luciferin can be determined.

Luciferase. Approximate assays can be performed using I_{max} of light emission on injection of a solution of the sample of enzyme in 1 ml of 0.1 *M* phosphate buffer, pH 7.0, containing 0.5 *M* NaCl into a cuvette containing a standard amount of luciferin in 2 ml of the same buffer.[8] However, owing to the complicated nature of the enzymic reaction, I_{max} is not strictly proportional to the concentration of luciferase.[6] Hence purification of the enzyme is routinely followed quantitatively using the peroxidasic activity of luciferase with ascorbic acid and H_2O_2 as substrates.[8] This activity appears to be a true property of the enzyme and is always totally associated with the luciferase activity.

The standard assay mixture contains 0.15 μmole of H_2O_2, 0.1 μmole of ascorbic acid, 10 μmole of Tris · HCl (pH 8.5) and the sample of luciferase in a volume of 1 ml, using stock solutions of 10 m*M* H_2O_2 in H_2O and 10 m*M* ascorbate in 10 m*M* Tris·HCl, pH 7.5, to prepare the mixture of H_2O_2 and ascorbate just before use. Variation of absorption at 265 nm to measure oxidation of ascorbic acid was followed in a Cary 15 spectrophotometer at 25°, the initial velocity being used for the determination of activity. Before addition of the enzyme, the nonenzymic oxidation was recorded, and this rate was used as a blank. A unit of activity is defined as the amount of enzyme that induces a decrease of 0.1 OD unit at 265 nm/min at 25°. The assay is linear between 0.2 and 10 units.

The activity can also be expressed in arbitrary light units. Calibration of the light-measuring apparatus[6] according to Lee *et al.*[9] or with the light standard of Hastings and Weber[10] gave results that differ by a factor of about 3. One standard light unit is thus equivalent to 6.1×10^8 photons/ml if the Lee *et al.* calibration is used or to 20.7×10^8 photons/ml according to the Hastings and Weber standard source.

[6] J.-P. Henry, M. F. Isambert, and A. M. Michelson, *Biochim. Biophys. Acta* **205,** 437 (1970).

[7] J.-P. Henry and A. M. Michelson, *Biochimie* **55,** 75 (1973).

[8] J.-P. Henry, C. Monny, and A. M. Michelson, *Biochemistry* **14,** 3458 (1975).

[9] J. Lee, A. S. Wesley, J. F. Ferguson, and H. H. Seliger, *in* "Bioluminescence in Progress" (F. H. Johnson and Y. Haneda, eds.), p. 35. Princeton Univ. Press, Princeton, New Jersey, 1966.

[10] J. W. Hastings and G. Weber, *J. Opt. Soc. Am.* **53,** 140 (1963).

Purification of Luciferase[8]

An acetone powder is first prepared,[6] which can be stocked at −20° without loss of activity. This powder is used for the preparation of both luciferase and luciferin.

About 100 pholades are dissected in ice or on a cold plate to remove the luminescent parts of the siphon and the poli triangles, avoiding excitation of the organs by cutting around, but not into, the specialized regions (visible as distinct white areas). This material (45 g) is then mixed at high speed with 800 ml of acetone precooled to −20° in a Waring blender for several minutes and then filtered, the residue being washed twice with cold acetone. After drying in a desiccator *in vacuo,* about 10 g of powder are obtained.

Subsequent operations are conducted at 4°. The acetone powder (10 g) is homogenized in 150 ml of 10 m*M* Tris · HCl buffer, pH 7.5, in a Polytron grinder (Kinematica GMBH, Lucerne) at full speed for 1 min then centrifuged at 28,000 *g* for 15 min (Sorvall). A second homogenization of the residue, in 80 ml of buffer, is effected and the combined supernatants are pooled. This extract (step 1) is adjusted with 2 *M* NaCl and 1 *M* acetate buffer, pH 4.8, to 0.1 *M* NaCl in 50 m*M* acetate. After 5 min, precipitated proteins are removed by centrifugation (15 min at 28,000 *g*) and discarded; the supernatant is adjusted to pH 6.5 with 0.6 *N* NH_4OH (step 2). This solution is applied to a column of DEAE-Sephadex A-50 (3 × 24 cm) preequilibrated with 0.1 *M* NaCl in 50 m*M* Tris · HCl, pH 7.5. The column is washed with 200 ml of the same buffer, then a gradient of NaCl (0.1 to 0.6 *M*) in 50 m*M* Tris · HCl, pH 7.5, is used to elute the luciferase. Two peaks of activity are eluted: the first corresponds to free luciferase, and the second to a stable complex of luciferase and inactivated luciferin (step 3). Both fractions can be further purified by the following processes, which describe a typical run for the second (major) activity.

Solid ammonium sulfate is added to the pooled activity of the second peak to a concentration of 210 g/liter; the mixture is stirred, then centrifuged for 15 min at 28,000 *g*. The residue is discarded, and solid ammonium sulfate (210 g/liter) again is slowly added to the supernatant with stirring. Precipitated protein is collected by centrifugation and resuspended in 50 m*M* Tris · HCl, pH 7.5, then dialyzed overnight against this same buffer (step 4).

Solid cesium chloride (400 g/liter) is added, and the clear solution is transferred to polycarbonate tubes and centrifuged at 50,000 rpm for 45 hr in a type 65 fixed-angle rotor in a Spinco L2-65 B ultracentrifuge. The contents of each tube is sucked off using a peristaltic pump and fractionated in a suitable fraction collector. Fractions (0.2 ml) are assayed for protein and activity after dilution, using the peroxidase activity of the enzyme (see assay techniques). Active fractions are pooled and poured directly onto a column (3 × 50 cm) of Sepharose 6B (or Sephadex G-100) equilibrated in 50 m*M* Tris·HCl, pH 7.5, and the enzyme is eluted with the same buffer. Active fractions are pooled, then concentrated by use of Diaflo PM-30 membranes; the pure enzyme is stored in small tubes at −20° (step 5).

The results are summarized in Table I. After the final step the enzyme shows a single component in ultracentrifugation and in polyacrylamide gel (5% and 7.5%) electrophoresis. Specific activity is constant throughout the peak of protein in Sepharose 6B chromatography.

Properties of Luciferase[8,11]

Pholad luciferase is a very asymmetric acidic metalloglycoprotein of relatively high molecular weight (310,000) containing 2 g-atoms of copper per molecule of enzyme, at least partially in cupric form as shown by the electron spin resonance spectra. Amino acid and sugar analyses are presented in Table II. The relative abundance of aspartic and glutamic acids undoubtedly contribute to the extremely acidic isoelectric point.

Physical properties of the enzyme are summarized in Table III. The high buoyant density is a result of the sugar content and provides a useful purification step, using centrifugation in a cesium chloride (or sucrose) gradient. Strong asymmetry of the molecule is shown by the frictional ratio of 1.8.

Purification of *Pholas* Luciferin and Oxyluciferin

Early purifications[6] of luciferin yielded material that was homogeneous with respect to protein but of low luminescent activity, since a thermal denaturation step to destroy luciferase activity was used. These conditions also denature about 90% of the luciferin activity. Since luciferase is a copper protein, activity is readily inhibited by diethyldithiocarbamate, and highly active luciferin is obtained by using this reagent in the preliminary steps.[11]

The acetone powder (4 g) obtained from dissected luminous organs (see second paragraph of purification of luciferase) is homogenized in 30 ml of an aqueous solution of diethyldithiocarbamate (1 m*M*) in a Polytron grinder at full speed for 1 min. The suspension is centrifuged at 28,000 *g* for 15 min at 4°, and the supernatants are adjusted to 50 m*M* NaCl–50 m*M* acetate, pH 4.8. After several minutes, precipitated proteins are removed by centrifugation and the supernatant is neutralized with 0.1 *M* phosphate buffer pH

[11] J.-P. Henry and C. Monny, *Biochemistry* **16**, 2517 (1977).

TABLE I
PURIFICATION PROCEDURE FOR *Pholas* LUCIFERASE

Step and fraction	Volume (ml)	Protein concentration (mg/ml)	Total protein (mg)	Activity (U/ml)	Total activity[a]	Specific activity (U/mg)
1. Crude extract	207	12.30	2550	32	6,620	2.6
2. Acid extract	231	9.43	2180	37.6	8,690	4.0
3. DEAE-Sephadex						
1st peak	107	1.26	135	48	5,136	38.1
2nd peak	58	1.86	108	184	10,670	99.0
4. Ammonium sulfate fractionation (2nd peak)						
0–30%	9.3	3.95	36.7	18.8	175	4.8
30–60%	32	2.32	74.2	423	13,540	182.3
5. Cesium chloride gradient and Sephadex G-100	50.6	0.30	15.2	203	10,270	676.7

[a] The increase in total activity has been observed in all preparations.

TABLE II
CHEMICAL ANALYSES OF *Pholas* LUCIFERIN, OXYLUCIFERIN, AND LUCIFERASE

Component	Luciferin (No. of residues)	Oxyluciferin (No. of residues)	Luciferase (No. of residues)
Amino acids			
Aspartic acid	31.2	31.3	338.3
Threonine	20.0	18.2	175.8
Serine	7.1	7.6	167.0
Glutamic acid	25.0	25.9	257.7
Proline	5.3	7.2	109.0
Glycine	13.2	13.7	257.1
Alanine	9.8	10.2	150.9
Valine	14.6	12.9	101.6
Methionine	8.3	10.4	76.9
Isoleucine	9.0	8.2	130.2
Leucine	9.9	9.1	173.1
Tyrosine	7.5	7.2	90.1
Phenylalanine	8.1	7.2	90.5
Histidine	0	0.3	53.4
Lysine	7.8	4.9	127.0
Arginine	8.6	7.9	152.1
Cysteine	8.5	6.8	61.6
Tryptophan	6.2	6.2	56.5
Sugars			
Glucosamine	11.1	11.3	39.4
Fucose	9.8	10.6	86.4
Mannose	7.1	6.6	57.0
Galactose	5.2	5.1	20.9
Sulfate	2.6	3.3	

7.0. Crude luciferin is separated from this solution by precipitation between 50 and 80% saturated ammonium sulfate (the precipitate at 50% saturation being discarded).

This material is further purified by filtration through Sephadex G-100. About 10 mg of protein in 3 ml of 10 m*M* phosphate buffer, pH 7.0, are applied to a column of Sephadex G-100 (50 × 2.5 cm), and the luciferin is eluted with the same buffer. Active fractions are collected and pooled. The final step involves purification on DEAE-Sephadex A-50. Material (25 mg of accumulated crude luciferin) after Sephadex G-100 filtration is adjusted to 0.1 *M* NaCl–50 m*M* phosphate, pH 7.0, and applied to a column of DEAE-Sephadex (23 × 1.5 cm) preequilibrated in the same buffer. After washing with 40 ml of buffer, a gradient of NaCl (0.1 to 0.8 *M*) is applied to elute the luciferin. Active fractions are pooled and dialyzed against 10 m*M* phosphate buffer, pH 7.0; the pure luciferin is stored frozen at −20°.

TABLE III
PHYSICAL PROPERTIES OF *Pholas* LUCIFERASE, LUCIFERIN, AND OXYLUCIFERIN

	Luciferase	Luciferin	Oxyluciferin
Molecular weight	310,000	34,600	34,600
Subunits	2(MW 150,000)	1	1
Copper atoms/molecule	2.18 ± 0.13	0	0
Sugars (%)	10.8	16.7	16.9
$E^{1\%}_{1\,cm}$ 280 nm	14.8	14.9	12.7
Isoelectric point	< 3.5	< 3.5	< 3.5
Partial specific volume	0.707	0.707	0.707
Sedimentation coefficient (S)	10.7	3.05	3.05
Stokes' radius a (Å)	83	29.8	29.8
Frictional ratio f/f_0	1.8	1.4	1.4
Diffusion constant ($\times 10^{-7}$/cm^2 sec^{-1})	—	7.2	7.2
Apparent buoyant density	1.340	—	—
Thermal stability	45°	60°	—
ΔH^* of thermal denaturation	55 kcal/mol	—	—
pH Stability	3–9	—	—
pH Optimum: luciferase	8	—	—
peroxidase	5 and 8	—	—

Oxyluciferin. Acetone powder (1 g) is dispersed in 30 ml of 10 mM phosphate buffer, pH 7.0, in a Polytron grinder, and the homogenate is clarified by centrifugation. The supernatant is kept at room temperature until essentially no luciferin activity (light emission) is present, and the oxyluciferin formed (by oxidation due to the luciferase also present in the extract) is purified by the above techniques for luciferin, using disc gel electrophoresis to follow the purification.

Properties of Luciferin and Oxyluciferin[8,11]

Like *Pholas* luciferase, the substrate luciferin is a glycoprotein. The molecule is a single unit of molecular weight 34,600. In addition to amino acids and sugars, three sulfate residues are present, which account for the strongly acidic nature of the luciferin. Chemical analyses of luciferin and oxyluciferin are given in Table II, and physical properties are summarized in Table III. The similarity in chemical composition with luciferase may be noted.

The quantum yield for purified luciferin is 0.09 using calibration as described by Lee *et al.*[9] or 0.27 according to the Hastings and Weber standard.[10] These figures represent lower limits since, under nondenaturing conditions, oxyluciferin cannot be separated from luciferin and in gel electrophoresis both migrate at the same rate. Hence even the most active preparations probably contain a certain percentage of oxy (or inactivated) luciferin. Under strong denaturing conditions (8 *M* urea, β-mercaptoethanol) oxyluciferin and luciferin can be separated by polyacrylamide gel electrophoresis.[11] The characteristics of migration in different concentrations of acrylamide indicate that this separation is due not to a difference in molecular weight, but rather to a charge difference. (It may be noted that the mobility of glycoproteins is not uniquely a function of molecular weight.) The characteristically faster band (R_m 0.62) of oxyluciferin (under conditions of denaturation) is only obtained when luciferin (R_m 0.55) is oxidized with concomitant light emission, for example with luciferase, horseradish peroxidase or chemically with Fe^{2+}/O_2/phosphate. If luciferin is denatured by heating at 65°, the fast band corresponding to oxyluciferin is not obtained, and this product is now completely inactive both with respect to light emission and conversion to oxyluciferin.[11]

Prosthetic Group. Since amino acid or sugar components of luciferin cannot give rise to a light emission centered about 490 nm, it is clear that the glycoprotein must contain a prosthetic group (probably of small molecular weight) that undergoes oxidation and is indeed the true luciferin. So far, attempts to separate this molecule have been unsuccessful,[6] in part owing to shortage of material. The integrity of the protein appears to be essential—thermal denaturation or proteolytic treatment yield inactive products.[6] However, this may well be due to an extreme instability of the luciferin prosthetic molecule when liberated from its protective glycoprotein carrier. That this group exists has been shown by various spectroscopic studies.[6,11,12] Thus apart from the typical protein absorption spectra characteristics, luciferin shows a band of absorption at 307 nm with an ϵ = 11,800. During oxidation to oxyluciferin catalyzed by luciferase, this band disappears and concurrently a band at longer wavelengths (355–360 nm) appears. The difference spectrum between oxyluciferin and luciferin is shown in Fig. 1. Similar changes can also be seen when luciferin is oxidized by horseradish peroxidase or by dialysis of luciferin against 5 *M* guanidine.[12] Fluorescence spectra of luciferin show a typical protein (tryptophan) fluorescence with excitation at ~290 nm and emission at ~345 nm.[6] In addition, a weaker second fluorescence can be observed at longer wavelengths[12] with excitation λ_{max} at 325 nm and emission λ_{max} at 400 nm. During conversion to oxyluciferin these fluorescence characteristics disappear and thus probably represent fluorescence of the prosthetic group. When luciferin is dialyzed against 5 *M* guanidine, or when oxyluciferin is

[12] J.-P. Henry, M. F. Isambert, and A. M. Michelson, *Biochimie* **55,** 83 (1973).

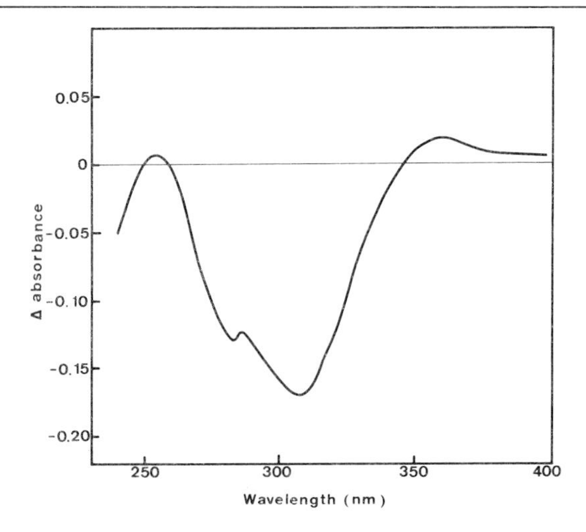

FIG. 1. Changes in absorption spectrum of luciferin (14.5 μM) during oxidation by luciferase (1.1 μM). The difference spectrum between a cuvette containing the mixture of luciferin and luciferase in 0.9 ml of 0.1 M Tris·HCl buffer, pH 8.5, and 0.5 M NaCl and two cuvettes (in the reference compartment) containing separated solutions of luciferin and luciferase at the same concentrations was recorded with a Cary 14. Four additions of ascorbate (3 μM) were made to the luciferin–luciferase cuvette. The spectrum was recorded after 120 min and corrected for variation of the base line.

denatured (5 M guanidine followed by dialysis against water), a much more intense fluorescence with excitation at 365 nm and emission at 460 nm is observed. This is possibly characteristic of the emitter molecule covalently linked to the glycoprotein. The relationship with enzymic bioluminescent emission (λ_{max} 490 nm) is unclear, particularly since this fluorescence can be seen only after denaturation of oxyluciferin, and the native protein appears to quench excited singlet emission.

Reaction Luciferin–Luciferase; Protein–Protein Interactions; Mechanism

Both components of the system, enzyme and substrate, are glycoproteins. The luciferin is readily identified since with a constant amount of this protein, addition of the second in varying amounts gives rise to a constant total light output, whereas, with a given quantity of luciferase, total light emitted is a linear function of the amount of luciferin added. Optimal pH of the reaction is at pH 8–9 and I_{max} (i.e., V_{max}), but not total light emitted, is dependent on ionic strength, with an optimum at 0.5 M NaCl.

The spectrum of emission[6,12] is broad, λ_{max} 490 nm, with a half-width of the band of 95 nm. Oxygen is necessary for this emission, since if luciferin and luciferase are mixed under strictly anaerobic conditions no luminescence is observed. However, formation of a stable enzyme–substrate complex during anaerobic incubation is suggested by the fact that admission of O_2 after several minutes gives rise to a high-intensity flash during the first minute followed by the normal slow reaction (which can last several days at 25°). The reaction can therefore be expressed:

$$LH_2 + E \rightarrow E - LH_2 \xrightarrow{O_2} \longrightarrow \text{product}^* \rightarrow \text{product} + h\nu$$

where LH_2 represents the luciferin.

This complex can also be accumulated in the presence of oxygen by preincubation at acid pH (4.8) at 0°. Injection of the solution into phosphate buffer (pH 7.0) at 20° again gives a biphasic emission, in which the second, slow reaction is unchanged but is preceded by a rapid flash emission. The intensity of this flash increases with the time of preincubation at pH 4.8, with a maximum at about 30 min at 0°; i.e., under these conditions formation of the complex enzyme–substrate is slow. It may be noted that decay of the fast emission is exponential and the rate constant is independent of the concentration of luciferase, in accord with the concept of a preformed intermediate. Indeed this intermediate can be isolated[6] by filtration of a mixture of luciferin and luciferase on Sephadex G-100 at pH 4.8. As mentioned earlier, purification of luciferase gives rise to a major fraction in which the enzyme is complexed with luciferin (shown by gel electrophoresis in sodium dodecyl sulfate; apart from the enzyme subunits of MW 150,000, a second band of MW 35,000 is observed). During purification this complex is converted to an oxyluciferin (or inactivated) complex in absence of dissociation. It is indeed formation of these enzyme–substrate, enzyme–product complexes (and the limited dissociation) which complicates the kinetics of light emission in terms of I_{max}, which is not directly proportional to the amount of luciferase with a constant amount of luciferin, but is somewhat sigmoid.[6] Addition of oxyluciferin inhibits the enzymic reaction due to formation of an inactive complex with luciferase.

Protein–Protein Interactions.[11] The stoichiometry and kinetic and thermodynamic characteristics of the interactions between luciferase and luciferin, oxyluciferin or the inactivated form of luciferin (as isolated as a 1:1 complex with luciferase during purification of the enzyme) are presented in Table IV. The strong interactions are probably due to the fact that associations between highly asymmetric glycoproteins are involved. In the case of studies with oxyluciferin, the molecule was radioactively labeled with ^{125}I, whereas results with luciferin were obtained at pH 4.8 (to prevent enzymic oxidation during the determinations). At this pH the complex luciferin–luciferase has a half-life of $>$ 120 min at 0°. The quantity of

TABLE IV
KINETIC AND THERMODYNAMIC CHARACTERISTICS OF *Pholas* PROTEIN–PROTEIN INTERACTIONS

	Luciferin–luciferase complex	Oxyluciferin–luciferase complex	Inactive luciferin–luciferase complex
Stoichiometry	2	2	1
Dissociation equilibrium constant k_d (M)	1.2×10^{-11}	1.7×10^{-8}	2.4×10^{-11}
Association rate constant k_a (M^{-1} sec^{-1})	1.6×10^{7}	1.6×10^{7}	1.6×10^{7}
Dissociation rate constant k_d (sec^{-1})	1.9×10^{-4}	0.27	3.8×10^{-4}
Half-lifetime $t_{1/2}$ (sec)	3600	2.6	1800
ΔG°_a (kcal mol^{-1})	—	−10.6	—
ΔH°_a (kcal mol^{-1})	—	+3.5	—
ΔS°_a (cal mol^{-1} deg^{-1})	—	+47.3	—

complex formed is readily estimated by measurement of the flash obtained when a pH jump to 8.5 is effected. Titration of a constant quantity of enzyme to saturation gives a stoichiometry of 2 molecules of luciferin per molecule of enzyme. Similar titration of the luciferase–inactive luciferin complex (isolated during purification of luciferase) with active luciferin gave a ratio of 1:1; i.e., in the final complex the luciferase is bound to one active luciferin and one inactive molecule when short (5 min) preincubation times are used. With longer times, the inactive luciferin is displaced and a 2:1 luciferin–luciferase complex is formed. The half-time for displacement of inactive luciferin from the second site by luciferin (at pH 4.8 and 0°) is about 30 min with a rate constant of 3.8×10^{-4} sec^{-1}.

Both luciferin and oxyluciferin form 2:1 complexes with luciferase with similar association constants. Indeed, when competition is observed using mixtures of luciferase, luciferin, and oxyluciferin (using preincubation at pH 4.8, a pH jump to 8.5 to measure the luciferin bound to luciferase giving rise to a flash, followed by chemiluminescent oxidation of residual free luciferin by subsequent addition of ferrous ions), the ratio of the two association rate constants is 0.94. Nevertheless, despite the very minor differences between luciferin and oxyluciferin (slight change in charge) the two complexes are quite distinct with respect to dissociation rate characteristics and dissociation equilibrium constants (Table IV). Confirmation of these differences is obtained by experimental determination of the ratio of the equilibrium dissociation constants using displacement of oxyluciferin from the preformed complex by luciferin. This gives a value k_d oxyluciferin/k_d luciferin of about 1300, equilibrium being attained after 8 hr.

Luciferase thus contains two equivalent and independent binding sites with equal affinity (Scatchard plots of the association luciferase–oxyluciferin, single discontinuity in titration curves of luciferin with luciferase, blockage of one site by an inactive form of luciferin which reduces the activity by a factor of 2). The two sites are probably related to the dimeric nature of the enzyme, each subunit possessing an active site containing one atom of copper and a binding site for one molecule of substrate or product. The complex between luciferase and the inactive form of luciferin closely resembles the enzyme–substrate complex rather than oxyluciferin–luciferase. Indeed, while luciferin can displace the inactive luciferin from the complex, considerable difficulty occurs when preparation of free luciferase is attempted by simple dissociation techniques in the absence of added luciferin.

The enzymic turnover sequences may be expressed

$$\begin{array}{lll}
E + LH_2 \longleftrightarrow & E - LH_2 & \longrightarrow E - \text{oxy L} \\
E - LH_2 + LH_2 \longleftrightarrow & E - 2\,LH_2 & \longrightarrow E - 2\text{ oxy L} \\
E - 2\text{ oxy L} \longleftrightarrow & E - \text{oxy L} + \text{oxy L} & \\
E - \text{oxy L} + LH_2 \longleftrightarrow & E - LH_2 - \text{oxy L} & \\
E - LH_2 - \text{oxy L} \longleftrightarrow & E - LH_2 + \text{oxy L} & \\
E - LH_2 - \text{oxy L} + LH_2 \longleftrightarrow & E - 2\,LH_2 + \text{oxy L} &
\end{array}$$

etc.

Since, in absence of other factors, the half-life of the luciferase–oxyluciferin complex is relatively long (though very much shorter than that of the enzyme–substrate complex), a rapid catalytic reaction is excluded. (Since I_{max} is a function of ionic strength it is likely that salt concentration has a considerable effect on the association and dissociation characteristics of the different complexes.) Thermodynamically the association between luciferase and oxyluciferin has positive enthalpy and entropy, and formation of the complex is an entropy-driven process. The association rate constant approaches diffusion rates, and thus the reaction may well be diffusion limited. It is likely that the sugar residues in both proteins play an important role. These interactions have been compared with other strong protein–protein complexes.[11]

Mechanism. Luciferase catalyzes the oxidation of luciferin by O_2 with emission of light. This luminescence is inhibited by superoxide dismutase[13] at pH 8 (but not at pH 7). The energy of activation ΔH^* of the enzymic oxidation of luciferin is about 8.6 kcal and closely resembles that obtained for chemiluminescence of luciferin on oxidation with $O_2^{\cdot-}$ (electrolytic or via Fe^{2+}/O_2/phosphate) or with horseradish peroxidase acting as an

[13] A. M. Michelson and M. F. Isambert, *Biochimie* **55**, 619 (1973).

oxidase.[13] This suggests that the same activated intermediate is involved and that the kinetics of decay under flash conditions represent degradation of this intermediate. The reaction may be written:

$$\begin{aligned}
E - Cu^{2+} + LH_2 &\longrightarrow E - Cu^{+} \ldots LH^{\cdot} + H^{+} \\
E - Cu^{+} \ldots LH^{\cdot} + O_2 &\longrightarrow E - Cu^{2+} \ldots LH^{\cdot} \ldots O_2^{\cdot -} \\
E - Cu^{2+} \ldots LH^{\cdot} \ldots O_2^{-} &\longrightarrow E - Cu^{2+} + LHOO^{-} \\
LHOO^{-} &\longrightarrow \text{product} + h\nu
\end{aligned}$$

At pH 7, the $O_2^{\cdot -}$ is enzyme bound since superoxide dismutase does not inhibit, but at alkaline pH some dissociation occurs.

In the absence of reducing agents, turnover is extremely slow, and indeed it can be shown that the enzyme oxidizes only two molecules of substrate, corresponding to the two active sites.[11] However, the reaction is very strongly stimulated on addition of reducing agents such as dihydroxyfumaric acid, pyrogallol, catechol, or ascorbic acid.[13] Addition of limiting amounts of ascorbate produces a limited amount of light, and this is correlated with a decrease in absorption of the 307-nm band characteristic of luciferin; these effects can be followed several times on successive additions of ascorbate, which is consumed in the reaction.[11] In the case of addition of dihydroxyfumaric acid, the stimulated emission is strongly inhibited by superoxide dismutase even at pH 7, indicating a continuous production of free $O_2^{\cdot -}$ (which can oxidize luciferin nonenzymically) despite the fact that dihydroxyfumarate is not a substrate for luciferase (as is ascorbate in presence of H_2O_2). In the case of ascorbate (and catechol) the stimulated light emission is inhibited by catalase (but not by superoxide dismutase), and it is probable that here the ascorbate plays a role as substrate for the luciferase acting as a peroxidase, facilitating reduction of the enzyme to the cuprous state. With pyrogallol neither $O_2^{\cdot -}$ nor H_2O_2 appear to play a role in the stimulation, which may be due to production of pyrogallol radicals which then react with luciferin. Apart from the induction of free-radical chain mechanisms, a possible function of these effectors may well be to modify the oxyluciferin complexed with luciferase so that binding characteristics are modified and dissociation occurs more readily. In this respect it may be noted that in total absence of reducing agents, oxidation of the first two molecules of luciferin by luciferase is not accompanied by spectral modification at 307 nm, which thus occurs after light emission and is a secondary event that precedes or accompanies dissociation of oxyluciferin from the enzyme product complex.

Luciferase as a Limited Peroxidase

Apart from its bioluminescent activity, *Pholas* luciferase also possesses a peroxidase activity with respect to a small number of substrates.[12] Although typical substrates of horseradish peroxidase such as guaiacol, hydroquinone, catechol, and *p*-phenylenediamine are not significantly oxidized by luciferase in the presence of H_2O_2, pyrogallol is converted to purpurogallin (at a rate some 300 times less than with peroxidase on a weight basis, or about 40 times less on a molar basis). Oxidation of ascorbate by H_2O_2 is also catalyzed by luciferase in a manner strictly comparable with that of horseradish peroxidase. Thus variation of rate with concentration of H_2O_2 gives a linear Lineweaver–Burk plot, and the K_m for H_2O_2 is 16 μM. As with peroxidases, the reciprocal of the square root of the reaction rate is proportional to the reciprocal of the concentration of ascorbic acid at constant H_2O_2. As shown by gel electrophoresis, both luciferase and peroxidase activities are manifested by the same protein band. Two pH optima are shown for the peroxidase reaction, at pH 5 and 8, whereas the luciferase activity shows only the second optimum.[8] Under optimal conditions, the turnover number for oxidation of ascorbate is about 43,400 molecules per minute per molecule of enzyme compared with 246,000 for ascorbic acid oxidase. Turnover of the enzyme as a luciferase, in the absence of H_2O_2, is enormously slower owing to product inhibition. Indeed the ratio of activities luciferase/peroxidase (in arbitrary units) is 1.21 for free luciferase, but this is reduced to 0.58 for the complex of luciferase with inactivated luciferin. Thus formation of the protein–protein complex of luciferase with luciferin, oxyluciferin, or inactivated luciferin does not inhibit the peroxidase activity with H_2O_2 and ascorbate whereas such stable complexes do interfere with the bioluminescent reaction.

Enzymic activity as a peroxidase is inhibited by metal-chelating agents such as diethyldithiocarbamate and 8-hydroxyquinoline and by high concentrations of cyanide (10 m*M*), but not by reagents that complex with cuprous copper (such as bathocuprein sulfonate) even in presence of ascorbate. In the case of diethyldithiocarbamate, the chelator remains attached to the protein after dialysis (as shown by absorption spectra) and can be only partially, and with difficulty, removed by exogenous cupric ions. The mechanism of luciferase as peroxidase probably involves reduction by ascorbate of a Cu^{2+} enzyme to Cu^{+}, followed by reoxidation to Cu^{2+} by H_2O_2 (in absence of formation of $O_2^{\cdot-}$, which is involved in the luciferase mechanism.)

Peroxidase

$$\mathrm{Cu(II) - E + AH_2 \longrightarrow Cu(I) - E + AH\cdot + H^+}$$

$$\mathrm{Cu(I) - E + H_2O_2 \longrightarrow Cu(II) - E + HO\cdot + HO^-}$$

Luciferase

$$\mathrm{Cu(II) - E + LH_2 \longrightarrow Cu(I) - E + LH\cdot + H^+}$$

$$\mathrm{Cu(I) - E + O_2 \longrightarrow Cu(II) - E + O_2^{\cdot-}}$$

$$\mathrm{LH\cdot + O_2^{\cdot-} \longrightarrow LHOO^- \rightsquigarrow P} + h\nu$$

Oxidation of Luciferin by Horseradish Peroxidase[12,13]

Horseradish peroxidase reacts with *Pholas* luciferin with concomitant light emission in the absence of H_2O_2 or mediators such as dihydroxyfumaric acid. Total light emitted is independent of pH between 4.8 and 8.0, but maximal velocity (I_{max}) decreases with increase in pH. High ionic strength is not necessary (in contrast with luciferase). Molecular oxygen is required, and it appears that with this particular substrate the peroxidase (both molecules are glycoproteins) acts as an oxidase. The emission spectrum is identical with that obtained using luciferase, and the first-order decay characteristics of the flash obtained with an excess of peroxidase relative to luciferin are similar ($k = 0.07$ sec^{-1}) to those seen on oxidation of luciferin by $O_2^{\cdot-}$ or the rapid reaction obtained by preformation of a luciferin–luciferase complex. In contrast with the luciferase reaction, oxidation of luciferin by peroxidase is strongly inhibited by ascorbate, catechol, pyrogallol, dihydroxyfumarate, low concentrations of cyanide (0.1 m*M*) and mercaptoethanol. Ferric ions inhibit the oxidation, whereas at low ionic strength manganous ions stimulate. Superoxide dismutase does not inhibit under any conditions attempted, and it is likely that luciferin is in fact oxidized by an enzyme-bound form of $O_2^{\cdot-}$ via a typical Yamazaki type of mechanism.[14]

Chemiluminescence of Luciferin[7,13,15,16]

Injection of ferrous ions or of reduced FMN into mixtures of luciferin and luciferase during the enzymic reaction produces a flash of light imposed on the normal slow reaction.[6] This is not due to regeneration of oxyluciferin, and the flash height is directly related to the quantity of luciferin present in the mixture. Indeed, luciferase is not necessary and the same effect is obtained with solutions of luciferin alone. This is due to production of superoxide radicals[17] by autoxidation of Fe^{2+} (in phosphate buffer) or of $FMNH_2$

$$Fe^{2+} + O_2 \longrightarrow Fe^{3+} \ldots O_2^{\cdot-}$$
$$FMNH_2 + 2\,O_2 \longrightarrow FMN + 2\,O_2^{\cdot-} + 2\,H^+$$

The same chemiluminescent oxidation is obtained when solutions of electrolytically prepared $O_2^{\cdot-}$ are injected, and all three reactions are inhibited by superoxide dismutase at pH 10, but not at pH 7; it is likely that the affinity of luciferin for $O_2^{\cdot-}$ at neutral pH is such that competition by the

[14] I. Yamazaki, L. H. Yokota, and R. Nakajuma, *in* "Oxidases and Related Redox Systems" (T. S. King, H. S. Mason, and M. Morrison, eds.), p. 485. Wiley, New York, 1965.

[15] J.-P. Henry and A. M. Michelson, *Biochim. Biophys. Acta* **205,** 451 (1970).

[16] A. M. Michelson, *Biochimie* **55,** 925 (1973).

[17] A. M. Michelson, *Biochimie* **55,** 465 (1972).

dismutase is much less effective. Indeed titration of the inhibitory effect of superoxide dismutase on light emission induced by Fe^{2+}/O_2/phosphate indicates a cooperative transition in the luciferin at about pH 9.15.

With respect to chemiluminescence of luciferin induced by ferrous ions and O_2 (in the presence of a suitable complexing anion, such as phosphate or pyrophosphate, but not acetate) the kinetics of decay of the flash are first order and independent of the concentration of Fe^{2+}, but the rate constant increases with increase in concentration of NaCl (from 0.055 sec^{-1} in 0.2 *M* NaCl to 0.115 sec^{-1} in 1.4 *M* NaCl at pH 7) as does I_{max}. Ionic strength thus affects the rate of decomposition of the luciferin peroxide intermediate. As mentioned earlier, the ΔH^* of this reaction (8.6 kcal) is constant whether chemical or enzymic oxidation of luciferin is used, as well as the rate of decay. Decomposition of this intermediate is accelerated at alkaline pH, and the rate constant increases markedly between pH 6 and pH 8. It may be noted that owing to the short lifetime of $O_2^{\cdot-}$ radicals, luminescence is not observed when luciferin is added to premixed Fe^{2+} and phosphate ions, but only when phosphate or ferrous ions are added as a final component. The sensitivity of the reaction is such that femtomole quantities (10^{-15} mol) of luciferin or 0.1 nmol quantities of Fe^{2+} can be conveniently estimated. As may be expected, light emission is strongly inhibited by $O_2^{\cdot-}$ scavenging reagents, such as ascorbate, cysteine, β-mercaptoethanol, and sulfite. Apart from the kinetics of emission, the same spectral distribution and quantum yield is obtained whether luciferin is oxidized chemically or enzymically in accord with the hypothesis that a common peroxide intermediate is involved.

Chemiluminescence of *Pholas* luciferin is induced by other metal ions and complexants, such as Fe^{2+}/O_2/dihydroxyfumarate, Co^{2+}/O_2/dihydroxyfumarate, or Co^{2+}/O_2/tetraglycine and related Ni^{2+} systems.[16] While certain of these systems resemble Fe^{2+}/O_2/phosphate in that $O_2^{\cdot-}$ is produced, in several cases the reaction involves direct reaction of a free radical (organic or hydroxyl radicals produced indirectly) with luciferin to give a luciferin free radical, which can then react with O_2 to give a luciferin peroxide.

Oxidation by hydroxyl radicals produced by H_2O_2 and ferrous ions (in the absence of phosphate or other complexing anion) also produces chemiluminescence of luciferin. For a given quantity of Fe^{2+}, I_{max} increases with increase in H_2O_2 concentration; similarly, with constant H_2O_2, intensity increases with the amount of Fe^{2+}. The kinetics are first order and the rate constant varies with H_2O_2 concentration.[15]

$$H_2O_2 + Fe^{2+} \longrightarrow Fe^{3+} + HO^{\cdot} + HO^{-}$$
$$LH_2 + HO^{\cdot} \longrightarrow LH^{\cdot} + HO^{-} + H^{+}$$
$$LH^{\cdot} + O_2 \longrightarrow LHOO^{\cdot}$$
$$LHOO^{\cdot} + LH_2 \longrightarrow LHOOH + LH^{\cdot}$$

In the absence of ferrous ions, using H_2O_2 alone, luminescence is reduced to about 2%. This residual activity is probably due to the presence of trace metals in the buffer system.

Luminescence of luciferin by enzymic systems of production of $O_2^{-\cdot}$, for example, xanthine oxidase in the presence of a suitable substrate such as hypoxanthine, has also been studied. Since in this case free $O_2^{-\cdot}$ is liberated in solution at a slow continuous rate, inhibition by superoxide dismutase (or by ferricytochrome *c*) is efficient at pH 7.8. Comparison of *Pholas* luciferin with luminol (also oxidized with resultant light emission) in this system indicates a more rapid reaction of $O_2^{-\cdot}$ with luciferin (much greater sensitivity, at least 100-fold) and also a stronger affinity, since 10-fold quantities of superoxide dismutase are necessary for equivalent inhibitions of luciferin luminescence compared with luminol.[7]

Luciferin chemiluminescence is obtained on oxidation with potassium permanganate. An optimum at pH 5 is observed; variation of ionic strength has no effect (in contrast with the reaction luciferin–luciferase). The system can be used to estimate luciferin concentrations as a function of total light emitted, or alternatively to measure permanganate (to about 5×10^{-11} mol) using I_{max}, which is a linear function of the concentration of oxidant. Similar results are obtained with potassium ferricyanide as oxidizing agent.

Electrochemical oxidation of luciferin[7] produces light emission at the anode, and, like activation by permanganate or ferricyanide, corresponds to a single electron transfer. Variation of intensity of light emission with change in applied voltage shows a plateau at high voltage; the peak current as indicated by the intensity of light emission is proportional to the amount of luciferin present, whereas the voltage corresponding to half-maximal light intensity ($E_{1/2}$) is constant, with a value of 770 mV using a calomel electrode as cathode. This corresponds to an anode potential of 1.02 V, that is, about 23 kcal for a one-electron oxidation. Reversibility of the anodic oxidation could not be demonstrated, hence $E_{1/2}$ cannot be assimilated to E_0. Nevertheless, since light emission at 490 nm requires at least 58 kcal, it is likely that a luciferin radical is formed, which then reacts with molecular oxygen to give a luciferin peroxide intermediate, degradation of which provides the necessary energy.

Thus two general mechanisms appear to operate in the chemiluminescence of *Pholas* luciferin. The first implies reaction with $O_2^{-\cdot}$

$$LH_2 + O_2^{-\cdot} \longrightarrow LH\cdot + HOO^-$$
$$LH\cdot + O_2 \longrightarrow LHOOH \rightsquigarrow \text{product} + h\nu$$

whereas the second involves formation of a luciferin free radical, which then reacts with molecular oxygen to yield the same peroxide intermediate.

Assay Systems for Other Molecules[18-21]

Light emission techniques using oxidation of luminol have been developed for microestimations of H_2O_2, $O_2^{\cdot-}$, glucose, glucose oxidase (or other oxidases and their substrates which produce H_2O_2), peroxidase, catalase, and superoxide dismutases. These methods are relatively sensitive, since estimations in the general range of 10^{-10} to 10^{-15} mol can be performed with precision (details are given below) and can be extended to enzyme-coupled antibodies in many cases. The general approach involves peroxidase-catalyzed oxidation of luminol by H_2O_2 and measurement of light emission (I_{max} and total light) using a suitable apparatus with which intensity of light emission and integrated total light are recorded continuously. Apart from the sensitivity obtained in static measurements of a given component, the techniques can often be modified to follow continuous variations or repeated fluxes of an enzymic activity or product *in vitro* or *in vivo* (since the reagents are generally nontoxic), for example, production of $O_2^{\cdot-}$ by polymorphonuclear neutrophiles during phagocytosis, since the rate-limiting step is not emission (and measurement) of light.

While such techniques involving luminol are convenient and undoubtedly of very general application, the sensitivity remains limited despite the enzymic amplification, which plays an important role. Replacement of luminol by *Pholas* luciferin (and perhaps by other luciferins) can increase sensitivity by several orders of magnitude in certain of the methods. In the case of horseradish peroxidase, the enzyme acts as an oxidase in the absence of H_2O_2 when *Pholas* luciferin is used as substrate, as described above. Light emitted owing to oxidation of the luciferin is correlated with specific quantities of peroxidase and can thus be used to estimate the enzyme in free solution, coupled to antibodies, or when such antibodies are subsequently bound to cell surface antigens. With this modification, peroxidase can be measured to a limit of about 100,000 molecules per milliliter with relative ease, and the number of cell-surface antigenic sites per cell (mouse lymphocytes) can be determined after interaction of a peroxidase-coupled antibody with the surface immunoglobulins. A further extension is the use of bacterial luciferase-coupled antibodies and subsequent determination of luciferase activity using appropriate methods.[20]

It is surprising that this approach has not been extensively developed in a more general manner. Given the specificity of antibodies for a wide range

[18] K. Puget and A. M. Michelson, *Biochimie* **58,** 757 (1976).

[19] A. M. Michelson, K. Puget, P. Durosay, and J. C. Bonneau *in* "Superoxide and Superoxide Dismutases" (A. M. Michelson, J. M. McCord, and I. Fridovich, eds.), p. 469. Academic Press, New York, 1977.

[20] K. Puget, A. M. Michelson, and S. Avrameas, *Anal. Biochem.* **79,** 447 (1977).

[21] K. Puget and A. M. Michelson, *Biochimie* **56,** 1255 (1974).

of molecules of biological or medical interest, coupling of such antibodies with a suitable luciferase, for example, *Cypridina* luciferase, followed by determination by light emission techniques after addition of the homologous luciferin, provides a convenient, rapid and general method of extremely high sensitivity often exceeding radioactive techniques, since turnover of the enzyme provides a preamplification of measurement of light emitted. Again the possibility exists of continuous measurement of pulses of activity since extraction processes are not necessary. In addition, use of light amplification techniques could permit the localization of certain cell components with a direct visual image.

Hydrogen Peroxide.[18] The cuvette contains 0.3 ml of 1 *M* K_2HPO_4 pH 7.8, 0.3 ml of 1 m*M* EDTA, 30 μl of 0.1 m*M* luminol, 10 μg of peroxidase, and water to 2 ml. Reaction is initiated by injection of 1 ml of the sample of H_2O_2 and both I_{max} and total light emitted are measured and compared with a standard plot. Satisfactory linearity is obtained with 10^{-7} to 10^{-11} mol of H_2O_2.

Glucose.[18] The cuvette contains 0.3 ml of 1 *M* K_2HPO_4, pH 7.8, 12 units of glucose oxidase, the sample of glucose to be estimated, and water to 2 ml. After incubation of this solution for 10 min at 20°, 1 ml of 0.3 m*M* luminol in H_2O containing 10 μg of horseradish peroxidase is injected to initiate light emission. Values of I_{max} and total light of the flash are measured and compared with a standard plot. The method is applicable to the estimation of 30 ng to 200 μg of glucose.

Glucose Oxidase.[18] An aliquot of the unknown sample of glucose oxidase is added to a solution of 1 mg/ml of glucose in 10 m*M* phosphate, 30 μ*M* NaF, pH 7.0, and incubated for 20 min at 25°. Samples of 1 ml are withdrawn and injected into the cuvette containing 0.3 ml of 1 *M* phosphate, pH 7.8, 30 μl of 10 m*M* luminol, 10 μg of horseradish peroxidase, and water to a final volume of 2 ml, to initiate light emission. The estimation is linear for 10^{-3} to 10^{-2} international units of glucose oxidase (i.e., approximately 10^{-14} to 10^{-13} mol of enzyme using a specific activity of 250 units/mg for the pure enzyme).

Essentially the same techniques can be applied to the estimation not only of a variety of substrates that produce H_2O_2 on oxidation, but also of a number of other oxidases that yield H_2O_2 in the presence of a specific substrate, the latter ensuring specificity of the oxidase estimated, even in the presence of other oxidases.

Glucose Oxidase-Coupled Antibodies.[20] A suspension of mouse lymphocyte cells (3×10^6/ml) which had been treated with glucose oxidase bound to antibody in 10 m*M* phosphate, pH 7.0, containing 30 μ*M* NaF and 1 mg of glucose per milliliter is incubated at 25° for 20 min; then 1 ml of the suspension is injected into 2 ml of 0.1 *M* phosphate, pH 7.8, containing 10

μg of horseradish peroxidase and 0.1 mM luminol, and the light emitted (I_{max} and total light) measured. The results are compared with a standard curve and indicate that about 5000 molecules of glucose oxidase are bound per cell. The control cells give a background of 10–15%. A variation of this technique involves continuous measurement of the glucose oxidase activity. The cells are suspended in 0.1 M phosphate, pH 7.8, containing 0.1 mM luminol, 3.3 μg of peroxidase and 0.33 mg of glucose per milliliter, and the light emitted owing to peroxidase-catalyzed oxidation of luminol by the H_2O_2 produced by the glucose oxidase is measured as a steady-state output, and the values are compared with those obtained for known amounts of glucose oxidase.

Bacterial Luciferase-Coupled Antibodies.[20] Bacterial luciferase oxidizes lauryl aldehyde in the presence of reduced FMN with emission of light. Measurement of light intensity allows determination of the luciferase activity and, hence, the amount of enzyme when compared with a standard.

To a suspension of 2 ml of cells (3×10^6/ml), treated with luciferase-bound antibody, in 0.1 M phosphate, pH 7, is added 0.15 μg of lauryl aldehyde (in 100 μl of water), followed by injection of 1 ml of 50 μM $FMNH_2$ in 10 mM phosphate, pH 7, containing 0.5 mM EDTA. A flash of light (I_{max} 3.1×10^6 photons/sec/ml) is observed, similar to that obtained with approximately 0.7 ng of luciferase per milliliter. This corresponds to about 260 molecules of active luciferase per cell.

Catalase.[19] The sample is added to 5 ml of 21 μM H_2O_2 in 10 mM K_2HPO_4, pH 7.8, and this solution is incubated at 25° for 20 min. Residual H_2O_2 (and hence catalase activity) is determined by measurement of light emitted by oxidation of luminol catalyzed by horseradish peroxidase, using the following technique. The cuvette contains 10 μg of horseradish peroxidase in 3 ml of 0.1 mM EDTA, 0.1 mM luminol in 0.1 M phosphate, pH 7.8. Reaction is initiated by injection of 1 ml of the above solution of H_2O_2 incubated 20 min with the sample, and values of I_{max} and total light emitted (integration with time) are measured. Calibration curves are established using 1 to 15 ng of crystalline beef liver catalase (50,000 units/mg) per milliliter of 21 μM H_2O_2. The two estimations (I_{max} and total light) should not differ by more than 5%; the results are correct within a range of 1–15 ng of catalase (i.e., femtomole quantities). Using this technique, the catalase content in human erythrocytes can be accurately measured with samples of 0.02–0.05 μl of total blood.[19] Catalase can be measured in presence of peroxidase if 10 μM 2-mercaptoimidazole is added to the initial incubation mixture; the peroxidase is completely inhibited.[20]

Superoxide Dismutases.[21] Superoxide dismutase activity is measured by inhibition of the chemiluminescence of luminol induced by $O_2^{\cdot-}$ produced by the system xanthine oxidase/O_2/hypoxanthine.

The cuvette contains 0.1 m*M* EDTA, 10 μM luminol in 0.1 *M* glycine–NaOH buffer, pH 9 (for a final volume of 3 ml), 5 μg of xanthine oxidase, and the aliquot of superoxide dismutase. Reaction is initiated by injection of 1 ml of 30 μM hypoxanthine in water. The maximum of light intensity (the peak occurs after several minutes) is compared with that of a blank in which the dismutase is absent. Inhibition of 50% corresponds to 0.1 unit (luminol) of the enzyme. Limits of about 0.2 ng (i.e., femtomole quantities) of erythrocyprein can be readily and precisely estimated.

Peroxidase.[20] The cuvette contains 0.3 ml of 1 *M* glycine–NaOH buffer, pH 9.0, 0.3 ml of 1 m*M* EDTA, 30 μl of 1 m*M* luminol (final concentration 10 μM), an aliquot of the peroxidase to be measured, and water to a volume of 2 ml. Reaction is initiated by injection of 1 ml of 0.3 m*M* H_2O_2, and the light intensity is measured 1 min after injection. In the absence of peroxidase an initial pulse of light emission is observed, but at 1 min the intensity is essentially zero.

Under these conditions light emission at 1 min is not strictly linear over the entire range of peroxidase examined (0.01–10 μg) since two slopes are present with a break at about 0.2 μg of peroxidase.

A modification of this method in which 1 ml of 30 μM H_2O_2 is injected gives results that are completely linear (log/log) with respect to $h\nu$ and the quantity of peroxidase. However, the sensitivity is reduced, and the method is valuable for the range 0.06 to 10 μg.

Microestimation of Peroxidase: Use of Pholas Luciferin.[20] The sample of peroxidase is added to a solution of *Pholas* luciferin (3 μg/ml; 85 n*M*) in 3.0 ml of 25 m*M* phosphate, pH 7.8, and the steady-state level of light emission is measured (photomultiplier 1500 V). The response is linear (on a log/log scale) over five orders of magnitude and provides a reliable measure of peroxidase between 10 pg/ml and 400 ng/ml (i.e., between about 2.5×10^{-16} and 10^{-11} mol/ml).

Application of an increased voltage (1650 V) to the photomultiplier to increase sensitivity permits estimation (under the same conditions) of femtogram quantities of peroxidase using the appropriate calibration curve, the lower limit observed being about 75,000 molecules/ml. The steady-state light output is linear between about 5 fg to 10 pg, giving a total range of measure of 4×10^{-7} to 5×10^{-15} g of peroxidase per milliliter.

Even at these low concentrations of enzyme and substrate, the kinetics of the reaction are completely normal. This is probably due to the strong interactions between luciferin and peroxidase (as for luciferase) and it is likely that with an excess of substrate free peroxidase is not present, but that oxyluciferin is displaced from an enzyme–product complex (the emitter) by luciferin during the reactions. However, such interactions have not as yet been studied.

In principle, the sensibility of the method can be further increased by measurement of total light emitted during a given time and by suitable modification of the light-measuring techniques. It is not impossible that with such refinements accurate estimations of several hundred molecules of peroxidase per milliliter could be effected.

Peroxidase-Coupled Antibodies Bound to Mouse Lymphocytes: Pholas Luciferin.[20] Antibodies coupled with peroxidase are incubated with spleen cells previously fixed for 1 hr at 4° with 4% formaldehyde in 0.1 *M* phosphate buffer, pH 7.5, and then washed. These cells (about 1 to 5×10^6 cells/ml) are then suspended in 2.8 ml of 0.125 *M* phosphate, 0.025 *M* NaCl, pH 7.8, containing 5 n*M* pholad luciferin, and light emission is measured, using steady-state output and total light emitted over a suitable time interval. The results indicate that approximately 1200 molecules of active peroxidase are bound per cell. Control cells give about 26% of the values observed. This background is much lower than when peroxidase fixed on cells is quantified by conventional chromogenic procedures. This background is due to endogenous peroxidase or pseudoperoxidase present in nonlymphocyte spleen cells. These enzymes seem much less active when their measurement is carried out with the present procedure, possibly because *Pholas* luciferin is a macromolecule and does not easily pass the cell membrane. This is a considerable advantage for all cellular measurements, since the preparation of pure cell populations is not necessary.

Acknowledgments

This work has received financial support from the CNRS (E.R. 103), the DGRST (Contract No. 77.7.02.80), INSERM (Contract No. 77.4.084.2) and the Fondation pour la Recherche Médicale Française. The efforts of various collaborators, in particular Dr. J.-P. Henry, largely responsible for the development of these studies, are acknowledged with appreciation.

Section VIII

Chemiluminescent Techniques

[36] The Chemiluminescence of Luminol and Related Hydrazides

By DAVID F. ROSWELL and EMIL H. WHITE

Chemiluminescence is the production of light by chemical reactions. Electronically excited states are formed in the chemical reaction, and light emission from these completes the process. The relationship between chemiluminescence and absorption/emission phenomena, such as fluorescence, are shown in Fig. 1.

A surprisingly large number of chemical reactions are chemiluminescent.[1–4] This paper, however, will deal only with a rather famous example, the chemiluminescence of luminol [5-amino-2,3-dihydrophthalazine-1,4-dione (compound I)]; some derivatives and analogs will also be discussed.

O
NH
NH
NH_2 O

(I)

The first report of chemiluminescence from luminol was made by Albrecht[5] in 1928. Since that time the reaction of luminol and other derivatives of the general hydrazide structure[6] have been studied extensively. A number of reviews on this subject have appeared over the last 15 years,[7–10] and hence the historical aspects of the problem will be treated only briefly.

[1] K.-D. Gundermann, *Top. Curr. Chem.* **46,** 61 (1974).

[2] E. H. White and D. F. Roswell, *Acc. Chem. Res.* **3,** 54 (1970).

[3] M. J. Cormier, D. M. Hercules, and J. Lee, eds., "Chemiluminescence and Bioluminescence." Plenum, New York, 1973.

[4] K.-D. Gundermann, "Chemilumineszenz Organischer Verbindungen." Springer-Verlag, Berlin and New York, 1968.

[5] H. O. Albrecht, *Z. Phys. Chem.* **136,** 321 (1928).

[6] Luminol as well as being named as a phthalazinedione is also referred to as 3-aminophthalic hydrazide.

[7] E. H. White, *in* "Light and Life" (W. D. McElroy and B. Glass, eds.), 1st ed., p. 183. Johns Hopkins Press, Baltimore, Maryland, 1961.

[8] K.-D. Gundermann, *Angew. Chem., Int. Ed.* **4,** 566 (1965).

[9] F. McCapra, *Quart. Rev. Chem. Soc.,* **20,** 485 (1966).

[10] J. W. Haas, *J. Chem. Educ.* **44,** 396 (1967).

ISBN 0-12-181957-4

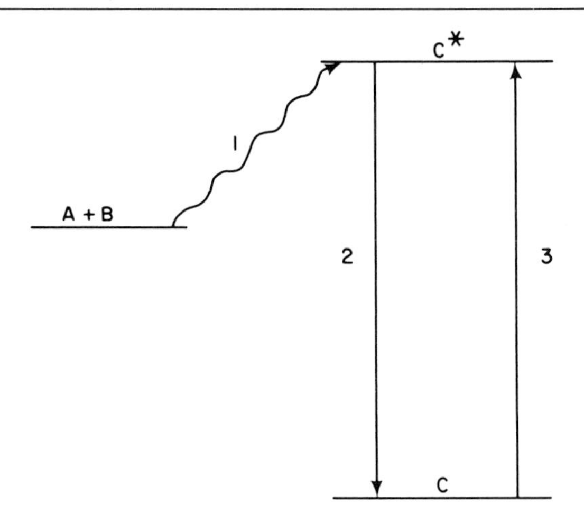

FIG. 1. Chemiluminescence, path 1 + 2; fluorescence, path 2; absorption, path 3.

Reaction Conditions

The chemiluminescent reactions of luminol are all oxidations [Eq. (1)]. A wide variety of reaction conditions have been employed; the reaction can be carried out either in protic media or in aprotic solvents such as dimethyl sulfoxide (DMSO), hexamethylphosphoric acid triamide (HMPT), or dimethylformamide (DMF). In aprotic media, only oxygen and a strong base are required for chemiluminescence.[7]

$$\text{R-C}_6\text{H}_3(\text{CO})_2(\text{NH})_2 + 2\,{}^{-}\text{OH} + [\text{O}_2] \longrightarrow \text{R-C}_6\text{H}_3(\text{CO}_2^-)_2 + 2\,\text{H}_2\text{O} + \text{N}_2 \qquad (1)$$

Water is the most common protic solvent, although the reaction has been studied in the lower alcohols as well. The reaction in these solvents usually requires base, an oxidizing agent, and either oxygen or a peroxide, depending on the system. Hypochlorite, ferricyanide, and persulfate are commonly used oxidizing agents. In place of the oxidizing agents, chelated transition metal complexes, pulse radiolysis,[11] or sonic waves[12] have been used in the presence of either oxygen or peroxides.

[11] J. H. Baxendale, *Chem. Commun.* **1971,** 1490 (1971).

[12] E. N. Harvey, *J. Am. Chem. Soc.* **61,** 2392 (1939).

One of the simplest systems involves adding luminol to a solution of hydrogen peroxide in dilute sodium carbonate that has been partially neutralized immediately before the luminol is added.[13]

Structural Effects

Early workers investigated the effect of gross structural modifications of luminol [Eq. (1)].[14–16] Compounds isomeric to (I) in the heterocyclic ring—compounds (II) and (III), as well as N or O methylation products (IV and V, respectively)—were nonchemiluminescent.

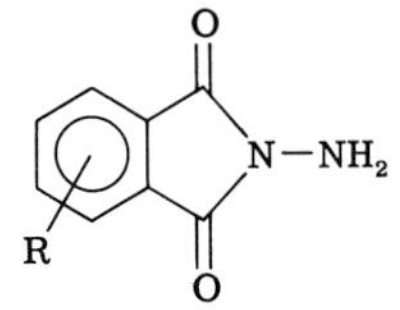

(II) 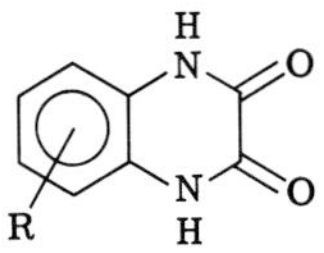 (III) 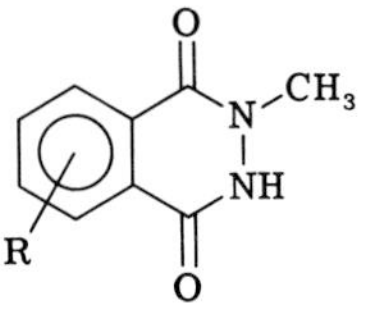 (IV) 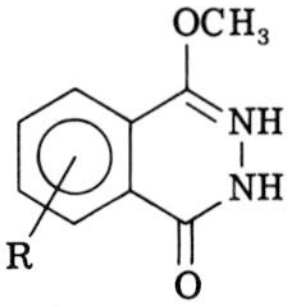 (V)

Analogs in which the amino group was replaced by other substituents, or analogs with substitution anywhere in the nonheterocyclic ring were generally found to be chemiluminescent. These compounds vary in their light-producing ability; e.g., compounds (VI)[1] and (VII)[13] are more chemiluminescent than luminol whereas compounds (VIII) and (IX) are of lower efficiency.[14]

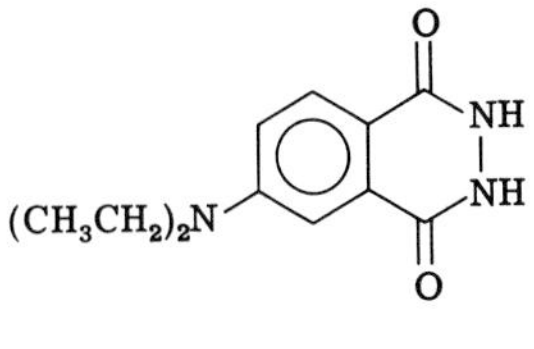

(VI) 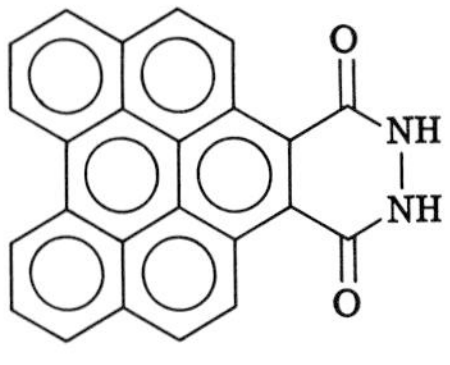 (VII)

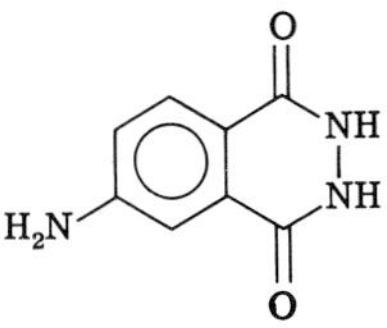

(VIII) 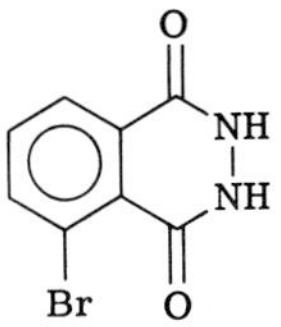 (IX)

[13] E. H. White and R. B. Brundrett, *in* "Chemiluminescence and Bioluminescence" (M. J. Cormier, D. M. Hercules, and J. Lee, eds.), p. 231. Plenum, New York, 1973.

[14] H. D. K. Drew and F. H. Pearman, *J. Chem. Soc. London* **1937,** 586 (1937).

[15] H. D. K. Drew and R. F. Garwood, *J. Chem. Soc. London* **1939,** 836 (1939).

[16] E. H. Huntress and J. V. K. Gladding, *J. Am. Chem. Soc.* **64,** 2644 (1942).

The chemiluminescence quantum yield (Φ_{ch}), where Φ_{ch} = photons produced/molecules reacted, is used as the measure of light-producing ability. For luminol, $\Phi_{ch} = 0.01$; and for compound (VII), $\Phi_{ch} = 0.08$. This quantum yield may also be defined as $\Phi_{ch} = \Phi_R \Phi_{ES} \Phi_{Fl}$, where Φ_R is the fraction of reacting molecules following the correct chemical path, Φ_{ES} is the fraction of excited states produced, and Φ_{Fl} is the fluorescent quantum yield of the emitter. The light-producing ability of luminol analogs is thus a function of three factors; of these, Φ_{ES} and Φ_{Fl} appear to be the most important.[17,18]

Products

Under the mild reaction conditions possible in aprotic solvents, the organic product of the luminol oxidation is 3-aminophthalate (X), which is

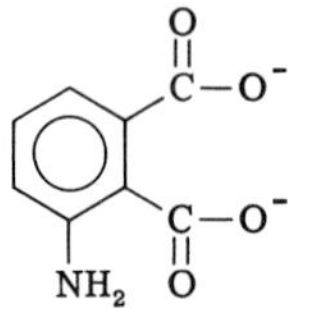

(X)

produced almost quantitatively.[19] Nitrogen is evolved and oxygen is consumed in essentially molar amounts.[19] In protic media the reaction conditions are so vigorous that degradation of the initially formed aminophthalate occurs, but the overall reaction is believed to be the same.

The Light Emitter

In both protic and aprotic media the product phthalate ion, (X), is found in an electronically excited state and is responsible for the observed emission. The identity of the fluorescence spectrum of the product with the chemiluminescence emission spectra of the reaction (see Fig. 2) establishes this point. As shown in Fig. 2a, the emission from the oxidation of luminol, (I), in aqueous media occurs at 425 nm and is superimposable on the fluorescence spectra of the total reaction products and also on the spectrum of 3-aminophthalate ion. In aprotic media the emission maxima are shifted to longer wavelengths. The long-wavelength emission has been attributed to the quininoid tautomer, (XI), of (X). This species would not be present in

[17] R. B. Brundrett, D. F. Roswell, and E. H. White, *J. Am. Chem. Soc.* **94**, 7536 (1972).
[18] R. B. Brundrett and E. H. White, *J. Am. Chem. Soc.*, **96**, 7497 (1974).
[19] E. H. White, O. C. Zafiriou, H. M. Kagi, and J. H. M. Hill, *J. Am. Chem. Soc.* **86**, 940 (1964).

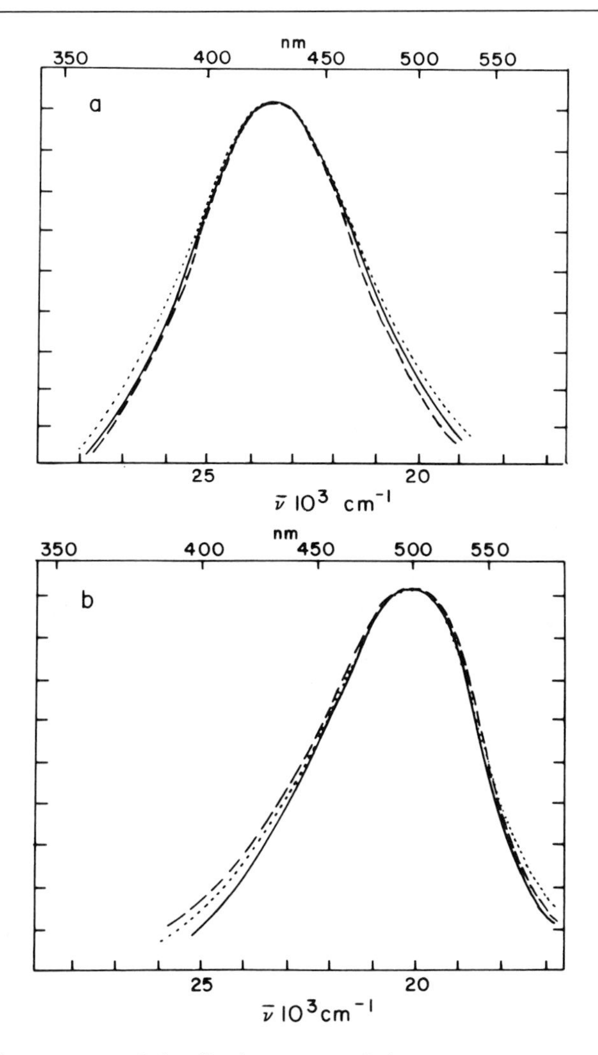

FIG. 2. Fluorescence and chemiluminescence emission spectra: (a) solvent is water; (b) dimethyl sulfoxide. ———, chemiluminescence of luminol; . . . . , fluorescence of 3-aminophthalate; -----, fluorescence of total reaction mixture. Permission to reprint is granted by the American Chemical Society. *J. Am. Chem. Soc.* **95,** 2610 (1973).

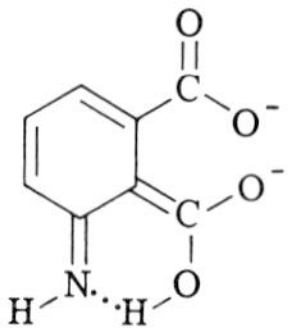

(XI)

highly aqueous media[20] because of the lowered basicity of a carboxylate group hydrogen-bonded to water molecules. In mixed solvent systems (DMSO–H_2O) both emissions occur, the 425-nm band increasing with water concentration. These reaction systems show a one-to-one correlation between chemiluminescence and the fluorescence of (X) as long as the base used is a quaternary ammonium hydroxide.

If other bases are used, this correspondence is lost; in a system of 90% dimethyl sulfoxide (DMSO)–water made basic with alkali metal hydroxides (MOH), the shorter-wavelength emission band is more pronounced in 3-aminophthalate fluorescence than in luminol chemiluminescence when M = Na^+ or K^+[20] (Fig. 3).

The fluorescence spectrum of 3-aminophthalate shows a marked increase in the shorter-wavelength emission band with increasing sodium ion concentration, and at high sodium ion concentrations the longer-wavelength band is almost lost. The absorption spectrum of 3-aminophthalate ion shifts with increasing sodium ion concentration, suggesting that there is considerable ion paring in the ground state. Because of the short lifetime of the excited state, this ion pairing persists in the excited state. The ion pair of aminophthalate ion, (X), would be expected to have a lower electron density at the carbonyl moiety of the carboxylate group, and thus be less able to bring about the intramolecular proton transfer required to form the quininoid form (XI), which is responsible for the long-wavelength emission. Thus, metal ions form ion pairs with (X), which fluoresce at about 425 nm, whereas the free ion can pass over to the quininoid form, (XI), which fluoresces at about 510 nm.

The differences in the fluorescence emission spectra of 3-aminophthalate ion and the chemiluminescence of luminol (Fig. 3) can also be explained by the ion-pair effect. The fraction of excited aminophthalate ions in the ion-pair form differs in fluorescence and chemiluminescence. The fraction in fluorescence is essentially determined by the ground-state equilibrium. The fraction of ion pairs in chemiluminescence is determined

[20] E. H. White, D. F. Roswell, C. C. Wei, and P. D. Wildes, *J. Am. Chem. Soc.* **94,** 6223 (1972); see also J. Lee and H. H. Seliger, *Photochem. Photobiol.* **15,** 227 (1972) and J. D. Gorsuch and D. M. Hercules, *Photochem. Photobiol.* **15,** 567 (1972).

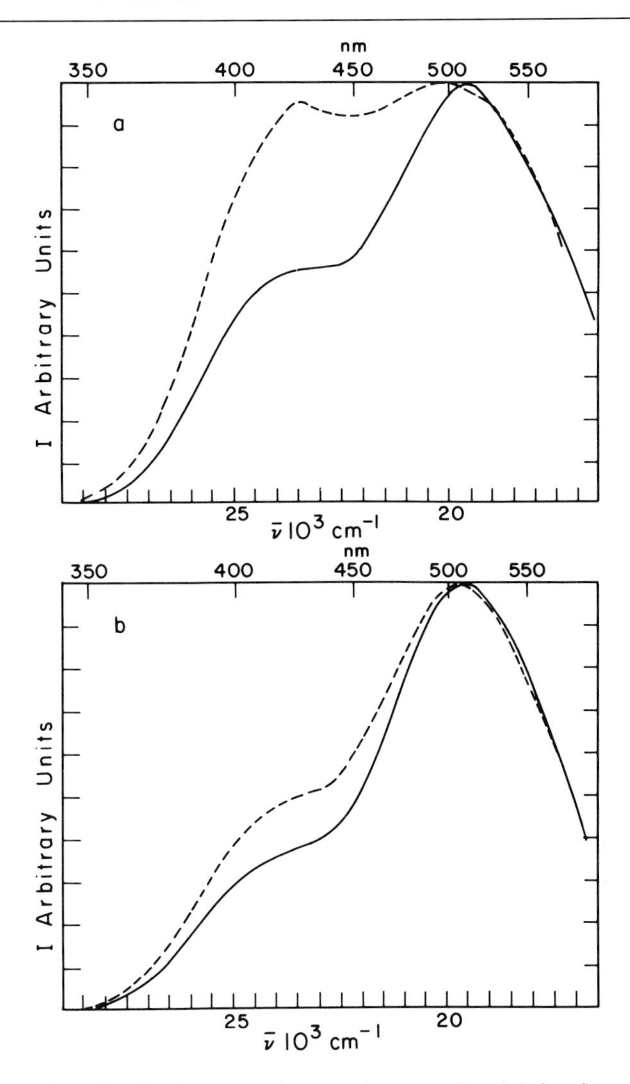

FIG. 3. Luminol chemiluminescence (————) and 3-aminophthalate fluorescence (----) in 30 mol % water, 70 mol % dimethyl sulfoxide with (a) 20 m*M* sodium hydroxide; (b) 20 m*M* potassium hydroxide. Permission to reprint is granted by the American Chemical Society. *J. Am. Chem. Soc.* **95,** 2610 (1973).

at the transition state of the reaction, and in the present case this value is lower than that in fluorescence. It should be pointed out that the above observations pertain to luminol and that usually the excited states formed in chemiluminescence and fluorescence are the same.

The general picture of the chemiluminescence of cyclic hydrazides—that light emission occurs from the corresponding phthalate ion [Eq. (2)]—appears to hold with only rare exceptions. The parent compound (R=H,

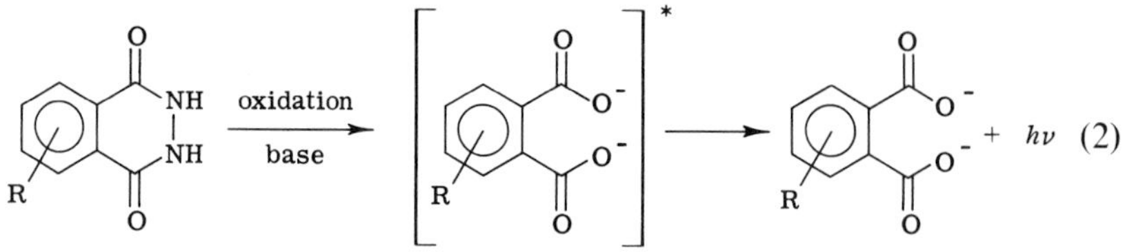

oxidation / base → [ ]* → + $h\nu$ (2)

phthalic hydrazide) is essentially nonchemiluminescent in the protic system;[21] this is not surprising, since phthalate ion under the reaction conditions is not fluorescent. In aprotic solvents such as DMSO, however, the reaction of phthalic hydrazide with base and oxygen does give a yellow emission of moderate intensity, yet under the same conditions phthalate ion is again not fluorescent. The emission spectrum of this chemiluminescence matches the fluorescence spectrum of the monoanion of the starting hydrazide, (XII), under the reaction conditions. Equation (3) has been

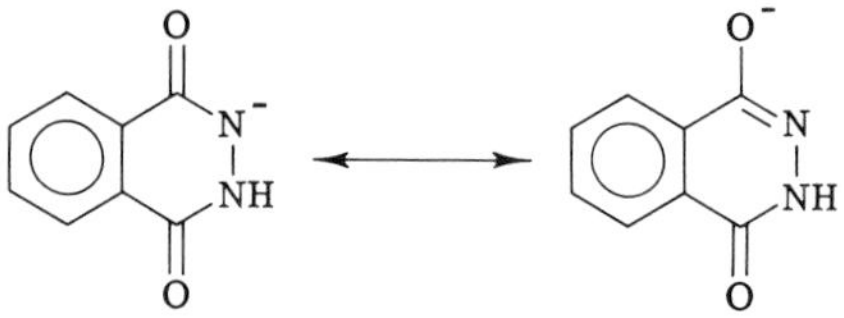

(XII)

proposed to account for this result.[22] This scheme requires production of excited phthalate and subsequent intermolecular energy transfer to the anion of the starting material from which emission occurs. The energy transfer apparently occurs between an excited phthalate ion hydrogen bonded to a phthalic hydrazide anion, since added fluorescers without acidic protons, such as acridone or the salt of *N*-methyl phthalic hydrazide, do not accept the excitation energy.

[21] J. Stauff and G. Hartmann, *Ber. Bunsenges. Phys. Chem.* **69**, 145 (1965).

[22] E. H. White, D. F. Roswell, and O. C. Zafiriou, *J. Org. Chem.* **34**, 2462 (1969).

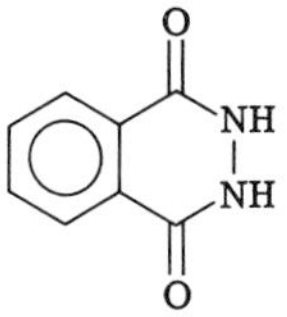

$$\text{(O, NH, NH, O)} \xrightarrow{\text{base}} \text{(XII)} \xrightarrow{\text{oxidation}} [\text{phthalate ion}]^* \tag{3}$$

$$[\text{Phthalate ion}]^* + \text{(XII)} \longrightarrow \text{phthalate ion} + \text{(XII)}^*$$

$$\text{(XII)}^* \longrightarrow h\nu + \text{(XII)}$$

Energy Transfer

Energy transfer has also been shown to occur in the chemiluminescence of hydrazides[23-26] which are linked to some highly fluorescent groups. Compound (XIII) is a good example of this type of structure. In the case of

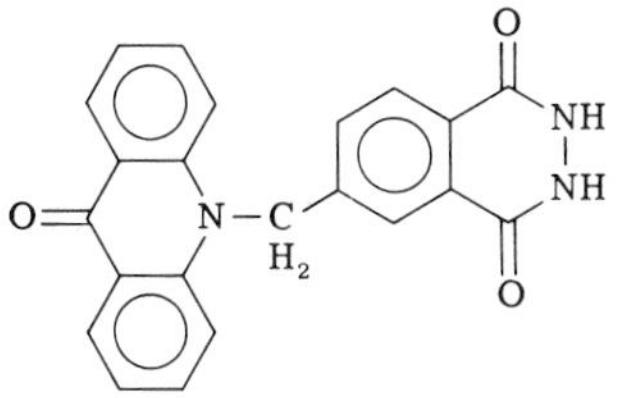

O= N–C H_2 NH NH O O

(XIII)

compound (XIII), oxidation of the hydrazide portion of the molecule (in protic or aprotic media) results in emission from the acridone moiety. This emission occurs because of rapid intramolecular singlet–singlet energy transfer from the electronically excited phthalate ion to the acridone portion of the molecule.

Reaction Mechanism

Information about the mechanism of oxidation of luminol and other hydrazides has been obtained in studies using both protic and aprotic media. In aprotic media (DMSO), the dinegative ion, (XIV), formed in a stepwise reaction with base is a crucial intermediate [Eq. (4)].[27] Subsequent

[23] E. H. White and D. F. Roswell, *J. Am. Chem. Soc.* **89,** 3944 (1967).

[24] E. H. White, D. R. Roberts, and D. F. Roswell, *in* "Molecular Luminescence" (E. C. Linn, ed.), 1st ed., p. 479. Benjamin, New York, 1969.

[25] D. F. Roswell, V. Paul, and E. H. White, *J. Am. Chem. Soc.* **92,** 4855 (1970).

[26] D. R. Roberts and E. H. White, *J. Am. Chem. Soc.* **92,** 4861 (1970).

[27] E. H. White and M. M. Bursey, *J. Am. Chem. Soc.* **86,** 941 (1964).

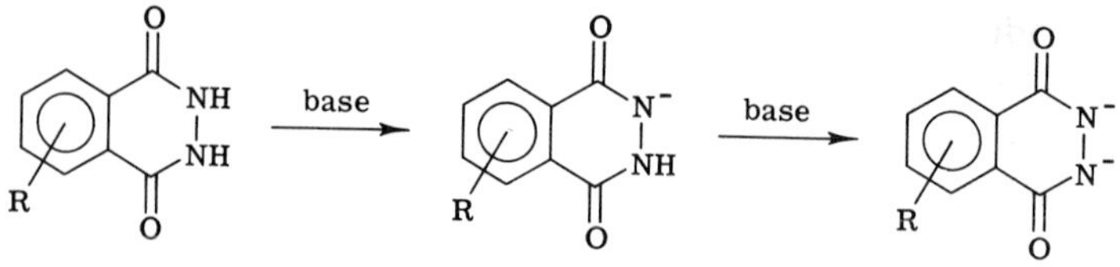

(4)

(XIV)

reaction of this intermediate with oxygen via some as yet undetermined pathway produces the phthalate ion in an excited state.

The reaction of the dinegative ion could follow several pathways, as indicated in Eqs. (5a)–(5d).

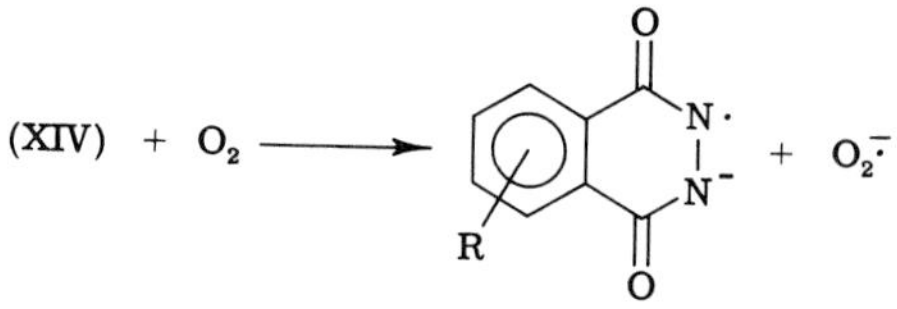

(XIV) + O_2 → (XV) + $O_2^{\cdot -}$ (5a)

(XIV) + O_2 → (XVI) (5b)

(XIV) + O_2 → (XVII) (5c)

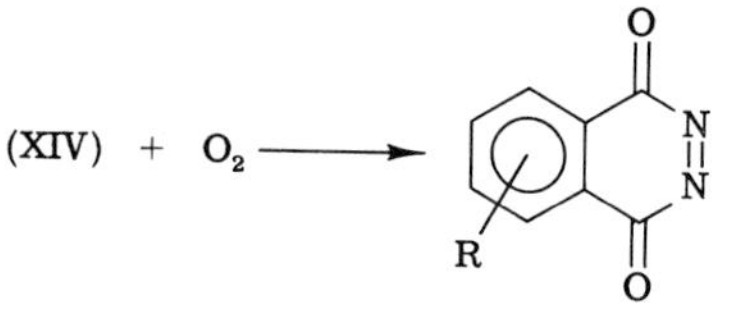

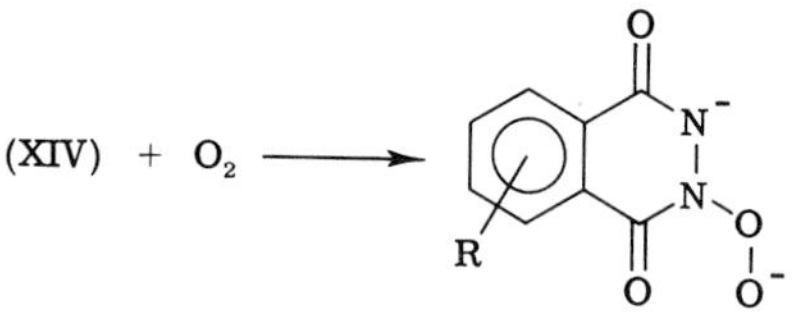

(XIV) ⟷ $\xrightarrow{O_2}$ (XVIII) (5d)

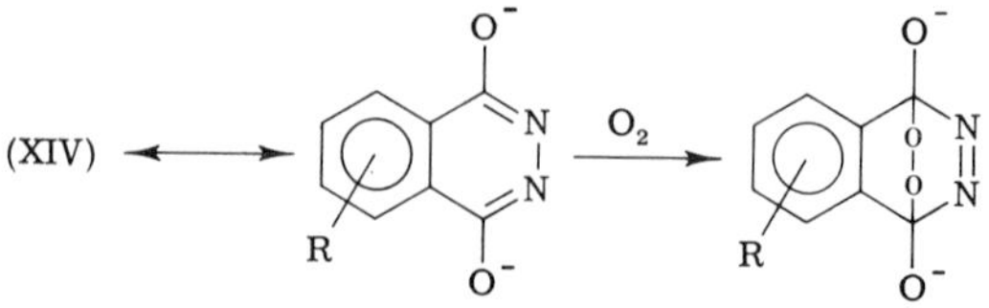

The radical anion intermediate, (XV), in Eq. (5a) has also been proposed for the reaction in protic media when transition metal complexes are used as catalysts, and for the electrochemical oxidation.[28] In protic media it could be formed by the pathway shown in Eq. (6).[7,29] Alternatively, it has

$$M^{2+} + H_2O_2 \longrightarrow M^{3+} + {}^-OH + \cdot OH$$

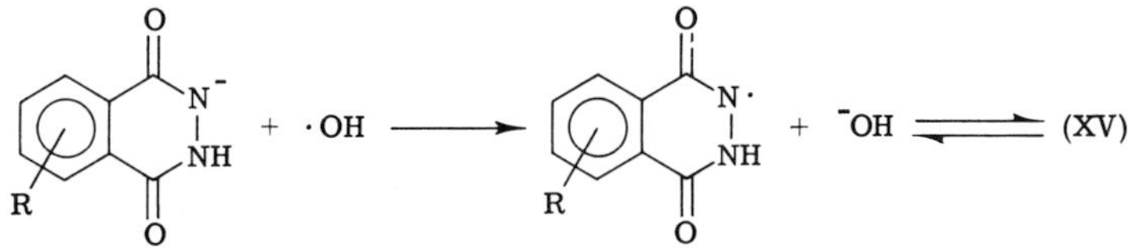

$$\text{(hydrazide anion, R-substituted)} + \cdot OH \longrightarrow \text{(hydrazide radical, R-substituted)} + {}^-OH \rightleftharpoons \text{(XV)} \quad (6)$$

been proposed[30] that the metal ion forms a complex with the hydrazide, which is then oxidized, and it has also been suggested that it is a $HO_2^- M^{2+}$ complex which produces the hydrazide radical anion.[31]

The azaquinone proposed in Eq. (5b) has received considerable attention as a possible reaction intermediate; it was proposed as an intermediate in the first paper published on the chemiluminescence of luminol.[5] Later workers, on the basis of kinetic evidence, have suggested azaquinones as intermediates in the persulfate/hydrogen peroxide-driven chemiluminescence of luminol as well as in the hypochlorite oxidation.[32,33] Azaquinones have been trapped as their cyclopentadiene adducts,[34] (XIX), from solution, which in the absence of the trapping agent normally exhibit

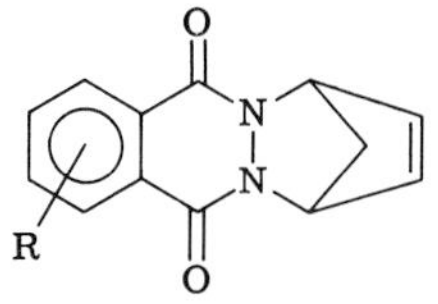

(XIX)

chemiluminescence. These adducts have been identified by thin-layer chromatography; the yield is low, accounting for only a small fraction of the starting hydrazide.

[28] B. Epstein and T. Kuwana, *Photochem. Photobiol.* **4,** 1157 (1965).
[29] P. B. Shevlin and H. A. Neufeld, *J. Org. Chem.* **35,** 2178 (1970).
[30] A. K. Babko and N. M. Lukovskaya, *Ukr. Khim. Zh.* **30,** 508 (1964).
[31] T. G. Burdo and W. R. Seitz, *Anal. Chem.* **47,** 1639 (1975).
[32] M. M. Rauhut, A. M. Semsel, and B. G. Roberts, *J. Org. Chem.* **34,** 2431 (1966).
[33] W. R. Seitz, *J. Phys. Chem.* **79,** 101 (1975).
[34] Y. Omote, T. Miyake, and N. Sugiyama, *Bull. Chem. Soc. Jpn.* **40,** 2446 (1967).

Diazaquinones are reactive but isolable compounds.[35-37] They chemiluminesce in the presence of alkaline hydrogen peroxide and, unlike the hydrazides, do not require peroxide-decomposing catalysts or oxidizing agents. Comparison of the emission spectra from the chemiluminescence of the hydrazide and the corresponding diazaquinone show that the emitter (the phthalate ion) is the same in the two cases. The diazaquinone has been detected by stop-flow spectroscopy during the hypochlorite-catalyzed oxidation of 2,3-naphthalic hydrazide.[38] However, its detection is a function of the hydrogen peroxide concentration. At high hydrogen peroxide concentrations it is not detectable. Rate studies are difficult to do and indicate that the steps prior to the production of the diazaquinone are rate determining. The few facts that are available support the hypothesis that azaquinones are intermediates in the light-producing oxidation of the hydrazides.

The intermediates proposed in Eqs. (5c) and (5d), instead of being formed directly, could also be thought of as arising from reaction of the azaquinone with hydroperoxide ion or superoxide ion [Eq. (7)]. Evidence

(XVI) + $^{-}$OOH ⟶ [$^{-}$O, OOH; N=N; R; O]

↓ base

[$^{-}$O, O—O^{-}; N=N; R; O] ↓ (XVIII)

↓ [O; N–OOH; N^{-}; R; O] ↓ base (XVII)

(7)

has been presented that superoxide may be involved in the reaction of luminol with base and oxygen in aqueous media, since superoxide dismutase quenches the chemiluminescent reaction.[39]

[35] K.-D. Gundermann, *Angew. Chem.* **80,** 494 (1968).

[36] E. H. White, E. G. Nash, D. R. Roberts, and O. C. Zafiriou, *J. Am. Chem. Soc.* **90,** 5932 (1968).

[37] K.-D. Gundermann, *in* "Chemiluminescence and Bioluminescence" (M. J. Cormier, D. M. Hercules, and J. Lee, eds.), p. 209. Plenum, New York, 1973.

[38] C. Lee, unpublished results, 1974.

[39] E. K. Hodgson and I. Fridovich, *Photochem. Photobiol.* **18,** 451 (1973).

Singlet oxygen, produced via dye sensitization, has been shown to cause luminol chemiluminescence in aqueous base.[40-42] Singlet oxygen produced by laser irradiation produces only very weak chemiluminescence from luminol, and 3-aminophthalate is not the major product. It has been suggested that emission arising in the dye sensitization experiments is due to radicals being produced concurrently with singlet O_2.[42]

The oxidation of hydrazides under mild conditions can produce products other than phthalate ion. Reaction of phthalic hydrazide with base and potassium ferricyanide or hydrogen peroxide–ammonium hydroxide–copper has been shown to produce phthalaldehydic acid, (XX), as well as phthalate ion itself.[43] Oxidation of the naphthalic hydrazide, (XXI), with

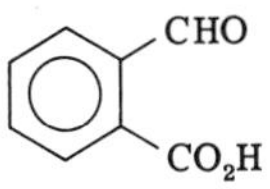

CHO
CO_2H
(XX)

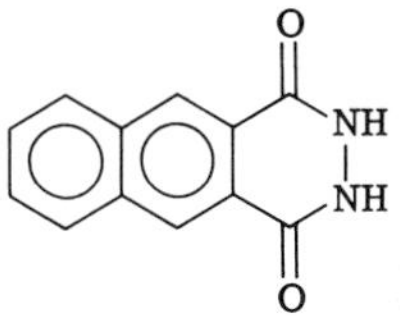

O
NH
NH
O
(XXI)

basic hypochlorite mixtures also produces the corresponding acid aldehyde.[38] The formation of the aldehyde is not surprising, since diacylimides are known to react with bases to produce one equivalent of acid and one of aldehyde [Eq. (8)].[44]

$$\text{R-}\overset{\text{O}}{\overset{\|}{\text{C}}}\text{-N=N-}\overset{\text{O}}{\overset{\|}{\text{C}}}\text{-R}' \longrightarrow \text{R-}\overset{\text{O}}{\overset{\|}{\text{C}}}\text{H} + \text{R-}\overset{\text{O}}{\overset{\|}{\text{C}}}\text{OH}$$

The yield of aldehyde in the hydrazide reactions decreases at the higher hydrogen peroxide concentrations that lead to enhanced light emission. Thus, the aldehyde may be a product of a dark reaction and the interception of species (XXII) may be directly on the light-emission pathway [Eq. (8)].

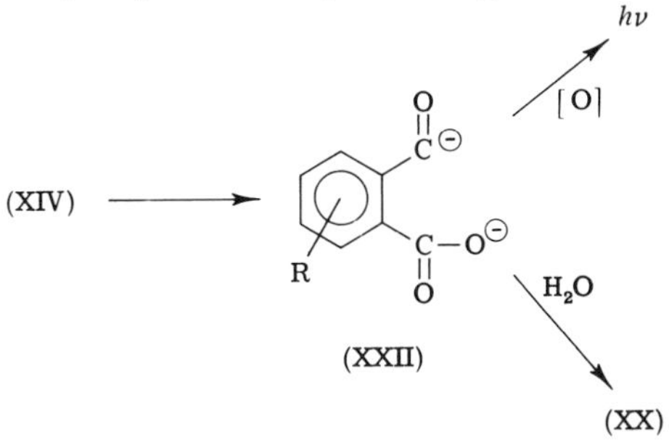

(XIV) ⟶ (XXII) $\xrightarrow{[O]}$ $h\nu$; (XXII) $\xrightarrow{H_2O}$ (XX) (8)

(XXII)

[40] K. Kuschnir and T. Kuwana, *Chem. Commun.* **1969,** 193 (1969).
[41] I. B. C. Matheson and J. Lee, *Photochem. Photobiol.* **12,** 9 (1970).
[42] I. B. C. Matheson and J. Lee, *Photochem. Photobiol.* **24,** 605 (1976).
[43] T. Huang, unpublished results, 1972.
[44] C. Niemann and J. T. Hays, *J. Am. Chem. Soc.* **74,** 5796 (1952).

In summary, it appears that dianion (XIV) is involved in aprotic oxidations, the radical anion (XV) in protic ones, and the azaquinones very probably are involved in the light pathway.

Monoacylhydrazides

Noncyclic nonoacylhydrazides, such as compounds (XXIII)–(XXV), have been reported to be chemiluminescent.[45,46] The oxidizing systems used are the same as those used for the cyclic hydrazides. The

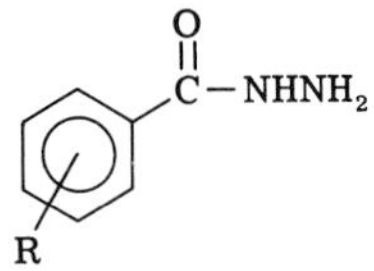

(XXIII) 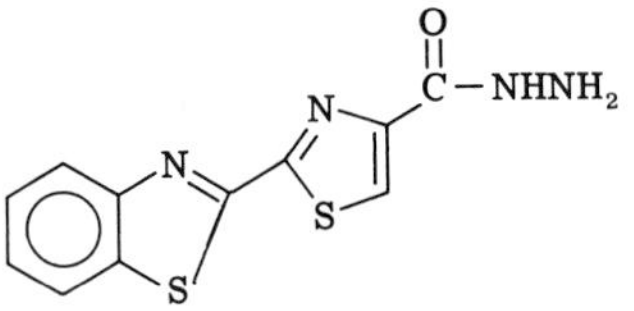 (XXIV) 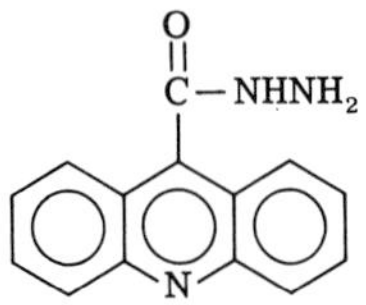 (XXV)

monoacylhydrazides are moderately chemiluminescent compounds; compounds (XXIV) and (XXV) have $\Phi_{Cl} = 0.003$ (in comparison, the value for luninol is 0.01).

The pathway leading to light emission appears to be different from that of luminol and the cyclic hydrazides,[47] and thus these compounds are not good models for the luminol chemiluminescence. The emitter in the case of compound (XXV) corresponds to the fluorescence of acridone and its anion. Thus the most efficient light-producing reaction requires a decarbonylation or decarboxylation. One crucial intermediate would be the acyl anion (XXVI), which could lose carbon monoxide and react with oxygen to

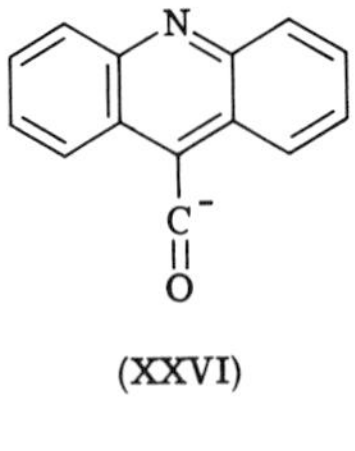

(XXVI)

[45] E. H. White, M. M. Bursey, D. F. Roswell, and J. H. M. Hill, *J. Org. Chem.* **32,** 1198 (1967).

[46] E. Rapaport, M. W. Cass, and E. H. White, *J. Am. Chem. Soc.* **94,** 3153 (1972).

[47] In an earlier report[46] it was erroneously proposed that the mechanism of oxidation of (XXIV) and (XXV) were the same as for the cyclic hydrazides, and that the corresponding carboxylate ion was the light emitter.

form the excited state of acridone anion[48,49] [(XXVII), Eq. (9)]. Alternatively, the acyl anion could react with oxygen and then decarboxylate via a dioxetane intermediate with formation of an excited state[50-52] [Eq. (10)]. Note that acridine carboxylic acid is only weakly fluorescent and that it may be formed via the luminol pathway also, but the low quantum yield to be expected precludes detection of this pathway.[47]

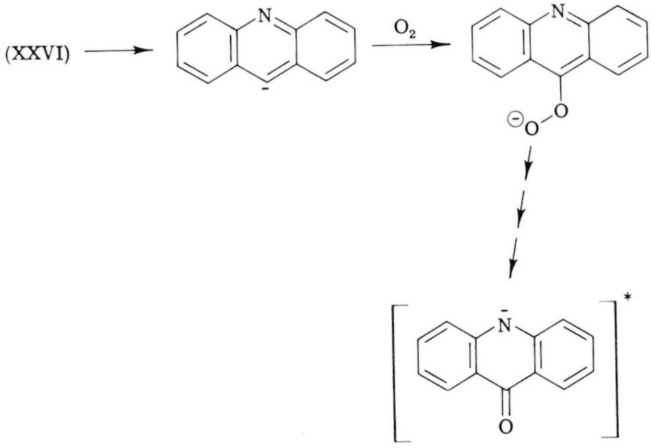

(XXVI) → N $\xrightarrow{O_2}$ N, O, ⊖O → [N, O]* (XXVII) (9)

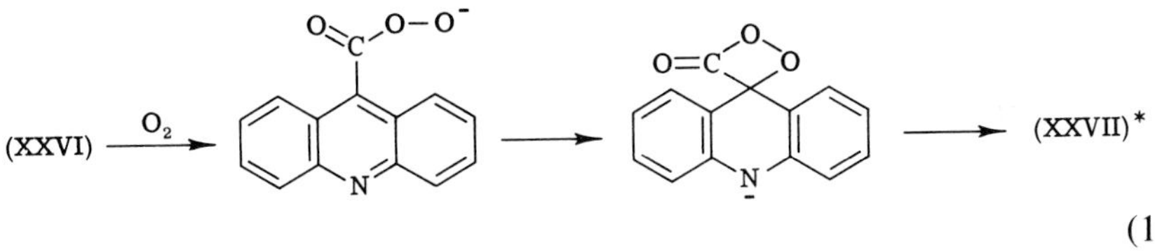

(XXVI) $\xrightarrow{O_2}$ O=C–O–O⁻, N → O=C, O, O, N → (XXVII)* (10)

The decarbonylation (decarboxylation) pathway does not appear to be operative in the case of the cyclic hydrazides, since naphthalic hydrazide, (XXI), yields naphthalate excited states, but not those of 3-hydroxyl-2-naphthoic acid.

[48] G. A. Russell and E. G. Janzen, *J. Am. Chem. Soc.* **84,** 4153 (1962).
[49] G. A. Russell and A. G. Bemis, *J. Am. Chem. Soc.* **88,** 5491 (1966).
[50] F. McCapra, *Acc. Chem. Res.* **9,** 201 (1976).
[51] N. J. Turro, P. Lechtken, N. E. Schore, G. Schuster, H. C. Steinmetzer, and A. Yekta, *Acc. Chem. Res.* **7,** 97 (1974).
[52] E. H. White, J. D. Miano, C. J. Watkins, and E. Breaux, *Angew. Chem., Int. Ed.* **13,** 229 (1974).

[37] Monitoring Specific Protein-Binding Reactions with Chemiluminescence

By HARTMUT R. SCHROEDER, ROBERT C. BOGUSLASKI, ROBERT J. CARRICO, and ROBERT T. BUCKLER

Low levels of hormones, drugs, and metabolites in complex biological fluids are frequently measured by competitive protein-binding methods. These assays take advantage of the ability of antibodies and certain other proteins to bind ligands with high affinity and specificity. Typically, the ligand of interest and a radiolabeled analog are allowed to compete for a limited number of protein-binding sites, and then the amount of bound or free radiolabeled ligand is determined. The result is related to the level of unlabeled ligand by means of a standard curve.

The use of radiolabeled ligands in these assays pose some convenience and stability problems. Therefore, nonradioactive materials that can be detected with high sensitivity are useful alternatives to radiolabels. Since some aminophthalhydrazides participate in simple oxidation reactions to produce light with high quantum efficiencies, they can be measured quantitatively at picomolar levels, and thus are useful labeling materials.[1] Ligands of interest are attached through an alkyl-bridging group to the amino residue of isoluminol (6-amino-2,3-dihydrophthalazine-1,4-dione). Isoluminol is employed because alkylation of the amino residue increases the quantum efficiency to about the same value as luminol, whereas similar substitution on luminol (5-amino-2,3-dihydrophthalazine-1,4-dione) decreases its efficiency substantially.[2-4]

The sensitivity of competitive binding assays is determined in part by the lower limit for detection of the label. Isoluminol derivatives with several different bridging groups coupled to various ligands have been synthesized. In addition, derivatives of 7-aminonaphthalene-1,2-dicarboxylic acid hydrazide have also been investigated as labels because such compounds have very high quantum efficiences. The lower limits for detection of these compounds using several oxidation systems are presented along with their use in protein-binding assays.

[1] H. R. Schroeder, P. O. Vogelhut, R. J. Carrico, R. C. Boguslaski, and R. T. Buckler, *Anal. Chem.* **48,** 1933 (1976).
[2] R. B. Brundrett, D. F. Roswell, and E. H. White, *J. Am. Chem. Soc.,* **94,** 7536 (1972).
[3] K.-D. Gundermann and M. Drawert, *Chem. Ber.* **95,** 2018 (1962).
[4] R. B. Brundrett and E. H. White, *J. Am. Chem. Soc.* **96,** 7497 (1974).

ISBN 0-12-181957-4

Light Measurements

Peak light intensities from chemiluminescent reactions described here were recorded automatically with a DuPont 760 luminescence biometer. Alternatively, a custom-built microprocessor connected to the biometer monitored the signal at 10-msec intervals and printed out the peak light intensity and total light production in arbitrary units. The microprocessor was activated when the signal from the photomultiplier exceeds a threshold that is five times greater than the electronic noise.

Other sensitive photometers for monitoring light production have been described.[5-7]

Preparation of Reagents

Water used for preparation of reagents is filtered through charcoal and distilled from a glass apparatus. All glass containers, except the 6 × 50 mm test tubes used for the chemiluminescent reactions, are exposed to concentrated nitric acid overnight and rinsed thoroughly with distilled water. Reagents are measured with disposable glass pipettes or pneumatic pipettes equipped with disposable plastic tips.

Chemiluminescent Compounds. One-millimolar stock solutions of isoluminol and its derivatives are prepared in 0.1 *M* Na_2CO_3, pH 10.5, and these are stable at 4° for several months. Naphthylhydrazides are dissolved in 0.1 *M* Na_2CO_3–0.15 *M* NaOH, pH 12.6. A few hours before use, further dilutions are prepared in water or an appropriate buffer.

Heme Catalysts. One milligram of microperoxidase[8] (Sigma Chemical, St. Louis, Missouri) is dissolved in 2.5 ml of 10 m*M* Tris · HCl, pH 7.4. This 200 μ*M* stock solution is stable at 4° for at least 1 month. A 2 μ*M* solution is prepared in 75 m*M* sodium barbital buffer, pH 8.6, within a few hours of use.

A 1 m*M* stock solution of hematin in 0.1 *M* Na_2CO_3, pH 10.5, is allowed to age at least 1 week at 4° in order to reduce variability of performance in the assay. This solution is diluted to 1 μ*M* just prior to use. The activity of the catalyst in chemiluminescent reactions decreases gradually and this effect is compensated by increasing the amount of catalyst per reaction. In this way, the stock solution is usable for at least 2 months.

Antibody to Thyroxine. The antibody is raised in rabbits immunized

[5] G. W. Mitchell and J. W. Hastings, *Anal. Biochem.* **39**, 243 (1971).

[6] H. Rasmussen and R. Nielson, *Acta Chem. Scand.* **22**, 1745 (1968).

[7] W. R. Seitz and M. P. Neary, *Methods Biochem. Anal.* **23**, 161 (1976).

[8] N. J. Feder, *Histochem. Cytochem.* **18**, 911 (1970).

with a thyroxine-bovine serum albumin conjugate.[9] The antibody is purified by a published method.[10]

Chemiluminescent Reactions

Reaction mixtures (150 μl) containing the chemiluminescent compound and the appropriate catalyst are measured into 6 × 50 mm test tubes (Kimble, Division of Owens-Illinois, No. 73500). Each test tube is mounted in the photometer and a 10-μl aliquot of the oxidant is rapidly injected from a 25-μl Hamilton syringe. (The bevel on the syringe needle is removed so that the stream of oxidant will be directed downward.)

Alternatively, the injection of oxidant is accomplished with a custom-built solenoid pump equipped with a sapphire piston in a Teflon cylinder. Flow of liquid to and from the pump is controlled by Teflon valves (Series 2, Model 1 Teflon solenoid valve, General Valve Corp., East Hanover, New Jersey) operated by solenoids. The pump and valves are coordinated electronically. Teflon tubing (0.95 mm i.d. and 0.3 mm wall thickness) carries the liquid to a glass orifice (0.78 mm i.d.) mounted over the reaction tube in front of the photomultiplier. The pump delivers 10.1 ± 0.06 μl of solution in 30 msec.

Oxidation Systems

H_2O_2–Microperoxidase. Reaction mixtures (150 μl) contain 50 m*M* NaOH, 0.27 μ*M* microperoxidase, and the chemiluminescent compound. In addition, these mixtures contain 59 m*M* sodium barbital, which is the buffer for diluting microperoxidase and chemiluminescent compounds. The final pH is about 12.6. After the mixture has stood for at least 10 min at room temperature, the chemiluminescent reaction is initiated by injection of 10 μl of 90 m*M* H_2O_2 in 10 m*M* Tris·HCl, pH 7.4.

Alternatively, chemiluminescent reactions are carried out in 75 m*M* sodium barbital, pH 8.6, containing 0.27 μ*M* microperoxidase and variable concentrations of chemiluminescent compounds. The mixture is left for at least 10 min at room temperature, then a 10-μl aliquot of 3 m*M* H_2O_2 in 10 m*M* Tris·HCl, pH 7.4, is injected to initiate chemiluminescence.

H_2O_2–Hematin. In addition to the chemiluminescent compounds, these reaction mixtures (150 μl) contain 50 m*M* NaOH and 0.07 μ*M* hematin. They are allowed to stand at room temperature for at least 10 min, and then 10 μl of 90 m*M* H_2O_2 is injected to initiate light production.

[9] N. N. Alexander and V. F. Jennings, *Clin. Chem.* **20,** 1353 (1974).
[10] D. Livingston, this series, Vol. 34, p. 725.

Hypochlorite–CoCl₂. Light production is initiated by addition of a 10-μl aliquot of 10 m*M* NaOCl to a solution of the chemiluminescent compound (150 μl) containing 50 m*M* NaOH and 333 n*M* $CoCl_2$.

H_2O_2–$K_3Fe(CN)_6$. A solution of the chemiluminescent compound (150 μl) is prepared in 50 m*M* Na_2CO_3, pH 10.5. A 10-μl aliquot of a solution containing 1 m*M* H_2O_2 and 5 μ*M* $K_3Fe(CN)_6$ is injected to initiate light emission.

H_2O_2–Lactoperoxidase. This reaction system (150 μl) contains the chemiluminescent compound and 10 μg of lactoperoxidase in 0.1 *M* Tris · HCl at either pH 7.4 or 8.8. After incubation at 25° for 2 min, 10 μl of 1 m*M* H_2O_2 in 10 m*M* Tris · HCl, pH 7.4, are injected.

Synthesis of 6-*N*-(3-Amino-2-hydroxypropyl)amino-2,3-Dihydro-1,4-phthalazine-dione [Table I (I)]

The synthesis of 4-amino-*N*-methylphthalimide is carried out according to the method of Drew and Pearman.[11] 4-Nitrophthalic acid, 211 g (1 mol), is stirred with 400 ml of acetic anhydride at reflux temperature; when complete solution is achieved, the temperature is maintained for an additional 15 min. The solvent is removed under vacuum and the residue is recrystallized from toluene to yield 103 g (53% yield) of 4-nitrophthalic anhydride, mp 119°–120°.

4-Nitrophthalic anhydride, 103 g (0.5 mol), is stirred with 200 ml of 20% methylamine at 0° for 30 min. When the solution is adjusted to pH 2.0 with concentrated HCl, a white precipitate forms which is collected by filtration, washed with water, and dried. The solid is then stirred with 200 ml of acetic anhydride at reflux temperature for 15 min after complete solution is obtained. As the reaction mixture cools, the product crystallizes. The crystals are collected by filtration and washed with acetic acid and water. The dried crystals provide 61 g (59% yield) of *N*-methyl-4-nitrophthalimide, mp 178°.

A solution of 281 g of $SnCl_2 \cdot 2H_2O$ (1.25 mol) is prepared in 500 ml of water and 1.5 liters of concentrated HCl. To this is added 67 g (0.3 mol) of *N*-methyl-4-nitrophthalimide with vigorous stirring. The reaction mixture becomes warm, and complete solution is obtained. Then it is allowed to stand overnight, and a pale yellow solid which forms is collected by filtration. This material is washed with copious amounts of water and dried to give 36 g (67% yield) of 4-amino-*N*-methylphthalimide, mp 247°.

Twenty-five grams (0.142 mol) of 4-amino-*N*-methylphthalimide and 20.7 g (0.21 mol) of 1-chloro-2,3-epoxypropane are combined with 150 ml of

[11] H. D. K. Drew and F. H. Pearman, *J. Chem. Soc. London* p. 26 (1937).

TABLE I

RELATIVE PEAK LIGHT INTENSITIES AND LIMITS FOR DETECTION OF CHEMILUMINESCENT COMPOUNDS[a]

Aminophthalhydrazides

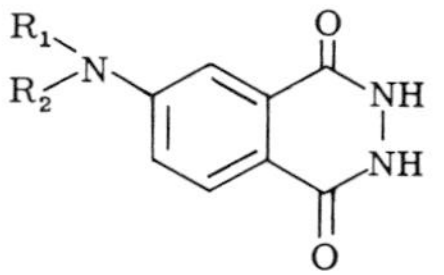

	R_1	R_2	Relative peak light intensity[b] (%)	Detection limit (p*M*)
Isoluminol	$-H$	$-H$	5	30
Diethyl-isoluminol[c]	$-CH_2CH_3$	$-CH_2CH_3$	100(100)[d]	1(1)[d]
(I)	$-H$	$-CH_2CH(OH)CH_2NH_2$	10(5)[d]	20(10)[d]
(II)	$-H$	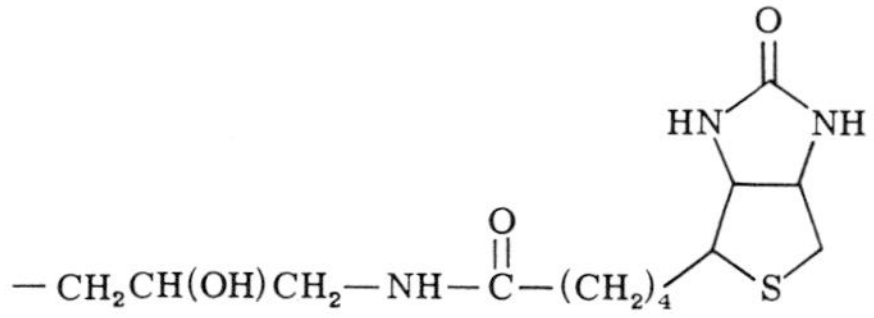	(5)[d]	(10)[d]
(III)	$-CH_2CH_3$	$-CH_2CH(OH)CH_2NH_2$	46	5
(IV)	$-CH_2CH_3$	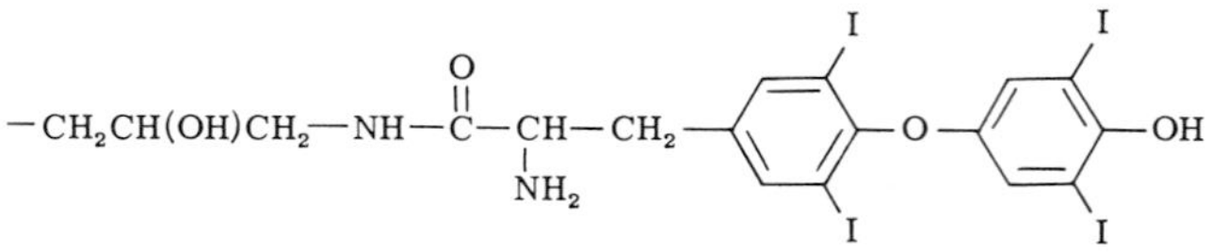	2	100
(V)	$-H$	$-(CH_2)_4NH_2$	14	20
(VI)	$-CH_2CH_3$	$-(CH_2)_4NH_2$	84	2

(VII)	$-CH_2CH_3$	$-(CH_2)_4NH-C(=O)-CH(NH_2)-CH_2-$	2	100
(VIII)	$CH_3CH_2-N-(CH_2)_4NH_2$		$(420)^d$ 120^e	$(0.1)^d$ 0.5^e
(IX)	$CH_3CH_2-N-(CH_2)_4NH-C(=O)-CH(NH_2)-CH_2-$		$(20)^d$ 3.2^e	$(2.0)^d$ 10^e

[a] Measurements were conducted with the H_2O_2–hematin system, pH 12.6.

[b] Plots of peak light intensities versus concentration were prepared for each chemiluminescent compound. The slopes were compared to that obtained with diethylisoluminol to calculate the relative peak light intensities.

[c] Synthesized according to K.-D. Gundermann and M. Drawert, *Chem. Ber.* **95,** 2018 (1962).

[d] Results in parentheses were obtained with the H_2O_2–microperoxidase system, pH 12.6.

[e] Measurements conducted with the H_2O_2–microperoxidase system, pH 8.6.

2,2,2-trifluoroethanol and stirred at reflux temperature for 48 hr. Part of the solvent, 70–80 ml, is removed by distillation, and the remaining solution is allowed to cool to room temperature. A yellow solid forms and is collected by filtration. It is digested with ethyl acetate and collected again by filtration to give 29.5 g (77% yield) of 4-*N*-(3-chloro-2-hydroxypropyl)amino-*N*-methylphthalimide, mp 136°–138°.

Potassium phthalimide, 15.7 g (0.085 mol), and 13.5 g (0.05 mol) of 4-*N*-(3-chloro-2-hydroxypropyl)amino-*N*-methylphthalimide are stirred in 150 ml of dimethylformamide at reflux temperature for 24 hr. The dimethylformamide is removed at low pressure, and the residue is washed with water and collected by filtration. The yellow solid is recrystallized from a glacial acetic acid–water solvent pair to give 12.8 g (67% yield) of *N*-methyl-4-*N*-[2-hydroxy-3-(*N*-phthalimido)propyl]-aminophthalimide, mp 247°–248°.

Thirty-five milliliters of 95% hydrazine and 5 g of *N*-methyl-4-*N*-[2-hydroxy-3-(*N*-phthalimido)propyl]aminophthalimide (13.2 mmol) are combined with 90 ml of ethanol and stirred at reflux temperature for 4 hr. The solvent is removed under reduced pressure, and the resulting solid is dried in vacuum at 120°. The dried material is stirred for 1 hr with 70 ml of 0.1 *N* HCl and then the insoluble by-product is removed by filtration. The filtrate is adjusted to pH 6.5 with a saturated solution of sodium bicarbonate. A white precipitate forms and is collected and dried to give 2.2 g (67% yield) of (I). Recrystallization from water provides analytically pure material that decomposes at 273°.

Synthesis of 6-*N*-[(3-Biotinylamido)-2-hydroxypropyl]amino-2,3-dihydrophthalazine-1,4-dione [Table I (II)]

Biotin, 0.29 g (1.2 mmol), and 0.17 ml of triethylamine are dissolved in 20 ml of dry dimethylformamide and cooled to −10° under anhydrous conditions. A solution containing 0.14 ml of ethyl chloroformate in 2.9 ml of diethylether is added slowly to the reaction mixture, which then is allowed to stand for 30 min. A precipitate of triethylammonium chloride is removed by filtration. One milliliter of dry pyridine and a suspension of 600 mg (2.4 mmol) of (I) in 20 ml of dry dimethylformamide is added to the filtrate. This mixture is stirred at −10° for 30 min and then allowed to warm to room temperature overnight. The solvent is removed at reduced pressure and the oily residue is stirred with 50 ml of 0.1 *N* HCl for 1 hr. A white solid which forms is collected by filtration and washed with 0.1 *N* HCl and with water. This product is dried under vacuum overnight at room temperature to give 0.55 g (97% yield) of (II), mp. 170°–173°.

Synthesis of 6-[*N*-(3-Amino-2-hydroxypropyl)-*N*-ethyl]amino-2,3-dihydro-1,4-phthalazine-dione [Table I (III)]

One hundred-six grams (0.7 mol) of diethyl sulfate and 61 g (0.35 mol) of 4-amino-*N*-methylphthalimide are combined in 500 ml of 2,2,2-trifluoroethanol and heated at reflux temperature for 48 hr. Then the reaction mixture is concentrated under reduced pressure and carefully neutralized with a saturated solution of Na_2CO_3. A precipitate which forms is collected by filtration and dried to yield 62 g of 4-(*N*-ethylamino)-*N*-methylphthalimide. Recrystallization from aqueous methanol provides 51 g (74% yield) of pale yellow product of analytical purity, mp 149°–151°.

A mixture containing 51 g of 4-(*N*-ethylamino)-*N*-methylphthalimide (0.25 mol) and 54 g of 1-chloro-2,3-epoxypropane (0.59 mol) is heated at reflux temperature in 500 ml of 2,2,2-trifluoroethanol for 48 hr. The resulting yellow solution is concentrated under vacuum, and the residue is chromatographed on a 7 × 76 cm column of silica gel 60 (1000 g) (E. Merck, Darmstadt, Germany) equilibrated with CCl_4:acetone (9:1). The chromatogram is developed with this solvent, and the desired product elutes between 4 and 6.9 liters of effluent. This material is recrystallized several times from aqueous methanol to yield 27 g (44%) of analytically pure 4-[*N*-(3-chloro-2-hydroxypropyl)-*N*-ethyl]amino-*N*-methylphthalimide, mp 122°.

Potassium phthalimide, 23 g (0.13 mol), and 25 g of 4-[*N*-(3-chloro-2-hydroxypropyl)-*N*-ethyl]amino-*N*-methylphthalimide (0.08 mol) are stirred at reflux temperature in 150 ml of dry dimethylformamide for 36 hr. The solvent is removed under vacuum, leaving a brown residue that is triturated with methanol to produce 19 g of yellow solid. Crystallization of the solid from aqueous acetic acid gives 16 g (49% yield) of 4-{*N*-ethyl-*N*-[2-hydroxy-3-(*N*-phthalimido)propyl]}amino-*N*-methylphthalimide. Recrystallization from aqueous methanol provides this compound in analytically pure form, mp 158°–160°.

Sixty milliliters of 95% hydrazine and 300 ml of absolute ethanol are combined with 15 g of 4-{*N*-ethyl-[*N*-2-hydroxy-3-(*N*-phthalimido)-propyl]}amino-*N*-methylphthalimide (0.037 mol) and heated at reflux for 3 hr. Then the solvent is removed under vacuum and the residue is dried overnight at 110° and 0.05 mm Hg. This material is stirred in dilute HCl and filtered to remove insoluble by-products. The filtrate is adjusted to pH 7.0 with a saturated solution of sodium carbonate, and the precipitate that forms is collected to provide 4.6 g (46% yield) of (III). Recrystallization from water provides an analytically pure sample, mp 208°–211°.

Synthesis of 6-{*N*-Ethyl-*N*-[2-hydroxy-3-(thyroxinylamido)propyl]}-amino-2,3-dihydro-1,4-phthalazine-dione [Table I (IV)]

Twenty grams of L-thyroxine (25.6 mmol) are stirred with 240 ml of anhydrous ethyl acetate at 0°, and 46 ml of trifluoroacetic acid and 7.6 ml of trifluoroacetic anhydride are added. The reaction mixture becomes clear while stirring for 1 hr; 200 ml of cold water (0°) are added, and this mixture is saturated with NaCl. The undissolved NaCl is removed by filtration, and the organic and aqueous layers are separated. The organic phase is extracted twice with saturated NaCl (100 ml each) and then dried over $MgSO_4$. The solvent is removed at 35° under vacuum and the white residue is washed twice with water and dried at 60° and 0.05 mm Hg to provide 21 g (95% yield) of *N*-trifluoroacetylthyroxine. Analytically pure material is obtained by recrystallization from ether–petroleum ether, mp 233°–235° (decomp.).

Triethylamine, 0.17 ml, and 1.05 g of *N*-trifluoroacetylthyroxine (1.2 mmol) are dissolved in 20 ml of dimethylformamide at −10° under anhydrous conditions. A solution composed of 0.14 ml of ethyl chloroformate (1.2 mmol) and 2.9 ml of anhydrous diethylether is added and allowed to react for 30 min. A precipitate of triethylammonium chloride is removed by filtration, and the *N*-trifluoroacetylthyroxinyl ethyl carbonic anhydride in the filtrate is used without isolation for the next step of the synthesis.

Compound (III) (670 mg; 2.4 mmol) is suspended in 20 ml of dimethylformamide and 1 ml of pyridine at −10° under anhydrous conditions and combined with the anhydride described above. The reaction mixture is allowed to stand at room temperature for 2 days. Then the solvent is removed under vacuum and the residue is washed with 10% HCl. The solid material is collected by filtration and dried.

The trifluoroacetyl blocking group is removed by treatment with alkali. The solid is stirred with 40 ml of 0.1 *M* Na_2CO_3, pH 10.5, and a small amount of dimethylformamide is added to obtain a clear solution. After 24 hr, the solvents are removed under vacuum. The residue is dissolved in 30 ml of water, and this solution is adjusted to pH 7.2 with dilute HCl. A white precipitate which forms is collected by filtration and dried at 60° and 0.05 mm Hg to give 600 mg (50% yield) of (IV), mp > 240° (d).

Synthesis of 6-*N*-(4-Aminobutyl)amino-2,3-dihydro-1,4-phthalazine-dione [Table I (V)]

Forty-two grams of *N*-(4-bromobutyl)phthalimide (0.15 mol) and 51.5 g of 4-amino-*N*-methylphthalimide (0.29 mol) are stirred with 300 ml dimethylformamide at reflux temperature for 24 hr. A yellow precipitate which forms on cooling to room temperature is collected by filtration and dried to give 38.5 g (68% yield) of *N*-methyl-4-*N*-[4-(*N*-phthalimido) butyl]aminophthalimide. Recrystallization from aqueous acetic acid provides a product of analytical purity, mp 217°–218°.

Fifteen milliliters of 95% hydrazine in 75 ml of absolute ethanol is combined with 4 g of *N*-methyl-4-*N*-[4-(*N*-phthalimido)butyl]aminophthalimide and heated to reflux temperature for 3 hrs. The solvent is removed under vacuum, and the residue is dried at 110° and 0.1 mm Hg. This material is stirred with 50 ml of 10% HCl and filtered to remove insoluble by-products. The filtrate is adjusted to pH 7 with sodium bicarbonate and the precipitate which forms is collected by filtration to give 400 mg (15% yield) of (V). A sample recrystallized from water decomposes at 335°. Chemical ionization mass spectrometry gives a P + 1 peak at 249 and P + 2 peak at 250. The calculated molecular weight is 248.

Synthesis of 6-[*N*-(4-Aminobutyl)-*N*-ethyl]amino-2,3-dihydro-1,4-phthalazine-1,4-dione [Table I (VI)]

One hundred milliliters of diethyl sulfate (0.77 mol) and 38 g (0.1 mol) of *N*-methyl-4-*N*-[4-(*N*-phthalimido)butyl]aminophthalimide is heated at 160° under anhydrous conditions in an oil bath for 45 min. The mixture is cooled to room temperature, then poured into 3 liters of ice water. A yellow precipitate which forms is collected by filtration and dried. Recrystallization from aqueous acetic acid gives 29 g (72% yield) of analytically pure 4-[*N*-ethyl-*N*-4-(*N*-phthalimido)butyl]amino-*N*-methylphthalimide, mp 164°–165°.

Eighty milliliters of 95% hydrazine in 300 ml of absolute ethanol is combined with 29 g of 4-[*N*-ethyl-*N*-(4-*N*-phthalimido)butyl]amino-*N*-methylphthalimide (0.07 mol) and heated at reflux temperature for 3 hr. Then the solvent is removed under vacuum, and the residue is dried at 110° and 0.05 mm Hg. The dried material is stirred with dilute HCl and filtered to remove insoluble by-products. A precipitate forms when the filtrate is adjusted to pH 9.0 with KOH. This solid is recrystallized from 50% aqueous dimethylformamide to give 6.5 g (33% yield) of analytically pure (VI), mp 255°–257°.

Synthesis of 6-[*N*-Ethyl-*N*-(4-thyroxinylamido)butyl]amino-2,3-dihydro-1,4-phthalazine-dione [Table I (VII)]

Carbonyldiimidazole, 0.8 g (5.0 mmol), is added to a solution of 4.4 g (5.0 mmol) of *N*-trifluoroacetylthyroxine in 50 ml of dry tetrahydrofuran, and this mixture is heated at reflux temperature for 10 min. The solvent is removed at low pressure. A suspension of 1.4 g of (VI) (5 mmol) in dry dimethylformamide is added and stirred at room temperature for 48 hr. The solvent is removed at reduced pressure, and the residue is stirred with 80 ml of 10% HCl. The insoluble material is collected by filtration and dried at room temperature and 0.1 mm Hg.

A portion of this material, 2.1 g, is dissolved in 35 ml of 0.5 *N* NaOH and allowed to react for 1 hr at room temperature to remove the trifluoroacetyl blocking group. The solution is adjusted to pH 5.0 with concentrated HCl, and a precipitate which forms is collected by filtration, washed with water, and dried at room temperature and at 0.1 mm Hg. This material is chromatographed on 200 g of silica gel 60, using ethanol/triethylammonium bicarbonate, pH 7.5 (7:3) as solvent. The product elutes between 490 and 650 ml of effluent, and the solvent is removed under vacuum to give 680 mg of cream-colored solid. This material is stirred with 50 ml of 50% dimethylformamide and filtered, and the product is precipitated by addition of water. The white precipitate is collected by centrifugation at 20,000 *g* for 25 min and dried at room temperature to give 200 mg of analytically pure (VII), mp 200° (decomp.).

Synthesis of 7-[*N*-(4-Aminobutyl)-*N*-ethyl]aminonaphthalene-1,2-dicarboxylic Acid Hydrazide [Table I (VIII)]

Initial steps in the synthesis of (VIII) are based on the method of Gundermann.[12] One liter of tetrachloroethane, 250 ml of nitrobenzene, 108 g (1 mol) of anisole, and 111 g (1.1 mol) of succinic anhydride are stirred at room temperature for 30 min and then cooled to 0°. Then 280 g (2.1 mol) of $AlCl_3$ are added, and the reaction mixture is stirred for 3 days at 0°. The reaction is quenched by addition of ice until evolution of HCl gas ceases. Two hundred milliliters of 10% hydrochloric acid are added, and the organic solvent is removed by steam distillation. On cooling, the aqueous phase deposits a tan solid, which is crystallized from aqueous ethanol to give 188 g (91% yield) of 4-(4-methoxyphenyl)-4-oxobutyric acid, mp 149°–150°.

A zinc amalgam is prepared by combining 374 g (5.7 g-atom) of mossy zinc, 37.4 g (0.18 mol) of mercuric chloride, 560 ml of H_2O, and 19 ml of concentrated hydrochloric acid. This mixture is stirred vigorously for 5 min. The liquid is decanted, and 280 ml of water, 655 ml of concentrated hydrochloric acid, 375 ml of toluene, 20 ml of glacial acetic acid, and 188 g of 4-(4-methoxyphenyl)-4-oxobutyric acid are added. The reaction mixture

[12] K.-D. Gundermann, W. Horstmann, and G. Bergmann, *Justus Liebigs Ann. Chem.* **684,** 127 (1965).

is heated at reflux temperature, and at 6-hr intervals during the next 24 hr 200 ml of concentrated hydrochloric acid are added. After cooling, the aqueous phase is separated and washed with three portions (500 ml each) of ether, and these extracts are combined with the organic phase from the reaction. The combined organic extracts are mixed with 500 ml of 2 *N* NaOH, and the organic solvent is removed by steam distillation. Then 88 g (0.7 mol) of dimethyl sulfate are added and the temperature is maintained at 80°–90°. Additional 2 *N* NaOH (approximately 300 ml) is added when the reaction becomes neutral. After 5 hr, the reaction is cooled, acidified with concentrated hydrochloric acid, and extracted with three portions (500 ml each) of ether. These extracts are combined and dried, and the solvent is removed. The residue is recrystallized from aqueous ethanol to give 90 g (52% yield) of 4-(4-methoxyphenyl)butyric acid as white crystals, mp 59°–61°.

A mixture of 90 g (0.47 mol) of 4-(4-methoxyphenyl)butyric acid, 124 g (0.94 mol) of boron trifluoride–methanol complex, and 400 ml of methanol is refluxed for 4 hr. After cooling, the reaction mixture is extracted with three portions (400 ml each) of ether, and the solvent is removed from the extracts to give 96 g (98% yield) of methyl 4-(4-methoxyphenyl)butyrate as a yellow oil.

To a dry, three-neck round-bottom flask, fitted with a mechanical stirrer, a reflux condenser, and an inlet for argon, are added 18.2 g (0.48 g-atom) of potassium, 800 ml of dry ether (distilled from $LiAlH_4$), and 28.3 ml (0.48 mol) of anhydrous ethanol. While the reaction mixture is refluxed for 5 hr, the potassium reacts to give a white suspension, and 94 ml (0.69 mol) of dimethyloxalate is added dropwise over 15 min to give a clear brown solution. Then 95.6 g (0.46 mol) of methyl 4-(4-methoxyphenyl)-butyrate is added, and a tan solid, which forms while the reaction refluxes overnight, is collected by filtration and washed with the ether. On drying, this solid provides 106 g (67% yield) of ethyl [2-hydroxy-3-methoxycarbonyl-5-(4-methoxyphenyl)]valerate potassium enolate, mp 120°–122°.

This product (0.31 mol) is added in portions to 326 ml of 60% H_2SO_4 at 5°–15°. When the addition is complete, the reaction mixture is heated on a steam bath for 15 min and then cooled to 20°. An additional 652 ml of 80% H_2SO_4 is added and the heating is resumed for 15 min. During this time a yellow precipitate forms, and the mixture is poured into 1 liter of cold water. The product is collected by filtration to give 24.9 g (35% yield) of 7-methoxy-3,4-dihydronaphthalene-1,2-dicarboxylic acid anhydride as yellow crystals, mp 163°–165°.

A mixture of 19.9 g (0.087 mol) of the 7-methoxy-3,4-dihydronaphthalene-1,2-dicarboxylic acid and 2.8 g (0.087 g-atom) of sulfur are heated at 250° for 3 hr under argon. On cooling, the brown solution solidifies and crystallization from benzene gives 15 g (80% yield) of 7-methoxynaphthalene-1,2-dicarboxylic anhydride as yellow crystals, mp 193°–194°.

A mixture of 15 g (0.07 mol) of 7-methoxynaphthalene-1,2-dicarboxylic anhydride, 240 ml of glacial acetic acid and 240 ml of 48% HBr are refluxed for 2 hr, and during this time a thick yellow precipitate forms. After cooling, the precipitate is collected by filtration to give 11.2 g (75% yield) of 7-hydroxynaphthalene-1,2-dicarboxylic anhydride, mp 285°.

A solution is prepared by dissolving 180 g of SO_2 in 240 ml of 20% aqueous ammonium hydroxide at 0°. To this solution is added 200 ml of 20% aqueous ammonium hydroxide and 25 g (0.12 mol) of 7-hydroxynaphthalene-1,2-dicarboxylic anhydride. This reaction mixture is placed in a steel autoclave and heated at 170° for 8 hr. After cooling, the reaction mixture is concentrated under reduced pressure to remove NH_3. Neutralization with acetic acid gives 7-aminonaphthalene-1,2-dicarboxylic acid as a red solid. This solid is heated in a steam bath for 2 hr in 70 ml of absolute methanol and 40 ml of concentrated H_2SO_4. Then the reaction mixture is cooled and poured into 1 liter of crushed ice containing excess Na_2CO_3. The neutralized mixture is extracted with three portions (250 ml each) of ether, and the combined extracts are dried over anhydrous $MgSO_4$. Removal of the solvent leaves 20 g (66% yield) of dimethyl 7-aminonaphthalene-1,2-dicarboxylate as a red oil.

Dimethyl 7-aminonaphthalene-1,2-dicarboxylate, 25.9 g (0.1 mol), and 14.1 g (0.05 mol) of *N*-(4-bromobutyl)phthalimide are refluxed in 100 ml of 2,2,2-trifluoroethanol under argon for 16 hr. The solvent is removed under vacuum and the residue is partitioned between 400 ml of ether and 300 ml of water. The ether is dried over anhydrous $MgSO_4$ and cleared by filtration, and then the solvent is removed under vacuum. The dark red residue was dissolved in 100 ml of benzene and chromatographed on 1200 g of silica gel 60 equilibrated with benzene: methanol (19: 1). The chromatogram is developed with this solvent, and the desired product elutes between 5.1 and 5.7 liters of effluent. The solvent is removed under vacuum to give 9 g (40% yield) of dimethyl 7-*N*-(4-phthalimidobutyl) aminonaphthalene-1,2-dicarboxylate as a red oil.

Nine grams (0.02 mol) of dimethyl 7-*N*-(4-phthalimidobutyl)aminonaphthalene-1,2-dicarboxylate dissolved in 20 ml of diethyl sulfate is heated at 130° for 2 hr under argon. Then the reaction mixture is poured into 200 ml of saturated $NaHCO_3$ containing crushed ice. The aqueous solution is extracted with 250 ml of ether, and this extraction is repeated twice. The combined ether extracts are dried, and the solvent is removed to give a red oil contaminated with diethyl sulfate. This oil is chromatographed on a 600-g column of silica gel 60 with benzene: methanol (19: 1) as eluting solvent. The yellow fractions are pooled and concentrated under vacuum to provide an oil that contains some diethyl sulfate. This residual impurity is removed by sublimation at 50° and 0.01 mm Hg to give 4.0 g (42% yield) of dimethyl 7-1(*N*-ethyl-4-[*N*-phthalimido)butyl]aminonaphthalene-1,2-dicarboxylate as a light red oil.

Fifteen milliliters of 85% hydrazine, 4 g (8 mmol) of dimethyl 7-[(*N*-ethyl)-4-(*N*-phthalimido)butyl]aminonaphthalene-1,2-dicarboxylate, and 50 ml of absolute methanol are refluxed for 3 hr. Then the solvent is removed under vacuum, and the crystalline residue is dried at 80° and 0.01 mm Hg overnight. This material is chromatographed on a 200 g column of silica gel 60 equilibrated with ethanol: 1 *M* triethylammonium bicarbonate, pH 7.5, (7:3) as solvent. The desired product elutes between 0.5 and 1.5 liters of the effluent, and the solvent is removed to provide a yellow solid. It is recrystallized twice from pyridine to provide 1.1 g (42% yield) of 7-[*N*-(4-aminobutyl)-*N*-ethyl]aminonaphthalene-1,2-dicarboxylic acid hydrazide (VIII) as fine yellow crystals, mp 246°–247°.

Synthesis of 7-[*N*-Ethyl-*N*-(4-thyroxinylamido)butyl]aminonaphthalene-1,2-Dicarboxylic Acid Hydrazide [Table I, (IX)]

A solution of 1.7 g of *N*-trifluoroacetylthyroxine (2 mmol) in 20 ml of dry pyridine is stirred at −10° under argon, and 450 mg of dicyclohexylcarbodiimide (2.2 mmol) are added. Forty-five minutes later, 980 mg of (VIII) is added and allowed to react for 3 hr. Then the reaction is allowed to warm to room temperature overnight. One hundred milliliters of pyridine and 10 g of silica gel 60 are added, and the pyridine is removed under vacuum. The impregnated silica gel is placed on top of a column of 200 g of silica gel 60 equilibrated with ethanol: 1 *M* triethylammonium bicarbonate, pH 7.5 (7:3). The chromatogram is developed with this solvent and the desired product elutes between 0.78 and 1.0 liter of effluent. The solvent is removed to give a yellow crystalline residue, which is dissolved in 300 ml of 1 *M* triethylammonium bicarbonate, pH 7.5, and refluxed for 3 hr to remove the trifluoroacetyl blocking group. The hot reaction mixture is filtered to remove some oily material. Then 15 g of silica gel 60 is added, and the slurry is dried on a rotary evaporator. The impregnated silica is placed on top of a 200 g column of silica gel 60 and the chromatogram is developed with ethanol: 1 *M* triethylammonium bicarbonate, pH 7.5 (7:3). The desired product elutes between 0.6 and 1.0 liter of effluent, and the solvent is removed to give 0.9 g of a yellow solid. Of this solid, 110 mg are chromatographed on a 3.2 × 45 cm column of Sephadex LH-20 equilibrated with methanol, and the desired product elutes between 450 and 530 ml of effluent to provide 60 mg (22% yield) of (IX) as a yellow solid, mp 218° (decomp.).

Performance of Oxidation Systems for Chemiluminescent Compounds

Sensitivity of Chemiluminescent Oxidation Systems. A number of oxidation systems can be used to produce chemiluminescence (Table II). Peak light intensities produced by several levels of luminol in the picomolar range were measured with each oxidizing system. Plots of peak light intensity versus luminol concentration were prepared, and the sensitivity of measurements with each oxidation system was estimated by determining the luminol concentration that gave a peak light intensity at 1.5 times the background level determined with a control without luminol. The results obtained from the various oxidation systems differ by more than 100-fold (Table II). The best sensitivity was achieved with the H_2O_2–microperoxidase and the H_2O_2–hematin systems, which could detect 1 p*M* (0.15 fmol) luminol. Measurements of total light production rather than peak light intensities did not change the limits of sensitivity.

Light was not produced when H_2O_2 and isoluminol were mixed in the absence of a catalyst.

Effects of Buffers on Chemiluminescence. The microperoxidase catalyst has the advantage that it can be used equally effectively at pH 8.6 and 12.6 (Fig. 1); whereas, hematin is active only at high pH. Peak light intensities increased linearly with the diethylisoluminol concentration at either pH (Fig. 1). When barbital buffer, pH 8.6, was replaced by 50 m*M* sodium phosphate, pH 8.6, the peak light intensities were unchanged. Chemiluminescent reactions conducted with this buffer at pH 8.0 and 7.5 gave 10 and 1%, respectively, the peak light intensities observed at pH 8.6. Reactions conducted in 50 m*M* Tris · HCl buffer, pH 8.6, gave one-third the

TABLE II

LIMITS FOR DETECTION OF LUMINOL WITH VARIOUS OXIDATION SYSTEMS[a]

Oxidation system	pH	Detection limit (p*M*)
H_2O_2–microperoxidase	8.6	1
	12.6	1
H_2O_2–hematin	12.6	1
NaOCl–$CoCl_2$	13.0	10
$NaIO_4$	13.0	20
H_2O_2–$K_3Fe(CN)_6$	10.5	40
H_2O_2–lactoperoxidase	8.8	50
	7.4	300

[a] Ten-microliter aliquots of the oxidants were added to 150 μl of reaction mixtures containing various concentrations of luminol. The detection limits are the luminol concentrations that gave peak light intensities at least 1.5 times those determined from controls without luminol.

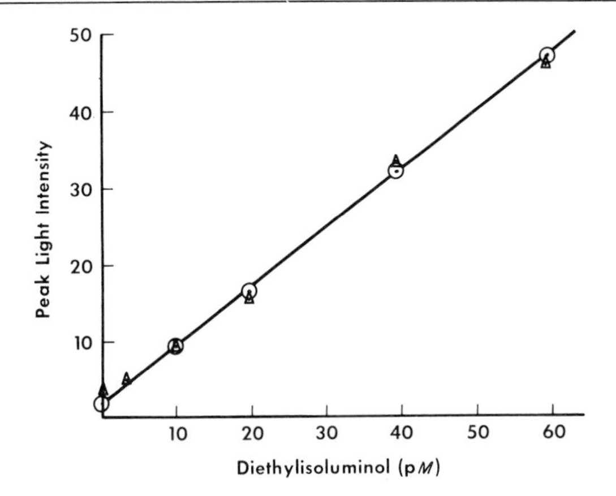

FIG. 1. Peak light intensities obtained on oxidation of diethylisoluminol with the H_2O_2–microperoxidase system. Reaction mixtures (150 μl) containing the indicated concentrations of diethylisoluminol were oxidized with the H_2O_2–microperoxidase system at pH 12.6 (○) and pH 8.6 (Δ). Each data point is the average of triplicate measurements.

peak light intensity measured with the barbital buffer, and 50 m*M* Bicine buffer, pH 8.6, completely quenched light production.

Kinetics of Light Emission. Light emission produced by diethylisoluminol and (VII) at pH 8.6 and 12.6, using microperoxidase as a catalyst, reached peak intensity within 1 sec after H_2O_2 was added. Then the light intensities decreased with half-lives of 0.5 and 4.5 sec at pH 8.6 and 12.6, respectively.

Effects of Proteins on Chemiluminescent Reactions. Competitive protein-binding reactions are often used to assay for substances in biological fluids containing proteins and other materials that might interfere with chemiluminescent reactions. To study the effects of proteins on the reactions, 30 p*M* (VI) was oxidized with the H_2O_2–microperoxidase system, pH 12.6, in the presence of various concentrations of proteins. Reaction mixtures containing 0.8 and 8.3 mg of bovine serum albumin per milliliter gave peak light intensities that were 86 and 48%, respectively, of control values measured in the absence of albumin. The peak light intensity produced by (VII) with the H_2O_2–microperoxidase system, pH 12.6, was enhanced about 2-fold in the presence of 8.3 mg of bovine serum albumin per milliliter; however, in the H_2O_2–hematin system, the light emission was quenched. These results indicate that the effects of proteins vary depending on the oxidation system.

In additional experiments, 3 n*M* (VII) was oxidized with the H_2O_2–microperoxidase system, pH 12.6, in the presence of 0.01 and 0.1% serum, and the peak light intensities were 80% and 50%, respectively, of the control value measured without the protein. In some cases, interference by endogenous materials in the specimen can be avoided by dilution. In other cases, it might be necessary to carry out a preliminary extraction of the ligand being assayed.

Limits for Quantitative Measurement of Chemiluminescent Labels and Their Ligand Conjugates

Modification of the chemical structure of aminophthalhydrazides can have a large effect on the chemiluminescent quantum yields.[3] Ligands are attached to the amino group of isoluminol through an alkyl bridging arm because alkylation at this position increases light production in chemiluminescent reactions (Table I).[2] The chemical constitution of the bridging arm has only a small effect on light emission as demonstrated by (I) and (V), which produced similar peak light intensities (Table I). Ethylation of the secondary amino residues of (I) and (V) leads to about 5-fold increase in the peak light intensities [compared to (III) and (VI), respectively].

Attachment of the ligand biotin to the primary amino residue of (I) did not change the peak light intensity produced in the H_2O_2–microperoxidase system (Table I). However, when thyroxine was coupled to (I), the peak light intensity was 10% of the value observed with (I) (data not shown). Similarly, thyroxine conjugates of (III) and (VI) produced peak light intensities that were less than 5% of the values obtained with the corresponding unconjugated aminophthalhydrazides (Table I). Since addition of exogenous thyroxine at levels 1000 times the label (VI) does not affect light emission, the losses observed with the conjugate (VII) may be due to quenching by iodine via an intramolecular route.[13,14] It is unlikely that most other ligands will have as large an effect on light production when they are coupled to isoluminol derivatives.

Aminonaphthylhydrazides can be measured quantitatively at lower levels than the aminophthalhydrazides. For instance, (VIII) can be determined at levels as low as 0.1 p*M* (0.015 fmol) with the H_2O_2–microperoxidase system, pH 12.6 (Table I). At pH 8.6 the limit is about 0.5 p*M*, and this difference might be due to the poorer solubility of (VIII) in aqueous solvents that are not strongly basic. The thyroxine–naphthylhydrazide conjugate (IX) can be measured at substantially lower levels than the corresponding aminophthalhydrazide conjugates (IV) and (VII).

[13] G. K. Radda, *in* "Methods in Membrane Biology" (E. D. Moru, ed.), Vol. 4, p. 140. Plenum, New York, 1975.

[14] G. R. Fleming, A. W. E. Knight, J. M. Morris, R. J. S. Morrison, and G. W. Robinson, *J. Am. Chem. Soc.* **99,** 4306 (1977).

Quenching of light also occurs with the thyroxine–naphthylhydrazide conjugate (IX), but to lesser extent.

In general, the relative peak light intensities obtained with compounds in Table I are consistent with quantum yields reported for similar chemiluminescent compounds.[2-4,12]

Reproducibility of Chemiluminescence Measurements

The reproducibility of chemiluminescence measurements was evaluated with the H_2O_2–microperoxidase system. Peak light intensities were measured for 15 identical reactions containing 1.67 n*M* (VII) at pH 12.6. The average peak light intensity was 29.0 ± 5.2 (SD), i.e., ±18%. The standard deviation decreased to ±5% when triplicate results were averaged. Similar results were obtained when the oxidant was added with the automatic injection device.

Reactions conducted with 11.3 p*M* luminol, a highly efficient chemiluminescent compound, gave a mean peak light intensity for triplicates of 29.3 ± 3.7 (SD), i.e., ±12.7%. When the automatic injector was employed, the SD decreased to ±5%. The reproducibility was virtually the same when reactions were conducted at pH 8.6 and, also, when total light production rather than peak light intensity was measured.

Example of a Homogeneous Competitive Protein-Binding Assay

Reaction mixtures containing 84 n*M* (II) and various levels of avidin up to 60 μg/ml are prepared in 0.1 *M* Tris · HCl, pH 7.4, and incubated at 25° for 15 min. A ligand–protein complex forms, and this is a substrate in the H_2O_2–lactoperoxidase system.[1] The peak light intensities measured with these reaction mixtures increase about 10-fold as the avidin level increases up to 14 μg/ml and then the light emission decreases at higher levels of avidin (Fig. 2). Total light production varies similarly. In control measurements, avidin does not affect peak light intensities produced with (I). This enhancement of light production by avidin provides a means for carrying out a homogeneous assay for biotin. (The term "homogeneous" indicates that separation of protein-bound biotin–isoluminol from the free conjugate is not required.) In a typical assay, reaction mixtures containing various levels of biotin, 90 n*M* (II) and 3.7 μg of avidin per milliliter are prepared in 0.1 *M* Tris · HCl, pH 7.4. These mixtures are incubated at 25° for 5 min and assayed with the H_2O_2–lactoperoxidase system, pH 7.4. As the biotin level increases, less (II) is bound by the avidin and the light intensity decreases. This competitive binding assay can measure 50–400 n*M* biotin.[1]

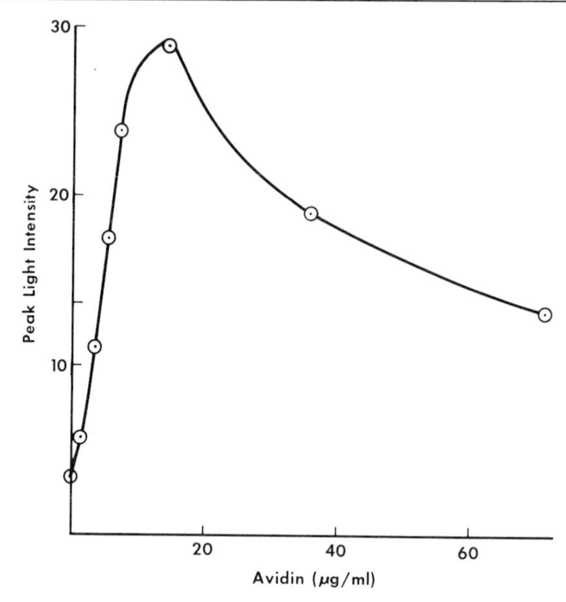

FIG. 2. Peak light intensities produced with the biotin–isoluminol conjugate (II) in the presence of various levels of avidin. Reaction mixtures containing 84 n*M* (II) and the indicated levels of avidin in 0.1 *M* Tris·HCl, pH 7.4, were incubated at 25° for 15 min. Then peak light intensities were measured during oxidation with the H_2O_2–lactoperoxidase system, pH 7.4.

Example of a Heterogeneous Competitive Binding Assay

A heterogeneous assay is one in which a separation of the protein-bound and free forms of the labeled ligand is required. For example, the antibody-bound and the free-thyroxine conjugate (VII) can be separated on small columns of Sephadex G-25. The free conjugate binds to the Sephadex, and the antibody-bound form washes directly through the column and is used for measurement of chemiluminescence.

Titration of (VII) with Antibody. Binding reactions (250 μl) containing 19.3 n*M* (VII) and various levels of rabbit antibody to thyroxine are prepared in 75 m*M* sodium barbital buffer, pH 8.6, and incubated at room temperature for 1 hr. Then a 200-μl aliquot of each binding reaction is applied to a separate 2-ml column of Sephadex G-25 equilibrated with 75 m*M* sodium barbital buffer, pH 8.6. The Sephadex columns are poured in 8 × 65 mm polyethylene barrels (Sherwood Medical, St. Louis, Missouri) with porous Vyon disks (ESB, Inc., Mertztown, Pennsylvania) at the top and bottom. The columns are washed with 2 ml of the same buffer and the effluents are collected. Aliquots (95 μl) of each effluent are assayed with the

H_2O_2–hematin system. Typical results are shown in Fig. 3. In this example, 5.6 μl of antibody binds 50% of the thyroxine conjugate (VII). Similar results are obtained with (IX).

Chemiluminescence Immunoassay for Thyroxine. Competitive binding reactions (250 μl) containing 19.3 n*M* (VII), various levels of thyroxine as indicated in Fig. 4 and 8.3 μl of antibody are prepared in 75 m*M* barbital buffer, pH 8.6. (The antibody is added last.) The reactions are incubated at room temperature for 1 hr and then a 200-μl aliquot from each one is applied to a separate 2-ml column of Sephadex G-25 as described above for titering the antibody. Again, triplicate 95-μl aliquots of the effluent are assayed with the H_2O_2–hematin system. The results shown in Fig. 4 demonstrate that the assay will quantitatively measure thyroxine in the clinically significant range, 10 to 60 n*M*.

The specificity of this competitive binding assay can be verified by substituting isoluminol for the thyroxine conjugate. Isoluminol, 3.4 n*M*, binds to Sephadex G-25 and is not eluted by the buffer or by antibody to thyroxine.

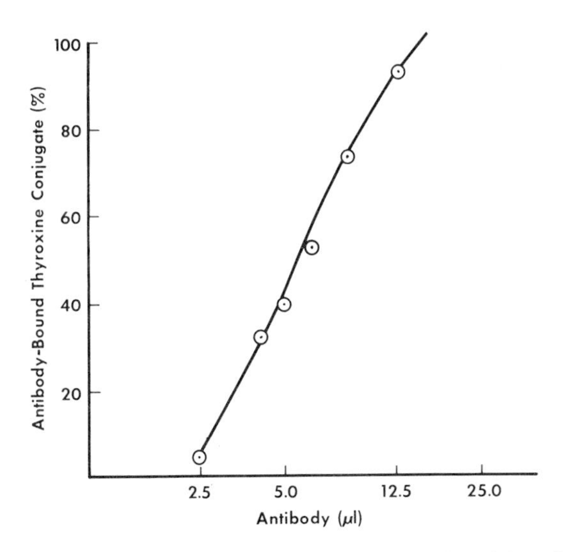

FIG. 3. Titration of the thyroxine–isoluminol conjugate (VII) with antibody. Duplicate reaction mixtures containing 19.3 n*M* (VII) and the indicated levels of antibody in 75 m*M* sodium barbital buffer, pH 8.6, were allowed to stand at room temperature for 1 hr. A 200-μl aliquot of each reaction was applied to a column of Sephadex G-25, and the antibody-bound form of (VII) was eluted with 2 ml of the barbital buffer. Triplicate aliquots (95 μl) of this effluent were assayed for antibody-bound (VII) with the H_2O_2–hematin system, and the results are presented as percent of (VII) bound by the antibody.

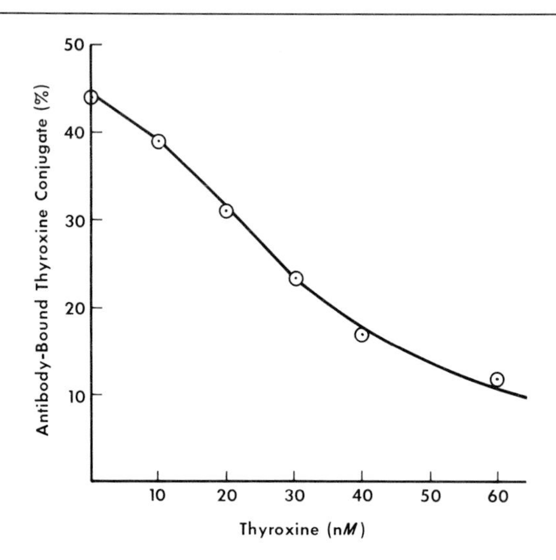

FIG. 4. A standard curve for a competitive protein-binding assay for thyroxine. Duplicate binding reactions (250 μl) containing 19.3 n*M* (VII), 8.3 μl of antibody to thyroxine and the indicated levels of thyroxine were allowed to stand at room temperature for 1 hr. Then the antibody-bound form of (VII) was isolated and assayed as outlined in Fig. 3.

Comments

Aminophthalhydrazides, aminonaphthylhydrazides, and their ligand conjugates can be measured at picomolar concentrations by chemiluminescence. Therefore, competitive protein binding assays monitored by chemiluminescence can measure some ligands at concentrations present in biological materials. The model assays described here were not optimized extensively; therefore, the lowest possible limits for quantitative detection of competing ligands were probably not achieved.

Light production by the biotin–isoluminol conjugate (II) is enhanced about 10-fold when the biotin moiety is bound to avidin. This enhancement provided a means for monitoring competitive binding assays for biotin without separation of the protein-bound and free forms. Light intensities produced by the antibody-bound forms of (VII) and (IX) are enhanced about 3-fold and 8-fold, respectively, compared to the intensities for the free conjugates.

Among the oxidation systems tested, the H_2O_2–microperoxidase and the H_2O_2–hematin systems provide the most sensitive assays for the chemiluminescent compounds. The H_2O_2–hematin system is more difficult to use because the activity of hematin solutions varies with age. This problem can be circumvented by using stock solutions of a stable heme-containing protein, such as catalase. When the protein is added to the reaction mixture at pH 12.6, the heme moiety provides a reproducible catalyst.

Light emission by chemiluminescent reactions can be influenced by contaminants in the reagents. Most of the work described here was carried out by injecting the oxidant from a Hamilton syringe into a mixture containing a catalyst and the chemiluminescent compound. After a syringe was used for an unpredictable period, the results became erratic, but this problem could be corrected by using a new syringe. Apparently, corrosion of the stainless steel plunger and needle on the syringe introduced metal(s) into the oxidant, and this altered the efficiency of the chemiluminescent reaction. The problem was avoided with the automatic injector because only glass and plastic surfaces came into contact with the oxidant.

Acknowledgments

The authors gratefully acknowledge the assistance of Edgar O. Snoke and Frances M. Yeager in carrying out this investigation. The authors also acknowledge the invaluable contributions of Dr. Paul O. Vogelhut, Robert W. Rogers, and Fotios F. Marmarinos, who designed and built the instrumentation for this investigation.

[38] Chemiluminescence Detection of Enzymically Generated Peroxide

By W. RUDOLF SEITZ

Enzyme-catalyzed processes that lead to the formation of hydrogen peroxide can be coupled to chemiluminescence (CL) detection. The conditions of the enzymic process are controlled so that the amount of hydrogen peroxide generated is proportional to the concentration or activity to be measured. The hydrogen peroxide is then allowed to react with excess CL-generating reagents so that the resulting CL intensity is proportional to peroxide concentration. Two CL systems have been evaluated for analytical peroxide detection: luminol and peroxyoxalate CL. Several other CL reactions such as lucigenin also require hydrogen peroxide as a reactant;

ISBN 0-12-181957-4

however, the analytical possibilities of these reactions have not been investigated, and it is unlikely that they would offer significant advantages.

The primary advantages of CL detection of peroxide are low detection limits, of the order of 10 m*M* H_2O_2, and wide linear dynamic ranges, roughly four orders of magnitude. The only instrumentation required is a light detector and a means of bringing the analyte and reactants together reproducibly. In these respects, CL detection of peroxide is similar to the well known, widely practiced firefly ATP assay. However, CL offers an important additional advantage in that the reagents are relatively inexpensive, stable, and available in large quantities. This permits considerable flexibility in the overall analytical system, since it need not be designed to conserve reagent. This will be illustrated in this chapter in the discussion of some of the analytical configurations possible with CL.

There are many possible applications of CL peroxide detection in enzymic analysis. Many oxidase enzymes catalyze reactions that lead directly to hydrogen peroxide. Important examples include the oxidation of glucose by oxygen to gluconolactone and peroxide catalyzed by glucose oxidase, the oxidation of uric acid by oxygen to allantoin and hydrogen peroxide catalyzed by uricase and the oxidation of cholesterol by oxygen to Δ^4-cholesterone and hydrogen peroxide catalyzed by cholesterol oxidase. In addition, processes generating NADH are amenable to CL detection, since NADH reduces molecular oxygen to hydrogen peroxide in the presence of a mediator.[1] Processes that consume NADH or H_2O_2 are also potentially coupled to CL detection by relating a decrease in CL intensity to an analytical parameter of interest. No assays of this sort have been developed, however. It would be very difficult, if not impossible, to design an assay of this sort that would retain the main advantages of CL detection, i.e., the low detection limit and the wide dynamic range.

This chapter will discuss in some detail the characteristics of luminol and peroxyoxalate CL as detection reactions for any process, enzymic or nonenzymic, giving rise to hydrogen peroxide. It should, however, be made clear that to date CL has been coupled only to assays involving relatively high concentrations of hydrogen peroxide, primarily glucose determination using glucose oxidase to catalyze peroxide formation. As a result, the samples have been diluted by a large factor prior to the analysis. For this reason, it is still unresolved whether CL detection can be used for interference-free peroxide determination in undiluted biological samples. This should be carefully checked by any investigator attempting to use CL detection in undiluted samples.

[1] D. C. Williams III and W. R. Seitz, *Anal. Chem.* **48,** 1478 (1976).

Luminol

The luminol reaction is shown below.

$$\text{Luminol (NH}_2\text{, C=O, NH, N}^{\ominus}\text{, C=O)} + 2\,H_2O_2 + OH^- \xrightarrow{\text{base, catalyst, or cooxidant}} \text{(1)}$$

$$\text{NH}_2\text{-C}_6\text{H}_3(\text{COO}^-)_2 + N_2 + 3\,H_2O \longrightarrow \text{NH}_2\text{-C}_6\text{H}_3(\text{COO}^-)_2 + \text{light}_{\lambda\,max} = 430\ \text{nm}$$

For peroxide analysis, luminol and catalyst/cooxidant must be in sufficient excess so that peroxide is the limiting reagent. In this case reaction kinetics and CL intensity will be proportional to peroxide concentration. General characteristics of luminol CL are considered by Roswell and White in this volume [36].

Choice of Catalyst/Cooxidant

Luminol and hydrogen peroxide will not react to any significant extent unless a third component is present. The third component can function either as a catalyst or a cooxidant or both. An example of a catalyst would be peroxidase, since it is unchanged in the overall reaction. An example of a cooxidant would be peroxydisulfate. Peroxydisulfate oxidizes luminol by two electrons to a diazaquinone, which then reacts with hydrogen peroxide in a one-to-one reaction yielding CL.[2] An example of a catalyst/cooxidant may be ferricyanide which oxidizes luminol and is reduced to ferrocyanide but is subsequently reoxidized to ferricyanide.[3]

The choice of catalyst/cooxidant is critical for analytical purposes. The catalyst/cooxidant influences the rate of the CL reaction, CL efficiency, the order of the reaction, and the stoichiometry. Since the overall oxidation of luminol to aminophthalate involves the loss of four electrons, it will

[2] M. M. Rauhut, A. M. Semsel, and B. G. Roberts, *J. Org. Chem.* **31,** 2431 (1966).

[3] P. B. Shevlin and H. A. Neufeld, *J. Org. Chem.* **35,** 2178 (1970).

require two hydrogen peroxide molecules for each luminol [as written in reaction (1), which is balanced] unless part of the oxidation is done by a cooxidant.

Hemin has been the usual catalyst/cooxidant for fundamental studies of luminol CL; however, it has not been evaluated for peroxide analysis. It is not very soluble and is more expensive than other cooxidants/catalysts.

Several transition metal ions catalyze luminol CL.[4] Co(II) and Cu(II) are the most efficient. The mechanism is thought to involve a metal ion–peroxide complex as the active species reacting with luminol.[5] Both Co(II) and Cu(II) have been investigated as catalysts for peroxide determination using luminol;[6,7] however, two problems are encountered. With Cu(II) as a catalyst, sensitive peroxide analysis is possible, but intensity is proportional to $[H_2O_2]^n$, where n is larger than one. The value of n is quite sensitive to conditions such as pH and luminol concentration. Probably this behavior is in some way due to the 2:1 stoichiometry of the peroxide–luminol reaction.

The principal problem with Co(II) is solubility over the pH range from 10 to 12, where luminol CL is most efficient. The solubility can be increased by adding ligands like NH_3 to form soluble complexes; however, this is self-defeating because complexing agents reduce catalytic efficiency by preventing the formation of metal ion–peroxide complex. Co(III) complexes have recently been reported as catalysts for peroxide determination, but the detection limit for peroxide was of the order of 30 μM, and response was nonlinear.[8]

Peroxydisulfate has also been evaluated as a cooxidant for peroxide analysis. Response to peroxide is linear; however, there is background CL from peroxydisulfate and luminol in the absence of peroxide. The background is not sufficiently constant so that it can be satisfactorily subtracted out.

Ferricyanide has been the cooxidant/catalyst of choice for peroxide analysis coupled to enzymic assays.[6,9,10] The rate of the ferricyanide-catalyzed luminol reaction is in the optimum range for analytical purposes, i.e., fast enough so that one gets appreciable intensity in the first several

[4] W. R. Seitz and D. M. Hercules, *in* "Chemiluminescence and Bioluminescence" (M. J. Cormier, D. M. Hercules, and J. Lee, eds.), p. 427. Plenum, New York, 1973.

[5] T. Burdo and W. R. Seitz, *Anal. Chem.* **47,** 1639 (1975).

[6] D. T. Bostick and D. M. Hercules, *Anal. Chem.* **47,** 447 (1975).

[7] W. A. Armstrong and W. G. Humphreys, *Can. J. Chem.* **43,** 2326 (1965).

[8] V. Patrovsky, *Talanta* **23,** 553 (1976).

[9] D. T. Bostick and D. M. Hercules, *Anal. Lett.* **7,** 347 (1974).

[10] J. P. Auses, S. L. Cook, and J. T. Maloy, *Anal. Chem.* **47,** 244 (1975).

seconds, but not so fast that it is highly sensitive to variations in mixing. Apparently, the rate is not highly temperature dependent since good precision and good agreement with other methods were obtained performing the ferricyanide-catalyzed reaction without provisions for temperature control.[6] Figure 1 shows the dependence of CL intensity measured in a flow system on ferricyanide concentration. The rapid decrease in CL at ferricyanide concentrations above 10 m*M* is at least partly due to the fact the ferricyanide absorbs luminol CL.

The problem with using ferricyanide for peroxide analysis is that ferricyanide, luminol, and oxygen produce CL in the absence of added peroxide. Thus it is necessary to have provisions for subtracting out this unwanted signal. An alternative solution to the background problem has been to work at pH 9, where the background CL is much lower in intensity.[10] However, the signal from peroxide is also much lower at this pH. In principle, it should be possible to deoxygenate the reagents prior to an analysis; however, this is not conveniently incorporated into an analytical procedure. Another consequence of the reaction between ferricyanide, luminol, and oxygen is that it is not possible to prepare a single stable ferricyanide–luminol reagent. The analytical system must be designed so that ferricyanide and luminol come in contact either when the CL reaction is initiated or in some controlled manner just before the reaction is started.

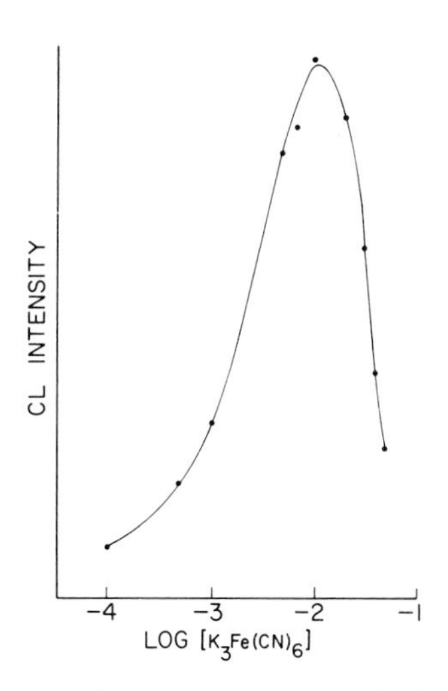

FIG. 1. Chemiluminescence (CL) intensity vs $K_3Fe(CN)_6$ concentration measured in a flow system [D. T. Bostick and D. M. Hercules, *Anal. Chem.* **47**, 447 (1975)]. Conditions: pH 10.8, 0.2 m*M* luminol.

Peroxidase catalysis is promising over the pH range from 7 to 9, since it gives greater CL intensity than other catalysts/cooxidants under these conditions. Peroxidase-catalyzed luminol CL has been studied in detail, although not from an analytical point of view.[11] While peroxidase is suitable for low pH, it has very low activity above pH 9 where luminol CL is most efficient.

pH

The luminol reaction is most efficient at high pH. Figure 2 shows intensity vs pH, using ferricyanide catalysis. Maximum CL is observed in the pH range from 10.4 to 10.8. The decrease as one goes to lower pH values is quite rapid. Since background CL behaves similarly with pH, however, the signal-to-background ratio does not decrease nearly as rapidly as the absolute response. The pH range from 10.4 to 10.8 should give best precision, since in this range small variations in pH will not significantly affect CL intensity. Other catalysts/cooxidants show similar but not identical CL intensity vs pH behavior.

The pH requirements of the luminol reaction can be used to advantage when coupling to enzymic processes, since the pH adjustment to conditions suitable for observing luminol CL can simultaneously serve as a means of stopping enzyme/catalyzed processes. However, pH is also a very serious limitation to applications of luminol CL. One problem is that the luminol reaction and the enzymic processes producing hydrogen peroxide cannot occur simultaneously without severe compromises with respect to luminol intensity and the rates of the enzyme-catalyzed reactions. For example, in principle a method for assaying glucose or glucose

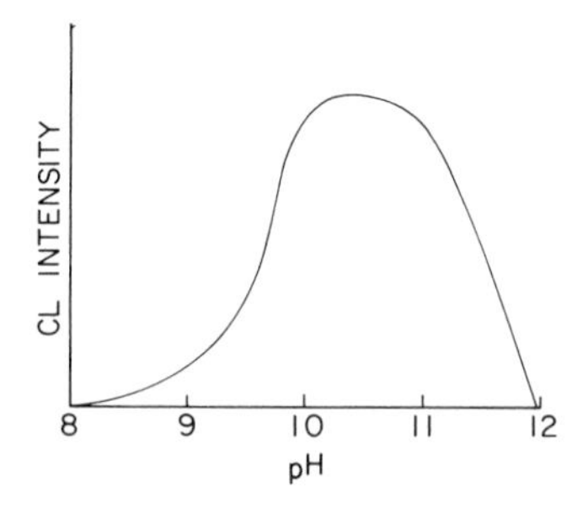

FIG. 2. Chemiluminescence (CL) intensity from luminol vs pH using ferricyanide as cooxidant/catalyst [D. T. Bostick and D. M. Hercules, *Anal. Chem.* **47,** 447 (1975); J. P. Auses, S. L. Cook, and J. T. Maloy, *Anal. Chem.* **47,** 244 (1975)].

[11] M. J. Cormier and P. M. Prichard, *J. Biol. Chem.* **243,** 4706 (1968).

oxidase activity could be developed by performing reactions (2) and (3) simultaneously under one set of conditions.

$$\beta\text{-}d\text{-Glucose} + O_2 \xrightarrow[\text{glucose oxidase}]{} d\text{-gluconolactone} + H_2O_2 \quad (2)$$

$$H_2O_2 + \text{luminol} \xrightarrow[\text{catalyst}]{} \text{CL} \quad (3)$$

By appropriately adjusting concentrations, it should be possible to achieve conditions where CL intensity would reflect the concentration of glucose or glucose oxidase activity as desired. In practice, the differences in pH optima severely limit the possibilities of this otherwise attractive approach. Instead, it is necessary to do the two reactions involved in a coupled scheme sequentially under two different sets of conditions.

The second problem associated with the pH requirements of luminol is even more serious. High pH may accelerate the rate of reaction between hydrogen peroxide and reducing components in biological samples. These reactions consume hydrogen peroxide before it can react with luminol, thus reducing observed CL intensity and interfering negatively in analytical procedures. Uric acid was observed to interfere in this manner when an attempt was made to develop a CL method for glucose in urine.[6] To get satisfactory results it was necessary to separate uric acid by adsorption on a $Zn(OH)_2$–$BaSO_4$ precipitate before doing the CL reaction.[12] Interferences of this sort have not been observed in studies involving serum samples; however, the procedures involve dilution by a factor of 100 or more.[6,10] Problems may be encountered if undiluted samples are to be coupled to assays involving luminol CL.

The problems with pH account for the interest in peroxidase as a catalyst, since it is best suited for low pH.

Luminol Concentration

The luminol concentration is not a critical variable in designing methods based on coupling to peroxide using ferricyanide. As luminol concentration increases, CL intensity increases. However, the signal due to peroxide and background CL from oxygen are affected similarly, so there is no gain in peroxide detectability by using high concentrations of luminol. In fact, at high levels of luminol the observed CL signals become more sensitive to variations in mixing, causing precision to deteriorate. Concentrations in the range from 0.1 m*M* to 1 m*M* luminol are suitable for peroxide analysis.

[12] D. C. Williams III, G. F. Huff, and W. R. Seitz, *Clin. Chem.* **22**, 372 (1976).

Analytical Configurations

Direct Injection

CL analysis for peroxide can be done by injecting reagents into the sample (or vice versa) in a container placed in front of a light detector and measuring the resulting intensity as a function of time. Either peak CL intensity or CL integrated over some time interval can be related to concentration. This approach is commonly used for the firefly ATP assay. It is inexpensive to implement and consumes relatively little reagent since commercial instruments of this sort are designed around volumes of 0.2–0.5 ml. On the other hand, this approach is subject to poor precision due to nonreproducible injection. Also this is necessarily a manual procedure and cannot be incorporated into an automated method.

For coupling purposes, the enzyme-catalyzed process would be allowed to proceed for a known time under controlled conditions. It would then be stopped and peroxide analysis performed. A procedure of this sort is described below. It was developed by Auses *et al.*[10] for glucose analysis; however, other enzymic processes could be substituted for the glucose oxidase-catalyzed oxidation of glucose. Two features of this procedure are noteworthy. First, the luminol is added along with the glucose so that it will not react with ferricyanide before initiating the CL reaction. Second, the enzyme-catalyzed reaction is stopped by adding a buffered ferricyanide solution, which simultaneously starts the analysis.

Procedure

A 4.5 m*M* stock solution of luminol is prepared. Initially, sufficient luminol is dissolved with the aid of a small quantity of base, but, before the solution is brought to volume, it is adjusted to pH 7. Similarly, a glucose oxidase stock solution containing 150 units/ml is adjusted to pH 7 before final dilution. Neither of these solutions is buffered. A stock solution of 2 m*M* $K_3Fe(CN)_6$ is prepared in 1 *M* ammonia buffer at pH 9.5. Aqueous glucose standards are prepared by dilution of a stock solution containing 200 mg of glucose per deciliter.

In a typical run, a 10-μl sample of glucose solution is introduced into a cuvette containing 1.6 ml of an unbuffered mixture of glucose oxidase and luminol at pH 7. After agitation to ensure mixing of the reagents, the cuvette is placed in the detector-cell compartment. A delay time of 5 min is allowed to elapse for hydrogen peroxide to build up in the system. While the light output is being monitored, 0.8 ml of 1 *M* ammonia buffer containing $K_3Fe(CN)_6$ is rapidly injected into the cuvette. The chemiluminescence

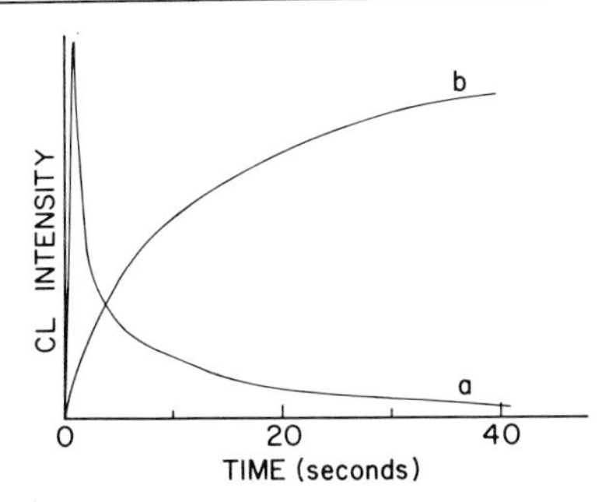

FIG. 3. Typical chemiluminescence (CL) vs time curve for luminol reaction with ferricyanide and peroxide initiated by injecting sample into reagents. Zero time corresponds to the time of initiating the reaction. Curve a shows the CL signal, and curve b the integrated signal. Taken from J. P. Auses, S. L. Cook, and J. T. Maloy, *Anal. Chem.* **47**, 244 (1975).

obtained immediately upon buffer–ferricyanide addition is recorded as a function of time. Simultaneously, the integrated CL intensity is obtained and recorded. Typical results are shown in Fig. 3; the integrated CL intensity (curve b) approaches a limiting value at times in excess of 40 sec.

Temperature control is not specified in the procedure but clearly is required for any enzyme process that does not go to completion. There is no reason not to buffer the enzymic process, provided that the buffer determining the pH of the luminol reaction is designed to have sufficient buffer capacity to overcome the effect of the enzymic buffer.

The detection limit for glucose was estimated to be 2 mg/dl. After dilution, this corresponds to less than 1 part per million or 6 μM glucose. Because a low pH was chosen for the procedure there was no detectable background CL from ferricyanide, luminol, and oxygen.

Commercial luminol (Eastman Kodak or Aldrich) is frequently recrystallized before use although there is no evidence that this is required for the analytical method. Luminol solutions show greater activity, i.e., give more light, when they are initially prepared. It is recommended that luminol solutions be given 2 or 3 days to stabilize prior to use.

Ferricyanide solutions should be stored in the dark and should be prepared fresh every few days.

Flow Systems

Flow systems offer better control of mixing than direct injection as well as being amenable to automation. Figure 4 shows a flow system designed for glucose analysis using ferricyanide-catalyzed luminol CL. An infusion pump drives three syringes containing 0.2 mM luminol in 0.1 M KOH–H_3BO_3 buffer, 10 mM potassium ferricyanide, and 4 mM acetate buffer as

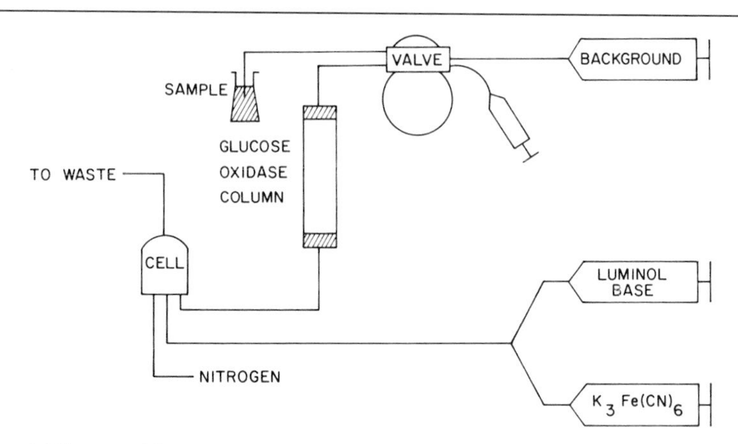

FIG. 4. Diagram of flow system for determining glucose coupled to luminol chemiluminescence. From D. T. Bostick and D. M. Hercules, *Anal. Lett.* **7**, 347 (1974).

the background. The amount of KOH in the buffer is adjusted so that the pH of the CL reaction is 10.5. The ferricyanide and luminol are mixed directly in the flow system to avoid decomposition. A small metal coil to promote mixing where the ferricyanide and luminol flows join is very helpful in keeping background CL from the luminol ferricyanide–oxygen reaction constant. A sampling valve is used to insert 1–2-ml slugs of sample into the background flow line.

The cell where the CL reaction occurs has a volume of about 1 ml and is stirred by bubbling a stream of nitrogen through it. The quality of the data is not highly dependent on the dimensions of the cell or the method of stirring.

The background solution would be the same buffer used for the enzymic process. When background solution flows through the cell, a constant level of CL is observed from the reaction between ferricyanide, luminol, and oxygen. When a peroxide sample flows through the cell, a new steady state is achieved. The data appear in the form of peaks with height proportional to peroxide concentration. One of the advantages of a flow system is that all measurements are relative to background, and it is not necessary to run a blank. This system has a detection limit of 7 n*M* hydrogen peroxide and is linear up to 0.1 m*M* peroxide.

Coupling to enzymic processes can be accomplished in several ways. The simplest approach is to generate peroxide in some controlled way and then stop the reaction and introduce peroxide sample to the flow system. A more elegant approach is to perform the enzymic process in the flow system itself. The apparatus in Fig. 4 is one example of this. Glucose samples are introduced to the flow system. Before they reach the CL cell, they pass over a column of glucose oxidase immobilized on a solid support. The glucose is quantitatively converted to peroxide in the column. The peroxide then reacts to produce CL. Some typical data from this system are shown in Fig. 5. Mixing between the sample and background causes the rise and fall times of the peaks to be fairly long. It is not practical to separate sample and background by air bubbles, because the bubbles get hung up in the enzyme column and interfere with its operation.

The glucose method was applied to serum. Samples were deproteinated and diluted by a factor of 2000. Deproteination was performed to avoid column clogging. This can be better accomplished by immobilizing the enzyme on a rigid material, such as porous glass. The performance of the CL method on six reference samples is summarized in Table I.

The use of an immobilized enzyme column is only one possible way of performing the enzyme-catalyzed process in the flow system. Alternatively, it would be possible to have two background flow lines, one containing enzyme and the other substrate. The lines would be joined and would flow through a delay coil, allowing time to generate peroxide. A system of this sort for coupling the determination of lactate dehydrogenase (LDH) activity to peroxide has been described,[1] but it was combined with peroxyoxalate CL rather than luminol.

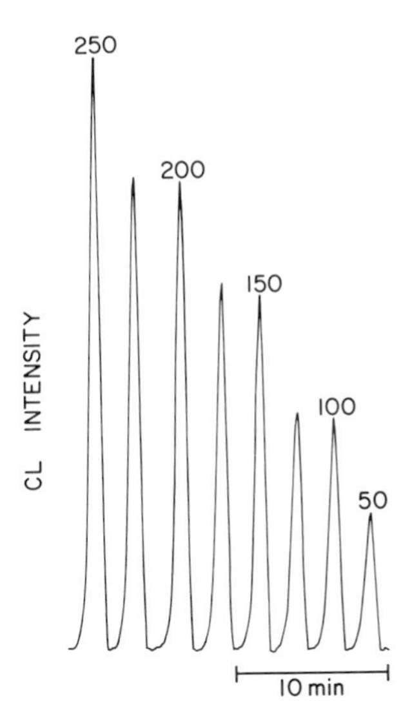

FIG. 5. Typical response for flow system of Fig. 4. The numbers above the peaks correspond to glucose concentrations in milligrams per deciliter (before dilution by a factor of 2000). Conditions, 20 mM $K_3Fe(CN)^{-3}_6$; 0.2 mM luminol in 0.1 M $KOH–H_3BO_3$ buffer, pH 10.5; pH 5.6 4 mM acetate buffer for sample and background; flow rate 2.76 ml/min per syringe. From D. T. Bostick and D. M. Hercules, *Anal. Chem.* **47,** 447 (1975).

Unfortunately the CL-flow system method for peroxide using ferricyanide has one serious deficiency. It requires that all flow rates be constant. Otherwise, there are severe fluctuations in background CL, causing the detection limit for peroxide to deteriorate by about two orders of magnitude. Most automated systems involve pulsating flow, which is unsatisfactory. The infusion pump system of Fig. 4 requires that the syringes be periodically refilled.

Immobilized Catalyst

The newest approach to doing CL analysis is to immobilize the catalyst. This has been done using immobilized peroxidase. Two approaches have been tried. Peroxidase has been immobilized by covalent attachment to the inside walls of transparent nylon tubing[13] and by entrapment in a polyacrylamide gel on the end of a fiber optic.[14] In this case, CL intensity will depend on the means of mass transfer to the immobilized enzyme. One advantage of this approach is that peroxidase is conserved. This approach is compatible with flow systems by simply positioning the immobilized catalyst opposite a light detector and flowing sample over it.

Peroxyoxalate Chemiluminescence

Peroxyoxalate chemiluminescence refers to a large class of reactions summarized below.

$$RO-\overset{\overset{O}{\|}}{C}-\overset{\overset{O}{\|}}{C}-OR + H_2O_2 \xrightarrow{\text{activator}} \begin{array}{c} O\;\;\;O \\ \|\;\;\;\| \\ C-C \\ |\;\;\;\;| \\ O-O \end{array} \tag{4}$$

1,4-Dioxetane dione

$$\left[\begin{array}{c} O\;\;\;O \\ \|\;\;\;\| \\ C-C \\ |\;\;\;\;| \\ O-O \end{array}\right] + \text{flr} \longrightarrow \text{flr}^* + 2\,CO_2 \tag{5}$$

$$\text{flr}^* \longrightarrow \text{flr} + \text{CL} \tag{6}$$

The first example of peroxyoxalate CL involved oxalyl chloride.[15] This discovery stimulated a major research effort at American Cyanamid, ulti-

[13] W. Hornby, presented at the Summer Symposium on Enzymes in Analytical Chemistry, Amherst, Massachusetts, June 14–17 (1977).

[14] W. R. Seitz, presented at the Summer Symposium on Enzymes in Analytical Chemistry, Amherst, Massachusetts, June 14–17 (1977).

[15] E. A. Chandross, *Tetrahedron Lett.* **1963,** 761 (1963).

TABLE I
CORRELATION STUDY BETWEEN REFERENCE AND THE CHEMILUMINESCENCE (CL) VALUES FOR SIX CONTROL SAMPLES OF GLUCOSE

Sample	Average CL value (mg/dl)	Standard deviation in Cl value (mg/dl)[a]	Control value (mg/dl)
1	42.8	1.45	42.0
2	79.6	1.72	78.0
3	79.8	1.66	81.0
4	106.4	1.26	110.0
5	136.1	3.89	135.8
6	212.4	2.31	210.0

[a] Based on 4 replicates.

mately leading to a practical chemiluminescence light source.[16] Several types of oxalates react with peroxide to give the necessary intermediate for CL; however, greatest efficiency is observed with electronegatively substituted oxalate esters. Characteristics of peroxyoxalate CL have been reviewed.[17]

Peroxyoxalate CL is the most efficient known nonbiological CL reaction, having an efficiency around 0.3 under optimum conditions. The fluorophor does not participate in the reaction but serves only to accept energy from 1,4-dioxetanedione decomposition. As a result, the wavelength and color of CL can be varied by appropriate choice of fluorophor. The reaction produces CL in the absence of "activator"; however, the efficiency is relatively low. The activator in most systems is a weak base, such as lithium salicylate or triethylamine, which prolongs the lifetime of chemiluminescence and increases efficiency. Peroxyoxalate reactions are most efficient in ethers and ester, although they are compatible with other solvents. The work at American Cyanamid was directed toward a practical light source, i.e., maximum efficiency and lifetime. The requirements for a useful analytical reaction are different and will be discussed below.

Solvent System

The most serious limitation of peroxide analysis using peroxyoxalate CL is the need for an organic solvent. Although peroxyoxalate CL is observed in mixed solvent systems involving water, oxalate esters as well as other types of oxalates that give efficient CL are insoluble in water.

[16] M. M. Rauhut, *in* "Chemiluminescence and Bioluminescence" (M. J. Cormier, D. M. Hercules, and J. Lee, eds.), p. 451. Plenum, New York, 1973.

[17] M. M. Rauhut, *Acc. Chem. Res.* **2,** 80 (1969).

Esters and ethers are the most suitable solvents for peroxyoxalate CL. Some ethers, such as dioxane, are miscible with water and would seem, therefore, to be good for determining peroxide in an aqueous solution. However, because ethers interact with oxygen to form peroxides, there is significant CL from an ether solution of oxalate, fluorophor, and activator in the absence of added peroxide. This leads to oxalate decomposition as well as high background CL that obscures signals from low levels of peroxide in the sample of interest. Most common esters are not very miscible with water. Poor precision results if one attempts to do CL analysis using two immiscible reagents.

One solution to this problem is to use a third solvent to create a ternary miscible solvent system. The solvent system chosen for analytical work was ethyl acetate, methanol, and water.[18] The methanol and ethyl acetate were mixed in a flow system immediately before the analysis because the oxalate ester gradually decomposes in the presence of primary and secondary alcohols.

Another potentially useful solvent is *tert*-butyl alcohol, since oxalate esters are stable in this solvent and it is miscible with water. The melting point for *tert*-butyl alcohol is 25.5°, just slightly above normal room temperature; therefore some form of temperature control is needed.

Reaction Conditions

Increasing oxalate and fluorophor concentrations lead to an increase in CL intensity and efficiency. The optimal concentrations are the maximum amount that can be conveniently and rapidly dissolved in the solvent to be used. The critical component of the reaction for analytical purposes is the activator concentration. Figure 6 illustrates qualitatively the effect of activator on CL vs time. In the absence of activator, there is a significant lag time before CL starts. The lag time decreases as activator level increases, so that ultimately the point is reached where CL is observed immediately upon mixing. The overall efficiency decreases at high activator concentrations, since CL is short lived. However, analytical measurements normally involve observing CL for a relatively short time, so the efficiency loss is not a problem.

For any analytical system, optimum activator level should be established empirically for the desired oxalate and fluorophor concentrations. As oxalate level increases, the amount of activator required to eliminate the lag time also increases. Therefore, any changes in oxalate level will require

[18] D. C. Williams III, G. F. Huff, and W. R. Seitz, *Anal. Chem.* **48,** 1003 (1976).

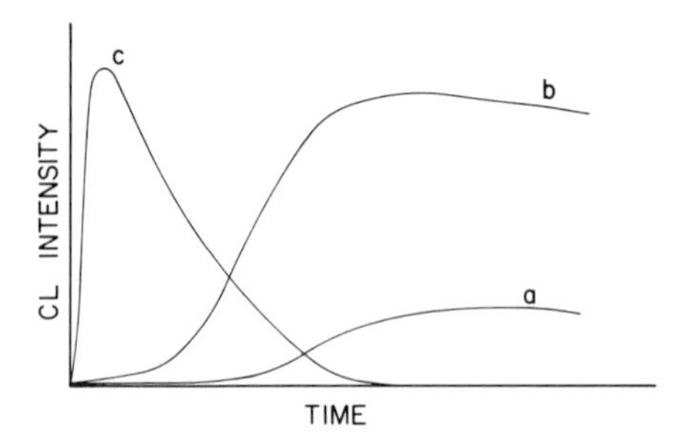

FIG. 6. Effect of activator on chemiluminescence (CL) from oxalate ester system. Curve a, no activator; curve b, low activator concentration; curve c, high activator concentration.

a compensating change in activator concentration. Triethylamine was used as the activator in the reported analytical work.[1,18]

Analytical Configurations

Peroxyoxalate CL is not suitable for analysis using injection techniques because the ternary solvent system does not mix readily. The analytical work to date has all involved flow systems. The initial study involved the flow system of Fig. 4 with different reagents.[18] The luminol solution is replaced by a solution of 600 mg of bis-2,4,6-trichlorophenyloxalate (TCPO) and 50 mg of perylene per liter as the fluorophor in ethyl acetate. TCPO was prepared according to a literature procedure.[19] Because pure oxalates are not available commercially, they must either be synthesized or begged from American Cyanamid. The ferricyanide solution is replaced by 50 μl of triethylamine per liter in methanol. The background solution contains aqueous buffer as before. Peroxide samples are introduced as before, and the data are in the form of peaks.

This system has a detection limit of 70 nM hydrogen peroxide with linear response up to 1 mM. The response is pH dependent; however, there is sufficient sensitivity for peroxide analysis from pH 4 to pH 10. Maximum CL is observed at pH 8. This is a major advantage of peroxyoxalate CL relative to luminol for H_2O_2 determination. By working at pH 6 in the aqueous solution, glucose analysis in urine was possible without interference from uric acid.[18] At the luminol pH, uric acid had to be physically removed prior to analysis.

In principle, the peroxyoxalate system should not produce CL in the absence of peroxide; in practice, however, using the flow system of Fig. 4

[19] A. G. Mohan and N. J. Turro, *J. Chem. Educ.* **51,** 528 (1974).

some background CL was observed. This background apparently arises from the mixing itself, since it has been subsequently found that the background vanishes if the mixing is done just before the reagents enter the area viewed by the detector. In this case, the detection limit for peroxide is 10 n*M* or lower.

The other major advantage of peroxyoxalate CL for analytical purposes is that it can be used in flow systems with pulsating flow without serious deterioration of the detection limit. Thus, it is more suitable for automated analysis than luminol. However, it is necessary to use latex rubber or some other type of tubing compatible with methanol and ethyl acetate since these solvents attack Tygon.

Peroxide detection by peroxyoxalate CL can be coupled to enzymic processes in the same way as luminol. Since peroxyoxalate CL is amenable to automation, complex coupling schemes can be carried out in a flow system. For example, peroxyoxalate CL has been coupled to the measurement of LDH activity via the following reaction scheme:

$$\mathrm{NAD^+ + lactate \rightleftarrows NADH + H^+ + pyruvate} \tag{7}$$

$$\mathrm{NADH + MB_{oxidized} \rightarrow NAD^+ + MB_{reduced}} \tag{8}$$

$$\mathrm{MB_{reduced} + O_2 \rightarrow MB_{oxidized} + H_2O_2} \tag{9}$$

where NADH is the reduced form of nicotinamide adenine dinucleotide and MB is methylene blue. Figure 7 shows the flow system for performing the analysis. The ethyl acetate solution contains 600 mg of TCPO and 50 mg of perylene per liter. The methanol solution contains 500 ml of

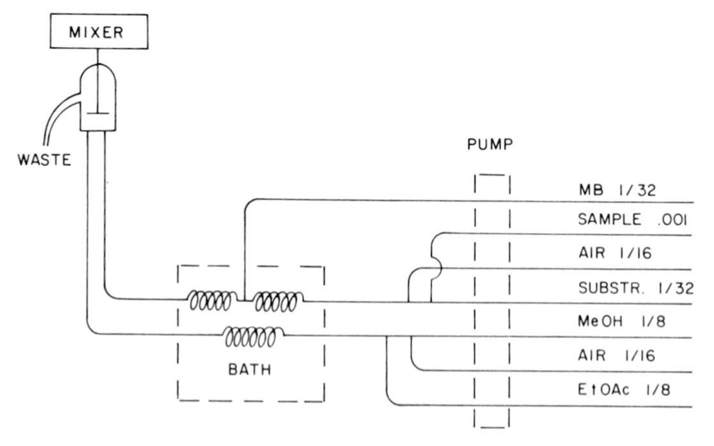

FIG. 7. Flow system for automated determination of lactate dehydrogenase activity coupled to peroxyoxalate chemiluminescence. Tubing diameters are given in inches. MB, methylene blue.

triethylamine per liter. The MB concentration was 0.15 m*M*. The substrate was 0.2 *M* tris(hydroxymethylamino)methane buffer at pH 9 containing 50 m*M* L-lactate and 1.5 m*M* NAD. The sample line was placed in water when no sample was being introduced into the system. The flow system contains three delay coils. One allows time for LDH to generate NADH, and the second allows the reaction between NADH and $MB_{oxidized}$ to go almost to completion. The bath temperature is 37°. The delay coil in the ethyl acetate–methanol line is for temperature equilibration. In the flow cell used in this study, mixing is accomplished by a small vibrating disk.

The correlation between the CL method and a standard spectrophotometric method for LDH in serum samples is shown in Fig. 8. The correlation coefficient is 0.9996, and the slope is 0.927. The CL method correlates closely with the standard method but gives slightly higher values. This is thought to reflect the fact that the CL method uses considerably less sample, and therefore LDH inhibitors in the sample are being diluted to a greater extent.

The above type of system extends the advantages of CL not only to peroxide generating systems, but also to NADH-generating systems.

Conclusion

Both luminol and peroxyoxalate reactions are suitable for coupling to enzymic processes that produce hydrogen peroxide. Luminol is better known and has been more extensively evaluated for analytical purposes. However, it has two fundamental limitations: the requirement of alkaline pH and not being suitable for flow systems requiring pulsating flow, i.e.,

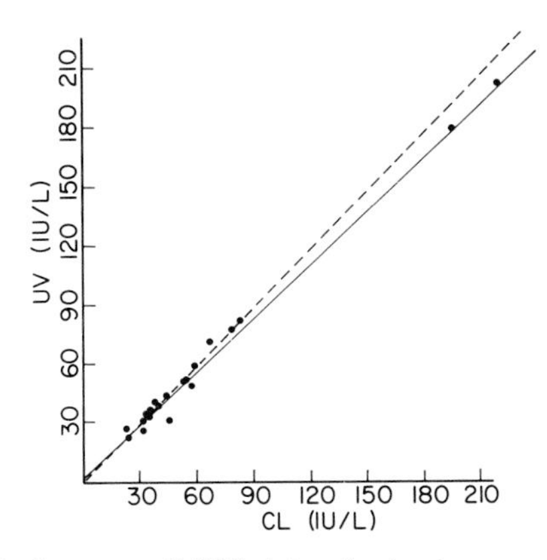

FIG. 8. Lactate dehydrogenase (LDH) determined using peroxyoxalate chemiluminescence (CL) vs LDH determined by spectrophotometry. Both axes are expressed in international units; ———, least-squares line.

autoanalyzers. The peroxyoxalate reaction is compatible with neutral pH values and pulsating flow systems; however, it requires the use of organic solvents. Both reactions can analyze peroxide down to 10 n*M* and both are linear over four orders of magnitude.

[39] The Generation of Chemiluminescence (CL) by Phagocytic Cells

By MICHAEL A. TRUSH, MARK E. WILSON, and KNOX VAN DYKE

The physical and biochemical processes whereby phagocytic cells such as polymorphonuclear leukocytes (PMNs) engulf and subsequently destroy (through both enzymic and nonenzymic reactions) bacteria have been the target of active research in recent years. After the initial isolation of the enzyme myeloperoxidase (MPO) from PMNs by Agner,[1] Klebanoff demonstrated that MPO exhibits a marked bactericidal activity when combined with hydrogen peroxide (H_2O_2) and various oxidizable cofactors (particularly halides).[2,3] These now classic experiments performed by Klebanoff provided the first substantial evidence supportive of an enzymic basis for the bactericidal activity of PMNs.

Hypochlorous acid (HOCl) formed during the course of the MPO–H_2O_2–halide reaction has been suggested to contribute to the bactericidal effects of MPO.[4] Mallet[5] in 1927 and subsequently Seliger[6] observed and reported that the interaction between H_2O_2 and hypochlorite (OCl^-) results in light emission. Kasha and Khan have since shown that this reaction between excess H_2O_2 and OCl^- results in the generation of singlet molecular oxygen ($^1O_2$), an electronically excited species that emits a photon upon relaxing to its triplet ground state (i.e., chemiluminescence).[7] On the basis of these earlier observations, Allen postulated that $^1O_2$ may be generated by phagocytosing PMNs and may, in part, contribute to the bactericidal activity of these cells as an intermediate of the MPO reaction.[8] The mecha-

[1] K. Agner, *Acta Physiol. Scand. Suppl.* **8,** 1 (1941).
[2] S. J. Klebanoff, *J. Exp. Med.* **126,** 1063 (1967).
[3] S. J. Klebanoff, *J. Bacteriol,* **95,** 2131 (1968).
[4] J. M. Zgliezynski, T. Stelmaszynska, W. Ostrowski, J. Naskalski, and J. Sznajd, *Biochim. Biophys. Acta* **235,** 419 (1971).
[5] L. Mallet, *C. R. Acad. Sci.* **185,** 352 (1927).
[6] H. H. Seliger, *Anal. Biochem.* **1,** 60 (1960).
[7] M. Kasha and A. U. Khan, *Ann. N. Y. Acad. Sci.* **171,** 5 (1970).
[8] R. C. Allen, Ph.D. dissertation, Tulane University, New Orleans, Louisiana, 1973.

ISBN 0-12-181957-4

nisms by which MPO-mediated reactions may exert bactericidal activity, including the generation of $^1O_2$, have been the subject of a recent review.[9] Further, Allen suggested that if $^1O_2$ is in fact generated via the MPO–H_2O_2–halide reaction, then chemiluminescence should accompany the onset of phagocytosis, a phenomenon that he subsequently demonstrated.[8,10]

Allen also proposed the formation of superoxide anions (O_2^-) by phagocytosing PMNs and suggested a role for these free radicals in the CL phenomenon.[11,12] Babior and co-workers subsequently provided experimental evidence for the generation of O_2^- by phagocytosing PMNs.[13] The interaction between O_2^- and H_2O_2 results in the production of still another toxic metabolite, the hydroxyl radical (OH·).[14] This reaction, referred to as the Haber–Weiss reaction, appears to be operative in PMNs, as evidenced by indirect measurements of OH· in phagocytic cells.[15,16] Evidence, in addition to the CL response, for the generation of $^1O_2$ by PMNs[17] and the participation of this species in cidal activity has been reported.[18] Figure 1 is a schematic representation of the sequence of events in the engulfment and destruction of bacteria by PMNs.

Whereas the exact light-emitting species has (have) not to date been identified (both $^1O_2$ and excited carbonyl groups have been proposed), nor has the precise relationship between CL and bactericidal activity been defined, the five following recent observations strongly indicate that the CL phenomenon is related to and dependent upon both cellular activation and intracellular metabolism.

1. Elicitation of CL by a phagocytosable particle is dependent upon a series of events occurring in normal serum and upon the subsequent activation of PMNs. Following the initial observation by Allen and co-workers that phagocytic cells generate a burst of chemiluminescence during activation by suitable particles, a number of subsequent studies have attempted to correlate the CL phenomenon with other parameters of phagocyte

[9] S. J. Klebanoff, *Semin. Hematol.* **12,** 117 (1975).
[10] R. C. Allen, R. L. Stjernholm, and R. H. Steele, *Biochem. Biophys. Res. Commun.* **47,** 679 (1972).
[11] R. C. Allen, R. L. Stjernholm, R. R. Benerito, and R. H. Steele, *in* "Chemiluminescence and Bioluminescence" (M. J. Cormier, D. M. Hercules, and J. Lee, eds.), p. 498. Plenum, New York, 1973.
[12] R. C. Allen, S. J. Yevich, R. W. Orth, and R. H. Steele, *Infec. Immun.* **60,** 909 (1974).
[13] B. M. Babior, R. S. Kipnes, and J. T. Curnutte, *J. Clin. Invest.* **52,** 741 (1973).
[14] F. Haber and J. Weiss, *Proc. R. Soc. Edinburgh Sect. A* **147,** 332 (1934).
[15] A. I. Tauber and B. M. Babior, *J. Clin. Invest.* **60,** 374 (1977).
[16] S. J. Weiss, G. W. King, and A. F. LoBuglio, *J. Clin. Invest.* **60,** 370 (1977).
[17] H. Rosen and S. J. Klebanoff, *J. Biol. Chem.* **252,** 4803 (1977).
[18] N. I. Krinsky, *Science* **186,** 364 (1974).

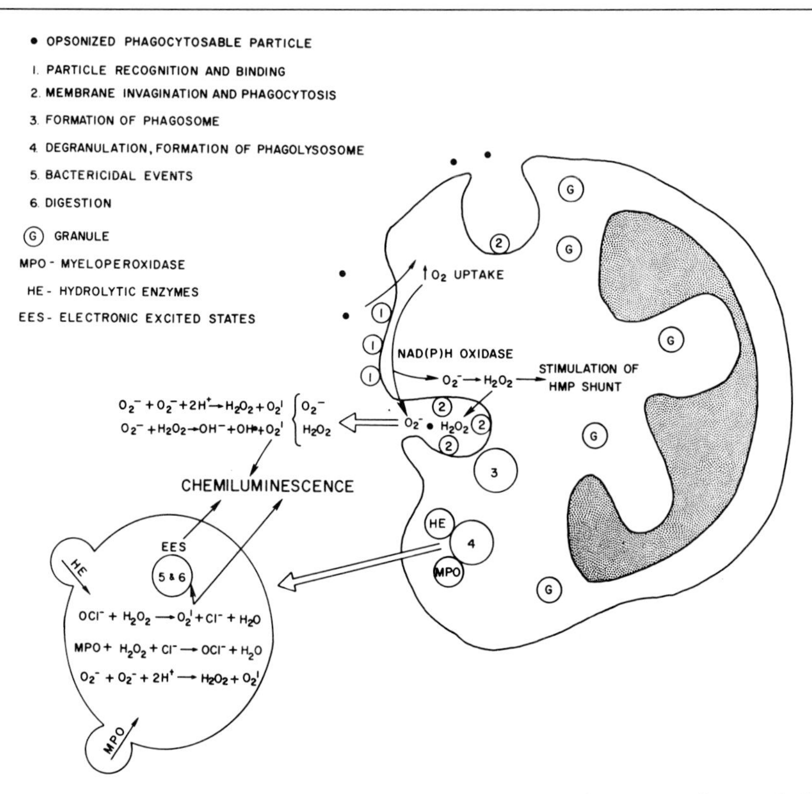

FIG. 1. Schematic representation of the endocytic events and accompanying metabolic processes that occur during phagocytosis by polymorphonuclear leukocytes (PMNs). For more detailed discussions of these processes see: (a) B. D. Cheson, J. T. Curnutte, and B. M. Babior, *in* "Progress in Clinical Immunology" (B. S. Schwartz, ed.), Vol. 3, p. 1. Grune & Stratton, New York, 1977. PMN biochemistry. (b) S. J. Klebanoff, *Semin. Hematol.* **12,** 117 (1975). PMN antimicrobial mechanisms. (c) T. P. Stossel, *N. Engl. J. Med.* **290,** 717, 774, 833 (1974). General review on phagocytosis.

activation, including oxygen consumption, particle uptake, oxidative metabolism, and *in vitro* killing of bacteria. One of the earliest studies to address itself to this matter was conducted by Stevens and Young, who reported a correlation between resistance of certain strains of *E. coli* to opsonization and decreased *in vitro* killing, oxygen consumption, visual phagocytosis and chemiluminescent response of human granulocytes.[19] Subsequently, Grebner *et al.* demonstrated parallel relationships between phagocytosis and chemiluminescence under a variety of conditions designed to alter opsonization of bacteria, and led these authors to suggest

[19] P. Stevens and L. S. Young, *Fed. Proc., Fed. Am. Soc. Exp. Biol.* **35,** 738 (1976).

that the biochemical processes that control phagocytosis and CL may be "closely related or interdependent."[20]

Whereas Grebner *et al.* were concerned primarily with amplification of the opsonization process by complement, Hemming *et al.* correlated phagocytic and neutrophile chemiluminescence assay results in a system that stressed the importance of type-specific (IgG) opsonins in serum in the efficient killing of group B streptococcal organisms.[21] Hence, opsonization of particles via type-specific (antibody) or nonspecific (complement) opsonins has been utilized in correlating phagocytosis and neutrophile chemiluminescence.

2. CL is dependent upon several intracellular metabolic events for the adequate generation and utilization of O_2^- and H_2O_2.

The importance of the burst of oxidative metabolism in the CL phenomenon is exemplified by reports of significantly depressed CL responses by PMNs of patients with chronic granulomatous disease of childhood[22] or myeloperoxidase deficiency.[23] The former disease is characterized by a failure of the patient's PMNs to generate O_2^- and H_2O_2 in significant amounts.[24] Myeloperoxidase-deficient PMNs exhibit normal generation of O_2^- and H_2O_2; however, the absence of the enzyme precludes the formation of $^1O_2$ via the MPO–H_2O_2-halide reaction.[23]

3. Agents that interfere with the activity of reactive oxygen species formed by PMNs (including superoxide dismutase, catalase, and sodium benzoate),[25] or that inhibit certain metabolic events (such as the MPO inhibitor sodium azide),[8,26] lead to decreased CL responses in otherwise normally functional PMNs.

4. The generation of CL is characteristic of phagocytic cells other than neutrophiles, including alveolar[27,28] and peritoneal macrophages,[29] monocytes,[30-32] and eosinophils.[33]

[20] J. V. Grebner, E. L. Mills, B. H. Gray, and P. G. Quie, *J. Lab. Clin. Med.* **89,** 153 (1977).

[21] V. G. Hemming, R. T. Hall, P. G. Rhodes, A. O. Shigeolea, and H. R. Hill, *J. Clin. Invest.* **58,** 1379 (1976).

[22] R. L. Stjernholm, R. C. Allen, R. H. Steele, W. W. Waring, and J. A. Harris, *Infect. Immun.* **7,** 313 (1973).

[23] H. Rosen and S. J. Klebanoff, *J. Clin. Invest.* **58,** 50 (1976).

[24] J. T. Curnutte, D. M. Whitten, and B. M. Babior, *N. Engl. J. Med.* **290,** 593 (1974).

[25] L. S. Webb, B. B. Keele, Jr., and R. B. Johnston, Jr., *Infect. Immun.* **9,** 1051 (1974).

[26] A. L. Sagone, Jr., D. S. Mendelson, and Earl N. Metz, *J. Lab. Clin. Med.* **89,** 1333 (1977).

[27] P. R. Miles, P. Lee, M. A. Trush, and K. Van Dyke, *Life Sci.* **20,** 165 (1977).

[28] G. D. Beall, J. E. Repine, J. R. Hoidal, and Fred L. Rasp, *Infect. Immun.* **17,** 117 (1977).

[29] E. C. Allen and L. D. Loose, *Biochem. Biophys. Res. Commun.* **69,** 245 (1976).

[30] A. L. Sagone, Jr., G. W. King, and Earl N. Metz, *J. Clin. Invest.* **57,** 352 (1976).

[31] R. D. Nelson, E. L. Mills, R. L. Simmons, and P. G. Quie, *Infect. Immun.* **14,** 129 (1976).

[32] R. B. Johnston, Jr., J. E. Lehmeyer, and L. A. Guthrie, *J. Exp. Med.* **143,** 1551 (1976).

[33] S. J. Klebanoff, D. T. Durack, H. Rosen, and R. A. Clark, *Infect. Immun.* **17,** 167 (1977).

5. Soluble agents that activate PMNs can induce CL. Recent studies in our laboratory have dealt with the effects of a soluble stimulant, the cation ionophore A23187, on neutrophile chemiluminescence.[34] We have demonstrated that this ionophore, in concentrations ranging from 0.5 to 10 μM, can activate chemiluminescence generation in resting neutrophiles in the presence (but not in the absence) of extracellular calcium. We have found that the CL response to the ionophore is a function of the extracellular calcium concentration, and that Mg^{2+} can antagonize but not substitute for Ca^{2+} in activating the neutrophile. Root and Metcalf have shown that the ionophore can also activate H_2O_2 and O_2^- production in resting granulocytes in a calcium-dependent manner, two species which are intimately involved in the generation of electronically excited states of oxygen.[35] Further, A23187 has been reported to activate NADPH oxidase activity in the 27,000 *g* fraction of human neutrophiles, an enzyme that has recently received much support as being the primary enzyme responsible for the respiratory burst of phagocytic cells.[36] Hence, our findings, together with those of Root and Metcalf, appear to indicate that a correlation exists between the activation of chemiluminescence and oxidative metabolism in resting neutrophiles exposed to the soluble agent, ionophore A23187. Another soluble stimulant, phorbol myristate acetate, has been shown to produce comparable results.[37]

From the above observations, it appears then that the generation of CL by PMNs is dependent upon a complex relationship between cellular activation and a series of interlinked enzymic and nonenzymic reactions, which ultimately result in the production of electronically excited states. If opsonized bacteria are the activating agent, then these excited states serve an important functional role in the bactericidal action of phagocytic cells. A more detailed representation of the mechanisms that contribute to the development of this phenomenon (CL) is provided in Fig. 2.

It is the purpose of this chapter to present the methods involved in generating chemiluminescence from phagocytic cells (emphasizing PMNs) and the procedures involved in utilizing CL as both an index to study defects of phagocytosis (of both cellular and humoral origin) and to study the interaction of pharmacologic and toxicologic agents with phagocytic cells.

[34] M. Wilson, M. A. Trush, and K. Van Dyke, *Fed. Proc., Fed. Am. Soc. Exp. Biol.* **36,** 1071 (1977).

[35] R. K. Root and J. Metcalf, *Clin. Res.* **24,** 318A (1976).

[36] L. C. McPhail, L. R. DeChatelet, J. E. Lehmeyer, and R. B. Johnston, Jr., *Clin. Res.* **25,** 343A (1977).

[37] L. R. DeChatelet, P. S. Shirley, and R. B. Johnston, Jr., *Blood* **47,** 545 (1976).

Cell Isolation Procedures

Methods for Isolation of Peripheral Blood Leukocytes (PMNs and Monocytes)

The initial obstacle in conducting *in vitro* chemiluminescence studies employing peripheral granulocytes rests in the separation of these cells from the other formed elements of whole blood, particularly erythrocytes. While no technique presently available completely satisfies this requirement, several methods have been reported that permit the isolation of leukocyte-rich fractions and further, the separation of mononuclear and polymorphonuclear leukocytes.

Erythrocyte contamination is conventionally removed through the use of aggregating agents, such as high molecular weight dextran,[38] methylcellulose, and Ficoll.[39] These agents promote the aggregation of erythrocytes in a fluid medium, such that the Stokes radius of these particles exceeds that of leukocytes and consequently leads to the more rapid sedimentation rate of the erythrocytes even under 1 *g* conditions. However, sedimentation of erythrocytes, using aggregating agents such as dextran, does not completely remove all erythrocyte contamination from the leukocyte-rich plasma. Removal of residual erythrocyte contaminants is particularly important in CL studies, as colored solutions quench the light emission. Erythrolysis of a leukocyte-rich pellet can be performed using ammonium chloride (0.83–0.87%) or osmotic lysis (in which the cells are briefly exposed to distilled water or hypotonic saline, followed by adjustment of the solution to isotonicity). These procedures effectively remove erythrocytes but must be done rapidly (<30 sec) to leave the leukocytes intact. Leukocyte viability can be assessed by trypan blue exclusion, lactate dehydrogenase release, or oxygen uptake studies.

Some studies may necessitate the use of more highly purified preparations of granulocytes ($>95\%$ pure). In such cases, for example in subcellular fractionation studies, the use of a combined Ficoll–Hypaque and dextran sedimentation procedure yields granulocyte-rich leukocyte preparations of 97–99% purity routinely.

In our laboratory, we routinely employ the following cell isolation procedure, obtaining typical yields of 5 to 7×10^7 PMNs from 30 ml of blood.

[38] W. A. Skoog and W. S. Beck, *Blood* **11,** 436 (1956).

[39] A. Boyum, *Scand. J. Clin. Lab. Invest.* **21,** 97 (1968).

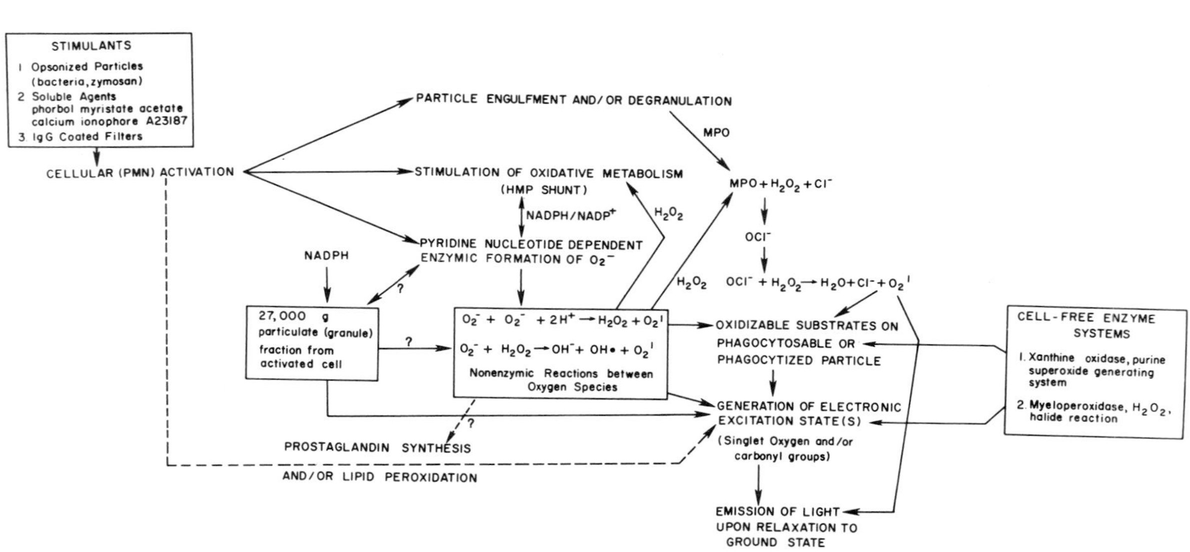

FIG. 2. Representation of the enzymic and nonenzymic mechanisms involved in the generation of chemiluminescence (CL) by polymorphonuclear leukocytes (PMNs) following cellular activation by both particulate and soluble stimulants. Dashed line represents two processes, prostaglandin synthesis and lipid peroxidation, known to generate CL; however, the contribution, if any, of these processes to the CL response of PMNs has yet to be defined. This figure was compiled from information presented in the following references.

1. *Cellular Studies*
   a. R. C. Allen, Ph.D. dissertation, Tulane University, New Orleans, Louisiana, 1973.
   b. R. C. Allen, R. L. Stjernholm, and R. H. Steele, *Biochem. Biophys. Res. Commun.* **47**, 679 (1972).

c. R. C. Allen, S. J. Yevich, R. W. Orth, and R. H. Steele, *Infect. Immun.* **60,** 909 (1974).
d. B. M. Babior, R. S. Kipnes, and J. T. Curnutte, *J. Clin. Invest.* **52,** 741 (1973).
e. L. R. DeChatelet, P. S. Shirley, and R. B. Johnston, Jr., *Blood* **47,** 545 (1976).
f. F. Haber and J. Weiss, *Proc. R. Soc. Edinburgh Sect. A* **147,** 332 (1934).
g. S. J. Klebanoff, *Semin. Hematol.* **12,** 117 (1975).
h. R. B. Johnston, Jr., B. B. Keele, Jr., H. P. Misra, L. S. Webb, J. E. Lehmeyer, and K. V. Rajagapalan, *in* "The Phagocytic Cell in Host Resistance" (J. A. Bellanti and D. H. Dayton, eds.), p. 61. Raven, New York, 1975.
i. R. B. Johnston, Jr., and J. E. Lehmeyer, *J. Clin. Invest.* **57,** 836 (1976).
j. H. Rosen and S. J. Klebanoff, *J. Biol. Chem.* **252,** 4803 (1977).
k. A. I. Tauber and B. M. Babior, *J. Clin. Invest.* **60,** 374 (1977).
l. L. S. Webb, B. B. Keele, Jr., and R. B. Johnston, Jr., *Infect. Immun.* **9,** 1051 (1974).
m. M. Wilson, M. A. Trush, and K. Van Dyke, *Fed. Proc., Fed. Am. Soc. Exp. Biol.* **36,** 1071 (1977).

2. *Subcellular Studies*
n. B. M. Babior, J. T. Curnutte, and B. J. McMurrich, *J. Clin. Invest.* **58,** 989 (1976).
o. K. Kakinuma, A. Boveris, and B. Chance, *FEBS Lett.* **74,** 295 (1977).
p. L. C. McPhail, L. R. DeChatelet, J. E. Lehmeyer, and R. B. Johnston, Jr., *Clin. Res.* **25,** 343A (1977).

3. *CL from Cell-Free Enzyme Systems*
q. R. C. Allen, *Biochem. Biophys. Res. Commun.* **63,** 675 (1975).
r. R. C. Allen, *Biochem. Biophys. Res. Commun.* **63,** 684 (1975).
s. B. D. Cheson, R. L. Christensen, R. Sperling, B. E. Kohler, and B. M. Babior, *J. Clin. Invest.* **58,** 789 (1976).
t. J. F. Piatt, A. S. Cheema, and P. J. O'Brien, *FEBS Lett.* **74,** 251 (1977).
u. H. Rosen and S. J. Klebanoff, *J. Clin. Invest.* **58,** 50 (1976).

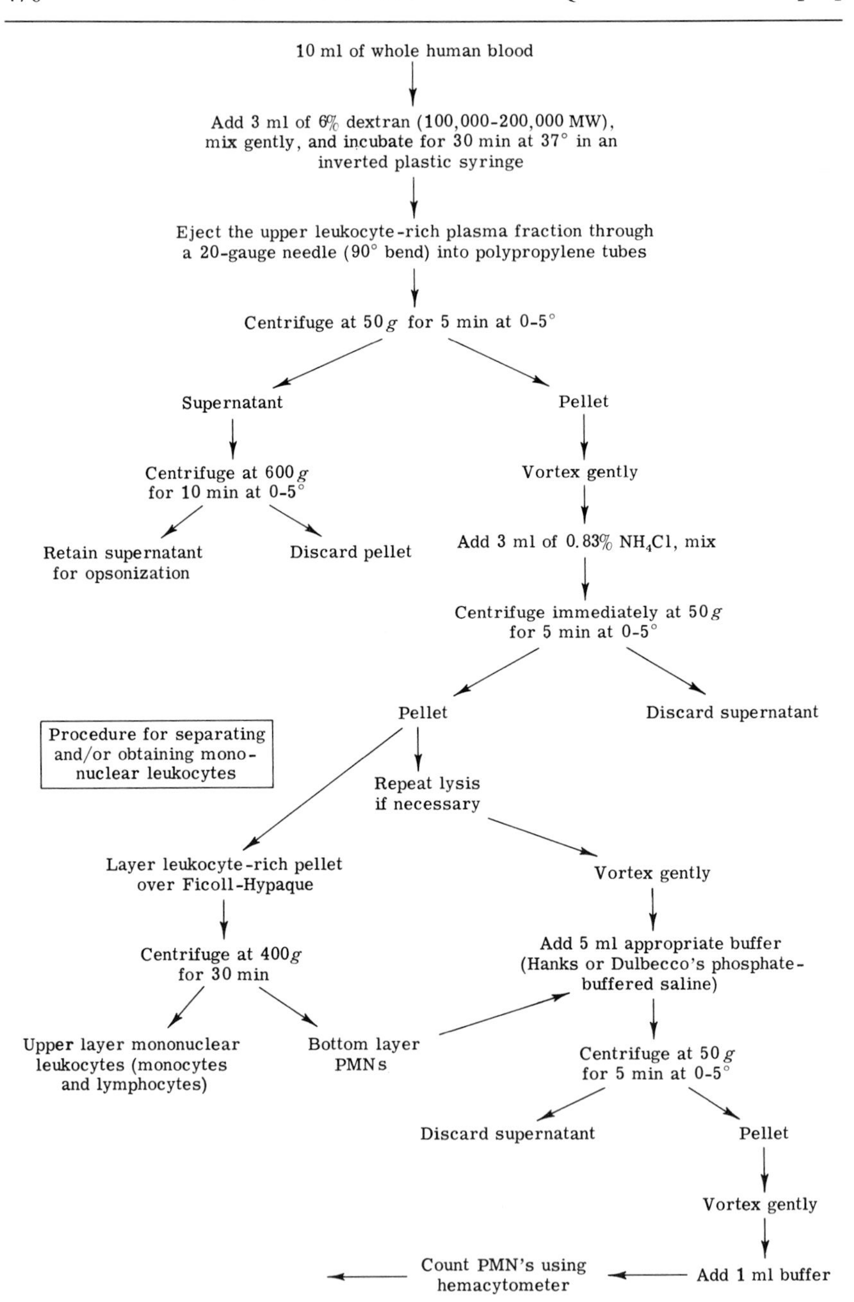

FIG. 3. Isolation of polymorphonuclear leukocytes (PMNs).

Methods for Isolation of Macrophages for Chemiluminescence Studies

While phagocytic cells (PMNs and monocytes) obtained from human blood have been utilized most extensively in CL studies, phagocytic cells (macrophages) harvested from the lungs and peritoneal cavity of experimental animals also demonstrate a CL response.[27,29] In a recent report by Beall *et al.*, the CL response of human alveolar macrophages was described.[28] The methods for obtaining macrophages from these sites are quite different from the procedure described for isolating PMNs from human blood.

Alveolar Macrophages. Lung alveolar macrophages for CL studies can be obtained in quite pure suspensions from rabbits, rats, and guinea pigs. In general, the procedures utilized to obtain these cells are (1) tracheal lavage of the lung or (2) isolation of macrophages from a minced lung preparation. For CL studies the lavage technique is preferred because contamination by red blood cells is unavoidable with the minced lung procedure, thus necessitating a lysis step.

The animal species used to obtain the macrophages will be dependent on the number of cells required. It has been our experience that 5×10^6 alveolar macrophages from either rats or rabbits and 4 mg of zymosan yields an adequate CL response (60,000–70,000 cpm above background). Naturally a larger yield of cells can be obtained from a rabbit as compared to a rat. The yield of alveolar macrophages can be increased by using a lavaging fluid (such as Hanks' Balanced Salt Solution) supplemented with EDTA[40] or lacking calcium and magnesium ions.[41] Another procedure that has been employed to increase cell yields is to inject animals with killed Bacillus Calmette-Guérin (BCG) bacteria[42] or complete Freund's adjuvant[43] to induce granulomatous lesions in the lung. However, one must realize that this population of cells may be already activated and therefore demonstrate increased phagocytic and metabolic capabilities.

Another important consideration when selecting the species of animal is the purity of the preparation. While the degree of red cell contamination is a matter of lavage technique and thus generally preventable, the amount of PMN contamination is related to the animal species. Preparations of alveolar macrophages from guinea pigs seem to be contaminated with PMNs to a much larger extent than preparations from rats and rabbits. Thus, suspensions of alveolar macrophages should be examined microscopically for PMN contamination.

[40] E. R. Pavillard, *Aust. J. Exp. Biol. Med. Sci.* **41,** 265 (1973).

[41] J. D. Brain and R. Frank, *J. Appl. Physiol.* **34,** 75 (1973).

[42] Q. N. Myrvik, E. S. Leak, and S. Oshima, *J. Immunol.* **89,** 745 (1962).

[43] B. M. Gesner and J. G. Howard, *in* "Handbook of Experimental Immunology" (D. M. Weir, ed.), p. 1009. Blackwell, Oxford, 1967.

When using alveolar macrophages for CL studies, a much better response will be obtained if the cells are incubated at 37° between counts and if zymosan rather than bacteria is used as the phagocytosable particle.

Lavage Procedures. RABBIT. The animal is anesthetized by an injection of pentobarbital into one of the ear veins. Wet the fur on the abdominal cavity, dissect the skin from the abdominal cavity, and exsanguinate the animal by cutting the abdominal aorta and inferior vena cava. Open the thorax as far as the thyroid cartilage, spread the rib cage, and then carefully dissect the tissue surrounding the trachea. Next, cut the trachea across and insert a piece of plastic tubing with an outside diameter that fits the trachea tightly. It is at this point that care must be taken to prevent any extraneous blood from entering the trachea. The lungs may be carefully dissected out of the chest cavity or left *in situ* for the lavage procedure. The lavage fluid, generally a buffer solution containing glucose, can be instilled into the lungs by means of a syringe or wash bottle (30–50 ml) being careful not to overdistend the lungs. Gently massage the lungs and then aspirate or pour out the lavage fluid. Repeat the washing several times, pool the lavages, and centrifuge (120 *g*).

RAT. Anesthetize the animal with an intraperitoneal injection. Using a scalpel, cut through the skin from the neck region to the abdominal cavity. Carefully, using blunt-tip scissors, dissect away the muscle exposing the trachea, being careful not to cut any of the small blood vessels in this region. Snip an opening in the exposed trachea, insert a piece of plastic tubing to serve as a cannula, and tie in place. Open the abdominal cavity and exsanguinate the animal by cutting the abdominal aorta and vena cava. Perforate the diaphragm (with blunt-tip scissors) to deflate the lungs, and carefully cut the diaphragm away, exposing the lungs. Lavage the lungs *in situ* using a syringe and needle fitted into the tracheal cannula. Inflate the lungs with 4 ml of buffer (without Ca^{2+} or Mg^{2+}), massage the lungs, and slowly aspirate the lavage fluid. Repeat this procedure 17–20 times, pool the washings, and centrifuge (120 *g*).

Peritoneal Macrophages. The yield of peritoneal macrophages from normal healthy animals (rabbits and guinea pigs) is generally too low for doing adequate studies. However, agents (for example glycogen, casein, and mineral oil) can be injected into the intraperitoneal cavity; this will elicit the accumulation of phagocytic cells (neutrophiles and/or macrophages) from the peripheral circulation into the peritoneum. Although the use of inducing agents increases the number of cells that can be obtained from the peritoneal cavity, this population should be looked upon as already being activated. Stuart *et al.* describe the composition of various inducing agents and possible drawbacks to their use.[44] The macrophages

[44] A. E. Stuart, J. A. Habeshaw, and A. E. Davidson, *in* "Handbook of Experimental Immunology" (D. M. Weir, ed.), p. 24.1. Blackwell, Oxford, 1973.

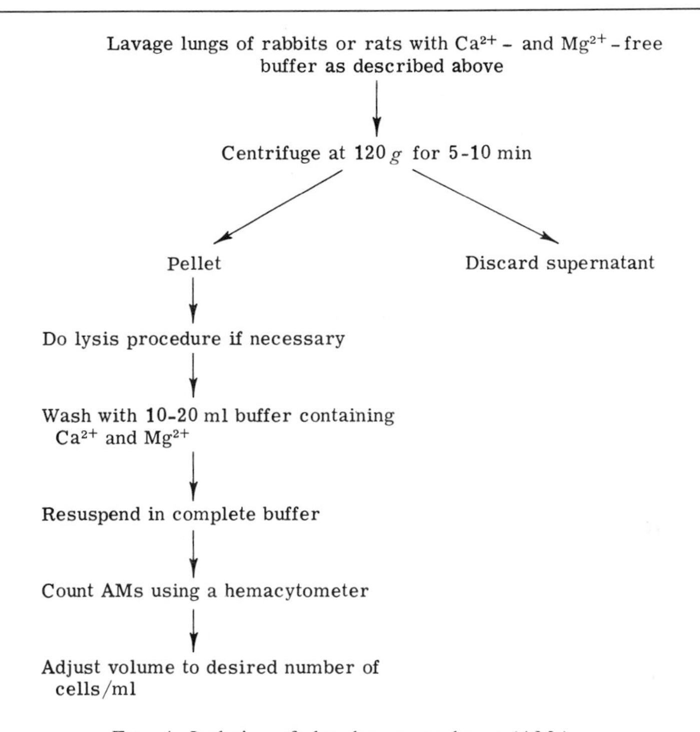

FIG. 4. Isolation of alveolar macrophages (AMs).

are lavaged from the peritoneal cavity 4–5 days after injection of the inducing agents (from 15 to 24 hr after administration of the inducing agent, the cellular population consists mainly of elicited PMNs). The lavage technique for obtaining peritoneal macrophages is described by Stuart *et al.*[44]

Instrumentation and Techniques to Measure CL Responses

In discussing the instrumentation capable of detecting the CL response of cells undergoing phagocytosis, it is necessary to realize, in relative terms, the amount of light being generated and the wavelength of this light. Basically, there are two systems that can be used: (1) measurement of the weak inherent cellular light or (2) an amplified-light system dependent on and reflective of the reactive oxygen species generated during phagocytosis.

Experiments using crude filters and a liquid scintillation counter (LSC) with bialkali photomultiplier tubes have demonstrated that the inherent cellular CL generated has a broad wavelength spectrum with the main

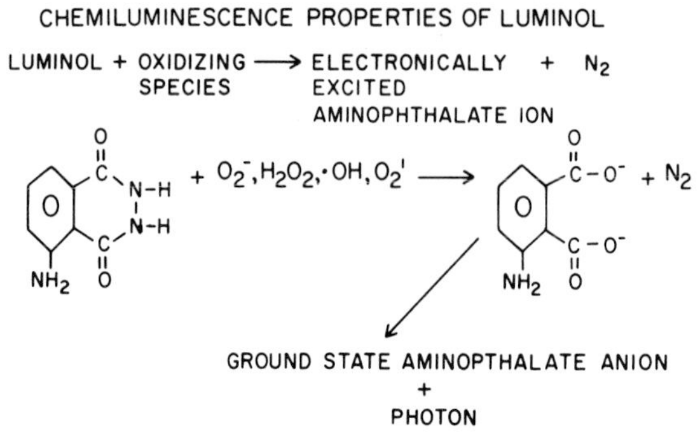

CHEMILUMINESCENCE PROPERTIES OF LUMINOL

LUMINOL + OXIDIZING SPECIES ⟶ ELECTRONICALLY EXCITED AMINOPHTHALATE ION + N_2

$+ O_2^-, H_2O_2, \cdot OH, O_2' \longrightarrow$ + N_2

GROUND STATE AMINOPTHALATE ANION + PHOTON

FIG. 5. Schematic representation of the oxidation of luminol which results in the generation of photons. The oxidizing species result from cellular activation.

wavelength in the red region.[45] Presently, the only instrument available to detect this CL response is the LSC operated in the out-of-coincidence mode. We have attempted to use other instruments, such as photon counters and the Chem-Glow photometer, but only on a few occasions (using high cell numbers) could inherent cellular CL be detected.

The amplified system uses the chemiluminescent compound luminol (5-amino-2,3-dihydro-1,4-phthalazinedione), which is converted to an excited aminophthalate ion in the presence of oxidizing species.[46] This reaction, which is illustrated in Fig. 5, emits blue light (425 nm).[47] Luminol has a high quantum efficiency and produces enough light so that almost any instrument equipped with a photomultiplier tube with an S-4 to S-20 response can be used for CL measurements.

Which system one chooses is dependent upon the instrumentation available and upon the experimental limitations one establishes (e.g., cell number).

Use of the Liquid Scintillation Counter

The original system developed by Allen utilized the LSC in the out-of-coincidence mode. One of the problems in using a counter in this manner is the high counter backgrounds observed. Allen used carefully matched bialkali photomultiplier (PM) tubes to achieve a machine background of a few thousand counts per minute at counter settings of 100% gain and

[45] T. F. Lint, A. M. Brendzel, and B. R. Anderson, *Fed. Proc., Fed. Am. Soc. Exp. Biol.* **36,** 1071 (1977).

[46] H. O. Albrecht, *Z. Phys. Chem.* **136,** 321 (1928).

[47] E. H. White, O. Zafiriou, H. H. Kagi, and J. H. M. Hill, *J. Am. Chem. Soc.* **86,** 940 (1964).

discriminators from 0 to 1000. Since in most cases it will not be possible to select matched PM tubes, background can be reduced by other means: (1) dark adapt the scintillation vials, preferably for 24 hr; (2) use plastic scintillation vials instead of glass vials; (3) store buffers in the dark and/or in dark containers; (4) do all manipulations related to the vials in a darkened room.

The out-of-coincidence mode is used because of the low energy of the light-emitting species. Scintillation counting (in the in-coincidence mode) for radioactivity differs from CL detection in that the light pulse from the higher energy radioactive source must be detected simultaneously by both PM tubes to be accurately registered as a count. In contrast, the out-of-coincidence mode allows the events from both PM tubes to be seen individually and summed. This statement applies only to Packard's LSC and the latest (1977) models of Beckman counters. In some counters the coincidence circuitry between the PM tubes cannot be eliminated, but rather one of the PM tubes must be switched off, leaving only one tube as the detector and thus reducing the efficiency of detection by 50%.

Phagocytic cells are sensitive to temperature and, as such, prolonged exposure to cold temperatures causes decreased metabolic activity and therefore a decreased CL response. If the only counter available for CL assay is refrigerated, shut the refrigeration off for at least 24 hr before doing CL experiments.

A minor problem with the LSC is the initial delay and loss of counts after the vial drops inside the counting chamber and activates the counting mechanism. If this presents a problem, it is possible to have this delay time reduced.

A major advantage of the LSC is that the sample changer is precisely timed. This permits the use of a multiple vial system whereby one can rotate a series of vials and study controls and experimentals at the same time. The number of vials in a cycle would be determined by how often one wishes to observe the light emission from a single vial. This system should be set up so that one can observe the peak response. Alternatively, a sample can be analyzed continuously by switching the counter to continuous repetition, thereby following the response every minute. This is a good procedure to use in screening for the peak response or in the construction of kinetic curves (slope analysis). However, there is less agitation and mixing of the cells in the vial by this procedure as opposed to the multiple vial system.

The settings of the LSC for detecting a CL response from phagocytizing cells can vary. The determinant of what settings are used is the amount of background noise one is willing to accept. In our laboratory a gain of 60% and discriminator range of 50–1000 generally gives us $\geq$ 200,000 cpm with

TABLE I
LIQUID SCINTILLATION COUNTER (LSC) SETTINGS USED TO MEASURE PHAGOCYTOSIS-ASSOCIATED CHEMILUMINESCENCE

1. Preset time	1 Minute
2. Counting channel (linear 0–60 to 1000 divisions)	Width 5 divisions Lower discriminator A, 0–60 Upper discriminator B, 1000
3. Gain	60–100%
4. Coincidence[a]	Off
5. Refrigeration[a]	For optimal activity, turn off for at least 24 hr prior to use
6. Input selector[a] (photomultiplier tube switch)	1 + 2

[a] Located on the back of LSC.

1×10^7 PMNs and 4 mg of zymosan. Obviously, more counts could be observed with a gain of 100%. Table I represents counter settings that can be used.

Luminol-Amplified Chemiluminescence Systems

Luminol has been demonstrated to chemiluminesce upon reacting with potent oxidizing agents. The high quantum efficiency of this reaction[48] led to the initial use of this compound in the detection of free-radical production by xanthine oxidase.[49] Subsequently, Allen and Loose described the use of luminol in detecting the formation of oxidizing agents by activated rabbit alveolar and peritoneal macrophages and human PMNs.[29] The high quantum efficiency of the luminol reaction suggested to us that the luminol-amplified CL system may be clinically useful in assessing opsonophagocytic defects in much the same way as native chemiluminescence has been used, but with the advantage of requiring smaller blood volumes. Figure 6 depicts the possible mechanisms by which luminol-amplified CL responses may be diminished.

There are certain difficulties associated with the use of luminol:

1. Solubility. Three approaches have been taken in dealing with the relative insolubility of luminol in aqueous solutions:

a. Choose a suitable solvent in which luminol will dissolve, but that will not itself interfere with the metabolic activity of PMNs or macrophages. In experiments employing the liquid scintillation counter, we prepare a stock luminol solution by dissolving 0.5 mg of luminol (Aldrich Chemical Co.) in

[48] U. Isacsson and G. Wettermark, *Anal. Chim. Acta* **68,** 339 (1974).

[49] J. R. Totter, E. Castro de Dugros, and C. Riveiro, *J. Biol. Chem.* **235,** 1839 (1960).

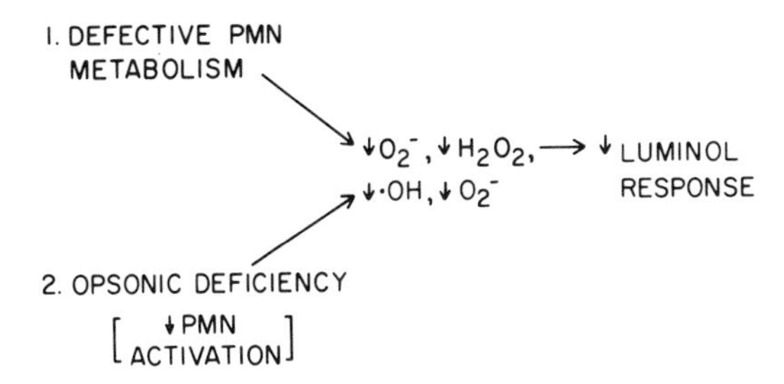

FIG. 6. Representation of how a deficiency in polymorphonuclear leukocyte (PMN) metabolism and/or activation would bring about a diminished luminol response (i.e., diminished amplification of the inherent or native response).

0.05 ml of dimethyl sulfoxide (DMSO) and then diluting to a concentration of 10 mg/100 ml in Dulbecco's phosphate buffer (PBS), pH 7.2. This stock solution was then diluted as indicated in PBS to concentrations of 1 : 100, 1 : 250, or 1 : 500 for use in various experiments. For experiments using the Chem-Glow photometer, 10 mg of luminol were dissolved in 1 ml of DMSO and diluted to a final volume of 1 liter using distilled water.

b. Convert the parent luminol compound to a more water-soluble salt. This may be accomplished by using sodium hydroxide, for example, to yield a sodium salt. Alternatively, Sheehan and Hercules used 1.42 g of luminol, 78 g of potassium hydroxide, and 61.8 g of boric acid in 1 liter of water to prepare an 8 m*M* stock solution of luminol.[50]

c. Add luminol directly to a specified volume of serum to achieve saturation, then filter the serum to remove any undissolved compound. The obvious difficulty inherent to this approach is batch-to-batch variation in luminol concentration.[29]

2. Luminol concentration. Preliminary studies indicated that to some extent the concentration of luminol relative to cell concentration is important. As a result of the reactivity of luminol with oxidizing agents generated by both resting and phagocytosing cells, addition of too concentrated a solution of luminol results in excessive light emission, such that the capacity of the liquid scintillation counter to accurately record such emission in out-of-coincidence mode is exceeded. Figure 7 demonstrates the relationship between luminol concentration and peak CL response. The native CL response of 10^6 PMNs upon addition of 3 mg of opsonized zymosan is included for comparative purposes. The liquid scintillation counter operated in the out-of-coincidence mode can record a maximum of 900,000 cpm, and so luminol concentrations that produce peak CL responses exceeding this value are not practical to use. Figure 8 depicts the relationship between

[50] T. L. Sheehan and D. M. Hercules, *Anal. Chem.* **49,** 446 (1977).

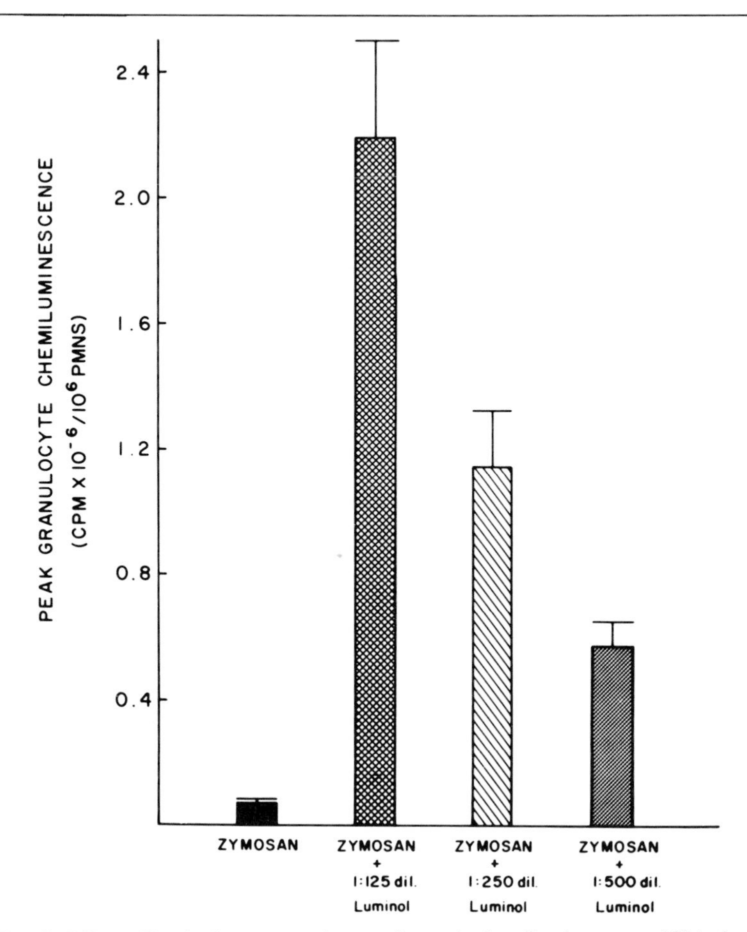

FIG. 7. Effect of luminol concentration on the peak chemiluminescence (CL) observed. The luminol (0.1 ml) was added simultaneously with zymosan (4 mg) to 1×10^6 PMNs. Each bar is the mean $\pm$ SE of 3 experiments. Peak CL occurred at 5 min for both native (inherent) and luminol-amplified CL.

PMN concentration and peak luminol-amplified CL, using 0.1 ml of a 1:250 luminol solution and 3 mg of opsonized zymosan. Whereas studies based upon measurement of native CL typically require at least 5×10^6 cells, the luminol-amplified system permits the use of at least an order of magnitude fewer cells (when using the liquid scintillation counter), which is a distinct advantage of the latter system. Cell (PMN or macrophage), particle, and luminol concentrations employed in studies using the Aminco Chem-Glow photometer have recently been reported.[51] Figures 9 and 10 depict the

[51] K. Van Dyke, M. Trush, M. Wilson, P. Stealey, and P. Miles, *Microchem. J.* **22,** 463 (1977).

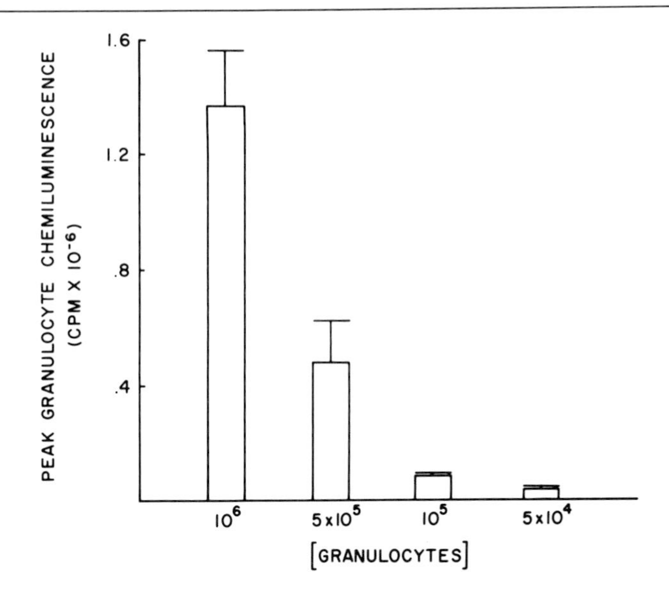

FIG. 8. Relationship between polymorphonuclear leukocyte concentration and peak luminol-amplified chemiluminescence (CL). Each bar is the mean ± SE of 4 experiments. Peak CL occurred at 10 min.

luminol-amplified CL responses of CGD PMNs and neonatal blood examples of cellular and opsonic defects, respectively. The luminol-amplified responses reflect defects similar to those seen with native CL (compare Fig. 9 with Fig. 14 in the opsonic defect section of this paper).

Use of the Aminco Chem-Glow Photometer

An alternative, less expensive instrument, the Chem-Glow photometer, can be used to assess, by CL measurement, phagocytic activity and opsonophagocytic defects.[51] Because of the sensitivity of this instrument, however, the luminol-amplified system is used. As demonstrated in Figs. 11 and 12, the CL responses observed are similar to the responses from a LSC. Thus similar events appear to be measured by both instruments.

Although this instrument lacks the sensitivity of the LSC in measuring inherent cellular CL, there are certain possible advantages to this instrument. These include the following: (1) The initial cost of the system is 1/10 of the LSC unit. Since this unit is modular, the Chem-Glow photometer, integrator-timer, and recorder can be purchased separately, depending upon the need of the user. (2) The cells can be incubated during the reaction by flowing 37° water through the aluminum head. (3) Initial events can be

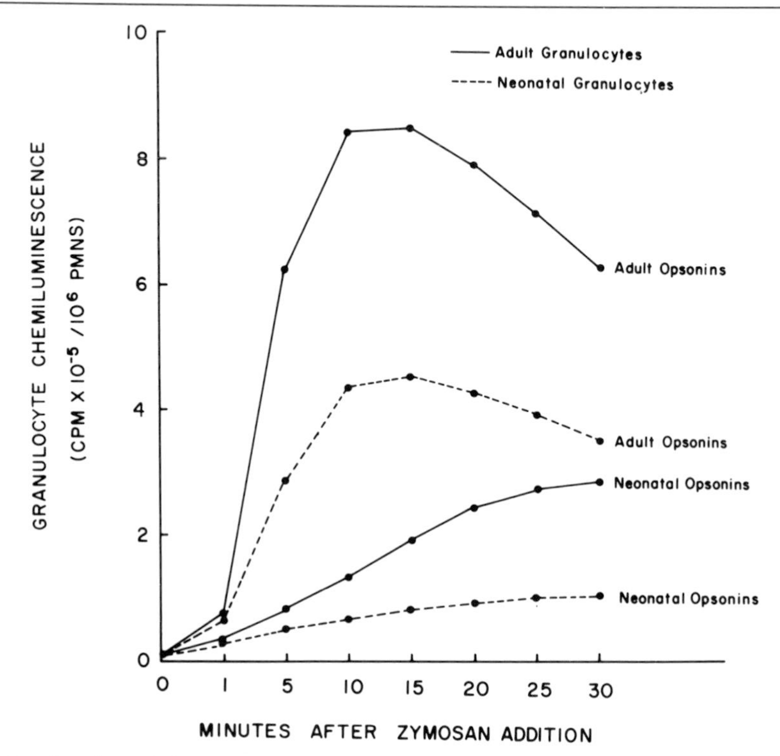

FIG. 9. Effect of the source of polymorphonuclear leukocytes (PMNs) and opsonins on the luminol-amplified chemiluminescence (CL) response. PMNs were isolated from adult and neonatal blood. Zymosan (10 mg) was opsonized with 1.0 ml of autologous neonatal or adult plasma, washed, and diluted to 3 mg/ml. Luminol (0.1 ml of the 1 : 250 dilution) and zymosan (3 mg) were added simultaneously to 1×10^6 PMNs.

easily measured. (4) A stirring apparatus is available that can agitate the cells and particles.

Methods of Data Presentation

Prior to the initiation of a CL response, the background of the vials and its contents should be determined. This background value represents a composite of the inherent electronic noise of the counter and the inherent CL of the vials, solutions added (buffer) and resting cell activity. As mentioned previously, this value can be controlled somewhat by dark-adapting the vials (storing them in the dark and/or wrapping them in aluminum foil) and by preventing the exposure of solutions to the room lights as much as possible. Further, all additions and handling of the vials

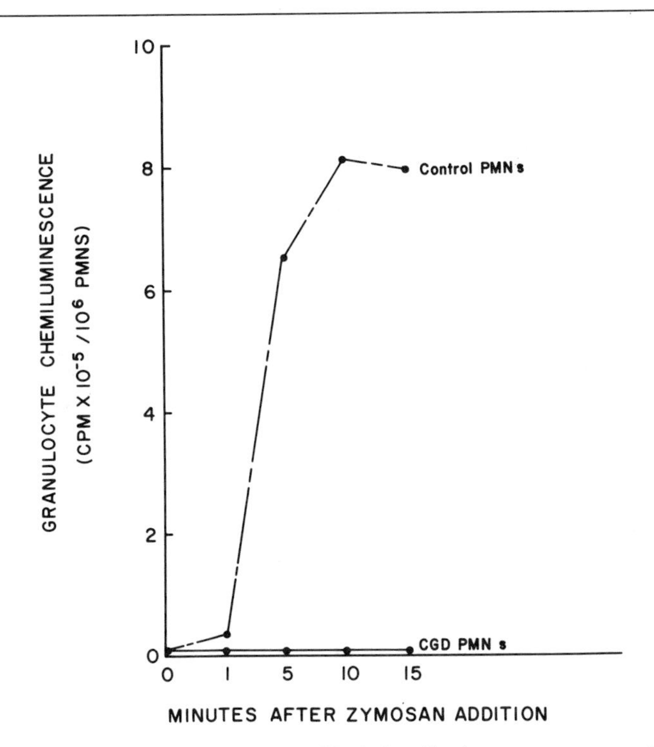

FIG. 10. Comparison of the luminol-amplified chemiluminescent response of polymorphonuclear leukocytes (PMN s) obtained from a normal adult donor (control) and an 11-year-old boy with chronic granulomatous disease (CGD). Luminol (0.1 ml of the 1:250 dilution) and zymosan (2 mg) were added simultaneously to 1×10^6 PMNs.

should be done in a darkened room. Minimizing the background is important because it is subtracted from the counts obtained from each response and as such can contribute to experimental variability. The data obtained after background subtraction can be expressed in several ways (Table II).

What method of data presentation is used will be determined, in part, by the response observed and the experimental conditions. Often under different experimental conditions the pattern of the temporal response between control and experimental differs and/or the time when the peak response occurs differs. The time of the response chosen to determine percent change from control then becomes an important consideration. While illustration of the temporal responses or comparison of peak responses are more than sufficient, the comparison of the response slopes may be more valuable. This method was initially utilized by Allen to compare differences in opsonic activity but is an interesting approach to

TABLE II
METHODS OF DATA PRESENTATION

Method	Description
1. Temporal response	Plot of the data (cpm $\times$ 10^{-n}) as a function of time
2. Integrated counts	A presentation of the integrated total counts at given time intervals (cpm $\times$ 10^{-n}) obtained by connecting a rate meter and a strip-chart recorder to the counter. Allows an easier comparison of chemiluminescence to metabolic parameters (i.e., HMP shunt activity), which are generally reported as activity per time unit (30 min)[a]
3. Peak or maximum response	Reporting the peak response obtained under the different conditions. If this method is used, some indication at what time of the response the peak occurred should be made
4. Comparison of response slopes	Calculation of the slope from the temporal response curves[b]

[a] R. C. Allen, Ph.D. dissertation, Tulane University, New Orleans, Louisiana, 1973.
[b] R. C. Allen, *Infect. Immun.* **15,** 828 (1977).

compare the decrease in activation of cells which have been exposed to pharmacologic agents.[52]

Assessing Opsonophagocytic Dysfunctions by CL Assay

Analysis of Opsonic Defects by Granulocyte CL Assay

The initiation of enhanced oxidative metabolism and accompanying chemiluminescence following the activation of phagocytic cells by particulate substances is dependent upon both cellular and humoral factors. The cellular factors are multifaceted and will be discussed in detail later in this section. The humoral factors consist of a heterogeneous group of serum proteins collectively referred to as "opsonins," whose function is to promote the attachment to and subsequent ingestion of phagocytosable material by phagocytic cells, principally neutrophilic leukocytes. Opsonization of particulate material is accomplished via one or more of the following routes: (1) generation of type-specific opsonins of the IgG and/or IgM classes of immunoglobulins;[53] (2) generation of opsonins of the C3-opsonin (C3o) type via the classical (antibody-dependent) complement pathway;

[52] R. C. Allen, *Infect. Immun.* **15,** 828 (1977).
[53] K. E. Christie, C. D. Solberg, B. Larsen, A. Grov, and O. Tonder, *Acta Pathol. Microbiol. Scand. Sect. C.* **184,** 119 (1976).

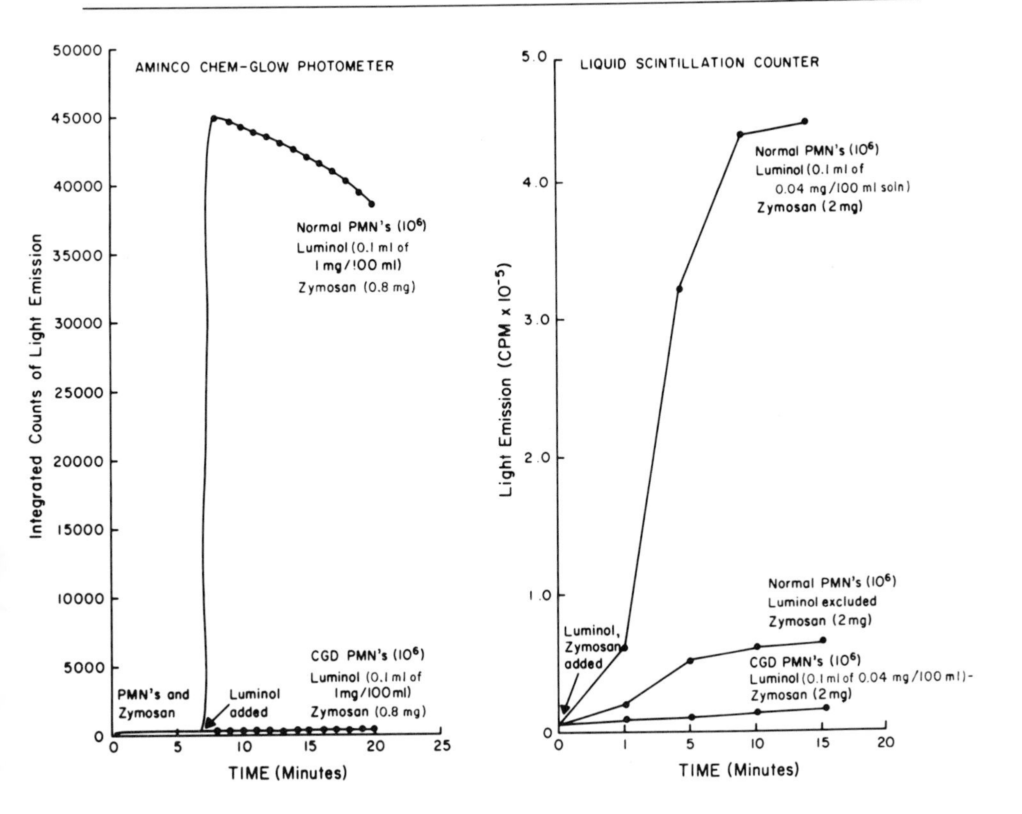

FIG. 11. Detection of a cellular polymorphonuclear leukocyte (PMN) defect using the luminol-amplified chemiluminescence system with the Aminco Chem-Glow photometer.

(3) generation of opsonins of the C3o type via the alternative, or properdin, pathway (antibody-independent) first described by Pillemer *et al.*[54]

Nonspecific opsonins of the C3o type consist of a cleavage product of complement protein C3 which is either C3b itself or a closely related molecule. Deposition of IgG and/or IgM antibodies (via their Fab, or antigen-binding, sites) as well as C3o onto the surface of microorganisms results in a marked enhancement of the rates of attachment to and ingestion of these foreign particles. This enhancement arises from the fact that phagocytic cells possess both Fc and C3b receptors on their plasma membranes which promote the association of these cells with particle-bound opsonins.[55]

[54] L. Pillemer, L. Blume, I. H. Lepow, A. O. Ross, E. W. Todd, and A. C. Wardlaw, *Science* **120,** 279 (1954).

[55] D. J. Scribner and D. Fahrney, *J. Immunol.* **116,** 892 (1976).

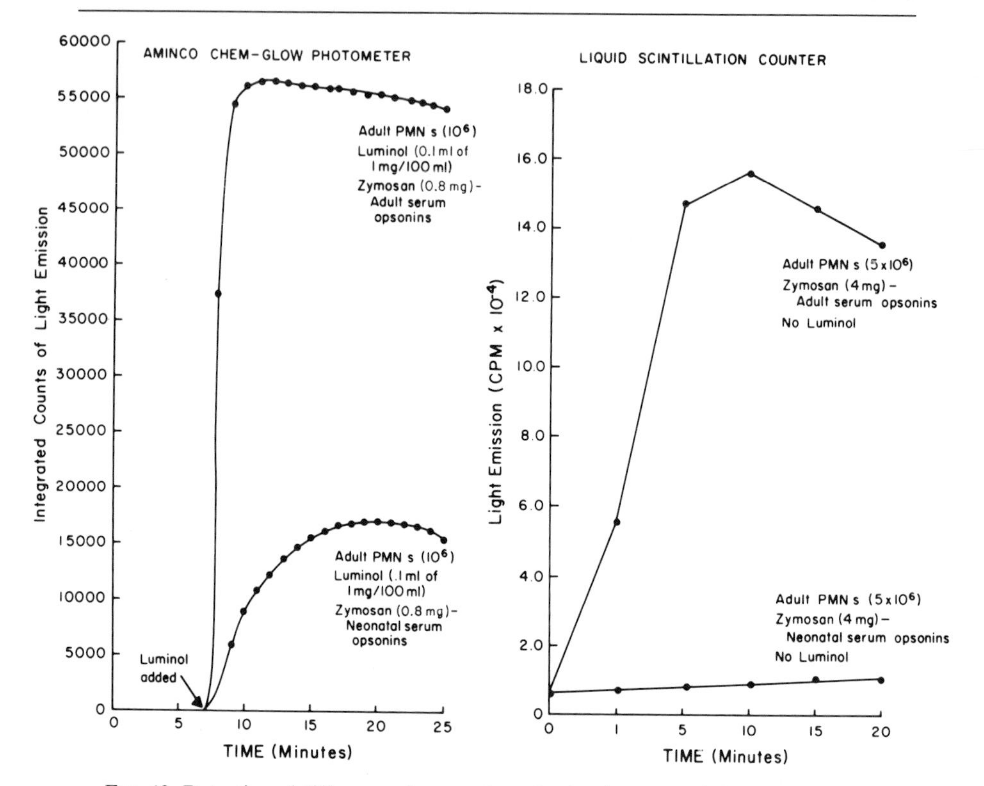

FIG. 12. Detection of differences in opsonic activation between adult and neonatal opsonins using the luminol-amplified chemiluminescence system with the Aminco Chem-Glow photometer.

Aberrations in the circulating levels and/or functional activities of serum opsonins may lead to defective activation of otherwise normal phagocytic cells. Several clinical disorders involving deficiencies of both type-specific and nonspecific opsonization are known, and they are frequently associated with increased susceptibility to bacterial infection. These disorders include agammaglobulinemia, congenital C3 deficiency, and hypercatabolic C3 deficiency.[56] Defects in opsonization may be assessed from a functional standpoint through the use of the granulocyte CL assay, and four examples of such an approach are provided below:

Neonatal Opsonic Defects. Newborn infants, particularly premature infants, represent a clinically important example of defective opsonization associated with increased incidence of sepsis (systemic bacterial infec-

[56] R. B. Johnston, Jr., and R. M. Stroud, *J. Pediatr.* **90,** 169 (1977).

tion).[57] Although the human fetus possesses functional B lymphocytes by week 20 of gestation,[58] neonates at birth carry circulating antibodies that are largely maternally derived through placental transport. This is presumably due to a relative lack of antigenic stimulation encountered by the fetus *in utero*. Whereas IgG antibodies are transported across the placenta with little difficulty, IgM antibodies are unable to cross this maternal–fetal barrier. Most antibodies possessing opsonic activity toward gram-negative bacteria fall into the IgM class, which explains in part the neonate's unusual susceptibility to gram-negative sepsis. However, recent studies suggest that defects in the alternative pathway of opsonization (which operates independently of antibody involvement) may be of greater importance in predisposing neonates to sepsis. Stossel and co-workers reported that the deficiency of opsonic activity in cord sera correlated with depressed levels of factor B (C3 proactivator) but not with immunoglobulin or C3 deficiencies.[59]

We have employed the granulocyte CL assay in assessing the functional status of the alternative pathway in neonatal serum by utilizing for opsonization a substrate (zymosan) that activates the alternative pathway. A small volume (approximately 3–6 ml) of peripheral venous blood was withdrawn from premature neonates of varying gestational age (26–38 weeks) by way of an indwelling umbilical venous catheter and was immediately transferred to heparinized vacutainers. Isolation of granulocytes and opsonization of zymosan were performed as discussed earlier. Immediately after obtaining neonatal blood samples, venous blood was drawn from healthy adult volunteers and processed in an identical manner. Zymosan previously opsonized with either neonatal or adult serum (refer to opsonization procedure as outlined in Fig. 13) was added to dark-adapted plastic scintillation vials containing neonatal or adult granulocytes. This "cross-matching" procedure not only permits an evaluation of neonatal serum opsonic capacity (by comparing the CL responses of adult PMNs activated by zymosan coated with neonatal or autologous serum opsonins), but also lends itself to an evaluation of neonatal PMN function (by comparing the CL responses of neonatal and adult PMNs activated by zymosan opsonized with adult serum). On the basis of this protocol, we recently reported that opsonic activity generated by the alternative complement pathway is markedly depressed in neonatal serum.[60] Further, our results indicated that neonatal PMNs generate consistently lower CL responses

[57] H. D. Wilson and H. F. Eichenwald, *Pediatr. Clinics N. Am.* **21,** 571 (1974).

[58] R. Van Furth, H. R. E. Schuit, and W. Hijmans, *Immunology* **11,** 1 (1966).

[59] T. P. Stossel, C. A. Alper, and F. S. Rosen, *Pediatrics* **52,** 134 (1973).

[60] M. E. Wilson, M. A. Trush, K. Van Dyke, M. D. Mullett, and W. A. Neal, *Pediatr. Res.* **11,** 496 (1977).

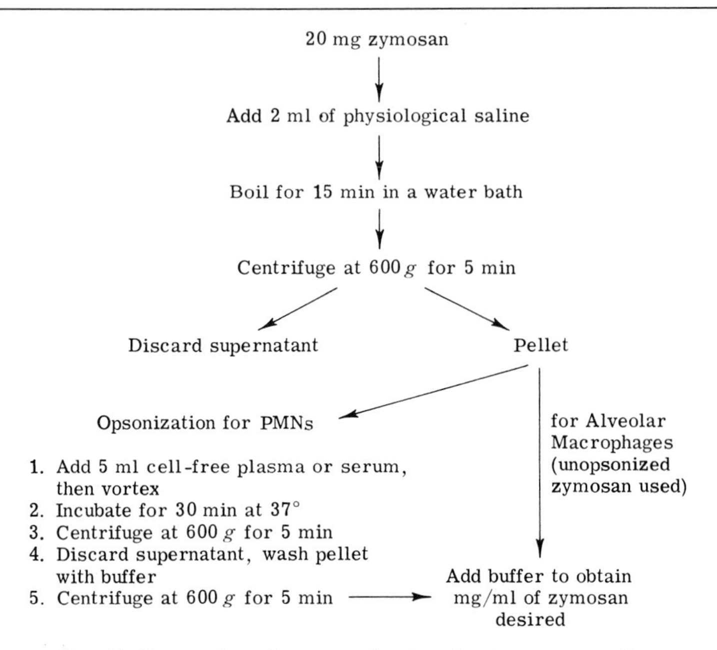

FIG. 13. Preparation of zymosan for chemiluminescence studies.

to zymosan opsonized with adult serum than do adult PMNs. The results of a typical CL assay performed with blood obtained from a 28-week-old neonate are shown in Fig. 14 and are consistent with the findings of other groups.[61,62] The defect in neonatal PMN activation does not appear to be attributable to a diminished population of membrane receptors for C3b and Fc, according to Pross *et al.*[63]

Defective Type-Specific Opsonization. Whereas our studies of neonatal defense mechanisms have focused primarily on opsonization via the alternative pathway, Hemming and co-workers have utilized the granulocyte CL assay to assess the importance of type-specific (IgG) opsonins in host defense against group B streptococci.[21] Similarly, Stevens and Young have demonstrated a correlation between resistance of certain clinically isolated strains of *Escherichia coli* to opsonization by nonimmune serum and decreased CL responses.[19,64]

[61] E. L. Mills, T. Thompson, B. Bjorkstein, J. Verhoef, P. K. Peterson, and P. G. Quie, *Abstracts of the Midwest Society of Pediatric Research,* Nov. 3–5, 1976, Chicago. Abst. No. 10.

[62] H. R. Hill, N. A. Hogan, and V. G. Hemming, *Clin. Res.* **24,** 178A (1976).

[63] S. H. Pross, J. A. Hallock, R. Armstrong, and C. W. Fishel, *Pediatr. Res.* **11,** 135 (1977).

[64] P. Stevens and L. S. Young, *Infec. Immun.* **16,** 796 (1977).

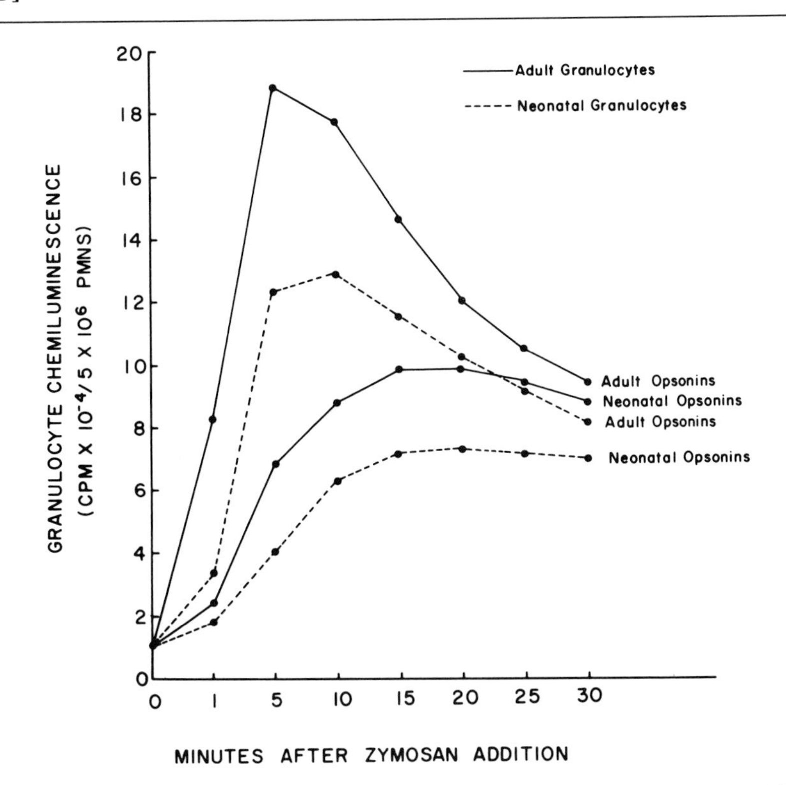

FIG. 14. Temporal chemiluminescence response of granulocytes (5×10^6) isolated from adult and neonatal blood to zymosan (4 mg) opsonized with 2 ml of autologous or neonatal plasma.

Differentiating Classical versus Alternative Pathway Opsonization. Grebner *et al.* utilized a basic difference in the ion requirements for activation of the classical and alternative pathways in order to determine the mode of opsonization of a number of particles.[20] In particular, the classical pathway requires calcium ions for activation whereas the alternative pathway requires only magnesium. Consequently, these workers added a chelating agent (EGTA, 10 m*M*) to serum in order to remove all divalent cations and then replaced the chelated magnesium with 10 m*M* $MgCl_2$ prior to opsonization. This approach makes it possible to discern between classical and alternative pathway opsonization using the granulocyte CL assay.

Quantitative Assessment of Serum Opsonic Capacity. Allen has recently stressed the utility of performing slope analyses on the initial CL responses of phagocytosing PMNs in quantitating serum opsonic capacity.[52] As was discussed in a prior section, this approach can be more useful than mere comparison of peak CL values.

Analysis of Cellular Defects by Granulocyte CL Assay

Following opsonization and binding of a phagocytosable particle to a phagocytic cell, such as a neutrophilic leukocyte, an early and dramatic increase in oxidative metabolic activity is normally observed, a phenomenon classically referred to as the "respiratory burst." Recent studies support the contention that the enzyme responsible for initiating the burst may be NADPH oxidase, a flavin enzyme.[65] As indicated in Fig. 2, the oxidase is responsible for catalyzing the monovalent reduction of molecular oxygen to superoxide anion. This reaction can undergo spontaneous (nonenzymic) dismutation to yield singlet oxygen, an electronically excited species that may contribute to the generation of chemiluminescence, either directly via relaxation to the triplet ground state, or via formation and subsequent relaxation of excited carbonyl moieties. Consequently, defective generation of superoxide anion can lead to defective generation of CL.

Stjernholm *et al.* first reported that granulocytes from patients with chronic granulomatous disease of childhood (CGD) generate little or no CL in the presence of particles opsonized by normal serum.[22] Many subsequent studies have shown that, although these cells phagocytose particles normally, CGD granulocytes do not exhibit the characteristic burst of oxidative metabolic activity, and are markedly deficient with regard to superoxide production.[24] The defect in superoxide production may be associated with a deficiency of NADPH oxidase activity in these cells.[66] Consequently, defective production of superoxide anion in CGD granulocytes appears to play an important role in diminished CL generation by these cells. Further, the granulocyte CL assay has received increasing support as a clinical tool in the initial diagnosis of CGD.

As indicated in Fig. 2, electronically excited molecules are generated via one of several reactions, including the reaction of myeloperoxidase with the hypochlorite ion and excess hydrogen peroxide. Consequently, granulocytes deficient in MPO would also be anticipated to yield diminished CL responses. Indeed, Rosen and Klebanoff reported defective generation of CL in MPO-deficient leukocytes.[23] The authors reported normal generation of superoxide anion by these cells, but noted that superoxide production and CL fall off less rapidly than in control leukocytes. Their findings support a role for MPO-catalyzed reactions in the generation of CL. That superoxide production and CL generation was prolonged in these cells might be explained by the fact that since H_2O_2 is not

[65] D. Iverson, L. R. DeChatelet, J. K. Spitznagel, and P. Wang, *J. Clin. Invest.* **59,** 282 (1977).

[66] L. C. McPhail, L. R. DeChatelet, P. S. Shirley, C. Wilfert, R. B. Johnston, Jr., and C. E. McCall, *J. Pediatr.* **90,** 213 (1977).

consumed in the MPO reaction, it is available to react with other molecules, such as superoxide anion, to generate electronically excited species via the Haber–Weiss reaction.

In view of the fact that CGD granulocytes fail to produce superoxide and hydrogen peroxide, whereas MPO-deficient granulocytes exhibit normal oxidative metabolic activity, it seems reasonable that the CL assay may be utilized to distinguish between these two diseases by adding exogenous hydrogen peroxide to both cell preparations during the course of the assay. Hydrogen peroxide readily diffuses across membranes and should augment intracellular levels of H_2O_2. Since MPO-deficient leukocytes generate adequate amounts of H_2O_2, exogenously added H_2O_2 should not appreciably alter CL generation by these cells. On the other hand, CGD granulocytes contain normal amounts of MPO but lack H_2O_2, and so exogenous H_2O_2 would be expected to markedly augment CGD CL responses. Such a response would be consistent with the clinical finding that CGD patients derive benefit from a high-dose ascorbate regimen, which would generate H_2O_2 via a quinone–hydroquinone reaction.

Whereas defects in the enzymes associated with the hexose monophosphate shunt may also contribute to diminished CL responses, no reports of such defects are presently available. However, patients with selective glucose-6-phosphate dehydrogenase deficiency (G-6-PDH catalyzes the first reaction in the HMP shunt) exhibit a CGD syndrome that is distinguishable from CGD disease. Murphy and co-workers have demonstrated defective neutrophile chemiluminescence generation by leukocytes of patients with recurrent pneumonias.[67] Defective chemotaxis appears to have contributed in some cases; however, further characterization of the defect(s) is required.

The results indicate that, in addition to assessing opsonic capacity, the granulocyte chemiluminescence assay has application in the initial diagnosis of a number of defects in oxidative metabolism and intracellular killing, which may be further characterized by subsequent biochemical studies.

Figure 15 summarizes the diagnostic potential of the "cross-match" method of performing CL assays. Using this approach it is possible to differentiate between defects in opsonization (classical or alternative pathway) and defects in cellular metabolic activation. Following the initial determination of an opsonic or cellular defect, it is then possible to characterize the defect(s) by instituting studies of other biochemical parameters, such as enzyme activities, immunoglobulin or complement protein titers.

[67] S. A. Murphy, A. H. Cushing, S. B. Witemeyer, and D. E. Van Epps, *Am. Rev. Respir. Dis.* **115,** 68 (1977).

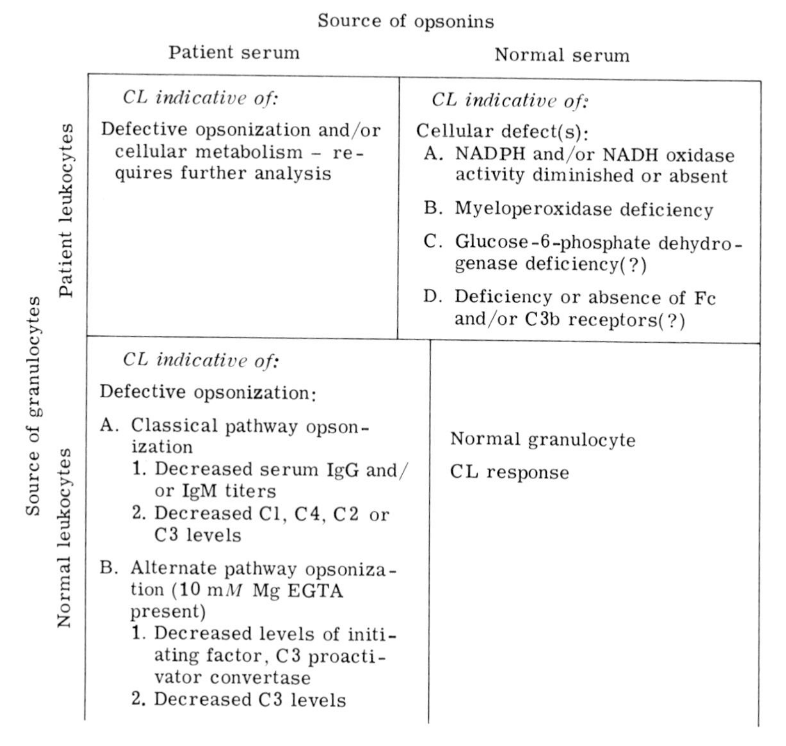

Source of granulocytes	Source of opsonins: Patient serum	Source of opsonins: Normal serum
Patient leukocytes	*CL indicative of:* Defective opsonization and/or cellular metabolism – requires further analysis	*CL indicative of:* Cellular defect(s): A. NADPH and/or NADH oxidase activity diminished or absent B. Myeloperoxidase deficiency C. Glucose-6-phosphate dehydrogenase deficiency(?) D. Deficiency or absence of Fc and/or C3b receptors(?)
Normal leukocytes	*CL indicative of:* Defective opsonization: A. Classical pathway opsonization 1. Decreased serum IgG and/or IgM titers 2. Decreased Cl, C4, C2 or C3 levels B. Alternate pathway opsonization (10 m*M* Mg EGTA present) 1. Decreased levels of initiating factor, C3 proactivator convertase 2. Decreased C3 levels	Normal granulocyte CL response

FIG. 15. Assessment of opsonophagocytic dysfunctions by granulocyte chemiluminescence (CL) assay.

The Use of Chemiluminescence as an Index to Study the Interaction and/or Effect of Pharmacologic and Toxicologic Agents on Phagocytic Cells

Since the chemiluminescent response of phagocytic cells is dependent on cellular metabolism and activation, the measurement of CL represents a potentially useful index by which to assess the effect of pharmacologic and toxicologic agents on phagocytic cells. The methodology for assessing such effects is similar to that described above for determining opsonophagocytic defects except that pharmacologic agents should be added to the cells under various conditions to obtain a more complete characterization of their action. These conditions include (1) preincubation of the cells with the agent for 15 min at 37° before addition of the phagocytosable particle; (2) simultaneous addition of the agent and particle (i.e., opsonized zymosan); and (3) addition of the agent after activation of the cells. [This experimental protocol was used by Klebanoff and Clark to distinguish between an agent

TABLE III
METHODS THAT CAN BE USED IN CONJUNCTION WITH CHEMILUMINESCENCE TO CHARACTERIZE THE EFFECT OF PHARMACOLOGIC AGENTS ON PHAGOCYTIC CELLS

Parameter	Method
1. Superoxide production	Cytochrome *c* reduction[a]
2. Hydrogen peroxide generation	a. Extinction of scopoletin fluorescence[b]
	b. Generation of labeled CO_2 from [^{14}C]formate oxidation[c]
3. HMP shunt activity	[1-^{14}C]glucose oxidation[d]
4. Oxygen uptake	Oxygen electrode
5. Particle ingestion	a. Microscopic examination of number of particles in cells
	b. Oil Red O paraffin procedure[e]
6. Cell viability	a. LDH release
	b. Trypan blue uptake

[a] B. M. Babior, R. S. Kipnes, and J. T. Curnutte, *J. Clin. Invest.* **52,** 741 (1973).
[b] R. K. Root, J. Metcalf, N. Oshino, and B. Chance, *J. Clin. Invest.* **55,** 945 (1975).
[c] G. Y. N. Iyer, D. M. F. Islam, and J. H. Quastel, *Nature (London)* **192,** 535 (1961).
[d] B. Holmes, P. G. Quie, D. B. Windhorst, and R. A. Good, *Lancet* **1, 1966,** 1225 (1966).
[e] T. P. Stossel, R. J. Mason, J. Hartwig, and M. Vaughan, *J. Clin. Invest.* **51,** 615 (1972).

that inhibits iodination (an MPO-dependent reaction) by altering particle ingestion (sodium polyantholesulfonate) from one (sodium azide) having a more direct effect on MPO.[68]]

The parameters one is looking for in determining whether an agent has an effect on phagocytosis-associated CL are a change in peak CL and/or an alteration in the temporal response (for example, the time when peak CL occurs). These data can be expressed as (1) percent inhibition by comparing peak responses, (2) illustration of the temporal response, or (3) calculation of slopes to reflect differences in cellular activation. A change in any or all of these parameters is reflective of a possible alteration in cellular metabolism brought about by the drug–cell interaction. As such, other methods should be utilized to determine (if possible) the basis for the alteration in CL (Table III). Since these parameters and ultimately CL are dependent upon the activation by, and the subsequent ingestion of, the phagocytosable particle, it becomes necessary to determine whether the alteration in CL results from the inhibition of particle uptake and/or the inhibition of some aspect of cellular metabolism upon which CL is dependent. This can be accomplished by inhibiting particle uptake with cytocholasin B (5 μg/ml) and studying the drug's effect on O_2^- produc-

[68] S. J. Klebanoff and R. A. Clark, *J. Lab. Clin. Med.* **89,** 675 (1977).

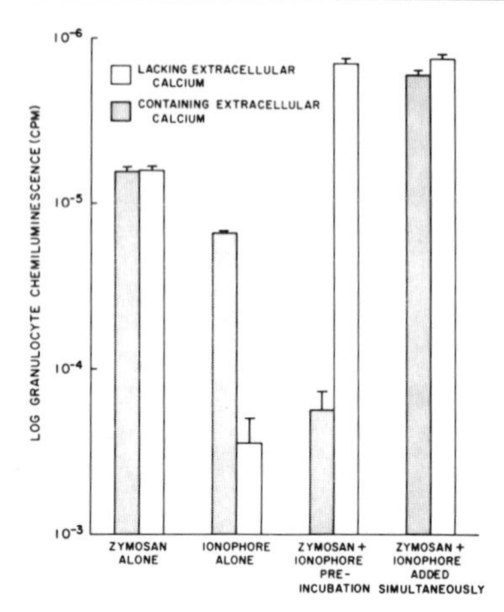

FIG. 16. The peak chemiluminescence responses of 5×10^6 PMNs stimulated by opsonized zymosan (4 mg) and/or ionophore A23187 (10 μM) in the presence or the absence of extracellular calcium.

tion.[69] As illustrated in Figs. 1 and 2, the physical and metabolic events of phagocytosis are very much interlinked, and as such it often becomes very difficult to interpret and dissect an agent's effect on phagocytic cell function. Nonetheless, CL measurement represents a rapid and easy assay to study the effects and/or interaction of agents with phagocytic cells.

Using CL generated by PMNs and AMs as the experimental parameter, the following approaches can be used to study the pharmacologic or toxicologic effect of agents on phagocytic cells:

1. Using pharmacologic agents as tools to study the contribution of ions, enzymes or oxygen species. Examples of this approach include: (a) the inhibition of CL by SOD, catalase, and benzoate[25]; (b) the inhibition of CL by the MPO inhibitor azide;[8,26] (c) the interaction of the calcium ionophore A23187 with PMNs in the presence or the absence of calcium.[34]

2. Studying the *in vitro* effect of agents that have been observed to affect some other aspect of phagocytic function when applied *in vitro*. Examples include: (a) the inhibition of CL by amphotericin B;[70] (b) the inhibition of CL by acrylamide.[71]

3. Studying the *in vitro* effect of agents that have been reported to affect some aspect of phagocytic function when administered *in vivo*. The interaction of promethazine with human PMNs is an example of this approach.[72,73]

[69] R. B. Zurier, S. Hoffstein, and G. Weissmann, *Proc. Natl. Acad. Sci. U.S.A.* **70,** 844 (1973).

[70] B. Bjorkstein, C. Ray, and P. G. Quie, *Infect. Immun.* **14,** 315 (1976).

[71] D. M. Mangan, Masters thesis, West Virginia University, Morgantown, 1977.

[72] M. A. Trush and K. Van Dyke, *Fed. Proc., Fed. Am. Soc. Exp. Biol.* **35,** 748 (1976).

[73] M. A. Trush and K. Van Dyke, *Pharmacology,* in press.

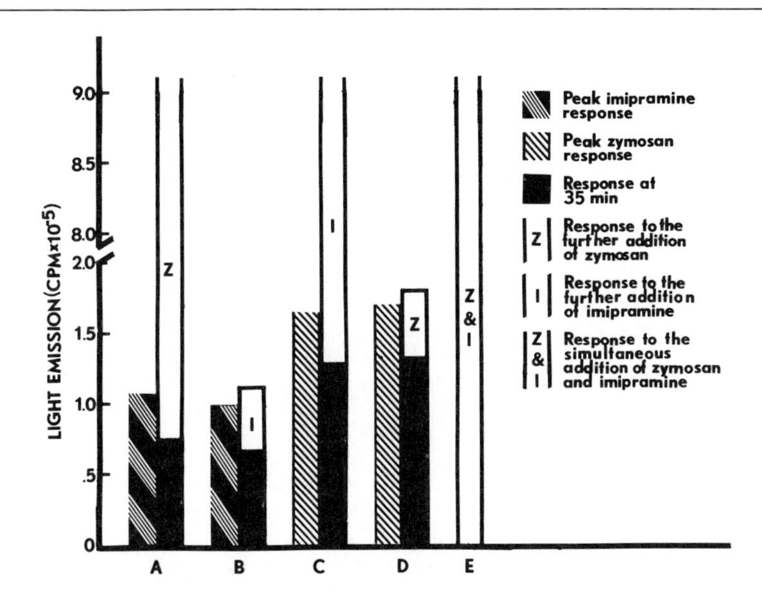

FIG. 17. This figure illustrates the chemiluminescence (CL) interactions that occur when 0.1 m*M* imipramine (a tricyclic antidepressant) or zymosan (4 mg) are added either separately, sequentially, or simultaneously to 1×10^7 PMNs. Bar A, sequential addition of zymosan to cells initially exposed to 0.1 m*M* imipramine; bar B, sequential addition of imipramine to cells initially exposed to the drug; bar C, sequential addition of 0.1 m*M* imipramine to cells initially stimulated by zymosan (4 mg); bar D, sequential addition of zymosan to PMNs initially stimulated by zymosan; bar E, peak CL response (5 min) to the simultaneous addition of 0.1 m*M* imipramine and 4 mg of zymosan. All sequential additions were made just prior to the 40-min reading. The open bars (A, C, and E) indicate that the magnitude of light emission was greater than 900,000 cpm.

4. Studying the response of phagocytic cells altered morphologically and/or biochemically as a result of the *in vivo* administration of drugs. Changes in the CL response of rat alveolar macrophages that are morphologically altered as a result of the administration of chlorphentermine represent an example of this approach.[74]

5. Using resting or phagocytosing human PMNs as a cellular model to study the interaction and possible activation of compounds by reactive oxygen species. Recently, estradiol[75] has been reported to be irreversibly bound to phagocytosing PMNs, and studies conducted in our laboratory have indicated a possible activation of certain tricyclic antidepressants upon addition to both resting and phagocytosing PMNs.[76,77]

[74] M. A. Trush and M. J. Reasor, *Fed. Proc., Fed. Am. Soc. Exp. Biol.* **37,** 501 (1978).

[75] S. J. Klebanoff, *J. Exp. Med.* **145,** 983 (1977).

[76] M. A Trush and K. Van Dyke, *Pharmacologist* **18,** 219 (1976).

[77] M. A. Trush, M. E. Wilson, and K. Van Dyke, *Pharmacologist* **19,** 163 (1977).

Obviously, the use of CL as an index to study the pharmacologic and/or toxicologic effects of agents on phagocytic cells has widespread application and potential.

Comments

We have found the measurement of CL by phagocytic cells to be a reliable and reproducible parameter by which to assess phagocytosis-associated events and as such feel that this system has widespread applicability to a variety of problems, both scientific and clinical in nature. The future application of this system will represent a tribute to the scientific creativity of R. C. Allen, who discovered this system.

Acknowledgments

We are grateful to the following individuals for their assistance in experiments reported by the authors: Drs. Martha Mullet and William E. Neal, Department of Pediatrics and Neonatal Intensive Care Unit, for help in obtaining blood from neonates; Dr. Ruth Phillips, Department of Pediatrics, for help in obtaining blood from the CGD child; Dr. Mark J. Reasor, Department of Pharmacology, for helpful discussions on procedures on the isolation of alveolar macrophages; and Mary Gallegly and Donna Canterbury, Outpatient Clinic Laboratory, for drawing blood from donors.

Work done by the authors has been supported, in part, by grants from the following sources: West Virginia University Senate Grant; West Virginia University Institutional Grant; West Virginia Heart Association; National Institute of Health Training Grant T 32 GM07039.

[40] Chemiluminescence from Electron-Transfer Processes

By LARRY R. FAULKNER

This volume is filled with stories about beautiful reactions that produce light. The phenomenon is striking. Yet its extraordinary aspect is not the appearance of the photons, but rather the production of electronically excited species in a chemical reaction. Why should any kind of chemistry yield excited products, when it can generate the same species in their ground states?

This question addresses the central issue of bioluminescence and chemiluminescence. Its answer would clarify the light-producing stages of many known systems, and it would permit an advanced strategy for design-

ISBN 0-12-181957-4

ing interesting new processes. The answer is also complex, elusive, and largely unknown.

Even so, important components can be set forth. For example, it is clear that significant generation of high-energy products rests on kinetic features, rather than thermodynamic aspects. Some special character in the reaction dynamics must enhance the attractiveness of the pathway to electronically excited products. Two critical elements seem to be a high reaction energy and a very short duration for the actual reaction event. In the context of chemiluminescence, a high energy is an obvious necessity. If there is to be any hope for light production, the exothermicity in forming ground-state products must be at least the energy required for electronic excitation. Speed seems to enter as a general agent encouraging the reaction to take the path to excited products, rather than the alternative route to ground-state species. Chemiexcitation can take place only in systems that find the latter pathway uncomfortable. Such a situation applies when a large energy is liberated over a very short time scale (e.g., less than a period of vibration). Then the system cannot readily absorb the energy in mechanical modes, and one reaches a kind of kinetic Franck–Condon limitation. The path to electronically excited products becomes relatively attractive, because its demand for mechanical accommodation is not nearly so great.

Many other factors certainly influence the yields of excited states from chemical processes. We overlook them in this introduction only to stress the seemingly general importance of speed, which itself implies a certain mechanistic simplicity.

Not many reaction types are simple enough and powerful enough to provide general chemical routes to excited products. Only two have emerged from the last decade of research into chemiluminescence from liquid solutions. One of them is electron transfer, which can be the simplest of bimolecular chemistries. The other is the thermal cleavage of annular peroxides, such as dioxetanes and dioxetanones. In a mechanical sense, those reactions are among the least complicated of unimolecular processes.

The small peroxides and their capacity for chemiluminescence have long been interesting to workers in bioluminescence because they appear to be useful models for natural luciferins.[1-6] By contrast, electron-transfer

[1] E. H. White, J. D. Miano, C. J. Watkins, and E. J. Breaux, *Angew. Chem., Int. Ed.* **13,** 229 (1974).

[2] N. J. Turro, P. Lechtken, N. E. Schore, G. Schuster, H. C. Steinmetzer, and A. Yekta, *Acc. Chem. Res.* **7,** 97 (1974).

[3] F. McCapra, *Prog. Org. Chem.* **8,** 231 (1973).

[4] F. McCapra, *Acc. Chem. Res.* **9,** 201 (1976).

[5] T. Wilson, *Int. Rev. Sci.: Phys. Chem. Ser. Two* **9,** 265 (1976).

[6] J. W. Hastings and T. Wilson, *Photochem. Photobiol.* **23,** 461 (1976).

chemiluminescence has seemed a rather separate domain. Very recently, the two worlds have been bridged by the results of Schuster and co-workers, who have shown that certain peroxides seem to favor an electron-transfer basis for chemiexcitation.[7-10] Their work alone justifies a discussion of electron-transfer chemiluminescence in this volume.

The subject has been considered in detail in several recent reviews,[11-16] and full coverage here is both unnecessary and undesirable. Instead, we shall examine very briefly the principles, the typical experimental techniques, and certain outstanding issues in the field. The following pages are intended only as an introduction to this fascinating phenomenon and its literature.

Basic Chemical Aspects

Prototypes

A typical example of redox chemiluminescence is the system in which the anion and cation radicals of rubrene annihilate to yield (ultimately) a rubrene molecule in its first excited singlet state,[17-22]

$$R^{\cdot +} + R^{\cdot -} \rightarrow {}^1R^* + R \tag{1}$$

Yellow chemiluminescence follows when the singlet ${}^1R^*$ emits,

$${}^1R^* \rightarrow R + h\nu \tag{2}$$

[7] J.-Y. Koo and G. B. Schuster, *J. Am. Chem. Soc.* **99,** 6107 (1977).
[8] J.-Y. Koo, S. P. Schmidt, and G. B. Schuster, *Proc. Natl. Acad. Sci. U.S.A.*, **75,** 30 (1978).
[9] S. P. Schmidt and G. B. Schuster, *J. Am. Chem. Soc.* **100,** 1966 (1978).
[10] J.-Y. Koo and G. B. Schuster, *J. Am. Chem. Soc.* **100,** in press.
[11] A. Weller and K. Zachariasse, *in* "Chemiluminescence and Bioluminescence" (M. Cormier, D. M. Hercules, and J. Lee, ed.), pp. 169–181. Plenum, New York, 1973.
[12] A. J. Bard, C. P. Keszthelyi, H. Tachikawa, and N. E. Tokel, *in* "Chemiluminescence and Bioluminescence" (M. Cormier, D. M. Hercules, and J. Lee, eds.), p. 193. Plenum, New York, 1973.
[13] K. A. Zachariasse, *in* "The Exciplex" (M. Gordon, W. R. Ware, P. De Mayo, and D. R. Arnold, eds.), p. 275. Academic Press, New York, 1975.
[14] A. J. Bard and S. M. Park, *in* "The Exciplex" (M. Gordon, W. R. Ware, P. De Mayo, and D. R. Arnold, eds.), p. 305. Academic Press, New York, 1975.
[15] L. R. Faulkner, *Int. Rev. Sci.: Phys. Chem. Ser. Two* **9,** 213 (1975).
[16] L. R. Faulkner and A. J. Bard, *Electroanal. Chem.* **10,** 1 (1977).
[17] R. E. Visco and E. A. Chandross, *Electrochim.* **13,** 1187 (1968).
[18] J. Chang, D. M. Hercules, and D. K. Roe, *Electrochim. Acta* **13,** 1197 (1968).
[19] L. R. Faulkner, H. Tachikawa, and A. J. Bard, *J. Am. Chem. Soc.* **94,** 691 (1972).
[20] R. Bezman and L. R. Faulkner, *J. Am. Chem. Soc.* **94,** 6324 (1972); **95,** 3083 (1973).
[21] H. Tachikawa and A. J. Bard, *Chem. Phys. Lett.* **26,** 246 (1974).
[22] N. Periasamy and K. S. V. Santhanam, *Proc. Indian Acad. Sci. Sect. A* **80,** 194 (1974).

This kind of process is quite general among aromatic molecules and their radical ions. The ions may be derived from a single precursor, as in the case of rubrene above, or they may come from two different parents. For example, the rubrene anion also gives yellow chemiluminescence when it reacts with the cation radical of *N,N,N′,N′*-tetramethyl-p-phenylene-diamine (TMPD),[19]

$$R^{\cdot-} + TMPD^{\cdot+} \rightarrow {}^1R^* + TMPD \tag{3}$$

and the rubrene cation does the same upon reaction with the anion radical of *p*-benzoquinone (BQ),[19]

$$R^+ + BQ^- \rightarrow {}^1R^* + BQ \tag{4}$$

These examples show that either ion precursor may emit. In fact, excited singlets of both can be generated, as they are in the reaction between the thianthrene (TM) cation radical and the anion of 2,5-diphenyl-1,3,4-oxadiazole (PPD),

$$TH^{\cdot+} + PPD^{\cdot-} \rightarrow \begin{cases} {}^1TH^* + PPD \\ TH + {}^1PPD^* \end{cases} \tag{5}$$

Dual emissions arise from the two singlets $^1TH^*$ and $^1PPD^*$.[23]

It is the practice of workers in this field to write reactions in terms of abbreviations and notations like those used in Eqs. (1)–(5), and that custom will be followed consistently here. The structures of most abbreviated compounds are shown in Fig. 1.

Hundreds of chemiluminescent reactions of radical ions have been catalogued, and work with them has dominated activity in the whole domain of electron-transfer luminescence.[11-16] Three factors underlie the popularity of radical ions: (a) They span an enormous range of potency as redox agents. It is easy to find reactions that are sufficiently energetic to populate an excited state; and one has the flexibility of allowing any given ion to react with a series of counter reactants that will provide a nearly continuous range of reaction energy. (b) Their redox reactions are simple and very fast.[24,25] Structural changes are usually not strongly tied to the electron-transfer process itself. (c) Their precursors (which are also the products of electron transfer) often have high quantum yields for fluorescence, so that chemiexcitation is well manifested by the appearance of luminescence.

Radical ion reactions cannot be carried out in the presence of significant amounts of protic impurities, such as water or alcohols, because anion

[23] C. P. Keszthelyi, H. Tachikawa, and A. J. Bard, *J. Am. Chem. Soc.* **94,** 1522 (1972).
[24] S. Arai, A. Kira, and M. Imamura, *J. Chem. Phys.* **54,** 5073 (1971).
[25] R. P. Van Duyne and S. F. Fischer, *Chem. Phys.* **5,** 183 (1974).

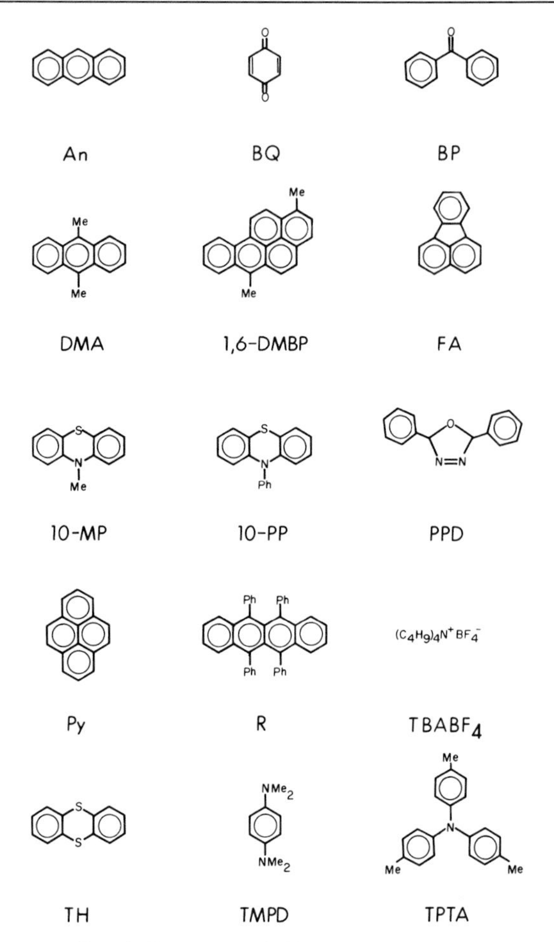

FIG. 1. Structures of abbreviated molecules.

radicals are fairly strong bases and they decay in the presence of proton donors. Cation radicals are similarly sensitive to nucleophiles. Again, water is a serious offender. Oxygen is also detrimental, because it reacts with most anion radicals. These considerations highlight the basis for the scrupulous attention that is paid to the purity of the solvent systems and to the exclusion of air.[16] Favorite solvents are acetonitrile, *N,N*-dimethylformamide (DMF), benzonitrile, tetrahydrofuran (THF), and 1,2-dimethoxyethane (DME). Samples are usually prepared and processed on vacuum lines, where deaeration is carried out by freeze–pump–thaw techniques. The radicals are generated chemically e.g., by reduction with an alkali metal,[26] or more commonly, by electrolysis of the precursors.[16] Experimental methods are covered in more detail in the next section.

Despite the attention that has been given to the radical ion annihilations, they do not encompass the whole domain of chemiexcitation in electron transfer. The phenomenon seems general for fast, simple redox processes that are sufficiently energetic to excite a product. For example, the 2,2′-bipyridine (bipy) complexes of ruthenium yield light as a consequence of the reaction[27]

$$\mathrm{Ru(biby)_3^{3+} + Ru(bipy)_3^{+} \rightarrow Ru(bipy)_3^{2+} + [Ru(bipy)_3^{2+}]^*} \qquad (6)$$

The luminescence has a spectrum that is characteristic of the well studied emitting charge-transfer state of trisbipyridylruthenium(II).

The same luminescence was observed by Martin *et al.* as an accompaniment to the reaction between aquated electrons and the tripositive Ru complex,[28]

$$\mathrm{Ru(bipy)_3^{3+}} + e^-_{aq} \rightarrow \mathrm{[Ru(bipy)_3^{2+}]^*} \qquad (7)$$

The solvated electrons were produced in their study by pulse radiolysis. Itaya *et al.* have seen similar reactions between various oxidants and solvated electrons in hexamethylphosphoramide (HMPA).[29,30] A straightforward example is the reaction with the cation radical of TMPD,

$$\mathrm{TMPD^{+}} + e^-_{solv} \rightarrow \mathrm{TMPD^*} \qquad (8)$$

Of interest to investigators with a biological inclination is the demonstration by Mayeda and Bard that singlet molecular oxygen is produced in electron-transfer processes involving superoxide ion.[31] An example is the reaction with ferricenium (Fc^+) in acetonitrile,

$$\mathrm{Fc^+ + O_2^- \rightarrow Fc + {}^1O_2^*} \qquad (9)$$

Singlet oxygen was detected by its reaction with the trapping agent 1,3-diphenylisobenzofuran,

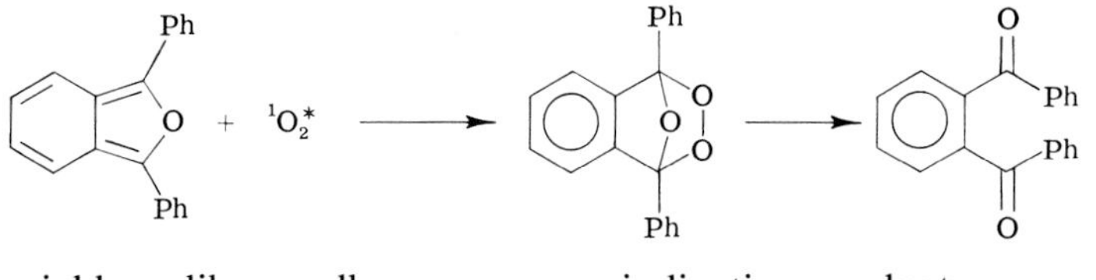

$$\text{1,3-diphenylisobenzofuran} + {}^1\mathrm{O}_2^* \longrightarrow \text{endoperoxide} \longrightarrow o\text{-dibenzoylbenzene} \qquad (10)$$

which yields *o*-dibenzoylbenzene as an indicative product.

[26] K. A. Zachariasse, Thesis, Vrije Universiteit te Amsterdam, 1972.

[27] N. E. Tokel-Takvoryan, R. E. Hemingway, and A. J. Bard, *J. Am. Chem. Soc.* **95,** 6582 (1973).

[28] J. E. Martin, E. J. Hart, A. W. Adamson, H. Gafney, and J. Halpern, *J. Am. Chem. Soc.* **94,** 9238 (1972).

[29] K. Itaya, M. Kawai, and S. Toshima, *Chem. Phys. Lett.* **42,** 179 (1976).

[30] K. Itaya, M. Kawai, and S. Toshima, submitted for publication.

[31] E. A. Mayeda and A. J. Bard, *J. Am. Chem. Soc.* **95,** 6223 (1973).

The examples of chemiexcitation that we have considered here all have the common property that they yield a locally excited state of a one-electron product. More complex cases also exist, but they will not be discussed until we have first examined the availability of energy in the chemiexcitation step itself. This aspect has an enormous influence on the mechanistic choices available to any particular system.

The Impact of Energy Balances

An accounting of energy in the electron-transfer reactions considered above shows immediately that they often cannot be as simple as the overall equations might suggest. Our concern now is whether or not the emitting state is accessible to the redox process. In other words, does the emitted photon carry more or less energy than a single redox event can supply? The answer is very important from a diagnostic standpoint.

Let us consider a general reaction between ions whose precursors are an acceptor molecule A and a donor species D. That is,

$$A^{\overline{\cdot}} + D^{\dot{+}} \rightarrow A + D \tag{11}$$

The free energy of this reaction is straightforwardly calculated from the standard electrochemical potentials for the couples,

$$A + e^- \rightleftharpoons A^{\overline{\cdot}} \qquad E^0(A/A^{\overline{\cdot}}) \tag{12}$$

$$D^+ + e^- \rightleftharpoons D \qquad E^0(D^{\dot{+}}/D) \tag{13}$$

and in electron volts it is simply,

$$\Delta G^0 = E^0(A/A^{\overline{\cdot}}) - E^0(D^{\dot{+}}/D) \tag{14}$$

For reactions of interest here, ΔG^0 is very large (1–4 eV) and negative. If the energy required to raise A or D to a given excited state, such as an emitting singlet, is less than $-\Delta G^0$, then the redox process could produce that excited species spontaneously, and chemiexcitation is at least a possibility. On the other hand, excited states with energies significantly greater than $-\Delta G^0$ are obviously not accessible to the electron transfer.

Figure 2 depicts the energy balances for three cases involving rubrene in the discussion above.[19] [see Eqs. (1), (3), and (4)]. The leftmost column indicates the energies of the free ions above their precursors. The heights of these levels above the base line are the available reaction energies $-\Delta G^0$. The remaining columns show the energies required for excitation of rubrene, BQ, and TMPD to local first excited singlet and lowest triplet states. All three reactions produce chemiluminescence having the same spectral distribution as rubrene's fluorescence, hence the level labeled $^1R^*$ is reached in each case.

Only the rubrene anion/cation reaction has even approximately the

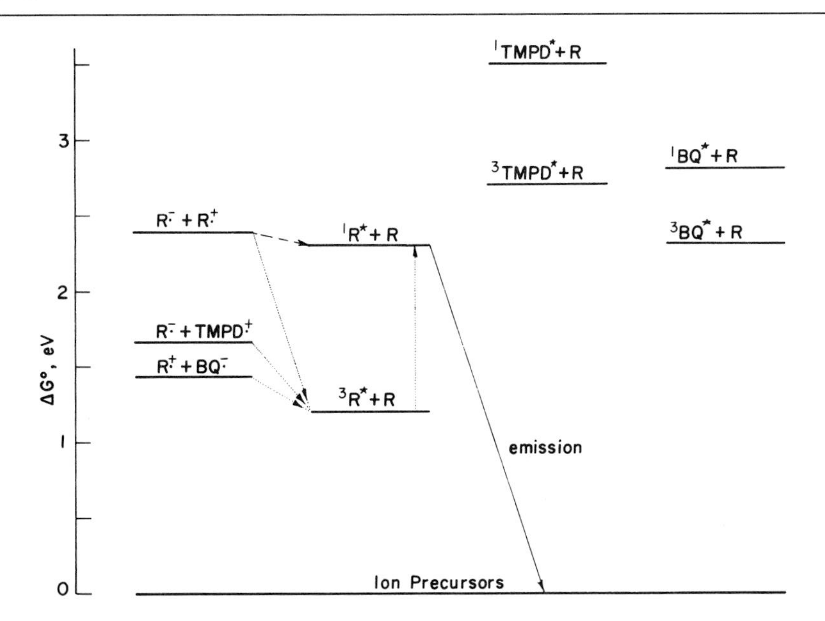

FIG. 2. Energy balances for chemiluminescent reactions of rubrene radical ions. All three reactions shown here give the same emission. Dashed arrow shows S route. Dotted arrows show T route; promotion from $^3R^* + R$ to $^1R^* + R$ requires another rubrene triplet. Data taken from L. R. Faulkner, H. Tachikawa, and A. J. Bard, *J. Am. Chem. Soc.* **94,** 691 (1972) and D. K. K. Liu and L. R. Faulkner, *J. Am. Chem. Soc.* **99,** 4594 (1977).

energy needed for a direct population of the emitting state. This system is termed energy-sufficient, and chemiluminescence from it can arise by the simple mechanism stated in Eqs. (1) and (2). On the other hand, the other two reactions are grossly energy-deficient. In either system, an emitted photon contains far more energy than a single redox reaction can provide; hence the overall effects indicated in Eqs. (3) and (4) must involve several mechanistic steps. Note also that the TMPD and BQ excited states are far too high to be populated in any of the reactions.

Light emission from energy-deficient systems is usually rationalized via a mechanism involving triplet intermediates.[11-16] For example, the reaction between rubrene anion and $TMPD^{\dagger}$ can populate the rubrene triplet directly,

$$TMPD^{\dagger} + R^{\bar{\cdot}} \rightarrow TMPD + {}^3R^* \tag{15}$$

then two of the triplets can undergo triplet–triplet annihilation[32,33] to produce the emitting rubrene singlet,

[32] C. A. Parker, "Photoluminescence of Solutions." Elsevier, Amsterdam, 1968.
[33] J. B. Birks, "Photophysics of Aromatic Molecules." Wiley, New York, 1970.

$$^3R^* + {}^3R^* \rightarrow {}^1R^* + R \tag{16}$$

The effect of Eq. (16) is to pool the energy from two redox events into a single photon. This two-step mechanism, which is widely known as the T route, is strongly supported by experimental evidence. We shall discuss some of it in later sections.

An energy-sufficient system, such as the rubrene anion/cation reaction, always has the dual possibilities of producing emitting singlets either directly as in Eq. (1) (S route) or indirectly via the T route. The factors that control the balance between the two pathways are not very well understood, but are of great interest. We shall examine the problem more fully below.

Excimers, Exciplexes, and Charge-Transfer States

The spectral distribution of emission from chemiluminescent electron transfer reactions is not always the same as the fluorescence from a local state of a product. Often there is a component at longer wavelengths. Sometimes it arises from a contaminant, such as a product of radical ion decay, but very frequently it is intrinsic to the system under study and manifests a more complex excited state in which two or more chromophores interact.

The classic case is the excimer emission that accompanies the reduction of $TMPD^{+}$ by the anion of pyrene (Py).[34,35] Curve a of Fig. 3 shows the spectral distribution of chemiluminescence. At wavelengths below 400 nm, there is a relatively sharp band that is readily ascribed to emission from $^1Py^*$, as one can see by comparison with the fluorescence spectrum of the same solution in curve b of Fig. 3. At longer wavelengths is the structureless band from the pyrene excimer Py_2^*

$$Py_2^* \longrightarrow 2Py + h\nu \tag{17}$$

This emission component is visible in the photoluminescence spectrum of curve b of Fig. 3, but it is not very strong, because the concentration of pyrene is too low to allow for effective formation of the excimer through the association

$$^1Py^* + Py \rightleftharpoons Py_2^* \tag{18}$$

At higher concentrations, the long-wavelength band dominates even the fluorescence.[36]

[34] A. Weller and K. Zachariasse, *J. Chem. Phys.* **46,** 4984 (1967).
[35] J. T. Maloy and A. J. Bard, *J. Am. Chem. Soc.* **93,** 5968 (1971).
[36] T. Förster and K. Kasper, *Z. Elektrochem.* **59,** 976 (1955).

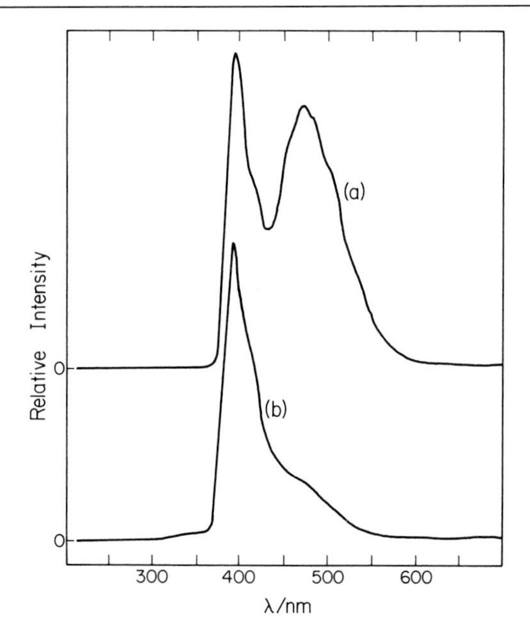

FIG. 3. Curve a: Chemiluminescence from the reaction between $Py^{\overline{\cdot}}$ and $TMPD^{\dagger}$ in dimethylformamide. Electrogeneration of ions was carried out in solutions containing 1 m*M* TMPD and 5 m*M* Py. Curve b: Fluorescence spectrum of the same solution under excitation at 350 nm. Original curves used with permission from J. T. Maloy and A. J. Bard, *J. Am. Chem. Soc.* **93,** 5968 (1971). Figure is reproduced with permission from L. R. Faulkner, *Int. Rev. Sci.: Phys. Chem. Ser. Two* **9,** 213 (1975).

Studies of excimer emissions in chemiluminescence have provided some important pieces of evidence supporting the T route's operation in energy-deficient systems. The reaction between $Py^{\overline{\cdot}}$ and $TMPD^{\dagger}$ is such a case. The interesting aspect of its chemiluminescence is the high relative strength of the excimer band. The ratio of dimer to monomer emission is very much greater than in prompt fluorescence,[34] and this aspect alone suggests that the excimers are not formed entirely from $^1Py^*$ via Eq. (18). Instead, there seems to be a path that generates them directly. Maloy and Bard demonstrated that point convincingly in quantitative studies based on effects of concentration.[35]

The simplest rationale for the results is that the excimers are formed by the annihilation of pyrene triplets,

$$Py^{\overline{\cdot}} + TMPD^{\dagger} \rightarrow {}^3Py^* + TMPD \tag{19}$$

$$^3Py^* + {}^3Py^* \rightarrow \begin{cases} {}^1Py^* + Py \\ Py_2^* \end{cases} \tag{20}$$

The parallel paths implied in Eq. (20) are known to exist from earlier studies of delayed fluorescence from pyrene.[32,33]

Weller and Zachariasse provided evidence that a mechanism like this one also applies to the system in which $TMPD^{+\cdot}$ oxidizes the anion radical of 9,10-dimethylanthracene (DMA).[37] The chemiluminescence spectrum from a THF solution shows excimer and monomer emissions from DMA, and the temperature dependence of their ratio behaves in a manner that is characteristic of systems that produce excimers by triplet–triplet annihilation.

Excimer bands are also prominent in chemiluminescence from several systems in which the reactant ions are derived from a single precursor. An interesting recent example is the annihilation of anion and cation radicals derived from 1,6-dimethylbenzo[*a*]pyrene (1,6-DMBP), which yields light almost exclusively from the excimer.[38] The $DMA^{+\cdot}/DMA^{-\cdot}$ reaction is more typical in producing a mixed spectrum.[39,40] These cases offer an intriguing prospect for direct excimer formation in the ion annihilation; e.g.,

$$DMA^{+\cdot} + DMA^{-\cdot} \rightarrow DMA_2^* \tag{21}$$

Whether this kind of process actually occurs is not yet clear, although the point has received some attention.[39-41]

On the other hand, there is little doubt that exciplexes do form directly in radical ion reactions.[11-16,26] In generalized notation, such a process is

$$A^{-\cdot} + D^{+\cdot} \rightarrow {}^1(A^-D^+)^* \tag{22}$$

where the exciplex ${}^1(A^-D^+)^*$ can be regarded as a kind of contact ion pair with radiative properties. When it emits, it returns to a dissociative ground state,

$${}^1(A^-D^+)^* \rightarrow A + D + h\nu \tag{23}$$

The properties of exciplexes have been extensively studied, mainly by photoluminescence techniques,[33,42] in which they are generated by reactions such as

$${}^1A^* + D \rightarrow {}^1(A^-D^+)^* \tag{24}$$

They are extremely sensitive to the polarity of the solvent. In nonpolar media, they emit with relatively high quantum yield, but in polar media they tend to dissociate into free ions.

[37] A. Weller and K. A. Zachariasse, *Chem. Phys. Lett.* **10,** 197 (1971).

[38] S. M. Park, *Photochem. Photobiol.,* submitted.

[39] T. C. Werner, J. Chang, and D. M. Hercules, *J. Am. Chem. Soc.* **92,** 763 (1970).

[40] C. A. Parker and G. D. Short, *Trans. Faraday Soc.* **63,** 2618 (1967).

[41] E. A. Chandross, J. W. Longworth, and R. E. Visco, *J. Am. Chem. Soc.* **87,** 3260 (1965).

[42] A. Weller, *in* "The Exciplex" (M. Gordon, W. R. Ware, P. De Mayo, and D. R. Arnold, eds.), p. 23. Academic Press, New York, 1975.

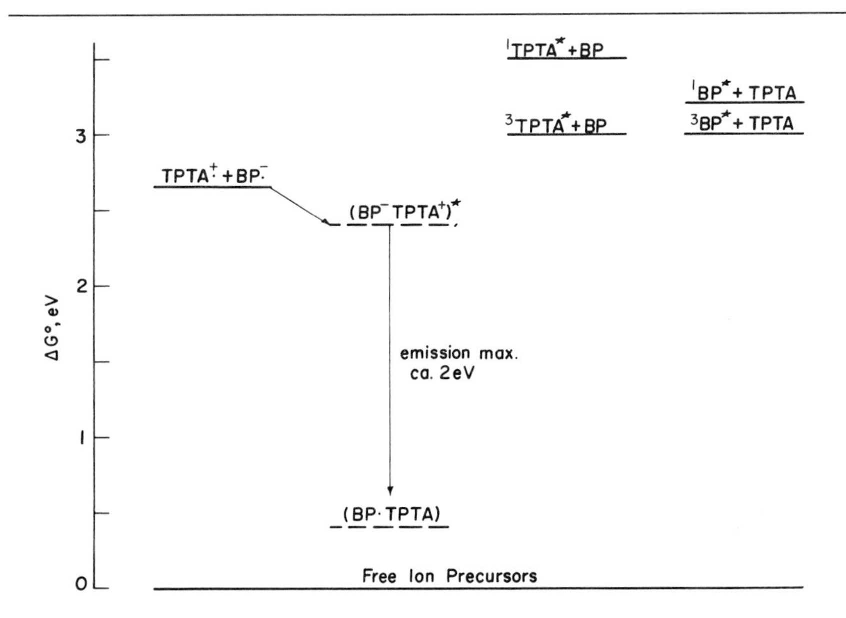

FIG. 4. Energetics of the reaction between $TPTA^{+}$ and BP^{-}. Exciplex level is dashed to indicate uncertainty. Emission energy is solvent-dependent and the transition takes place to a repulsive Franck–Condon state of unknown energy above the ground level. Radical ion energies apply to acetonitrile. Data taken from K. A. Zachariasse, Thesis, Vrije Universiteit te Amsterdam, 1972, and from H. Tachikawa and L. R. Faulkner, *J. Am. Chem. Soc.* **98,** 5569 (1976).

In a very comprehensive study,[11,13,26] Weller and Zachariasse demonstrated that Eq. (22) is a general reaction in low-polarity media such as THF. An interesting example is the oxidation of the benzophenone (BP) anion radical with the cation of tri-*p*-tolylamine (TPTA). This reaction shows a broad emission band in THF, even though it is not sufficiently energetic to populate any local state of TPTA or BP. Figure 4 shows the energetics in a graphical form, and it highlights the constraints on the system.

In favorable cases, exciplexes can also serve as intermediates in the generation of chemiluminescence from local states. Weller and Zachariasse made this point in a study of the reaction between $TPTA^{+}$ and DMA^{-} in THF.[43] This luminescence has components from the exciplex $^{1}(DMA^{-}TPTA^{+})^{*}$ and from $^{1}DMA^{*}$. Figure 5 shows that their relative strengths are quite dependent on temperature, and it suggests that $^{1}DMA^{*}$ might arise from the exciplex by thermally activated dissociation, e.g.

[43] A. Weller and K. A. Zachariasse, *Chem. Phys. Lett.* **10,** 590 (1971).

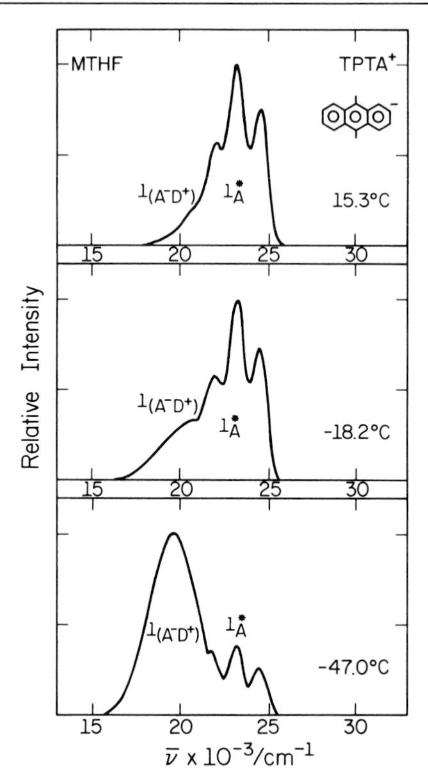

FIG. 5. Temperature dependence of the emission spectrum from the reaction between TPTA$^{\cdot+}$ and DMA$^{\cdot-}$ in 2-methyltetrahydrofuran. Reproduced by permission from A. Weller and K. A. Zachariasse, *Chem. Phys. Lett.* **10,** 590 (1971).

$$\begin{array}{ccccc} DMA^{\cdot -} + TPTA^{\cdot +} \rightarrow & {}^1(DMA^-TPTA^+)^* & \overset{\Delta}{\rightleftarrows} & {}^1DMA^* + TPTA & (25) \\ & \downarrow & & \downarrow & \\ & DMA + TPTA + h\nu' & & DMA + TPTA + h\nu & \end{array}$$

Strong support for this idea comes from the fact that the monomer-to-exciplex emission ratio is exponentially dependent on $1/T$. The apparent activation energy agrees well with the estimated energy deficit between $^1(DMA^-TPTA^+)^*$ and $^1DMA^*$.

Given the apparent instability of exciplexes in polar solvents, it has been surprising to see many recent reports of exciplex emission during radical ion reactions in such media.[14,38,44–49] Park and Bard first noted the effect in a

[44] S. M. Park and A. J. Bard, *J. Am. Chem. Soc.* **97,** 2978 (1975).
[45] R. E. Hemingway, S. M. Park, and A. J. Bard, *J. Am. Chem. Soc.* **97,** 200 (1975).
[46] H. Tachikawa and L. R. Faulkner, *J. Am. Chem. Soc.* **98,** 5569 (1976).
[47] S. M. Park, M. T. Paffett, and G. H. Daub, *J. Am. Chem. Soc.* **99,** 5393 (1977).
[48] S. M. Park and R. A. Caldwell, *J. Electrochem. Soc.* **124,** 1859 (1977).
[49] K. Itaya and S. Toshima, *Chem. Phys. Lett.* **51,** 447 (1977).

series of studies in acetonitrile.[14,44] Many of the phenomena observed earlier by Weller and Zachariasse with THF solutions were shown to apply also in the more polar medium. For example, a readily identifiable exciplex emission was seen[44] during the oxidation of $BP^{\cdot -}$ with $TPTA^{\cdot +}$. The energetics of the case remained virtually as depicted in Fig. 4, so that all locally excited states were inaccessible to the system. Park and Bard were also able to observe an instance in which local singlets were produced by thermal dissociation of an exciplex according to a scheme like Eq. (25).[44]

Both singlet and triplet exciplexes can be formed, in principle, but the usual emissions in chemiluminescence must be ascribed to the singlets on the basis of spin conservation. Triplet exciplexes have been discussed frequently, but are not readily detected. Only recently has a report seemed to manifest their presence in chemiluminescent systems. Tachikawa and Faulkner found evidence that $^3(BP^-TPTA^+)^*$ could transfer energy to any of several anthracenes in acetonitrile.[46] This process produced triplets of the anthracenes, which then annihilated to yield a characteristic fluorescence,

$$^3(BP^-TPTA^+)^* + An \rightarrow TPTA + BP + {}^3An^* \quad (26)$$

$$^3An^* + {}^3An^* \rightarrow {}^1An^* + An \quad (27)$$

Itaya and Toshima have added still another segment to the story of exciplexes in polar media by examining several systems in which donor and acceptor moieties were bound together by methylene chains.[49] The molecules had the structures,

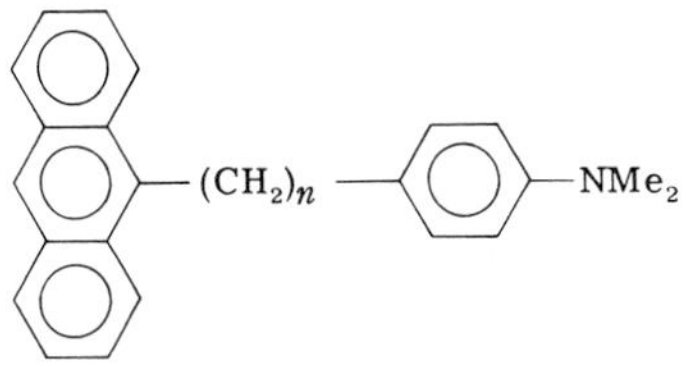

$(CH_2)_n$ — NMe_2

with $n = 1$ or 2. In acetonitrile, these species reduce much like anthracene to ions that can be abbreviated by $A^{\cdot -} - D$, whereas they oxidize much like *N,N*-dimethylaniline to cations, which are written as $A - D^{\cdot +}$. Annihilation of the ions gives emission from the intramolecular exciplex,

$$\begin{array}{c} A^{\cdot -} - D + A - D^{\cdot +} \rightarrow {}^1(A^- - D^+)^* + A - D \\ \downarrow \\ A - D + h\nu \end{array} \quad (28)$$

There is an interesting contrast here with results from the reaction in which $A^{\cdot -}$ and $D^{\cdot +}$ are separated centers. Itaya and Toshima found, for example, that the anion of 9-methylanthracene and the cation of *N,N*-dimethyl-*p*-anisidine gave emission only from the local singlet of 9-methylanthracene. Since the reaction is energy-deficient, this luminescence apparently arises via the T route. The linked systems either suppress this component or overwhelm it with an unusually efficient exciplex emission.

A related, and extremely interesting, species was examined by Itaya and Toshima in this same study.[49] It could be written in the format given above with $n = 0$; however, it differs fundamentally from the cases of $n = 1$ and 2, in that the A and D moieties are conjugatively linked through a twisted double bond. As a result, the excited state $^1(A^- - D^+)^*$ is not an exciplex, but instead it is a charge-transfer state of a single chromophore. Note also that it is stabilized in a polar medium, rather than being destabilized as true exciplexes are. It is not surprising that its chemiluminescence is very much brighter than that from the usual exciplex systems in polar solvents.

In general, those components are very weak and are often overwhelmed if significant emissions from local states are generated by the S or T routes. Nevertheless, interest in exciplexes is rather high, because they may be general intermediates in high-energy electron-transfer processes.[14,44] If so, they would be intimately implicated in the determination of excited-state yields. The work of Weller and Zachariasse shows that they are unquestionably important for reactions in relatively nonpolar media.[11-16,26,43] However, it is not yet possible to evaluate their role in polar solvents. Weakness in emission may indicate a negligible role. On the other hand, it may only reflect low visibility for species that are abundant, but possess very low quantum yields for fluorescence.[46]

Theoretical Expectations about Chemiexcitation Yields

The efficiencies with which various excited states are formed in electron-transfer reactions are fundamentally interesting because they offer valuable quantitative insight into the mechanisms for accommodating large amounts of energy in a kinetic system. Vast areas of science would benefit from an understanding of those mechanisms, and the hope of progress there underlies much current work in electron-transfer chemiluminescence. A number of plausible theoretical positions have been set forth, and it is useful to consider them here as a context for our later discussion of experimental findings.

Marcus, who employed an extension of his theory of electron transfer,[50] was first to attack the problem.[51] His theory has the fundamental presumption that every actual transfer event occurs on a Franck–Condon time scale. The zone in which the potential surfaces for reactants and products

[50] R. A. Marcus, *J. Chem. Phys.* **43,** 679 (1965).
[51] R. A. Marcus, *J. Chem. Phys.* **43,** 2654 (1965); **52,** 2803 (1970).

intersect is therefore of primary interest because it is the only region of configuration space in which the reactants and products have common conformations. The electron-transfer rate ought to be proportional to the likelihood that a pair of reactants reaches that intersection region. Marcus evaluates the probability as the classical chance that nuclear motion will yield the proper distorted configuration. Thus, the redox rate constant can be expressed as,

$$k_{et} \propto \kappa \exp(-\Delta G^*/kT) \tag{29}$$

where ΔG^* is an activation free energy, and κ is the adiabaticity.

Marcus has indicated that ΔG^* can be evaluated from the standard free energy of reaction, ΔG^o, by the expression,

$$\Delta G^* = \frac{w^r + w^p}{2} + \frac{\lambda}{4} + \frac{\Delta G^o}{2} + \frac{(\Delta G^o + w^p + w^r)^2}{4\lambda} \tag{30}$$

The variable λ is a "reorganization parameter," whose value is accessible from electron spin resonance (ESR) measurements and other experiments. It normally has a value of 0.4 to 0.8 eV. The work terms w^r and $-w^p$, which describe the energy used in creating the reactant pair and separating the product pair, are usually minor.

A consequence of Eq. (30) is that the maximum electron-transfer rate should be observed when $\Delta G^o \cong -\lambda$; i.e., for a mildly exothermic process. If the reaction is more or less exothermic, an activation barrier develops and the transfer rate declines. For a very exothermic redox process the reaction also becomes nonadiabatic ($\kappa \to 0$), so extremely low rate constants are predicted.[51]

The last conclusion is important for chemiluminescence because its implication is that a reaction will yield almost exclusively the highest accessible product state, unless λ is very large or several high-energy states are closely situated. This rather rigid forecast has been modified in more recent work by Marcus,[52] Hoytink,[53,54] and Van Duyne and Fischer,[25,55] who have all introduced nonclassical effects, such as nuclear tunneling, into the problem.

The effect of spin conservation has also received much discussion.[11-16,25,55,53,54] Most experimental systems involve doublet–doublet annihilations which produce singlets and triplets. If random spins react, then strict

[52] R. A. Marcus, *in* "Proceedings of the International Summer School on Quantum Mechanical Aspects of Electrochemistry" (P. Kirkov, ed.). Center for Radioisotope Applications in Science and Industry, Skopje, Yugoslavia, 1972.

[53] G. J. Hoytink, *Discuss. Faraday Soc.* **45,** 14 (1968).

[54] G. J. Hoytink, *in* "Chemiluminescence and Bioluminescence" (M. Cormier, D. M. Hercules, and J. Lee, eds.), p. 147. Plenum, New York, 1973.

[55] S. F. Fischer and R. P. Van Duyne, *Chem. Phys.* **26,** 9 (1977).

conservation dictates that 75% of the products will be triplet and 25% will be singlet, even though the distribution of products within the triplet and singlet manifolds might be governed by Franck–Condon factors as expressed in a relation like Eq. (29). As an alternative possibility, Hoytink suggested that spin factors of 3/4 and 1/4 might enter only the preexponential factors in the rate expressions.[53] If so, spin conservation would not be an important determinant of the product distribution, because the exponents primarily rule the rates.

Park and Bard,[14,44] building on the work of Zachariasse,[11,26] have advanced another possibility, which is largely unrelated to existing theories of electron transfer. Evidence that exciplexes form directly in ion annihilations, even in polar solvents, has evoked the suggestion that they may be general intermediates. If so, the distributions of localized product states would be hard to predict because they would depend on radiationless processes within the exciplexes themselves.

Unfortunately, the available experimental measurements of product yields are far too scarce to allow even tentative evaluations of these hypotheses. Current evidence has been considered in detail in a recent review,[15] and the interested reader may wish to pursue the subject there. More data are needed, and measurements in this area are likely to be a focus of research for some time to come.

Experimental Techniques

Only a brief introduction to experimental aspects will be provided here, because the whole subject has recently been reviewed in detail.[16] Our goal is simply to outline the types of experiments that are commonly performed and to indicate the kind of information that they provide.

Sometimes the reactive species involved in these studies can be isolated (e.g., as cation radical salts, such as $TMPD^{+}ClO_4^{-}$), or they can be produced chemically in large quantities in a convenient manner (e.g., by reduction with sodium or potassium). In these instances, reactions can be carried out in bulk, and spectroscopic measurements can be made on a reaction zone.[26]

Most of the time, it is more convenient to produce the radical ions in small quantities on a short, controlled time scale by electrochemical means.[16] *Electrochemiluminescence* (or *electrogenerated chemiluminescence,* ECL) is a generic term that describes the stimulation of light-producing bulk reactions by electrode processes[56]; however, usage alone has caused ECL to become practically synonymous with the more specific topic of electron-transfer chemiluminescence.

[56] T. Kuwana, *Electroanal. Chem.* **1,** 197 (1966).

Cyclic Voltammetry

The basic electrochemical measurement in this field is cyclic voltammetry,[57] which characterizes the potential regions in which the ions can be generated and yields estimates of their stabilities.

One begins with a cell containing three electrodes immersed in a quiescent solution of the ion precursors.[16] We will consider an example of 1 m*M* thianthrene (TH) and 1 m*M* 2,5-diphenyl-1,3,4-oxadiazole (PPD) in acetonitrile. A supporting electrolyte, such as 0.1 *M* tetra-*n*-butylammonium fluoborate ($TBABF_4$), is added in order to impart conductivity. The electrode of interest is the working electrode, which might be a platinum disk having an area of perhaps 0.05 cm^2. Current flow at this electrode is compensated by a counter electrode, so that these two are in series with the power source driving the current. That device is usually a programmable potentiostat, which uses feedback circuitry to force current through the working and counter electrodes in whatever fashion is required to maintain the working electrode at a programmed potential with respect to the third, or reference, electrode. The voltage sensor that measures the potential of the working electrode vs the reference has a very high impedance, hence current does not flow through the reference.

The reference electrode itself might be a conventional aqueous saturated calomel electrode (SCE), which has a fixed potential on a thermodynamic scale. Alternatively, it may be nothing more than an exposed wire, which will maintain a fixed potential over the course of an experiment, even though it is not fixed thermodynamically. Such devices are called quasi-reference electrodes (QRE). They are widely used, because they are convenient for sealed, vacuum-tight, water-free cells.

In cyclic voltammetry, the potential of the working electrode is scanned linearly with time, and the current flow is recorded as a function of potential. Figure 6 shows a typical record for the TH/PPD system.[23,58,59] Cathodic currents are positive, and anodic currents are negative.

The scan starts at 0.0 V vs QRE and moves positively at 100 mV/sec. Little current flows until the electrode reached a potential near 0.9 V, where TH is oxidized to TH^{+}. The current then rises sharply (in a negative direction) because the potential is near $E^{\circ}(TH^{+}/TH)$, and the balance in the interfacial equilibrium,

$$TH^{+} + e \rightleftharpoons TH \tag{31}$$

begins to shift rapidly in favor of the cation radical. However, this current flow eventually depletes the region near the electrode of TH, and the current falls, because of supply limitations. When the electrode reaches a

[57] R. S. Nicholson and I. Shain, *Anal. Chem.* **36,** 706 (1964).

[58] P. R. Michael, Ph.D. thesis, University of Illinois at Urbana-Champaign, 1976.

[59] P. R. Michael and L. R. Faulkner, *J. Am. Chem. Soc.* **99,** 7754 (1977).

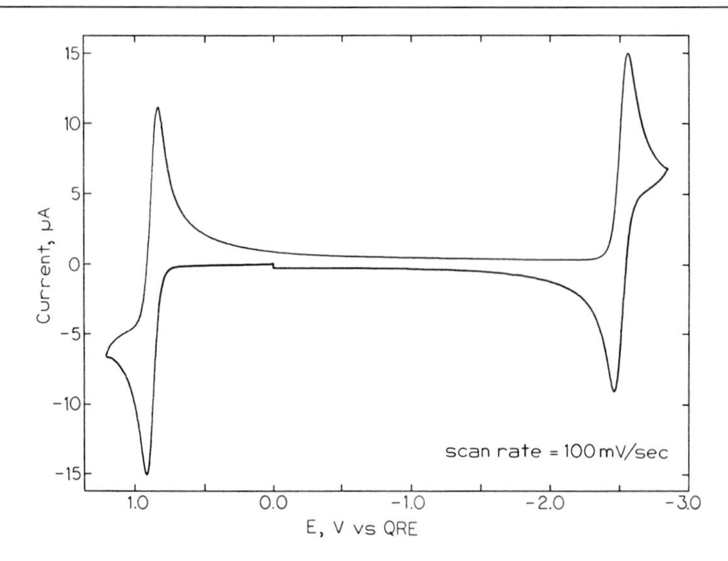

FIG. 6. Cyclic voltammogram of 1 m*M* thianthrene (TH) and 1 m*M* 2,5-diphenyl-1,3,4-oxadiazole (PPD) in acetonitrile containing 0.1 *M* tetra-*n*-butylammonium fluoborate. Reproduced with permission from P. R. Michael, Ph.D. thesis, University of Illinois at Urbana-Champaign, 1976.

potential near 1.2 V, the scan is reversed and a cathodic peak is recorded as the cation radical is reduced back to the neutral species. The height of this reversal peak, relative to the initial (or forward) peak, is a sensitive indicator of the stability of the cation on the time scale required to traverse the active potential region.[57] By varying the scan rate, this time scale can be changed over several orders of magnitude. The average potential of the two peaks is usually a very good estimate of the standard potential for the couple.[57]

After passing the reduction peak for TH^{+} in Fig. 6, the scan continues uneventfully in a negative direction until PPD can be reduced. Then there is a cathodic forward peak corresponding to the formation of $PPD^{\overline{\cdot}}$, and upon reversal, an anodic peak reflects its reoxidation to PPD. The scan finally returns to the starting point and stops.

Generating Light Electrochemically

The most widely used techniques for ECL involve the sequential production of the reactants at a single microelectrode.[16] The case of TH/PPD, which was considered above, is typical.[23,58,59]

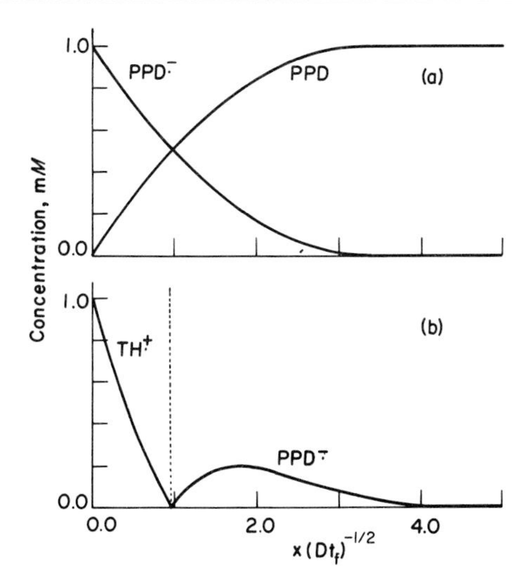

FIG. 7. Concentration profiles near the electrode during an electrochemiluminescence (ECL) step experiment. Data apply to 1 mM TH and 1 mM PPD. The parameter x measures distance from the electrode, and D is the diffusion coefficient of PPD. (a) Profiles of PPD and $PPD^{-\cdot}$ at the end of the forward step. (b) Concentrations of $TH^{+\cdot}$ and $PPD^{-\cdot}$ during the second step. Reaction boundary is shown by dotted line. Curves apply for a time $0.4t_f$ into the second step.

A three-step sequence is frequently employed to generate a single pulse of light.[15-18,20,25,58,59] At the outset, the working electrode is held at some potential where no electrolysis occurs. Operating from the voltammogram in Fig. 6, one might choose a value of 0.0 V vs QRE. At $t = 0$, the potential is changed abruptly (i.e., "stepped") to a value where one reactant is generated at the diffusion-limited rate. To produce $PPD^{-\cdot}$, one could use a value of -2.7 V vs QRE. The potential is held there for a period t_f, which might range from 1 μsec to 5 sec in length.[16,25] During this forward step, the reactant diffuses into the solution and forms a zone adjacent to the electrode, as shown in Fig. 7a. At $t = t_f$, the potential is stepped to a value at which the second reactant is electrogenerated at the diffusion-limited rate. In our example, thianthrene would be oxidized conveniently to $TH^{+\cdot}$ at $+1.2$ V vs QRE. It would then diffuse into the solution and form a reaction boundary with the previously deposited zone of $PPD^{-\cdot}$.[60-62] Figure 7b shows the concentration distributions in the diffusion layer at a time during this

[60] S. W. Feldberg, *J. Am. Chem. Soc.* **88,** 390 (1966).
[61] L. R. Faulkner, *J. Electrochem. Soc.* **124,** 1724 (1977).
[62] J. L. Morris, Jr., and L. R. Faulkner, *J. Electrochem. Soc.* **125,** in press.

pulse. Since $PPD^{\cdot-}$ and $TH^{\cdot+}$ both have concentration gradients toward the reaction boundary, they diffuse to it continuously. As $PPD^{\cdot-}$ is consumed, the boundary moves farther into the solution, and the rate of reaction declines. The light is therefore most intense at the start of the second step and decays monotonically thereafter. After a second period t_f, the generation of $TH^{\cdot+}$ is terminated by a third step back to the initial potential, where both reactants are destroyed electrolytically, and the solution is soon returned to its initial state.

There are two sources of information here, if one can accurately relate detector photocurrents to the absolute emission rate in quanta per second. The problems with calibrations of absolute photometers are examined by Seliger in this volume [44]. Other discussions with particular reference to ECL are available in the literature.[15,16,63–65]

The most straightforward way to use absolute data is to integrate the transient and to compare the total quantum output to the number of redox events. The latter figure can be derived easily from the charge injected in the forward step, hence a total output measurement determines the mean emission efficiency ($\bar{\phi}_{ecl}$, photons emitted per redox event). This parameter, which is of obvious practical interest, is significant from a fundamental standpoint too, for it reflects both the efficiency of excited state generation and the competition between the elementary processes that decide the fates of excited species in the chain to luminescence.

Such mechanistic and yield information is also contained in the shape of the light decay. The most direct and reliable way to glean it is to compare the total rate of light emission I (einstein/sec) to the total rate of the redox reaction N (mol/sec) at any instant. The ratio of these quantities I/N is the instantaneous chemiluminescence efficiency ϕ_{ecl}, i.e., the probability that an electron-transfer event leads to emission. Relative intensities proportional to the absolute rate I are directly recorded in the experiment. If an absolute detector is used, the relative rates can be converted to absolute ones, but this step is not usually necessary for diagnostic purposes. The reaction rate N must be calculated from theory, but it can be verified via the electrochemical current transients.[61] In most experiments, it depends only on the diffusion patterns of electrogenerated reactants in solution, because the rate constants for the reactions under study are extremely large. Very good approximate analytical relations for the calculation of N are available.[61]

The functional linkage between I and N differs greatly for S and T

[63] A. Weller and K. A. Zachariasse, *Chem. Phys. Lett.* **10,** 424 (1971).

[64] C. P. Keszthelyi, N. E. Tokel-Takvoryan, and A. J. Bard, *Anal. Chem.* **47,** 249 (1975).

[65] P. R. Michael and L. R. Faulkner, *Anal. Chem.* **48,** 1188 (1976).

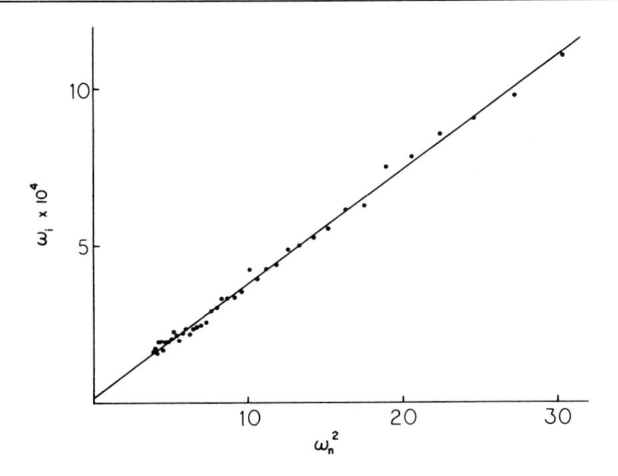

FIG. 8. Light intensity vs reaction rate for a step-generated transient from the reaction between TH^{+} and PPD^{-}. The parameters ω_i and ω_n are I and N, respectively, multiplied by the same experimental constant. The reductant was generated first in this case. Although emission comes from both $^1TH^*$ and $^1PPD^*$, the contribution from $^1PPD^*$ is negligible. Reproduced by permission from P. R. Michael and L. R. Faulkner, *J. Am. Chem. Soc.* **99,** 7754 (1977). Copyright by the American Chemical Society.

systems.[62] For the latter case, I is strictly proportional to N. The factor linking them is the product of the probability that a redox event will produce a singlet and the probability that the short-lived singlet will emit. In contrast, the T route forces I to depend on a higher power of N, because it contains a second-order reaction of the chemiexcited species.

The diagnostic value of this decay curve is highlighted in Fig. 8, which shows an analyzed transient from the TH/PPD system.[58,59] The parameters ω_i and ω_n used there are values of I and N that have been multiplied by a constant in order to render them dimensionless. The figure clearly demonstrates that I is proportional to N^2. Thus it rules out the S pathway and strongly supports the T route to luminescence. Such a conclusion is most intriguing, because the system is energy-sufficient and might be expected to pursue the S route.[23,58,59] The basis for this anomaly is not understood. We will return to it below in a consideration of magnetic field effects.

Step techniques are also used routinely for generating a steady train of pulses for spectral analysis and other purposes. One simply alternates the working electrode between the potentials required for producing the two reactants, and a pulse is emitted after each transition in potential.

A different approach is to generate the reactants at separate electrodes and flow them together.[16,35,64] Such a plan is implemented elegantly with the rotating ring-disk electrode (RRDE). This device has a planar face, which is rotated around an axis at perhaps 2000 rpm. Centered on the axis is a disk of platinum, and surrounding the disk, but separated from it by a thin insulating annulus, is a ring also made of platinum. Both the ring and the disk are working electrodes, and their potentials are controlled independently. The rotation causes the solution on the faces of the electrodes to be forced outward by a centrifugal effect; thus there is a pumping action in which fresh solution is brought along the axis to the face of the disk, where it is then redirected radially across the face of the ring. If one reactant is produced at the disk and the other is generated at the ring, they will meet and react over the surface of the ring. These devices are widely used in electrochemical studies, and the hydrodynamics governing their performance are well understood.[66] The advantage in ECL is that steady light levels are achieved. The studies described in the discussion surrounding Fig. 3 were carried out with an RRDE.[35]

Interception

The behavior of certain intermediates, particularly triplets, can often be illuminated by trapping them with interceptors. A dramatic example is the effect of added anthracene on chemiluminescence from the oxidation of the fluoranthene (FA) radical anion by the cation of 10-methylphenothiazine (10-MP).[67] Curve a of Fig. 9 shows the normal ECL spectrum, which corresponds to fluorescence from $^1FA^*$. The reaction is energy-deficient, so one suspects that the T route applies. The effect of anthracene provides strong support. This species can be added innocuously with respect to the electrochemistry, yet it transforms the emission spectrum entirely to that of anthracene (see curve b of Fig. 9).

The result is readily explained by invoking the exothermic energy transfer,

$$^3FA^* + An \rightarrow {}^3An^* + FA \tag{32}$$

which would then be followed by the annihilation of anthracene triplets.

Triplet interception has also been used to evaluate chemiexcitation yields.[68,69] Such experiments have featured reactions of the fluoranthene anion radical with one of several oxidants. Both reactants are generated simultaneously in equal amounts under bulk electrolytic conditions. Independent, large-area electrodes are used, and the solution is stirred. Also

[66] W. J. Albery and M. L. Hitchman, "Ring-Disc Electrodes." Oxford Univ. Press (Clarendon), London and New York, 1971.

[67] D. J. Freed and L. R. Faulkner, *J. Am. Chem. Soc.* **93,** 2097 (1971).

[68] D. J. Freed and L. R. Faulkner, *J. Am. Chem. Soc.* **93,** 3565 (1971).

[69] D. J. Freed and L. R. Faulkner, *J. Am. Chem. Soc.* **94,** 4790 (1972).

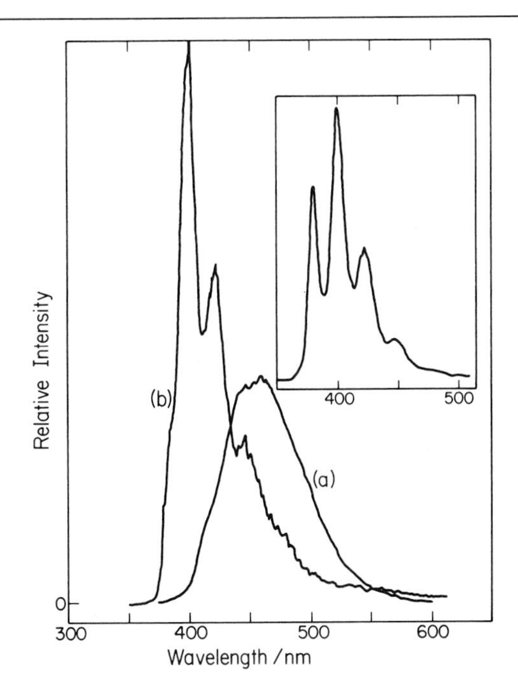

FIG. 9. Curve a: Chemiluminescence from the reaction between $FA^{\overline{\cdot}}$ and $10\text{-}MP^{\dot{+}}$ in dimethylformamide, (DMF). Reactants were electrogenerated from a solution 1 m*M* in each precursor. Curve b: Spectrum obtained upon addition of 1 m*M* anthracene. Anthracene ions were not produced by electrolysis. Inset shows fluorescence from anthracene at 10 μM and DMF. Reabsorption removes the 0,0 transition in curve b. Original data used with permission from D. J. Freed and L. R. Faulkner, *J. Am. Chem. Soc.* **93,** 2097 (1971). Figure reproduced with permission from L. R. Faulkner, *Int. Rev. Sci.: Phys. Chem. Ser. Two* **9,** 213 (1975).

included is *trans*-stilbene, which intercepts triplets via energy transfer; then the stilbene triplets undergo a partitioned decay to a cis-trans mixture. Gas chromatographic analysis of the resulting *cis*-stilbene provides the basis for calculating the probability of triplet formation in the redox reaction.

The results for reactions featuring 10-methyl- and 10-phenylphenothiazine cations ($10\text{-}MP^{\dot{+}}$ and $10\text{-}PP^{\dot{+}}$) gave an intriguing contrast.[68,69] The phenylated species oxidized $FA^{\overline{\cdot}}$ to produce triplets in nearly unit yield, whereas the $10\text{-}MP^{\dot{+}}$ showed only a 2–3% efficiency. The basis for this difference is really not understood, but it seems to indicate that the electron-transfer event has a fundamental sensitivity to substitution patterns in the oxidant.

Magnetic Field Effects

The study of magnetic effects on ECL grew out of earlier studies by Merrifield and co-workers,[70,71] who examined the annihilation of triplet excitons in molecular crystals. They observed that the rate constant for the process was dependent on field strength in the range below 10 kG. Merrifield later provided a theory to explain the effect.[72] His ideas have since been elaborated and tested widely. The theory is complex in detail, but simple in concept. Its essence is that spin-selection rules govern the reaction.

Let us consider two triplets that encounter each other with the prospect of annihilating. Each species is in one of three triplet sublevels, and the pair of triplets is one of nine randomly selected combinations. When the pair begins to interact, it is regarded as a single entity, which is described by a "pair state." Since two triplets were combined to create it, the pair state has a mixed singlet, triplet, and quintet character. If delayed fluorescence is to be produced, then the pair state must convert to a pure singlet. Merrifield's postulate[72] is that the probability for the conversion is proportional to the singlet character of the pair state. The rate constant for producing excited singlets from two triplets is a statistical quantity reflecting the distribution of singlet character among the nine possible pair states.

At low field the three triplet wave functions are determined by the dipolar interaction of unpaired electron spins, which leads to the zero-field splitting. At high field, the wave functions are determined by the Zeeman interactions. Thus, the angular momentum characteristics of the wave functions are continuously dependent on the field until the high-field limit is reached. Since the wave functions change with field strength, the distribution of singlet character among the nine pair states also changes, and one observes a field-dependent macroscopic rate constant. In solution-phase triplet–triplet annihilation, the rate constant generally declines monotonically with increasing strength until a saturation limit is reached above 5 kG.[73,74]

Magnetic effects on ECL were first investigated as a test of the T route

[70] R. C. Johnson, R. E. Merrifield, P. Avakian, and R. B. Flippen, *Phys. Rev. Lett.* **19,** 285 (1967).

[71] R. E. Merrifield, *Acc. Chem. Res.* **1,** 131 (1968).

[72] R. E. Merrifield, *J. Chem. Phys.* **48,** 4318 (1968).

[73] L. R. Faulkner and A. J. Bard, *J. Am. Chem. Soc.* **91,** 6495 (1969).

[74] P. Avakian, R. P. Groff, R. E. Kellogg, R. E. Merrifield, and A. Suna, *in* "Organic Scintillators and Liquid Scintillation Counting" (D. L. Horrocks and C.-T. Peng, eds.), p. 499. Academic Press, New York, 1971.

in energy-deficient systems,[19,75] and they have yielded an intriguing dichotomy[12,14-16,19,21-23,75]. Every energy-deficient system examined to date shows an enhancement of chemiluminescence in the presence of the field, whereas every chemically clean system with a reaction energy within a few kT of energy-sufficiency, save one (the rubrene anion/cation reaction),[19,21-23] shows no field effect.

The enhancements observed for energy-deficient systems were surprising in view of the established inhibition of triplet–triplet annihilation, but they have been rationalized by invoking a field-inhibited quenching of triplets by the radical ions.[19,76] Such reactions involve angular-momentum changes, hence they are also subject to the constraints of spin selection. A few experimental trials have verified the predicted inhibitions, which are usually large enough to cause increases in delayed fluorescence intensity as the field strength rises.[77,78] The lengthened triplet lifetime in the applied field therefore overrides the decreased rate constant for triplet–triplet annihilation.

The acceptance of this rationale for the intensity enhancements in ECL is based completely on the consistent correlation of the field effects and the energy balances. Unfortunately, quantitative tests have not been practical, because the chemiluminescent reaction zone is too complex to allow for tractable description.[62] If the qualitative explanation is correct, magnetic effects can be a powerful aid in mechanistic diagnosis. In fact, there is already some tendency to accept them in such a role.

However, caution is well advised. Not only is the rationale unconfirmed by direct tests, but several pieces of experimental evidence raise suspicion about it. First are the surprisingly large magnitudes of many enhancements. They can approach 30%, whereas the field effects in studies of delayed fluorescence rarely exceed 10%. In addition, there is an anomaly in results for the energy-sufficient reaction between TH^+ and PPD^-. The magnetic field does not affect its ECL intensity,[23] and this fact, taken together with the energy-sufficiency, could imply that light arises by the S route. However, the discussion surrounding Fig. 8 shows that the S mechanism cannot apply.[58,59] The discrepancy is not understood. It may be a fortuitous result that can be accommodated within the accepted rationale,[59,78] or it may be more significant. In either case, it demonstrates that mechanistic diagnosis cannot be made unambiguously on the basis of behavior in a magnetic field.

[75] L. R. Faulkner and A. J. Bard, *J. Am. Chem. Soc.* **91,** 209 (1969).
[76] P. W. Atkins and G. T. Evans, *Mol. Phys.* **29,** 921 (1975).
[77] L. R. Faulkner and A. J. Bard, *J. Am. Chem. Soc.* **91,** 6497 (1969).
[78] H. Tachikawa and A. J. Bard, *Chem. Phys. Lett.* **26,** 10 (1974).

However, research in this domain is active, and one can justify optimism that some of the larger uncertainties will soon be removed.

Chemically Initiated Electron-Exchange Luminescence

Schuster's group has just reported some intriguing results that point to a general role for electron-transfer processes in chemiluminescence from certain peroxides.[7-10] The central molecule in their work is diphenoyl-peroxide (DPP), which thermolyzes at room temperature in solution to give mostly benzocoumarin (BC) and CO_2.[7,10]

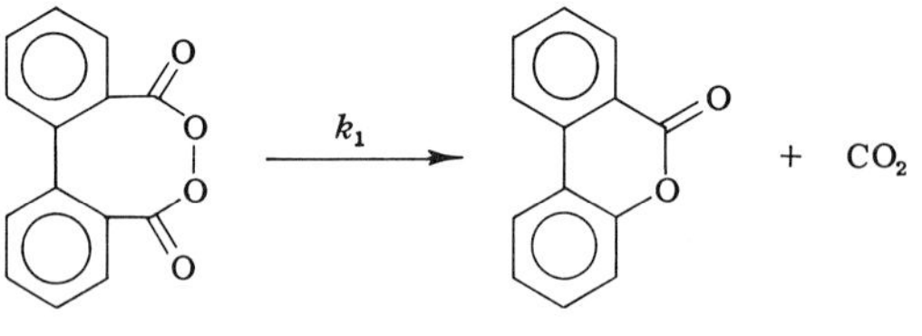

$$\text{DPP} \xrightarrow{k_1} \text{BC} + CO_2 \tag{33}$$

DPP BC

Interception experiments in which 9,10-dibromoanthracene and biacetyl were added as trapping agents indicated that $^1BC^*$ and $^3BC^*$ are not formed significantly in this reaction. However, strong chemiluminescence appeared when more easily oxidized molecules, such as rubrene or perylene, were substituted for the interceptors. Usually the light had the spectrum of fluorescence from the additive.

These species were not merely interceptors, because they stimulated the decay of DPP in a truly catalytic fashion. The rate constants governing the strictly exponential declines in chemiluminescence followed the form

$$k_{obs} = k_1 + k_2 [D] \tag{34}$$

where D is the added molecule, which is called the activator. Figure 10 displays kinetic data for several different systems. The rate constant k_2 depends on the chemical nature of the activator, and it clearly reflects the existence of a bimolecular decay channel parallel to Eq. (33). The thermolysis itself is manifested by k_1, which is independent of the activator. The chemiluminescence intensity rises with the concentration of D, and the measured activation energy for producing emission in a given system is identical to the measured activation energy of k_2. These facts strongly imply that chemiluminescence arises entirely from the bimolecular process.

The mechanism proposed by Schuster and Koo involves an activated charge transfer to produce a caged ion pair,[7,10]

$$D + DPP \rightarrow (D^{\pm} + DPP^{\mp})_{cage} \tag{35}$$

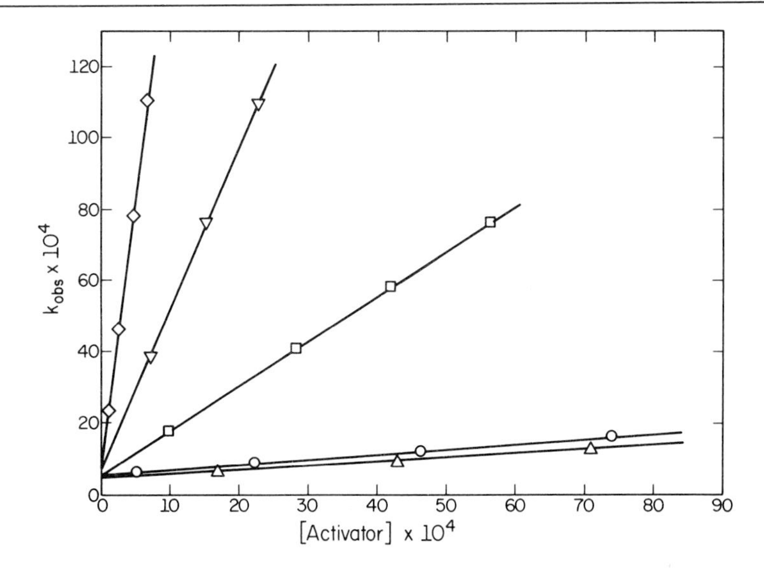

FIG. 10. Effects of the structure and concentration of the activator on the decay of chemiluminescence from diphenoylperoxide (DPP). In clockwise order, data are for rubrene, naphthacene, perylene, 9,10-diphenylanthracene, and pyrene. Reproduced by permission from J.-Y. Koo and G. B. Schuster, *J. Am. Chem. Soc.*, **100** (in press). Copyright by the American Chemical Society.

The anion $DPP^{\overline{\cdot}}$ is demonstrably unstable, and it is presumed here to lose CO_2 immediately. Recyclization forms the BC anion radical,

$$(D^{\dot{+}} + DPP^{\overline{\cdot}})_{cage} \rightarrow (D^{\dot{+}} + BC^{\overline{\cdot}} + CO_2)_{cage} \tag{36}$$

The species $BC^{\overline{\cdot}}$ is a powerful reductant [$E^0(BC/BC^{\overline{\cdot}}) = -1.92$ V vs SCE], and can produce an emitting activator singlet by charge annihilation,

$$(D^{\dot{+}} + BC^{\overline{\cdot}} + CO_2)_{cage} \rightarrow (^1D^* + BC + CO_2)_{cage} \tag{37}$$

This sequence has been named chemically initiated electron exchange luminescence (CIEEL).[7,10] Its fascinating aspect is the way in which a chemical process (elimination of CO_2) transforms the weak reductant $DPP^{\overline{\cdot}}$ into the strong reducing agent $BC^{\overline{\cdot}}$. A similar idea was introduced earlier by McCapra to account for chemiluminescence from oxalate esters,[3] but it has not been tested experimentally. On the other hand, the evidence that this mechanism actually occurs with DPP is extensive. Several points can be cited here. Most important is the strong correlation between k_2 and the oxidation potential of D. Figure 11 emphasizes this point. Spectroscopic properties of the activators generally fail to produce equivalent correlations.

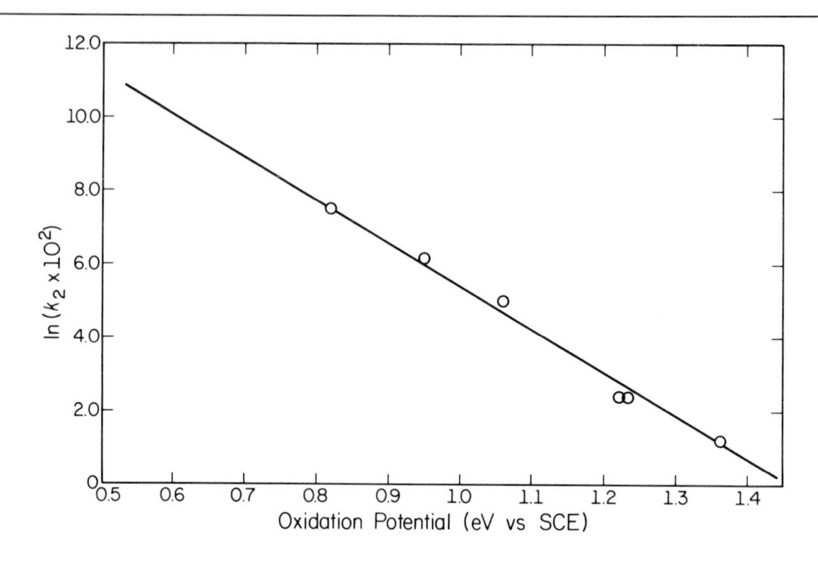

FIG. 11. Logarithmic correlation of k_2 with the standard potentials for oxidation of the activators. In order of increasing E° (D^{+}/D), the points represent rubrene, naphthacene, perylene, 9,10-diphenylanthracene, coronene, and pyrene. Reproduced by permission from J.-Y. Koo and G. B. Schuster, *J. Am. Chem. Soc.*, **100** (in press). Copyright by the American Chemical Society.

The effect of solvent is also suggestive. As its dielectric constant increases, the value of k_2 usually rises, while the efficiency of light production drops. The former effect is consistent with the generation of ionic intermediates in the rate-determining step, and the drop in yield suggests that improved solvation of those intermediates interferes with excited state production via charge recombination. It is possible that the interference takes the form of a facilitated escape of free ions from the cage. The idea that escape is competitive with recombination is supported by the additional observation that increased viscosity leads to improved emission yields.[10]

It is also instructive to consider results from the system in which triphenylamine (TPA) serves as the activator.[10] Its chemiluminescence comes wholly from the exciplex $^1(BC^{-}TPA^{+})^*$, which can be identified in independent studies of fluorescence. Figure 12 shows the spectroscopic data. The results there cannot be explained by exciplex formation via diffusive encounters involving $^1BC^*$ or $^1TPA^*$, hence it implies that the partners are together before the state is created. Such a conclusion obviously strengthens the premise that all the chemistry leading to excitation occurs within a cage.

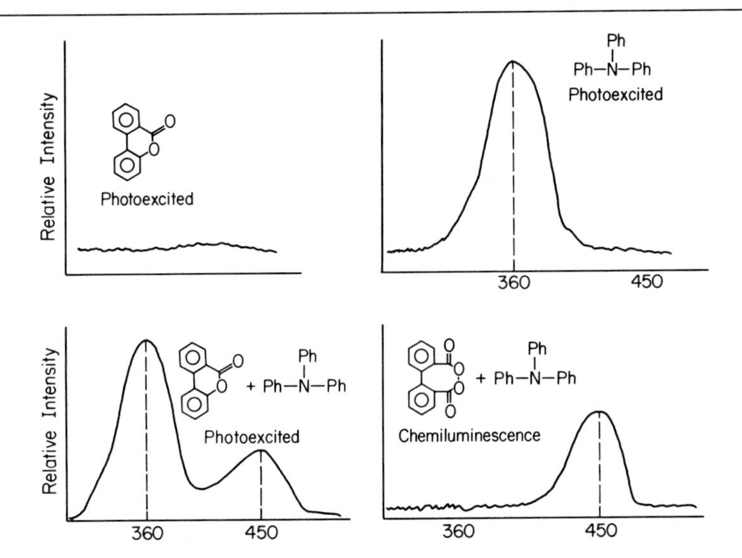

FIG. 12. Fluorescence and chemiluminescence spectra from the diphenoylperoxide/benzocoumarin/triphenylamene (DPP/BC/TPA) system. Chemiluminescence is typical of solutions containing 0.12 m*M* DPP and 0.38–1.4 m*M* TPA in benzene. The fluorescence spectra were taken for solutions containing 10 m*M* BC and/or 2.7 m*M* TPA in benzene. Excitation was carried out below 360 nm. Reproduced by permission from J.-Y. Koo and G. B. Schuster, *J. Am. Chem. Soc.*, **100**(in press). Copyright by the American Chemical Society.

Initial measurements indicate that excited states are produced in CIEEL rather efficiently ($\lesssim 10\%$), and there seems to be a surprisingly high favor toward singlet excitation. The latter feature may manifest spin conservation in a geminate recombination of doublets formed initially from a purely singlet system.

Of perhaps more general interest as a chemiluminescent species is dimethyldioxetanone (DMD), which is a model for luciferins active in bioluminescence, including that from the firefly.[3-6] DMD was first synthesized by Adam and co-workers, who also studied its chemiluminescence properties.[79,80] In a pure solution, it behaves like a dioxetane by undergoing thermolysis to produce excited acetone,

$$\text{DMD} \xrightarrow{\Delta} CO_2 + \text{Me}_2\text{C}{=}\text{O}^* \qquad (38)$$

[79] W. Adam and J.-C. Liu, *J. Am. Chem. Soc.* **94,** 2894 (1972).
[80] W. Adam, G. A. Simpson, and F. Yany, *J. Phys. Chem.* **78,** 2559 (1974).

However, it is also subject to activated decay by the addition of rubrene and other donors. The kinetic properties of the activated systems are virtually identical to those described earlier for DPP, thus it is likely that CIEEL also applies to DMD.[9]

The discovery of CIEEL may turn out to be an extremely important development for bioluminescence. Koo *et al.* have reinforced this point recently by noting that a proposed precursor to bioluminescence from the firefly is a dioxetanone which incorporates a peroxidic acceptor and a donor moiety within the same molecule.[8] It is therefore conceivable that chemiexcitation comes about in an intramolecular CIEEL process. Unfortunately, the structure of this key intermediate is not certain, and it is the focus of a vigorous debate.[81,82] Other chapters in this volume provide authoritative information on the subject. The alternative structure is a linear peroxide,[81] which could conceivably also undergo CIEEL. These ideas about chemiexcitation in the firefly system are speculative, of course, but they usefully underscore the experimental possibilities in this exciting new domain.

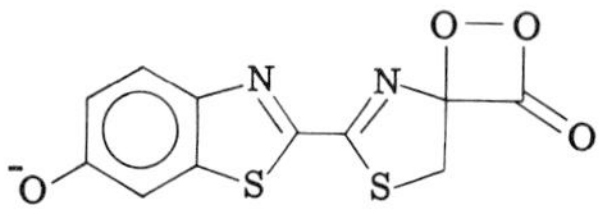

More Unusual Chemistry

Schuster's work with reductants that grow stronger via chemical changes[7-10] tends to parallel studies of several ECL systems in which reduction actually produces strong oxidants or oxidation yields potent reductants. Knowledge of these particularly unusual twists has improved considerably in the past few months, and it is interesting to examine, in this closing section, some of the late developments.

Chandross appears to have been first to understand the phenomenon.[83] An example comes from his work with *N*-tosylated carbazole, which reacts with solvated electrons in HMPA in two stages. The first electron forces the tosylate moiety to depart,

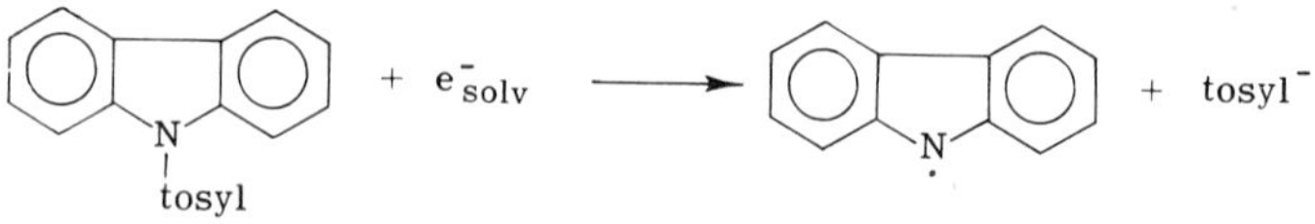

$$\text{carbazole-N-tosyl} + e^-_{solv} \longrightarrow \text{carbazolyl-N}^{\bullet} + \text{tosyl}^- \quad (39)$$

[81] F. I. Tsuji, M. DeLuca, P. D. Boyer, S. Endo, and M. Akutagawa, *Biochem. Biophys. Res. Commun.* **74**, 606 (1977).

[82] O. Shimomura, T. Goto, and F. H. Johnson, *Proc. Natl. Acad. Sci. U.S.A.* **74**, 2799 (1977).

[83] E. A. Chandross, *Trans. N. Y. Acad. Sci. Ser. 2* **31**, 571 (1969).

Left behind is a radical that is quite a strong oxidant. It then reacts with a second electron to produce a carbazole anion in its fluorescent singlet state,

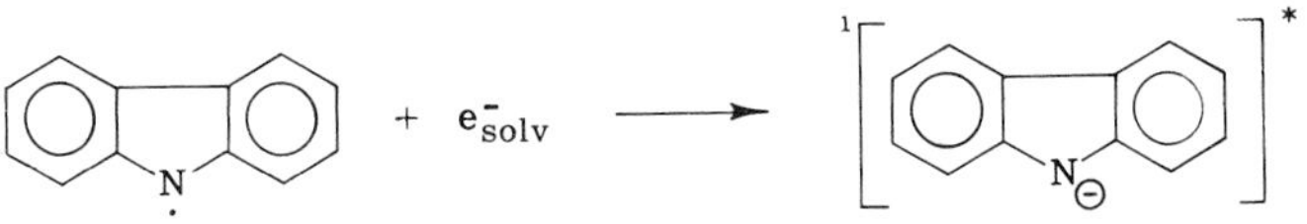

$$\text{carbazolyl radical} + e^-_{solv} \longrightarrow {}^1[\text{carbazole anion}]^* \qquad (40)$$

Itaya *et al.* have recently characterized this process in more detail, and they have provided many examples of similar cases involving leaving groups.[29,30]

A fascinating study with a comparable basis was just reported by Chang *et al.*,[84] who examined the oxidation of oxalate ion in an acetonitrile. They were hoping to generate dioxetanedione,

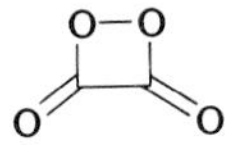

which has been of interest as a possible intermediate in chemiluminescence from oxalate esters.[6] Interceptors such as rubrene were added in order to trap excited states produced by thermolysis of the dioxetanedione. Interestingly, chemiluminescence was not observed when the electrode potential was set so as to oxidize only the oxalate, but it did appear at more extreme potentials where the interceptor was also oxidized. The emission spectrum always corresponded to fluorescence from the interceptor.

The proposed mechanism for this effect involves a homogeneous reaction between the interceptor's cation radical and $C_2O_4^{2-}$, i.e.,

$$R^{+\cdot} + C_2O_4^{2-} \rightarrow C_2O_4^{-\cdot} + R \qquad (41)$$

The resulting anion radical eliminates CO_2 and forms the powerful reductant $CO_2^{-\cdot}$,

$$C_2O_4^{-\cdot} \rightarrow CO^2 + CO_2^{-\cdot} \qquad (42)$$

which then reduces the neutral interceptor,

$$CO_2^{-\cdot} + R \rightarrow CO_2 + R^{-\cdot} \qquad (43)$$

Thus, the annihilation of $R^{+\cdot}$ and $R^{-\cdot}$ becomes possible, and light appears.

Strong evidence for this scheme comes from cases in which thianthrene (TH) and 2,5-diphenyl-1,3,4-oxadiazole (PPD) are added to the system. If only TH is added, no luminescence appears, even when it is oxidized in the presence of oxalate. However, light is emitted if one merely adds PPD. The effect is striking, because PPD is not oxidizable and cannot participate in the oxidative chemistry. Its role seems to be in the interception of $CO_2^{-\cdot}$,

$$CO_2^{-\cdot} + PPD \rightarrow CO_2 + PPD^{-\cdot} \qquad (44)$$

[84] M.-M. Chang, T. Saji, and A. J. Bard, *J. Am. Chem. Soc.* **99**, 5399 (1977).

to produce anions that undergo a known chemiluminescent reaction[23,58,59] with TH^{+}. In the absence of PPD, there is no light because TH is not reducible, and anions capable of electron-transfer excitation in a reaction with TH^{+} are not available.

Reports in the literature describe other systems with similar properties,[83,85-88] but they are not as well characterized as these examples. The cases considered here will suffice for our purpose, which is to show that surprisingly powerful electron-transfer reactions can arise in systems where milder redox agents undergo favorable chemical rearrangements or eliminations.

Acknowledgments

The author wishes to thank Professor G. B. Schuster for aiding this review by his stimulating discussions and by supplying preprints of his work. Appreciation is also extended to the National Science Foundation for financial support under Grant MPS-75-05361.

[85] T. M. Siegel and H. B. Mark, Jr., *J. Am. Chem. Soc.* **93,** 6281 (1971).
[86] T. M. Siegel and H. B. Mark, Jr., *J. Am. Chem. Soc.* **94,** 9020 (1972).
[87] K. G. Boto and A. J. Bard, *J. Electroanal. Chem.* **65,** 945 (1975).
[88] D. L. Akins and R. L. Birke, *Chem. Phys. Lett.* **29,** 428 (1974).

Section IX

Instrumentation and Methods

[41] Construction of Instrumentation for Bioluminescence and Chemiluminescence Assays[1]

By JAMES MICHAEL ANDERSON, GEORGE J. FAINI,[2] and JOHN E. WAMPLER

With the variety of bioluminescent and chemiluminescent assays described in this volume, the need for convenient and reliable light-measuring instruments is obvious. Several commercial instruments have been designed specifically to fill this need, and some of this instrumentation has been discussed by Picciolo *et al.* [43]. Alternatively many researchers fabricate their own instrument, an approach that is often desirable because of specialized measurement requirements or economics. The intent of this chapter is to provide the basic information necessary to utilize this alternative.

The major considerations in designing a radiometer or a photometer are the selection of components and their proper use. Design and construction are greatly simplified when components are selected from the wide variety of prepackaged electronic modules now available. In this chapter, all the major analog components are prepackaged electronic modules that require very few connecting wires and a minimum number of discrete passive components (resistors, capacitors, potentiometers, etc.). With these modules, the selection criteria are usually reduced to a few, easily understood parameters. Foremost among these considerations, however, are the reliability and sensitivity of the overall instrument, and user convenience.

Two photometers will be described—a simple unit and a more complex multifunctional instrument. The first is cheap and easily constructed (see block diagram of Fig. 1A). It reproducibly and reliably carries out the basic light-measuring functions and has good sensitivity and low noise. A more complex version of this instrument, which includes a variety of circuits to facilitate measurements and add to user convenience, is also described (Fig. 1B).

Both of these instruments can be fabricated without using printed circuit cards. In fact, since the current transducer (amplifier in circuit 1A or preamplifier in circuit 1B) should be mounted on a socket with Teflon insulators, many of the discrete components used in this part of the circuit

[1] This work was supported by grants from the National Science Foundation, PCM-74-08663 and PCM-76-15842 to J.E.W. and PCM-76-10573 to J.M.A. Contribution No. 368 from the University of Georgia, Marine Institute, Sapelo Island, Georgia.

[2] Currently Senior Applications Specialist, Automation and Control, Merck and Co., Westpoint, Pennsylvania.

ISBN 0-12-181957-4

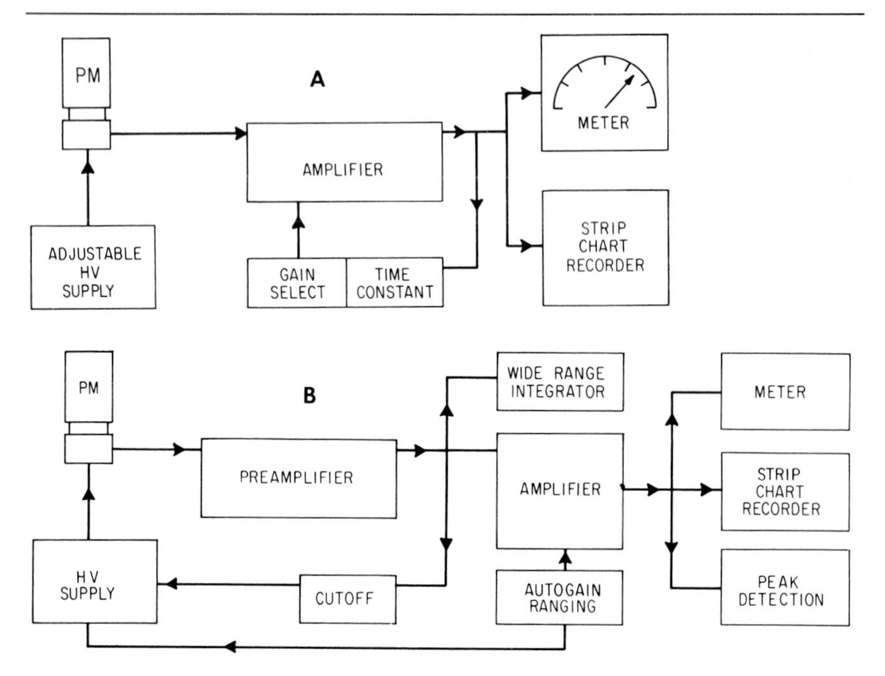

FIG. 1. Block diagram of photometers (A) Simple photometer. (B) Photometer with integrator, peak detector, autogain ranging, and phototube protection circuits.

can be mounted directly on the socket pins. Direct mounting of this transducer on a printed circuit board can lead to leakage currents at the inputs and reduce performance. The integrated circuits that carry out the various functions of Fig. 1B are most easily wired by using wirewrap techniques.

Preliminary Considerations

Light Detection. A variety of light-detecting components are available, but of these the photomultiplier tube (PM) offers the highest sensitivity in terms of amplified light signal equivalent to total instrument noise.[3] In addition to converting a light signal into current, the PM also amplifies this current and thus requires less subsequent amplification. A single photon event at the photocathode can produce an anode current of 10^6 electrons. By the very nature of its operation, a photomultiplier is a current source and, as such, should ideally be amplified by a current amplifier.

The major selection criteria for a photomultiplier for low-light measurements are the sensitivity, quantum efficiency, dark current, and

[3] A. T. Young, *Appl. Opt.* **8,** 2431 (1969).

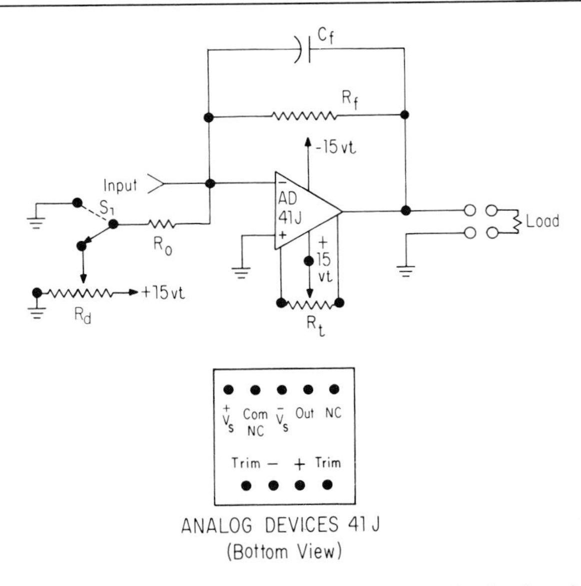

FIG. 2. Current to voltage transducer. R_t, trim adjust; R_f, feedback resistor; R_o, offset resistor; R_d, offset adjust; S_1, grounding switch, and C_f, the time constant capacitor. Also shown is pin function diagram for Analog Devices (AD) 41J operational amplifier.

wavelength response. For general use in an uncooled housing, red response can be sacrificed for lower dark current. An end-window tube with a large number of dynodes should be selected to obtain the most amplification and sensitivity. Potted dynode chains and Teflon sockets will help reduce leakage currents and improve noise performance.

Current to Voltage Conversion. The current produced by a PM is very conveniently amplified using an operational amplifier current transducer (Fig. 2).[4] In this circuit the current from the anode flows into a virtual ground at the amplifier's minus (inverting) input. In fact the amplifier's function here is to maintain the minus input at or near the same potential (zero) as that of the plus (noninverting) input. This is achieved by summing the currents into the minus input to zero; i.e., the output positive voltage develops a matching positive current into the summing junction (minus input) through the feedback resistor, R_f. The current-to-voltage gain is therefore determined by R_f. The additional components of the circuit of Fig. 2 gives the circuit complete function as a simple photometer. The trim pot, R_t, with resistance value and pin connections as specified by the amplifier manufacturer, is used to zero the amplifier's output when the

[4] "Applications Manual for Operational Amplifiers," Philbrick/Nexus Research, Dedham, Massachusetts. Nimrod Press, Boston, Massachusetts, 1968.

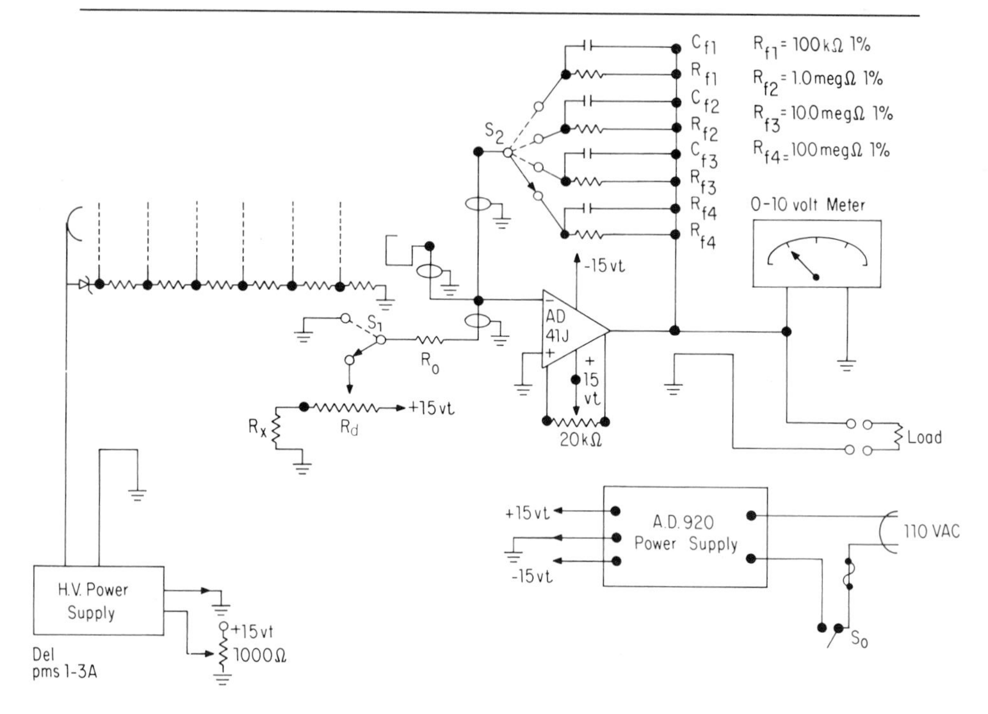

FIG. 3. Simple photometer. The circuit of Fig. 2 combined with power supplies, a photomultiplier, a meter, and gain selection via S_2 can effectively be used as a simple photometer for routine measurements.

input is shorted to ground via the resistor, R_0, and switch, S_1. This adjustment should be performed with the maximum value of R_f. It can also be implemented with a temporary clip lead in place of S_1, no high voltage on the PM and no offset of photomultiplier dark current. For switching gains, the gain resistor and its associated capacitor can be changed by a switch (S_2 in Fig. 3). The amplifier's time constant is determined by the product $R_f \times C_f$.

A Simple Photometer

By combining modular power supplies, voltmeter, and a simple, functional housing (Fig. 4) with the circuit of Fig. 2, the photometer pictured in Fig. 3 can be easily fabricated. The housing and cell compartment shown in Fig. 4 was designed for routine use in the Bioluminescence Laboratory. It has a large aperture shutter and can accommodate various cell holders; these, in turn, can include integral masks or filters as required.

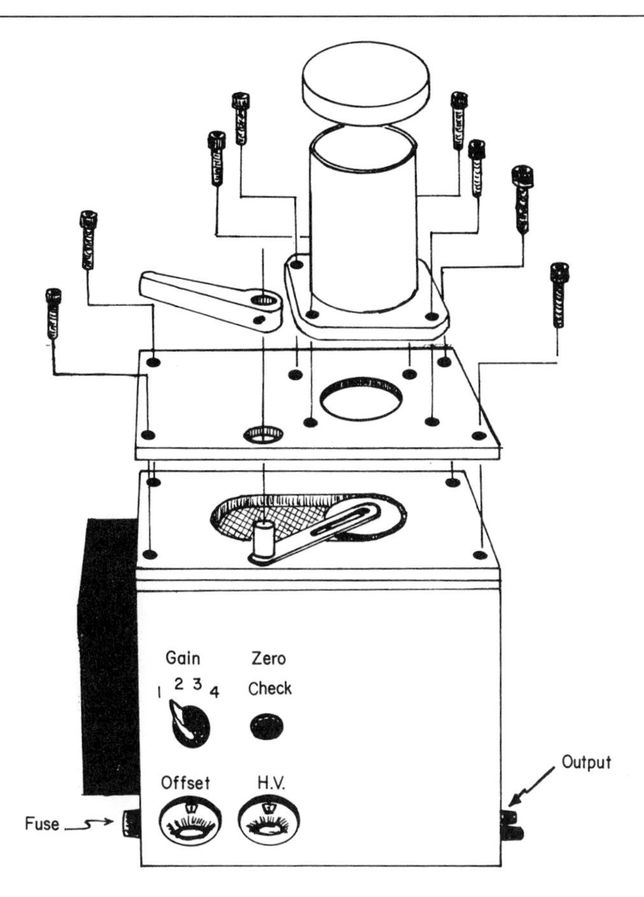

FIG. 4. Photometer housing. The circuit of Fig. 3 is easily packaged in a simple housing. The power supply (AD920) is mounted on the left side of the box. Output via the banana jacks can be used to drive a variety of readout devices.

These cell holders are made of easily machined plastic, the remainder of the housing being constructed of aluminum. A specialized version of this housing has been constructed for dual-wavelength measurements.

To improve the low light capabilities of the instrument, the PM housing also contains the current transducer circuit. This housing is light-tight and dry. Humidity variations in any circuit involving high voltages and high impedences can result in nonlinear performance and significantly increased noise. Therefore the photomultiplier dynode chain is potted in epoxy. Silica gel can also be enclosed in the housing as a drying agent, but it should be positioned in a container that the PM cannot "see," since these compounds often luminesce (Young, 1969).[3] The housing also includes a PM shield, as recommended by the manufacturer.

Criteria for Component Selections. The FET operational amplifier module (Table I) for the current transducer was chosen for its low input bias current and low input current noise. The amplifier chosen, as with most similar amplifiers, includes input and output overvoltage protection, and output short circuit protection. Its power supply requirements (±15 V DC at 20 mA) and the requirements of the DC-to-DC converter used to generate the PM high voltage (+15 at about 100 mA) are easily obtained from the 200 mA of the ±15 modular power supply (Table I). The ±15 power supply module is short-circuit protected and can be obtained with either socket mounting (as used here) or terminal strip input and output connections.

Fabrication. Well insulated 16- or 18-gauge stranded wire is used for all leads, with solid wire for interconnections. All connections are kept short and direct. The leads connected to the amplifier's summing junction from the PM, the switch, and the offset resistor, R_0, are shielded and kept as short as possible. The trim pot is epoxied to the amplifier socket. Ground and shield connections as shown in Fig. 3 are all connected by 16-gauge stranded wire to the power supply common.

Postconstruction Adjustments. Check all power connections and cross-check for shorting before applying any power to the circuit. Then,

TABLE I
PHOTOMETER PARTS LIST

Number	Manufacturer	Device
AD41J	Analog Devices	Operational amplifier
AD521J	Analog Devices	Instrumentation amplifier
AD920	Analog Devices	Power Supply, ±15 V
4709	Teledyne/Philbrick	Voltage to frequency converter, 100 KHz
PMS1-3A	Del Electronics	DC/DC converter
LM308N	Fairchild	Operational amplifier
LM310N	Fairchild	Voltage follower
LM339N	Fairchild	Quad comparator
LM555	Fairchild	Monostable
LM723	Fairchild	Voltage regulator
DL750	Litronics	LED display
MM74C926	National Semiconducter	CMOS counters
CD4052A	National Semiconducter	CMOS switch
CD4029A	National Semiconducter	CMOS counter
CD4049A	National Semiconductor	CMOS buffer/inverter
CD4011A	National Semiconductor	Nand gate
SN75492	Texas Instruments	Display driver
SN7400	Texas Instruments	Quad nand gate

mechanically zero the readout meter (Fig. 3) and set the gain selector switch to the lowest value feedback resistor. Before installing the photomultiplier tube, turn the trim pot, R_t, to zero offset and apply power to the circuit. Adjust the trim pot, with S ~ S_2 depressed, until the meter reads zero. Repeat this adjustment at each successively higher gain setting. Turn off the power and install the PM. Turn the power back on and set the dark current offset pot, R_D, so that the meter again reads zero. The photometer is now ready for use.

Multifunction Photometer

The circuit of Figs. 2 and 3 is also used for the more complex instrument. Minor changes are removal of switch S_2 and installation of a single 100 K feedback resistor on the current transducer as shown in the circuit of Fig. 5A. The 0.01 μf feedback capacitor gives a limiting time constant of 10^{-2} sec. The output of this amplifier (Fig. 5A) is then connected to an autoranging instrument amplifier (Fig. 5B), a digital integrator (Fig. 6), a peak-detector circuit (Fig. 7), and the high-voltage shutdown circuit (Fig. 8).

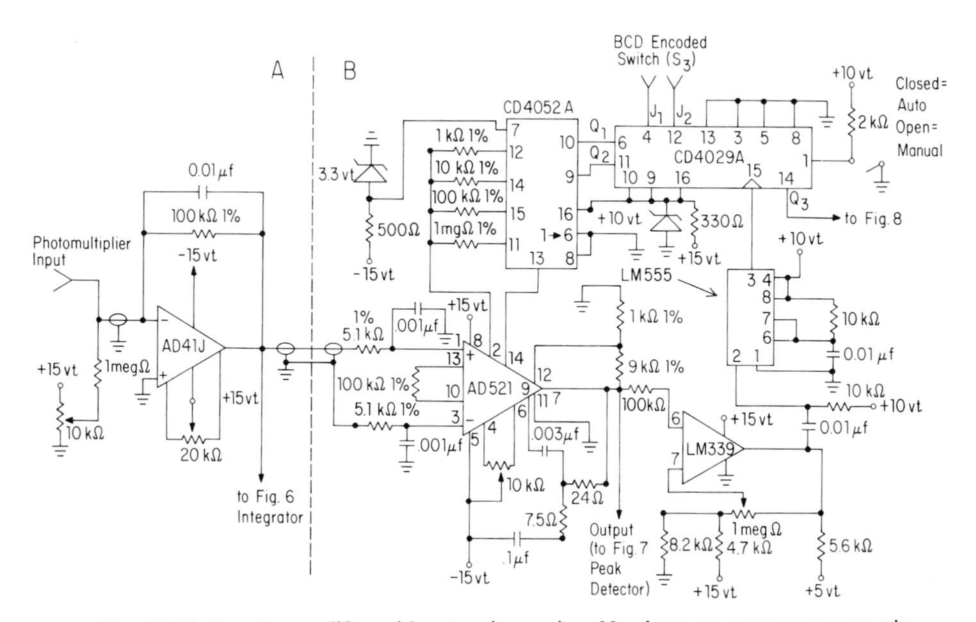

FIG. 5. Photometer amplifier with autogain ranging. Numbers on components are pin numbers. The preamp stage is mounted in a housing similar to that in Fig. 4, while the other circuitry is mounted in a separate readout chassis. See text for details.

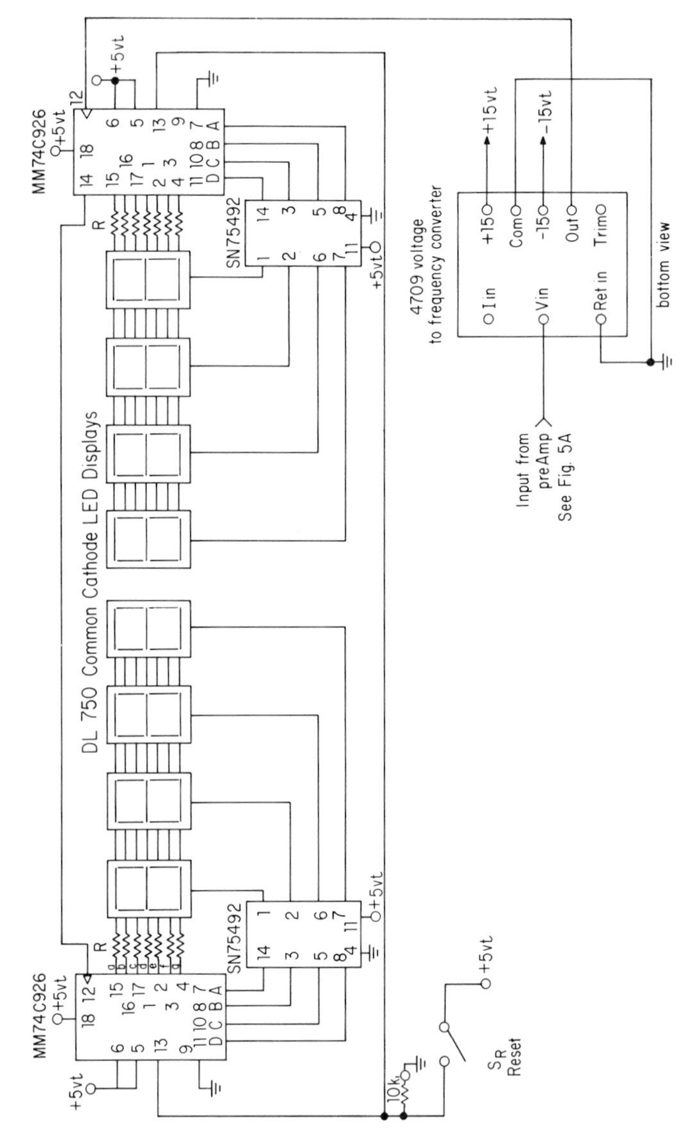

FIG. 6. Digital integrator. Numbers on components are pin numbers, and pins on 4709 are shown from bottom view. The resistors (R) are needed only under extreme environments. Their values are determined by the specifications of the display multiplexer.

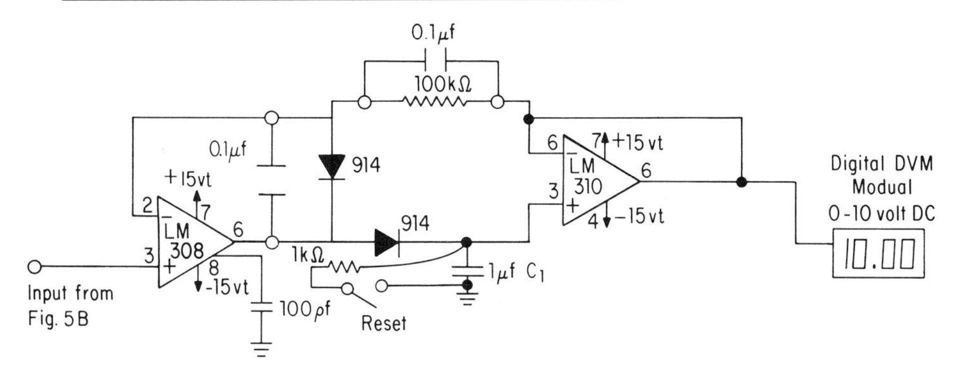

FIG. 7. Peak-detecting circuit. Capacitor C_1 should be a low-leakage type (such as Mylar) and burned-in prior to use.

Autoranging Amplifier. One problem with changing the gain at the OP-AMP current transducer, as in the simple instrument, is that high gain imposes a requirement for large resistor values and with long leads that can significantly contribute to noise pickup by the circuit. The simple solution to these problems is to use a fixed-gain preamplifier as in Fig. 5A and to feed the output of this circuit into a second amplifier for gain selection. Modular instrumentation amplifiers are ideal for this purpose, since they are noninverting and gain selection is easily achieved by switching a single resistor. Figure 5 shows this type of second stage.

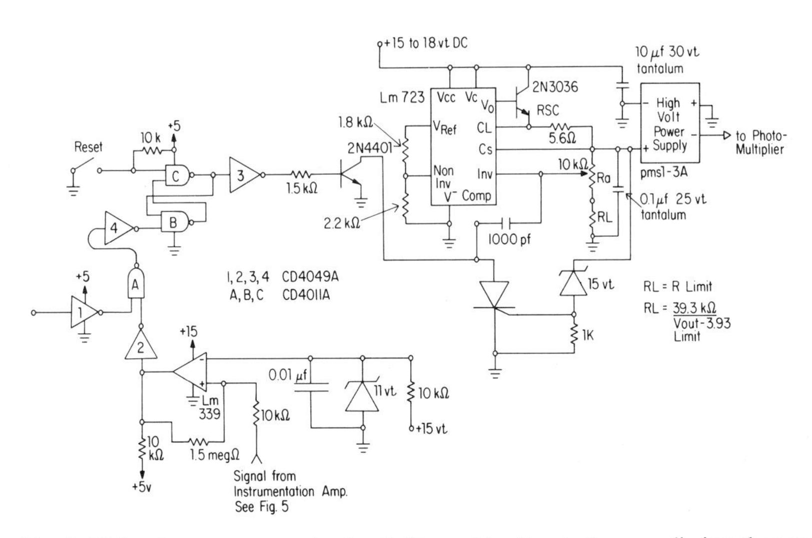

FIG. 8. High-voltage power supply circuit. The resistor R_L sets the upper limit to the power supply output. The PMS1–3A has a DC to DC voltage gain of 100.

This circuit switches gain automatically, a desirable feature when doing repetitive assays involving highly variable light intensity. Gain is changed by changing the gain resistors using a CMOS switch (CD4052A), which in turn is controlled by a counter (CD4029A). The gain in this circuit is manually set using the BCD encoded rotary switch, S_3, or, in the autoranging mode, is controlled by the monostable (LM555). An operational amplifier acts as a comparator to sense when 9.2 V is reached, and at this point fires the monostable to lower the gain. The circuit shown in Fig. 5B provides for switching four gains, but could be expanded to eight gains by using a CD4051A CMOS switch. The unused output of the counter (either Q3 or Q4) can be used to deactivate the photomultiplier's high-voltage supply for protection from excessively bright sources (see Fig. 8).

Integrator. A measure of the total amount of light produced during a bioluminescent event is often needed, for example, in quantum yield determinations. Analog integration of the output signal can be accomplished with a resistor and a capacitor in series. This simple integrator will work adequately with short transient signals, but with longer-lived output such a circuit will tend to give low values due to leakage of the capacitor. Various analog operational amplifier integrators could also be used,[4] but leakage from the integrating capacitor will still occur. More accurate integration, however, can be accomplished for all types of signal durations by use of digital, or counting, integrators. These have the added advantage of producing a numeric display of the result. The heart of a digital integrator is the voltage-to-frequency converter, which produces an output frequency directly proportional to the input voltage. Highly accurate (linear from 0.01 V or lower to 10 V input) and modular voltage to frequency converters are available at low cost (for example the Model 4709, Teledyne/Philbrick). The simplest digital integrator consists of a series of counters and display components as described previously.[5] This simple integrator, however, has high current consumption; in addition, the large number of connecting wires required for this type of integrator makes construction using printed circuit cards difficult.

The integrator shown in Fig. 6 solves both these problems and also allows for easy separation of counting and display components within the photometer housing. The display is multiplexed within the counters so that only one digit is on at any one time, and as the scan frequency is high, the display appears to be on continuously.

There are a few applications, such as firefly-ATP assays, where integration for a set period of time is desirable. The simplest way to produce the

[5] J. E. Wampler, *in* "Analytical Applications of Bioluminescence and Chemiluminescence" (E. W. Chappelle and G. L. Picciolo, eds.), p. 105 (NASA Publ. No. SP-388). National Aeronautics and Space Administration, Washington, D.C., 1975.

timed interval is to use a monostable multivibrator with variable timing resistance (Fig. 9). The desired interval is set by adjusting the potentiometer using a stop watch. A second monostable can be used if a delay is desired before integration commences (Fig. 9). More accurate timing is possible by using clock circuits, but discussion of these is beyond the scope of this chapter.

Peak Determinations. Since the maximum of light emission is often directly related to the overall rate of the reaction producing the light, a means for measuring this peak intensity is useful. The simplest and most straightforward measurement is obtained by use of a strip-chart recorder. Analog peak acquisition for later readout by a voltmeter is also possible (Fig. 7). Analog peak-hold circuits usually consist of one or more operational amplifiers and a holding capacitor. In the circuit in Fig. 7, the first OP-AMP pumps the capacitor up to the peak while the second OP-AMP acts as a buffer to help prevent leakage from the capacitor. The main problem with the analog peak-hold circuit is that the peak-holding capacitor leaks off the charge after the peak has been reached. Careful selection of the holding capacitor and use of an FET input OP-AMP can help diminish, but not eliminate this leakage. With noisy signals, this circuit will sense the peak transient of the noise on top of the signal maximum; this can produce very large errors. The circuit is, however, quite useful for measurements of rapid flashes of relatively high intensity, where the signal-to-noise ratio is high. Selection of the capacitor for this type of circuit is critical as is a burning-in period. Polystyrene or Mylar[4] capacitors should be used and the circuit left on with a constant high signal (10 V) at its input for several hours before use.

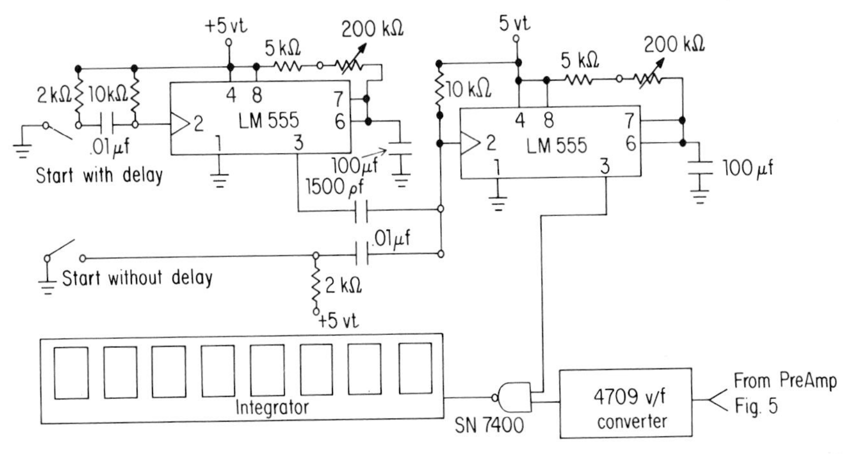

FIG. 9. Timed integration control circuit. The delay times are adjusted via the 200°kΩ variable resistors.

Power Supplies. The high voltage output of the DC-to-DC converter module can be adjusted either with a potentiometer as in the simple circuit (Fig. 3) or by use of the circuit shown in Fig. 8. In this circuit, the input voltage to the converter is controlled by a LM 723 integrated circuit power supply. This circuit also protects the radiometer from excessive signals, since overranging of the autoranging circuit or overvoltage of the preamplifier's output can both shut off the high-voltage supply by triggering the R/S flip-flop (CD4011).

The + 18 V D.C. required by the PM power supply and the + 5 V for the logic components can easily be produced by using integrated-circuit power supplies, for example LM340-18 and LM340-5, respectively (National Semiconductor).

[42] The Simultaneous Determination of Heat Changes and Light Production[1]

By NEAL LANGERMAN

The simultaneous automatic measurement of any two dissimilar physical parameters requires a combination of instrumentation. To appreciate the qualitative and quantitative power available from such a combination, one must understand each of the instruments. Traditionally, heat changes are measured using calorimeters, and light production using photomultiplier tubes. The latter are well understood by most scientists. Since this is not the case for calorimeters, a portion of this chapter will be devoted to explaining the principles of calorimetry.

Heat changes are ubiquitous to all physical and chemical processes. This is the source of the generality and strength of calorimetry. However, it is also the source of a major problem. All processes that produce heat within the calorimeter are detected by the sensors, often resulting in a thermogram that is complex and confusing. The preferred solution to this problem is to repeat the experiment using more well-defined conditions. Unfortunately, often this is not possible for complex chemical or biological processes. Under these conditions, monitoring two or more physical parameters simultaneously is exceedingly useful. In this chapter, a combined calorimeter–photometer system, and several of its different applications, will be described.

[1] Supported in part by Grants GM 22049 and GM 22676 (NIH) and 75-03030 (NSF).

ISBN 0-12-181957-4

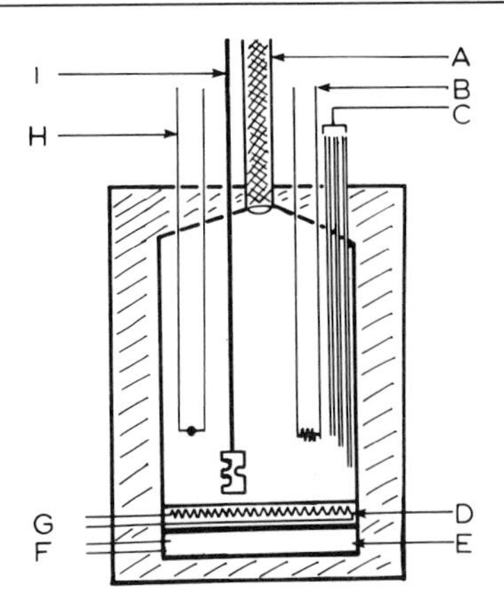

FIG. 1. Schematic outline of Tronac 550 Isothermal calorimeter vessel modified with fiber optical bundle. A, Fiber optics; B, calibration heater; C, sample inlet tubes; D, controlled heater; E. Peltier cooler; F,G, electrical connections; H, control thermister; I, stirrer.

Calorimetry: the Detection of Energy Utilization[2]

In the simplest terms, a calorimeter consists of an environmental control chamber and a temperature sensor. A polystyrene cup with a mercury thermometer is a trivial example of this device. Modern calorimeters are designed to detect exceedingly small heat effects, on the order of several microcalories, and thus require sophisticated temperature control and detection circuitry. Several commercial instruments are currently available, of which two are quite suitable for biochemical experimentation. One is manufactured by Tronac Corporation (Orem, Utah) and the other by LKB, Inc. (Rockville, Maryland). Of these, the Tronac Model 550 is by far more easily modified for dual sensor capability and is the instrument discussed throughout this article[3] (see Fig. 1).

A calorimeter measures energy changes in the form of heat occurring within the confines of the instrument. The heat can be measured in three distinct ways. The first of these is to measure the temperature rise associated with the heat production using a thermistor or similar device. In a purely adiabatic system, the temperature rise will equal the heat produced divided by the heat capacity of the system. Thus, the heat effect associated with a process can be readily determined from the temperature rise and knowledge of the apparent heat capacity of the system.

[2] In general, exothermic processes will be considered. Endothermic processes are treated in an identical manner.

[3] A modification to the LKB batch isothermal calorimeter will be discussed near the end of this article.

The other two methods of measuring heat production involve the measurement of heat flow. This can be done in a passive manner using a thermopile or in an active manner using a Peltier device. Thermopile sensors produce a voltage proportional to the temperature difference between the reaction cell and a heat sink. The Peltier device measures the power required to maintain constant temperature within the reaction cell. A more detailed treatment of these instruments is given by Langerman and Biltonen.[4,5]

The Tronac calorimeter is nominally designed as a titration device; i.e., to one reactant, in a reaction vessel, the second reactant (the titrant) is slowly and continuously added, initiating a reaction. This causes a heat effect, which, when graphed as a function of time, produces a thermogram. Several examples of thermograms will be found in this chapter. The Tronac titration calorimeter is available in two models, which are readily interchanged, providing a high level of versatility. The isothermal (Model 550) calorimeter maintains a constant temperature within the reaction vessel using a Peltier device; it is well suited for studying processes requiring long periods for completion. The Isoperibol (Model 450)[6] calorimeter maintains approximately adiabatic conditions within the calorimeter and is best applied to fast reactions. The choice of a particular instrument depends greatly on the experimental questions being asked and has been discussed in detail elsewhere.[4]

One particularly important distinction to note is the material from which the reaction vessels are made. The isothermal vessel is metal (either stainless steel or tantalum), and the Isoperibol vessel is glass. The latter is actually a glass Dewar and is highly reflective. As will be seen below, this has important consequences for detection of light produced within the calorimeter.

Either calorimeter can be used in a "batch" mode such that the reaction vessel is filled with one reactant and the second reactant is added either all at once or incrementally.

The precision of the measurements obtainable using the Tronac system varies from better than 0.1% to 10%, depending on the absolute magnitude of the heat effect. If large quantities of material are available, and the total effect is in excess of several millicalories, then extremely high precision is possible. On the other hand, if the total heat effect is on the order of

[4] N. Langerman and R. L. Biltonen, this series, Vol. 61 [13].

[5] R. L. Biltonen and N. Langerman, this series, Vol. 61 [14].

[6] The term "Isoperibol" is used by the manufacturer to reflect the constant environment of the interior of this type of calorimeter. In reality, the unit is "pseudo-adiabatic," that is, although heat flow does occur through the walls of the vessel, proper experimental design will minimize this effect.

microcalories, then a precision of 5–10% may have to be tolerated. This is often the case in biological problems because of the difficulty of obtaining adequate amounts of sample.

Adaptation of the Calorimeter for a Second Sensor

Our comments will be directed toward the simultaneous detection of light and heat. However, many other additional sensors could also be installed in the Tronac calorimeter insert. For example, D. J. Eatough (personal communication) has used a redox electrode system, and L. D. Hansen (personal communication) has used both a pH and PO_2 electrode system.

The light-detection system installed in the Tronac calorimeter was a 7-mm quartz optical fiber bundle placed in the calorimeter insert shaft and used to conduct the light to a photomultiplier tube. The position of the bundle in the isothermal vessel is shown in Fig. 1 and in the Isoperibol vessel in Fig. 2. The light-detection circuitry currently being used was designed by Mitchell and Hastings,[7] although a commercial device, such as a Pacific Photometrics, Inc. system (Model 110), has also been used. The modification of the calorimeter insert is not difficult but is best left to the manufacturer.

The simultaneous recording of the heat signal and the light signal can be handled either digitally or on a two-pen analog recorder. We have used the latter method because of cost and convenience.

Calibration of Heat- and Light-Detection Signals

Both the heat-detection and the light-detection circuits must be calibrated. Calorimeters are routinely calibrated electrically by passing a known current through a known resistor and comparing the observed signal to the known power expressed as calories (or joules) per second. Alternatively, the heat signal may be calibrated using a chemical reaction with a documented enthalpy change, such as the neutralization of a strong acid. Both procedures have been discussed in detail elsewhere.[4] The calibration of the optical system of the instrument will be discussed here.

Prior to describing the actual method for calibrating the instrument we use, it is necessary to define the actual information desired. The calibration should provide information as to the number of photons required to sweep out a unit of area on a strip chart or the number of photons per second that gives a specific response on the photometer. The calibration experiment

[7] G. W. Mitchell and J. W. Hastings, *Anal. Biochem.* **39,** 243 (1971).

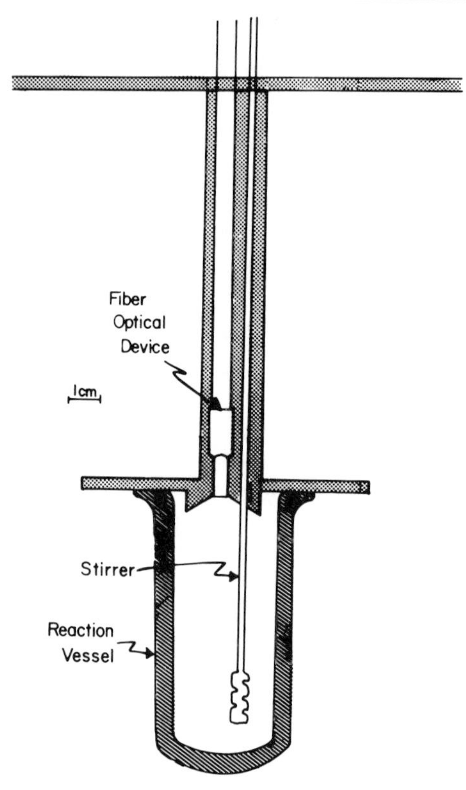

FIG. 2. Schematic outline of Tronac 450 Isoperibol calorimeter vessel modified with fiber optical bundle. Details of sample inlet tubes, calibration heaters, thermisters, and sample inlet tubes have been omitted for clarity. The highly polished interior walls and curved bottom contribute to the superior light detection of this reaction vessel over that shown in Fig. 1.

may also be used to estimate what fraction of the light generated within the reaction vessel is actually being delivered to the photomultiplier tube by the fiber optics. In practice, this last parameter is the most readily obtained. Ideally, the calibration method should closely reproduce the expected experimental system. Thus, if the light is being generated in a total volume of 15 ml, the calibration method should also generate light in this volume.

Use of the light standard of Hastings and Weber[8] is not adequate for this purpose. Because it is sealed within a vial, it does not have the appropriate geometry. The luminol reaction, as described by Lee and Seliger[9] and by Michael and Faulkner[10] was selected as the most appropriate method. The most accurate determination of the number of photons per unit area is

[8] J. W. Hastings and G. Weber, *J. Opt. Soc. Am.* **53,** 1410 (1963).

[9] J. Lee and H. H. Seliger, *Photochem. Photobiol.* **15,** 227 (1972).

[10] P. R. Michael and L. R. Faulkner, *Anal. Chem.* **48,** 1188 (1976).

obtained using the hemoglobin catalyzed–H_2O_2 oxidation of luminol. However, this reaction is relatively slow (several minutes) and it is useful to employ a much faster reaction for determining the relative efficiency of the optical system. The ferricyanide oxidation, although of lower quantum yield, is a better choice.

The reaction vessel of the calorimeter is filled with 15 ml of luminol such that the total absorbance of the solution at 347 nm is 0.013. The luminol solution is prepared as previously described.[9] The reaction vessel is placed on the calorimeter, the stirrer is turned on, and the unit is set to record the light signal. In practice, we find that black cloth must be placed over the top of the calorimeter and over the fiber optics to prevent light leaks. A 10-fold molar excess of ferricyanide in 1 ml total volume is then injected into the reaction vessel. Under these conditions, the light-emitting reaction is instantaneous and readily detected. This procedure is repeated several times, and the total area under each of the light signals is determined.

A similar experiment, properly scaled so as not to exceed the response capacity of the photometer, is now performed, placing the reactants in close proximity to the photomultiplier tube. These latter experiments are repeated several times, and the areas under the resulting recorder traces are again determined. The direct comparison of these signals, properly corrected for any scale changes, provides an estimate of the efficiency of the optical system.

If the actual response of the photomultiplier is known, in quanta per second per unit area, then the actual number of quanta released within the reaction vessel may be calculated from the previously determined efficiency. A summary of these various factors for the present version of the optical system is presented in Table I.

TABLE I
CALIBRATION OF OPTICAL SYSTEM RESPONSE

	Isoperibol calorimeter	Isothermal calorimeter	Light machine[c]
Quanta/area[a]	3.83×10^{11}	4.29×10^{12}	3.64×10^{10}
Efficiency[b] (%)	9.5	0.85	100

[a] Determined using the chemiluminescence of the ferricyanide oxidation of luminol as described by J. Lee and H. H. Seliger, *Photochem. Photobiol.* **15,** 227 (1972).

[b] The efficiency is defined as 100 times the ratio of the area:quanta observed via the optical system to the area: quanta observed when the reaction is initiated in close proximity to the photomultiplier tube.

[c] The light detector used in these experiments consisted of a Pacific Photometrics Instruments Laboratory photometer, Model 110, and a 1P21 photomultiplier tube. The latter was enclosed in a housing that would accept either the end of the optical fiber bundle from the calorimeter or a glass scintillation vial.

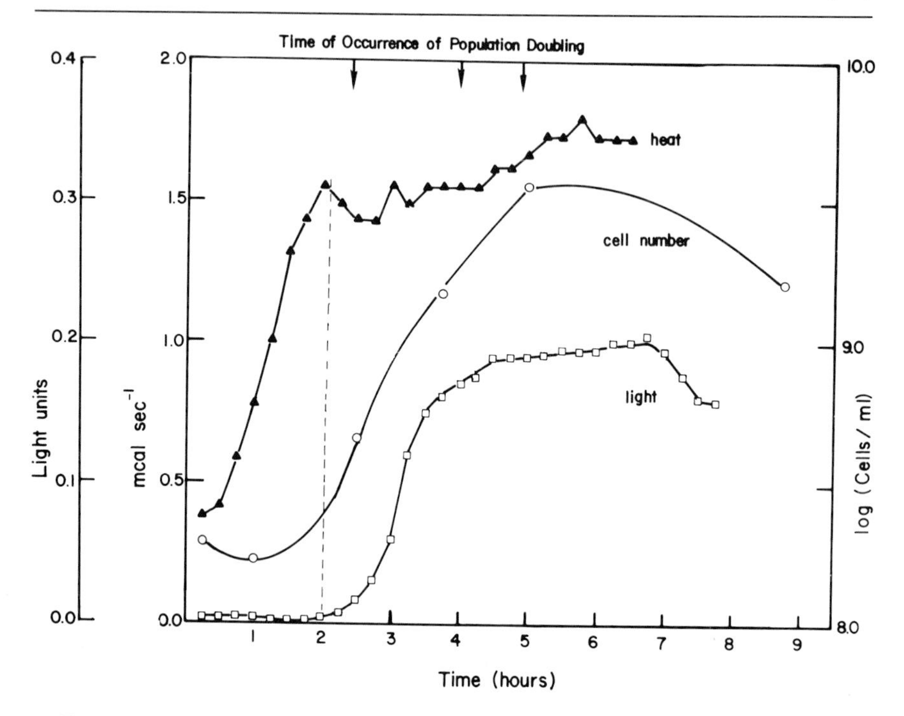

FIG. 3. Thermogram combining light production, expressed in light units (1 light unit = 1.44×10^{12} photons sec^{-1}), and viable cells (per milliliter) for *Beneckea harveyi* grown in 80% seawater containing 5 g of Bacto-tryptone, 0.5 g of Bacto Yeast Extract, and 3 ml of glycerol per liter. The scale at the top indicates when each population doubling occurred.

Applications

Several specific applications of this dual sensor calorimeter are examined below. Two of the examples illustrate applications in which light production was monitored, and the third example illustrates an application in which the process studied was driven by light.

Metabolic Studies of Luminescent Bacteria. McIlvaine and Langerman[11] have reported the application of this dual-sensor calorimeter to the study of the growth of several species of luminescent bacteria. The Tronac isothermal calorimeter was used in these studies. One thermogram resulting from this study is shown in Fig. 3. McIlvaine and Langerman also periodically sampled the contents of the calorimeter for viable cells. These data are plotted on the same graph; the data suggest that biosynthetic regulatory mechanisms manifest themselves in changes in the efficiency of

[11] P. McIlvaine and N. Langerman, *Biophys. J.* **17,** 17 (1977).

energy utilization and that the more efficiently a bacterium is growing, the less heat it releases to the environment. This hypothesis was tested by growing the luminescent bacteria in minimal media, where they are known to grow with higher energy efficiency, and observing a markedly reduced heat production. The optical system confirmed the growth of the bacteria by recording changes in the light production by the bacteria.

Luminol Chemiluminescence. The usefulness of the dual-sensor approach to calorimetry may be illustrated by examining an experiment that did not work as expected. Figure 4 presents the thermogram and light data obtained during the titration of luminol by ferricyanide. The experiment was done in conjunction with developing the luminol reaction for calibrating the optical system. Luminol was titrated in the Isoperibol calorimeter to evaluate ΔH for the reaction. When the observed data were used to estimate the stoichiometry, a ratio of ~ 15 mol of ferricyanide per mole of

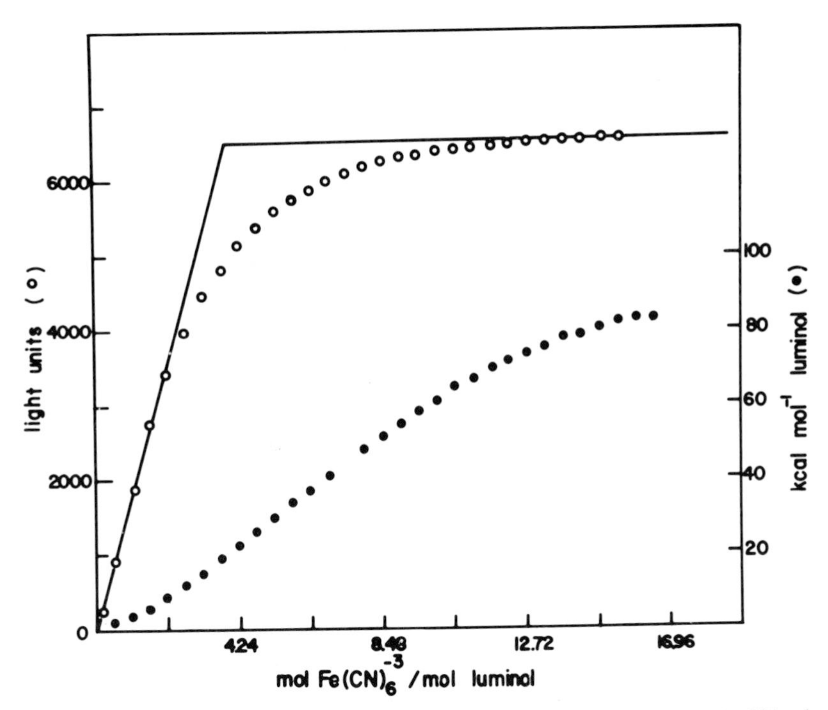

FIG. 4. Combined thermogram and light signal for the titration of 15 ml of 0.178 mM luminol with 0.38 M $Fe(CN)_6^{-3}$ in pH 12.2 carbonate buffer. The solid curve is drawn for the expected light signal to indicate the expected stoichiometry of 4:1. The heat signal, which should also reflect this stoichiometry, indicates the presence of a second process. In this case, the dual detector capability showed that the calorimetric data contained an artifact.

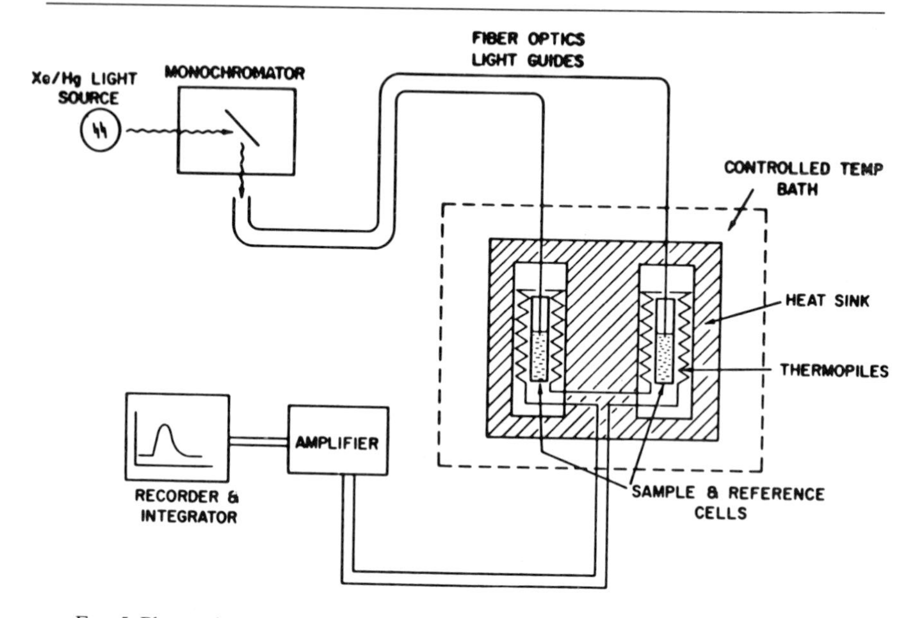

FIG. 5. Photocalorimeter schematic used by A. Cooper and C. A. Converse [*Biochemistry* **15**, 2970 (1976)]. This unit is designed around an LKB Batch Isothermal Calorimeter. Not shown are the shutter and movable fiber optics mount at the exit slit of the monochromator, which allow either the reference or sample cell to be illuminated independently.

luminol was obtained. When the light data were analyzed in the same way, however, the expected stoichiometry (4:1) was obtained. This immediately suggested that the ferricyanide was further reacting with a species in solution and that the observed heat contained an artifact not related to the reaction of interest. The availability of the simultaneously obtained light data prevented misinterpretation of the data.

Photocalorimetry of Rhodopsin in Rod Outer-Segment Membranes. Calorimetry in which the reaction of interest is initiated by light has been called "photocalorimetry." The only application of this method to date was by Cooper and Converse[12] in their study of the reaction:

$$\text{Rhodopsin} + h\nu \rightarrow \text{metarhodopsin I}$$

They used an LKB batch calorimeter as shown in Fig. 5. A thermogram showing their data is presented in Fig. 6; the figure legend explains the conditions of the experiment. The data they obtained indicate that the reaction of interest occurs with a ΔH of $+16.8$ kcal mol^{-1} at pH 8.0. They

[12] A. Cooper and C. A. Converse, *Biochemistry* **15**, 2970 (1976).

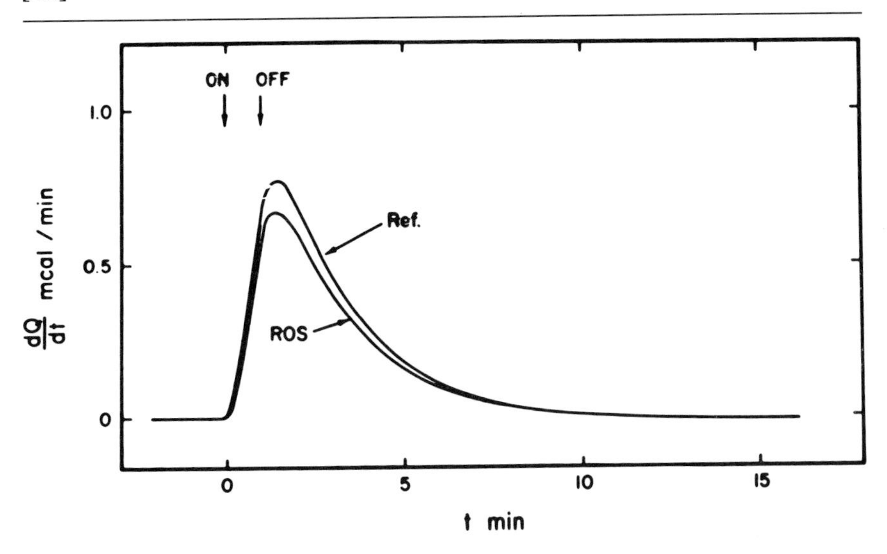

FIG. 6. Superimposed photocalorimeter response for 1-min irradiation at 546 nm of rod outer segment (ROS, 0.1 *M* Tris · acetate, pH 8.0, 3°) and of an inert dye solution (in reference cell). The reaction is rhodopsin→ meta-I and is endothermic. The noise level in this case is about twice the linewidths of the traces. Note that the shape of the curves is determined solely by the thermal response of the calorimeter and does not reflect the kinetics of the reaction, which is essentially instantaneous on this time scale.

further determined that no net proton exchange occurred concomitant with the reaction.

These few examples and the description of the instrument illustrate the potential available by combining calorimetry with other techniques. As previously pointed out,[4,5] calorimetry has now left the realm of the specialist and become a tool available to anyone interested in energy changes occurring during chemical or physical processes. If these processes should happen to emit light, then this light may be simultaneously detected and used to assist in the interpretation of the thermodynamic data.

[43] Characteristics of Commercial Instruments and Reagents for Luminescent Assays

By GRACE LEE PICCIOLO, JODY W. DEMING, DAVID A. NIBLEY, and EMMETT W. CHAPPELLE

Principles of Design

The basic design of a photometric instrument suitable for the detection of low-level light emissions, as encountered in the firefly luciferase ATP assay and other bioluminescent and chemiluminescent reactions, should include the following components, which are diagrammed schematically in Fig. 1:

- High-gain, low-noise photomultiplier tube with a well stabilized, high-voltage power supply to ensure sensitive, reproducible measurements of the light emissions
- High-gain, high-impedance, direct current amplifier with a wide range of sensitivity and linear response
- Means to quantitate the amplified electrical signal, i.e., meter, recorder, or oscilloscope
- Means to null background and endogenous light
- Reaction chamber to allow adequate mixing of reagents while protected from extraneous sources of light

Although suitable instrumentation is available commercially, an appropriate instrument can be fabricated by the investigator at minimal cost if tools and expertise in electrical engineering are available. A detailed description of design and fabrication is described elsewhere.[1,2]

Commercially available instrumentation specifically designed for luminescent assays (and evaluated by this laboratory) includes the ATP photometer, Model 2000, SAI Technology Company, San Diego, California; the Chem-Glow photometer, American Instrument Company, Silver Spring, Maryland; and the luminescent biometer, E. I. DuPont Company, Wilmington, Delaware. Vitatect Company has recently marketed a photometer that allows sample concentration and reaction with luciferase in a moving filter tape within the instrument. This unique photometer has been evaluated elsewhere.[3] Scintillation counters operated in the noncoinci-

[1] J. E. Wampler, *in* "Analytical Applications of Bioluminescence and Chemiluminescence" (E. W. Chappelle and G. L. Picciolo, eds.), p. 105 (NASA Publ. No. SP-388). Washington, D.C., 1975.

[2] E. W. Chappelle and G. V. Levin, *Biochem. Med.* **2,** 49 (1968).

[3] G. L. Picciolo, E. W. Chappelle, R. R. Thomas, and M. A. McGarry, *Appl. Environ. Microbiol.* **34,** 6 (1977).

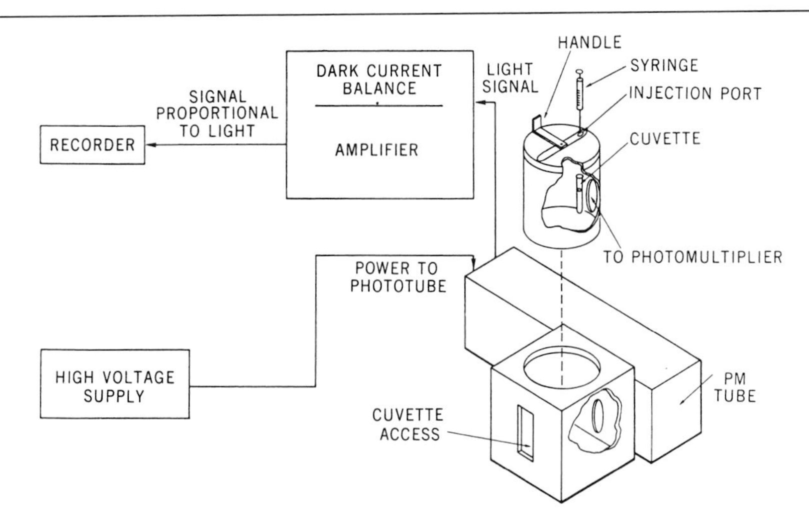

FIG. 1. Diagram of a basic photometer for luminescent measurements.

dence mode may be used to measure luminescent reactions via integration of the delayed light reaction curve. Barring major modifications of the instrument, however, it is not possible to obtain a value for the peak light flash that occurs in less than 1 sec in the case of the firefly luciferase ATP reaction. We have not considered scintillation counters in this evaluation of commercially available instrumentation, although others have addressed the use of this instrument in some detail.[4,5]

General Characteristics of Three Photometers

The three commercial photometers evaluated in this laboratory meet the basic requirements as set forth above. All are equipped with suitable photomultiplier tubes (PMT): the SAI photometer (Model 2000) has a 13-stage PMT with a bialkali cathode; the DuPont biometer has an 11-stage PMT with an S-11 cathode; and the Aminco Chem-Glow has a 9-stage PMT with an S-4 cathode. Replacement with tubes of different characteristics is possible in all the instruments, within certain limitations of size and shape. Each of the cathodes provides an operational spectral response of 350–600 nm.

The dark current of the PMT and the endogenous light emitted by a small volume of the luciferase–luciferin enzyme preparation can be nulled adequately by each instrument. If the endogenous light is too high to be

[4] S. Addanke, J. F. Sotos, and P. D. Rearick, *Anal. Biochem.* **14,** 261 (1966).

[5] R. Johnson, J. H. Gentile, and S. Cheer, *Anal. Biochem.* **60,** 115 (1974).

TABLE I
DECAY OF ENDOGENOUS LIGHT WITH TIME

Minutes after hydration[a]	Relative light units	Percent initial light emission remaining
0	4.71×10^6	100
1	3.90×10^6	68
2	3.06×10^6	41
3	2.37×10^6	30
4	1.98×10^6	20
5	1.62×10^6	14
10	8.16×10^5	4
15	5.49×10^5	2
20	4.11×10^5	1

[a] Using 1.5 ml of 0.25 *M* Tris, pH 8.2, 10 m*M* $MgSO_4$, 1 m*M* Cleland's per vial of DuPont enzyme.

zeroed by the instrument electronic system, the enzyme preparation itself should be incubated until the endogenous light level is manageable. Light emitted by the particular enzyme preparation used in the evaluation of these instruments (purified luciferase obtained from DuPont) decays significantly after rehydration over a 20-min incubation period at room temperature as demonstrated by Table I.

Endogenous light is thought to be a function of the amount of ATP present in the enzyme solution without the injection of a sample or blank solution. The blank solution consists of the sample fluid without any particulate sources of ATP (sterile media, filter-sterilized water sample, etc.), processed in the same fashion as the actual sample so as to contain equivalent volumes of extractant and diluent. Light produced as a result of injecting the blank solution into the enzyme preparation may be indicative of the dissociation of bound endogenous ATP. The activity of a particular enzyme preparation is that light produced when a known concentration of exogenous ATP is injected. Functional sensitivity is the difference between the blank and the activity. All three light measurements (endogenous, blank, and activity) decay at different rates, depending on conditions. To optimize the signal-to-noise ratio, removal of endogenous ATP by purification of the enzyme mixture is recommended, as well as equilibration of the ionic strength of the enzyme solution and the solution injected, stabilization of the purified enzyme by addition of Cleland's reagent, which protects the sulfhydryl bonds, and storage of the enzyme solution at 4° in the dark[6,7]

[6] J. W. Deming, G. L. Picciolo, and E. W. Chappelle, *in* "Proceedings of ASTM Symposium on Native Aquatic Bacteria," in press.

[7] G. L. Picciolo, E. W. Chappelle, J. W. Deming, M. A. McGarry, D. A. Nibley, H. Okrend, and R. R. Thomas, *in* "Handbook on Infectious Disease Methodology" (D. Amsterdam, ed.), Vol. 1. Dekker, New York, in press.

Each instrument differs in amplification range and means of quantitating the electrical signal produced as a result of light sensing. The SAI Model 2000 provides an amplification range of six decades with digital display of the signal amplitude. The DuPont biometer also provides a digital display, but amplifies over a range of 5 decades. The Chem-Glow amplifies over 4 decades and provides a meter display of the signal. All instruments can be used in conjunction with a strip-chart recorder for a permanent record of the light reaction curve. The Chem-Glow can be attached to an integrator–timer unit, made by Aminco, for an expanded range and digital display.

Both the SAI Model 2000 and the Chem-Glow, used with the integrator–timer, offer the option to integrate the area under the light reaction curve. The biometer registers only maximum light intensity after a 3-sec accumulation period and does not provide an integration mode, although the area on the strip chart could be measured to obtain a value for integrated area. With the SAI Model 2000, a 6- or 60-sec integration period is possible. An automatic 15 sec delay allows for manual external mixing of enzyme and sample and placement of the reaction cuvette into the reaction chamber; i.e., the initial light flash is not measured. (SAI's newer Model 3000 provides variable integration times.) To measure the peak of the initial flash, however, a means must be provided for injection of the sample into the luciferase cuvette, already positioned within the light-tight reaction chamber. A special adaptor kit equipped with an automatic injector can be purchased for use with the SAI Model 2000 to enable simultaneous injection and measurement of peak light response.

The Aminco integrator–timer, used in conjunction with the Chem-Glow photometer, provides 6 possible integration time intervals including a 3-sec interval as a measure of the initial light flash. The Chem-Glow is designed to accommodate light-tight injection of sample into luciferase and requires no further adaptations; however, we have added a hollow metal cylinder as an injection tower which holds the syringe and decreases stray light at very low emission levels (Fig. 2). Aminco also makes available an automatic injector.

Before deciding on the mode of light measurement, the investigator should be aware that crude preparations of the luciferase may contain phosphorylating enzymes that convert precursor molecules to ATP. These reactions occur at a slower rate than the reaction of *in situ* ATP with the luciferase enzyme. Consequently, the most accurate measure of total *in situ* ATP will be the maximum light response obtained immediately upon mixing sample with enzyme preparation, not the integrated value taken many seconds later. The potential problem of contaminating enzymes can be avoided by using purified luciferase preparations. In the evaluation of commercially available photometers, we have used only purified luciferase (from DuPont). Purified preparations also provide greater sensitivity, as

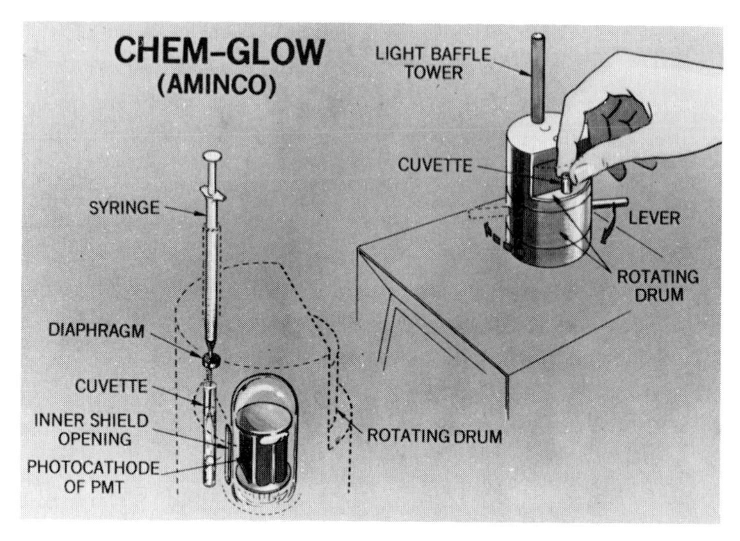

FIG. 2. Illustration of reaction chamber of Chem-Glow (Aminco) photometer. PMT, photomultiplier tube.

demonstrated in Fig. 3. A comparison of commercially available crude and purified enzymes, hydrated according to the manufacturers' instructions, and purified enzyme prepared in our laboratory from firefly lanterns[8] indicates that the purified extracts provide superior sensitivity. Holm-Hansen has shown that adding synthetic luciferin to crude extracts greatly enhances sensitivity.[9]

The feature that distinguishes all ATP photometers from other light-measuring devices, including spectrophotometers and scintillation counters, is the unique design of the reaction chamber. In the SAI Model 2000, the reaction cuvette is lowered into the chamber and protected from extraneous light by a rubber-sealed lid. A shutter that separates the cuvette from the photocathode surface beneath it opens 15 sec after the cuvette containing the reaction mixture has been lowered into position and the assay button has been depressed. For peak measurements, the shutter must be open at the time of injection.

The biometer and the Chem-Glow utilize a rotating drum for reaction chamber. The luciferase cuvette is loaded into the drum on the outside of the instrument and rotated in position in front of the photocathodes surface without exposing the photocathode to external light. A small rubber-sealed

[8] D. A. Nibley, G. L. Picciolo, E. W. Chappelle, and M. A. McGarry, *in* "Proceedings of the Second Biannual ATP Methodology Symposium," p. 441. SAI Technology Company, San Diego, California, 1977.

[9] D. M. Karl and O. Holm-Hansen, *Anal. Biochem*, **75**, 100 (1976).

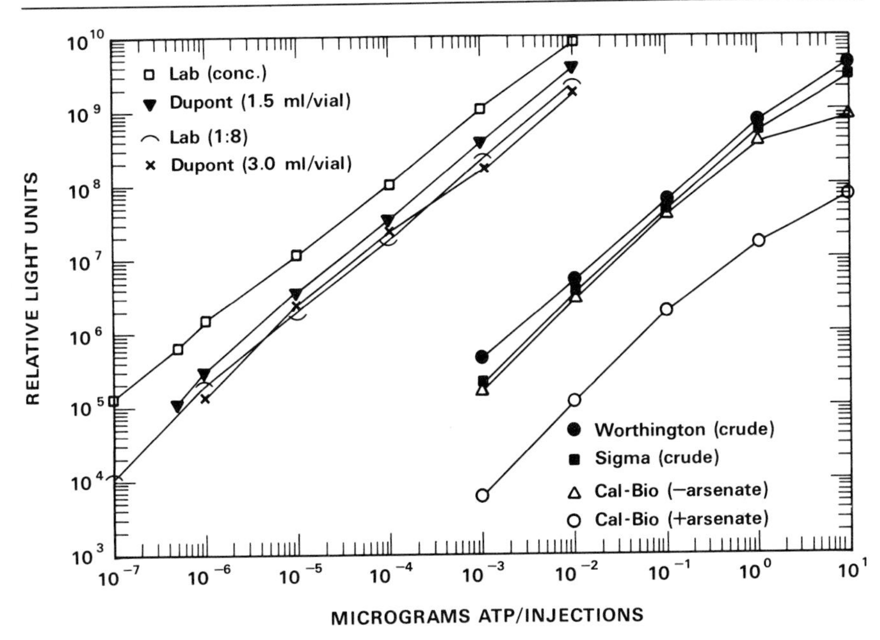

FIG. 3. ATP concentration curves using crude and purified luciferase preparations.

orifice in the top of the chamber permits injection of the sample into the luciferase without affecting the light-tight integrity of the chamber. The chambers of both these instruments accept a 6 × 50-mm cuvette. The biometer allows light emission from a maximum reaction volume of 0.2 ml to reach the photocathode, while the Chem-Glow allows the full volume of the cuvette (0.8 ml) to reach the sensor. When the SAI Model 2000 is fitted with a conical reflectorized cuvette holder, the 6 × 50-mm cuvette can be used, and 0.8 ml volume emission will reach the photocathode. Without this modification, only larger volumes from 2.0 to 10.0 ml contained within standard scintillation vials can be used. Aminco provides two alternative reaction chambers for the Chem-Glow. The "fat head" adapter accepts a larger cuvette allowing use of a maximum reaction volume of 1.75 ml. The "flow head" adapter houses a glass coil instead of an open cuvette. With this coil and a system of tubing and peristaltic pumps, a continuous flow can be constructed for luminescent assays. An example of a semiautomated flow system developed to assay levels of bacterial ATP in water samples is diagrammed schematically in Fig. 4. This flow system has been shown to be equivalent in sensitivity to standard injection methodology.[10]

[10] G. L. Picciolo, R. R. Thomas, E. W. Chappelle, R. E. Taylor, E. J. Jeffers, and M. A. McGarry, *Proc. Inst. Symp. Rapid Methods and Automation Microbiol., 2nd*, pp. 186–192 (1976).

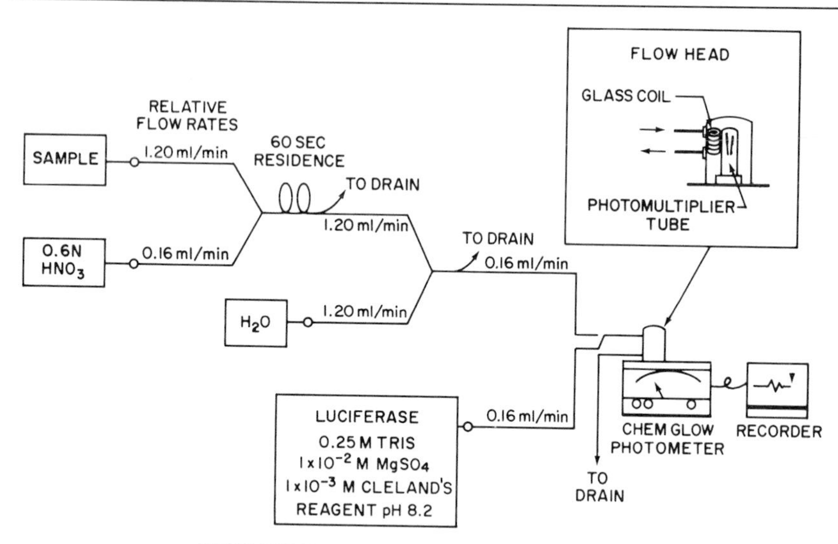

FIG. 4. Flow system for firefly luciferase ATP assay.

Evaluation Methods

Modification of Instruments. All three photometers were modified by the addition of a hollow, cylindrical metal tower designed to house a disposable 1-ml tuberculin syringe. Sample injection into a 6 × 50-mm cuvette containing 0.1 ml of the luciferase preparation, already positioned in front of the PMT, was accomplished by piercing with a 21-gauge needle a soft black rubber diaphragm at the junction of the base of the tower and the reaction chamber. These modifications prevented light leakage during the injection. In the case of the SAI photometer, the changes permitted peak-height measurements using disposable syringes. In addition, the reflectorized conical cuvette holder designed to hold a 6 × 50-mm cuvette was used.

Radioactive Light Standards. A series of four radioactive light standards, supplied by Johnston Laboratories, Cockeysville, Maryland, were used to calibrate the photometric instruments. Each standard consists of ^{14}C-labeled glucose in a liquid scintillator in concentrations that increase by factors of 10. Beckman Readi-solve VI is the phosphor that emits at a 395 nm peak. The highest standard, referred to as number 5 in the figures, contains 10 μCi of the labeled compound. The scintillation mixture was purged with N_2 and glass-sealed with a slightly positive pressure in the same type of 6 × 50 mm glass cuvette used for the luciferase measurements. Since these light standards have a constant light output, a long half-life, and a known numerical relationship to each other, they can be used to measure sensitivity, reproducibility, and linearity of photometric instrumentation without the complication of variables present in the luciferase–ATP reaction mixture.

For all tests, the PMT gain was adjusted so that a given light standard registered the same percentage of the total range for each instrument. This adjustment allowed comparison of the three instruments regardless of numerical values, and allowed full use of the range of each one independently of the others. In order to show all curves on the same graph, arbitrary *y* values were chosen for each graph. As a result, the figures do not show absolute relationships between the instruments.

The relative light units registered by each of the three photometers using the series of radioactive light standards are plotted in Fig. 5. All

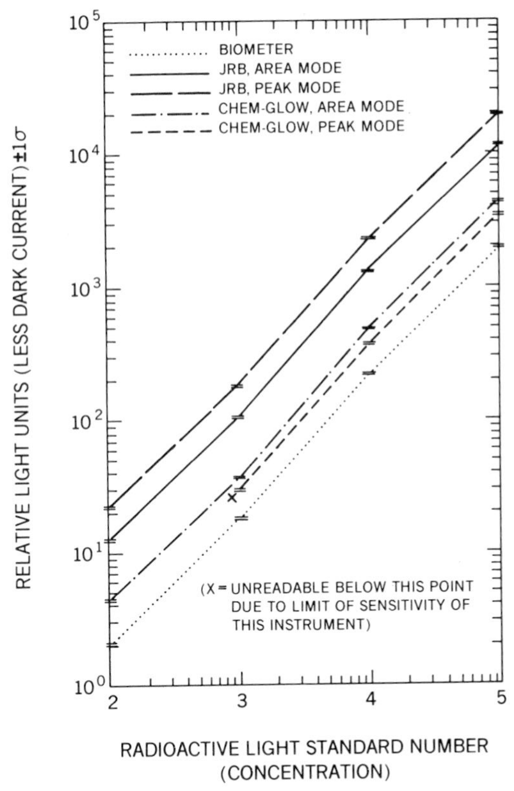

FIG. 5. Light standard concentration curves on three photometers.

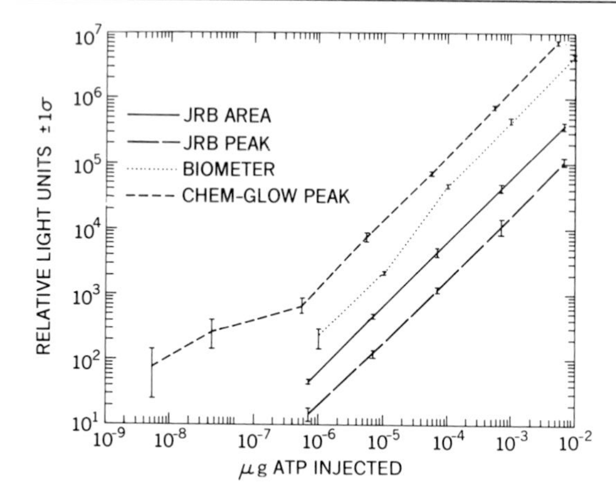

FIG. 6. ATP concentration curves (without nitric acid) on three photometers.

instruments gave linear results with small standard deviations at each point. Each demonstrated resolution of the lowest light standard at a value approaching dark current. The Chem-Glow must be used with the integrator–timer in order to resolve the lowest standard.

ATP Concentration Curves. Light emission in response to various concentrations of commercial ATP (Sigma) is compared on the three photometers in Fig. 6. The instruments are comparable and demonstrate good reproducibility and linearity with a working range from 10^{-6} to 10^{-2} μg of ATP injected. At lower concentrations, the curve flattens out and reproducibility is poor with all instruments. (Data shown are only for Chem-Glow used in the peak mode. The Chem-Glow operated in area mode, now shown, resulted in a parallel curve.)

Bacterial Concentration Curves. The three instruments were also used to measure levels of bacterial ATP (relative numbers of bacteria per milliliter), extracted with 0.1 *N* nitric acid, from dilutions of an overnight culture of *Escherichia coli* in trypticase soy broth. Direct microscopic counts versus light units are shown in Fig. 7. The results indicate comparable linearity among the instruments and good reproducibility, with larger variations occurring below a concentration of 10^4 bacteria per milliliter.

Summary

In selection of an instrument for use with liquid–liquid light reactions, several commercial photometers are available, varying in options for mode of operation, sample presentation, sample volume, readout, etc. Sensitiv-

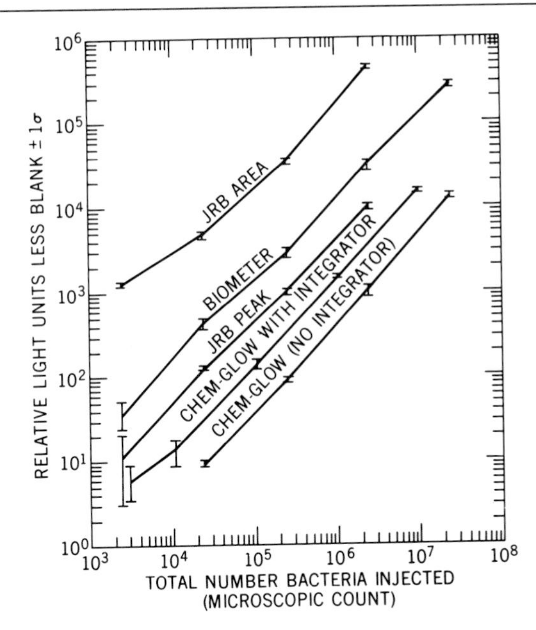

FIG. 7. Bacterial concentration curves (using nitric acid) on three photometers.

ity, reproducibility, and linearity are comparable for the three evaluated. Additionally, scintillation counters may be adapted for these types of light measurements. Another instrument, which incorporates a moving filter tape for sample concentration and processing, was evaluated separately. Further automated sample processing can be achieved with a flow system that uses peristaltic pumps, tubing, and a glass coil as the reaction "cuvette."

Commercially available crude luciferase preparations were found to be comparable in sensitivity, activity, and linearity, excluding one that included arsenate. Purified luciferase was found to be 1000 times more sensitive than the crude preparations; however, it is considerably more expensive.

[44] Excited States and Absolute Calibrations in Bioluminescence[1]

By H. H. Seliger

I. Introduction

Aside from the heating of the earth's crust by the decay of naturally occurring radioactive elements, volcanism, and cosmic rays, all of the energy for life on earth derives from a narrow spectral region of electromagnetic radiation from the sun, incident on the surface of the earth. Throughout the prebiological period of direct chemical synthesis of organic molecules by ultraviolet light and electrical discharges, to the present day where chlorophyll-sensitized photochemistry is the source of the free energy for all biological chemistry, an irradiance of 0.09 W cm^{-2} has driven life on earth.

Therefore, a major selection process for the structure of all molecules involved in the efficient capture of the free energy of photons, pigment molecules, pigment–protein complexes, and electron transport molecules has been directed to the production of excited states and the coupling of these electronically excited states to chemical reactions. The advantage of photochemistry is in its selectivity. Specific pigment molecules at ambient temperature can be raised to energy levels corresponding to 20,000–50,000°K, permitting selective reactions to occur that are not accessible otherwise.

Evolution is an expression of the Third Law of Thermodynamics, not only on the molecular level, but also in the most elaborate life forms. Were this not the case the biochemistry of the lowly bacterium might have little relationship to the dinosaur or to man. In anthropometric analogy, when Nature selected chemical reactions or sequences of chemical reactions that worked for the original single-celled, self-replicating life forms it was because they provided the most efficient thermodynamics. When environmental stress over and over again selected for those most efficient mutant combinations for structure and function, Nature always built on the original set of basic chemical steps.

Thus when it became time for functional bioluminescence, that is the

[1] Contribution No. 954 of the McCollum-Pratt Institute. This work has been supported under contracts from AEC, ERDA, and DOE EY-76-S-02-3277. The author expresses his deep appreciation to Mr. William H. Biggley for his invaluable assistance in carrying out the measurements described in these standardization techniques and for the preparation of the radioactive tertiary luminous standards.

ISBN 0-12-181957-4

emission of light for other organisms to "see," it is not surprising that a slight modification of the already-present monooxygenase biochemistry would be used. In retrospect it should not have been surprising that the quantum yield of firefly bioluminescence was close to 100%[2,3]; after all, the quantum yield of photosensitized one-electron reduction of ferridoxin or of abstraction of an electron from hydrated chlorophyll is also close to 100%.

So far as we know the only function for chemically produced excited states in biological systems is for bioluminescence, the emission of light for signaling. There are, however, a large number of apparently nonfunctional chemiluminescences that have been observed from biological systems, beginning with the mitogenetic radiation of Gurwitsch[4] and most recently with the observations of the microsomal chemiluminescence of metabolites of carcinogenic polycyclic aromatic hydrocarbons.[5,6] It is the purpose of this chapter to describe the methods for the absolute measurement of the intensities of these light emissions and for the measurement of their spectral distributions, for these measurements form the basis for the assay, for the stoichiometry, and for determination of the molecular mechanisms of chemiluminescence in biological systems.

II. Electronically Excited States of Molecules

When a system in an energy state E_2 decays to a lower state E_1 in a radiative transition, the frequency of the emitted photon is given by:

$$\nu = (E_2 - E_1)/h \tag{1}$$

where h is Planck's constant. The above equation also expresses the conditions for absorption of light. In the polyatomic molecule, vibrational and rotational interactions result in a large number of closely spaced energy levels giving rise to essentially continuous bands for absorption and for emission.

There are many ways besides the absorption of photons in which ground-state molecules may be raised to excited states. Examples are high temperature, electrical discharges, electroluminescence, collisions with corpuscular radiation, rupturing of crystalline bonds, as in crystalloluminescence and triboluminescence, frictional forces which produce high, local electrostatic fields, as in sonoluminescence and the "peeling" of

[2] H. H. Seliger and W. D. McElroy, *Biochem. Biophys. Res. Commun.* **1,** 21 (1959).

[3] H. H. Seliger and W. D. McElroy, *Arch. Biochem. Biophys.* **88,** 136 (1960).

[4] A. Gurwitsch, "Die Mitogenetische Strahlung." Springer Verlag, Berlin and New York, 1932.

[5] H. H. Seliger, *Photochem. Photobiol.* **21,** 355 (1975).

[6] J. P. Hamman and H. H. Seliger, *Biochem. Biophys. Res. Commun.* **70,** 675 (1976).

cellophane tape, and exergonic chemical reactions, including ion and free radical recombinations and reactions involving oxygen or peroxides. The delayed luminescence of preilluminated chloroplasts at times (10^{-5} to 10^{4} seconds) too long for the process to be a primary fluorescence or a slow fluorescence[7,8] is the result of a back reaction (recombination) of reactants within the thylacoid that have undergone charge separation due to the primary illumination.[9] In all of the functional bioluminescent reactions and in all of the nonfunctional or Adventitious Biological Chemiluminescence reactions,[10] oxygen is a specific requirement. Bioluminescence emission colors have been reported ranging from the blue of marine bacteria to the red of the railroad worm. This variety of colors is perfectly reasonable when one considers that in all of the cases studied the substrate molecules and the enzyme molecules are quite different. The color range observed is therefore no more unusual than the color range of fluorescences of conjugated organic molecules. In bioluminescent reactions the enzyme molecule, usually called a luciferase, in addition to catalyzing the chemical reaction of the substrate luciferin, can be demonstrated in certain cases to interact with the excited state to modify the energy-level distribution and therefore the color of the emitted light. In other cases a change in the color of the *in vivo* bioluminescence is the result of a "wavelength shifter," a lumiphore–protein complex that binds to the luciferin–luciferase complex and as the result of efficient energy transfer produces a sensitized chemiluminescence. Thus even here Nature shows her conservative nature. The spectral shift is to a region of the spectrum that has the greatest transmission in the environment in which the particular bioluminescent species has adapted. In the case of *Renilla*, the sea pen, the transmission of green light from the green fluorescent protein in the shallow eutrophic waters in which *Renilla* grows is much greater than that of the blue-emitting luciferin–luciferase reaction in the absence of this wavelength shifter. In the case of the luminous bacteria there appears to be an additional factor. Although the bioluminescent emissions of the bacteria in the absence and the presence of the blue fluorescent protein (BFP) are both in the blue region, characteristic of the maximum spectral transmission in ocean water, the fluorescence quantum yield of the BFP is higher, resulting in a significantly higher bioluminescent quantum yield in the presence of the native or *in vivo* BFP.

[7] B. Strehler and W. Arnold, *J. Gen. Physiol.* **34,** 809 (1951).

[8] W. Arnold and J. Azzi, *Photochem. Photobiol.* **14,** 233 (1971).

[9] G. P. B. Kraan, J. Amesz, B. R. Velthuys, and R. G. Steemers, *Biochim. Biophys. Acta* **223,** 129 (1970).

[10] H. H. Seliger and J. P. Hamman, *in* "10th Jerusalem Symposium on Excited States in Bio- and Organic Chemistry" (B. Pullman and N. Goldblum, eds.) p. 345. North-Holland Publ., Amsterdam, 1977.

Thus we can have two subclasses of chemiluminescent reactions, direct and sensitized chemiluminescence. In direct chemiluminescence a chemical product of the monooxygenation can be identified with the same fluorescence characteristics as the observed chemiluminescence or bioluminescence. Such a molecule has been found for the chemiluminescence of luminol[11-13] and lucigenin.[14] In sensitized chemiluminescence the emitter molecule is not a participant in the chemical reaction leading to the product excited state. Examples of this class of sensitized chemiluminescent reactions are the oxidations of oxalyl chloride in the presence of H_2O_2 and catalysts plus acceptor molecules such as anthracene, *N*-methylacridine, and 9,10-diphenylanthracene.[15-19] This mechanism of chemiluminescence was first suggested by Kautsky and Zocher[20] and Kautsky and Thiele[21] for the oxidation of calcium silicide by HCl.

The environment of the excited chromophore can affect the energy of the transition. In the case of the chemiluminescence of luminol, the color changes from violet in aqueous systems to blue in dimethyl sulfoxide (DMSO). In both cases the product is 3-aminophthalic acid,[22] and the chemiluminescence emission spectrum corresponds with the fluoresence emission spectrum of the product in the particular solvent system.[13] Similarly, the color of the *in vitro* bioluminescence emission of *Photinus pyralis* changes from the normal yellow-green to red as the pH of the reaction solution is lowered.[2,23] In addition there is a spread in the colors of the light emitted by the different firefly species. *In vitro* experiments with firefly reactants have demonstrated that the crucial variable is the luciferase

[11] H. H. Seliger, *in* "Light and Life" (W. D. McElroy and B. Glass, eds.), p. 200. Johns Hopkins Press, Baltimore, Maryland, 1961.

[12] E. H. White, *in* "Light and Life" (W. D. McElroy and B. Glass, eds.), p. 183. Johns Hopkins Press, Baltimore, Maryland, 1961.

[13] E. H. White and M. M. Bursey, *J. Am. Chem. Soc.* **86,** 941 (1964).

[14] J. R. Totter, *in* "Bioluminescence in Progress" (F. H. Johnson and Y. Haneda, eds.), p. 22. Princeton Univ. Press, Princeton, New Jersey, 1966.

[15] M. M. Rauhut and R. C. Hirt, "Chemiluminescent Materials" Tech. Rept. No. 1, August. American Cyanamid Co., Stamford, Connecticut, 1963.

[16] M. M. Rauhut and R. C. Hirt, "Chemiluminescent Materials" Tech. Rept. No. 2, November. American Cyanamid Co., Stamford, Connecticut, 1963.

[17] M. M. Rauhut and R. C. Hirt, "Chemiluminescent Materials" Tech. Rept. No. 3, February. American Cyanamid Co., Stamford, Connecticut, 1964.

[18] M. M. Rauhut and R. C. Hirt, "Chemiluminescent Materials" Tech. Rept. No. 4, May. American Cyanamid Co., Stamford, Connecticut, 1964.

[19] E. A. Chandross, *Tetrahedron Lett.* **12,** 761 (1963).

[20] H. Kautsky and H. Zocher, *Z. phys.* **9,** 267 (1922).

[21] H. Kautsky and H. Thiele, *Z. Anorg. Allgem. Chem.* **144,** 197 (1925).

[22] E. H. White, O. Zafiriou, H. H. Kagi, and J. H. M. Hill, *J. Am. Chem. Soc.* **86,** 940 (1964).

[23] H. H. Seliger and W. D. McElroy, *Radiat. Res. Suppl.* **2,** 528 (1960).

molecule.[24,25] Thus there can be an important difference between bioluminescence and chemiluminescence. The enzyme molecule, which functions as a catalyst, may also be a participant in the light-emission step and an excited enzyme–chromophore complex can be the light emitter. In bioluminescence we shall therefore be concerned with the electronic properties of enzyme–chromophore complexes. There is a diversity of luciferins and luciferases in the various phyla. On the other hand, there is a well established similarity among biochemical pathways in different organisms. It is therefore tempting to search for such similarities among bioluminescent reactions. One such similarity might be found in the chemical nature of the excitation step.[5]

The spectral distribution of emission in the transition from the first excited state to the ground state (or absorption in the reverse) depends on the individual transition dipole moments between the various levels of the upper and lower states. These transition dipole moments may vary from vibrational level to vibrational level because of the Franck–Condon principle,[26] which states that during an electronic transition the nuclear configurations do not change appreciably. It is for this reason that transitions from the lowest vibrational level of one state to higher vibrational levels of the final state are most probable, these being the ones that more nearly match the initial nuclear configuration. Thus the center of gravity of the fluorescence ($S_1 \rightarrow S_0$) is at a lower energy than the center of gravity of the absorption ($S_0 \rightarrow S_1$). The 0–0 vibrational level transition, which ideally should be the same in both absorption and fluorescence, can also be shifted in energy.

Excited states may be populated by light absorption and by chemical reaction (including collisions if enough energy is available). The rapid relaxation of molecules to thermal equilibrium means that whatever the initial distribution of excited states that are chemically formed, emission will be observed from only the first excited singlet or triplet states. In fact at room temperature most emission is from the lowest vibrational level of the first excited state. Thus, if a reaction were to occur which formed a product molecule in a triplet state, we would expect rapid relaxation to the lowest levels (assuming intersystem crossing does not occur). If the reaction were in fluid solution, it would probably be nonluminescent because of quenching processes. If the reaction occurred in the solid state, however, it would be another matter entirely, for diffusional quenching is inhibited and phosphorescence or α-delayed fluorescence is often observed. Trip-

[24] H. H. Seliger, J. B. Buck, W. G. Fastie, and W. D. McElroy, *J. Gen. Physiol.* **48,** 95 (1964).
[25] H. H. Seliger and W. D. McElroy, *Proc. Natl. Acad. Sci.* U.S.A. **52,** 75 (1964).
[26] H. H. Jaffé and M. Orchin, "Theory and Application of Ultraviolet Spectroscopy," p. 134. Wiley, New York, 1962.

let chemiluminescence in the solid state has been observed. The chemiluminescence observed under ordinary conditions in fluid media (and probably also bioluminescence occurring on a protein molecule) is, on the other hand, expected to come from singlet transitions.

This particular requirement has presented some theoretical problems in the understanding of the high bioluminescent quantum yields of the firefly reaction. It is now generally agreed, and the experimental evidence demonstrates, that a dioxetane intermediate is formed by the addition of oxygen to the luciferin substrate. This dioxetanone has been proposed to dissociate to produce CO_2 and the product excited state of luciferin, the decarboxy-keto-luciferin. However, from symmetry rules this excited state was predicted to be a triplet excited state, which should be nonfluorescent, in obvious contradiction to the experimentally observed high bioluminescent quantum yield and therefore the inferred high fluorescence yield of the product excited state. This apparent dilemma has been resolved very neatly by Koo and Schuster[27] and by Koo *et al.*,[28] who showed that the dioxetane reaction is a Chemically Initiated Electron Exchange Luminescence (CIEEL). In the case of the firefly luciferin dioxetanone, an intramolecular CIEEL from the easily oxidized polycyclic heterocycle portion of the luciferin to the high energy dioxetane moiety, followed by rapid decarboxylation, will generate a charge-transfer resonance structure of the product, which will be a *singlet* state and therefore no longer in contradiction with the high quantum yields observed. The reaction is proposed to be

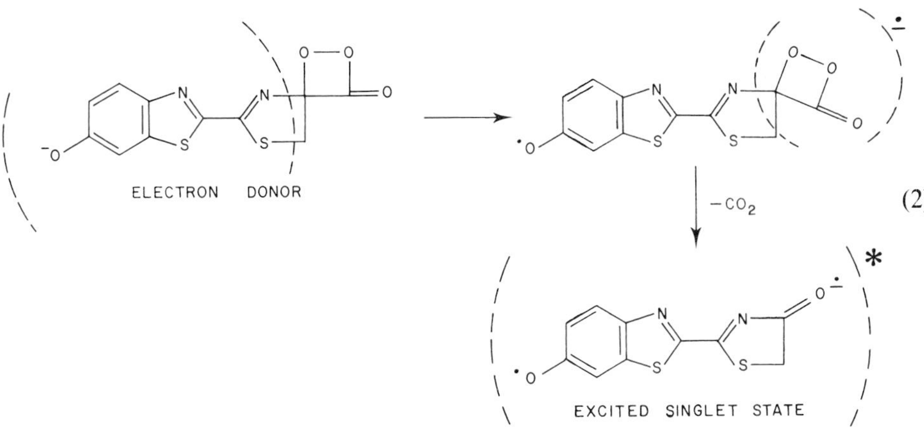

ELECTRON DONOR

$-CO_2$

EXCITED SINGLET STATE

(2)

Another result of the rapid relaxation to the lowest vibrational levels of S_1 is that the emission spectrum of chemiluminescence is predicted to be

[27] J.-Y. Koo and G. B. Schuster, *J. Am. Chem. Soc.* **99,** 6107 (1977).

[28] J.-Y. Koo, S. P. Schmidt, and G. B. Schuster, *Proc. Natl. Acad. Sci. U.S.A.* **75,** 30 (1978).

exactly the same as the fluorescence emission spectrum of the product molecule which has been formed by chemical reaction.

III. Perturbations of Electronically Excited States

Perturbations involve a relatively small change in the state of a system caused by change in some external parameters. Both the transition dipole moment leading to a change in the fluorescence quantum yield, and the energy of the transition, leading to a change in the spectral distribution, may occur.

Into this class we place weak interactions between molecules when one of the pairs has no absorption band in the region under consideration. An important example is solute–solvent interactions, and these will illustrate the principles involved. The intermolecular forces are mainly electric dipole forces caused by the following interactions: (1) two permanent dipole moments (not to be confused with the transition dipole moment): polar solvent and polar solute; (2) permanent dipole moment with induced dipole moment: polar solvent–nonpolar solute or vice versa; for these the polarizability of the molecule in which the dipole moment is being induced is a factor; (3) London or dispersion forces arising from temporary dipole moments in a molecule having, on the longer time average, no net dipole moment. These forces are always present but become the most important for nonpolar solvents and solutes.

If the perturbation is considered relative to an isolated molecule (i.e., the molecules in a dilute gas), then the energies of the electronic states are lowered by the solvation energy. For transitions induced by absorption of radiation, the ground state of a molecule is lowered (relative to a gas) by the solvation energy. However, since the excitation takes place within about 10^{-15} sec, the solvent molecules do not have time to reorient and therefore the solvation energy required for the excited states is not that for the equilibrium state but for a transient, "Franck–Condon" (FC) state. This FC state may be under considerable strain (i.e., higher energy) relative to the equilibrium state, and its solvation energy may even be negative. Equilibration with respect to the solvent structure about the excited state is expected within the time required for molecular vibrations or reorientations (10^{-13} to 10^{-11} second). A similar situation will hold for the transition from excited to ground states, accounting for fluorescence emission. This phenomenon can account for the shift between the 0–0 vibrational level transition of absorption and fluorescence.

Perturbation effects are strongest for the strongest forces, i.e., permanent dipole interactions. It is for this reason that the vibrational structure of a transition is often blurred for polar solutes, especially in a polar solvent.

Each vibrational level is perturbed by a solvent interaction which varies from solute molecule to solute molecule, the net result being a complete loss of vibrational level identity. For the same reasons it will be readily imagined that the breadth of an emission band can be affected by solvent interaction. Since the solute–solvent interaction energy depends on the solute dipole moment and since the dipole moment may be different in ground and excited states, the maximum effect of solvent perturbation is expected when a solute in a polar medium gains or loses a large dipole moment when excited. A representative value for a transition energy change between a nonpolar and a polar solvent may be 1000 cm^{-1} or 2.86 kcal/mol. This corresponds to a shift of about 24 nm at 500 nm. For a nonpolar solute (e.g., benzene) the energy shift would be much less, but it could be greater for a larger change in dipole moment between ground and excited states. It is inferred that the spectral shifts in the bioluminescence emission spectra of different species of fireflies[24,25,29,30] are the result of enzyme–substrate "solvent"-type interactions described above, similar to the case for the spectral changes in luminol chemiluminescence in DMSO and in aqueous solutions.

Once the possibility of protein perturbation of energy levels is accepted for the product chromophore in bioluminescent reactions, an effect similar to that producing the Franck–Condon state must be considered. The FC state represents a strained configuration produced by a nonequilibrium orientation of solvent molecules about the excited state. Because relaxation occurs so rapidly (10^{-13} to 10^{-11} second), fluorescence emission occurs from the equilibrium excited state. However, the relaxation time of a chromophore bound to an enzyme will be much slower. That is, the excited state of the product chromophore may be formed with the protein conformation appropriate to the substrate. Relaxation of the protein to the conformation appropriate to the product may not occur within the excited state lifetime (10^{-8} to 10^{-9} second for a $\pi^* \rightarrow \pi$ transition). Emission then occurs from a transient complex of enzyme and product not obtainable under equilibrium conditions. If the protein–chromophore interaction is strong enough to cause a spectral shift, the fluorescence emission from the protein–product complex may be different from the chemiluminescence emission spectrum of the same chemically produced complex.

When two molecules with similar energy levels interact, the possibility of energy transfer must be considered. When the two molecules are far apart,

[29] W. D. McElroy, H. H. Seliger, and M. DeLuca, *in* "Evolving Genes and Proteins" (V. Bryson and H. J. Vogel, eds.), p. 319. Academic Press, New York, 1965.

[30] W. D. McElroy and H. H. Seliger, *in* "Molecular Architecture in Cell Physiology" (T. Hayashi and A. Szent-Györgyi, eds.), p. 63. Prentice-Hall, Englewood Cliffs, New Jersey, 1966.

the interaction is weak and there is little effect on the transition energies. If they are close together and one molecule is in an excited state, it can transfer its energy to the other if the overlap between the emission spectrum of the one and the absorption spectrum of the other is favorable. Förster[31] has discussed this type of energy transfer. In the case of a stronger interaction, as for example when two monomers complex to form a dimer, the energy levels will be perturbed. For a dimer, one model based on excitation energy transfer[32] predicts a splitting of the original transition into two, the separation between them depending on the interaction energy and the oscillator strength. The total oscillator strength of the dimer transitions is the same as for the isolated molecules, but one or the other of the dimer transitions may be forbidden.

We have recently demonstrated chemiluminescence from excimers in our own laboratory. We have found that the microsomal chemiluminescence of benzo[*a*]pyrene-7,8-diol, the presumed proximate carcinogenic metabolite of benzo[*a*]pyrene that is epoxidized to the carcinogenic 9,10-epoxybenzo[*a*]pyrene-7,8-diol, has an unstructured emission spectrum that agrees with the fluorescence of the 7,8-diol excimer. This has provided a major step in the elucidation of the molecular mechanism for this reaction.

The effect of chemical substitution on the spectral properties of a molecule can be treated as a perturbation. An example of a chemical perturbation might be the difference between benzene and chloro- or nitrobenzene. This method works best when the groups added are not themselves strong chromophores and do not destroy the essential properties of the initial chromophore (e.g., the π-electron system). This approach seems useful when considering chemiluminescence of aromatic molecules if the product is unknown. The substrate molecule provides the basic chromophore and the reaction leading to an excited product produces a perturbation of its energy levels and transition probabilities. Because the nonradiative transition probabilities are more sensitive to chemical substitution than transition energies and absorption oscillator strengths, the quantum yields in a series of chemical analogs do not usually show any predictable relationship. Therefore, we would not expect to be able to predict the fluorescence efficiency of the product molecule. Examples of the effects of these substitutions on the luciferin molecule of the firefly *Photinus pyralis* have been given by White *et al.*[33,34]

[31] T. Förster, "Fluoreszenz Organischer Verbindungen." Vandenhoecht Ruprecht, Göttingen, 1951.

[32] M. Kasha, *Rev. Mod. Phys.* **31,** 162 (1959).

[33] E. H. White, H. Wörther, H. H. Seliger, and W. D. McElroy, *J. Am. Chem. Soc.* **88,** 2015 (1966).

[34] E. H. White, E. Rapaport, H. H. Seliger, and T. A. Hopkins, *Bioorg. Chem.* **1,** 92 (197[illegible]).

IV. Fluorescence and Chemiluminescence

The rate of deexcitation of the singlet excited state, S_1, to the ground state, S_0, is given by

$$\frac{d(S_1)}{dt} = -(k_f + k_{fq} + k_{isc})(S_1) + k_\alpha (T_1) + k_0 (S_0) \tag{3}$$

where k_f is the rate constant for fluorescence emission, k_{fq} is the rate constant for internal quenching or nonradiative transition, k_{isc} is the rate constant for intersystem crossing to the triplet excited state, k_α is the temperature-dependent rate constant for crossing back from the triplet state to the S_1 level, and k_0 is the rate of excitation to the singlet state. In chemiluminescent reactions the term $k_0\,(S_0)$ is replaced by the rate of the chemical reaction.

If we consider room-temperature reactions in solution, we can neglect k_α. In steady-state fluorescence the quantum intensity of fluorescence, I_f^0, in the absence of external quenchers is given by

$$I_f^0 = k_f (S_1) = \frac{k_f}{k_f + k_{fq} + k_{isc}} k_0 (S_0) \tag{4}$$

so that the quantum yield for fluorescence, ϕ_F, is

$$\phi_F = \frac{k_f}{k_f + k_{fq} + k_{isc}} \tag{5}$$

and the observed mean lifetime for fluorescence, τ_F, is

$$\tau_F = \frac{1}{k_f + k_{fq} + k_{isc}} = \phi_F\, k_f^{-1} \tag{6}$$

In the presence of an external quencher concentration, Q, the intensity of fluorescence is reduced to

$$I_F = \frac{k_f}{k_f + k_{fq} + k_{isc} + k_Q (Q)} \cdot k_0 (S_0) \tag{7}$$

Therefore from Eqs. (4) and (7) we obtain

$$\frac{I_F}{I_F^0} = \frac{1}{1 + \tau_F\, k_Q (Q)} \tag{8}$$

the Stern–Volmer equation. This relation holds when quenching occurs by collision of the quencher molecule with the excited state. If the quencher forms a nonfluorescent complex with the ground state molecule, the Stern–Volmer equation is not obeyed.

If $[(I_F^0/I_F) - 1]$ is plotted against Q, then both $k_Q\,\tau_F$ (called the quenching constant) and the association constant of the complex can be determined. It is the former in which we are interested because of its possible application to chemiluminescence. If the lifetime of the excited state can be experimentally determined or estimated, the rate constant for quenching can be determined, and used to gain information about the quenching process. Alternatively, if k_Q can be estimated, the lifetime of the excited state can be calculated, and it was this purpose for which Eq. (8) was used by Stern and Volmer. In practice, it would appear to be impossible to calculate k_Q except in the limit that the rate of the quenching reaction is limited by the rate of diffusional encounters between Q and S_1. Such a limit gives the largest possible value of k_Q, and thus the minimum lifetime. For aqueous solutions at room temperature (300°K, and viscosity 0.01 poise), the rate of quenching is given by Debye[35] as:

$$k_Q = \frac{1}{4}\left(2 + R_1/R_2 + R_2/R_1\right)(0.7 \times 10^{10})\ M^{-1}\ \text{sec}^{-1} \quad (9)$$

where R_1 and R_2 are the radii of chromophore and quencher. As can easily be seen from Eq. (9), the rate of collisional quenching is not very sensitive to molecular radii, and thus for molecules of reasonably similar size k_Q can be taken as about 10^{10} sec^{-1} M^{-1}. This leads to some qualitative conclusions about quenching. For fluorescence lifetimes of the order of 10^{-8} to 10^{-9} sec, we do not expect an effect on fluorescence yield unless $(Q) = 10^{-2}$ to 10^{-1} M. These values of Q reduce the fluorescence by one half. However, for long lifetimes involving a triplet state $\tau = 10^{-2}$ to 10^{-4} sec and $(Q)_{1/2} = 10^{-8}$ to 10^{-6} M for collisional quenchers. Thus, even trace amounts of an efficient collisional quencher can completely quench phosphorescence.

If we assume that the chemiluminescent reaction is

$$A + X \xrightarrow{k_e} S_1 \rightarrow S_0 + h\nu \quad (10)$$

the intensity of chemiluminescence is given by

$$I_{ch} = k_f(S_1) = \frac{k_f}{k_f + k_{fq} + k_{isc}} k_e(A)(X) \quad (11a)$$

which simplifies, from Eq. (5) to

$$I_{ch} = \phi_F k_e(A)(X) \quad (11b)$$

The quantum yield for chemiluminescence is defined in an analogous manner to that for fluorescence, except that it relates to the rate of substrate molecules reacting. Since it is also possible that the substrate molecule A may undergo other chemical reactions by competing pathways that do not

[35] P. Debye, *Trans. Electrochem. Soc.* **82**, 265 (1942).

lead to the excited state product molecule S_1, the quantum yield of chemiluminescent reactions can be defined as

$$\phi_{ch} = \phi_F \, \phi_{ex} \tag{12}$$

where ϕ_{ex} is the fraction of the substrate molecules going through the chemiluminescent pathway.[36] The equation corresponding to Eq. (8) is

$$\frac{\phi_{ch}}{\phi_{ch}^0} = \frac{\phi_{ex}}{\phi_{ex}^0} \frac{1}{1 + \tau_F k_Q (Q)} \tag{13}$$

V. Units of Light Intensities

In all of photobiology the absorption or emission of light is a photon process. Unfortunately, the historical development of the theory of radiation, based on the physics of the energy density properties of electromagnetic radiation, has perpetuated the use of energy units for the specification of light intensities in biology. These are not *wrong*; they are imprecise and often confusing. For example, the absorption spectrum of chlorophyll in the visible region exhibits a peak in the red region of the spectrum, ca. 680 nm, corresponding to its first excited state, and a second peak in the blue region, ca. 440 nm, corresponding to its second excited state. Blue or red photons absorbed and producing either of these excited states are equally effective in producing the first excited state through which energy transfer to the reaction-center chlorophyll occurs. This same argument applies to other blue- or green- or yellow-light absorbing accessory pigments—all of which transfer electronic excitation energy with essentially 100% efficiency to reaction-center chlorophyll either in system I or system II. Therefore the efficiency of a photochemical reaction is *not* an energy-stored to energy-absorbed ratio, but products formed or molecules reduced or oxidized or isomerized divided by *photons* absorbed. The conceptual distinction was explained many years ago in Einstein's treatment of the Photoelectric Effect. It was shown that the energy of the individual photon rather than the energy density of the photon beam was involved in overcoming the work function or binding energy of electrons in various atoms in order for photoelectrons to be released.

A second impreciseness relates to the historical development of spectroscopy and the interference properties of photon beams. Whether by means of prisms or gratings the experimental means of separating the spectral components of a light source have made use of these wave inter-

[36] H. H. Seliger, *in* "Chemiluminescence and Bioluminescence" (M. J. Cormier, D. M. Hercules, and J. Lee, eds.), p. 461. Plenum, New York, 1973.

ference properties, and the separation is made therefore as a function of wavelength. However, the proper physical description of the states of an atom or a molecule are in terms of its allowed *energy* levels. Therefore the *shape* of an absorption spectrum and the *shape* of a fluorescence emission spectrum, and properties related to the overlap of absorption and emission spectra or to the overlap of spectra in general are dependent upon the energy levels, not the wavelengths. Again, representation of spectra as functions of wavelength is not *wrong,* but as in the intensity argument it is imprecise and can be misleading.

It is no wonder that we see all combinations of these units and methods of plotting data in the photobiological and photochemical literature. Let us begin by stating that the most precise and universally applicable description of spectral distributions of light intensities is

$I_{\bar{\nu}}d\bar{\nu} \equiv$ number of photons passing unit area of 1 cm^2 sec^{-1} having energies corresponding to the wavenumber interval between $\bar{\nu}$ and $\bar{\nu} + d\bar{\nu}$. (14)

The relationships between energy, wavelength, and wavenumber of a photon are given by

$$E = h\nu = hc/\lambda = hc\bar{\nu} \tag{15a}$$

E = energy in ergs or Joules; 1 Joule = 10^7 ergs (15b)
h = Planck's constant; 6.625×10^{-27} erg (15c)
ν = frequency of the electromagnetic radiation in Hz (15d)
c = speed of light $\simeq 3 \times 10^{10}$ cm sec^{-1} (15e)
λ = wavelength in cm; 1 cm = 10^7 nm (nanometer) (15f)
$\bar{\nu}$ = wavenumber cm^{-1} with the unit Kayser (K) (15g)

A photon with a wavelength of 400 nm has $\bar{\nu}$ = 25,000 K or 25 kK.

The relationship between the irradiance, the power input per unit area, to the photon irradiance, the monoenergetic photon flux per unit area, is

$$\underset{\text{at wavelength } \lambda_{\text{nm}}}{1\ \mu\text{W cm}^{-2}} = \frac{\lambda\ [\text{nm}]}{198.7} \times 10^{+12} \underset{\text{at wavelength } \lambda_{\text{nm}}}{\text{photons cm}^{-2}\ \text{sec}^{-1}} \tag{16}$$

For the chemist interested in the identification and assay of pure compounds by their absorption spectra or in elemental analysis by identification of sharp atomic line spectra, the above discussion presents no problems. In the first case absorption spectra are determined by the logarithm of the reciprocal transmission per unit pathlength through the solid or solution to be identified or assayed. Thus,

$$\text{Optical density per centimeter at wavelength } \lambda \equiv (\text{OD})_\lambda = \log\ (I_o/I)_\lambda \tag{17}$$

is independent of intensity units. Therefore if $(OD)_\lambda$ is plotted as a function of λ, with a peak absorption at a wavelength λ_{max}, the wavenumber for peak absorption, $\bar{\nu}_{max} = 1/\lambda_{max}$. The shape of the absorption curve plotted as a linear function of λ will differ from the shape of the absorption curve plotted as a linear function of $\bar{\nu}$, but the peak positions for maximum absorption will coincide. Thus, for identification purposes, where consistency is important and compounds are identified by the absolute values of their extinction coefficients E_λ at their peak absorption wavelengths, it makes little difference whether the extinctions are reported as E_λ or $E_{\bar{\nu}}$. However, for the analysis of fluorescence or bioluminescence emission spectra consisting of broad continuous structureless bands, a significant discrepancy arises.

In most cases a plane-grating spectrometer is used, the slit widths are maintained constant throughout the measurement, and the corrected data are in units proportional to photons per second per unit wavelength interval ($dN/d\lambda$). For the purposes of identification and for internal consistency this method has been quite satisfactory, and there exists in the literature a very large number of fluorescence, chemiluminescence, and bioluminescence emission spectra, all plotted as functions of wavelength. However, for the examination of the detailed shape of the emission bands, for the calculation of lifetimes and transition probabilities, and for the accurate determination of energy differences between various levels, it is physically more significant to report the observed photon intensity distribution as photons per second per unit wavenumber interval ($dN/d\bar{\nu}$) against wavenumber $\bar{\nu}$, since $\bar{\nu}$ is directly proportional to the energy of the photon. In this case the energy width $\Delta\bar{\nu}$ is not constant with the wavelength setting and we can write

$$|\Delta\bar{\nu}| = \bar{\nu}_a - \bar{\nu}_b = \frac{\Delta\lambda}{\lambda_a\lambda_b} \quad (18)$$

If λ_0 is the spectrometer wavelength setting,and λ_a and λ_b the wavelengths at the ends of the bandwidth $\Delta\lambda$:

$$\lambda_b = \lambda_0 + \tfrac{1}{2}\,\Delta\lambda, \quad \lambda_a = \lambda_0 - \tfrac{1}{2}\,\Delta\lambda \quad (19)$$

Neglecting terms in $(\Delta\lambda)^2$,

$$\Delta\bar{\nu} = \Delta\lambda/\lambda_0^2 \quad (20)$$

and therefore:

$$\left.\frac{dN}{d\bar{\nu}}\right|_{\bar{\nu}_0} = \lambda_0^2 \left.\frac{dN}{d\lambda}\right|_{\lambda_0} \quad (21)$$

Equation (21) indicates how the photon distribution function $dN/d\bar{\nu}$ is calculated from the experimentally obtained function $dN/d\lambda$.

When spectral data are plotted as $dN/d\bar{\nu}$ against $\bar{\nu}$, the shapes of the curves are very closely Gaussian. In that case we can write:

$$dN/d\bar{\nu} = \text{const. } e^{-\left(\frac{\bar{\nu}_m - \bar{\nu}}{2\sigma}\right)^2} \tag{22}$$

where $\bar{\nu}_m$ is the peak position, and σ is the standard deviation.

From Eq. (22) Seliger and McElroy[37] obtained an expression relating $\bar{\nu}_m$ to λ_m, the wavelength of the peak photon emission of a $dN/d\lambda$ distribution as a function of λ. This expression is given as:

$$\lambda_m = \frac{1}{\bar{\nu}_m}\left\{1 - 0.36\left(\frac{\text{FWHM}}{\bar{\nu}_m}\right)^2\right\} \tag{23}$$

where FWHM (the full width at half maximum) is equal to 2.352σ.

The difference between a $dN/d\bar{\nu}$ and a $dN/d\lambda$ plot is illustrated in Fig. 1 for the fluorescence emission spectrum of products of the firefly reaction.

Not only do the peak wavelengths obtained by the two methods differ (by as much as 7 nm), but the relative intensities of the two fluorescence bands are actually reversed. For the $dN/d\bar{\nu}$ plot the ratio of peak heights is more nearly equal to the ratio of areas under each band.

Therefore, if fluorescence and bioluminescence emission spectra are to be used for identification of emitting molecules, the simple relationship valid for absorption spectra no longer holds and λ_m is no longer simply related to ν_m, as shown by Eq. (23). The distinction disappears when sharp-line atomic emission spectra are to be analyzed. From Eq. (23), as FWHM $\rightarrow 0$, $\lambda_m \rightarrow 1/\nu_m$.

The distinction becomes important for the broad emission bands of highly conjugated molecules and when solvent and binding interactions are to be studied. This encompasses all bioluminescent emission spectra.

A third, more confusing, *incorrect* and unnecessary complication in the specification of light intensities in photobiology has been introduced by the illuminating engineer. In this case the spectral intensity distribution is multiplied by an arbitrary human visual spectral response function and integrated over the wavelength limits of the human response function, with the result expressed in lumens or luminous flux. This has given rise to candelas, lamberts, stilbs, lux, phots, and footcandles. These anthropometric units are of use only to light bulb manufacturers and have no place in quantitative biology. Only for blackbody light sources can these units of physiological photometry be related to the true photon spectral intensity distributions (see Seliger and McElroy[38]).

[37] H. H. Seliger and W. D. McElroy, *in* "Bioluminescence in Progress" (F. H. Johnson and Y. Haneda eds.), p. 405. Princeton Univ. Press, New Jersey, 1966.

[38] H. H. Seliger and W. D. McElroy, "Light: Physical and Biological Action." Academic Press, New York, 1965.

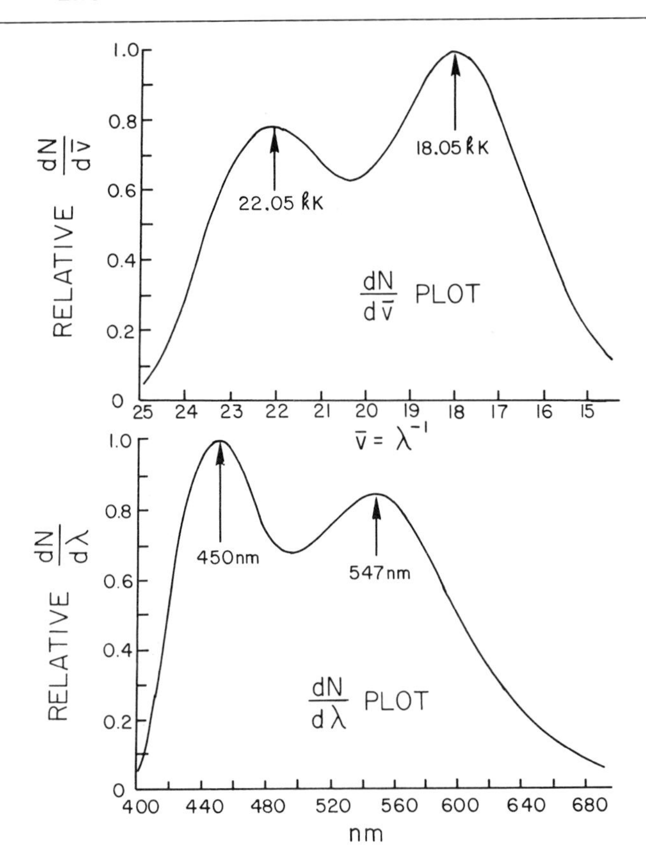

FIG. 1. Difference between a $dN/d\bar{v}$ and a $dN/d\lambda$ plot for the fluorescence emission spectrum of products of the firefly reaction.

VI. Measurement of Light

The earliest method for the measurement of light was by visual comparison of the illumination of a diffusing surface by an unknown source with the illumination by a known standard source. The eye was the detector and standard sources were actually very carefully constructed wax or spermaceti candles. The presences of ultraviolet (UV) light and infrared (IR) light were first detected by their blackening of a photographic plate and heating effects, respectively. The names given to these regions refer to the fact that they were outside the spectral sensitivity range of the normal human eye.

At the present time there are two major types of detectors for the quantitative measurement of light. The first and most universally applicable detector is the blackened thermopile, which has uniform absorption properties from the far UV to the far IR and measures the absorbed *energy* of incident photons by their heating effect on junctions of dissimilar metals or semiconductors, the Seebeck Effect. A more complete description can be found in Smith *et al.*[39] and Seliger and McElroy.[38] Since thermopiles are wavelength independent, they must be used in conjunction with a spectrometer in order to specify the photon irradiance. Their limitation, however, is in their intrinsic low sensitivity and the fact that they must be of small area in order to achieve optimum sensitivity. The best evacuated thermopiles that can be constructed have a Noise Equivalent Power (NEP) such that a light beam with an irradiance of 5×10^{-9} W cm^{-2} will produce a signal just equal to the fluctuation noise. From Eq. (16) this corresponds to 10^{10} photons sec^{-1} cm^{-2} at a wavelength of 400 nm and 18×10^{10} photons sec^{-1} cm^{-2} at 700 nm. The best commercially available thermopiles have a practical limit of around 10^{-8} W cm^{-2}, corresponding to 2×10^{10} photons sec^{-1} cm^{-2} at $\lambda = 400$ nm.

The thermopile detector is calibrated by measuring its response in a known radiation field. The output is then given in terms of incident energy flux, in units of μV per μW/cm^2. The radiation field from an ideal blackbody is calculable from Planck's Radiation Law, and it is found in practice that the radiation distribution from a properly constructed blackbody agrees very well with this law. Therefore, the thermopile can be calibrated by placing it at a fixed distance from a standard blackbody cavity or from a previously calibrated standard lamp operating at a specified voltage and current. With assumed equal absorptivity at all wavelengths, this calibrated thermopile can be used to determine any light energy flux regardless of the spectral distribution. Unfortunately, this rather simple procedure is limited to fairly strong light sources, fluxes greater than 10^{-1} μW/cm^2. In addition, the most sensitive thermopiles do not have a completely uniform response across their sensitive areas, requiring their use in uniform, large radiation fields to obtain an average response.

The thermopile can be used to characterize spectral photon irradiances of bright sources used for fluorescence excitation and for calibrating the spectral response curves for monochromators and spectrometers. It cannot be used for routine measurement of fluorescence or bioluminescence. It remains, however, as the basis for all derived calibrations using the more sensitive photon detectors, the photocell and the electron multiplier phototube.

The electron multiplier phototube is the most sensitive photon detector from the UV through the near IR. For comparison with the thermopile, there are now available specially selected phototubes with blue region

[39] R. A. Smith, F. E. Jones, and R. P. Chasmar, "The Detection and Measurement of Infrared Radiation." Oxford Univ. Press, London and New York, 1957.

response, 10 mm diameter photocathodes, that, at room temperature, have a photoelectron noise counting rate of between 30 and 50 sec^{-1}. The photocathode efficiency at $\lambda = 400$ nm is close to 20%. Thus, the noise equivalent photon irradiance at 400 nm corresponds to

$$\text{NE photon irradiance at 400 nm} = 2000 \text{ sec}^{-1} \text{ cm}^{-2} \qquad (24)$$

This is a factor of 10^7 greater than for the thermopile and therefore extends the range of light detection to levels where chemiluminescent reactions with extremely low quantum yields can be detected.

A. Relative Spectral Sensitivity of Spectrometer–Phototube Detector

The most basic characteristics to be associated with fluorescence and bioluminescence are the precise shapes of the emission spectra.

In order to obtain the experimental data in the desired form, certain corrections must be applied to take into account the overall spectral response of the instruments. The source of this spectral response is the variation in transmission of the spectrometer and the selectivity of the electron multiplier phototube detector with wavelength. There are many papers available in the literature which discuss these spectral corrections and describe other sources of error.[23,40-61]

[40] R. J. Argauer and C. E. White, *Anal. Chem.* **36,** 368 (1964).
[41] J. B. Birks, "Photophysics of Aromatic Molecules." Wiley, New York, 1970.
[42] E. J. Bowen, *Proc. R. Soc. London Ser. A* **154,** 349 (1936).
[43] R. L. Christensen and I. Ames, *J. Opt. Soc. Am.* **51,** 224 (1961).
[44] J. N. Demas and G. A. Crosby, *J. Phys. Chem.* **75,** 991 (1971).
[45] G. G. Guilbault, "Practical Fluorescence, Theory, Methods, and Techniques." Dekker, New York, 1973.
[46] D. M. Hercules, *Science* **125,** 1242 (1957).
[47] F. H. Johnson, O. Shimomura, Y. Saiga, L. C. Gershman, G. J. Reynolds, and J. R. Waters, *J. Cell. Comp. Physiol.* **60,** 85 (1962).
[48] J. Lee and H. H. Seliger, *Photochem. Photobiol.* **4,** 1015 (1965).
[49] E. Lippert, W. Nagele, I. Siebold-Blackenstein, U. Steiger, and W. Voss, *Z. Anal. Chem.* **170,** 1 (1959).
[50] F. R. Lipsett, *J. Opt. Soc. Am.* **49,** 673 (1959).
[51] W. H. Melhuish, *J. Phys. Chem.* **64,** 762 (1960).
[52] W. H. Melhuish, *J. Opt. Soc. Am.* **52,** 1256 (1962).
[53] C. A. Parker, *Anal. Chem.* **34,** 502 (1962).
[54] C. A. Parker, "Photoluminescence of Solutions." Elsevier, Amsterdam, 1968.
[55] C. A. Parker and W. J. Barnes, *Analyst* **82,** 606 (1957).
[56] C. A. Parker and W. T. Rees, *Analyst* **85,** 587 (1960).
[57] C. A. Parker and W. T. Rees, *Analyst* **87,** 83 (1962).
[58] H. Sprince and G. R. Rowley, *Science* **125,** 25 (1957).
[59] S. Udenfriend, "Fluorescence Assay in Biological Medicine." Academic Press, New York, 1962.
[60] G. Weber and F. W. J. Teale, *Trans. Faraday Soc.* **53,** 646 (1957).
[61] C. E. White, M. Ho, and E. Q. Weomer, *Anal. Chem.* **32,** 438 (1960).

The most direct method for determining the spectral efficiency of the spectrometer–phototube combination is with a National Bureau of Standards Color Temperature Standard Lamp.[48,62] In the case of tungsten, the emissivity over the visible spectrum (400–700 nm), while lower than that of a black body, changes in such a way that for any true temperature of tungsten there is a color match with a black body at a higher true temperature. The true temperature of the black body which would have the same color as the tungsten filament is defined as the color temperature of the tungsten filament. Therefore one can calculate the relative spectral intensity for the visible region of a Standard Lamp of Color Temperature by assuming that it radiates as a black body. The spectral distribution of radiant energy from a black body can be obtained from Planck's equation (see Forsythe,[63] p. 79ff).

When the spectral region extends below 400 nm and above 700 nm, there is a Standard Lamp of Spectral Radiant Intensity whose absolute radiant intensity or brightness is specified at wavelength steps from the UV through the IR in microwatts per unit wavelength interval per steradian per unit area of the strip tungsten filament.

A third type of standard lamp is available, a Standard Lamp of Spectral Irradiance. This is a tungsten halogen lamp where the irradiance at a fixed distance from the lamp is specified at wavelength steps, also from the UV through the IR, in microwatts per unit wavelength interval per square centimeter at the given distance.

The Spectral Irradiance Lamp and the Radiant Intensity Lamp have low emissions in the UV region below 350 nm, although they can be used, with care to avoid scattering effects, down to 250 nm. A deuterium lamp has recently been developed that can extend the calibration down to 200 nm and that will overlap with the tungsten filament lamps in the 350 nm region.[64] The deuterium lamp provides approximately 200 times more intensity in the UV than the tungsten filament lamps.

The Radiant Intensity Lamp and the Spectral Radiance Lamp can be used in the same manner as the Color Temperature Standard Lamp for the relative spectral calibration of a spectrometer–phototube. They have, however, major secondary advantages in that they can be used for absolute standardization as well.

All of these Standard Lamps deteriorate with use. Therefore it is advisable to purchase much less expensive commercial models of these lamps,

[62] D. W. Ellis, *in* "Fluorescence and Phosphorescence Analysis" (D. M. Hercules, ed.), p. 41. Wiley, New York, 1966.

[63] W. E. Forsythe, "Smithsonian Physical Tables," 9th rev. ed. Smithsonian Inst., Washington, D.C.

[64] R. D. Saunders, W. R. Ott, and J. M. Bridges, *Appl. Opt.* **17**, 593 (1978).

which are not standardized, to be used while setting up the geometries for calibration and for initial testing and then to substitute the Standard Lamps during the actual period of the calibration. This will prolong the absolute calibrations of the Standard Lamps for many years.

1. Measuring Geometries

There are a number of geometries in which these lamps may be used for spectrometer–phototube calibration:

a. Direct Focusing. For this method and for other calibrations to be treated in later sections it is advisable to obtain a matched pair of achromatic positive lenses of approximately 10 cm diameter and focal lengths of approximately 25 cm so that when an image of a filament source is focused on the entrance slit the f number of the lens system, in this case, f/2.5, will be faster than the f number of the spectrometer. In my own laboratory the spectrometers range from f/3 to f/6 so that the condition is always fulfilled. This ensures that the grating will always be filled and that there will be no vignetting. For work in the UV region, where light absorption by the lenses at the shorter wavelengths may be a problem, a front-surface aluminized spherical mirror of diameter 20 cm and focal length 50 cm is used. With either one of the lenses or with the mirror a 1:1 image of the filament is projected onto the plane of the entrance slit. The entrance slit may be closed down to the width of a razor blade and the slit height can be reduced by masking so that the image of the filament is much larger than the slit dimensions. Thus the spectrometer "sees" only the central section of the filament. The phototube voltage is reduced so that the maximum anode current does not exceed 1 μamp. The Color Temperature Lamp consists of a lateral array of six filaments. I have found that the outer two filaments may give results 1–2% less than the inner four filaments, and the inner four are identical within the precision of the measurements ($\pm$0.5%). Thus for all but the most demanding calibrations any one of the inner filaments will be entirely satisfactory. The Radiant Intensity Lamp is a single strip tungsten filament, and the Spectral Irradiance Lamp is a single coiled tungsten filament, so the problem does not arise with these lamps.

b. Diffuse Reflection—No Focusing Optics. An equally acceptable and possibly simpler method that does not require focusing optics is to place a square white diffusing plate near and facing the entrance slit so that the ratio of the distance of the screen from the slit to the side of the square is again 2.5 or less. Any one of the three standard lamps can be used to illuminate the diffusing surface facing the entrance slit. The diffuser is adjusted to be normal to the line joining the entrance slit to the center of the diffuser. The light source should illuminate the same face of the diffuser close to the normal without the detector housing obscuring illumination of any portion of the diffuser. The angle formed by the lamp–diffuser face–spectrometer entrance slit should be as small as possible. Care must be taken so that scattered and reflected light from the lamp, other than that directly incident on the diffusing plate, does not reach the entrance slit of the spectrometer. This is easily accomplished by the use of circular or square diaphragms and separators made of cardboard painted with a good grade of nonreflecting photographic black paint. A simple test for this is to interpose between the lamp and diffuser a black-painted square slightly larger than the diaphragm and to adjust the light-absorbing separators until no measurable signal appears at the phototube.

The lamp-to-diffuser distance should be greater than 1 meter so that the diffuser is illuminated uniformly. The distance is completely arbitrary and can be adjusted so that the slit opening and the phototube gain are approximately the same as would be used for the actual measurement of fluorescence and bioluminescence spectra. If the lamp-to-diffuser distance is measured accurately, the use of the Spectral Irradiance Lamp will also give the absolute as well as the relative spectral sensitivity of the spectrometer–phototube combination.

Diffusing screens can be prepared with MgO smoke in a chemical hood or can be sprayed with Eastman White Reflecting $BaSO_4$ paint.[65,66] A diffusing plate of multiply sprayed $BaSO_4$ has a reflectance of 0.991 ± 0.001 from 400 to 900 nm.[67] From 400 nm down to 300 nm the reflectance drops linearly to 0.968. Even at 250 nm the reflectance is extremely high, 0.950.

c. Diffuse Reflection—Focusing Optics. In those cases where the spectrometer–phototube or spectrometer–photodiode has insufficient sensitivity to be used in the nonfocused geometry, the strip tungsten filament of the Radiant Intensity Lamp can be focused by an achromatic lens with a circular stop onto the diffusing plate. In this case the lamp-to-focusing lens distance is adjusted so that a *magnified image* of the strip filament covers the entire diffusing plate. The brightness of the illuminated diffusing plate will depend only on the lamp-to-lens distance and the diameter of the circular stop. This method will also provide the absolute as well as the relative spectral sensitivity of the spectrometer–phototube combination.

d. Thermopile—Neutral Density Filter Substitution. The direct focusing method described above can be modified so that the spectral response curve of the spectrometer–phototube combination can be determined in the 400 to 800 nm region without the necessity of using any of the Standard

[65] F. Grum and G. W. Luckey, *Appl. Opt.* **7**, 2289 (1968).
[66] W. G. Egan and T. Hilgeman, *Appl. Opt.* **14**, 1137 (1975).
[67] F. Grum and T. E. Wightman, *Appl. Opt.* **16**, 2775 (1977).

Lamps. Any commercial tungsten filament–halogen lamp will suffice provided that the current through the lamp can be maintained constant throughout the calibration. In this case both achromatic lenses are used; the first with the filament at its focal length to make a parallel beam stopped down to slightly less than 5 cm diameter, and the second to focus this parallel beam onto the entrance slit. A commercially available calibrated direct-current or chopped thermopile is substituted for the phototube at the exit slit, and the entrance slit width is adjusted so that the thermopile and its associated microvoltmeter provide a set of readings of μW cm^{-2} as a function of wavelength setting of the spectrometer. This can be a continuous recorder trace of the output of the thermopile as a function of time when the grating cam is motor driven or can be a table of thermopile outputs as a function of manually set wavelengths. Next the phototube is substituted in the exit beam of the spectrometer and a suitable combination of neutral density filters is *placed in the parallel beam between the two achromatic lenses*. By closing down of the slit and adjustment of the high voltage which determines the gain, G, of the phototube, a similar set of readings of phototube current versus wavelength setting is obtained.

The thermopile may be calibrated absolutely in the laboratory by a simple procedure using a *fifth* type of standard lamp, a Total Radiation Standard Lamp. This Total Radiation Lamp is calibrated relative to a standard Blackbody and is specified in *total* μW cm^{-2} at a fixed distance from the lamp.

There are two advantages to using a thermopile as the basis for all absolute light measurements in the laboratory. A direct-current air (unevacuated) thermopile, if handled carefully and kept in a desiccator over Drierite or blue silica gel will retain its calibration for many years. It does not deteriorate with use and therefore can be used as an absolute check on all of the other Standard Lamps, which deteriorate with the length of time they are used. It is also used in the Band Emission Average (BEA) technique to be described in a later section.

2. *Details of Calibration*

The current measured by the spectrometer–phototube combination for any wavelength setting λ and slit width $d\lambda$ will be

$$I_\lambda = B(\lambda) d\lambda \Omega G T_\lambda \epsilon_\lambda \times \frac{\lambda\ [\text{nm}]}{198.7} \times 10^{12} \times 1.60 \times 10^{-19} \qquad (25)$$

where B (λ) is the power emitted by the Standard Lamp per unit area per unit wavelength per steradian in μW cm^{-2}

$d\lambda$ is the slit width or bandwidth of the slit system

Ω is the optical geometry factor of the experimental set up

T_λ is the transmission function of the grating-mirror combination of the spectrometer

ϵ_λ is the photoelectric efficiency function of the photocathode of the phototube

$\lambda/198.7 \times 10^{+12}$ from Eq. (16) converts microwatts at the wavelength to photons sec^{-1}

1.60×10^{-19} converts electrostatic units (esu) of electron charge sec^{-1} to amperes

If the geometry, the slit width, the phototube high voltage (gain, G), and the current through the Standard Lamp are kept constant during the measurement, all of these factors can be lumped into a single constant, K, and

$$I_\lambda = KB(\lambda)\mathrm{T}_\lambda\epsilon_\lambda \times \frac{\lambda\ [\mathrm{nm}]}{198.7} \times 10^{12} \tag{26}$$

The spectrometer–phototube spectral sensitivity function, S_λ, is just the product of $T_\lambda\epsilon_\lambda$.

In the case of the Color Temperature Lamp, the values of $B(\lambda)$ as functions of λ can be obtained from tables of Planck Blackbody spectral power emission (Forsythe[63] or any handbook of chemistry and physics). In the cases of the Radiant Intensity Lamp and the Spectral Irradiance Lamp the values of $B(\lambda)$ or irradiance, in microwatts as functions of wavelength are given with the Certificates of Standardization that come with the lamps.

In the case of the thermopile substitution method (Section VI,A,1,d) to measure the energy $E(\lambda)$ in W cm^{-2} in the beam at the exit slit and the neutral attenuation α by the neutral density filters, we have, analogous to Eq. (26) for the phototube current,

$$I_\lambda = K'E(\lambda)\ \alpha\ T_\lambda\epsilon_\lambda \times \frac{\lambda}{198.7} \times 10^{12} \tag{27}$$

where the constant K' takes into account that the thermopile at the exit slit beam may subtend a different solid angle than the substituted phototube.

From Eq. (26)

$$S_\lambda = \frac{I_\lambda}{KB(\lambda)} \times \frac{198.7}{\lambda\ [\mathrm{nm}]} \times 10^{-12} \tag{28}$$

If we simplify still further by collecting constants, it is seen that for either Eq. (26) or (27)

$$S_\lambda = \frac{I_\lambda}{B\ (\lambda)\ \lambda} K_1 \tag{29a}$$

or

$$S_\lambda = \frac{I_\lambda}{E\ (\lambda)\ \lambda} K_2 \tag{29b}$$

If we now normalize these data to unity at the maximum value of S_λ we obtain the relative spectral sensitivity function for the spectrometer–phototube combination.

The fully corrected spectrum of any fluorescent or bioluminescent solution can now be calculated by dividing each measured value I'_λ by the corresponding value of the normalized spectral sensitivity function S_λ and again normalizing to unity at the peak value of the corrected spectrum.

B. Absolute Measurement of Photon Emissions and Quantum Yields of Bioluminescent or Chemiluminescent Reactions

1. *Step-by-Step Absolute Calibration of Phototubes*

In this technique the calibration geometry should duplicate the experimental geometry as closely as possible. The phototube is placed at a known distance, d, from the bioluminescent or chemiluminescent solution and its photocathode area is defined by a diaphragm of diameter D, usually around 25 mm. The distance d is made large enough, and the volume of the emitting solution is made small enough, so that the emitting solution can be considered to be a point source with inverse square geometry.

We measure separately the corrected emission spectrum of the reaction, $F(\lambda)$, and we now renormalize $F(\lambda)$ such that

$$\int_\lambda F(\lambda)\, d\lambda = 1 \tag{30}$$

The total photons emitted per second into a 4π solid angle from the source is assumed to be N_0.

The energy flux incident on the phototube photocathode due to the source will then be

$$E\,[\mu\text{W}] = N_0 \times \frac{D^2}{16d^2}\left[\int_\lambda F(\lambda) \times \frac{198.7}{\lambda} \times 10^{-12}\, d\lambda\right] \tag{31}$$

The integral in Eq. (31) is the mean energy in μJoules of a photon emitted by the bioluminescent source with a spectral distribution $F(\lambda)$.

Now by use of a tungsten filament lamp at constant current as a source for a monochromator, or by use of the achromatic lens with the filament at the focal length to make a parallel beam and a series of narrow-band interference filters placed in the parallel beam, we produce monochromatic beams at a series of wavelengths over the entire range of $F(\lambda)$. Care is taken so that the width of the monochromatic beam is larger than the diameter D of the phototube diaphragm. By a substitution technique, interchanging first a calibrated thermopile and then a combination of filters of known neutral attenuation α with the phototube at a fixed gain G_1, the current I'_λ measured at any central wavelength λ in the calibration beam is

$$I'_\lambda = H_\lambda \frac{\pi D^2}{4} \alpha\, \epsilon'_\lambda G_1 \times 1.60 \times 10^{-19} \text{ amperes} \tag{32}$$

where H_λ is the energy in μW cm^{-2} measured with the calibrated thermopile and ϵ'_λ is the photoelectric efficiency of the photocathode, in units of photoelectrons per microwatt of incident on the total photocathode at the wavelength λ.

The absolute values of the function ϵ'_λ are determined from Eq. (32):

$$\epsilon'_\lambda = \frac{I'_\lambda}{H_\lambda \alpha G_1} \times \frac{4}{\pi D^2} \times \frac{10^{19}}{1.60} \tag{33}$$

The experimental measurement of the current, I^S, due to the bioluminescent source at a distance d will be

$$I^S = N_0 \times \frac{D^2}{16d^2} \int_\lambda [F(\lambda) \times \frac{198.7}{\lambda} \times 10^{-12} \times \epsilon'_\lambda d\lambda] \times G_2 \times 1.60 \times 10^{-19} \tag{34}$$

where the integral is now the mean photoelectric efficiency of the photocathode over the emission spectrum of the bioluminescent reaction and G_2 is the gain of the phototube in the experiment, which will usually be different from G_1. If we combine Eqs. (33) and (34) we obtain:

$$N_0 = 4\pi d^2\, I^S \frac{G_1}{G_2} \alpha \times \frac{10^{12}}{198.7} \left[\int_\lambda \frac{F(\lambda)}{\lambda} \frac{I'_\lambda}{H\lambda} d\lambda \right]^{-1} \tag{35}$$

From Eq. (35) it is seen that

a. If the diaphragm diameter D is the same for both the calibration and the experiment, it does not enter into the calculations.

b. The conversion factor for esu of charge to the Practical System drops out.

c. The ratio, G_2/G_1, rather than the knowledge of their absolute values, is required. This is a major simplification.

The value of N_0 can then be calculated by numerical integration of Eq. (35).

Thus far the technique has assumed that N_0 is a constant. For a bioluminescent reaction, initiated at $t = 0$ and going to completion at $t = t_f$, the total number of photons emitted from A substrate molecules over the entire emission spectrum will be simply

$$\text{Total light} = A\phi_{ch} \tag{36}$$

Since only N_0 and I^S in Eq. (35) are functions of time, it follows that

$$\text{Total light} = \int_{t_0}^{t_f} I dt \times 4\pi d^2 \frac{G_1}{G_2} \alpha \times \frac{10^{12}}{198.7} \left[\int_\lambda \frac{F(\lambda)}{\lambda} \frac{I'_\lambda}{H\lambda} d\lambda \right]^{-1} \tag{37}$$

If we know the total number of substrate molecules, A, initially in the reaction solution, the quantum yield for bioluminescence or chemiluminescence, ϕ_{ch}, is given by

$$\phi_{ch} = \frac{\text{Total light}}{A} \tag{38}$$

2. The BEA Technique

This Band Emission Average (BEA) technique is a simplification of the Step-by-Step Calibration. In this case both achromatic lenses are used to focus a tungsten filament, at a constant current, on the entrance slit of the spectrometer. The lamp used is a 6-V 18-ampere coiled filament, 1 mm × 2 mm, normally operated at 15.0 amperes. Since lifetime is a very critical function of undervoltage as well as overvoltage, the lamp can serve for hundreds of hours without deterioration. In the parallel beam of the first lens are inserted, *empirically*, Corning glass color filters such that the observed relative spectral distribution of the lamp–filter combination focused on the entrance slit by the second lens matches as closely as possible the relative spectral distribution of the bioluminescence whose absolute emission is to be determined.

There is a simplification to this empirical filter selection procedure. In this *pseudo-source* method it is only necessary that the uncorrected relative spectral emissions coincide. Since the detector is the same, the corrected spectral emissions will also coincide.

When this is accomplished the calibrated thermopile is placed in the parallel beam. Since the parallel beam is uniform and has the same relative spectral distribution as the bioluminescent source to be measured, the energy flux measured by the thermopile will correspond to the mean flux in $\mu W\ cm^{-2}$ of that particular spectral distribution.

Next the phototube with its diaphragm D and its neutral density attenuation filters as in Section VI, B, 1 a. is substituted for the thermopile.

Corresponding to Eq. (32) the calibration current for the pseudo-source beam measured by the phototube is:

$$I' = \overline{H}\,\frac{\pi D^2}{4}\,\alpha\bar{\epsilon}' G_1 \times 1.60 \times 10^{-19} \text{ amperes} \tag{39}$$

The subscripts are no longer used since we have carried out the calibration measurement over the complete emission spectrum.

The expression for the current measured by the phototube at the distance d from the experimental source is:

$$I = N_0 \times \frac{D^2}{16d^2}\left[\int_\lambda F(\lambda) \times \frac{198.7}{\lambda} \times 10^{-12}\, d\lambda\right] \times \bar{\epsilon}' G_2 \times 1.60 \times 10^{-19} \tag{40}$$

where ϵ' is the mean photoelectric efficiency over the emission spectrum.

From Eqs. (39) and (40) we obtain

$$N_0 = 4\pi d^2 I \frac{G_1}{G_2} \alpha \times \frac{10^{12}}{198.7} \frac{\bar{H}}{I'} \left[\int_\lambda \frac{F(\lambda)}{\lambda} d\lambda \right]^{-1} \quad (41)$$

completely analogous to Eqs. (35) and (37).

The BEA technique has a significant advantage over the step-by-step method. Once the colored glass filter combination has been selected, a number of phototubes may be calibrated by a single substitution measurement. The calibration can be preserved by storing the filters and the tungsten lamp.

The most important source of error in these calculations is the determination of the attenuation factor α, a factor of 10^{-3} to 10^{-4}. Only reflectance attenuation filters have a flat transmittance across the visible spectrum. Kodak Wratten filters are especially poor in the blue region. One must also be careful in using more than one reflection filter for attenuation. When reflection attenuation filters are used in tandem it no longer holds exactly that the total transmission, T, is the product of the individual transmissions (T_i). It can be shown that in multiple reflection if β_1 is the reflection coefficient of filter 1, and β_2 is the reflection coefficient of filter 2 ($T_1 = 1 - \beta_1$ etc.), the total transmission of the filter combination is given by

$$T = \frac{I}{I_0} = (1 - \beta_1)(1 - \beta_2)[1 + \beta_1\beta_2 + \beta_1^2\beta_2^2 + \beta_1^3\beta_2^3 + \ldots] \quad (42)$$

If one assumes a perfectly parallel beam and no internal absorption, the infinite series of Eq. (42) can be represented as

$$T = \frac{T_1 T_2}{1 - \beta_1\beta_2} \quad (43)$$

As an example, if the individually measured transmissions T_1 and T_2 are 0.01, that is, $\beta = 0.99$, instead of the expected total transmission of 10^{-4}, the transmission will be $T = 10^{-4}/(1 - 0.98) = 50 \times 10^{-4}$. Owing to the facts that neither is the beam exactly parallel, nor are the filters placed exactly parallel, one obtains experimentally some intermediate value, greater than 10^{-4} but less than 50×10^{-4}. It is therefore possible in even such a simple attenuation technique to be incorrect by an order of magnitude. In the present measurements this effect is eliminated by interposing at 45° between the two reflection attenuation filters the flat, partially reflecting base of a prism so that the singly attenuated beam is reflected at 90° and then passes through the second reflection attenuation filter before striking the phototube.

In this case a similar derivation, where R is the reflection by the flat face of the prism base gives

$$T = \frac{T_1 R T_2}{1 - R^2 \beta_1 \beta_2} \tag{44}$$

As an example, for T_1 and $T_2 = 0.05$ and $R = 0.05$, the true transmission differs from the calculated product transmission by the factor $1/(1 - R^2\beta_1\beta_2) = 1.002$. In this simple step all multiple reflection corrections have been eliminated.

3. Comparison with Chemiluminescent Reactions for which ϕ_{ch} Is Known

The techniques described thus far have been used to determine the absolute quantum yields for the firefly bioluminescent reaction,[2,3] for the bacterial bioluminescent reaction[48] for the luminol chemiluminescent reaction,[11,45,68-71] for the quantum yield of the ferrioxalate actinometer,[72] and for the determination of the total photons emitted by bioluminescent dinoflagellates upon complete mechanical stimulation.[73]

It is possible to relate the absolute bioluminescent or chemiluminescent quantum yields of any other reactions to these absolute quantum yields by direct comparison, using a spectrometer–phototube combination for which the precise relative spectral response functions, S_λ, have been measured.

Consider a bioluminescent reaction containing initially A^X substrate molecules which are reacted at the entrance slit of the spectrometer. From separate measurements it is necessary to determine $F^X(\lambda)$, the relative spectral distribution of the emission. Since the same argument relative to the time dependence of the emission holds here as in Section VI, B, 1 [Eq. (36)], we can use a variant of Eq. (37) for the derivation.

If we use superscripts X and S for unknown and standard chemiluminescent reactions, respectively, we can write Eqs. (45) through (51) for the phototube current integrals as functions of wavelength.

For the unknown quantum yield reaction, we set the spectrometer to the peak wavelength of the emission λ^X_{max} and we measure $\int_0^{t_f} I^X_{\lambda_{max}}\, dt$ at this fixed wavelength.

In a separate series of measurements we measure the linear dispersion

[68] H. H. Seliger, *Anal. Biochem.* **1,** 60 (1960).
[69] J. Lee and H. H. Seliger, *Photochem. Photobiol.* **11,** 247 (1970).
[70] J. Lee and H. H. Seliger, *Photochem. Photobiol.* **15,** 227 (1972).
[71] J. Lee, A. S. Wesley, J. F. Ferguson III, and H. H. Seliger, *in* "Bioluminescence in Progress" (F. A. Johnson and Y. Haneda, eds.), p. 35. Princeton Univ. Press, Princeton, New Jersey, 1966.
[72] J. Lee and H. H. Seliger, *J. Chem. Phys.* **40,** 519 (1964).
[73] H. H. Seliger, W. H. Biggley, and E. Swift, *Photochem. Photobiol.* **10,** 227 (1969).

of the slit system so that the exact wavelength band pass, $\Delta\lambda$, is known. This is done by measuring the half width, in nanometers, of the triangular response versus wavelength of the spectrometer–phototube to one of the sharp Hg lines from a low pressure Hg wavelength calibration lamp.

In another separate series of measurements we determine the corrected, renormalized emission spectrum $F^X(\lambda)$ of the chemiluminescent reaction whose quantum yield is unknown as well as that for the known luminol reaction, $F^S(\lambda)$

$$\int_\lambda F^X(\lambda)d\lambda = 1 \tag{45}$$

The renormalized emission spectrum $F^X(\lambda)$ is then plotted on coordinate paper as a function of λ and the fractional emission at λ_{max} is determined by:

$$f^X_{\lambda_{max}} = \frac{F^X(\lambda_{max})\Delta\lambda}{\int_\lambda F^X(\lambda)d\lambda} \tag{46}$$

This is calculated by integrating the area under the portion of the curve at $F^X(\lambda_{max})$ bounded by $\pm\ \Delta\lambda/2$ and dividing it by the total area under the curve.

At any time, t, during the experimental determination of $\int_0^{t_f} I^X_{\lambda_{max}}\,dt$, the corrected current is given by

$$\frac{I^X_{\lambda_{max}}(t)}{S^X_{\lambda_{max}}} = [A^X(t)\ \phi^X_{ch} f^X_{\lambda_{max}}\ \Omega^X\ G^X]\ 1.60 \times 10^{-19} \text{ amperes} \tag{47}$$

where $S^X_{\lambda_{max}}$ is the spectrometer–photometer response function at the maximum wavelength setting for the solution of unknown chemiluminescence quantum yield, Ω^X is the geometry factor for the isotropic emission, and G^X is the phototube gain used in the experimental determination of $\int_0^{t_f} I^X_{\lambda_{max}}\,dt$.

We now perform a similar measurement for the same volume of the known chemiluminescent reaction in the same cuvette at the entrance slit, so that $\Omega^S = \Omega^X$. Then

$$\frac{I^S_{\lambda_{max}}(t)}{S^S_{\lambda_{max}}} = [A^S(t)\ \phi^S_{ch} f^S_{\lambda_{max}}\ \Omega^X\ G^S]\ 1.60 \times 10^{-19} \text{ amperes} \tag{48}$$

We record the values of $I^X_{\lambda_{max}}$ and $I^S_{\lambda_{max}}$ as functions of time and integrate numerically. Next we divide the time integrals [Eq. (47) by Eq. (48)] and solve for ϕ^X_{ch}:

$$\phi^X_{ch} = \frac{\int_0^{t_f} I^X_{\lambda_{max}}\,dt}{\int_0^{t_f} I^S_{\lambda_{max}}\,dt}\ \frac{S^S_{\lambda_{max}}}{S^X_{\lambda_{max}}}\ \frac{A^S f^S_{\lambda_{max}}}{A^X f^X_{\lambda_{max}}}\ \frac{G^S}{G^X}\ \phi^S_{ch} \tag{49}$$

where

$\int_0^{t_f} I_\lambda dt = q_\lambda$, the time integral of the measured phototube current (50)

From Eq. (49) we draw the following conclusions:

i. It is not necessary for the emission spectrum of the unknown (ϕ^X_{ch}) reaction to overlap that of the known (ϕ^S_{ch}) reaction.

ii. If there is significant overlap we can set the measurement wavelengths equal. In this case Eq. (49) reduces to

$$\phi^X_{ch} = \frac{q^X_{\lambda_{max}}}{q^S_{\lambda_{max}}} \frac{A^S}{A^X} \frac{f^S_{\lambda_{max}}}{f^X_{\lambda_{max}}} \frac{G^S}{G^X} \phi^S_{ch} \tag{51}$$

iii. Again only the gain *ratio* G^S/G^X need be measured.

iv. Since $\Omega^X = \Omega^S$, the geometry need not be known and refractive index effects cancel out.

v. The accurate measurement of $I_\lambda(t)$ depends upon measuring the voltage drop across a precision resistor. In this case, if the same resistor is used for both sets of current measurements, its value need not be known accurately.

Equation (49) represents an extremely versatile and general method for determining the chemiluminescence or bioluminescence quantum yield of any reaction so long as $f^X_{\lambda_{max}}$ as given by Eq. (46), can be determined. Therefore, it is unnecessary to have a series of standard chemiluminescent reactions in various regions of the spectrum. Quantum yields of all bioluminescent reactions can be related back to the accurately known luminol chemiluminescent quantum yield in dimethyl sulfoxide (DMSO) at 25°.[70]

$$\phi_{ch} \text{ (luminol in DMSO)} = 0.0125$$

C. Absolute Measurement of Fluorescence Quantum Yields

In many cases we would like to determine the separate products ϕ_F and ϕ_{ex} of Eq. (12). In particular, the value

$$\phi_{ex} = \phi_{ch}/\phi_F \tag{52}$$

the fraction of the substrate molecules A, reacting through the pathway to produce excited state products, gives important information in terms of mechanism. In the case of sensitized bioluminescence, as in *Renilla* and the luminous bacteria

$$\phi_{ch} = \phi_{ex}\phi_{tr}\phi'_F \text{ (fluorescent protein)} \tag{53}$$

where ϕ_{tr} is the efficiency of energy transfer between the product excited state and the fluorescent protein, which has a fluorescence quantum yield ϕ'_F. It then becomes important to know ϕ_{tr}. Therefore, a separate measurement of ϕ_F, either of the product molecule in the case of direct bioluminescence or of the fluorescent protein in the case of sensitized bioluminescence becomes necessary.

It is possible that in some chemical reactions resulting in electronically excited states, the free energy released as the result of the chemical reaction may not always be greater than that necessary to leave the product molecule in a highly excited vibrational level of the first excited state, but may correspond to excitation on the long-wavelength side of the photoexcitation spectrum, in the region of anti-Stokes fluorescence. Valentiner and Rössiger[74] showed that the fluorescence quantum yield of a solution of fluorescein dropped rapidly as the exciting wavelength increased above 510 nm, on the long-wavelength side of the absorption spectrum, where the fluorescence band and the excitation band begin to overlap. This was verified by Vavilov,[75] Jablonski,[76] and a number of other careful investigators. These results have been interpreted as an artifact due to the contribution to the long-wavelength absorption of fluorescein solutions by nonfluorescent dimers of fluorescein where peak absorption is red shifted relative to the absorption of normally fluorescent monomers.[77,78] We have found a similar anti-Stokes drop in ϕ_F by a factor of 10 for Rhodamine B solutions in ethylene glycol and in ethyl alcohol solutions at concentrations as low as 10^{-6} M where dimer formation is negligible.

This presents an interesting theoretical problem because it implies that ground-state molecules in high vibrational ground states, when raised to the first excited state, may have different solvent cages that result in strong quenching before vibrational equilibrium is established. The implications to bioluminescence and chemiluminescence are that possibly the solvent cages of chemically formed, electronically excited product molecules can be different from photoexcited ground state molecules and therefore that in some cases the photoexcited value of k_f of Eq. (5) may not be the same as the chemiexcited rate constant.

1. Step-by-Step Phototube Calibration

This straightforward technique was first developed by Vavilov[79] using a MgO diffusing plate on which the image of a standard lamp filament was

[74] S. Valentiner and M. Rössiger, *Z. Phys.* **36,** 81 (1926).
[75] S. I. Vavilov, *Z. Phys.* **42,** 311 (1927).
[76] A. Jablonski, *Acta Phys. Pol.* **2,** 97 (1933).
[77] A. Jablonski, *Acta Phys. Pol.* **13,** 239 (1954).
[78] R. Drabent and D. Frackowiak, *Acta Phys. Pol.* **14,** 447 (1955).
[79] S. I. Vavilov, *Z. Phys.* **22,** 266 (1924).

focused in a geometry similar to that described in Section VI, B, 1. The spectrometer entrance slit was replaced by the photon detector. Narrow-band interference filters between the lamp and the focusing lens served to define the wavelength bands for which the phototube was calibrated. In later work a monchromator was used. Next the diffuser was substituted for by the front face of a cuvette containing an optically dense solution of the fluorescent material so that greater than 99% of the exciting light was absorbed within a few millimeters of the solution. The cuvette then served as a front face emitting source with the same geometry relative to the detector as the original diffusing screen. Polarization corrections were introduced much later. In this case the value of N_0 was

$$N_0 = P_0 \phi_F \tag{54}$$

where P_0 was the number of incident exciting quanta per second which were assumed to be completely absorbed in the solution.

A correction for refractive index is important. When light is emitted from a solution of refractive index η_{sol} through a plane glass cuvette face of refractive index η_g into air of refractive index 1, there is a change in the effective solid angle subtended by the detector and less light reaches the detector. Thus the fraction of light received by the detector from the fluorescent solution will be reduced by the factor[31,80,81] of

$$(\eta_g/\eta_{sol})^2 \, (1/\eta_g)^2 = 1/\eta_{sol}^2 \tag{55}$$

Therefore the calculated value of ϕ_F in fluorescence yield measurements involving rectangular cuvettes must be multiplied by η_{sol}^2 for the peak wavelength of the emission spectrum.

A complete treatment of the Vavilov method, modified by Melhuish[82, 83] is given by Demas and Crosby.[44]

2. *The Bowen Quantum Counter*

In place of the step-by-step calibration technique, Bowen and Sawtell[84] introduced the concept of an optically dense fluorescing material that would absorb *all* incident light essentially at the surface. Since the ϕ_F of the quantum counter was assumed constant, the measured fluorescence intensity from the quantum counter was directly proportional to the number of incident photons per second independent of their wavelength. Bowen and Sawtell first used solid uranyl nitrate. The Rhodamine B solution quantum

[80] B. A. Brice, M. Halwer, and R. Speiser, *J. Opt. Soc. Am.* **40,** 768 (1950).
[81] J. J. Hermans and S. Levinson, *J. Opt. Soc. Am.* **41,** 460 (1951).
[82] W. H. Melhuish, *N. Z. J. Sci. Technol.* **37,** 142 (1955).
[83] W. H. Melhuish, *J. Opt. Soc. Am.* **54,** 183 (1964).
[84] E. J. Bowen and J. W. Sawtell, *Trans. Faraday Soc.* **33,** 1425 (1937).

counter[82] consisting of 4 g l^{-1} of Rhodamine B in glycerol or 3 g l^{-1} of Rhodamine B in ethylene glycol[52] has been found to have a constant ϕ_F from 350 to 600 nm. Therefore, instead of a phototube detector whose photoelectric spectral efficiency function must be determined as in the original Vavilov technique (cf. Section VI,C,1) a front-face Rhodamine B quantum counter receives the incident radiation and is viewed through a red transmitting filter by its own phototube.

In the Vavilov technique, with the Rhodamine B cuvette subtending a small solid angle at the normal to the plane of the diffuser, the quantum counter phototube current due to the image of a monochromator slit on the diffuser would be

$$I_L = B_\lambda R_1 \frac{\Omega_1}{\pi}(1 - R_2)\, \phi_F^{\text{Rh B}} \frac{1}{\eta_1^2}\, \Omega_2 T_f\, \bar{\epsilon} G_1 \times 1.60 \times 10^{-19} \tag{56}$$

where B_λ is the photons $\sec^{-1}$ of exciting light of wavelength λ incident on the screen, R_1 is the reflectivity of the diffuser (0.991), Ω_1 is the solid angle subtended by the Rhodamine B cuvette, R_2 is the reflection from the glass or quartz face of the Rhodamine B cuvette, $\phi_F^{\text{Rh B}}$ is the fluorescence quantum yield of the Rhodamine B, η_1 is the index of refraction of the Rhodamine B solution for the fluorescence of Rhodamine B, Ω_2 is the solid angle subtended by the quantum counter phototube, T_f is the transmission of the red-transmitting filter for the Rhodamine B fluorescence, $\bar{\epsilon}$ is the mean photoelectric efficiency of the photocathode of the quantum counter phototube for the spectrum transmitted by the red filter, and G_1 is the gain of the phototube.

When the diffuser is replaced by the optically dense cuvette containing the fluorescent solution whose ϕ_F^X is to be determined, the phototube current is given by

$$I_X = B_\lambda (1 - R_3) \frac{\phi_F^X}{\eta_2^2}\, \Omega_1 (1 - R_2)\, \phi_F^{\text{Rh B}} \frac{1}{\eta_1^2}\, \Omega_2 T_f \bar{\epsilon} G_2 \times 1.60 \times 10^{-19} \tag{57}$$

where R_3 is the reflection from the cuvette containing the solution of unknown fluorescence quantum yield ϕ_F^X, and η_2 is the index of refraction of the unknown ϕ_F^X solution for its fluorescence.

If we divide Eq. (57) by Eq. (56) we obtain

$$\frac{I_X}{I_L} = \frac{(1 - R_3)}{R_1} \frac{\phi_F^X \pi}{\eta_2^2} \frac{G_2}{G_1} \tag{58}$$

from which

$$\phi_F^X = \frac{I_X}{I_L} \frac{R_1}{(1 - R_3)} \frac{\eta_2^2}{\pi} \frac{G_1}{G_2} \tag{59}$$

From Eq. (59) the following points emerge:

i. The measurement is independent of all of the geometries and efficiencies associated with the quantum counter, provided $\phi_F^{\mathrm{Rh\,B}}$ is constant for both the exciting light and the fluorescence.

ii. Only the phototube gain *ratio* need be measured.

iii. The technique is independent of the exciting wavelength and of the absolute intensity of the exciting light, just so long as B_λ remains constant, is within the absorption band of the unknown ϕ_F^X solution, and is of sufficient intensity to produce measurable fluorescence.

A more detailed treatment of the technique dealing with absorption-reemission due to the high optical density of the fluorescing solution is given by Melhuish[52] and Demas and Crosby.[44]

3. *Weber and Teale Solution Scatterers*

In 1957 Weber and Teale introduced dilute colloidal solutions as ideal dipole scatterers.[60] In their treatment the Rhodamine B solution is still used as the quantum counter. The colloidal scattering solution replaces the diffusing screen. The main advantage of the Weber and Teale method is that it permits the same geometry factors to be used (and therefore to cancel out) for fluorescent and scattering solutions of very low optical densities, eliminating self-quenching, absorption-reemission and absorption by nonfluorescing dimers, which can introduce significant errors in fluorescing solutions of high optical density (high concentrations).

Using monochromatic unpolarized exciting light, B_λ, the quantum counter phototube current for the scattering solution will be:

$$I_S = \frac{2\,B_{\lambda_{ex}}}{(3 + p_S)}(1 - e^{-E_S d})(e^{-E_S d})\frac{K_{\mathrm{Rh\,B}}}{\eta_S^2}G_1 \tag{60}$$

where E_S is the absorption coefficient of the scattering solution for the exciting light of wavelength λ_{ex}, p_S is the polarization ratio for the scattered light,[60, 85] and K_{RhB} contains all of the geometry factors and efficiency terms relating to the detection of the scattered light by the Rhodamine B quantum counter as in Eqs. (56) and (57). The factor $(1 - e^{-E_S d})$ is the fractional absorption of the exciting beam in the scatterer, and $e^{-E_S d}$ is the attenuation of the scattered radiation in the solution. η_S is the index of refraction of the colloidal solution for the exciting light.

If we differentiate I_S with respect to E_S (the concentration of the scattering solution) and let $E_S \rightarrow 0$

[85] P. P. Feofilov, "The Physical Basis of Polarized Emission" p. 274. Consultants Bureau, New York, 1961.

$$\left[\frac{(dI_S)}{(dE_S)}\right]_{E_S \to 0} = \frac{2B_{\lambda_{ex}}}{(3 + p_S)} d \frac{K_{RhB}}{\eta_S^2} G_1 \quad (61)$$

A similar equation for the unknown ϕ_F^X fluorescing solution will be

$$I_X = \frac{2B_{\lambda_{ex}}}{(3 + p_X)} (1 - e^{-E_X d}) (e^{-E_X d}) \frac{K_{RhB}}{\eta_X^2} \phi_F^X G_2 \quad (62)$$

where E_X is the absorption coefficient of the fluorescing solution for the exciting light of wavelength λ_{ex}, p_X is the polarization ratio of the fluorescent radiation and the fractional absorption and attenuation terms have the same meaning. η_X is the index of refraction of the fluorescing solution for its fluorescence.

Again, differentiating Eq. (62) with respect to the concentration of fluorescing solution and letting $E_X \to 0$

$$\left(\frac{dI_X}{dE_X}\right)_{E_X \to 0} = \frac{2B_{\lambda_{ex}}}{(3 + p_X)} d \frac{K_{RhB}}{\eta_X^2} \phi_F^X G_2 \quad (63)$$

Dividing Eq. (63) by Eq. (61) and solving for ϕ_F^X:

$$\phi_F^X = \frac{\left(\frac{dI_X}{dE_X}\right)_{E_X \to 0}}{\left(\frac{dI_S}{dE_S}\right)_{E_S \to 0}} \frac{(3 + p_X)}{(3 + p_S)} \frac{\eta_X^2}{\eta_S^2} \frac{G_1}{G_2} \quad (64)$$

From Eq. (64) we draw the following conclusions:

i. The slopes of the variation of I_X with E_X and of the variation of I_S with E_S need to be measured. Therefore only one absolute measurement of optical density is necessary. At low concentrations of the ideal scatterer and of the fluorescing solution, the plot of I versus concentration will be a straight line passing through the origin and therefore can be determined simply by measurements at known dilutions based on one absolute measurement.

ii. The absolute measurement of ϕ_F^X is again independent of the geometries and the efficiencies associated with the quantum counter, provided $\phi_F^{Rh\,B}$ is constant for both the exciting light and the fluorescence. Only the gain *ratio* G_1/G_2, as in Eq. (59), need be determined.

iii. The technique is independent of the exciting wavelength and of the absolute intensity of B_λ provided it is the same for both the scattering and the fluorescence measurements.

The polarization p of the fluorescent or scattered beams is defined as

$$p = \frac{I_\parallel - I_\perp}{I_\parallel + I_\perp} \quad (65)$$

Where $I_{\parallel}$ and $I_{\perp}$ are the intensities of the components of the fluorescent beam resolved in directions parallel and perpendicular to the normal to the direction of observation. For dipolar scattering p_S will be nearly 1. For fluorescent solutions excited by unpolarized light $-1/7 < p_X < 1/3$. For low viscosity solutions p_X will be at most a few percent. Therefore, only moderate precision is needed for the determination of p_S and p_X since $(3 + p_X)/(3 + p_S)$ will be approximately equal to 0.7 within narrow limits.

4. *Comparisons with Fluorescent Solutions for which Φ_F Is Known*

If we have a fluorescent solution of known fluorescence yield ϕ_F it is possible to measure the fluorescence yield of any other solution by a direct relative comparison, using a calibrated spectrometer–phototube combination. The technique is a combination of the Weber and Teale method and the method described in Section VI,B,3, for the measurement of ϕ^X_{ch} with respect to the luminol chemiluminescence quantum yield. For this purpose we can consider that the absolute fluorescence quantum yield of quinine sulfate (less than 10^{-4} *M*) in 0.1 *N* sulfuric acid, is constant over the excitation region 200–350 nm (see Chen[86]) and is 0.546 at 25°C[87].

The simplifying condition here is that the solution of unknown ϕ^X_F must be able to be excited to fluorescence by UV excitation so that the same exciting light beam can be used for both the unknown ϕ^X_F solution and the quinine sulfate solution. If the unknown ϕ^X_F solution has an absorption band outside of the quinine absorption, a freshly prepared solution of fluorescein, less than 10^{-6} *M* in 0.1 *N* sodium hydroxide can be used as a standard with excitation in the 400–500 nm region. Weber and Teale[60] reported a value of ϕ_F of 0.92 at 25°C.

As before, the corrected fluorescence emission spectra $F^S(\lambda)$ and $F^X(\lambda)$ must be determined and normalized such that

$$\int_\lambda F^S(\lambda)\, d\lambda = 1 \tag{66}$$

and

$$\int_\lambda F^X(\lambda)\, d\lambda = 1 \tag{67}$$

Next $\Delta\lambda$, the wavelength bandpass of the spectrometer, must be determined. It may be that the fluorescence emission spectra of the standard and unknown solutions overlap sufficiently for the comparisons of their fluorescence intensities to be measured at the same spectrometer wavelength setting. However, since this is not necessary the derivation will assume that each will be measured at its respective wavelength of maximum intensity.

[86] R. F. Chen, *Anal. Biochem.* **19,** 374 (1967).
[87] W. H. Melhuish, *J. Phys. Chem.* **65,** 229 (1961).

Therefore let us assume that the signal due to fluorescence of the standard will be measured at λ^{S}_{max} and that of the unknown at λ^{X}_{max}

Again we calculate graphically

$$f^{S}_{\lambda_{max}} = \frac{F^{S}(\lambda^{S}_{max})\,\Delta\lambda}{\int_{\lambda} F^{S}(\lambda)\,d\lambda} \tag{68}$$

and

$$f^{X}_{\lambda_{max}} = \frac{F^{X}(\lambda^{X}_{max})\,\Delta\lambda}{\int_{\lambda} F^{X}(\lambda)\,d\lambda} \tag{69}$$

Next we place a solution of known optical density of the standard at a concentration below $10^{-4}\,M$ at the entrance slit of the spectrometer so that when irradiated the fluorescent beam fills the entrance slit. Irradiation is normal to the entrance optics. It is usually inconvenient to place the cuvette directly at the entrance slit. The fluorescent image produced by the irradiation beam can be focused on the entrance slit by means of the front-surface aluminized mirror. The irradiation beam, from a monochromator or by isolation of one of the Hg lines at a wavelength λ_{ex}, is common to both the standard and the unknown solutions. Neither the wavelength λ_{ex} nor its absolute intensity $B_{\lambda_{ex}}$ need be known precisely provided there is sufficient intensity to excite fluorescence.

The spectrometer–photometer current at λ^{S}_{max} due to the irradiation of the standard solution with $B_{\lambda_{ex}}$ will be

$$I^{S}_{\lambda_{max}} = B_{\lambda_{ex}}(1 - e^{-E_{S}d})(e^{-E_{S}d})\frac{(1 - R^{S})}{(3 + p_{S})} \times \phi^{S}_{F}\frac{\Omega^{S}}{\eta^{2}_{S}} f^{S}_{\lambda_{max}} S^{X}_{\lambda_{max}} G^{S} \times 1.60 \times 10^{-19} \tag{70}$$

As in the Weber and Teale derivation

$$\left(\frac{dI^{S}_{\lambda_{max}}}{dE_{S}}\right)_{E_{S}\to 0} = B_{\lambda_{ex}} d\frac{(1 - R^{S})}{(3 + p_{s})} \times \Phi^{S}_{F}\frac{\Omega^{S}}{\eta^{2}_{S}} f^{S}_{\lambda_{max}} S^{S}_{\lambda_{max}} G^{S} \times 1.60 \times 10^{-19} \tag{71}$$

The same irradiation of the same volume of the unknown ϕ^{X}_{F} solution gives

$$I^{X}_{\lambda_{max}} = B_{\lambda_{ex}}(1 - e^{-E_{X}d})(e^{-E_{X}d})\frac{(1 - R^{X})}{(3 + p_{X})} \times \phi^{X}_{F}\frac{\Omega^{X}}{\eta^{2}_{X}} f^{X}_{\lambda_{max}} S^{X}_{\lambda_{max}} G^{X} \times 1.60 \times 10^{-19} \tag{72}$$

and

$$\left(\frac{dI^{X}_{\lambda_{max}}}{dE_X}\right)_{E_X\to 0} = B_{\lambda_{ex}} d\frac{(1-R^X)}{(3+p_X)} \phi^X_F \frac{\Omega^X}{\eta^2_X} f^X_{\lambda_{max}} S^X_{\lambda_{max}} G^X \times 1.60 \times 10^{-19} \quad (73)$$

Solving for ϕ^X_F we obtain

$$\phi^X_F = \frac{\left(\dfrac{dI^X_{\lambda_{max}}}{dE_X}\right)_{E_X\to 0}}{\left(\dfrac{dI^S_{\lambda_{max}}}{dE_S}\right)_{E_S\to 0}} \frac{(3+p_X)}{(3+p_S)} \frac{(1-R^S)}{(1-R^X)} \times \frac{\eta^2_X}{\eta^2_S} \frac{\Omega^S}{\Omega^X} \frac{f^S_{\lambda_{max}}}{f^X_{\lambda_{max}}} \frac{S^S_{\lambda_{max}}}{S^X_{\lambda_{max}}} \frac{G^S}{G^X} \phi^S_F \quad (74)$$

From the conditions of the experiment, $R^S = R^X$, $\Omega^S = \Omega^X$ and $\eta^2_S \cong \eta^2_X$. For small fluorescent molecules in low-viscosity solvents $p_S \cong p_X$. However, this should be verified separately. In most cases the optical densities can be made extremely small by precise dilution of carefully measured, more concentrated solutions, so that a single pair of measurements will suffice. Then

$$\phi^X_F = \frac{I^X_{\lambda_{max}}}{I^S_{\lambda_{max}}} \frac{(OD)^S_{\lambda_{max}}}{(OD)^X_{\lambda_{max}}} \frac{f^S_{\lambda_{max}}}{f^X_{\lambda_{max}}} \frac{S^S_{\lambda_{max}}}{S^X_{\lambda_{max}}} \frac{G^S}{G^X} \frac{(3+p_X)}{(3+p_S)} \phi^S_F \quad (75)$$

VII. Conclusions and Summary

In this presentation only phototube techniques have been presented for absolute and relative spectral measurements of light, since these are the most sensitive means for detection. All absolute determinations are based on the Planck formula for the brightness of a blackbody radiator at a given temperature and a primary standard blackbody radiator maintained at the National Bureau of Standards. Tungsten filament lamps, compared against this primary blackbody can then serve in individual laboratories as secondary standards to be operated at carefully specified voltages and currents.

Total Power: total μW cm^{-2} at a fixed distance from the lamp

Spectral Brightness: μW per unit wavelength interval per steradian per unit area of the filament

Spectral Irradiance: μW per unit wavelength interval per square centimeter incident on an illuminated surface at a fixed distance from the lamp

Color Temperature: the effective blackbody temperature that produces the same relative spectral energy distribution as the tungsten filament lamp in the 400–700 nm region

The Total Power Lamp can be used to standardize a linear thermopile or to standardize any of the other lamps for total irradiance.

The high reflectance and near-perfect diffusivity (see Egan and Hilgeman[88]) of $BaSO_4$ powders are the basis of the Vavilov technique.

The total absorption quantum counter was a major conceptual contribution, eliminating the necessity for measuring the absolute spectral response function of phototubes. This is the converse of the BEA technique, where a pseudosource achieves the same simplification.

The measurement techniques presented are quite straightforward. There is no reason therefore for any spectral data to be presented in uncorrected form. Since a large body of spectra in the literature has been presented in the form of $dN/d\lambda$ as a function of wavelength and the data are not incorrect, it is difficult to make an abrupt transition to the more physically meaningful presentation of spectra as $dN/d\bar{\nu}$ as a function of $\bar{\nu}$. Perhaps during the transition *both* curves should be presented. It is not sufficient to superimpose a wavenumber scale on a wavelength plot since the wavenumber scale will be nonlinear and the ordinates will still be $dN/d\lambda$. It must be remembered that $\bar{\nu}_{max}$ for a band emission spectrum is not equal to $1/\lambda_{max}$ and that only $\bar{\nu}_{max}$ has a physical significance in terms of excited-state energy levels.

There are available commercially, and it is possible to prepare, phosphors mixed uniformly with ^{14}C-labeled barium carbonate in transparent epoxy in standard test tubes. These long-lived self-luminous sources can be used as permanent tertiary emission standards for chemiluminescent reactions.

Once the ϕ_{ch} for a chemiluminescent reaction has been determined, the same reaction can be carried out in any light-measuring instrument normally used in the laboratory for routine assays. This geometry consists essentially of a tube support that holds the solution at a fixed distance from a phototube in a small, light-tight box.

If we react a known concentration (optical density) and volume of the chemiluminescent solution containing A substrate molecules, in this geometry the integrated current measured by the phototube will be:

$$\int_0^{t_f} I_{ch} dt = A\phi_{ch}\Omega_1\bar{S}_1G_1 \times 1.60 \times 10^{-19} \text{ coulombs} \tag{76}$$

where Ω_1 is the physical geometry of the light box, $\bar{S}_1$ is the average photocathode photoelectric efficiency for the chemiluminescent emission,

[88] W. G. Egan and T. Hilgeman, *Appl. Opt.* **15,** 1845 (1976); **16,** 2861 (1977).

and G_1 is the gain of the phototube, determined by the negative high voltage on the cathode.

If we replace the chemiluminescent reaction with the radioactive luminous source

$$I \text{ (source)} = B \text{ (source)} \, \Omega_2 \, \bar{S}_2 \, G_2 \times 1.60 \times 10^{-19} \text{ amperes} \tag{77}$$

where B is the photons $\sec^{-1}$ emitted by the source, Ω_2 is the source–phototube geometry, and $\bar{S}_2$ is the average photocathode photoelectric efficiency for the emission spectrum of the phosphor. *This spectrum does not need to be identical with the chemiluminescence emission spectrum. The only requirement is that it be within the range of spectral sensitivity of the phototube.*

From Eq. (77) we see that B (source) $\Omega_2\bar{S}_2$ can be equated to $B(\text{CL})\Omega_1\bar{S}_1$, where $B(\text{CL})$ is the *equivalent* chemiluminescent emission in photons $\sec^{-1}$ of the radioactive source. Thus

$$B(\text{CL}) = \frac{B(\text{source}) \, \Omega_2\bar{S}_2}{\Omega_1\bar{S}_1} \tag{78}$$

From Eq. (77)

$$B(\text{CL}) = \frac{I(\text{source})}{G_2 \times 1.60 \times 10^{-19}} \times \frac{1}{\Omega_1\bar{S}_1} \tag{79}$$

If we now substitute in Eq. (76) for $\Omega_1\bar{S}_1$ we obtain

$$B(\text{CL}) = \frac{I(\text{source})}{\int_0^{t_f} I_{\text{ch}} \, dt} A \phi_{\text{ch}} \frac{G_1}{G_2} \tag{80}$$

where the right side of the equation contains only measured quantities and known constants. Again only the *ratio* of phototube gains need be measured. It can be seen from this derivation that it is not necessary to know $\bar{S}_1$ or $\bar{S}_2$ or the emission spectra of the bioluminescent reaction or that of the luminous source or any of the geometries or even to be able to measure the current exactly by using high-accuracy resistors, so long as the same resistors are used for the measurement of I(source) and $\int I_{\text{ch}} \, dt$. All these unknowns cancel by virtue of the substitution technique.

This same radioactive phosphor can be used as a reference standard for *all* bioluminescent reactions within the range of spectral sensitivity of the phototube. However, since $B(\text{CL})$ is *equivalent* photons $\sec^{-1}$ it will have a different value for each bioluminescent reaction. It will be necessary to initially compare each different bioluminescent reaction with the luminous radioactive source in the same geometry. If we use the superscript i to denote the bioluminescent reaction for which the source is to be calibrated, the general form of Eq. (80) becomes

$$B^{i}(\mathrm{CL}) = \frac{I(\text{source})}{\int_{0}^{t_f} I^{i}_{ch}\, dt} \frac{A^{i}\phi^{i}_{ch}G^{i}_{1}}{G_2} \quad (81)$$

It is recommended that the luminol chemiluminescent reaction at less than 10^{-6} M in DMSO be used as the basis for all bioluminescent and chemiluminescent quantum yields. The best value for the molar extinction in DMSO at λ_{max} = 359 nm is 7756 ± 191, based on purified samples from seven independent sources. From measurements made on these 7 samples, 1 cm^3 of a luminol solution of OD 1 cm^{-1} emits $9.74 \pm 0.28 \times 10^{14}$ photons. The best estimate for the absolute chemiluminescent quantum yield of the luminol reaction at 25°C is therefore 0.0125 ± 6%, taking into account all sources of error of the phototube absolute standardization technique[48] and the internal precision of the luminol measurements.[70]

It is recommended that quinine sulfate solution (0.1 N H_2SO_4 at less than 10^{-5} M be used as the basis for all fluorescent quantum yield comparisons. For unpolarized monochromatic exciting light from 250 to 350 nm, there is general agreement that at 25°C the absolute fluorescence quantum yield of quinine sulfate is 0.55 ± 5%.[44, 87]

Author Index

Numbers in parentheses are footnote reference numbers and indicate that an author's work is referred to although his name is not cited in the text.

C

D

H

I

J

K

L

M

N

Q

R

S

Subject Index

E

I

O

Q

A
B
C 8
D 9
E 0
F 1
G 2
H 3
I 4
J 5